STUDENT'S SOLUTIONS MANUAL

CHRISTINE VERITY

INTERMEDIATE ALGEBRA: CONCEPTS & APPLICATIONS

EIGHTH EDITION

Marvin L. Bittinger

Indiana University Purdue University Indianapolis

David J. Ellenbogen

Community College of Vermont

Addison-Wesley
is an imprint of

PEARSON

Addison-Wesley
is an imprint of

PEARSON

www.pearsonhighered.com

Table of Contents

Chapter 1

Algebra and Problem Solving

Exercise Set 1.1

1. A letter representing a specific number that never changes is called a <u>constant</u>.

3. In the expression $7y$, the multipliers 7 and y are called <u>factors</u>.

5. When all variables in a variable expression are replaced by numbers and a result is calculated, we say that we are <u>evaluating</u> the expression.

7. A number that can be written in the form a/b, where a and b are integers (with $b \neq 0$) is said to be a <u>rational</u> number.

9. Division can be used to show that $\frac{7}{40}$ can be written as a <u>terminating</u> decimal.

11. Five less than some number

    Let n represent the number; $n - 5$

13. Twice a number

    Let x represent the number; $2x$

15. Twenty-nine percent of some number

    Let x represent the number; $0.29x$, or $\frac{29}{100}x$

17. Six less than half of a number

    Let y represent the number; $\frac{1}{2}y - 6$

19. Seven more than ten percent of some number

    Let s represent the number; $0.1s + 7$, or $\frac{10}{100}s + 7$

21. One less than the product of two numbers

    Let m and n represent the numbers; $mn - 1$

23. Ninety miles per every four gallons of gas

    We have
    $$90 \div 4, \text{ or } \frac{90}{4}.$$

25. The area of a square is given by $A = s^2$. We substitute $s = 6$ and solve.
    $$A = s^2 = 6^2 = 36 \text{ ft}^2$$

27. The area of a square is given by $A = s^2$. We substitute $s = 0.5$ and solve.
    $$A = s^2 = 0.5^2 = 0.25 \text{ m}^2$$

29. The area of a triangle is given by $A = \frac{1}{2}bh$. We substitute $b = 3$, $h = 5$ and solve.
    $$A = \frac{1}{2}bh = \frac{1}{2}(3)(5) = 7.5 \text{ ft}^2$$

31. The area of a triangle is given by $A = \frac{1}{2}bh$. We substitute $b = 3$, $h = 2.4$ and solve.
    $$A = \frac{1}{2}bh = \frac{1}{2}(3)(2.4) = 3.6 \text{ m}^2$$

33. Substitute and carry out the operations indicated.
    $$\begin{aligned} 4x + y &= 4 \cdot 2 + 3 \\ &= 8 + 3 \\ &= 11 \end{aligned}$$

35. Substitute and carry out the operations indicated.
    $$\begin{aligned} 20 + r^2 - s &= 20 + (5)^2 - 10 \\ &= 20 + 25 - 10 \\ &= 35 \end{aligned}$$

37. Substitute and carry out the operations indicated.
    $$\begin{aligned} 2c \div 3b &= 2 \cdot 6 \div 3 \cdot 2 \\ &= 12 \div 3 \cdot 2 \\ &= 4 \cdot 2 \\ &= 8 \end{aligned}$$

39. $3n^2p - 3pn^2 = 3 \cdot 5^2 \cdot 9 - 3 \cdot 9 \cdot 5^2$

    Observe that $3 \cdot 5^2 \cdot 9$ and $3 \cdot 9 \cdot 5^2$ represent the same number, so their difference is 0.

41. Substitute and carry out the operations indicated.
    $$\begin{aligned} 5x \div (2 + x - y) &= 5 \cdot 6 \div (2 + 6 - 2) \\ &= 5 \cdot 6 \div (8 - 2) \\ &= 5 \cdot 6 \div 6 \\ &= 30 \div 6 \\ &= 5 \end{aligned}$$

43. Substitute and carry out the operations indicated.

$$\left[10 - (a - b)\right]^2 = \left[10 - (7 - 2)\right]^2$$
$$= \left[10 - 5\right]^2$$
$$= 5^2$$
$$= 25$$

45. Substitute and carry out the operations indicated.

$$\left[5(r + s)\right]^2 = \left[5(1 + 2)\right]^2$$
$$= \left[5(3)\right]^2$$
$$= 15^2$$
$$= 225$$

47. Substitute and carry out the operations indicated.

$$x^2 - \left[3(x - y)\right]^2 = 6^2 - \left[3(6 - 4)\right]^2$$
$$= 6^2 - \left[3(2)\right]^2$$
$$= 6^2 - 6^2$$
$$= 0$$

49. Substitute and carry out the operations indicated.

$$(m - 2n)^2 - 2(m + n) = (8 - 2 \cdot 1)^2 - 2(8 + 1)$$
$$= (8 - 2)^2 - 2(9)$$
$$= 6^2 - 2(9)$$
$$= 36 - 2(9)$$
$$= 36 - 18$$
$$= 18$$

51. List the letters in the set: {a, l, g, e, b, r}

53. List the numbers in the set: {1, 3, 5, 7, ... }

55. List the numbers in the set: {10, 20, 30, 40, ... }

57. Specify the conditions under which a number is in the set: $\{x | x$ is an even number between 9 and 99$\}$

59. Specify the conditions under which a number is in the set: $\{x | x$ is whole number less than 5$\}$

61. Specify the conditions under which a number is in the set: $\{x | x$ is an odd number between 10 and 20$\}$

63. a) 0 and 6 are whole numbers.

b) –3, 0, and 6 are integers.

c) –8.7, –3, 0, $\frac{2}{3}$, and 6 are rational numbers.

d) $\sqrt{7}$ is an irrational number.

e) All of the given numbers are real numbers.

65. a) 0 and 8 are whole numbers.

b) –17, 0, and 8 are integers.

c) –17, –0.01, 0, $\frac{5}{4}$, and 8 are rational numbers.

d) $\sqrt{77}$ is an irrational number.

e) All of the given numbers are real numbers.

67. Since 196 is a natural number, the statement is true.

69. Since every whole number is an integer, the statement is true.

71. Since $\frac{2}{3}$ is a rational number, the statement is false.

73. Since $\sqrt{10}$ is an irrational number, and every member of the set of irrational numbers is a member of the set of real numbers, the statement is true.

75. Since the set of integers includes some numbers that are not natural numbers, the statement is true.

77. Since the set of rational numbers includes some numbers that are not integers, the statement is false.

79. *Writing Exercise.* Rational numbers can be expressed as p/q, where p and q are integers and $q \neq 0$. The set of integers is a subset of the set of rational numbers in which p is a multiple of q.

81. *Writing Exercise.* The statement is true because every member of the set {2, 4, 6} is also a member of the set {2, 4, 6}.

83. The quotient of the sum of two numbers and their difference

Let a and b represent the numbers. Then we have $\dfrac{a + b}{a - b}$.

85. Half of the difference of the squares of two numbers

Let r and s represent the numbers. Then we have

$$\frac{1}{2}\left(r^2 - s^2\right), \text{ or } \frac{r^2 - s^2}{2}.$$

87. The only whole number that is not also a natural number is 0. Using roster notation to name the set, we have {0}.

89. List the numbers in the set: {5, 10, 15, 20, ...}

91. List the numbers in the set: {1, 3, 5, 7, ...}

93. Recall from geometry that when a right triangle has legs of length 2 and 3, the length of the hypotenuse is $\sqrt{2^2 + 3^2} = \sqrt{4 + 9} = \sqrt{13}$. We draw such a triangle:

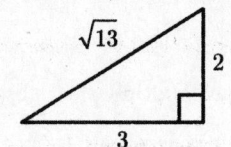

Exercise Set 1.2

1. It is true that the sum of the two negative number is always negative.

3. It is true that the product of a negative number and a positive number is always negative.

5. The statement is false. Consider $-5 + 2 = -3$, for example.

7. The statement is false. Let $a = 2$ and $b = 6$. Then $2 < 6$ and $|2| < |6|$.

9. It is true that the associative law of multiplication states that for all real numbers a, b, and c, $(ab)c$ is equivalent to $a(bc)$.

11. $|-10| = 10$ -10 is 10 units from 0

13. $|7| = 7$ 7 is 7 units from 0

15. $|-46.8| = 46.8$ -46.8 is 46.8 units from 0

17. $|0| = 0$ 0 is 0 units from itself

19. $\left|1\frac{7}{8}\right| = 1\frac{7}{8}$ $1\frac{7}{8}$ is $1\frac{7}{8}$ units from 0

21. $|-4.21| = 4.21$ -4.21 is 4.21 units from 0

23. $-5 \leq -4$

-5 is less than or equal to -4, a true statement since -5 is to the left of -4.

25. $-9 > 1$

-9 is greater than 1, a false statement since -9 is to the left of 1.

27. $0 \geq -5$

0 is greater than or equal to -5, a true statement since -5 is to the left of 0.

29. $-8 < -3$

-8 is less than -3, a true statement since -8 is to the left of -3.

31. $-4 \geq -4$

-4 is greater than or equal to -4. Since $-4 = -4$ is true, $-4 \geq -4$ is true.

33. $-5 < -5$

-5 is less than -5, a false statement since -5 does not lie to the left of itself.

35. $4 + 8$

Two positive numbers: Add the numbers getting 12. The answer is positive 12.

37. $(-3) + (-9)$

Two negative numbers: Add the absolute values, getting 12. The answer is negative, -12.

39. $-5.3 + 2.8$

A negative number and a positive number: The absolute values are 5.3 and 2.8. Subtract 2.8 from 5.3 to get 2.5. The negative number is farther from 0, so the answer is negative, -2.5.

41. $\frac{2}{7} + \left(-\frac{3}{5}\right) = \frac{10}{35} + \left(-\frac{21}{35}\right)$

A positive and negative number. The absolute values are $\frac{10}{35}$ and $\frac{21}{35}$. Subtract $\frac{10}{35}$ from $\frac{21}{35}$ to get $\frac{11}{35}$. The negative number is farther from 0, so the answer is negative, $-\frac{11}{35}$.

43. $-3.26 + (-5.8)$

Two negative numbers: Add the absolute values, getting 9.06. The answer is negative, -9.06.

45. $-\frac{1}{9} + \frac{2}{3} = -\frac{1}{9} + \frac{6}{9}$

A negative and positive number. The absolute values are $\frac{1}{9}$ and $\frac{6}{9}$. Subtract $\frac{1}{9}$ from $\frac{6}{9}$ to get $\frac{5}{9}$. The positive number is farther from 0, so the answer is positive, $\frac{5}{9}$.

47. $-6.25 + 0$

One number is zero: The sum is the other number, -6.25.

49. $4.19 + (-4.19)$

A negative and a positive number: The numbers have the same absolute value, 4.19, so the answer is 0.

51. $-18.3 + 22.1$

A negative and a positive number; The absolute values are 18.3 and 22.1. Subtract 18.3 from 22.1 to get 3.8. The positive number is farther from 0, so the answer is positive, 3.8.

53. The opposite of 2.37 is -2.37, because $2.37 + (-2.37) = 0$.

55. The opposite of -56 is 56, because $-56 + 56 = 0$.

57. The opposite of 0 is 0, because $0 + 0 = 0$.

59. If $x = 8$, then $-x = -8$. (The opposite of 8 is -8.)

61. If $x = -15$, then $-x = -(-15) = 15$. (The opposite of -15 is 15.)

63. If $x = -4.67$, then $-x = -(-4.67) = 4.67$. (The opposite of -4.67 is 4.67.)

65. If $x = 0$, then $-x = 0$. (The opposite of 0 is 0.)

67. $\begin{aligned}10 - 4 &= 10 + (-4) \qquad \text{Change the sign and add.}\\ &= 6\end{aligned}$

69. $\begin{aligned}4 - 10 &= 4 + (-10) \qquad \text{Change the sign and add.}\\ &= -6\end{aligned}$

71. $\begin{aligned}-5 - (-12) &= -5 + 12 \qquad \text{Change the sign and add.}\\ &= 7\end{aligned}$

73. $-5 - 14 = -5 + (-14) = -19$

75. $2.7 - 5.8 = 2.7 + (-5.8) = -3.1$

77. $\begin{aligned}-\frac{3}{5} - \frac{1}{2} &= -\frac{3}{5} + \left(-\frac{1}{2}\right)\\ &= -\frac{6}{10} + \left(-\frac{5}{10}\right) \qquad \text{Finding a common denominator}\\ &= -\frac{11}{10}\end{aligned}$

79. $\begin{aligned}0 - (-5.37) &= 0 + 5.37 \qquad \text{Change the sign and add.}\\ &= 5.37\end{aligned}$

81. $(-3)8$

Two numbers with unlike signs: Multiply their absolute values, getting 24. The answer is negative, -24.

83. $(-2)(-11)$

Two numbers with the same sign: Multiply their absolute values, getting 22. The answer is positive, 22.

85. $(4.2)(-5)$

Two numbers with unlike signs: Multiply their absolute values, getting 21. The answer is negative, -21.

87. $\frac{3}{7}(-1)$

Two numbers with unlike signs: Multiply their absolute values, getting $\frac{3}{7}$. The answer is negative, $-\frac{3}{7}$.

89. $(-17.45) \cdot 0 = 0$

91. $-\frac{2}{3}\left(\frac{3}{4}\right)$

Two numbers with unlike signs: Multiply their absolute values, getting $\frac{1}{2}$. The answer is negative, $-\frac{1}{2}$.

93. $\frac{-28}{-7}$

Two numbers with same sign: Divide their absolute values, getting 4. The answer is positive, 4.

95. $\frac{-100}{25}$

Two numbers with unlike signs: Divide their absolute values getting 4. The answer is negative, -4.

97. $\frac{73}{-1}$

Two numbers with unlike signs: Divide their absolute values, getting 73. The answer is negative, -73.

99. $\frac{0}{-7} = 0$

101. The reciprocal of 8 is $\frac{1}{8}$, because $8 \cdot \frac{1}{8} = 1$.

103. The reciprocal of $-\frac{5}{7}$ is $-\frac{7}{5}$, because $-\frac{5}{7} \cdot \left(-\frac{7}{5}\right) = 1$.

105. Does not exist

107. $\begin{aligned}\frac{3}{5} \div \frac{6}{7}&\\ &= \frac{3}{5} \cdot \frac{7}{6} \qquad \text{Multiplying the reciprocal of } \frac{6}{7}\\ &= \frac{21}{30}, \text{ or } \frac{7}{10}\end{aligned}$

109. $\begin{aligned}\left(-\frac{3}{5}\right) \div \frac{1}{2}&\\ &= -\frac{3}{5} \cdot \frac{2}{1} \qquad \text{Multiplying the reciprocal of } \frac{1}{2}\\ &= -\frac{6}{5}\end{aligned}$

111. $\left(-\dfrac{2}{9}\right) \div (-8)$

$= -\dfrac{2}{9} \cdot \left(-\dfrac{1}{8}\right)$ Multiplying the reciprocal of -8

$= \dfrac{2}{72}$, or $\dfrac{1}{36}$

113. $-\dfrac{12}{7} \div \left(-\dfrac{12}{7}\right)$

This is a number divided by itself so the quotient is 1. We would also do this exercise as follows.

$-\dfrac{12}{7} \div \left(-\dfrac{12}{7}\right) = -\dfrac{12}{7} \cdot \left(-\dfrac{7}{12}\right)$ Multiplying the reciprocal of $-\frac{12}{7}$

$= 1$

115. $-10^2 = -(10 \cdot 10) = -100$ Squaring 10 and then taking the opposite

117. $-(-3)^2 = -(-3)(-3) = -9$ Squaring (-3) and then taking the opposite

119. $(2-5)^2 = (-3)^2$ Working within the

$= 9$ parentheses first

121. $9 - (8 - 3 \cdot 2^3) = 9 - (8 - 3 \cdot 8)$ Working within the

$= 9 - (8 - 24)$ parentheses first

$= 9 - (-16)$

$= 9 + 16$

$= 25$

123. $\dfrac{5 \cdot 2 - 4^2}{27 - 2^4} = \dfrac{5 \cdot 2 - 16}{27 - 16} = \dfrac{10 - 16}{11} = \dfrac{-6}{11}$, or $-\dfrac{6}{11}$

125. $\dfrac{3^4 - (5-3)^4}{8 - 2^3} = \dfrac{3^4 - 2^4}{8 - 8} = \dfrac{81 - 16}{0}$

Since division by 0 is undefined, this quotient is undefined.

127. $\dfrac{(2-3)^3 - 5|2-4|}{7 - 2 \cdot 5^2} = \dfrac{(-1)^3 - 5|-2|}{7 - 2 \cdot 25} = \dfrac{-1 - 5(2)}{7 - 50}$

$= \dfrac{-1 - 10}{-43} = \dfrac{-11}{-43} = \dfrac{11}{43}$

129. $\left|2^2 - 7\right|^3 + 4 = |4 - 7|^3 + 4 = |-3|^3 + 4 = 3^3 + 4$

$= 27 + 4 = 31$

131. $32 - (-5)^2 + 15 \div (-3) \cdot 2$

$= 32 - 25 + 15 \div (-3) \cdot 2$ Evaluating the

                             exponential expression

$= 32 - 25 - 5 \cdot 2$ Dividing

$= 32 - 25 - 10$ Multiplying

$= -3$ Subtracting

133. Using the commutative law of addition, we have

$6 + xy = xy + 6.$

Using the commutative law of multiplication, we have

$6 + xy = 6 + yx.$

135. Using the commutative law of multiplication, we have

$-9(ab) = (ab)(-9)$

or $-9(ab) = -9(ba)$

137. $(3x)y$

$= 3(xy)$ Associative law of multiplication

139. $(3y + 4) + 10$

$= 3y + (4 + 10)$ Associative law of addition

141. $7(x+1) = 7 \cdot x + 7 \cdot 1$ Using the distributive law

$= 7x + 7$

143. $5(m-n) = 5 \cdot m - 5 \cdot n$ Using the distributive law

$= 5m - 5n$

145. $-5(2a + 3b)$

$= -5 \cdot 2a + (-5) \cdot 3b$

$= -10a - 15b$

147. $9a(b - c + d)$

$= 9a \cdot b - 9a \cdot c + 9a \cdot d$

$= 9ab - 9ac + 9ad$

149. $8a + 8b = 8 \cdot a + 8 \cdot b = 8(a + b)$

151. $9p - 3 = 3 \cdot 3p - 3 \cdot 1 = 3(3p - 1)$

153. $7x - 21y + 14z = 7 \cdot x - 7 \cdot 3y + 7 \cdot 2z = 7(x - 3y + 2z)$

155. $255 - 34b = 17 \cdot 15 - 17 \cdot 2b = 17(15 - 2b)$

157. *Writing Exercise.* The sum has the same sign as the number that is farther from 0.

159. Substitute and carry out the indicated operations.

$2(x + 5) = 2(3 + 5) = 2 \cdot 8 = 16$

$2x + 10 = 2 \cdot 3 + 10 = 6 + 10 = 16$

161. *Writing Exercise.* The quotient 7/0 is defined to be the number that gives a result of 7 when multiplied by 0. There is no such number, so we say the quotient is undefined.

163. $(8 - 5)^3 + 9 = 36$

165. $5 \cdot 2^3 \div (3 - 4)^4 = 40$

167. $17 - \sqrt{11 - (3 + 4)} \div \left[-5 - (-6)\right]^2$

$= 17 - \sqrt{11 - 7} \div \left[-5 + 6\right]^2$

$= 17 - \sqrt{4} \div \left[1\right]^2$

$= 17 - 2 \div 1$

$= 17 - 2$

$= 15$

169. Any value of a such that $a \leq -6.2$ satisfies the given conditions. The largest of these values is -6.2.

171. *Writing Exercise.* No; $5 - 3 = 2$, but $3 - 5 = -2$; also $8 \div 2 = 4$, but $2 \div 8 = 0.25$.

Exercise Set 1.3

1. By the distributive law, the expression $2(x + 7)$ is equivalent to the expression $2x + 14$, so they are equivalent expressions.

3. $3(t - 5) = 18$ and $3t - 15 = 18$ are equations and they have the same solution, 11, so they are equivalent equations.

5. $4x - 9 = 7$ and $4x = 16$ are equations and they have the same solution, 4, so they are equivalent equations.

7. Combining like terms in the expression $8t + 5 - 2t + 1$, we get the expression $6t + 6$, so these are equivalent expressions.

9. $6x - 3 = 10x + 5$ and $-8 = 4x$ are equations and they have the same solution, -2, so they are equivalent equations.

11. $3t = 21$ and $t + 4 = 11$

Each equation has only one solution, the number 7. Thus, the equations are equivalent.

13. $12 - x = 3$ and $2x = 20$

When x is replaced by 9, the first equation is true, but the second equation is false. Thus the equations are not equivalent.

15. $5x = 2x$ and $\dfrac{4}{x} = 3$

When x is replaced by 0, the first equation is true, but the second equation is not defined. Thus the equations are not equivalent.

17. $\quad x - 2.9 = 13.4$

$x - 2.9 + 2.9 = 13.4 + 2.9 \quad$ Adding princple; adding 2.9

$x + 0 = 13.4 + 2.9 \quad$ Law of opposites

$x = 16.3$

Check: $\quad \dfrac{x - 2.9 = 13.4}{16.3 - 2.9 \mid 13.4}$

$13.4 \overset{?}{=} 13.4 \quad$ TRUE

The solution is 16.3.

19. $\quad 8t = 72$

$\dfrac{1}{8} \cdot 8t = \dfrac{1}{8} \cdot 72 \quad$ Multiplication principle; multiplying by $\dfrac{1}{8}$, the reciprocal of 8

$1t = 9$

$t = 9$

Check: $\quad \dfrac{8t = 72}{8 \cdot 9 \mid 72}$

$72 \overset{?}{=} 72 \quad$ TRUE

The solution is 9.

21. $\quad \dfrac{2}{3}x = 30$

$\dfrac{3}{2} \cdot \dfrac{2}{3}x = \dfrac{3}{2} \cdot 30$

$1x = 45$

$x = 45$

Check: $\quad \dfrac{2}{3}x = 30$

$\dfrac{\dfrac{2}{3} \cdot 45 \mid 30}{}$

$30 \overset{?}{=} 30 \quad$ TRUE

The solution is 45.

23. $\quad 4a + 25 = 9$

$4a + 25 - 25 = 9 - 25$

$4a = -16$

$\dfrac{1}{4} \cdot 4a = \dfrac{1}{4}(-16)$

$1a = -4$

$a = -4$

Check: $\quad \dfrac{4a + 25 = 9}{4(-4) + 25 \mid 9}$

$-16 + 25 \mid$

$9 \overset{?}{=} 9 \quad$ TRUE

The solution is -4.

25.
$$2y - 8 = 9$$
$$2y - 8 + 8 = 9 + 8$$
$$2y = 17$$
$$\frac{1}{2} \cdot 2y = \frac{1}{2} \cdot 17$$
$$1y = \frac{17}{2}$$
$$y = \frac{17}{2}$$

Check:
$$\begin{array}{c|c} 2y - 8 = 9 \\ \hline 2\left(\dfrac{17}{2}\right) - 8 & 9 \\ 17 - 8 & 9 \\ & \overset{?}{9 = 9} \quad \text{TRUE} \end{array}$$

The solution is $\frac{17}{2}$.

27. $3x + 7x = (3 + 7)x = 10x$

29. $9t^2 + t^2 = (9 + 1)t^2 = 10t^2$

31. $16a - a = (16 - 1)a = 15a$

33. $n - 8n = (1 - 8)n = -7n$

35. $5x - 3x + 8x = (5 - 3 + 8)x = 10x$

37.
$$4x - 2x^2 + 3x$$
$$= 4x + 3x - 2x^2 \quad \text{Commutative law of addition}$$
$$= (4 + 3)x - 2x^2$$
$$= 7x - 2x^2$$

39.
$$18p - 12 + 3p + 8$$
$$= 18p + 3p - 12 + 8 \quad \text{Commutative law of addition}$$
$$= (18 + 3)p + (-12 + 8)$$
$$= 21p - 4$$

41.
$$-7t^2 + 3t + 5t^3 - t^3 + 2t^2 - t$$
$$= (-7 + 2)t^2 + (3 - 1)t + (5 - 1)t^3$$
$$= -5t^2 + 2t + 4t^3$$

43.
$$2x + 3(5x - 7)$$
$$= 2x + 15x - 21$$
$$= 17x - 21$$

45.
$$7a - (2a + 5)$$
$$= 7a - 2a - 5$$
$$= 5a - 5$$

47.
$$m - (6m - 2)$$
$$= m - 6m + 2$$
$$= -5m + 2$$

49.
$$3d - 7 - (5 - 2d)$$
$$= 3d - 7 - 5 + 2d$$
$$= 5d - 12$$

51.
$$2(x - 3) + 4(7 - x)$$
$$= 2x - 6 + 28 - 4x$$
$$= -2x + 22$$

53.
$$3p - 4 - 2(p + 6)$$
$$= 3p - 4 - 2p - 12$$
$$= p - 16$$

55.
$$-2(a - 5) - [7 - 3(2a - 5)]$$
$$= -2a + 10 - [7 - 6a + 15]$$
$$= -2a + 10 - [22 - 6a]$$
$$= -2a + 10 - 22 + 6a$$
$$= 4a - 12$$

57.
$$5\{-2x + 3[2 - 4(5x + 1)]\}$$
$$= 5\{-2x + 3[2 - 20x - 4]\}$$
$$= 5\{-2x + 3[-20x - 2]\}$$
$$= 5\{-2x - 60x - 6\}$$
$$= 5\{-62x - 6\}$$
$$= -310x - 30$$

59.
$$8y - \{6[2(3y - 4) - (7y + 1)] + 12\}$$
$$= 8y - \{6[6y - 8 - 7y - 1] + 12\}$$
$$= 8y - \{6[-y - 9] + 12\}$$
$$= 8y - \{-6y - 54 + 12\}$$
$$= 8y - \{-6y - 42\}$$
$$= 8y + 6y + 42$$
$$= 14y + 42$$

61.
$$4x + 5x = 63$$
$$9x = 63$$
$$\frac{1}{9} \cdot 9x = \frac{1}{9} \cdot 63$$
$$x = 7$$

Check:
$$\begin{array}{c|c} 4x + 5x = 63 \\ \hline 4 \cdot 7 + 5 \cdot 7 & 63 \\ 28 + 35 & \\ & \overset{?}{63 = 63} \quad \text{TRUE} \end{array}$$

The solution is 7.

63.
$$\frac{1}{4}y - \frac{2}{3}y = 5$$
$$\frac{3}{12}y - \frac{8}{12}y = 5$$
$$-\frac{5}{12}y = 5$$
$$\left(-\frac{12}{5}\right)\left(-\frac{5}{12}y\right) = \left(-\frac{12}{5}\right) \cdot 5$$
$$1y = -12$$
$$y = -12$$

Check:
$$\begin{array}{c|c} \dfrac{1}{4}y - \dfrac{2}{3}y = 5 \\ \hline \dfrac{1}{4}(-12) - \dfrac{2}{3}(-12) & 5 \\ -3 + 8 & 5 \\ & \overset{?}{5 = 5} \quad \text{TRUE} \end{array}$$

The solution is -12.

65. $4(t-3)-t=6$

$4t-12-t=6$

$3t-12=6$

$3t-12+12=6+12$

$3t=18$

$\dfrac{1}{3}\cdot 3t=\dfrac{1}{3}\cdot 18$

$t=6$

Check: $\dfrac{4(t-3)-t=6}{\begin{array}{c|c} 4(6-3)-6 & 6 \\ 4(3)-6 & \\ 12-6 & \\ & \overset{?}{6=6} \quad \text{TRUE} \end{array}}$

The solution is 6.

67. $3(x+4)=7x$

$3x+12=7x$

$3x+12-3x=7x-3x$

$12=4x$

$\dfrac{1}{4}\cdot 12=\dfrac{1}{4}\cdot 4x$

$3=x$

Check: $\dfrac{3(x+4)=7x}{\begin{array}{c|c} 3(3+4) & 7\cdot 3 \\ 3\cdot 7 & \\ & \overset{?}{21=21} \quad \text{TRUE} \end{array}}$

The solution is 3.

69. $70=10(3t-2)$

$70=30t-20$

$70+20=30t-20+20$

$90=30t$

$\dfrac{1}{30}\cdot 90=\dfrac{1}{30}\cdot 30t$

$3=t$

Check: $\dfrac{70=10(3t-2)}{\begin{array}{c|c} 70 & 10(3\cdot 3-2) \\ & 10(9-2) \\ & 10\cdot 7 \\ & \overset{?}{70=70} \quad \text{TRUE} \end{array}}$

The solution is 3.

71. $1.8(2-n)=9$

$3.6-1.8n=9$

$3.6-1.8n-3.6=9-3.6$

$-1.8n=5.4$

$\left(-\dfrac{1}{1.8}\right)(-1.8n)=\left(-\dfrac{1}{1.8}\right)\cdot 5.4$

$n=-3$

Check: $\dfrac{1.8(2-n)=9}{\begin{array}{c|c} 1.8(2-(-3)) & 9 \\ 1.8(2+3) & \\ 1.8(5) & \\ & \overset{?}{9=9} \quad \text{TRUE} \end{array}}$

The solution is –3.

73. $5y-(2y-10)=25$

$5y-2y+10=25$

$3y+10=25$

$3y+10-10=25-10$

$3y=15$

$\dfrac{1}{3}\cdot 3y=\dfrac{1}{3}\cdot 15$

$y=5$

Check: $\dfrac{5y-(2y-10)=25}{\begin{array}{c|c} 5\cdot 5-(2\cdot 5-10) & 25 \\ 25-(10-10) & \\ 25-0 & \\ & \overset{?}{25=25} \quad \text{TRUE} \end{array}}$

The solution is 5.

75. $\dfrac{9}{10}y-\dfrac{7}{10}=\dfrac{21}{5}$

$\dfrac{9}{10}y-\dfrac{7}{10}+\dfrac{7}{10}=\dfrac{21}{5}+\dfrac{7}{10}$

$\dfrac{9}{10}y=\dfrac{42}{10}+\dfrac{7}{10}$

$\dfrac{9}{10}y=\dfrac{49}{10}$

$\dfrac{10}{9}\cdot\dfrac{9}{10}y=\dfrac{10}{9}\cdot\dfrac{49}{10}$

$y=\dfrac{49}{9}$

Check: $\dfrac{\dfrac{9}{10}y-\dfrac{7}{10}=\dfrac{21}{5}}{\begin{array}{c|c} \dfrac{9}{10}\cdot\dfrac{49}{9}-\dfrac{7}{10} & \dfrac{21}{5} \\ \dfrac{49}{10}-\dfrac{7}{10} & \\ \dfrac{42}{10} & \\ & \overset{?}{\dfrac{21}{5}=\dfrac{21}{5}} \quad \text{TRUE} \end{array}}$

The solution is $\dfrac{49}{9}$.

77. $7r - 2 + 5r = 6r + 6 - 4r$

$$12r - 2 = 2r + 6$$
$$12r - 2 - 2r = 2r + 6 - 2r$$
$$10r - 2 = 6$$
$$10r - 2 + 2 = 6 + 2$$
$$10r = 8$$
$$\frac{1}{10} \cdot 10r = \frac{1}{10} \cdot 8$$
$$r = \frac{8}{10}$$
$$r = \frac{4}{5}$$

Check: $7r - 2 + 5r = 6r + 6 - 4r$

$$\begin{array}{c|c} 7 \cdot \dfrac{4}{5} - 2 + 5 \cdot \dfrac{4}{5} & 6 \cdot \dfrac{4}{5} + 6 - 4 \cdot \dfrac{4}{5} \\[2mm] \dfrac{28}{5} - 2 + 4 & \dfrac{24}{5} + 6 - \dfrac{16}{5} \end{array}$$

$$\overset{?}{\dfrac{38}{5} = \dfrac{38}{5}} \quad \text{TRUE}$$

The solution is $\dfrac{4}{5}$.

79. $\dfrac{2}{3}(x - 2) - 1 = \dfrac{1}{4}(x - 3)$

$$\frac{2}{3}x - \frac{4}{3} - 1 = \frac{1}{4}x - \frac{3}{4}$$
$$\frac{2}{3}x - \frac{7}{3} = \frac{1}{4}x - \frac{3}{4}$$
$$\frac{2}{3}x - \frac{7}{3} - \frac{1}{4}x = \frac{1}{4}x - \frac{3}{4} - \frac{1}{4}x$$
$$\frac{5}{12}x - \frac{7}{3} = -\frac{3}{4}$$
$$\frac{5}{12}x - \frac{7}{3} + \frac{7}{3} = -\frac{3}{4} + \frac{7}{3}$$
$$\frac{5}{12}x = \frac{19}{12}$$
$$\frac{12}{5} \cdot \frac{5}{12}x = \frac{12}{5} \cdot \frac{19}{12}$$
$$x = \frac{19}{5}$$

The check is left to the student. The solution is $\dfrac{19}{5}$.

81. $2(t - 5) - 3(2t - 7) = 12 - 5(3t + 1)$

$$2t - 10 - 6t + 21 = 12 - 15t - 5$$
$$-4t + 11 = -15t + 7$$
$$-4t + 11 + 15t = -15t + 7 + 15t$$
$$11t + 11 = 7$$
$$11t + 11 - 11 = 7 - 11$$
$$11t = -4$$
$$\frac{1}{11} \cdot 11t = \frac{1}{11}(-4)$$
$$t = -\frac{4}{11}$$

Check:

$$2(t - 5) - 3(2t - 7) = 12 - 5(3t + 1)$$

$$\begin{array}{c|c} 2\left(-\dfrac{4}{11} - 5\right) - 3\left(2\left(-\dfrac{4}{11}\right) - 7\right) & 12 - 5\left(3\left(-\dfrac{4}{11}\right) + 1\right) \\[2mm] 2\left(-\dfrac{59}{11}\right) - 3\left(-\dfrac{8}{11} - 7\right) & 12 - 5\left(-\dfrac{12}{11} + 1\right) \\[2mm] -\dfrac{118}{11} - 3\left(-\dfrac{85}{11}\right) & 12 - 5\left(-\dfrac{1}{11}\right) \\[2mm] -\dfrac{118}{11} + \dfrac{255}{11} & 12 + \dfrac{5}{11} \end{array}$$

$$\overset{?}{\dfrac{137}{11} = \dfrac{137}{11}} \quad \text{TRUE}$$

The solution is $-\dfrac{4}{11}$.

83. $3[2 - 4(x - 1)] = 3 - 4(x + 2)$

$$3[2 - 4x + 4] = 3 - 4x - 8$$
$$3[6 - 4x] = -4x - 5$$
$$18 - 12x = -4x - 5$$
$$18 - 12x + 12x = -4x - 5 + 12x$$
$$18 = 8x - 5$$
$$18 + 5 = 8x - 5 + 5$$
$$23 = 8x$$
$$\frac{1}{8} \cdot 23 = \frac{1}{8}(8x)$$
$$\frac{23}{8} = x$$

Check: $3[2 - 4(x - 1)] = 3 - 4(x + 2)$

$$\begin{array}{c|c} 3\left[2 - 4\left(\dfrac{23}{8} - 1\right)\right] & 3 - 4\left(\dfrac{23}{8} + 2\right) \\[2mm] 3\left[2 - 4\left(\dfrac{15}{8}\right)\right] & 3 - 4\left(\dfrac{39}{8}\right) \\[2mm] 3\left[2 - \dfrac{15}{2}\right] & 3 - \dfrac{39}{2} \\[2mm] 3\left(-\dfrac{11}{2}\right) & \dfrac{6}{2} - \dfrac{39}{2} \end{array}$$

$$\overset{?}{-\dfrac{33}{2} = -\dfrac{33}{2}} \quad \text{TRUE}$$

The solution is $\dfrac{23}{8}$.

85. $7x - 2 - 3x = 4x$

$$4x - 2 = 4x$$
$$4x - 2 - 4x = 4x - 4x$$
$$-2 = 0$$

Since the original equation is equivalent to the false equation $-2 = 0$, there is no solution. The solution set is $\varnothing$. The equation is a contradiction.

87.
$$2 + 9x = 3(4x + 1) - 1$$
$$2 + 9x = 12x + 3 - 1$$
$$2 + 9x = 12x + 2$$
$$2 + 9x - 2 = 12x + 2 - 2$$
$$9x = 12x$$
$$9x - 9x = 12x - 9x$$
$$0 = 3x$$
$$\frac{1}{3} \cdot 0 = \frac{1}{3} \cdot 3x$$
$$0 = x$$

The solution set is {0}. The equation is a conditional equation.

89. $-9t + 2 = -9t - 7(6 \div 2(49) + 8)$

Observe that $-7(6 \div 2(49) + 8)$ is a negative number. Then on the left side we have $-9t$ plus a positive number and on the right side we have $-9t$ plus a negative number. This is a contradiction, so the solution set is $\varnothing$.

91.
$$2\{9 - 3[-2x - 4]\} = 12x + 42$$
$$2\{9 + 6x + 12\} = 12x + 42$$
$$2\{21 + 6x\} = 12x + 42$$
$$42 + 12x = 12x + 42$$
$$42 + 12x - 12x = 12x + 42 - 12x$$
$$42 = 42$$

The original equation is equivalent to the equation $42 = 42$, which is true for all real numbers. Thus the solution set is the set of all real numbers. The equation is an identity.

93. *Writing Exercise.* The statement, "The equation has no solution: says that there is no number that makes the equation true. The statement, "The solution of the equation is zero" says that there is a number, the number zero that makes the equation true.

95. Let n represent the number. Then we have
$$2n - 9.$$

97. *Writing Exercise.*
$$3x + 6y + 4x + 2y$$
$$= 3x + 4x + 6y + 2y \qquad \text{Commutative law of addition}$$
$$= (3 + 4)x + (6 + 2)y \qquad \text{Distributive law}$$
$$= 7x + 8y$$

99.
$$-0.00458y + 1.7787 = 13.002y - 1.005$$
$$-13.00658y + 1.7787 = -1.005$$
$$-13.00658y = -2.7837$$
$$y = \frac{-2.7837}{-13.00658}$$
$$0.2140224409 \approx y$$

The check is left to the student. The solution is approximately 0.2140224409.

101.
$$6x - \{5x - [7x - (4x - (3x + 1))]\} = 6x + 5$$
$$6x - \{5x - [7x - (4x - 3x - 1)]\} = 6x + 5$$
$$6x - \{5x - [7x - (x - 1)]\} = 6x + 5$$
$$6x - \{5x - [7x - x + 1]\} = 6x + 5$$
$$6x - \{5x - [6x + 1]\} = 6x + 5$$
$$6x - \{5x - 6x - 1\} = 6x + 5$$
$$6x - \{-x - 1\} = 6x + 5$$
$$6x + x + 1 = 6x + 5$$
$$7x + 1 = 6x + 5$$
$$x = 4$$

103.
$$23 - 2\{4 + 3[x - 1]\} + 5\{x - 2(x + 3)\}$$
$$= 7\{x - 2[5 - (2x + 3)]\}$$
$$23 - 2\{4 + 3x - 3\} + 5\{x - 2x - 6\}$$
$$= 7\{x - 2[5 - 2x - 3]\}$$
$$23 - 2\{3x + 1\} + 5\{-x - 6\} = 7\{x - 2[-2x + 2]\}$$
$$23 - 6x - 2 - 5x - 30 = 7\{x + 4x - 4\}$$
$$-11x - 9 = 7\{5x - 4\}$$
$$-11x - 9 = 35x - 28$$
$$-9 = 46x - 28$$
$$19 = 46x$$
$$\frac{19}{46} = x$$

The check is left to the student. The solution is $\frac{19}{46}$.

105. *Writing Exercise.* Answers may vary. One such equation is $\frac{2}{5}x + \frac{1}{10} = \frac{9}{5}x - \frac{3}{10}$. If we first multiply both sides by 10, we avoid having to add fractions when we use the addition principle.

Connecting the Concepts

1. Expression;
$$3x - 5 - x + 12 = 2x + 7$$

3. Expression;
$$4t - (3t - 1)$$
$$= 4t - 3t + 1$$
$$= t + 1$$

5. Equation;
$$n - 7n - 2n = 3$$
$$-8n = 3$$
$$-\frac{1}{8}(-8n) = -\frac{1}{8} \cdot 3$$
$$n = -\frac{3}{8}$$

7. Expression;

$$8x + 2[x - (x - 1)]$$
$$= 8x + 2[x - x + 1]$$
$$= 8x + 2[1]$$
$$= 8x + 2$$

9. Equation;

$$2x - 6 = 3x + 5$$
$$2x - 6 - 2x = 3x + 5 - 2x$$
$$-6 = x + 5$$
$$-6 - 5 = x + 5 - 5$$
$$-11 = x$$

11. Expression;

$$-(p - 4) - [3 - (9 - 2p)] + p$$
$$= -p + 4 - [3 - 9 + 2p] + p$$
$$= -p + 4 - [-6 + 2p] + p$$
$$= -p + 4 + 6 - 2p + p$$
$$= -2p + 10$$

13. Equation;

$$3(x - 1) - 2(2x + 1) = 5(x - 1)$$
$$3x - 3 - 2x - 2 = 5x - 5$$
$$x - 5 = 5x - 5$$
$$x - 5 - x = 5x - 5 - x$$
$$-5 = 4x - 5$$
$$-5 + 5 = 4x - 5 + 5$$
$$0 = 4x$$
$$\frac{1}{4} \cdot 0 = \frac{1}{4} \cdot 4x$$
$$0 = x$$

15. Equation;

$$4n - (5 - n) - 2 = 3n + n + 8$$
$$4n - 5 + n - 2 = 4n + 8$$
$$5n - 7 = 4n + 8$$
$$5n - 7 - 4n = 4n + 8 - 4n$$
$$n - 7 = 8$$
$$n = 15$$

17. Expression;

$$t - 3t - 2(t + 3 - 2t) + 6 + 4t$$
$$= t - 3t - 2(3 - t) + 6 + 4t$$
$$= t - 3t - 6 + 2t + 6 + 4t$$
$$= 4t$$

19. Equation;

$$x = 2 - \{x - 2[3 - 2(x - 7) + 1]\}$$
$$x = 2 - \{x - 2[3 - 2x + 14 + 1]\}$$
$$x = 2 - \{x - 2[18 - 2x]\}$$
$$x = 2 - \{x - 36 + 4x\}$$
$$x = 2 - \{5x - 36\}$$
$$x = 2 - 5x + 36$$
$$x = 38 - 5x$$
$$x + 5x = 38 - 5x + 5x$$
$$6x = 38$$
$$\frac{1}{6} \cdot 6x = \frac{1}{6} \cdot 38$$
$$x = \frac{19}{3}$$

Exercise Set 1.4

1. *Familiarize*. We want to find two numbers. We can let x represent the first number and note that the second number is 9 more than the first. Also the sum of the numbers is 91.

Translate. The second number can be named $x + 9$. We translate to an equation:

First number　plus　second number　is　91.

$$\begin{array}{ccccc} \downarrow & \downarrow & \downarrow & \downarrow & \downarrow \\ x & + & (x + 9) & = & 91 \end{array}$$

3. *Familiarize*. Let $t =$ the time, in hours, it will take Stella to paddle in 6 mi. We will use the formula Distance = Speed × Time. Stella's speed paddling against the current is 3.5 – 1.9, or 1.6 mph.

Translate. We substitute in the formula

$$6 = 1.6t.$$

5. *Familiarize*. There are three angle measures involved, and we want to find all three. We can let x represent the smallest angle measure and note that the second is one more than x and the third is one more than the second, or two more than x. We also note that the sum of the three angle measures must be 180°.

Translate. The three angle measures are x, $x + 1$, and $x + 2$. We translate to an equation:

First　plus　second　plus　third　is　180°.

$$\begin{array}{ccccccc} \downarrow & \downarrow & \downarrow & \downarrow & \downarrow & \downarrow & \downarrow \\ x & + & (x + 1) & + & (x + 2) & = & 180 \end{array}$$

7. *Familiarize*. Since the escalator's speed is 90 ft/min and Dominik's walking speed is 100 ft/min, Dominik will move at a speed of $90 + 100$, or 190 ft/min on the escalator. Let $t =$ the time, in minutes, it takes to reach the top.

Translate. We will use the formula Distance = Speed × Time.

$$
\begin{array}{ccccc}
\underline{\text{Distance}} & = & \underline{\text{Speed}} & \times & \text{Time} \\
\downarrow & & \downarrow & \downarrow & \downarrow \\
230 & = & 190 & \times & t
\end{array}
$$

9. *Familiarize*. Let w represent the wholesale price, in dollars. Then the wholesale price raised by 50% is $w + 0.5w$, or $1.5w$.

Translate.

$$
\begin{array}{ccccc}
\underline{\substack{\text{Wholesale price} \\ \text{raised 50\%}}} & \text{plus} & \underline{\$1.50} & \text{is} & \underline{\text{selling price.}} \\
\downarrow & & \downarrow & \downarrow & \downarrow \\
1.5w & + & 1.50 & = & 22.50
\end{array}
$$

11. *Familiarize*. Let $t =$ the time, in hours, required for the plane to reach 29,000 ft. Since Distance = Speed × Time, the plane will travel $3500 \times t$ ft in t hr. Note that the plane starts at an altitude of 8000 ft.

Translate.

$$
\begin{array}{ccccc}
\substack{\text{Current} \\ \text{attitude}} & \text{plus} & \underline{\substack{\text{distance} \\ \text{climbed}}} & \text{is} & \underline{\text{29,000 ft.}} \\
\downarrow & & \downarrow & \downarrow & \downarrow \\
8000 & + & 3500t & = & 29{,}000
\end{array}
$$

13. *Familiarize*. Let x represent the measure of the second angle. Then the first angle is four times x, and the third angle is 5° more than twice x. The sum of the three angles is 180°.

Translate. The first angle is $4x$, the second angle is x, and the third angle is $2x + 5$. Translate to an equation:

$$
\begin{array}{ccccccc}
\underline{\text{First}} & \text{plus} & \underline{\text{second}} & \text{plus} & \underline{\text{third}} & \text{is} & 180°. \\
\downarrow & & \downarrow & & \downarrow & & \downarrow \\
4x & + & x & + & (2x+5) & = & 180
\end{array}
$$

15. *Familiarize*. Note that each odd integer is 2 more than the one preceding it. If we let n represent the first odd integer then $n + 2$ represents the next odd integer and $(n + 2) + 2$, or $n + 4$ is the third odd integer.

Translate.

$$
\begin{array}{ccccccc}
\underline{\text{First}} & \text{plus} & \underline{\substack{\text{twice the} \\ \text{second}}} & \text{plus} & \underline{\substack{\text{three times} \\ \text{the third}}} & \text{is} & 70. \\
\downarrow & & \downarrow & & \downarrow & & \downarrow\ \ \downarrow \\
n & + & 2(n+2) & + & 3(n+4) & = & 70
\end{array}
$$

17. *Familiarize*. The perimeter of an equilateral triangle is 3 times the length of a side. Let $s =$ the length of a side of the smaller triangle. Then $2s =$ the length of a side of the larger triangle. The sum of the two perimeters is 90 cm.

Translate.

$$
\begin{array}{ccccc}
\underline{\substack{\text{Perimeter of} \\ \text{smaller triangle}}} & \text{plus} & \underline{\substack{\text{perimeter of} \\ \text{larger triangle}}} & \text{is} & \underline{\text{90 cm.}} \\
\downarrow & & \downarrow & \downarrow & \downarrow \\
3s & + & 3 \cdot 2s & = & 90
\end{array}
$$

19. *Familiarize*. Let c represent the number of calls Cody will need on his next shift if he is to average 3 calls per shift. We find the average by adding the number of calls on each of the 5 shifts and then dividing by the number of addends.

Translate.

$$
\begin{array}{ccc}
\underline{\text{Average number of calls per shift}} & \text{is} & 3. \\
\downarrow & \downarrow & \downarrow \\
\dfrac{5+2+1+3+c}{5} & = & 3
\end{array}
$$

21. *Familiarize*. Let p represent the price Tony paid for his graphing calculator.

Translate.

$$
\begin{array}{ccccc}
\underline{\substack{\text{Price Tess} \\ \text{paid}}} & \text{is} & \underline{\substack{\text{price Tony} \\ \text{paid}}} & \text{less} & \$13. \\
\downarrow & \downarrow & \downarrow & \downarrow & \downarrow \\
84 & = & p & - & 13
\end{array}
$$

Carry out. We solve the equation.

$$84 = p - 13$$
$$97 = p \qquad \text{Adding 13 to both sides}$$

Check. The price Tess paid, \$84, is \$13 less than \$97, so the answer checks.

State. Tony paid \$97 for his graphing calculator.

23. *Familiarize*. Let d represent the cost of daycare in Billings, in dollars.

Translate.

$$
\begin{array}{ccccc}
\underline{\text{Cost in Boston}} & \text{is} & \dfrac{11}{4} & \text{of} & \underline{\text{cost in Billings}} \\
\downarrow & \downarrow & \downarrow & & \downarrow \\
1089 & = & \dfrac{11}{4} & \cdot & d
\end{array}
$$

Carry out. We solve the equation.

$$1089 = \frac{11}{4}d$$

$$\frac{4}{11} \cdot 1089 = \frac{4}{11} \cdot \frac{11}{4}d$$

$$396 = d$$

Check. Since $\frac{11}{4}$ of $396 is $1089, the answer checks.

State. The cost of child care in Billings is $396.

25. *Familiarize*. Let p represent the number of pictures taken by Robbin. Then $p + 8$ represents the number of pictures taken by Michelle.

Translate.

Total number of pictures	is	40.
↓	↓	↓
$p + (p + 8)$	$=$	40

Carry out. We solve the equation.

$$p + p + 8 = 40$$
$$2p + 8 = 40$$
$$2p = 32 \qquad \text{Subtracting 8 from both sides}$$
$$p = 16$$

If $p = 16$, then $p + 8 = 16 + 8$, or 24.

Check. 16 is 8 less than 24. Also, $16 + 24 = 40$ pictures. The answer checks.

State. Robbin photographed 16 seniors.

27. *Familiarize*. Let w represent the width of the mirror, in cm. Then $3w$ represents the length. Recall that the formula for the perimeter P of a rectangle with length l and width w is $P = 2l + 2w$.

Translate.

Perimeter	is	120 cm.
↓	↓	↓
$2 \cdot 3w + 2 \cdot w$	$=$	120

Carry out. We solve the equation.

$$2 \cdot 3w + 2 \cdot w = 120$$
$$6w + 2w = 120$$
$$8w = 120$$
$$w = 15$$

When $w = 15$, then $3w = 3 \cdot 15 = 45$.

Check. If the length is 45 cm and the width is 15 cm, then the length is three times the width. Also

$P = 2 \cdot 45 + 2 \cdot 15 = 90 + 30 = 120$ cm. The answer checks.

State. The length of the mirror is 45 cm, and the width is 15 cm.

29. *Familiarize*. Let l represent the length of the greenhouse, in meters. Then $\frac{1}{4}l$ represents the width. Recall that the formula for the perimeter P, of a rectangle with length l and width w is $P = 2l + 2w$.

Translate.

Perimeter	is	130 m.
↓	↓	↓
$2l + 2 \cdot \frac{1}{4}l$	$=$	130

Carry out. We solve the equation.

$$2l + 2 \cdot \frac{1}{4}l = 130$$
$$2l + \frac{1}{2}l = 130$$
$$\frac{5}{2}l = 130$$
$$\frac{2}{5} \cdot \frac{5}{2}l = \frac{2}{5} \cdot 130$$
$$l = 52$$

When $l = 52$, then $\frac{1}{4}l = 13$.

Check. If the length is 52 m and the width is 13 m, then the width is $\frac{1}{4}$ of the length. Also,

$P = 2 \cdot 52 + 2 \cdot 13 = 104 + 26 = 130$ m. The answer checks.

State. The length of the greenhouse is 52 m, and the width is 13 m.

31. The Familiarize and Translate steps were done in Exercise 3.

Carry out. We solve the equation.

$$1.6t = 6$$
$$t = \frac{1}{1.6}(6)$$
$$t = 3.75$$

Check. At a speed of 1.6 mph, in 3.75 hr, Stella paddles 1.6(3.75), or 6 mi. Our answer checks.

State. Stella takes 3.75 hr to paddle 6 mi.

33. The Familiarize and Translate steps were done in Exercise 13.

Carry out. We solve the equation.

$$4x + x + (2x + 5) = 180$$
$$7x + 5 = 180$$
$$7x = 180 - 5$$
$$7x = 175$$
$$\frac{1}{7} \cdot 7x = \frac{1}{7} \cdot 175$$
$$x = 25$$

When $x = 25$, then $4x = 4 \cdot 25 = 100$, and

$2x + 5 = 2 \cdot 25 + 5 = 55$.

Check. One angle, 100°, is 4 times the 25° angle, and the

third angle, 55°, is 5 more than twice the 25° angle. Also $100° + 25° + 55° = 180°$. The answer checks.

State. The measures of the angles are 100°, 25°, and 55°.

35. The Familiarize and Translate steps were done in Exercise 10.

Carry out. We solve the equation.

$$c - 0.05c = 142.50$$
$$0.95c = 142.50$$
$$\frac{0.95c}{0.95} = \frac{142.50}{0.95}$$
$$c = 150$$

Check. 5% of $150 is $7.50 and $150 - $7.50 = $142.50. The answer checks.

State. The cost would have been $150 if the bill had not been paid promptly.

37. The Familiarize and Translate steps were done in Exercise 9.

Carry out. We solve the equation.

$$1.5w + 1.50 = 22.50$$
$$1.5w = 21$$
$$w = \frac{1}{1.5} \cdot 21$$
$$w = 14$$

Check. If a wholesale price of $14 is raised by 50%, we have $14 + 0.5($14) = $14 + $7 = $21. When $1.50 is added to this figure, we have $21 + $1.50 = $22.50. The answer checks.

State. The wholesale price is $14.

39. *Writing Exercise.* Answers may vary. One possibility is: Jason has $\frac{1}{3}$ of a pizza. He offers half to Beth. What part of the pizza did Beth receive?

41. $30 = 5(6x)$
$30 = 30x$
$1 = x$
The solution is 1.

43. $$3 = \frac{x}{-2}$$
$$-2(3) = (-2)\left(\frac{x}{-2}\right)$$
$$-6 = x$$
The solution is –6.

45. *Writing Exercise.* The manner in which a guess or estimate is manipulated can give insight into the form of the equation to which the problem will be translated.

47. *Familiarize.* The average score on the first four tests is $\frac{83 + 91 + 78 + 81}{4}$, or 83.25. Let $x =$ the number of points above this average that Tico scores on the next test. Then the score on the fifth test is $83.25 + x$.

Translate.

Average score on 5 tests	is	2	more than	average score on 4 tests.
$\frac{83+91+78+81+(83.25+x)}{5}$	$=$	2	$+$	83.25

Carry out. Carry out some algebraic manipulation.

$$\frac{83 + 91 + 78 + 81 + (83.25 + x)}{5} = 2 + 83.25$$
$$\frac{416.25 + x}{5} = 85.25$$
$$416.25 + x = 426.25$$
$$x = 10$$

Check. If Tico scores 10 points more than the average of the first four tests on the fifth test, his score will be $83.25 + 10$, or 93.25. Then the five-test average will be $\frac{83 + 91 + 78 + 81 + 93.25}{5}$, or 85.25. This is 2 points above the four-test average, so the answer checks.

State. Tico must score 10 points above the four-test average in order to raise the average 2 points.

49. *Familiarize.* Let $p =$ the price of the house in 2004. From 2004 to 2005 real estate prices increased 6%, so the house was worth $p + 0.06p$, or $1.06p$. From 2006 to 2008 prices increased 2%, so the house was then worth $1.06p + 0.02(1.06p)$, or $1.02(1.06p)$. From 2006 to 2008 prices dropped 1%, so the value of the house became $1.02(1.06p) - 0.01(1.02)(1.06p)$, or $0.99(1.02)(1.06p)$.

Translate.

The price of the house in 2004	was	$117,743.
$0.99(1.02)(1.06p)$	$=$	117,743

Carry out. We solve the equation.

$$0.99(1.02)(1.06p) = 117,743$$
$$p = \frac{117,743}{0.99(1.02)(1.06)}$$
$$p \approx 110,000$$

Check. If the price of the house in 2004 was $110,000, then in 2005 it was worth 1.06($110,000), or $116,600. In 2006 it was worth 1.02($116,600), or $118,932, and in 2008 it was worth 0.99($118,932), or $117,734. Our answer checks.

State. The house was worth \$110,000 in 2004.

Exercise Set 1.5

1. A formula is an <u>equation</u> that uses letters to represent a relationship between two or more quantities.

3. The formula $C = \pi d$ is used to calculate the <u>circumference</u> of a circle.

5. The formula <u>$A = bh$</u> is used to calculate the area of a parallelogram of height h and base length b.

7. In the formula for the area of a trapezoid,

$A = \dfrac{h}{2}\left(b_1 + b_2\right)$ the numbers 1 and 2 are referred to as <u>subscripts</u>.

9.
$$E = wA$$
$$\frac{1}{w} \cdot E = \frac{1}{w} \cdot wA \quad \text{Multiplying both sides by } \frac{1}{w}$$
$$\frac{E}{w} = A \quad \text{Simplifying}$$

11.
$$d = rt$$
$$\frac{1}{t} \cdot d = \frac{1}{t} \cdot rt \quad \text{Multiplying both sides by } \frac{1}{t}$$
$$\frac{d}{t} = r \quad \text{Simplifying}$$

13.
$$V = lwh$$
$$\frac{1}{lw} \cdot V = \frac{1}{lw} \cdot lwh \quad \text{Multiplying both sides by } \frac{1}{lw}$$
$$\frac{V}{lw} = h \quad \text{Simplifying}$$

15.
$$L = \frac{k}{d^2}$$
$$d^2 \cdot L = d^2 \cdot \frac{k}{d^2} \quad \text{Multiplying both sides by } d^2$$
$$d^2 L = k \quad \text{Simplifying}$$

17.
$$G = w + 150n$$
$$G - w = 150n \quad \text{Subtracting } w \text{ from both sides}$$
$$\frac{1}{150}(G - w) = \frac{1}{150} \cdot 150n \quad \text{Multiplying both sides by } \frac{1}{150}$$
$$\frac{G - w}{150} = n \quad \text{Simplifying}$$

19. $2w + 2h + l = p$
$$l = p - 2w - 2h \quad \text{Adding } -2w - 2h \text{ to both sides}$$

21. $2x + 3y = 4$
$$3y = 4 - 2x \quad \text{Subtracting } 2x \text{ from both sides}$$
$$\frac{1}{3} \cdot 3y = \frac{1}{3}(4 - 2x) \quad \text{Multiplying both sides by } \frac{1}{3}$$
$$y = \frac{4 - 2x}{3} \quad \text{Simplifying}$$

23. $Ax + By = C$
$$By = C - Ax \quad \text{Subtracting } Ax \text{ from both sides}$$
$$\frac{1}{B} \cdot By = \frac{1}{B}(C - Ax) \quad \text{Multiplying both sides by } \frac{1}{B}$$
$$y = \frac{C - Ax}{B} \quad \text{Simplifying}$$

25.
$$C = \frac{5}{9}(F - 32)$$
$$\frac{9}{5} \cdot C = \frac{9}{5} \cdot \frac{5}{9}(F - 32) \quad \text{Multiplying both sides by } \frac{9}{5}$$
$$\frac{9}{5}C = F - 32 \quad \text{Simplifying}$$
$$\frac{9}{5}C + 32 = F$$

27.
$$V = \frac{4}{3}\pi r^3$$
$$\frac{3}{4\pi} \cdot V = \frac{3}{4\pi} \cdot \frac{4}{3}\pi r^3 \quad \text{Multiplying both sides by } \frac{3}{4\pi}$$
$$\frac{3V}{4\pi} = r^3 \quad \text{Simplifying}$$

29. $np + nm = t$
$$n(p + m) = t \quad \text{Factoring}$$
$$n = \frac{t}{p + m} \quad \text{Dividing both sides by } p + m$$

31. $uv + wv = x$
$$v(u + w) = x \quad \text{Factoring}$$
$$v = \frac{x}{u + w} \quad \text{Dividing both sides by } u + w$$

33.
$$A = \frac{q_1 + q_2 + q_3}{n}$$
$$n \cdot A = n \cdot \frac{q_1 + q_2 + q_3}{n} \quad \text{Clearing the fraction}$$
$$nA = q_1 + q_2 + q_3$$
$$nA \cdot \frac{1}{A} = (q_1 + q_2 + q_3) \cdot \frac{1}{A} \quad \text{Multiplying both sides by } \frac{1}{A}$$
$$n = \frac{q_1 + q_2 + q_3}{A}$$

35.
$$v = \frac{d_2 - d_1}{t}$$
$$t \cdot v = t \cdot \frac{d_2 - d_1}{t} \quad \text{Clearing the fraction}$$
$$tv = d_2 - d_1$$
$$tv \cdot \frac{1}{v} = (d_2 - d_1) \cdot \frac{1}{v} \quad \text{Multiplying both sides by } \frac{1}{v}$$
$$t = \frac{d_2 - d_1}{v}$$

37.
$$v = \frac{d_2 - d_1}{t}$$
$$t \cdot v = t \cdot \frac{d_2 - d_1}{t} \quad \text{Clearing the fraction}$$
$$tv = d_2 - d_1$$
$$tv - d_2 = -d_1 \quad \begin{array}{l}\text{Subtracting } d_2 \text{ from}\\ \text{both sides}\end{array}$$
$$-1 \cdot (tv - d_2) = -1(-d_1) \quad \begin{array}{l}\text{Multiplying both}\\ \text{sides by } -1\end{array}$$
$$-tv + d_2 = d_1,$$
$$\text{or } d_2 - tv = d_1$$

39.
$$bd = c + ba$$
$$bd - ba = c \quad \text{Adding } -ba \text{ to both sides}$$
$$b(d - a) = c \quad \text{Factoring}$$
$$b = \frac{c}{d - a} \quad \text{Dividing both sides by } d - a$$

41.
$$v - w = uvw$$
$$v = uvw + w \quad \text{Adding } w \text{ to both sides}$$
$$v = w(uv + 1) \quad \text{Factoring}$$
$$\frac{v}{uv + 1} = w \quad \text{Dividing both sides by } uv + 1$$

43.
$$n - mk = mt^2$$
$$n = mt^2 + mk \quad \text{Adding } mk \text{ to both sides}$$
$$n = m(t^2 + k) \quad \text{Factoring}$$
$$\frac{n}{t^2 + k} = m \quad \text{Dividing both sides by } t^2 + k$$

45. *Familiarize*. In Example 2, we find the formula for simple interest, $I = Prt$, when I is the interest, P is the principal, r is the interest rate, and t is the time, in years.

***Translate*.** We want to find the interest rate, so we solve the formula for r.
$$I = Prt$$
$$\frac{1}{Pt} \cdot I = \frac{1}{Pt} \cdot Prt$$
$$\frac{I}{Pt} = r$$

***Carry out*.** The model $r = \frac{I}{Pt}$ can be used to find the rate of interest at which an amount (the principal) must be invested in order to earn a given amount. We substitute $2600 for P, $\frac{1}{2}$ for t (6 months $= \frac{1}{2}$ yr), and $104 for I.

$$\frac{I}{Pt} = r$$
$$\frac{\$104}{\$2600\left(\frac{1}{2}\right)} = r$$
$$\frac{104}{1300} = r$$
$$0.08 = r$$
$$8\% = r$$

***Check*.** Since $\$2600(0.08)\left(\frac{1}{2}\right) = \$104$, the answer checks.

***State*.** The interest rate must be 8%.

47. *Familiarize*. On page 45 of the text we find the formula for the area of a parallelogram, $A = bh$, where b is the base and h is the height.

***Translate*.** We solve the formula for h.
$$A = bh$$
$$\frac{1}{b} \cdot A = \frac{1}{b} \cdot bh$$
$$\frac{A}{b} = h$$

***Carry out*.** The model $h = \frac{A}{b}$ can be used to find the height of any parallelogram for which the area and base are known. We substitute 96 for A and 6 for b.
$$h = \frac{A}{b}$$
$$h = \frac{96}{6}$$
$$h = 16$$

***Check*.** We repeat the calculation. The answer checks.

***State*.** The height is 16 cm.

49. *Familiarize and Translate*. In Example 5, the formula for body mass index is solved for W:
$$W = \frac{IH^2}{704.5}$$

***Carry out*.** We substitute 30.2 for I and 74 for H (6 ft 2 in. is 74 in.) and calculate W.
$$W = \frac{30.2(74)^2}{704.5} \approx 235$$

***Check*.** We could repeat the calculations or substitute in the original formula and then solve for W. The answer checks.

***State*.** Arnold Schwarzenegger weighs about 235 lb.

51. *Familiarize and Translate*. We will use the model developed in Example 6, $m = \pi r^2 hD$ to find the weight of the salt. Then we will add 28 g, the weight of the empty canister, to find the weight of the filled canister.

***Carry out*.** We substitute 4 for r, 13.6 for h, and 2.16 for

D and calculate m.

$$m = \pi r^2 hD$$
$$= \pi (4)^2 (13.6)(2.16)$$
$$\approx 1476.6$$

Add the weight of the empty canister:

$$1476.6 + 28 = 1504.6$$

Check. We repeat the calculations. The answer checks.

State. The filled canister weighs about 1504.6 g.

53. Familiarize. The formula for the area of a trapezoid is $A = \frac{1}{2}h(b_1 + b_2)$, where A is the area, h is the height, and b_1 and b_2 are the bases.

Translate. The unknown dimension is the height, so we solve the formula for h. We have $h = \dfrac{2A}{b_1 + b_2}$.

Carry out. We substitute.

$$h = \frac{2A}{b_1 + b_2}$$
$$h = \frac{2 \cdot 90}{8 + 12}$$
$$h = \frac{180}{20}$$
$$h = 9$$

Check. We repeat the calculation. The answer checks.

State. The unknown dimension, the height of the trapezoid, is 9 ft.

55. Observe that 4% of $1000 is $40, so $40 is the amount of simple interest that would be earned in 1 yr. Thus, it will take 1 yr for the investment to be worth $1040.

57. Familiarize. We use the formula given in the text,

$$E = w \cdot A.$$

Translate. The unknown quantity is the estimated blood volume E.

Carry out. We substitute 10.2 for w and 80 for A and calculate E.

$$E = 10.2(80)$$
$$E = 816$$

Check. We repeat the calculations. The answer checks.

State. The estimated blood volume is 816 ml.

59. Familiarize. We will use the formula given in the text,

$$R = r + \frac{400(W - L)}{N}.$$

Translate. We solve the formula for r.

$$R = r + \frac{400(W - L)}{N}$$
$$R - \frac{400(W - L)}{N} = r$$

Carry out. We substitute 1305 for R, 5 for w, 3 for L, and 5 + 3, or 8, for N and calculate r.

$$1305 - \frac{400(5 - 3)}{8} = r$$
$$1205 = r$$

Check. We can repeat the calculation or substitute in the original formula and then solve for r. The answer checks.

State. The average rating of Ulana's opponents was 1205.

61. Familiarize. We will use the formula given in the text, $K = 917 + 6(w + h - a)$.

Translate. We solve the formula for h.

$$K = 917 + 6(w + h - a)$$
$$K = 917 + 6w + 6h - 6a$$
$$K - 917 - 6w + 6a = 6h$$
$$\frac{K - 917 - 6w + 6a}{6} = h$$

Carry out. We substitute 1901 for K, 120 for w, and 23 for a and calculate h.

$$\frac{1901 - 917 - 6 \cdot 120 + 6 \cdot 23}{6} = h$$
$$67 = h$$

Check. We can repeat the calculation or substitute in the original formula and then solve for h. The answer checks.

State. Julie is 67 in., or 5 ft 7 in., tall.

63. Familiarize. We use the formula given in the text.

$$m = 0.32810w + 0.33929h - 29.5336$$

Translate. We solve the formula for w.

$$m = 0.32810w + 0.33929h - 29.5336$$
$$w = \frac{m - 0.33929h + 29.5336}{0.32810}$$

Carry out. We substitute 62 for m and 185 for h and calculate w.

$$w = \frac{62 - 0.33929 \cdot 185 + 29.5336}{0.32810}$$
$$w \approx 88$$

Check. We can repeat the calculations or substitute in the original formula and solve for w. The answer checks.

State. Marv has total body mass of about 88 kg.

65. Familiarize. We use the formula given in the text.

$$r = \frac{tmap}{hs}$$

Translate. We solve the formula for t.

$$r = \frac{tmap}{hs}$$
$$t = \frac{rhs}{map}$$

Carry out. We substitute 30 for s, 4 for h, 0.05 for m, 100 for a, 0.15 for p, 3.2 for r and solve for t.

$$t = \frac{3.2 \cdot 4 \cdot 30}{0.05 \cdot 100 \cdot 0.15}$$
$$t = 512$$

Check. We repeat the calculation or substitute in the original formula and solve for t. The answer checks.

State. The average daily blog traffic is 512 visits per day.

67. *Familiarize*. We will use Goiten's model,

$I = 1.08(T/N)$. Note that 8 hr = 8 × 1 hr = 8 × 60 min = 480 min.

Translate. We solve the formula for N.

$$I = 1.08\left(\frac{T}{N}\right)$$
$$N \cdot I = N(1.08)\left(\frac{T}{N}\right)$$
$$NI = 1.08T$$
$$N = \frac{1.08T}{I}$$

Carry out. We substitute 480 for T and 15 for I.

$$N = \frac{1.08T}{I}$$
$$N = \frac{1.08(480)}{15}$$
$$N = 34.56$$
$$N \approx 34 \quad \text{Rounding down}$$

Check. We repeat the calculations. The answer checks.

State. Dr. Cruz should schedule 34 appointments in one day.

69. *Familiarize*. We will use Thurnau's model,

$P = 9.337da - 299$.

Translate. Since we want to find the diameter of the fetus' head, we solve for d.

$$P = 9.337da - 299$$
$$P + 299 = 9.337da$$
$$\frac{P + 299}{9.337a} = d$$

Carry out. Substitute 1614 for P and 24.1 for a in the formula and calculate:

$$\frac{1614 + 299}{9.337(24.1)} = d$$
$$8.5 \approx d$$

Check. We repeat the calculation. The answer checks.

State. The diameter of the fetus' head at 29 weeks is about 8.5 cm.

71. *Writing Exercise*. No; a trapezoid has *exactly* one pair of parallel sides, but a rectangle has two pairs of parallel sides.

73. $2(c-1) - 5[3 - (c - 5)]$
$= 2c - 2 - 5[3 - c + 5]$
$= 2c - 2 - 5[8 - c]$
$= 2c - 2 - 40 + 5c$
$= 7c - 42$

75. *Writing Exercise*. Given equal volumes of cork and steel, it is reasonable to assume the steel would have greater mass. Since $\text{Density} = \dfrac{\text{mass}}{\text{Volume}}$, the steel would be expected to have greater density.

77. *Familiarize*. First we find the volume of the ring. Note that the inner diameter is 2 cm, so the inner radius is 2/2 or 1 cm. Then the volume of the ring is the volume of a right circular cylinder with height 0.5 cm and radius $1 + 0.15$, or 1.15 cm, less the volume of a right circular cylinder with height 0.5 cm and radius 1 cm. Recall that the formula for the volume of a right circular cylinder is $V = \pi r^2 h$. Then the volume of the ring is

$\pi(1.15)^2(0.5) - \pi(1)^2(0.5) = 0.16125\pi$ cm^3.

Translate. To find the weight of the ring we will use the formula $D = \dfrac{m}{V}$. Solving for m, we get

$$D = \frac{m}{V}$$
$$V \cdot D = V \cdot \frac{m}{V}$$
$$V \cdot D = m$$

Carry out. We substitute in the formula $m = V \cdot D$

$$m = 0.16125\pi(21.5)$$
$$m \approx 10.9$$

Check. We repeat the calculations. The answer checks.

State The ring will weigh about 10.9 g.

79. *Familiarize*. We use the formula given in the text.

$r = 0.3b - 0.6c$

Translate. We solve the formula for c.

$$r = 0.3b - 0.6c$$
$$r - 0.3b = -0.6c$$
$$\frac{r - 0.3b}{-0.6} = c$$

Carry out. We substitute -2.7 for r and 11 for b and calculate c.

$$c = \frac{-2.7 - 0.3 \cdot 11}{-0.6}$$
$$c = 10$$

Check. We repeat the calculations. The answer checks.

State. Alex Sanchez was caught stealing 10 times.

81.
$$s = v_i t + \frac{1}{2} a t^2$$
$$s - v_i t = \frac{1}{2} a t^2$$
$$2\left(s - v_i t\right) = a t^2$$
$$\frac{2\left(s - v_i t\right)}{t^2} = a, \text{ or}$$
$$\frac{2s - 2v_i t}{t^2} = a$$

83.
$$b = \frac{h + w + p}{a + w + p + f}$$
$$b\left(a + w + p + f\right) = h + w + p \quad \text{Multiplying both sides by } a + w + p + f$$
$$ab + bw + bp + bf = h + w + p$$
$$bw - w = h + p - ab - bp - bf$$
$$w\left(b - 1\right) = h + p - ab - bp - bf$$
$$w = \frac{h + p - ab - bp - bf}{b - 1} \quad \text{Multiplying both sides by } \frac{1}{b - 1}$$

85.
$$\frac{b}{a - b} = c$$
$$b = c\left(a - b\right)$$
$$b + bc = ac$$
$$b\left(1 + c\right) = ac$$
$$b = \frac{ac}{1 + c}$$

87. $s + \dfrac{s + t}{s - t} = \dfrac{1}{t} + \dfrac{s + t}{s - t}$

Observe that if we subtract $\dfrac{s + t}{s - t}$ from both sides we are left with an equivalent equation, $s = \dfrac{1}{t}$. We solve this equation for t.

$$s = \frac{1}{t}$$
$$st = 1 \quad \text{Multiplying both sides by } t$$
$$t = \frac{1}{s} \quad \text{Multiplying both sides by } \frac{1}{s}$$

89. *Writing Exercise*. The trapezoids, positioned as shown, form a parallelogram with base $b_1 + b_2$ and height h. The area of a parallelogram is base × height, so the area of the given parallelogram is $\left(b_1 + b_2\right)h$, or $h\left(b_1 + b_2\right)$. Since the area of each trapezoid is one-half the area of the parallelogram, the area of a trapezoid is $\frac{1}{2} \cdot h\left(b_1 + b_2\right)$, or $\frac{h}{2}\left(b_1 + b_2\right)$.

Exercise Set 1.6

1. The power rule

3. Raising a product to a power

5. The product rule

7. Raising a quotient to a power

9. The quotient rule

11. $6^4 \cdot 6^7 = 6^{4+7} = 6^{11}$

13. $m^0 \cdot m^8 = m^{0+8} = m^8$

15. $5x^4 \cdot 4x^3 = 5 \cdot 4 \cdot x^4 \cdot x^3 = 20x^{4+3} = 20x^7$

17. $\left(-3a^2\right)\left(-8a^6\right) = (-3)(-8)\, a^2 \cdot a^6 = 24a^{2+6} = 24a^8$

19. $\left(m^5 n^2\right)\left(m^3 n p^0\right) = \left(m^5 m^3\right)\left(n^2 n\right)\left(p^0\right) = m^{5+3} n^{2+1} \cdot 1$
$$= m^8 n^3$$

21. $\dfrac{t^8}{t^3} = t^{8-3} = t^5$

23. $\dfrac{15a^7}{3a^2} = \dfrac{15}{3} a^{7-2} = 5a^5$

25. $\dfrac{m^7 n^9}{m^2 n^5} = m^{7-2} \cdot n^{9-5} = m^5 n^4$

27. $\dfrac{32x^8 y^5}{8x^2 y} = \dfrac{32}{8} \cdot x^{8-2} \cdot y^{5-1} = 4x^6 y^4$

29. $\dfrac{28x^{10} y^9 z^8}{-7x^2 y^3 z^2} = \dfrac{28}{-7} \cdot x^{10-2} \cdot y^{9-3} \cdot z^{8-2} = -4x^8 y^6 z^6$

31. $-x^0 = -(-2)^0 = -(1) = -1$

33. $\left(4x\right)^0 = \left(4(-2)\right)^0 = (-8)^0 = 1$

35. $t^{-9} = \dfrac{1}{t^9}$

37. $6^{-2} = \dfrac{1}{6^2} = \dfrac{1}{36}$

39. $(-3)^{-2} = \dfrac{1}{(-3)^2} = \dfrac{1}{9}$

41. $-3^{-2} = -\dfrac{1}{3^2} = -\dfrac{1}{9}$

43. $-1^{-10} = -\dfrac{1}{1^{10}} = -1$

45. $\dfrac{1}{10^{-3}} = 10^3 = 1000$

47. $6x^{-1} = 6 \cdot \dfrac{1}{x} = \dfrac{6}{x}$

49. $3a^8 b^{-6} = 3a^8 \cdot \dfrac{1}{b^6} = \dfrac{3a^8}{b^6}$

51. $\dfrac{2z^{-3}}{x^5} = \dfrac{2}{x^5} \cdot \dfrac{1}{z^3} = \dfrac{2}{x^5 z^3}$

53. $\dfrac{3y^2}{z^{-4}} = 3y^2 \cdot z^4 = 3y^2 z^4$

55. $\dfrac{ab^{-1}}{c^{-1}} = a \cdot \dfrac{1}{b} \cdot c = \dfrac{ac}{b}$

57. $\dfrac{pq^{-2}r^{-3}}{2u^5 v^{-4}} = \dfrac{p}{2u^5} \cdot \dfrac{1}{q^2} \cdot \dfrac{1}{r^3} \cdot v^4 = \dfrac{pv^4}{2q^2 r^3 u^5}$

59. $\dfrac{1}{x^3} = x^{-3}$

61. $\dfrac{1}{(-10)^3} = (-10)^{-3}$

63. $8^{10} = \dfrac{1}{8^{-10}}$

65. $4x^2 = 4 \cdot \dfrac{1}{x^{-2}} = \dfrac{4}{x^{-2}}$

67. $\dfrac{1}{(5y)^3} = (5y)^{-3}$

69. $\dfrac{1}{3y^4} = \dfrac{1}{3} \cdot \dfrac{1}{y^4} = \dfrac{1}{3} \cdot y^{-4} = \dfrac{y^{-4}}{3}$

71. $6^{-3} \cdot 6^{-5} = 6^{-3+(-5)} = 6^{-8}$, or $\dfrac{1}{6^8}$

73. $a \cdot a^{-8} = a^{1+(-8)} = a^{-7}$, or $\dfrac{1}{a^7}$

75. $x^{-7} \cdot x^2 \cdot x^5 = x^{-7+2+5} = x^0 = 1$

77. $(4mn^3)(-2m^3 n^2) = 4(-2) \cdot m \cdot m^3 \cdot n^3 \cdot n^2$
$ = -8m^{1+3} n^{3+2}$
$ = -8m^4 n^5$

79. $(-7x^4 y^{-5})(-5x^{-6} y^8) = (-7)(-5) \cdot x^4 \cdot x^{-6} \cdot y^{-5} \cdot y^8$
$\phantom{(-7x^4 y^{-5})(-5x^{-6} y^8)} = 35x^{4+(-6)} y^{-5+8}$
$\phantom{(-7x^4 y^{-5})(-5x^{-6} y^8)} = 35x^{-2} y^3$, or $\dfrac{35y^3}{x^2}$

81. $(5a^{-2} b^{-3})(2a^{-4} b) = 5 \cdot 2 \cdot a^{-2} \cdot a^{-4} \cdot b^{-3} \cdot b$
$\phantom{(5a^{-2} b^{-3})(2a^{-4} b)} = 10a^{-2+(-4)} b^{-3+1}$
$\phantom{(5a^{-2} b^{-3})(2a^{-4} b)} = 10a^{-6} b^{-2}$, or $\dfrac{10}{a^6 b^2}$

83. $\dfrac{10^{-3}}{10^6} = 10^{-3-6} = 10^{-9}$, or $\dfrac{1}{10^9}$

85. $\dfrac{2^{-7}}{2^{-5}} = 2^{-7-(-5)} = 2^{-7+5} = 2^{-2}$, or $\dfrac{1}{2^2}$, or $\dfrac{1}{4}$

87. $\dfrac{y^4}{y^{-5}} = y^{4-(-5)} = y^{4+5} = y^9$

89. $\dfrac{24a^5 b^3}{-8a^4 b} = \dfrac{24}{-8} a^{5-4} b^{3-1} = -3ab^2$

91. $\dfrac{15m^5 n^3}{10m^{10} n^{-4}} = \dfrac{15}{10} m^{5-10} n^{3-(-4)} = \dfrac{3}{2} m^{-5} n^7$, or $\dfrac{3n^7}{2m^5}$

93. $\dfrac{-6x^{-2} y^4 z^8}{-24x^{-5} y^6 z^{-3}} = \dfrac{-6}{-24} x^{-2-(-5)} y^{4-6} z^{8-(-3)}$
$\phantom{\dfrac{-6x^{-2} y^4 z^8}{-24x^{-5} y^6 z^{-3}}} = \dfrac{1}{4} x^3 y^{-2} z^{11}$, or $\dfrac{x^3 z^{11}}{4y^2}$

95. $(x^4)^3 = x^{4 \cdot 3} = x^{12}$

97. $(9^3)^{-4} = 9^{3(-4)} = 9^{-12}$, or $\dfrac{1}{9^{12}}$

99. $(t^{-8})^{-5} = t^{-8(-5)} = t^{40}$

101. $(-5xy)^2 = (-5)^2 x^2 y^2 = 25x^2 y^2$

103. $(-2a^{-2} b)^{-3} = (-2)^{-3} \cdot (a^{-2})^{-3} \cdot b^{-3} = (-2)^{-3} a^6 b^{-3}$
$\phantom{(-2a^{-2} b)^{-3}} = \dfrac{1}{(-2)^3} \cdot a^6 \cdot \dfrac{1}{b^3} = -\dfrac{a^6}{8b^3}$

105. $\left(\dfrac{m^2 n^{-1}}{4}\right)^3 = \dfrac{m^6 n^{-3}}{4^3} = \dfrac{m^6 n^{-3}}{64}$, or $\dfrac{m^6}{64n^3}$

107. $\dfrac{(2a^3)^3 4a^{-3}}{(a^2)^5} = \dfrac{2^3 a^{3 \cdot 3} 4a^{-3}}{a^{2 \cdot 5}} = \dfrac{8a^9 4a^{-3}}{a^{10}}$
$\phantom{\dfrac{(2a^3)^3 4a^{-3}}{(a^2)^5}} = 8 \cdot 4a^{9+(-3)-10} = 32a^{-4}$, or $\dfrac{32}{a^4}$

109. $(8x^{-3} y^2)^{-4} (8x^{-3} y^2)^4 = (8x^{-3} y^2)^{-4+4} = (8x^{-3} y^2)^0 = 1$

111. $\dfrac{(5a^3 b)^2}{10a^2 b} = \dfrac{5^2 (a^3)^2 b^2}{10a^2 b} = \dfrac{25a^6 b^2}{10a^2 b}$
$\phantom{\dfrac{(5a^3 b)^2}{10a^2 b}} = \dfrac{25}{10} a^{6-2} b^{2-1} = \dfrac{5}{2} a^4 b$, or $\dfrac{5a^4 b}{2}$

113. $\left(\dfrac{2x^3 y^{-2}}{3y^{-3}}\right)^3 = \left(\dfrac{2}{3}x^3 y^{-2+3}\right)^3 = \left(\dfrac{2}{3}x^3 y\right)^3 = \dfrac{2^3}{3^3}\left(x^3\right)^3 y^3$

$\qquad\qquad = \dfrac{8x^9 y^3}{27}$

115. $\left(\dfrac{21x^5 y^{-7}}{14x^{-2}y^{-6}}\right)^0 = 1$

(Any nonzero real number raised to the zero power is 1.)

117. $\left(\dfrac{5x^0 y^{-7}}{2x^{-2}y^4}\right)^{-2} = \left(\dfrac{5}{2}x^{0+2}y^{-7-4}\right)^{-2} = \left(\dfrac{5x^2 y^{-11}}{2}\right)^{-2}$

$\qquad\qquad = \left(\dfrac{2}{5x^2 y^{-11}}\right)^2 = \dfrac{2^2}{5^2\left(x^2\right)^2\left(y^{-11}\right)^2}$

$\qquad\qquad = \dfrac{4}{25x^4 y^{-22}} = \dfrac{4}{25}x^{-4}y^{22}, \text{ or } \dfrac{4y^{22}}{x^4}$

119. *Writing Exercise* For any even number n.

$\quad (-1)^n = \underbrace{\left[(-1)(-1)\right]\left[(-1)(-1)\right]\cdots\left[(-1)(-1)\right]}_{n/2 \text{ pairs of factors}}$

$\qquad\quad = \underbrace{1\cdot 1\cdots 1}_{n/2 \text{ factors of } 1}$

$\qquad\quad = 1$

121. $4.9t^2 + 3t = 4.9(-3)^2 + 3(-3)$

$\qquad\qquad = 4.9(9) + 3(-3)$

$\qquad\qquad = 44.1 - 9$

$\qquad\qquad = 35.1$

123. *Writing Exercise.* In the expression $3 - (-2)^{-1}$, the first "$-$" after the 3 indicates subtraction. The second "$-$" inside the parentheses indicates the number opposite of two, or negative two. The third "$-$", found in the exponent, indicates that the expression $(-2)^{-1} = \dfrac{1}{(-2)}$.

125. $\dfrac{8a^{x-2}}{2a^{2x+2}} = \dfrac{8}{2}\cdot a^{x-2-(2x+2)} = 4a^{x-2-2x-2} = 4a^{-x-4}$

127. $\dfrac{\left(2^{-2}\right)^a \cdot \left(2^b\right)^{-a}}{\left(2^{-2}\right)^{-b}\left(2^b\right)^{-2a}} = \dfrac{2^{-2a}\cdot 2^{-ab}}{2^{2b}\cdot 2^{-2ab}} = \dfrac{2^{-2a-ab}}{2^{2b-2ab}}$

$\qquad\qquad = 2^{-2a-ab-(2b-2ab)} = 2^{-2a-ab-2b+2ab} = 2^{-2a-2b+ab}$

129. $\left(3^{a+2}\right)^a = 3^{(a+2)a} = 3^{a^2+2a}$

131. $\dfrac{4x^{2a+3}y^{2b-1}}{2x^{a+1}y^{b+1}} = \dfrac{4}{2}\cdot x^{2a+3-(a+1)}y^{2b-1-(b+1)}$

$\qquad\qquad = 2x^{2a+3-a-1}y^{2b-1-b-1} = 2x^{a+2}y^{b-2}$

133. $\dfrac{3^{q+3} - 3^2\left(3^q\right)}{3\left(3^{q+4}\right)} = \dfrac{3^{q+3} - 3^{q+2}}{3^{q+5}} = \dfrac{3^{q+2}(3-1)}{3^{q+2}\left(3^3\right)} = \dfrac{2}{3^3} = \dfrac{2}{27}$

135. $\left[\left(\dfrac{a^{-2c}}{b^{7c}}\right)^{-3}\left(\dfrac{a^{4c}}{b^{-3c}}\right)^2\right]^{-a} = \left(\dfrac{a^{6c}}{b^{-21c}}\cdot\dfrac{a^{8c}}{b^{-6c}}\right)^{-a} = \left(\dfrac{a^{14c}}{b^{-27c}}\right)^{-a}$

$\qquad\qquad = \dfrac{a^{-14ac}}{b^{27ac}}$

Exercise Set 1.7

1. The length of an Olympic marathon, in centimeters, is a large number so its representation in scientific notation would include a positive power of 10.

3. The mass of a hydrogen atom, in grams, is a small number so its representation in scientific notation would include a negative power of 10.

5. The time between leap years, in seconds, is a large number so its representation in scientific notation would include a positive power of 10.

7. $64,000,000,000 = \dfrac{64,000,000,000}{10^{10}}\cdot 10^{10}$ Multiplying by 1: $10^{10}/10^{10} = 1$

$\qquad\qquad = 6.4\times 10^{10}$ This is scientific notation.

9. $1,091,000,000 = \dfrac{1,091,000,000}{10^9}\cdot 10^9$ Multiplying by 1: $10^9/10^9 = 1$

$\qquad\qquad = 1.091\times 10^9$ This is scientific notation.

11. $0.0000013 = \dfrac{0.0000013}{10^6}\cdot 10^6$ Multiplying by 1: $10^6/10^6 = 1$

$\qquad\qquad = \dfrac{1.3}{10^6}$

$\qquad\qquad = 1.3\times 10^{-6}$ This is scientific notation.

13. $0.00009 = \dfrac{0.0009}{10^5}\cdot 10^5$ Multiplying by 1: $10^5/10^5 = 1$

$\qquad\qquad = \dfrac{9}{10^5}$

$\qquad\qquad = 9\times 10^{-5}$ This is scientific notation.

15. $803,000,000,000 = \dfrac{803,000,000,000}{10^{11}}\cdot 10^{11}$

$\qquad\qquad = 8.03\times 10^{11}$

17. $0.000000904 = \dfrac{0.000000904}{10^7}\cdot 10^7$

$\qquad\qquad = \dfrac{9.04}{10^7}$

$\qquad\qquad = 9.04\times 10^{-7}$

19. $431,700,000,000 = \dfrac{431,700,000,000}{10^{11}}\cdot 10^{11}$

$\qquad\qquad = 4.317\times 10^{11}$

21. $4 \times 10^5 = 400,000$ Moving the decimal point 5 places to the right.

23. $1.2 \times 10^{-4} = 0.00012$ Moving the decimal point 4 places to the left.

25. $3.76 \times 10^{-9} = 0.00000000376$ Moving the decimal point 9 places to the left.

27. $8.056 \times 10^{12} = 8,056,000,000,000$ Moving the decimal point 12 places to the right.

29. $7.001 \times 10^{-5} = 0.00007001$ Moving the decimal point 5 places to the left.

31. $9.06 \times 10^9 = 9,060,000,000$ Moving the decimal point 9 places to the right.

33. $\left(3.4 \times 10^{-8}\right)\left(2.6 \times 10^{15}\right)$
$= (3.4 \times 2.6)\left(10^{-8} \times 10^{15}\right)$
$= 8.84 \times 10^7$
$= 8.8 \times 10^7$ Rounding to 2 significant digits

35. $\left(2.36 \times 10^6\right)\left(1.4 \times 10^{-11}\right)$
$= (2.36 \times 1.4)\left(10^6 \times 10^{-11}\right)$
$= 3.304 \times 10^{-5}$
$= 3.3 \times 10^{-5}$ Rounding to 2 significant digits

37. $\left(5.2 \times 10^6\right)\left(2.6 \times 10^4\right) = (5.2 \times 2.6)\left(10^6 \times 10^4\right)$
$= 13.52 \times 10^{10}$
$= (1.352 \times 10) \times 10^{10}$
$= 1.352 \times \left(10 \times 10^{10}\right)$
$= 1.352 \times 10^{11}$
$= 1.4 \times 10^{11}$ (2 significant digits)

39. $\left(7.01 \times 10^{-5}\right)\left(6.5 \times 10^{-7}\right)$
$= (7.01 \times 6.5)\left(10^{-5} \times 10^{-7}\right)$
$= 45.565 \times 10^{-12}$
$= (4.5565 \times 10) \times 10^{-12}$
$= 4.5565 \times 10^{-11}$
$= 4.6 \times 10^{-11}$ (2 significant digits)

41. $\left(2.0 \times 10^6\right)\left(3.02 \times 10^{-6}\right)$

Observe that $10^6 \times 10^{-6} = 1$, so the product is $2.0(3.02)$, or 6.04, or 6.0 rounded to two significant digits.

43. $\dfrac{6.5 \times 10^{15}}{2.6 \times 10^4} = \dfrac{6.5}{2.6} \times \dfrac{10^{15}}{10^4}$
$= 2.5 \times 10^{11}$ (2 significant digits)

45. $\dfrac{9.4 \times 10^{-9}}{4.7 \times 10^{-2}} = \dfrac{9.4}{4.7} \times \dfrac{10^{-9}}{10^{-2}}$
$= 2.0 \times 10^{-7}$ (2 significant digits)

47. $\dfrac{3.2 \times 10^{-7}}{8.0 \times 10^8} = \dfrac{3.2}{8.0} \times \dfrac{10^{-7}}{10^8}$
$= 0.40 \times 10^{-15}$ (2 significant digits)
$= \left(4.0 \times 10^{-1}\right) \times 10^{-15}$
$= 4.0 \times 10^{-16}$

49. $\dfrac{9.36 \times 10^{-11}}{3.12 \times 10^{11}} = \dfrac{9.36}{3.12} \times \dfrac{10^{-11}}{10^{11}}$
$= 3.00 \times 10^{-22}$ (3 significant digits)

51. $\dfrac{6.12 \times 10^{19}}{3.06 \times 10^{-7}} = \dfrac{6.12}{3.06} \times \dfrac{10^{19}}{10^{-7}}$
$= 2.00 \times 10^{26}$ (3 significant digits)

53. $4.6 \times 10^{-9} + 3.2 \times 10^{-9} = (4.6 + 3.2) \times 10^{-9}$
$= 7.8 \times 10^{-9}$

55. $5.9 \times 10^{23} + 6.3 \times 10^{23} = (5.9 + 6.3) \times 10^{23}$
$= 12.2 \times 10^{23}$
$= (1.22 \times 10) \times 10^{23}$
$= 1.22 \times 10^{24}$
$= 1.2 \times 10^{24}$ (2 significant digits)

57. *Familiarize.* Let $n =$ the number of kg of municipal solid waster per person.

Translate. We divide.

$$n = \frac{2.02 \times 10^{12}}{6.5 \text{ billion}}$$

Carry out. We perform the calculations and write scientific notation for the answer.

$$n = \frac{2.02 \times 10^{12}}{6.5 \times 10^9}$$
$$= \frac{2.02}{6.5} \times \frac{10^{12}}{10^9}$$
$$\approx 0.3108 \times 10^3$$
$$= \left(3.108 \times 10^{-1}\right) \times 10^3$$
$$= 3.108 \times 10^2$$
$$= 3.1 \times 10^2 \quad \text{(2 significant digits)}$$

Check. Recheck the translation and calculation. The answer checks.

State. Each person generates an average of 3.1×10^2 kg of municipal solid waste.

59. *Familiarize*. We have a cylinder with diameter 4.0×10^{-10} in. and length 100 yd. We will use the formula for the volume of a cylinder $V = \pi r^2 h$. The radius is $\frac{4.0 \times 10^{-10}}{2}$, or 2.0×10^{-10} in. We convert 100 yd to inches: $100 \text{ yd} = 100 \times 1 \text{ yd} = 100 \times 36 \text{ in.} = 3600 \text{ in.}$, or 3.6×10^3 in.

Translate. We substitute in the formula.

$$V = \pi r^2 h$$
$$V = \pi \left(2.0 \times 10^{-10}\right)^2 \left(3.6 \times 10^3\right)$$

Carry out. We do the calculation.

$$V = \pi \left(2.0 \times 10^{-10}\right)^2 \left(3.6 \times 10^3\right)$$
$$= \pi \times 4.0 \times 10^{-20} \times 3.6 \times 10^3$$
$$= \left(\pi \times 4.0 \times 3.6\right) \times \left(10^{-20} \times 10^3\right)$$
$$\approx 45.2 \times 10^{-17}$$
$$\approx \left(4.52 \times 10\right) \times 10^{-17}$$
$$\approx 4.52 \times 10^{-16}$$
$$\approx 4.5 \times 10^{-16} \quad \text{(2 significant digits)}$$

Check. Recheck the translation and the calculations. The answer checks.

State. The volume of a 100-yd carbon nanotube is about 4.5×10^{-16} in^3.

61. *Familiarize*. Let $w =$ the weight of each sheet. 1 ream $= 500$ sheets.

Translate. We divide.

$$w = \frac{2.25}{500}$$

Carry out.

$$w = \frac{2.25}{5 \times 10^2}$$
$$= 0.450 \times 10^{-2}$$
$$= 4.50 \times 10^{-3}$$

Check. Recheck the translation and calculations. The answer checks.

State. Each sheet of copier paper weighs 4.50×10^{-3} kg, or 4.50 g.

63. *Familiarize*. We know that 1 light year $= 5.88 \times 10^{12}$ mi. Let $y =$ the number of light years from one and of the Milky Way galaxy to the other.

Translate. The distance is $\left(5.88 \times 10^{12}\right) y$ mi. It is also given by 5.88×10^{17} mi. We write the equation:

$$\left(5.88 \times 10^{12}\right) y = 5.88 \times 10^{17}$$

Carry out. We solve the equation.

$$\left(5.88 \times 10^{12}\right) y = 5.88 \times 10^{17}$$
$$y = \frac{5.88 \times 10^{17}}{5.88 \times 10^{12}}$$
$$y = \frac{5.88}{5.88} \times \frac{10^{17}}{10^{12}}$$
$$y = 1.00 \times 10^5$$

Check. Since light travels 5.88×10^{12} mi in one light year, in 1.00×10^5 yr it will travel $\left(1.00 \times 10^5\right) \times \left(5.88 \times 10^{12}\right) = 5.88 \times 10^{17}$ mi. The answer checks.

State. The distance from one end of the galaxy to the other is 1.00×10^5 light years.

65. *Familiarize*. We are told that 1 Angstrom $= 1 \times 10^{-10}$ m, 1 parsec ≈ 3.26 light years, and 1 light year $= 9.46 \times 10^{15}$ m. Let a represent the number of Angstroms in one parsec.

Translate The length of one parsec is $a \times 10^{-10}$ m. It can also be expressed as 3.26 light years, or $3.26 \times 9.46 \times 10^{15}$ m. Since these quantities represent the same number, we can write the equation.

$$a \times 10^{-10} = 3.26 \times 9.46 \times 10^{15}.$$

Carry out. Solve the equation:

$$a \times 10^{-10} = 3.26 \times 9.46 \times 10^{15}$$
$$a \times 10^{-10} \times \frac{1}{10^{-10}} = 3.26 \times 9.46 \times 10^{15} \times \frac{1}{10^{-10}}$$
$$a = \frac{3.26 \times 9.46 \times 10^{15}}{10^{-10}}$$
$$= \left(3.26 \times 9.46\right) \times \frac{10^{15}}{10^{-10}}$$
$$= 30.8396 \times 10^{25}$$
$$= \left(3.08396 \times 10\right) \times 10^{25}$$
$$= 3.08396 \times \left(10 \times 10^{25}\right)$$
$$= 3.08 \times 10^{26} \quad \text{(Rounding to 3 significant digits)}$$

Check. We recheck the translation and calculation.

State. There are about 3.08×10^{26} Angstroms in one parsec.

67. *Familiarize*. We have a very long cylinder. Its length is the average distance from Earth to the sun, 1.5×10^{11} m, and the diameter of its base is 3 Å. We will use the formula for the volume of a cylinder, $V = \pi r^2 h$ (See Example 8.)

Translate. We will express all distances in Angstroms.

Height (length): 1.5×10^{11} m $= \dfrac{1.5 \times 10^{11}}{10^{-10}}$ Å, or

$\qquad 1.5 \times 10^{21}$ Å

Diameter: 3 Å

The radius is half the diameter:

Radius: $\dfrac{1}{2} \times 3$ Å $= 1.5$ Å

Now substitute into the formula (using 3.14 for π):

$\qquad V = \pi r^2 h$

$\qquad V = 3.14 \times 1.5^2 \times 1.5 \times 10^{21}$

Carry out. Do the calculations.

$\qquad V = 3.14 \times 1.5^2 \times 1.5 \times 10^{21}$

$\qquad\quad = 10.5975 \times 10^{21}$

$\qquad\quad = 1.05975 \times 10^{22}$

$\qquad\quad = 1 \times 10^{22}$ Rounding to 1 significant digit

We can convert this result to cubic meters, if desired.

1 Å $= 10^{-10}$ m, So 1 cu Å $= \left(10^{-10}\right)^3$ m$^3 = 10^{-30}$ m^3.

Then 1×10^{22} cu Å $= 1 \times 10^{22}$ cu Å $\times \dfrac{10^{30} \text{ m}^3}{1 \text{ cu Å}}$

$= 1 \times 10^{22} \times 10^{-30} \times \dfrac{\text{cu Å}}{\text{cu Å}} \times \text{m}^3 = 1 \times 10^{-8}$ m^3.

Check. We recheck the translation and the calculations.

State. The volume of the sunbeam is about

1×10^{22} cu Å, or 1×10^{-8} m^3.

69. *Familiarize.* First we will find d, the number of drops in a pound. Then we will find b, the number of bacteria in a drop of U.S. mud.

Translate. To find d we convert 1 pound to drops:

$\qquad d = 1 \text{ lb} \cdot \dfrac{16 \text{ oz}}{1 \text{ lb}} \cdot \dfrac{6.0 \text{ tsp}}{1 \text{ oz}} \cdot \dfrac{60.0 \text{ drops}}{1 \text{ tsp}}$.

Then we divide to find b:

$\qquad b = \dfrac{4.55 \times 10^{11}}{d}$.

Carry out. We do the calculations.

$\qquad d = 1 \text{ lb} \cdot \dfrac{16 \text{ oz}}{1 \text{ lb}} \cdot \dfrac{6.0 \text{ tsp}}{1 \text{ oz}} \cdot \dfrac{60.0 \text{ drops}}{1 \text{ tsp}}$

$\qquad\quad = 5760 \text{ drops}$

Now we find b.

$b = \dfrac{4.55 \times 10^{11}}{d} = \dfrac{4.55 \times 10^{11}}{5.760 \times 10^3} \approx 0.790 \times 10^8$

$\approx \left(7.90 \times 10^{-1}\right) \times 10^8 = 7.9 \times 10^7$

(Our answer must have 2 significant digits.)

Check. If there are about 7.9×10^7 bacteria in a drop of U.S. mud, then in a pound there are about

$\dfrac{7.9 \times 10^7}{1 \text{ drop}} \cdot \dfrac{60.0 \text{ drops}}{1 \text{ tsp}} \cdot \dfrac{6.0 \text{ tsp}}{1 \text{ oz}} \cdot \dfrac{16 \text{ oz}}{1 \text{ lb}}$

$= \dfrac{45,504 \times 10^7}{1 \text{ lb}} \approx 4.55 \times 10^{11}$ bacteria per pound. The

answer checks.

State. About 7.9×10^7 bacteria live in a drop of U.S. mud.

71. *Familiarize.* First we will find the distance C around Jupiter at the equator, in km. Then we will use the formula Speed × Time = Distance to find the speed s at which Jupiter's equator is spinning.

Translate. We will use the formula for the circumference of a circle to find the distance around Jupiter at the equator:

$\qquad C = \pi d = \pi\left(1.43 \times 10^5\right)$.

Then we find the speed s at which Jupiter's equator is spinning:

Speed	×	Time	=	Distance
↓	↓	↓	↓	↓
s	×	10	=	C

Carry out. First we find C.

$\qquad C = \pi\left(1.43 \times 10^5\right) \approx 4.49 \times 10^5$

Then we find s.

$\qquad s \times 10 = C$

$\qquad s \times 10 = 4.49 \times 10^5$

$\qquad\quad s = \dfrac{4.49 \times 10^5}{10}$

$\qquad\quad s = 4.49 \times 10^4$

Check. At 4.49×10^4 km/h. in 10 hr, Jupiter's equator travels $4.49 \times 10^4 \times 10$, or 4.49×10^5 km. A circle with circumference 4.49×10^5 km has a diameter of

$\dfrac{4.49 \times 10^4}{\pi} \approx 1.43 \times 10^5$ km. The answer checks.

State. Jupiter's equator spins at a speed of about 4.49×10^4 km/h.

73. *Writing Exercise.* Answers may vary.

The height of a section of a laser's light beam is 10.0 mm. The radius is 1.00×10^2 mm. Find the volume of the light beam. (Use 3.14 for π.)

75. $3x - y \div 2z$
$= 3(-2) - (-12) \div 2(3)$
$= -6 + 12 \div 2 \cdot 3$
$= -6 + 6 \cdot 3$
$= -6 + 18$
$= 12$

77. *Writing Exercise.* $5 million in $20 bills contains

$\dfrac{5 \times 10^6}{20} = 0.25 \times 10^6 = 2.5 \times 10^5$ bills. In Exercise 62 we

found that a $5 bill weighs about 2.20×10^{-3} lb.

Assuming that a $20 bill weighs the same as a $5 bill,

2.5×10^5 bills would weigh

$2.5 \times 10^5 \times 2.20 \times 10^{-3} = 5.5 \times 10^2$, or 550 lb. Thus, it is

not possible that a criminal is carrying $5 million in $20

bills in a briefcase.

79. *Familiarize.* Let $d = $ the average density, $v = $ the

volume and $m = $ the mass.

Translate. We divide.

$$d = \frac{m}{v}$$

To convert the volume from km^3 to cm^3,

$$v = 1.08 \times 10^{12} \text{ km}^3 \cdot \left(\frac{1000 \text{ m}}{1 \text{ km}}\right)^3 \cdot \left(\frac{100 \text{ cm}}{1 \text{ m}}\right)^3$$

To convert the mass from kg to g

$$m = 5.976 \times 10^{24} \text{ kg} \cdot \frac{1000 \text{ g}}{1 \text{ kg}}$$

Carry out. We do the calculations.

$$v = 1.08 \times 10^{12} \cdot (1000)^3 \cdot (100)^3$$
$$= 1.08 \times 10^{27}$$
$$m = 5.976 \times 10^{24} \cdot 1000$$
$$= 5.976 \times 10^{27}$$
$$d = \frac{m}{v} = \frac{5.976 \times 10^{27}}{1.08 \times 10^{27}}$$
$$= \frac{5.976}{1.08} \times \frac{10^{27}}{10^{27}}$$
$$= 5.53 \quad \text{(3 significant digits)}$$

Check. We recheck the translation and calculations.

State. The average density of Earth is approximately

5.53 g/cm^3.

81. The larger number is the one in which the power of ten

has the larger exponent. Since –90 is larger than –91,

8×10^{-90} is larger than 9×10^{-91}.

$$8 \times 10^{-90} - 9 \times 10^{-91} = 10^{-90}\left(8 - 9 \times 10^{-1}\right)$$
$$= 10^{-90}\left(8 - 0.9\right)$$
$$= 7.1 \times 10^{-90}$$

Thus, 8×10^{-90} is larger by 7.1×10^{-90}.

83. $(4096)^{0.05}(4096)^{0.2} = 4096^{0.25}$
$$= \left(2^{12}\right)^{0.25}$$
$$= 2^3$$
$$= 8$$

85. *Familiarize.* Observe that there are 2^{n-1} grains of sand

on the nth square of the chessboard. Let g represent this

quantity. Recall that a chessboard has 64 squares. Note

also that $2^{10} \approx 10^3$.

Translate. We write the equation

$g = 2^{n-1}$.

To find the umber of grains of sand on the last (or 64·)

square, substitute 64 for n: $g = 2^{64-1}$.

Carry out. Recheck the translation and the calculations.

$$g = 2^{64-1} = 2^{63} = 2^3\left(2^{10}\right)^6$$
$$\approx 2^3\left(10^3\right)^6 \approx 8 \times 10^{18}$$

State. Approximately 8×10^{18} grains of sand are required

for the last square.

Chapter 1 Review

1. The correct choice is (e), since the equation $2x - 1 = 9$ is

equivalent to $2x = 10$.

3. The correct choice is (j), since the equation $\dfrac{3}{4}x = 5$ is

equivalent to $\dfrac{4}{3} \cdot \dfrac{3}{4}x = \dfrac{4}{3} \cdot 5$.

5. The correct choice is (i), since the expression $2(x + 7)$ is

equivalent to $2x + 14$.

7. The correct choice is (f), since the equation

$4x - 3 + 2x = 5$ is equivalent to $6x - 3 = 5$.

9. The correct choice is (d), since the expression $6 + 2x$ is

equivalent to $2(3 + x)$.

11. Eight less than the quotient of two numbers

Let x and y represent the numbers.

Then we have $\frac{x}{y} - 8$.

13. Name the set consisting of the first five odd natural numbers

$\{1, 3, 5, 7, 9\}$

or $\{x \mid x \text{ is an odd natural number less than } 10\}$

15. $\mid -19 \mid = 19$ -19 is 19 units from 0

17. $\mid 6.08 \mid = 6.08$ 6.08 is 6.08 units from 0

19. $\left(-\frac{2}{5}\right) + \frac{1}{3} = -\frac{6}{15} + \frac{5}{15}$

A positive and a negative. The absolute values are $\frac{6}{15}$

and $\frac{5}{15}$. Subtract $\frac{5}{15}$ from $\frac{6}{15}$ to get $\frac{1}{15}$. The negative

number is farther from 0, so the answer is negative,

$-\frac{1}{15}$.

21. $-13 - 12 = -13 + (-12) = -25$

23. $12.3 - 16.1 = 12.3 + (-16.1) = -3.8$

25. $\left(-\frac{2}{3}\right)\left(\frac{5}{8}\right)$

Two numbers with unlike signs: Multiply their absolute

values to get $\frac{10}{24}$, or $\frac{5}{12}$. The answer is negative, $-\frac{5}{12}$.

27. $\frac{-24}{-6}$

Two numbers with like signs. Divide their absolute

values, getting 4. The answer is positive, 4.

29. $-7 \div \frac{4}{3} = -7 \cdot \left(\frac{3}{4}\right)$ Multiplying by the reciprocal of $\frac{4}{3}$

$= -\frac{21}{4}$

31. $12 + x = x + 12$ Using the commutative law of addition

33. $5x + y = x \cdot 5 + y$ Using the commutative

                            law of multiplication

or $= y + 5x$ Using the commutative

                            law of addition

35. $x(yz) = (xy)z$ Using the associative

                            law of multiplication

37. $3x^3 - 6x^2 + x^3 + 5$

$= (3 + 1)x^3 - 6x^2 + 5$

$= 4x^3 - 6x^2 + 5$

39. $3(t + 1) - t = 4$

$3t + 3 - t = 4$

$2t + 3 = 4$

$2t + 3 - 3 = 4 - 3$

$2t = 1$

$\frac{1}{2} \cdot 2t = \frac{1}{2} \cdot 1$

$t = \frac{1}{2}$

The solution is $\frac{1}{2}$.

41. $-9x + 4(2x - 3) = 5(2x - 3) + 7$

$-9x + 8x - 12 = 10x - 15 + 7$

$-x - 12 = 10x - 8$

$-x - 12 + x = 10x - 8 + x$

$-12 = 11x - 8$

$-12 + 8 = 11x - 8 + 8$

$-4 = 11x$

$\frac{1}{11}(-4) = \frac{1}{11} \cdot 11x$

$-\frac{4}{11} = x$

The solution is $-\frac{4}{11}$.

43. $5t - (7 - t) = 4t + 2(9 + t)$

$5t - 7 + t = 4t + 18 + 2t$

$6t - 7 = 6t + 18$

$-7 = 18$ False equation

The solution set is $\varnothing$. The equation is a contradiction.

45. *Familiarize.* Let $x =$ one number, then $x - 19 =$ the

other number.

Translate.

<u>First number</u>	plus	<u>second number</u>	is	115.
↓	↓	↓	↓	↓
x	$+$	$(x - 19)$	$=$	115

Carry out. We solve the equation.

$x + x - 19 = 115$

$2x - 19 = 115$

$2x - 19 + 19 = 115 + 19$

$2x = 134$

$x = 67$

When $x = 67$, $x - 19 = 67 - 19 = 48$

Check. 48 is 19 less than 67. Also $48 + 67 = 115$. The

answer checks.

State. The smaller number is 48.

47. $x = \frac{bc}{t}$

$tx = bc$ Multiplying both sides by t

$\frac{tx}{b} = c$ Multiplying both sides by $\frac{1}{b}$

49. *Familiarize*. Recall the formula for the volume of a right circular cylinder is $V = \pi r^2 h$.

Translate. We solve the formula for h.

$$V = \pi r^2 h$$
$$h = \frac{V}{\pi r^2}$$

Carry out. We substitute 3.5 for r, 538.51 for V, and 3.14 for π and calculate.

$$h = \frac{538.51}{3.14(3.5)^2} = 14$$

Check. We can repeat the calculation or substitute into the original formula and then solve for h. The answer checks.

State. The height of the candle is 14 cm.

51. $\dfrac{12x^3y^8}{3x^2y^2} = \dfrac{12}{3}x^{3-2}y^{8-2} = 4xy^6$

53. $3^{-5} \cdot 3^7 = 3^{-5+7} = 3^2$, or 9

55. $\left(-5a^{-3}b^2\right)^{-3} = (-5)^{-3}\left(a^{-3}\right)^{-3}\left(b^2\right)^{-3}$

$$= -\frac{1}{5^3} \cdot a^9 \cdot b^{-6}$$
$$= -\frac{a^9}{125b^6}$$

57. $\left(\dfrac{3m^{-5}n}{9m^2n^{-2}}\right)^4 = \left(\dfrac{1}{3}m^{-5-2}n^{1-(-2)}\right)^4$

$$= \left(\frac{1}{3}m^{-7}n^3\right)^4$$
$$= \frac{m^{-28}n^{12}}{81}$$
$$= \frac{n^{12}}{81m^{28}}$$

59. $1 - (2-5)^2 + 5 \div 10 \cdot 4^2$

$$= 1 - (-3)^2 + 5 \div 10 \cdot 16$$
$$= 1 - 9 + \frac{1}{2} \cdot 16$$
$$= 1 - 9 + 8$$
$$= 0$$

61. $30{,}860{,}000{,}000{,}000$

$$= \frac{30{,}860{,}000{,}000{,}000}{10^{13}} \times 10^{13}$$
$$= 3.086 \times 10^{13}$$

63. $\dfrac{1.2 \times 10^{-12}}{6.1 \times 10^{-7}} = \dfrac{1.2}{6.1} \times \dfrac{10^{-12}}{10^{-7}}$

$$= 0.1967 \times 10^{-5}$$
$$= \left(1.967 \times 10^{-1}\right) \times 10^{-5}$$
$$= 1.967 \times 10^{-6}$$
$$\approx 2.0 \times 10^{-6} \qquad \text{(2 significant digits)}$$

65. *Writing Exercise*. To write an equation that has no solution, begin with a simple equation that is false for any value of x, such as $x = x + 1$. Then add or multiply by the same quantities on both sides of the equation to construct a more complicated equation with no solution.

67. First we express the quotient, 3 parts per billion, as the fraction and then express as a percent.

$$\frac{3}{1{,}000{,}000{,}000}$$
$$= \frac{3}{1{,}000{,}000{,}000} \cdot \frac{100}{100}$$
$$= \frac{300}{1{,}000{,}000{,}000}\%$$
$$= 0.0000003\%$$

69. *Familiarize*. First we find the area of each pizza. Then we find the price per square inch of each pizza. Recall, the formula for the area of a circle is $A = \pi r^2$ and the radius is $\dfrac{d}{2}$.

Translate. For each pizza, the price per square inch is the cost of each pizza divided by the area of each pizza.

$$P = \frac{C}{A}$$

Carry out. We substitute in the formula $A = \pi r^2$. For the first pizza, which costs \$8, $r = \dfrac{13}{2}$, or 6.5. We use 3.14 for π. $\qquad A = 3.14(6.5)^2 = 132.665$

So the price per square inch is

$$P = \frac{8}{132.665} \approx 0.06$$

For the second pizza, which costs \$11, $r = \dfrac{17}{2}$, or 8.5.

We use 3.14 for π. $\qquad A = 3.14(8.5)^2 = 226.865$

So the price per square inch is

$$P = \frac{11}{226.865} \approx 0.05$$

Check. Repeat the calculations. The answer checks.

State. The 17-inch pizza is about 5¢ per square inch, while the 13-inch pizza is about 6¢ per square inch. The 17-inch pizza is a better deal.

71.
$$m = \frac{x}{y - z}$$
$$m(y - z) = x$$
$$my - mz = x$$
$$-mz = -my + x$$
$$z = \frac{-my + x}{-m}, \text{ or } \frac{my - x}{m}$$

73. *Familiarize*. Let $x =$ the number of points on a quiz, then $3x$ is the number of points on the test.

Translate.

$$\underbrace{\text{Average score on 1 test and four quizzes}}_{\dfrac{3x + 4(82.5)}{7}} \quad \underset{=}{\text{is}} \quad \underset{85}{85.}$$

Carry out. We solve the equation.

$$\frac{3x + 330}{7} = 85$$
$$3x + 330 = 595$$
$$3x = 265$$
$$x = 88.\overline{3}$$

Check. Since $\dfrac{3(88.\overline{3}) + 4(82.5)}{7}$ is 85, the answer checks.

State. Garry must score 88.3 or greater to raise is average to 85.

75.
$$20 - 7[3(2x + 4) - 10] = 9 - 2(x - 5) + \underline{\quad}$$
$$20 - 7[6x + 12 - 10] = 9 - 2x + 10 + \underline{\quad}$$
$$20 - 7[6x + 2] = 19 - 2x + \underline{\quad}$$
$$20 - 42x - 14 = 19 - 2x + \underline{\quad}$$
$$6 - 42x = 19 - 2x + \underline{\quad}$$
$$6 - 40x = 19 + \underline{\quad}$$
$$-40x = \underline{\quad}$$

77. Answers will vary; $\sqrt{5} / 4$ is one example.

Chapter 1 Test

1. Let m and n represent the numbers; $mn - 4$

3. We substitute 7.8 for b and 46.5 for h and multiply.
$$A = \frac{1}{2} \cdot b \cdot h = \frac{1}{2}(7.8)(46.5) = 181.35 \text{ sq m}$$

5. $-7.5 + 3.8$

A negative and positive number: The absolute values are 7.5 and 3.8. Subtract 3.8 from 7.5 to get 3.7. The negative number is farther from 0, so the answer is negative, -3.7.

7. $29.5 - 43.7 = 29.5 + (-43.7)$
$= -14.2$ Change the sign and add.

9. $-6.4(5.3)$

Two numbers with unlike signs: Multiply their absolute values getting 33.92. The answer is negative, -33.92.

11. $-\dfrac{2}{7}\left(-\dfrac{5}{14}\right)$

Two numbers with the same sign: Multiply their absolute values getting $\dfrac{10}{98}$, or $\dfrac{5}{49}$. The answer is positive, $\dfrac{5}{49}$.

13. $\dfrac{2}{5} \div \left(-\dfrac{3}{10}\right) = \dfrac{2}{5} \cdot \left(-\dfrac{10}{3}\right)$

Two numbers with unlike signs: Divide their absolute values getting $\dfrac{4}{3}$. The answer is negative, $-\dfrac{4}{3}$.

15. Using the commutative law of addition, we have
$3 + x = x + 3$.

17. $6a^2b - 5ab^2 + ab^2 - 5a^2b + 2$
$= (6 - 5)a^2b + (-5 + 1)ab^2 + 2$
$= a^2b - 4ab^2 + 2$

19. $13t - (5 - 2t) = 5(3t - 1)$
$$13t - 5 + 2t = 15t - 5$$
$$15t - 5 = 15t - 5$$
$$-5 = -5 \quad \text{True}$$

All real numbers are solutions. The equation is an identity.

21. *Familiarize*. Let $x =$ the number of points on the sixth test.

Translate

$$\underbrace{\text{Average score on 6 tests}}_{\dfrac{84 + 80 + 76 + 96 + 80 + x}{6}} \quad \underset{=}{\text{is}} \quad \underset{85}{\underline{85.}}$$

Carry out. We solve the equation.

$$\frac{416 + x}{6} = 85$$
$$416 + x = 510$$
$$x = 94$$

Check. $\dfrac{84 + 80 + 76 + 96 + 80 + 94}{6}$, or 85. The answer checks.

State. Linda must score 94 so her average will be 85.

23. $3x - 7 - (4 - 5x)$
$= 3x - 7 - 4 + 5x$
$= 8x - 11$

25. $\left(7x^{-4}y^{-7}\right)\left(-6x^{-6}y\right)$

$\quad = (7)(-6)\left(x^{-4}x^{-6}\right)\left(y^{-7}y\right)$

$\quad = -42x^{-4+(-6)}y^{-7+1}$

$\quad = -42x^{-10}y^{-6}$, or $-\dfrac{42}{x^{10}y^6}$

27. $\left(-5x^{-1}y^3\right)^3 = (-5)^3\left(x^{-1}\right)^3\left(y^3\right)^3$

$\qquad\qquad = -125x^{-3}y^9$

$\qquad\qquad = -\dfrac{125y^9}{x^3}$

29. $\left(7x^3y\right)^0 = 1$

31. $\dfrac{1.8\times10^{-4}}{4.8\times10^{-7}} = \dfrac{1.8}{4.8}\times\dfrac{10^{-4}}{10^{-7}}$

$\quad = 0.375\times10^3$

$\quad = \left(3.75\times10^{-1}\right)\times10^3$

$\quad = 3.75\times10^2$

$\quad \approx 3.8\times10^2 \qquad \text{(2 significant digits)}$

33. $\left(2x^{3a}y^{b+1}\right)^{3c} = 2^{3c}\cdot\left(x^{3a}\right)^{3c}\cdot\left(y^{b+1}\right)^{3c} = 2^{3c}x^{9ac}y^{3bc+3c}$

35. $\dfrac{\left(-16x^{x-1}y^{y-2}\right)\left(2x^{x+1}y^{y+1}\right)}{\left(-7x^{x+2}y^{y+2}\right)\left(8x^{x-2}y^{y-1}\right)}$

$\quad = \dfrac{-16\cdot2\cdot x^{(x-1)+(x+1)}y^{(y-2)+(y+1)}}{-7\cdot8\cdot x^{(x+2)+(x-2)}y^{(y+2)+(y-1)}}$

$\quad = \dfrac{-32x^{2x}y^{2y-1}}{-56x^{2x}y^{2y+1}}$

$\quad = \dfrac{4}{7}x^{2x-2x}y^{(2y-1)-(2y+1)}$

$\quad = \dfrac{4}{7}y^{-2}$, or $\dfrac{4}{7y^2}$

Chapter 2

Graphs, Functions, and Linear Equations

Exercise Set 2.1

1. The two perpendicular number lines that are used for graphing are called <u>axes</u>.

3. In the <u>third</u> quadrant, both coordinates of a point are negative.

5. To graph an equation means to make a drawing that represents all <u>solutions</u> of the equation.

7. A is 5 units right of the origin and 3 units up, so its coordinate are (5, 3).

   B is 4 units left of the origin and 3 units up, so its coordinates are (−4, 3).

   C is 0 units right or left of the origin and 2 units up, so its coordinates are (0, 2).

   D is 2 units left of the origin and 3 units down, so its coordinates are (−2, −3).

   E is 4 units right of the origin and 2 units down, so its coordinates are (4, −2).

   F is 5 units left of the origin and 0 units up or down, so its coordinates are (−5, 0).

9.

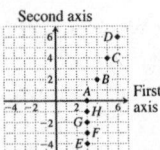

   $A(3, 0)$ is 3 units right and 0 units up or down.

   $B(4, 2)$ is 4 units right and 2 units up.

   $C(5, 4)$ is 5 units right and 4 units up.

   $D(6, 6)$ is 6 units right and 6 units up.

   $E(3, −4)$ is 3 units right and 4 units down.

   $F(3, −3)$ is 3 units right and 3 units down.

   $G(3, −2)$ is 3 units right and 2 units down.

   $H(3, −1)$ is 3 units right and 1 unit down.

11.

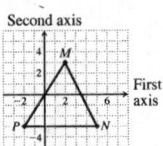

A triangle is formed. The area of a triangle is found by using the formula $A = \frac{1}{2}bh$. In this triangle the base and height are 7 units and 6 units, respectively.

$$A = \frac{1}{2}bh = \frac{1}{2} \cdot 7 \cdot 6 = \frac{42}{2} = 21 \text{ square units}$$

13. The first coordinate is positive and the second negative, so the point (3, −5) is in quadrant IV.

15. Both coordinates are negative, so the point (−3, −12) is in quadrant III.

17. Both coordinates are positive, so the point $\left(11, \frac{1}{4}\right)$ is in quadrant I.

19. The first coordinate is negative and the second positive, so the point (−1.2, 46) is in quadrant II.

21. $y = 3x - 7$

   $$\begin{array}{c|c} -1 & 3 \cdot 2 - 7 \\ & 6 - 7 \end{array}$$ Substituting 2 for x and −1 for y (alphabetical order of variables)

   $$-1 \overset{?}{=} -1$$

   Since −1 = −1 is true, (2, −1) is a solution of $y = 3x - 7$.

23. $2x - y = 5$

   $$\begin{array}{c|c} 2 \cdot 3 - 2 & 5 \\ 6 - 2 & \end{array}$$ Substituting 3 for x and 2 for y (alphabetical order of variables)

   $$4 \overset{?}{=} 5$$

   Since 4 = 5 is false, (3, 2) is not a solution of $2x - y = 5$.

25. $a - 5b = 8$

   $$\begin{array}{c|c} 3 - 5(-1) & 8 \\ 3 + 5 & \end{array}$$ Substituting 3 for a and −1 for b

   $$8 \overset{?}{=} 8$$

   Since 8 = 8 is true, (3, −1) is a solution of $a - 5b = 8$.

27.
$$6x + 8y = 4$$

$$\begin{array}{c|c} 6 \cdot \dfrac{2}{3} + 8 \cdot 0 & 4 \\ 4 + 0 & \end{array}$$ Substituting $\dfrac{2}{3}$ for x and 0 for y

$$4 \overset{?}{=} 4$$

Since $4 = 4$ is true, $\left(\dfrac{2}{3},\ 0\right)$ is a solution of $6x + 8y = 4$.

29.
$$r - s = 4$$

$$\begin{array}{c|c} 6 - (-2) & 4 \\ 6 + 2 & \end{array}$$ Substituting 6 for r and -2 for s

$$8 \overset{?}{=} 4$$

Since $8 = 4$ is false, $(6, -2)$ is not a solution of $r - s = 4$.

31.
$$y = 2x^2$$

$$\begin{array}{c|c} 1 & 2(2)^2 \end{array}$$ Substituting 2 for x and 1 for y

$$1 \overset{?}{=} 8$$

Since $1 = 8$ is false, $(2, 1)$ is not a solution of $y = 2x^2$.

33.
$$x^3 + y = 1$$

$$\begin{array}{c|c} (-2)^3 + 9 & 1 \\ -8 + 9 & \end{array}$$ Substituting -2 for x and 9 for y

$$1 \overset{?}{=} 1$$

Since $1 = 1$ is true, $(-2, 9)$ is a solution of $x^3 + y = 1$.

35. $y = 3x$

To find an ordered pair, we choose any number for x and then determine y by substitution.

When $x = 0$, $y = 3 \cdot 0 = 0$.

When $x = 1$, $y = 3 \cdot 1 = 3$.

When $x = -2$, $y = 3 \cdot (-2) = -6$.

x	y	$(x,\ y)$
0	0	$(0,\ 0)$
1	3	$(1,\ 3)$
-2	-6	$(-2, -6)$

Plot these points, draw the line they determine, and label the graph $y = 3x$.

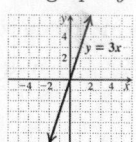

37. $y = x + 4$

To find an ordered pair, we choose any number for x and then determine y by substitution.

When $x = 0$, $y = 0 + 4 = 4$.

When $x = 1$, $y = 1 + 4 = 5$.

When $x = -2$, $y = -2 + 4 = 2$.

x	y	$(x,\ y)$
0	4	$(0,\ 4)$
1	5	$(1,\ 5)$
-2	2	$(-2,\ 2)$

Plot these points, draw the line they determine, and label the graph $y = x + 4$.

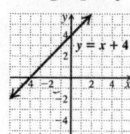

39. $y = x - 4$

To find an ordered pair, we choose any number for x and then determine y by substitution. For example, if we choose 1 for x, then $y = 1 - 4 = -3$. We find several ordered pairs, plot them and draw the line.

x	y	$(x,\ y)$
1	-3	$(1, -3)$
0	-4	$(0, -4)$
-2	-6	$(-2, -6)$

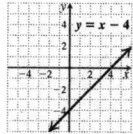

41. $y = -2x + 3$

To find an ordered pair, we choose any number for x and then determine y. For example, if $x = 1$, then $y = -2 \cdot 1 + 3 = -2 + 3 = 1$. We find several ordered pairs, plot them, and draw the line.

x	y
1	1
3	-3
-1	5
0	3

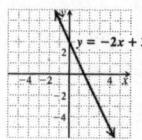

43. $y + 2x = 3$

$$y = -2x + 3 \quad \text{Solving for } y$$

Observe that this is the equation that was graphed in Exercise 41. The graph is shown above.

45. $y = -\dfrac{3}{2}x$

To find an ordered pair, we choose any number for x and then determine y by substitution. For example, if $x = 2$, then $y = -\dfrac{3}{2} \cdot 2 = -3$. We find several ordered pairs, plot them and draw the line.

x	y
2	-3
0	0
-4	6

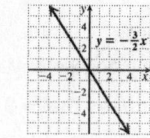

47. $y = \dfrac{3}{4}x - 1$

To find an ordered pair, we choose any number for x and then determine y by substitution. For example, if $x = 4$, then $y = \dfrac{3}{4} \cdot 4 - 1 = 3 - 1 = 2$. We find several ordered

pairs, plot them and draw the line.

x	y
4	2
0	−1
−8	−7

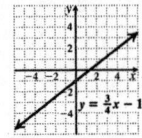

49. $y = |x| + 2$

We select x-values and find the corresponding y-values. The table lists some ordered pairs. We plot these points.

x	y
3	5
1	3
0	2
−1	3
−3	5

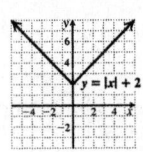

Note that the graph is V-shaped, centered at (0, 2).

51. $y = |x| - 2$

We select x-values and find the corresponding y-values. The table lists some ordered pairs. We plot these points.

x	y
3	1
1	−1
0	−2
−1	−1
−4	2

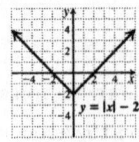

53. $y = x^2 + 2$

To find an ordered pair, we choose any number for x and then determine y. For example, if $x = 2$, then $y = 2^2 + 2 = 6$. We find several ordered pairs, plot them, and connect them with a smooth curve.

x	y
2	6
1	3
0	2
−1	3
−2	6

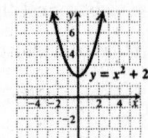

55. $y = x^2 - 2$

To find an ordered pair, we choose any number for x and then determine y. For example, if $x = 2$, then $y = 2^2 - 2 = 4 - 2 = 2$. We find several ordered pairs, plot

them, and connect them with a smooth curve.

x	y
2	2
1	−1
0	−2
−1	−1
−2	2

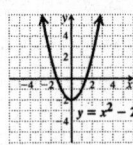

57. *Writing Exercise.* Suppose, for instance, we had plotted only the points (−3, 3) and (−1, 1). Drawing a line through these points would yield a straight line rather than the correct V-shaped graph.

59. $5t - 7 = 5 \cdot 10 - 7 = 50 - 7 = 43$

61. $(3-x)^2(1-2x)^3 = \left(3 - \dfrac{1}{2}\right)^2 \left(1 - 2 \cdot \dfrac{1}{2}\right)^3$

$$= \left(\dfrac{6}{2} - \dfrac{1}{2}\right)^2 (1-1)^3 = \left(\dfrac{5}{2}\right)^2 (0)^3 = 0$$

63. $\dfrac{2x+3}{x-4} = \dfrac{2(0)+3}{0-4} = \dfrac{0+3}{-4} = -\dfrac{3}{4}$

65. $x + 4 = 0$

    $x = -4$

The solution is −4.

67. $1 - 2x = 0$

    $1 = 2x$

    $\dfrac{1}{2} = x$

The solution is $\dfrac{1}{2}$.

69. *Writing Exercise.*

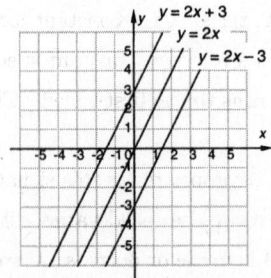

When −3 is added, the graph of $y = 2x$ is moved down 3 units. When 3 is added, the graph of $y = 2x$ is moved up 3 units. The three lines are parallel.

71. *Writing Exercise.* When $x < 0$, then $y < 0$ and the graph contains points in quadrant III. When $0 < x < 30$, then $y < 0$ and the graph contains points in quadrant IV. When $x > 30$, then $y > 0$ and the graph contains points in quadrant I. Thus, the graph passes through three quadrants.

73. a) Graph III seems most appropriate for this situation. It reflects less than 40 hours of work per week until September, 40 hours per week from September until December, and more than 40 hours per week in December.

  b) Graph II seems most appropriate for this situation. It reflects 40 hours of work per week until September, 20 hours per week from September until December, and 40 hours per week again in December.

  c) Graph I seems most appropriate for this situation. It reflects more than 40 hours of work per week until September, 40 hours per week from September until December, and more than 40 hours per week again in December.

  d) Graph IV seems most appropriate for this situation. It reflects less than 40 hours of work per week until September, approximately 20 hours per week from September until December, and about 40 hours per week in December.

75. a) Graph III seems most appropriate for this situation. It reflects a constant speed for a lakeshore loop.

  b) Graph II seems most appropriate for this situation. In climbing a monster hill, the speed gradually decreases, the speed bottoms out at the top of the hill, and increases for the downhill ride.

  c) Graph IV seems most appropriate for this situation. For interval training, the speed is constant for an interval, then increases to a new constant speed for an interval, then returns to the first speed. The sequence repeats.

  d) Graph I seems most appropriate for this situation. For a variety of intervals, the speed varies with no particular pattern of increase or decrease in speed.

77. Plot $(-10,-2)$, $(-3, 4)$, and $(6, 4)$, and sketch a parallelogram.

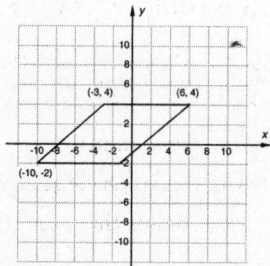

Since $(6, 4)$ is 9 units directly to the right of $(-3, 4)$, a

fourth vertex could lie 9 units directly to the right of $(-10,-2)$. Then its coordinates are $(-10+9,-2)$, or $(-1,-2)$.

If we connect the points in a different order, we get a second parallelogram.

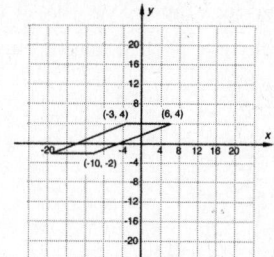

Since $(-3, 4)$ is 9 units directly to the left of $(6, 4)$, a fourth vertex could lie 9 units directly to the left of $(-10,-2)$. Then its coordinates are $(-10-9,-2)$, or $(-19,-2)$.

If we connect the points in yet a different order, we get a third parallelogram.

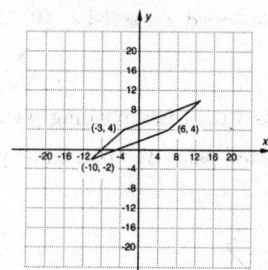

Since $(6, 4)$ lies 16 units directly to the right of and 6 units above $(-10, -2)$, a fourth vertex could lie 16 units to the right of and 6 units above $(-3, 4)$. Its coordinates are $(-3+16, 4+6)$, or $(13, 10)$.

79. $y = \dfrac{1}{x^2}$

Choose x-values from -4 to 4 and use a calculator to find the corresponding y-values. (Note that we cannot choose 0 as a first coordinate since $1/0^2$, or $1/0$, is not defined.) Plot the points and draw the graph. Note that it has two branches, one on each side of the y-axis.

x	y
-4	0.0625
-3	$0.\overline{1}$
-2	0.25
-1	1
-0.5	4
0.5	4
1	1
2	0.25
3	$0.\overline{1}$
4	0.0625

81. $y = \dfrac{1}{x-2}$

Choose x-values from -2 to 6, and use a calculator to find the corresponding y-values. (Note that we cannot choose 2 as a first coordinate since $\dfrac{1}{2-2}$, or $\dfrac{1}{0}$, is not defined.) Plot the points, and draw the graph. Note that it has two branches, one on each side of a vertical line through $(2, 0)$.

x	y
-2	-0.25
0	-0.5
1	-1
1.9	-10
2.1	10
2.5	2
3	1
4	0.5
6	0.25

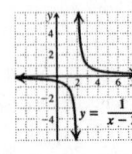

83. $y = \sqrt{x} + 1$

Choose x-values from 0 to 10 and use a calculator to find the corresponding y-values. Plot the points and draw the graph.

x	y
0	1
1	2
2	2.414
3	2.732
4	3
5	3.236
6	3.449
7	3.646
8	3.828
9	4
10	4.162

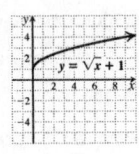

85. $y = x^3$

Choose x-values from -2 to 2 and use a calculator to find the corresponding y-values. Plot the points and draw the graph.

x	y
-2	-8
-1.5	-3.375
-1	-1
-0.5	-0.125
-0.25	-0.016
0	0
0.5	0.125
1	1
1.5	3.375
2	8

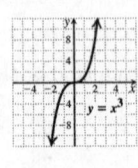

87. $y = \dfrac{1}{x} + 3$

Choose x-values from 3.5 to 4 and use a calculator to

find the corresponding y-values. (Note that we cannot choose 0 as a first coordinate since $1/0$ is not defined.) Plot the points and draw the graph. Note that it has two branches, one in the first quadrant and the other in the third quadrant.

x	y
-4	2.75
-3	2.67
-2	2.5
-1	2
-0.25	-1
0.25	7
1	4
2	3.5
3	3.33
4	3.25

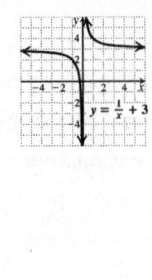

89. a)

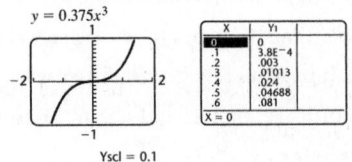

$y = 0.375x^3$

Yscl = 0.1

b)

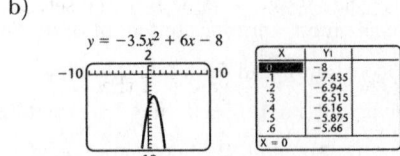

$y = -3.5x^2 + 6x - 8$

c)

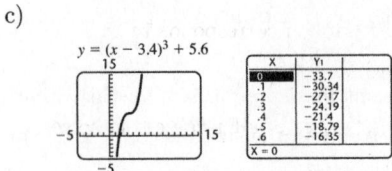

$y = (x - 3.4)^3 + 5.6$

Exercise Set 2.2

1. A function is a special kind of <u>correspondence</u> between two sets.

3. For any function, the set of all inputs, or first values, is called the <u>domain</u>.

5. When a function is graphed, members of the domain are located on the <u>horizontal</u> axis.

7. The notation $f(3)$ is read <u>"f of 3," "f at 3," or "the value of f at 3."</u>

9. The correspondence is a function, because each member of the domain corresponds to exactly one member of the range.

11. The correspondence is a function, because each member of the domain corresponds to exactly one member of the range.

13. The correspondence is not a function because a member of the domain (June 9 or October 5) corresponds to more than one member of the range.

15. The correspondence is a function, because each member of the domain corresponds to exactly one member of the range.

17. The correspondence is a function, because each flash drive has only one storage capacity.

19. This correspondence is a function, because each player has only one uniform number.

21. a) The domain is the set of all x-values of the set. It is $\{-3, -2, 0, 4\}$.

   b) The range is the set of all y-values of the set. It is $\{-10, 3, 5, 9\}$.

   c) The correspondence is a function, because each member of the domain corresponds to exactly one member of the range.

23. a) The domain is the set of all x-values of the set. It is $\{1, 2, 3, 4, 5\}$.

   b) The range is the set of all y-values of the set. It is $\{1\}$.

   c) The correspondence is a function, because each member of the domain corresponds to exactly one member of the range.

25. a) The domain is the set of all x-values of the set. It is $\{-2, 3, 4\}$.

   b) The range is the set of all y-values of the set. It is $\{-8, -2, 4, 5\}$.

   c) The correspondence is not a function, because a member of the domain (4) corresponds to more than one member of the range.

27. a) Locate 1 on the horizontal axis and then find the point on the graph for which 1 is the first coordinate. From that point, look to the vertical axis to find the

corresponding y-coordinate, -2. Thus, $f(1) = -2$.

   b) The set of all x-values in the graph extends from -2 to 5, so the domain is $\{x \mid -2 \leq x \leq 5\}$.

   c) To determine which member(s) of the domain are paired with 2, locate 2 on the vertical axis. From there look left and right to the graph to find any points for which 2 is the second coordinate. One such point exists. Its first coordinate is 4. Thus, the x-value for which $f(x) = 2$ is 4.

   d) The set of all y-values in the graph extends from -3 to 4, so the range is $\{y \mid -3 \leq y \leq 4\}$.

29. a) Locate 1 on the horizontal axis and then find the point on the graph for which 1 is the first coordinate. From that point, look to the vertical axis to find the corresponding y-coordinate, -2. Thus, $f(1) = -2$.

   b) The set of all x-values in the graph extends from -4 to 2, so the domain is $\{x \mid -4 \leq x \leq 2\}$.

   c) To determine which member(s) of the domain are paired with 2, locate 2 on the vertical axis. From there look left and right to the graph to find any points for which 2 is the second coordinate. One such point exists. Its first coordinate is -2. Thus, the x-value for which $f(x) = 2$ is -2.

   d) The set of all y-values in the graph extends from -3 to 3, so the range is $\{y \mid -3 \leq y \leq 3\}$.

31. a) Locate 1 on the horizontal axis and then find the point on the graph for which 1 is the first coordinate. From that point, look to the vertical axis to find the corresponding y-coordinate, 3. Thus, $f(1) = 3$.

   b) The set of all x-values in the graph extends from -4 to 3, so the domain is $\{x \mid -4 \leq x \leq 3\}$.

   c) To determine which member(s) of the domain are paired with 2, locate 2 on the vertical axis. From there look left and right to the graph to find any points for which 2 is the second coordinate. One such point exists. Its first coordinate is -3. Thus, the x-value for which $f(x) = 2$ is -3.

   d) The set of all y-values in the graph extends from -2 to 5, so the range is $\{y \mid -2 \leq y \leq 5\}$.

33. a) Locate 1 on the horizontal axis and then find the point on the graph for which 1 is the first coordinate. From that point, look to the vertical axis to find the corresponding y-coordinate, 3. Thus, $f(1) = 3$.

b)　The domain is the set of all x-values in the graph. It is $\{-4,-3,-2,-1,\ 0,\ 1,\ 2\}$.

c)　To determine which member(s) of the domain are paired with 2, locate 2 on the vertical axis. From there look left and right to the graph to find any points for which 2 is the second coordinate. There are two such points, $(-2, 2)$ and $(0, 2)$. Thus, the x-values for which $f(x)=2$ are -2 and 0.

d)　The range is the set of all y-values in the graph. It is $\{1, 2, 3, 4\}$.

35. a)　Locate 1 on the horizontal axis and then find the point on the graph for which 1 is the first coordinate. From that point, look to the vertical axis to find the corresponding y-coordinate, 4. Thus, $f(1)=4$.

b)　The set of all x-values in the graph extends from -3 to 4, so the domain is $\{x\,|-3\le x\le 4\}$.

c)　To determine which member(s) of the domain are paired with 2, locate 2 on the vertical axis. From there look left and right to the graph to find any points for which 2 is the second coordinate. There are two such points, $(-1, 2)$ and $(3, 2)$. Thus, the x-values for which $f(x)=2$ are -1 and 3.

d)　The set of all y-values in the graph extends from -4 to 5, so the range is $\{y\,|-4\le y\le 5\}$.

37. a)　Locate 1 on the horizontal axis and then find the point on the graph for which 1 is the first coordinate. From that point, look to the vertical axis to find the corresponding y-coordinate, 2. Thus, $f(1)=2$.

b)　The set of all x-values in the graph extends from -4 to 4, so the domain is $\{x\,|-4\le x\le 4\}$.

c)　To determine which member(s) of the domain are paired with 2, locate 2 on the vertical axis. From there look left and right to the graph to find any points for which 2 is the second coordinate. All points in the set $\{x\,|\,0<x\le 2\}$ satisfy this condition. These are the x-values for which $f(x)=2$.

d)　The range is the set of all y-values in the graph. It is $\{1, 2, 3, 4\}$.

39. The domain of f is the set of all x-values that are used in the points on the curve.

The domain is $\{x\,|\,x$ is a real number$\}$ or $\mathbb{R}$.

The range of f is the set of all y-values that are used in the points on the curve.

The range is $\{y\,|\,y$ is a real number$\}$ or $\mathbb{R}$.

41. The domain of f is the set of all x-values that are used in the points on the curve.

The domain is $\{x\,|\,x$ is a real number$\}$ or $\mathbb{R}$.

The range of f is the set of all y-values that are used in the points on the curve.

The range is $\{4\}$.

43. The domain of f is the set of all x-values that are used in the points on the curve.

The domain is $\{x\,|\,x$ is a real number$\}$ or $\mathbb{R}$.

The range of f is the set of all y-values that are used in the points on the curve.

The range is $\{y\,|\,y\ge 1\}$.

45. The domain of f is the set of all x-values that are used in the points on the curve.

The domain is $\{x\,|\,x$ is a real number and $x\ne -2\}$.

The range of f is the set of all y-values that are used in the points on the curve.

The range is $\{y\,|\,y$ is a real number and $y\ne -4\}$.

47. The domain of f is the set of all x-values that are used in the points on the curve. The domain is $\{x\,|\,x\ge 0\}$.

The range of f is the set of all y-values that are used in the points on the curve. The range is $\{y\,|\,y\ge 0\}$.

49. We can use the vertical-line test:

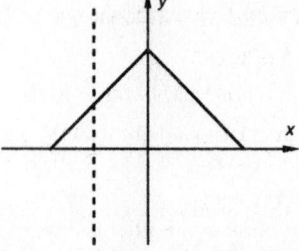

Visualize moving this vertical line across the graph. No vertical line will intersect the graph more than once. Thus, the graph is a graph of a function.

51. We can use the vertical-line test:

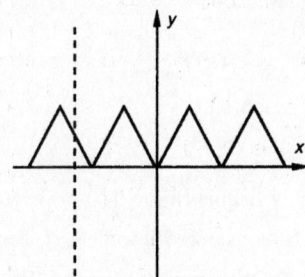

Visualize moving this vertical line across the graph. No vertical line will intersect the graph more than once. Thus, the graph is a graph of a function.

53. We can use the vertical line test.

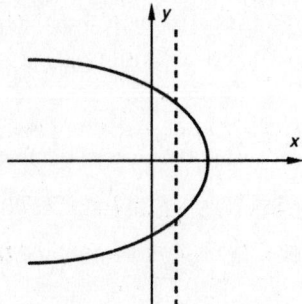

It is possible for a vertical line to intersect the graph more than once. Thus this is not the graph of a function.

55. $g(x) = 2x + 5$

   a) $g(0) = 2(0) + 5 = 0 + 5 = 5$

   b) $g(-4) = 2(-4) + 5 = -8 + 5 = -3$

   c) $g(-7) = 2(-7) + 5 = -14 + 5 = -9$

   d) $g(8) = 2(8) + 5 = 16 + 5 = 21$

   e) $g(a + 2) = 2(a + 2) + 5 = 2a + 4 + 5 = 2a + 9$

   f) $g(a) + 2 = 2(a) + 5 + 2 = 2a + 7$

57. $f(n) = 5n^2 + 4n$

   a) $f(0) = 5 \cdot 0^2 + 4 \cdot 0 = 0 + 0 = 0$

   b) $f(-1) = 5(-1)^2 + 4(-1) = 5 - 4 = 1$

   c) $f(3) = 5 \cdot 3^2 + 4 \cdot 3 = 45 + 12 = 57$

   d) $f(t) = 5t^2 + 4t$

   e) $f(2a) = 5(2a)^2 + 4 \cdot 2a = 5 \cdot 4a^2 + 8a = 20a^2 + 8a$

   f) $f(3) - 9 = 5 \cdot 3^2 + 4 \cdot 3 - 9 = 5 \cdot 9 + 4 \cdot 3 - 9$
                   $= 45 + 12 - 9 = 48$

59. $f(x) = \dfrac{x-3}{2x-5}$

   a) $f(0) = \dfrac{0-3}{2 \cdot 0 - 5} = \dfrac{-3}{0-5} = \dfrac{-3}{-5} = \dfrac{3}{5}$

   b) $f(4) = \dfrac{4-3}{2 \cdot 4 - 5} = \dfrac{1}{8-5} = \dfrac{1}{3}$

   c) $f(-1) = \dfrac{-1-3}{2(-1)-5} = \dfrac{-4}{-2-5} = \dfrac{-4}{-7} = \dfrac{4}{7}$

   d) $f(3) = \dfrac{3-3}{2 \cdot 3 - 5} = \dfrac{0}{6-5} = \dfrac{0}{1} = 0$

   e) $f(x+2) = \dfrac{x+2-3}{2(x+2)-5} = \dfrac{x-1}{2x+4-5} = \dfrac{x-1}{2x-1}$

   f) $f(a+h) = \dfrac{a+h-3}{2(a+h)-5} = \dfrac{a+h-3}{2a+2h-5}$

61. $f(x) = \dfrac{5}{x-3}$

Since $\dfrac{5}{x-3}$ cannot be computed when the denominator is 0, we find the x-value that causes $x-3$ to be 0:

      $x - 3 = 0$

         $x = 3$ Adding 3 to both sides

Thus, 3 is not in the domain of f, while all other real numbers are. The domain of f is

$\{x \mid x \text{ is a real number and } x \neq 3\}$.

63. $g(x) = 2x + 1$

Since we can compute $2x + 1$ for any real number x, the domain is the set of all real numbers.

65. $h(x) = |6 - 7x|$

Since we can compute $|6 - 7x|$ for any real number x, the domain is the set of all real numbers.

67. $f(x) = \dfrac{3}{8 - 5x}$

Solve: $8 - 5x = 0$

           $x = \dfrac{8}{5}$

The domain is $\left\{ x \mid x \text{ is a real number and } x \neq \dfrac{8}{5} \right\}$.

69. $h(x) = \dfrac{x}{x+1}$

Solve: $x + 1 = 0$

          $x = -1$

The domain is $\{x \mid x \text{ is a real number and } x \neq -1\}$.

71. $f(x) = \dfrac{3x+1}{2}$

Since we can compute $\dfrac{3x+1}{2}$ for any real number x, the domain is the set of all real numbers.

73. $g(x) = \dfrac{1}{2x}$

Solve: $2x = 0$

         $x = 0$

The domain is $\{x \mid x \text{ is a real number and } x \neq 0\}$.

75. $A(s) = s^2 \dfrac{\sqrt{3}}{4}$

$A(4) = 4^2 \dfrac{\sqrt{3}}{4} = 4\sqrt{3} \approx 6.93$

The area is $4\sqrt{3}$ cm$^2 \approx 6.93$ cm^2.

77. $V(r) = 4\pi r^2$

$V(3) = 4\pi(3)^2 = 36\pi$

The area is 36π in$^2 \approx 113.10$ in^2.

79. $H(x) = 2.75x + 71.48$

$H(34) = 2.75(34) + 71.48 = 164.98$

The predicted height is 164.98 cm.

81. $F(C) = \dfrac{9}{5}C + 32$

$F(-5) = \dfrac{9}{5}(-5) + 32 = -9 + 32 = 23$

The equivalent temperature is 23°F.

83. Locate the point that is directly above 225. Then estimate its second coordinate by moving horizontally from the point to the vertical axis. The rate is about 75 heart attacks per 10,000 men.

85. Locate the point on the graph that is directly above '00. Then estimate its second coordinate by moving horizontally from the point to the vertical axis. In 2000, about 500 movies were released. That is $F(2000) \approx 500$.

87. Plot and connect the points, using the wattage of the incandescent as the first coordinate and the wattage of the CFL as the second coordinate.

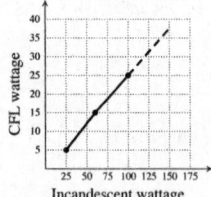

To estimate the wattage of the CFL bulb that creates light equivalent to a 75-watt incandescent bulb, first locate the point directly above 75. Then estimate the second coordinate by moving horizontally from the point to the vertical axis. Read the approximate function value there. The wattage is about 19 watts.

To predict the wattage of the CFL bulb that creates light equivalent to a 120-watt incandescent bulb, extend the graph and extrapolate. The wattage is about 30 watts.

89. Plot and connect the points, using body weight as the first coordinate and the corresponding number of drinks as the second coordinate.

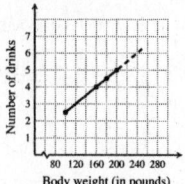

To estimate the number of drinks that a 140-lb person would have to drink to be considered intoxicated, first locate the point that is directly above 140. Then estimate its second coordinate by moving horizontally from the point to the vertical axis. Read the approximate function value there. The estimated number of drinks is 3.5.

To predict the number of drinks it would take for a 230-lb person to be considered intoxicated, extend the graph and extrapolate. It appears that it would take about 6 drinks.

91. Plot and connect the points, using the year as the first coordinate and the total sales as the second coordinate.

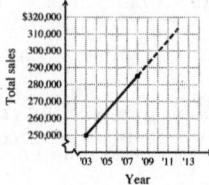

To estimate the total sales for 2004, first locate the point directly above 2004. Then estimate its second coordinate by moving horizontally to the vertical axis. Read the approximate function value there. The estimated 2004 sales total is about $257,000.

To estimate the sales for 2011, extend the graph and extrapolate. We estimate the sales for 2011 to be about $306,000.

93. *Writing Exercise.* You could measure time in years, because it seems reasonable that each graph represents a ten-year span. Answers may vary.

95. $\dfrac{6-3}{-2-7} = \dfrac{3}{-9} = -\dfrac{1}{3}$

97. $\dfrac{-5-(-5)}{3-(-10)} = \dfrac{-5+5}{3+10} = \dfrac{0}{13} = 0$

99. $2x - y = 8$

$-y = -2x + 8$

$y = 2x - 8$

101. $2x + 3y = 6$

$3y = 6 - 2x$

$y = \dfrac{6-2x}{3}$, or $2 - \dfrac{2}{3}x$, or $-\dfrac{2}{3}x + 2$

103. *Writing Exercise.* The independent variable should be chosen as the number of fish in an aquarium, since the survival of the fish is dependent upon an adequate food supply.

105. To find $f(g(-4))$, we first find $g(-4)$:

$g(-4) = 2(-4) + 5 = -8 + 5 = -3$.

Then

$f(g(-4)) = f(-3) = 3(-3)^2 - 1 = 3 \cdot 9 - 1 = 27 - 1 = 26$.

To find $g(f(-4))$, we first find $f(-4)$:

$f(-4) = 3(-4)^2 - 1 = 3 \cdot 16 - 1 = 48 - 1 = 47$.

Then $g(f(-4)) = g(47) = 2 \cdot 47 + 5 = 94 + 5 = 99$.

107. $f(\text{tiger}) = \text{dog}$
$f(\text{dog}) = f(f(\text{tiger})) = \text{cat}$
$f(\text{cat}) = f(f(f(\text{tiger}))) = \text{fish}$
$f(\text{fish}) = f(f(f(f(\text{tiger})))) = \text{worm}$

109. Locate the highest point on the graph. Then move vertically to the horizontal axis and read the corresponding time. It is about 2 min, 50 sec.

111. The two largest contractions occurred at about 2 minutes, 50 seconds and 5 minutes, 40 seconds. The difference in these times, is 2 minutes, 50 seconds, so the frequency is about 1 every 3 minutes.

113.

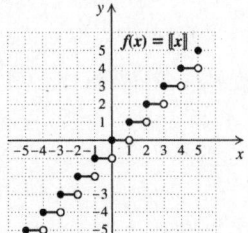

Exercise Set 2.3

1. A y-intercept is the form (f), $(0, b)$.

3. Rise is (e), the difference in y.

5. Slope-intercept form is (a), $y = mx + b$.

7. Graph: $f(x) = 2x - 1$.

We make a table of values. Then we plot the corresponding points and connect them.

x	$f(x)$
1	1
2	3
0	-1
-1	-3

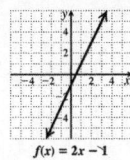

9. Graph: $g(x) = -\frac{1}{3}x + 2$.

We make a table of values. Then we plot the corresponding points and connect them.

x	$g(x)$
-3	3
0	2
3	1
6	0

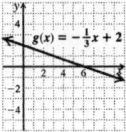

11. Graph: $h(x) = \frac{2}{5}x - 4$.

We make a table of values. Then we plot the corresponding points and connect them.

x	$h(x)$
-5	-6
0	-4
5	-2

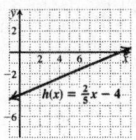

13. $y = 5x + 3$

The y-intercept is $(0, 3)$, or simply 3.

15. $g(x) = -x - 1$

The y-intercept is $(0, -1)$, or simply -1.

17. $y = -\frac{3}{8}x - 4.5$

The y-intercept is $(0, -4.5)$, or simply -4.5.

19. $f(x) = 1.3x - \frac{1}{4}$

The y-intercept is $\left(0, -\frac{1}{4}\right)$, or simply $-\frac{1}{4}$.

21. $y = 17x + 138$

The y-intercept is $(0, 138)$, or simply 138.

23. Slope $= \dfrac{\text{difference in } y}{\text{difference in } x} = \dfrac{11 - 3}{10 - 8} = \dfrac{8}{2} = 4$

25. Slope $= \dfrac{\text{difference in } y}{\text{difference in } x} = \dfrac{-5 - 3}{-4 - (-8)} = \dfrac{-8}{4} = -2$

27. Slope $= \dfrac{\text{difference in } y}{\text{difference in } x} = \dfrac{4 - (-7)}{13 - (-20)} = \dfrac{11}{33} = \dfrac{1}{3}$

29. Slope $= \dfrac{\text{difference in } y}{\text{difference in } x} = \dfrac{\dfrac{2}{3} - \dfrac{1}{6}}{\dfrac{1}{2} - \dfrac{1}{6}} = \dfrac{\dfrac{5}{6}}{\dfrac{2}{6}} = -\dfrac{5}{2}$

31. $\text{Slope} = \dfrac{\text{difference in } y}{\text{difference in } x} = \dfrac{43.6 - 43.6}{4.5 - (-9.7)} = \dfrac{0}{14.2} = 0$

33. $y = \dfrac{5}{2}x - 3$

Slope is $\dfrac{5}{2}$; y-intercept is $(0, -3)$.

From the y-intercept, we go *up* 5 units and to the *right* 2 units. This gives us the point $(2, 2)$. We can now draw the graph.

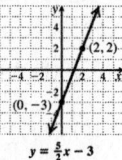

As a check, we can rename the slope and find another point.

$$\dfrac{5}{2} = \dfrac{5}{2} \cdot \dfrac{-1}{-1} = \dfrac{-5}{-2}$$

From the y-intercept, we go *down* 5 units and to the *left* 2 units. This gives us the point $(-2, -8)$. Since $(-2, -8)$ is on the line, we have a check.

35. $f(x) = -\dfrac{5}{2}x + 2$

Slope is $-\dfrac{5}{2}$, or $\dfrac{-5}{2}$; y-intercept is $(0, 2)$.

From the y-intercept, we go *down* 5 units and to the *right* 2 units. This gives us the point $(2, -3)$. We can now draw the graph.

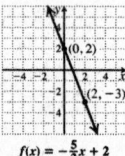

As a check, we can rename the slope and find another point.

$$\dfrac{-5}{2} = \dfrac{5}{-2}$$

From the y-intercept, we go *up* 5 units and to the *left* 2 units. This gives us the point $(-2, 7)$. Since $(-2, 7)$ is on the line, we have a check.

37. $F(x) = 2x + 1$

Slope is 2, or $\dfrac{2}{1}$; y-intercept is $(0, 1)$.

From the y-intercept, we go *up* 2 units and to the *right* 1 unit. This gives us the point $(1, 3)$. We can now draw the graph.

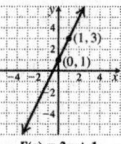

As a check, we can rename the slope and find another point.

$$2 = \dfrac{2}{1} \cdot \dfrac{3}{3} = \dfrac{6}{3}$$

From the y-intercept, we go *up* 6 units and to the *right* 3 units. This gives us the point $(3, 7)$. Since $(3, 7)$ is on the line, we have a check.

39. Convert to a slope-intercept equation.

$$\begin{aligned} 4x + y &= 3 \\ y &= -4x + 3 \end{aligned}$$

Slope is -4, or $\dfrac{-4}{1}$; y-intercept is $(0, 3)$.

From the y-intercept, we go *down* 4 units and to the *right* 1 unit. This gives us the point $(1, -1)$. We can now draw the graph.

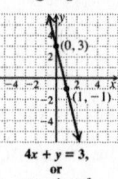

As a check, we can rename the slope and find another point.

$$\dfrac{-4}{1} = \dfrac{-4}{1} \cdot \dfrac{-1}{-1} = \dfrac{4}{-1}$$

From the y-intercept, we go *up* 4 units and to the *left* 1 unit. This gives us the point $(-1, 7)$. Since $(-1, 7)$ is on the line, we have a check.

41. Convert to a slope-intercept equation.

$$\begin{aligned} 6y + x &= 6 \\ 6y &= -x + 6 \\ y &= -\dfrac{1}{6}x + 1 \end{aligned}$$

Slope is $-\dfrac{1}{6}$, or $\dfrac{-1}{6}$; y-intercept is $(0, 1)$.

From the y-intercept, we go *down* 1 unit and to the *right* 6 units. This gives us the point $(6, 0)$. We can now draw the graph.

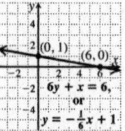

As a check, we choose some other value for x, say -6, and determine y:

$$y = -\frac{1}{6}(-6) + 1 = 1 + 1 = 2$$

We plot the point (–6, 2) and see that it *is* on the line.

43. $g(x) = -0.25x$

Slope is –0.25, or $\frac{-1}{4}$; y-intercept is (0, 0).

From the y-intercept, we go *down* 1 unit and to the *right* 4 units. This gives us the point (4, –1). We can now draw the graph.

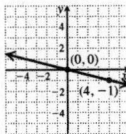

$g(x) = -0.25x$

As a check, we can rename the slope and find another point.

$$\frac{-1}{4} = \frac{-1}{4} \cdot \frac{-1}{-1} = \frac{1}{-4}$$

From the y-intercept, we go *up* 1 unit and to the *left* 4 units. This gives us the point (–4, 1). Since (–4, 1) is on the line, we have a check.

45. Convert to a slope-intercept equation.
$$4x - 5y = 10$$
$$-5y = -4x + 10$$
$$y = \frac{4}{5}x - 2$$

Slope is $\frac{4}{5}$; y-intercept is (0, –2).

From the y-intercept, we go *up* 4 units and to the *right* 5 units. This gives us the point (5, 2). We can now draw the graph.

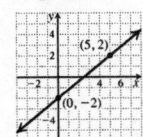

$4x - 5y = 10$, or $y = \frac{4}{5}x - 2$

As a check, we choose some other value for x, say –5, and determine y:

$$y = \frac{4}{5}(-5) - 2 = -4 - 2 = -6$$

We plot the point (–5, –6) and see that it *is* on the line.

47. Convert to a slope-intercept equation.
$$2x + 3y = 6$$
$$3y = -2x + 6$$
$$y = -\frac{2}{3}x + 2$$

Slope is $-\frac{2}{3}$; y-intercept is (0, 2).

From the y-intercept, we go *down* 2 units and to the *right* 3 units. This gives us the point (3, 0). We can now draw

the graph.

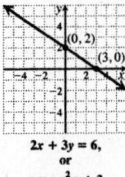

$2x + 3y = 6$,
or
$y = -\frac{2}{3}x + 2$

As a check, we choose some other value for x, say –3, and determine y:

$$y = -\frac{2}{3}(-3) + 2 = 2 + 2 = 4$$

We plot the point (–3, 4) and see that it *is* on the line.

49. Convert to a slope-intercept equation.
$$5 - y = 3x$$
$$-y = 3x - 5$$
$$y = -3x + 5$$

Slope is –3, or $\frac{-3}{1}$; y-intercept is (0, 5).

From the y-intercept, we go *down* 3 units and to the *right* 1 unit. This gives us the point (1, 2). We can now draw the graph.

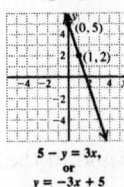

$5 - y = 3x$,
or
$y = -3x + 5$

As a check, we choose some other value for x, say –1, and determine y:

$$y = -3(-1) + 5 = 3 + 5 = 8$$

We plot the point (–1, 8) and see that it *is* on the line.

51. $g(x) = 4.5 = 0x + 4.5$

Slope is 0; y-intercept is (0, 4.5).

From the y-intercept, we go up or down 0 units and any number of nonzero units to the left or right. Any point on the graph will lie on a horizontal line 4.5 units above the x-axis. We draw the graph.

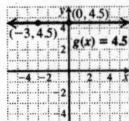

53. Use the slope-intercept equation, $f(x) = mx + b$, with $m = 2$ and $b = 5$.
$$f(x) = mx + b$$
$$f(x) = 2x + 5$$

55. Use the slope-intercept equation, $f(x) = mx + b$,

with $m = -\dfrac{2}{3}$ and $b = -2$.

$$f(x) = mx + b$$
$$f(x) = -\dfrac{2}{3}x - 2$$

57. Use the slope-intercept equation, $f(x) = mx + b$,

with $m = -7$ and $b = \dfrac{1}{3}$.

$$f(x) = mx + b$$
$$f(x) = -7x + \dfrac{1}{3}$$

59. We can use the coordinates of any two points on the line. Let's use (0, 5) and (4, 6).

$$\text{Rate of change} = \frac{\text{change in } y}{\text{change in } x} = \frac{6-5}{4-0} = \frac{1}{4}$$

The distance from home is increasing at a rate of $\dfrac{1}{4}$ km per minute.

61. We can use the coordinates of any two points on the line. We'll use (0, 100) and (9, 40).

$$\text{Rate of change} = \frac{\text{change in } y}{\text{change in } x} = \frac{40-100}{9-0} = \frac{-60}{9} = -\frac{20}{3},$$

or $-6\dfrac{2}{3}$

The distance from the finish line is decreasing at a rate of $6\dfrac{2}{3}$ m per second.

63. We can use the coordinates of any two points on the line. We'll use (3, 2.5) and (6, 4.5).

$$\text{Rate of change} = \frac{\text{change in } y}{\text{change in } x} = \frac{2.5-4.5}{3-6} = \frac{-2}{-3} = \frac{2}{3}$$

The number of bookcases stained is increasing at a rate of $\dfrac{2}{3}$ bookcase per quart of stain used.

65. We can use the coordinates of any two points on the line. We'll use (35, 490) and (45, 500).

$$\text{Rate of change} = \frac{\text{change in } y}{\text{change in } x} = \frac{490-500}{35-45} = \frac{-10}{-10}, \text{ or } 1$$

The average SAT math score is increasing at a rate of 1 point per thousand dollars of family income.

67. a) Graph II indicated that 200 ml of fluid was dripped in the first 3 hr, a rate of $\frac{200}{3}$ ml/hr. It also indicates that 400 ml of fluid was dripped in the next 3 hr, a rate of $\frac{400}{3}$ ml/hr, and that this rate continues until the end of the time period shown. Since the rate of $\frac{400}{3}$ ml/hr is double the rate of $\frac{200}{3}$ ml/hr, this

graph is appropriate for the given situation.

b) Graph IV indicates that 300 ml of fluid was dripped in the first 2 hr, a rate of 300/2, or 150 ml/hr. In the next 2 hr, 200 ml was dripped. This is a rate of 200/2, or 100 ml/hr. Then 100 ml was dripped in the next 3 hr, a rate of 100/3, or $33\frac{1}{3}$ ml/hr. Finally, in the remaining 2 hr, 0 ml of fluid was dripped, a rate of 0/2, or 0 ml/hr. Since the rate at which the fluid was given decreased as time progressed and eventually became 0, this graph is appropriate for the given situation.

c) Graph I is the only graph that shows a constant rate for 5 hours, in this case from 3 PM to 8 PM. Thus, it is appropriate for the given situation.

d) Graph III indicates that 100 ml of fluid was dripped in the first 4 hr, a rate of 100/4, or 25 ml/hr. In the next 3 hr, 200 ml was dripped. This is a rate of 200/3, or $66\frac{2}{3}$ ml/hr. Then 100 ml was dripped in the next hour, a rate of 100 ml/hr. In the last hour 200 ml was dripped, a rate of 200 ml/hr. Since the rate at which the fluid was given gradually increased, this graph is appropriate for the given situation.

69. The skier's speed is given by $\dfrac{\text{change in distance}}{\text{change in time}}$. Note that the skier reaches the 12-km mark 45 min after the 3-km mark was reached or after $15 + 45$, or 60 min. We will express time in hours: $15\,\text{min} = 0.25\,\text{hr}$ and $60\,\text{min} = 1\,\text{hr}$. Then

$$\frac{\text{change in distance}}{\text{change in time}} = \frac{12-3}{1-0.25} = \frac{9}{0.75} = 12.$$

The speed is 12 km/h.

71. The work rate is given by

$$\frac{\text{change in portion of house painted}}{\text{change in time}}.$$

$$\frac{\text{change in portion of house painted}}{\text{change in time}} = \frac{\frac{2}{3} - \frac{1}{4}}{8-0} = \frac{\frac{5}{12}}{8}$$

$$= \frac{5}{12} \cdot \frac{1}{8} = \frac{5}{96}$$

The painter's work rate is $\dfrac{5}{96}$ of the house per hour.

73. The rate at which the number of hits is increasing is given by $\dfrac{\text{change in number of hits}}{\text{change in time}}$.

$$\frac{\text{change in number of hits}}{\text{change in time}} = \frac{430{,}000 - 80{,}000}{2009 - 2007}$$
$$= \frac{350{,}000}{2} = 175{,}000$$

The number of hits is increasing at a rate of 175,000 hits/yr.

75. $C(d) = 0.75d + 30$

0.75 signifies the cost per mile is $0.75; 30 signifies that the minimum cost to rent a truck is $30.

77. $L(t) = \frac{1}{2}t + 5$

$\frac{1}{2}$ signifies that Lauren's hair grows $\frac{1}{2}$ in. per month; 5 signifies that her hair is 5 in. long immediately after she gets it cut.

79. $A(t) = \frac{1}{8}t + 75.5$

$\frac{1}{8}$ signifies that the life expectancy of American women increases $\frac{1}{8}$ yr per year, for years after 1970; 75.5 signifies that the life expectancy in 1970 was 75.5 yr.

81. $P(t) = 0.89t + 16.63$

0.89 signifies that the average price of a ticket increases by $0.89 per year, for years after 2000; 16.63 signifies that the cost of a ticket is $16.63 in 2000.

83. $C(t) = 849t + 5960$

849 signifies that the number of acres of organic cotton increases by 849 acres per year, for years after 2006; 5960 signifies that 5960 acres were planted in organic cotton in 2006.

85. $F(t) = -5000t + 90{,}000$

a) -5000 signifies that the truck's value depreciates $5000 per year; 90,000 signifies that the original value of the truck was $90,000.

b) We find the value of t for which $F(t) = 0$.
$$0 = -5000t + 90{,}000$$
$$5000t = 90{,}000$$
$$t = 18$$
It will take 18 yr for the truck to depreciate completely.

c) The truck's value goes from $90,000 when $t = 0$ to $0 when $t = 18$, so the domain of F is $\{x \mid 0 \le t \le 18\}$.

87. $v(n) = -200n + 1800$

a) -200 signifies that the depreciation is $200 per year; 1800 signifies that the original value of the bike was $1800.

b) We find the value of n for which $v(n) = 600$.
$$600 = -200n + 1800$$
$$-1200 = -200n$$
$$6 = n$$
The trade-in value is $600 after 6 yrs of use.

c) First we find the value of n for which $v(n) = 0$.
$$0 = -200n + 1800$$
$$-1800 = -200n$$
$$9 = n$$
The value of the bike goes from $1800 when $n = 0$, to $0 when $n = 9$, so the domain of v is $\{n \mid 0 \le n \le 9\}$.

89. *Writing Exercise.* As president, it would be a good idea for the federal debt to decrease, so m should be negative.

91. $\dfrac{-8 - (-8)}{6 - (-6)} = \dfrac{-8 + 8}{6 + 6} = \dfrac{0}{12} = 0$

93.
$$3 \cdot 0 - 2y = 9$$
$$0 - 2y = 9$$
$$-2y = 9$$
$$y = -\frac{9}{2}$$

95. $f(x) = 2x - 7$
$$f(0) = 2(0) - 7 = 0 - 7 = -7$$

97. *Writing Exercise.* Yes, the population can be modeled as a linear function with –10%, or –0.10 as the slope and the current population for the y-intercept.

99. a) Graph III indicates that the first 2 mi and the last 3 mi were traveled in approximately the same length of time and at a fairly rapid rate. The mile following the first two miles was traveled at a much slower rate. This could indicate that the first two miles were driven, the next mile was swum and the last three miles were driven, so this graph is most appropriate for the given situation.

b) The slope in Graph IV decreases at 2 mi and again at 3 mi. This could indicate that the first two miles were traveled by bicycle, the next mile was run, and the last 3 miles were walked, so this graph is most appropriate for the given situation.

c) The slope in Graph I decreases at 2 mi and then increases at 3 mi. This could indicate that the first two miles were traveled by bicycle, the next mile was hiked, and the last three miles were traveled by bus, so this graph is most appropriate for the given situation.

d) The slope in Graph II increases at 2 mi and again at 3 mi. This could indicate that the first two miles were hiked, the next mile was run, and the last three miles were traveled by bus, so this graph is most appropriate for the given situation.

101. The longest uphill climb is the widest rising line. It is the trip from Sienna to Castellina in Chianti.

103. Reading from the graph the trip from Castellina in Chianti to Ponte sul Pesa is downhill, then to Panzano is uphill and then to Creve in Chianti is downhill. All sections are about the same grade. So the trip began at Castellina in Chianti.

105. $rx + py = s - ry$
$ry + py = -rx + s$
$y(r + p) = -rx + s$
$$y = -\frac{r}{r+p}x + \frac{s}{r+p}$$

The slope is $-\frac{r}{r+p}$, and the y-intercept is $\left(0, \frac{s}{r+p}\right)$.

107. See the answer section in the text.

109. Let $c = 1$ and $d = 2$. Then
$f(c+d) = f(1+2) = f(3) = 3m + b$, but
$f(c) + f(d) = (m+b) + (2m+b) = 3m + 2b$.
The given statement is false.

111. Let $k = 2$. Then $f(kx) = f(2x) = 2mx + b$, but
$kf(x) = 2(mx + b) = 2mx + 2b$. The given statement is false.

113. a) $\dfrac{-c-(-6c)}{b-5b} = \dfrac{5c}{-4b} = -\dfrac{5c}{4b}$

b) $\dfrac{(d+e)-d}{b-b} = \dfrac{e}{0}$

Since we cannot divide by 0, the slope is undefined.

c) $\dfrac{(-a-d)-(a+d)}{(c-f)-(c+f)} = \dfrac{-a-d-a-d}{c-f-c-f}$
$$= \frac{-2a-2d}{-2f}$$
$$= \frac{-2(a+d)}{-2f}$$
$$= \frac{a+d}{f}$$

115.

$y_1 = 1.4x + 2, \quad y_2 = 0.6x + 2,$
$y_3 = 1.4x + 5, \quad y_4 = 0.6x + 5$

117. *Writing Exercise.* Using algebra, we find that the slope-intercept form of the equation is $y = \frac{5}{2}x - \frac{3}{2}$. This indicates that the y-intercept is $\left(0, -\frac{3}{2}\right)$, so a mistake has been made. It appears that the student graphed $y = \frac{5}{2}x + \frac{3}{2}$.

Exercise Set 2.4

1. Every <u>horizontal</u> line has a slope of 0.

3. The slope of a vertical line is <u>undefined</u>.

5. To find the x-intercept, we let $y = \underline{0}$ and solve the original equation for $\underline{x}$.

7. To solve $3x - 5 = 7$, we can graph $f(x) = 3x - 5$ and $g(x) = 7$ and find the x-value at the point of <u>intersection</u>.

9. Only <u>linear</u> equations have graphs that are straight lines.

11. $y - 2 = 6$
$\quad y = 8$

The graph of $y = 8$ is a horizontal line. Since $y - 2 = 6$ is equivalent to $y = 8$, the slope of the line $y - 2 = 6$ is 0.

13. $8x = 6$
$\quad x = \dfrac{3}{4}$

The graph of $x = \dfrac{3}{4}$ is a vertical line. Since $8x = 6$ is equivalent to $x = \dfrac{3}{4}$, the slope of the line $8x = 6$ is undefined.

15. $3y = 28$

$y = \dfrac{28}{3}$

The graph of $y = \dfrac{28}{3}$ is a horizontal line. Since $3y = 28$ is equivalent to $y = \dfrac{28}{3}$, the slope of the line $3y = 28$ is 0.

17. $5 - x = 12$

$x = -7$

The graph of $x = -7$ is a vertical line. Since $5 - x = 12$ is equivalent to $x = -7$, the slope of the line $5 - x = 12$ is undefined.

19. $2x - 4 = 3$

$2x = 7$

$x = \dfrac{7}{2}$

The graph of $x = \dfrac{7}{2}$ is a vertical line. Since $2x - 4 = 3$ is equivalent to $x = \dfrac{7}{2}$, the slope of the line $2x - 4 = 3$ is undefined.

21. $5y - 4 = 35$

$5y = 39$

$y = \dfrac{39}{5}$

The graph of $y = \dfrac{39}{5}$ is a horizontal line. Since $5y - 4 = 35$ is equivalent to $y = \dfrac{39}{5}$, the slope of the line $5y - 4 = 35$ is 0.

23. $4y - 3x = 9 - 3x$

$y = \dfrac{9}{4}$

The graph of $y = \dfrac{9}{4}$ is a horizontal line. Since $4y - 3x = 9 - 3x$ is equivalent to $y = \dfrac{9}{4}$, the slope of the line $4y - 3x = 9 - 3x$ is 0.

25. $5x - 2 = 2x - 7$

$5x = 2x - 5$

$3x = -5$

$x = -\dfrac{5}{3}$

The graph of $x = -\dfrac{5}{3}$ is a vertical line. Since $5x - 2 = 2x - 7$ is equivalent to $x = -\dfrac{5}{3}$, the slope of the line $5x - 2 = 2x - 7$ is undefined.

27. $y = -\dfrac{2}{3}x + 5$

The equation is written in slope-intercept form. We see that the slope is $-\dfrac{2}{3}$.

29. Graph $y = 4$.

This is a horizontal line that crosses the y-axis at $(0,\ 4)$.

If we find some ordered pairs, note that, for any x-value chosen, y must be 4.

x	y
-2	4
0	4
3	4

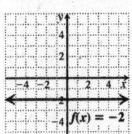

31. Graph $x = 3$.

This is a vertical line that crosses the x-axis at $(3, 0)$. If we find some ordered pairs, note that, for any y-value chosen, x must be 3.

x	y
3	-1
3	0
3	2

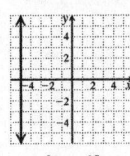

33. Graph $f(x) = -2$.

This is a horizontal line that crosses the y-axis at $(0, -2)$.

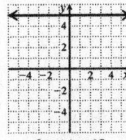

35. Graph $3x = -15$.

Since y does not appear, we solve for x.

$3x = -15$

$x = -5$

This is a vertical line that crosses the x-axis at $(-5, 0)$.

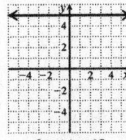

$3x = -15$

37. Graph $3 \cdot g(x) = 15$.

First solve for $g(x)$.

$3 \cdot g(x) = 15$

$g(x) = 5$

This is a horizontal line that crosses the vertical axis at $(0,\ 5)$.

$3 \cdot g(x) = 15$

39. Graph $x + y = 4$.

To find the y-intercept, let $x = 0$ and solve for y.

$0 + y = 4$

$y = 4$

The y-intercept is $(0, 4)$.

To find the x-intercept, let $y = 0$ and solve for x.

$$x + 0 = 4$$
$$x = 4$$

The x-intercept is $(4, 0)$.

Plot these points and draw the line. A third point could be used as a check.

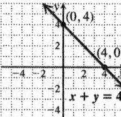

41. Graph $f(x) = 2x - 6$.

Because the function is in slope-intercept form, we know that the y-intercept is $(0, -6)$. To find the x-intercept, let $f(x) = 0$ and solve for x.

$$0 = 2x - 6$$
$$6 = 2x$$
$$3 = x$$

The x-intercept is $(3, 0)$.

Plot these points and draw the line. A third point could be used as a check.

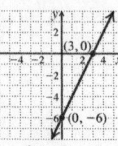

43. Graph $3x + 5y = -15$.

To find the y-intercept, let $x = 0$ and solve for y.

$$3 \cdot 0 + 5y = -15$$
$$5y = -15$$
$$y = -3$$

The y-intercept is $(0, -3)$.

To find the x-intercept, let $y = 0$ and solve for x.

$$3x + 5 \cdot 0 = -15$$
$$3x = -15$$
$$x = -5$$

The x-intercept is $(-5, 0)$.

Plot these points and draw the line. A third point could be used as a check.

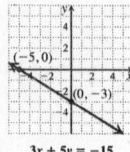

45. Graph $2x - 3y = 18$.

To find the y-intercept, let $x = 0$ and solve for y.

$$2 \cdot 0 - 3y = 18$$
$$-3y = 18$$
$$y = -6$$

The y-intercept is $(0, -6)$.

To find the x-intercept, let $y = 0$ and solve for x.

$$2x - 3 \cdot 0 = 18$$
$$2x = 18$$
$$x = 9$$

The x-intercept is $(9, 0)$.

Plot these points and draw the line. A third point could be used as a check.

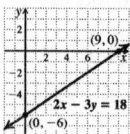

47. Graph $3y = -12x$.

To find the y-intercept, let $x = 0$ and solve for y.

$$3y = -12 \cdot 0$$
$$3y = 0$$
$$y = 0$$

The y-intercept is $(0, 0)$. This is also the x-intercept.

We find another point on the line. Let $x = 1$ and find the corresponding value of y.

$$3y = -12 \cdot 1$$
$$3y = -12$$
$$y = -4$$

The point $(1, -4)$ is on the graph.

Plot these points and draw the line. A third point could be used as a check.

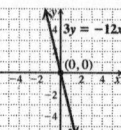

49. Graph $f(x) = 3x - 7$.

Because the function is in slope-intercept form, we know that the y-intercept is $(0, -7)$. To find the x-intercept, let $f(x) = 0$ and solve for x.

$$0 = 3x - 7$$
$$7 = 3x$$
$$\frac{7}{3} = x$$

The x-intercept is $\left(\frac{7}{3}, 0\right)$.

Plot these points and draw the line. A third point could be used as a check.

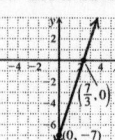

51. Graph $5y - x = 5$.

To find the y-intercept, let $x = 0$ and solve for y.
$$5y - 0 = 5$$
$$5y = 5$$
$$y = 1$$

The y-intercept is $(0, 1)$.

To find the x-intercept, let $y = 0$ and solve for x.
$$5 \cdot 0 - x = 5$$
$$-x = 5$$
$$x = -5$$

The x-intercept is $(-5, 0)$.

Plot these points and draw the line. A third point could be used as a check.

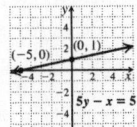

53. $0.2y - 1.1x = 6.6$
$\quad 2y - 11x = 66 \qquad$ Multiplying by 10

Graph $2y - 11x = 66$.

To find the y-intercept, let $x = 0$.
$$2y - 11x = 66$$
$$2y - 11 \cdot 0 = 66$$
$$2y = 66$$
$$y = 33$$

The y-intercept is $(0, 33)$.

To find the x-intercept, let $y = 0$.
$$2y - 11x = 66$$
$$2 \cdot 0 - 11x = 66$$
$$-11x = 66$$
$$x = -6$$

The x-intercept is $(-6, 0)$.

Plot these points and draw the line. A third point could be used as a check.

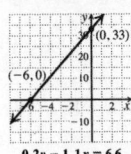

55. $x + 2 = 3$

Graph $f(x) = x + 2$ and $g(x) = 3$ on the same grid.

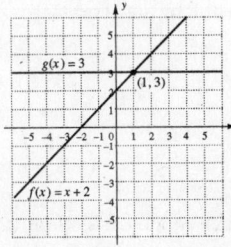

The lines appear to intersect at $(1, 3)$, so the solution is

apparently 1.

Check: $x + 2 = 3$

$\dfrac{}{1 + 2 \mid 3}$

$\quad \overset{?}{3 = 3} \qquad$ TRUE

The solution is 1.

57. $2x + 5 = 1$

Graph $f(x) = 2x + 5$ and $g(x) = 1$ on the same grid.

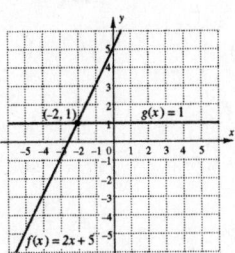

The lines appear to intersect at $(-2, 1)$, so the solution is

apparently -2.

Check: $\quad 2x + 5 = 1$

$\dfrac{}{2(-2) + 5 \mid 1}$

$\quad -4 + 5 \mid$

$\quad \overset{?}{1 = 1} \qquad$ TRUE

The solution is -2.

59. $\frac{1}{2}x + 3 = 5$

Graph $f(x) = \frac{1}{2}x + 3$ and $g(x) = 5$ on the same grid.

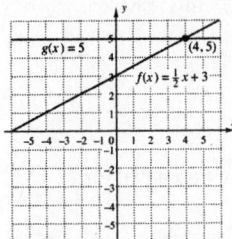

The lines appear to intersect at $(4, 5)$, so the solution is

apparently 4.

Check:

$$\frac{1}{2}x + 3 = 5$$

$$\dfrac{}{\frac{1}{2}(4) + 3 \mid 5}$$

$$2 + 3 \mid$$

$$\overset{?}{5 = 5} \qquad \text{TRUE}$$

The solution is 4.

61. $x - 8 = 3x - 5$

Graph $f(x) = x - 8$ and $g(x) = 3x - 5$ on the same grid.

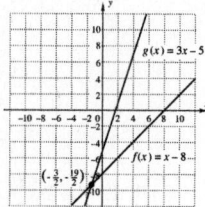

The lines appear to intersect at $\left(-\dfrac{3}{2},\ -\dfrac{19}{2}\right)$, so the solution is apparently $-\dfrac{3}{2}$.

Check:

$$
\begin{array}{c|c}
\multicolumn{2}{c}{x - 8 = 3x - 5}\\
\hline
-\dfrac{3}{2} - 8 & 3\left(-\dfrac{3}{2}\right) - 5\\
-\dfrac{19}{2} & -\dfrac{9}{2} - 5\\
& \overset{?}{\ }\\
-\dfrac{19}{2} &= -\dfrac{19}{2} \quad \text{TRUE}
\end{array}
$$

The solution is $-\dfrac{3}{2}$.

63. $4x + 1 = -x + 11$

Graph $f(x) = 4x + 1$ and $g(x) = -x + 11$ on the same grid.

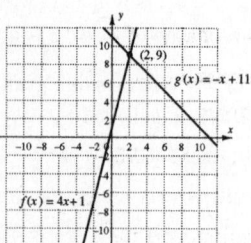

The lines appear to intersect at $(2, 9)$, so the solution is apparently 2.

Check:

$$
\begin{array}{c|c}
\multicolumn{2}{c}{4x + 1 = -x + 11}\\
\hline
4 \cdot 2 + 1 & -2 + 11\\
8 + 1 & \\
& \overset{?}{\ }\\
9 &= 9 \quad \text{TRUE}
\end{array}
$$

The solution is 2.

65. *Familiarize*. After an initial fee of $75, an additional fee of $35 is charge each month. After one month, the total cost is $75 + $35 = $110. After two months, the total cost is $75 + 2 \cdot 35 = \$145$. We can generalize this with a model, letting $C(t)$ represent the total cost, in dollars, for t months of membership.

Translate. We reword the problem and translate.

Total Cost	is	initial cost	plus	\$35 per month
↓	↓	↓	↓	↓
$C(t)$	$=$	75	$+$	$35t$

Carry out. First we write the model in slope-intercept form. $C(t) = 35t + 75$. The vertical intercept is $(0, 75)$, and the slope, or rate, is $35 per month. Plot $(0, 75)$ and from there go up $35 and to the right 1 month. This takes us to $(1, 110)$. Draw a line passing through both points.

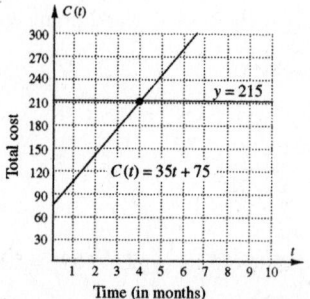

To estimate the time required for the total cost to reach $215, we are estimating the solution of $35t + 75 = 215$. We do this by graphing $y = 215$ and finding the point of intersection of the graphs. This point appears to be $(4, 215)$. Thus, we estimate that it takes 4 months for the total cost to reach $215.

Check. We evaluate.

$$
\begin{aligned}
C(4) &= 35 \cdot 4 + 75\\
&= 140 + 75\\
&= 215
\end{aligned}
$$

The estimate is precise.

State. It takes 4 months for the total cost to reach $215.

67. *Familiarize*. After paying the first $250, the patient must pay $\dfrac{1}{5}$ of all additional charges. For a $1000 bill, the patient pays $250 + \dfrac{1}{5}(\$1000 - \$250)$, or $400. We can generalize this with a model, letting $C(b)$ represent the total cost to the patient, in dollars, for a hospital bill of b dollars.

Translate. We reword the problem and translate.

Total Cost to patient	is	\$250	plus	$\dfrac{1}{5}$	of	Cost in excess of \$250
↓	↓	↓	↓	↓	↓	↓
$C(b)$	$=$	250	$+$	$\dfrac{1}{5}$	$\times$	$(b - 250)$

Carry out. First we rewrite the model in slope-intercept form.

$$C(b) = 250 + \frac{1}{5}(b - 250)$$
$$= 250 + \frac{1}{5}b - 50$$
$$= \frac{1}{5}b + 200$$

The vertical intercept is $(0, 200)$ and the slope is $\frac{1}{5}$. We plot $(0, 200)$ and, from there, to up 1 unit and right 5 units to $(5, 201)$. Draw a line through both points.

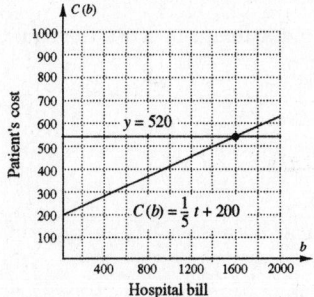

First we will find the value of b for which $C(b) = 520$. Then we will subtract 250 to find by how much Gerry's bill exceeded \$250. We find the solution of $\frac{1}{5}b + 200 = 520$. We graph $y = 520$ and find the point of intersection of the graphs. This point appears to be $(1600, 520)$. Thus, Gerry's total hospital bill was \$1600 and it exceeded \$250 by \$1600 − \$250, or \$1350.

Check. We evaluate.

$$C(1600) = \frac{1}{5}(1600) + 200 = 320 + 200 = 520$$

The estimate is precise.

State. Gerry's hospital bill exceeded \$250 by \$1350.

69. Familiarize. After an initial \$5.00 parking fee, an additional 50¢ fee is changed for each 15-min unit of time. After one 15-min unit of time, the cost is \$5.00 + \$0.50, or \$5.50. After two 15-min units, or 30 min, the cost is \$5.00 + 2(\$0.50), or \$6.00. We can generalize this with a model if we let $C(t)$ represent the total cost, in dollars, for t 15-min units of time.

Translate. We reword the problem and translate.

Total Cost	is	initial cost	plus	\$0.50 per 15-min time unit.
$\downarrow$	$\downarrow$	$\downarrow$	$\downarrow$	$\downarrow$
$C(t)$	$=$	5.00	$+$	$0.50t$

Carry out. First write the model in slope-intercept form: $C(t) = 0.50t + 5$. The vertical intercept is $(0, 5)$ and the slope, or rate, is 0.50, or $\frac{1}{2}$. Plot $(0, 5)$ and from there go

up \$1, and to the right 2 15-min units of time. This takes us to $(2, 6)$. Draw a line passing through both points.

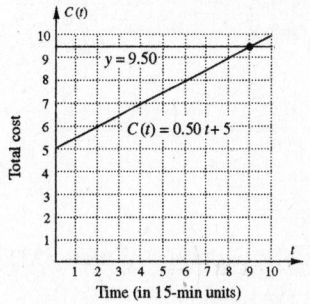

To estimate how long someone can park for \$9.50, we are estimating the solution of $0.50t + 5 = 9.50$. We do this by graphing $y = 9.50$ and finding the point of intersection of the graphs. The point appears to be $(9, 9.50)$. Thus, we estimate that someone can park for nine 15-min units of time, or 2 hr, 15 min, for \$9.50.

Check. We evaluate.

$$C(9) = 0.50(9) + 5$$
$$= 4.50 + 5$$
$$= 9.50$$

The estimate is precise.

State. Someone can park for 2 hr 15 min for \$9.50.

71. Familiarize. In addition to a charge of \$173, FedEx charges \$1.25 per pound in excess of 25 lb. It costs \$173 + \$1.25(30 − 25), or \$179.25 to ship a 30-lb package. It costs \$173 + \$1.25(50 − 25), or \$204.25 to ship a 50-lb package. We can generalize this with a model if we let $C(w)$ represent the cost of shipping a package weighing w lb, where $26 \le w \le 70$.

Translate. We reword the problem and translate.

Shipping Cost	is	\$173 charge	plus	\$1.25 per pound over 25 lbs.
$\downarrow$	$\downarrow$	$\downarrow$	$\downarrow$	$\downarrow$
$C(w)$	$=$	173	$+$	$1.25(w - 25)$

Carry out. First write the model in slope-intercept form: $C(w) = 1.25w + 141.75$.

The vertical intercept is $(0, 141.75)$ and the slope is 1.25, or $\frac{125}{100}$ or $\frac{5}{4}$. Plot $(0, 141.75)$ and from there to up 5 units and to the right 4 units to $(4, 146.75)$. Draw a line passing through both points.

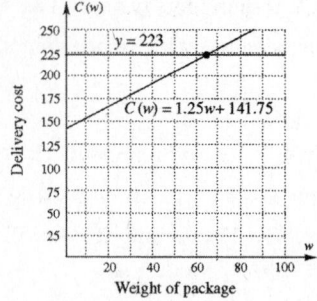

To estimate the weight of a package that costs $223 to ship, we are estimating the solution of $1.25w + 141.75 = 223$. We do this by graphing $y = 223$ and finding the point of intersection of the graphs. The point appears to be $(65, 223)$. Thus, we estimate that a package that costs $223 to ship, weighs 65 lbs.

Check. We evaluate.

$$C(65) = 1.25(65) + 141.75$$
$$= 81.25 + 141.75$$
$$= 223$$

The estimate is precise.

State. A package that costs $223 to ship weighs 65 lbs.

73. $5x - 3y = 15$

This equation is in the standard form for a linear equation, $Ax + By = C$, with $A = 5$, $B = -3$, and $C = 15$. Thus, it is a linear equation.

Solve for y to find the slope.

$$5x - 3y = 15$$
$$-3y = -5x + 15$$
$$y = \frac{5}{3}x - 5$$

The slope is $\frac{5}{3}$.

75. $8x + 40 = 0$

$$8x = -40$$
$$x = -5$$

The equation is linear. Its graph is a vertical line.

77. $4g(x) = 6x^2$

Replace $g(x)$ with y and attempt to write the equation in standard form.

$$4y = 6x^2$$
$$-6x^2 + 4y = 0$$

The equation is not linear, because it has an x^2-term.

79.
$$3y = 7(2x - 4)$$
$$3y = 14x - 28$$
$$-14x + 3y = -28$$

The equation can be written in the standard form for a linear equation, $Ax + By = C$, with $A = -14$, $B = 3$, and

$C = -28$. Thus, it is a linear equation. Solve for y to find the slope.

$$-14x + 3y = -28$$
$$3y = 14x - 28$$
$$y = \frac{14}{3}x - \frac{28}{3}$$

The slope is $\frac{14}{3}$.

81. $f(x) - \frac{5}{x} = 0$

Replace $f(x)$ with y and attempt to write the equation in standard form.

$$y - \frac{5}{x} = 0$$
$$xy - 5 = 0 \qquad \text{Multiplying by } x$$
$$xy = 5$$

The equation is not linear because it has an xy-term.

83.
$$\frac{y}{3} = x$$
$$y = 3x$$
$$3x - y = 0$$

The equation can be written in standard form for the linear equation $Ax + By = C$, with $A = 3$, $B = -1$, and $C = 0$. Thus, it is a linear equation. Solve for y to find the slope.

$$y = 3x$$

The slope is 3.

85. $xy = 10$

The equation is not linear, because it has an xy-term.

87. *Writing Exercise.*

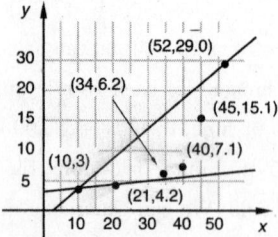

The lines obtained by connecting different pairs of points vary greatly. For example, the line obtained by connecting $(10, 3)$ and $(21, 4.2)$ is quite different from the line obtained by connecting $(10, 3)$ and $(52, 29.0)$. There is no line that can be drawn close to all of the points plotted. Therefore, a linear function does not give a good fit.

89. $-\frac{3}{10}\left(\frac{10}{3}\right) = -\frac{30}{30} = -1$

91. $-3[x - (-1)] = -3[x + 1] = -3x - 3$

93. $\frac{2}{3}\left[x-\left(-\frac{1}{2}\right)\right]-1=\frac{2}{3}\left[x+\frac{1}{2}\right]-1=\frac{2}{3}x+\frac{1}{3}-1=\frac{2}{3}x-\frac{2}{3}$

95. *Writing Exercise.* Consider the equation $Ax+By=C$. Writing this equation in slope-intercept form we have $y=-\frac{A}{B}x+\frac{C}{B}$. If we choose a value for x for which $-Ax+C$ is not a multiple of B, the corresponding y-value will be a fraction. Similarly, if $-Ax$ is a multiple of B but C is not then the corresponding y-value will be a fraction. Under these conditions, Jim will avoid fractions if he graphs using intercepts.

97. The line contains the points $(5,0)$ and $(0,-4)$. We use the points to find the slope.

$$\text{Slope}=\frac{-4-0}{0-5}=\frac{-4}{-5}=\frac{4}{5}$$

Then the slope-intercept equation is $y=\frac{4}{5}x-4$. We rewrite this equation in standard form.

$$
\begin{aligned}
y&=\frac{4}{5}x-4\\
5y&=4x-20 \qquad \text{Multiplying by 5 on both sides}\\
-4x+5y&=-20 \qquad \text{Standard form}
\end{aligned}
$$

This equation can also be written as $4x-5y=20$.

99. $rx+3y=p^2-s$

The equation is in standard form with $A=r$, $B=3$, and $C=p^2-s$. It is linear.

101. Try to put the equation in standard form.

$$
\begin{aligned}
r^2x&=py+5\\
r^2x-py&=5
\end{aligned}
$$

The equation is in standard form with $A=r^2$, $B=-p$, and $C=5$. It is linear.

103. Let equation A have intercepts $(a,0)$ and $(0,b)$. Then equation B has intercepts $(2a,0)$ and $(0,b)$.

Slope of $A=\frac{b-0}{0-a}=-\frac{b}{a}$

Slope of $B=\frac{b-0}{0-2a}=-\frac{b}{2a}=\frac{1}{2}\left(-\frac{b}{a}\right)$

The slope of equation B is $\frac{1}{2}$ the slope of equation A.

105. First write the equation in standard form.

$$
\begin{aligned}
ax+3y&=5x-by+8\\
ax-5x+3y+by&=8 \qquad \text{Adding } -5x+by\\
&\qquad\qquad \text{on both sides}\\
(a-5)x+(3+b)y&=8 \qquad \text{Factoring}
\end{aligned}
$$

If the graph is a vertical line, then the coefficient of y is 0.

$$
\begin{aligned}
3+b&=0\\
b&=-3
\end{aligned}
$$

Then we have $(a-5)x=8$.

If the line passes through $(4,0)$, we have:

$$
\begin{aligned}
(a-5)4&=8 \qquad \text{Substituting 4 for } x\\
a-5&=2\\
a&=7
\end{aligned}
$$

107. We graph $C(t)=0.50t+3$, where t represents the number of 15-min units of time, as a series of steps. The cost is constant within each 15-min unit of time. Thus,

for $0<t\leq1$, $C(t)=0.5(1)+3=\$3.50$;
for $1<t\leq2$, $C(t)=0.5(2)+3=\$4.00$;
for $2<t\leq3$, $C(t)=0.5(3)+3=\$4.50$;

and so on. We draw the graph. An open circle at a point indicates that the point is not on the graph.

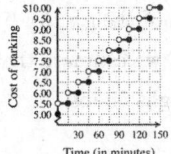

109. Graph $y_1=4x-1$ and $y_2=3-2x$ in the same window and use the Intersect feature to find the first coordinate of the point of intersection, 0.66666667. We check by solving the equation algebraically.

$$
\begin{aligned}
4x-1&=3-2x\\
6x-1&=3\\
6x&=4\\
x&=\frac{2}{3}=0.\overline{6}\approx0.66666667
\end{aligned}
$$

111. Graph $y_1=8-7x$ and $y_2=-2x-5$ in the same window and use the Intersect feature to find the first coordinate of the point of intersection, 2.6. We check by solving the equation algebraically.

$$
\begin{aligned}
8-7x&=-2x-5\\
8-5x&=-5\\
-5x&=-13\\
x&=2.6
\end{aligned}
$$

113. Graph $y=38+4.25x$. Use the Intersect feature to find the x-coordinate that corresponds to the y-coordinate 671.25. We find that 149 shirts were printed.

Exercise Set 2.5

1. False; see page 119 in the text.

3. False; given just one point, there is an infinite number of lines that can be drawn through it.

5. True; see Example 2 in the text.

7. False; see page 122 in the text.

9. True; see page 122 in the text.

11. $y - y_1 = m(x - x_1)$ — Point-slope equation
   $y - 2 = 3(x - 5)$ — Substituting 3 for m, 5 for x_1, and 2 for y_1

To graph the equation, we count off a slope of $\frac{3}{1}$,

starting at $(5, 2)$, and draw the line.

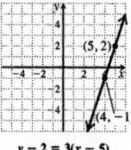

$y - 2 = 3(x - 5)$

13. $y - y_1 = m(x - x_1)$ — Point-slope equation
   $y - 2 = -4(x - 1)$ — Substituting -4 for m, 1 for x_1, and 2 for y_1

To graph the equation, we count off a slope of $-\frac{4}{1}$,

starting at $(1, 2)$, and draw the line.

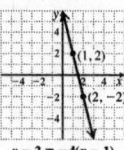

$y - 2 = -4(x - 1)$

15. $y - y_1 = m(x - x_1)$ — Point-slope equation
   $y - (-4) = \frac{1}{2}[x - (-2)]$ — Substituting $\frac{1}{2}$ for m, -2 for x_1, and -4 for y_1

To graph the equation, we count off a slope of $\frac{1}{2}$,

starting at $(-2, -4)$, and draw the line.

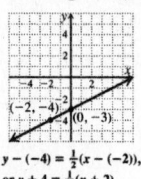

$y - (-4) = \frac{1}{2}(x - (-2))$,
or $y + 4 = \frac{1}{2}(x + 2)$

17. $y - y_1 = m(x - x_1)$ — Point-slope equation
   $y - 0 = -1(x - 8)$ — Substituting -1 for m, 8 for x_1, and 0 for y_1

To graph the equation, we count off a slope of $\frac{-1}{1}$,

starting at $(8, 0)$, and draw the line.

$y - 0 = -1(x - 8)$, or
$y = -(x - 8)$

19. $y - 3 = \frac{1}{4}(x - 5)$
   $y - y_1 = m(x - x_1)$

$m = \frac{1}{4}$, $x_1 = 5$, $y_1 = 3$, so the slope is $\frac{1}{4}$ and a point

(x_1, y_1) on the graph is $(5, 3)$.

21. $y + 1 = -7(x - 2)$
   $y - (-1) = -7(x - 2)$
   $y - y_1 = m(x - x_1)$

$m = -7$, $x_1 = 2$, $y_1 = -1$, so the slope is -7 and a point

(x_1, y_1) on the graph is $(2, -1)$.

23. $y - 6 = -\frac{10}{3}(x + 4)$
   $y - 6 = -\frac{10}{3}(x - (-4))$
   $y - y_1 = m(x - x_1)$

$m = -\frac{10}{3}$, $x_1 = -4$, $y_1 = 6$, so the slope is $-\frac{10}{3}$ and a

point (x_1, y_1) on the graph is $(-4, 6)$.

25. $y = 5x$
   $y - 0 = 5(x - 0)$
   $y - y_1 = m(x - x_1)$

$m = 5$, $x_1 = 0$, $y_1 = 0$, so the slope is 5 and a point

(x_1, y_1) on the graph is $(0, 0)$.

27. $y - y_1 = m(x - x_1)$ — Point-slope equation
   $y - (-4) = 2(x - 1)$ — Substituting 2 for m, 1 for x_1, and -4 for y_1
   $y + 4 = 2x - 2$ — Simplifying
   $y = 2x - 6$ — Subtracting 4 from both sides
   $f(x) = 2x - 6$ — Using function notation

To graph the equation, we count off a slope of $\frac{2}{1}$,

starting at $(1, -4)$ or $(0, -6)$, and draw the line.

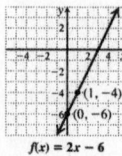

$f(x) = 2x - 6$

29. $y - y_1 = m(x - x_1)$ — Point-slope equation
   $y - 8 = -\frac{3}{5}[x - (-4)]$ — Substituting $-\frac{3}{5}$ for m, -4 for x_1, and 8 for y_1

   $y - 8 = -\frac{3}{5}(x + 4)$

   $y - 8 = -\frac{3}{5}x - \frac{12}{5}$ — Simplifying

   $y = -\frac{3}{5}x + \frac{28}{5}$ — Adding 8 to both sides

   $f(x) = -\frac{3}{5}x + \frac{28}{5}$ — Using function notation

To graph the equation, we count off a slope of $\frac{-3}{5}$, starting at $(-4, 8)$, and draw the line.

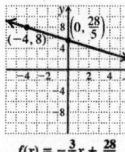

$f(x) = -\frac{3}{5}x + \frac{28}{5}$

31. $\quad y - y_1 = m(x - x_1)$ Point-slope equation

$y - (-4) = -0.6[x - (-3)]$ Substituting -0.6 for m,
$\qquad\qquad\qquad\qquad\qquad -3$ for x_1, and -4 for y_1

$\begin{aligned} y + 4 &= -0.6(x + 3) \\ y + 4 &= -0.6x - 1.8 \\ y &= -0.6x - 5.8 \\ f(x) &= -0.6x - 5.8 \qquad \text{Using function notation} \end{aligned}$

To graph the equation, we count off a slope of -0.6, or

$\dfrac{-3}{5}$, starting at $(-3, -4)$, and draw the line.

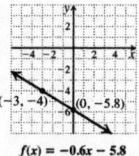

$f(x) = -0.6x - 5.8$

33. $m = \dfrac{2}{7}$; $(0, -6)$

Observe that the slope is $\dfrac{2}{7}$ and the y-intercept is

$(0, -6)$. Thus, we have $f(x) = \dfrac{2}{7}x - 6$.

To graph the equation, we count off a slope of $\dfrac{2}{7}$,

starting at $(0, -6)$, and draw the line.

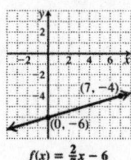

$f(x) = \frac{2}{7}x - 6$

35. $\quad y - y_1 = m(x - x_1)$ Point-slope equation

$y - 6 = \dfrac{3}{5}[x - (-4)]$ Substituting $\dfrac{3}{5}$ for m,
$\qquad\qquad\qquad\qquad\qquad -4$ for x_1, and 6 for y_1

$\begin{aligned} y - 6 &= \dfrac{3}{5}(x + 4) \\ y - 6 &= \dfrac{3}{5}x + \dfrac{12}{5} \\ y &= \dfrac{3}{5}x + \dfrac{42}{5} \\ f(x) &= \dfrac{3}{5}x + \dfrac{42}{5} \qquad \text{Using function notation} \end{aligned}$

To graph the equation, we count off a slope of $\dfrac{3}{5}$,

starting at $(-4, 6)$, and draw the line.

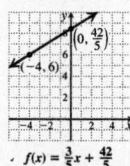

$f(x) = \frac{3}{5}x + \frac{42}{5}$

37. First find the slope of the line:

$$m = \frac{7 - 3}{3 - 2} = \frac{4}{1} = 4$$

Use the point-slope equation with $m = 4$ and

$(2, 3) = (x_1, y_1)$. (We could let $(3, 7) = (x_1, y_1)$ instead

to obtain an equivalent equation.)

$\begin{aligned} y - 3 &= 4(x - 2) \\ y - 3 &= 4x - 8 \\ y &= 4x - 5 \\ f(x) &= 4x - 5 \qquad \text{Using function notation} \end{aligned}$

39. First find the slope of the line:

$$m = \frac{5 - (-4)}{3.2 - 1.2} = \frac{5 + 4}{2} = \frac{9}{2} = 4.5$$

Use the point-slope equation with $m = 4.5$ and

$(1.2, -4) = (x_1, y_1)$.

$\begin{aligned} y - (-4) &= 4.5(x - 1.2) \\ y + 4 &= 4.5x - 5.4 \\ y &= 4.5x - 9.4 \\ f(x) &= 4.5x - 9.4 \qquad \text{Using function notation} \end{aligned}$

41. First find the slope of the line:

$$m = \frac{-1 - (-5)}{0 - 2} = \frac{-1 + 5}{0 - 2} = \frac{4}{-2} = -2$$

Use the point-slope equation with $m = -2$ and

$(0, -1) = (x_1, y_1)$.

$\begin{aligned} y - (-1) &= -2(x - 0) \\ y + 1 &= -2x \\ y &= -2x - 1 \\ f(x) &= -2x - 1 \qquad \text{Using function notation} \end{aligned}$

43. First find the slope of the line:

$$m = \frac{-5 - (-10)}{-3 - (-6)} = \frac{-5 + 10}{-3 + 6} = \frac{5}{3}$$

Use the point-slope equation with $m = \dfrac{5}{3}$ and

$(-3, -5) = (x_1, y_1)$.

$\begin{aligned} y - (-5) &= \dfrac{5}{3}(x - (-3)) \\ y + 5 &= \dfrac{5}{3}(x + 3) \\ y + 5 &= \dfrac{5}{3}x + 5 \\ y &= \dfrac{5}{3}x \\ f(x) &= \dfrac{5}{3}x \qquad \text{Using function notation} \end{aligned}$

45. a) Let t represent the number of years after 2000. We

form the pairs $(2, 14.5)$ and $(7, 19)$. First we find the

slope of the function that fits the data:

$$m = \frac{19 - 14.5}{7 - 2} = \frac{4.5}{5} = 0.9.$$

Use the point-slope equation with $m = 0.9$ and

$(7, 19) = (t_1, a_1)$.

$$a - 19 = 0.9(t - 7)$$
$$a - 19 = 0.9t - 6.3$$
$$a = 0.9t + 12.7$$
$$a(t) = 0.9t + 12.7$$

b) In 2013, $t = 2013 - 2000 = 13$

$$a(13) = 0.9(13) + 12.7 = 24.4 \text{ million cars}$$

c) We substitute 25 for $a(t)$ and solve for t.
$$25 = 0.9t + 12.7$$
$$12.3 = 0.9t$$
$$13.\overline{6} = t \text{ or about 14}$$

There will be 25 million cars produced approximately 14 yrs after 2000 or 2014.

47. a) Let t represent the number of years since 1990, and form the pairs (4, 79.0) and (14, 80.4). First we find the slope of the function that fits the data:
$$m = \frac{80.4 - 79.0}{14 - 4} = \frac{1.4}{10} = 0.14.$$

Use the point-slope equation with $m = 0.14$ and $(4, 79.0) = (t_1, E_1)$.

$$E - 79.0 = 0.14(t - 4)$$
$$E - 79.0 = 0.14t - 0.56$$
$$E = 0.14t + 78.44$$
$$E(t) = 0.14t + 78.44$$

b) In 2012, $t = 2012 - 1990 = 22$

$$E(22) = 0.14(22) + 78.44 = 81.52$$

The life expectancy of females in 2012 is 81.52 yrs.

49. a) Let t represent the number of years since 2000, and form the pairs (2, 282) and (6, 372.1). First we find the slope of the function that fits the data:
$$m = \frac{372.1 - 282}{6 - 2} = \frac{90.1}{4} = 22.525.$$

Use the point-slope equation with $m = 22.525$ and $(2, 282) = (t_1, A_1)$.

$$A - 282 = 22.525(t - 2)$$
$$A - 282 = 22.525t - 45.05$$
$$A = 22.525t + 236.95$$
$$A(t) = 22.525t + 236.95$$

b) In 2010, $t = 2010 - 2000 = 10$

$$A(10) = 22.525(10) + 236.95 = 462.2$$

In 2010, the PAC contributions will be approximately \$462.2 million.

51. a) Let t represent the number of years since 2000, and form the pairs (0, 52.7) and (5, 58.4). First we find the slope of the function that fits the data:
$$m = \frac{58.4 - 52.7}{5 - 0} = \frac{5.7}{5} = 1.14.$$

We know the y-intercept is (0, 52.7), so we write a function in slope-intercept form.

$$N(t) = 1.14t + 52.7$$

b) In 2012, $t = 2012 - 2000 = 12$

$$N(12) = 1.14(12) + 52.7 = 66.38$$

In 2012, approximately 66.38 million tons will be recycled.

53. a) Let t represent the number of years since 2000, and form the pairs (0, 16) and (5, 63). First we find the slope of the function that fits the data:
$$m = \frac{63 - 16}{5 - 0} = \frac{47}{5} = 9.4.$$

We know the y-intercept is (0, 16), so we write a function in slope-intercept form.

$$N(t) = 9.4t + 16$$

b) In 2010, $t = 2010 - 2000 = 10$

$$N(10) = 9.4(10) + 16 = 110$$

In 2010, 110 million Americans will use online banking.

c) We substitute 157 for $N(t)$ and solve for t.
$$157 = 9.4t + 16$$
$$141 = 9.4t$$
$$15 = t$$

There will be 157 million Americans using online banking 15 years after 2000, or 2015.

55. a) Let t represent the number of years after 1990, and form the pairs (4, 74.9) and (15, 79). First we find the slope of the function that fits the data:
$$m = \frac{79 - 74.9}{15 - 4} = \frac{4.1}{11} = \frac{41}{110}.$$

Use the point-slope equation with $m = \frac{41}{110}$ and $(4, 74.9) = (t_1, A_1)$.

$$A - 74.9 = \frac{41}{110}(t - 4)$$
$$A - 74.9 = \frac{41}{110}t - \frac{82}{55}$$
$$A(t) = \frac{41}{110}t + \frac{1615}{22}$$

b) In 2010, $t = 2010 - 1990 = 20$

$$A(20) = \frac{41}{110}(20) + \frac{1615}{22} \approx 80.9 \text{ million acres}$$

In 2010, the amount of land in the National Park system will be approximately 80.9 million acres.

57. We first solve for y and determine the slope of each line.
$$x + 2 = y$$
$$y = x + 2 \qquad \text{Reversing the order}$$

The slope of $y = x + 2$ is 1.

$$y - x = -2$$
$$y = x - 2$$

The slope of $y = x - 2$ is 1.

The slopes are the same; the lines are parallel.

59. We first solve for y and determine the slope of each line.

$$y + 9 = 3x$$
$$y = 3x - 9$$

The slope of $y = 3x - 9$ is 3.

$$3x - y = -2$$
$$3x + 2 = y$$
$$y = 3x + 2 \quad \text{Reversing the order}$$

The slope of $y = 3x + 2$ is 3.

The slopes are the same; the lines are parallel.

61. We determine the slope of each line.

The slope of $f(x) = 3x + 9$ is 3.

$$2y = 8x - 2$$
$$y = 4x - 1$$

The slope of $y = 4x - 1$ is 4.

The slopes are not the same; the lines are not parallel.

63. First solve the equation for y and determine the slope of the given line.

$$x - 2y = 3 \quad \text{Given line}$$
$$-2y = -x + 3$$
$$y = \frac{1}{2}x - \frac{3}{2}$$

The slope of the given line is $\frac{1}{2}$.

The slope of every line parallel to the given line must also be $\frac{1}{2}$. We find the equation of the line with slope $\frac{1}{2}$ and containing the point $(2, 5)$.

$$y - y_1 = m(x - x_1) \quad \text{Point-slope equation}$$
$$y - 5 = \frac{1}{2}(x - 2) \quad \text{Substituting}$$
$$y - 5 = \frac{1}{2}x - 1$$
$$y = \frac{1}{2}x + 4$$

65. First solve the equation for y and determine the slope of the given line.

$$x + y = 7 \quad \text{Given line}$$
$$y = -x + 7$$

The slope of the given line is -1.

The slope of every line parallel to the given line must also be -1. We find the equation of the line with slope -1 and containing the point $(-3, 2)$.

$$y - y_1 = m(x - x_1) \quad \text{Point-slope equation}$$
$$y - 2 = -1(x - (-3)) \quad \text{Substituting}$$
$$y - 2 = -1(x + 3)$$
$$y - 2 = -x - 3$$
$$y = -x - 1$$

67. The slope of $y = 4x + 3$ is 4. The given point $(0, -5)$ is the y-intercept, so we substitute in the slope-intercept equation.

$$y = 4x - 5.$$

69. First solve the equation for y and determine the slope of the given line.

$$2x + 3y = -7 \quad \text{Given line}$$
$$3y = -2x - 7$$
$$y = -\frac{2}{3}x - \frac{7}{3}$$

The slope of the given line is $-\frac{2}{3}$.

The slope of every line parallel to the given line must also be $-\frac{2}{3}$. We find the equation of the line with slope $-\frac{2}{3}$ and containing the point $(-2, -3)$.

$$y - y_1 = m(x - x_1) \quad \text{Point-slope equation}$$
$$y - (-3) = -\frac{2}{3}[x - (-2)] \quad \text{Substituting}$$
$$y + 3 = -\frac{2}{3}(x + 2)$$
$$y + 3 = -\frac{2}{3}x - \frac{4}{3}$$
$$y = -\frac{2}{3}x - \frac{13}{3}$$

71. First solve the equation for y and determine the slope of the given line.

$$3x - 9y = 2 \quad \text{Given line}$$
$$3x - 2 = 9y$$
$$\frac{1}{3}x - \frac{2}{9} = y$$

The slope of the given line is $\frac{1}{3}$.

The slope of every line parallel to the given line must also be $\frac{1}{3}$. We find the equation of the line with slope $\frac{1}{3}$ and containing the point $(-6, 2)$.

$$y - y_1 = m(x - x_1) \quad \text{Point-slope equation}$$
$$y - 2 = \frac{1}{3}[x - (-6)] \quad \text{Substituting}$$
$$y - 2 = \frac{1}{3}(x + 6)$$
$$y - 2 = \frac{1}{3}x + 2$$
$$y = \frac{1}{3}x + 4$$

73. $x = 2$ is a vertical line. A line parallel to it that passes through $(5, -4)$ is the vertical line 5 units to the right of the y-axis, or $x = 5$.

75. We determine the slope of each line.

$$x - 2y = 3$$
$$-2y = -x + 3$$
$$y = \frac{1}{2}x - \frac{3}{2}$$

The slope of $x - 2y = 3$ is $\frac{1}{2}$.

$$4x + 2y = 1$$
$$2y = -4x + 1$$
$$y = -2x + \frac{1}{2}$$

The slope of $4x + 2y = 1$ is -2.

The product of their slopes is $\left(\frac{1}{2}\right)(-2)$, or -1; the lines are perpendicular.

77. We determine the slope of each line.

The slope of $f(x) = 3x + 1$ is 3.

$$6x + 2y = 5$$
$$2y = -6x + 5$$
$$y = -3x + \frac{5}{2}$$

The slope of $6x + 2y = 5$ is -3.

The product of their slopes is $3(-3)$, or $-9 \neq -1$, so the lines are not perpendicular.

79. First solve the equation for y and determine the slope of the given line.

$$2x - 3y = 4 \qquad \text{Given line}$$
$$-3y = -2x + 4$$
$$y = \frac{2}{3}x - \frac{4}{3}$$

The slope of the given line is $\frac{2}{3}$.

The slope of perpendicular line is given by the opposite of the reciprocal of $\frac{2}{3}$, $-\frac{3}{2}$. We find the equation of the line with slope $-\frac{3}{2}$ and containing the point $(3, 1)$.

$$y - y_1 = m(x - x_1) \qquad \text{Point-slope equation}$$
$$y - 1 = -\frac{3}{2}(x - 3) \qquad \text{Substituting}$$
$$y - 1 = -\frac{3}{2}x + \frac{9}{2}$$
$$y = -\frac{3}{2}x + \frac{11}{2}$$

81. First solve the equation for y and determine the slope of the given line.

$$x + y = 6 \qquad \text{Given line}$$
$$y = -x + 6$$

The slope of the given line is -1.

The slope of perpendicular line is given by the opposite of the reciprocal of -1, 1. We find the equation of the line with slope 1 and containing the point $(-4, 2)$.

$$y - y_1 = m(x - x_1) \qquad \text{Point-slope equation}$$
$$y - 2 = 1(x - (-4)) \qquad \text{Substituting}$$
$$y - 2 = x + 4$$
$$y = x + 6$$

83. First solve the equation for y and determine the slope of the given line.

$$3x - y = 2 \qquad \text{Given line}$$
$$-y = -3x + 2$$
$$y = 3x - 2$$

The slope of the given line is 3.

The slope of perpendicular line is given by the opposite of the reciprocal of 3, $-\frac{1}{3}$.. We find the equation of the line with slope $-\frac{1}{3}$ and containing the point $(1, -3)$.

$$y - y_1 = m(x - x_1) \qquad \text{Point-slope equation}$$
$$y - (-3) = -\frac{1}{3}(x - 1) \qquad \text{Substituting}$$
$$y + 3 = -\frac{1}{3}x + \frac{1}{3}$$
$$y = -\frac{1}{3}x - \frac{8}{3}$$

85. First solve the equation for y and find the slope of the given line.

$$3x - 5y = 6$$
$$-5y = -3x + 6$$
$$y = \frac{3}{5}x - \frac{6}{5}$$

The slope of the given line is $\frac{3}{5}$. The slope of a perpendicular line is given by the opposite of the reciprocal of $\frac{3}{5}$, $-\frac{5}{3}$.

We find the equation of the line with slope $-\frac{5}{3}$ and containing the point $(-4, -7)$.

$$y - y_1 = m(x - x_1) \qquad \text{Point-slope equation}$$
$$y - (-7) = -\frac{5}{3}[x - (-4)] \qquad \text{Substituting}$$
$$y + 7 = -\frac{5}{3}(x + 4)$$
$$y + 7 = -\frac{5}{3}x - \frac{20}{3}$$
$$y = -\frac{5}{3}x - \frac{41}{3}$$

87. The slope of a line perpendicular to $2x - 5 = y$ is $-\frac{1}{2}$ and we are given the y-intercept of the desired line, $(0, 6)$. Then we have $y = -\frac{1}{2}x + 6$.

89. $y = 5$ is a horizontal line, so a line perpendicular to it must be vertical. The equation of the vertical line containing $(-3, 7)$ is $x = -3$.

91. *Writing Exercise.* First the slope formula would be used to determine the slope. Then using the slope-intercept formula, the slope and y-intercept are substituted. The result is the equation of the line in slope-intercept form.

93. $(2x^2 - x) + (3x - 5) = 2x^2 - x + 3x - 5$
$$= 2x^2 + (-1 + 3)x - 5$$
$$= 2x^2 + 2x - 5$$

95. $(2t - 1) - (t - 3) = 2t - 1 - t + 3$
$$= (2 - 1)t + (-1 + 3)$$
$$= t + 2$$

97. $f(x) = \dfrac{x}{x - 3}$

Since $\dfrac{x}{x - 3}$ cannot be computed when the denominator

is 0, we find the x-value that causes $x - 3$ to be 0.

$$x - 3 = 0$$
$$x = 3$$

Thus, 3 is not in the domain of f, while all other real numbers are. The domain of f is

$\{x | x$ is a real number and $x \neq 3\}$.

99. $g(x) = |6x + 11|$

Since we can compute $|6x + 11|$ for any real number x, the domain is the set of all real numbers.

101. *Writing Exercise.* In 2004, contributions were \$310.5 million. However, suing the model from Exercise 49

$A(4) = 22.525(4) + 236.95$, or \$327.05 million.

This indicates that the answer to the exercise might be high.

103. *Familiarize.* Celsius temperature C corresponding to a Fahrenheit temperature F can be modeled by a line that contains the points (32, 0) and (212, 100).

Translate. We find an equation relating C and F.

$$m = \frac{100 - 0}{212 - 32} = \frac{100}{180} = \frac{5}{9}$$
$$C - 0 = \frac{5}{9}(F - 32)$$
$$C = \frac{5}{9}(F - 32)$$

Carry out. Using function notation we have

$C(F) = \frac{5}{9}(F - 32)$. Now we find $C(70)$:

$C(70) = \frac{5}{9}(70 - 32) = \frac{5}{9}(38) \approx 21.1$.

Check. We can repeat the calculations. We could also graph the function and determine that (70, 21.1) is on the graph.

State. A temperature of about 21.1°C corresponds to a temperature of 70°F.

105. *Familiarize.* The total cost C of the phone in dollars, after t months, can be modeled by a line that contains the points (5, 410) and (9, 690).

Translate. We find an equation relating C and t.

$$m = \frac{690 - 410}{9 - 5} = \frac{280}{4} = 70$$
$$C - 410 = 70(t - 5)$$
$$C - 410 = 70t - 350$$
$$C = 70t + 60$$

Carry out. Using function notation, we have

$C(t) = 70t + 60$. To find the costs already incurred when the service began, we find $C(0)$:

$$C(0) = 70 \cdot 0 + 60 = 60.$$

Check. We can repeat the calculations. We could also graph the function and determine that (0, 60) is on the graph.

State. Tam had already incurred \$60 in costs when her service just began.

107. We find the value of p for which $A(p) = -2.5p + 26.5$ and $A(p) = 2p - 11$ are the same.

$$-2.5p + 26.5 = 2p - 11$$
$$26.5 = 4.5p - 11$$
$$37.5 = 4.5p$$
$$8.\overline{3} = p$$

Supply will equal demand at a price of about \$8.33 per pound.

109. The price must be a positive number, so we have

$p > 0$. Furthermore, the amount of coffee supplied must be a positive number. We have:

$$2p - 11 > 0$$
$$2p > 11$$
$$p > 5.5$$

Thus, the domain is $\{p | p > 5.5\}$.

111. Find the slope of $5y - kx = 7$:

$$5y - kx = 7$$
$$5y = kx + 7$$
$$y = \frac{k}{5}x + \frac{7}{5}$$

The slope is $\frac{k}{5}$.

Find the slope of the line containing $(7, -3)$ and $(-2, 5)$:

$$m = \frac{5 - (-3)}{-2 - 7} = \frac{5 + 3}{-9} = -\frac{8}{9}$$

If the lines are parallel, their slopes must be equal:

$$\frac{k}{5} = -\frac{8}{9}$$
$$k = -\frac{40}{9}$$

113. *Graphing Calculator and Writing Exercise.*

a) Following the instructions for entering data and using the linear regression option on a graphing calculator, we find the following function:

$f(x) = 0.256x - 1.746.$

b) $f(75) = 0.256(75) - 1.746 \approx 17$

The result we found in Exercise 87 was 19W. The answer found using the linear regression seems more reliable because it was found using a function that is based on more data points than the function in Section 2.2.

113. *Graphing Calculator and Writing Exercise.*

a) Following the instructions for entering data and using the linear regression option on a graphing calculator, we find the following function:

$f(x) = 0.256x - 1.746.$

b) $f(75) = 0.256(75) - 1.746 \approx 17$

The result we found in Exercise 87 was 19W. The answer found using the linear regression seems more reliable because it was found using a function that is based on more data points than the function in Section 2.2.

115. *Graphing Calculator Exercise*

Connecting the Concepts

1. $2x + 5y = 8$ is in standard form.

3. $x - 13 = 5y$ is none of these forms.

5. $x - y = 1$ is in standard form.

7.
$$y = \frac{2}{5}x + 1$$
$$5y = 2x + 5$$
$$-2x + 5y = 5 \quad \text{or} \quad 2x - 5y = -5$$

9. $3x - 5y = 10$
$$-5y = -3x + 10$$
$$y = \frac{3}{5}x - 2$$

11. Graph $y = 2x - 1$.

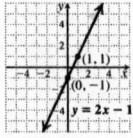

13. Graph $y - 2 = \frac{1}{2}(x - 1)$.

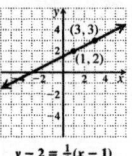

15. Graph $f(x) = -\frac{3}{4}x + 5$.

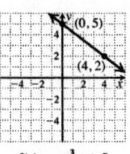

17. $f(x) = -3x + 7$

19. $y = \frac{2}{3}x - 8$

Exercise Set 2.6

1. sum; see page 130 in the text.

3. evaluate; see page 130 in the text.

5. excluding; see page 133 in the text.

7. Since $f(3) = -2 \cdot 3 + 3 = -3$ and $g(3) = 3^2 - 5 = 4$, we have $f(3) + g(3) = -3 + 4 = 1$.

9. Since $f(1) = -2 \cdot 1 + 3 = 1$ and $g(1) = 1^2 - 5 = -4$, we have $f(1) - g(1) = 1 - (-4) = 5$.

11. Since $f(-2) = -2 \cdot (-2) + 3 = 7$ and
$g(-2) = (-2)^2 - 5 = -1$ we have
$f(-2) \cdot g(-2) = 7(-1) = -7$.

13. Since $f(-4) = -2 \cdot (-4) + 3 = 11$ and
$g(-4) = (-4)^2 - 5 = 11$, we have $\dfrac{f(-4)}{g(-4)} = \dfrac{11}{11} = 1$.

15. Since $g(1) = 1^2 - 5 = -4$ and $f(1) = -2 \cdot 1 + 3 = 1$, we have $g(1) - f(1) = -4 - 1 = -5$.

17. $(f + g)(x) = f(x) + g(x) = (-2x + 3) + (x^2 - 5)$
$$= x^2 - 2x - 2$$

19. $(F + G)(x) = F(x) + G(x)$
$$= x^2 - 2 + 5 - x$$
$$= x^2 - x + 3$$

21. $(F-G)(x) = F(x) - G(x)$
$$= x^2 - 2 - (5-x)$$
$$= x^2 - 2 - 5 + x$$
$$= x^2 + x - 7$$

Then we have
$$(F-G)(3) = 3^2 + 3 - 7$$
$$= 9 + 3 - 7$$
$$= 5.$$

23. $(F \cdot G)(x) = F(x) \cdot G(x)$
$$= (x^2 - 2)(5 - x)$$
$$= 5x^2 - x^3 - 10 + 2x$$

Then we have
$$(F \cdot G)(-3) = 5(-3)^2 - (-3)^3 - 10 + 2(-3)$$
$$= 5 \cdot 9 - (-27) - 10 - 6$$
$$= 45 + 27 - 10 - 6$$
$$= 56.$$

25. $(F/G)(x) = F(x)/G(x)$
$$= \frac{x^2 - 2}{5 - x}, \ x \neq 5$$

27. Using our work in Exercise 25, we have
$$(G/F)(-2) = \frac{5 - (-2)}{(-2)^2 - 2} = \frac{5 + 2}{4 - 2} = \frac{7}{2}.$$

29. $(F+F)(x) = F(x) + F(x) = (x^2 - 2) + (x^2 - 2) = 2x^2 - 4$
$(F+F)(x) = 2(1)^2 - 4 = -2$

31. $N(2004) = (C+B)(2004) = C(2004) + B(2004)$
$$\approx 1.2 + 2.9 = 4.1 \text{ million}$$
We estimate the number of births in 2004 to be 4.1 million.

33. $(P-L)(2) = P(2) - L(2) \approx 26.5\% - 22.5\% \approx 4\%$

35. $(p+r)('05) = p('05) + r('05)$
$$\approx 25 + 70 = 95 \text{ million}$$
This represents the number of tons of municipal solid waste that was composted or recycled in 2005.

37. $F('96) \approx 215$ million

This represents the number of tons of municipal solid waste in 1996.

39. $(F-p)('04) = F('04) - p('04)$
$$\approx 260 - 30 = 230 \text{ million}$$
This represents the number of tons of municipal solid waste that was not composted in 2004.

41. The domain of f and of g is all real numbers. Thus,
$$\text{Domain of } f + g = \text{Domain of } f - g = \text{Domain of } f \cdot g$$
$$= \{x \mid x \text{ is a real number}\}.$$

43. Because division by 0 is undefined, we have
$$\text{Domain of } f = \{x \mid x \text{ is a real number } and \ x \neq -5\},$$
and Domain of $g = \{x \mid x \text{ is a real number}\}$.
Thus,
$$\text{Domain of } f + g = \text{Domain of } f - g = \text{Domain of } f \cdot g$$
$$= \{x \mid x \text{ is a real number } and \ x \neq -5\}.$$

45. Because division by 0 is undefined, we have
$$\text{Domain of } f = \{x \mid x \text{ is a real number } and \ x \neq 0\},$$
and Domain of $g = \{x \mid x \text{ is a real number}\}$.
Thus, Domain of
$$f + g = \text{Domain of } f - g = \text{Domain of } f \cdot g$$
$$= \{x \mid x \text{ is a real number } and \ x \neq 0\}.$$

47. Because division by 0 is undefined, we have
$$\text{Domain of } f = \{x \mid x \text{ is a real number } and \ x \neq 1\},$$
and Domain of $g = \{x \mid x \text{ is a real number}\}$.
Thus,
$$\text{Domain of } f + g = \text{Domain of } f - g = \text{Domain of } f \cdot g$$
$$= \{x \mid x \text{ is a real number } and \ x \neq 1\}.$$

49. Because division by 0 is undefined, we have
$$\text{Domain of } f = \left\{x \mid x \text{ is a real number } and \ x \neq -\frac{9}{2}\right\},$$
and Domain of $g = \{x \mid x \text{ is a real number } and \ x \neq 1\}$.
Thus, Domain of $f + g = $ Domain of $f - g = $ Domain of $f \cdot g$
$$= \left\{x \mid x \text{ is a real number } and \ x \neq -\frac{9}{2} \ and \ x \neq 1\right\}.$$

51. Domain of $f = $ Domain of $g = \{x \mid x \text{ is a real number}\}$.
Since $g(x) = 0$ when $x - 3 = 0$, we have $g(x) = 0$ when $x = 3$. We conclude that
$$\text{Domain of } f/g = \{x \mid x \text{ is a real number } and \ x \neq 3\}.$$

53. Domain of $f = $ Domain of $g = \{x \mid x \text{ is a real number}\}$.
Since $g(x) = 0$ when $2x + 8 = 0$, we have $g(x) = 0$ when $x = -4$. We conclude that
$$\text{Domain of } f/g = \{x \mid x \text{ is a real number } and \ x \neq -4\}.$$

55. Domain of $f = \{x \mid x \text{ is a real number and } x \neq 4\}$.
Domain of $g = \{x \mid x \text{ is a real number}\}$.
Since $g(x) = 0$ when $5 - x = 0$, we have $g(x) = 0$ when $x = 5$. We conclude that Domain of
$$f/g = \{x \mid x \text{ is a real number and } x \neq 4 \text{ and } x \neq 5\}.$$

57. Domain of $f = \{x \,|\, x \text{ is a real number and } x \neq -1\}$.

Domain of $g = \{x \,|\, x \text{ is a real number}\}$.

Since $g(x) = 0$ when $2x + 5 = 0$, we have $g(x) = 0$

when $x = -\dfrac{5}{2}$. We conclude that Domain of f/g

$= \left\{x \,\Big|\, x \text{ is a real number and } x \neq -1 \text{ and } x \neq -\dfrac{5}{2}\right\}$.

59. $(F+G)(5) = F(5) + G(5) = 1 + 3 = 4$

$(F+G)(7) = F(7) + G(7) = -1 + 4 = 3$

61. $(G-F)(7) = G(7) - F(7) = 4 - (-1) = 4 + 1 = 5$

$(G-F)(3) = G(3) - F(3) = 1 - 2 = -1$

63. From the graph we see that Domain of $F = \{x \,|\, 0 \leq x \leq 9\}$

and Domain of $G = \{x \,|\, 3 \leq x \leq 10\}$. Then

Domain of $F + G = \{x \,|\, 3 \leq x \leq 9\}$. Since $G(x)$ is never 0,

Domain of $F/G = \{x \,|\, 3 \leq x \leq 9\}$.

65. We use $(F+G)(x) = F(x) + G(x)$.

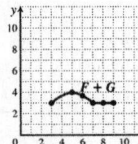

67. *Writing Exercise.* For the years from 1985 through 2000,

Americans consumed more soft drinks than juice, bottled

water and milk combined. We are using the

approximation of $S(t) - [M(t) + J(t) + W(t)]$.

69. $x - 6y = 3$

$\quad -6y = -x + 3$

$\qquad y = \dfrac{1}{6}x - \dfrac{1}{2}$

71. $5x + 2y = -3$

$\quad\ 2y = -5x - 3 \qquad$ Subtracting $5x$

$\quad \dfrac{1}{2} \cdot 2y = \dfrac{1}{2}(-5x - 3) \quad$ Multiplying by $\dfrac{1}{2}$

$\qquad y = -\dfrac{5}{2}x - \dfrac{3}{2}$

73. Let n represent the number; $2n + 5 = 49$.

75. Let n represent the first integer; $x + (x+1) = 145$.

77. *Writing Exercise.* First draw four graphs with the number

of hours after the first dose is taken on the horizontal axis

and the amount of Advil absorbed, in mg, on the vertical

axis. Each graph would show the absorption of one dose

of Advil. Then superimpose the four graphs and, finally,

add the amount of Advil absorbed to create the final

graph.

79. Domain of $F = \{x \,|\, x \text{ is a real number and } x \neq 4\}$.

Domain of $G = \{x \,|\, x \text{ is a real number and } x \neq 3\}$.

$G(x) = 0$ when $x^2 - 4 = 0$, or when $x = 2$ or

$x = -2$. Then Domain of $F/G = \{x \,|\, x \text{ is a real}$

number and $x \neq 4$ and $x \neq 3$ and $x \neq 2$ and $x \neq -2\}$.

81. Answers may vary.

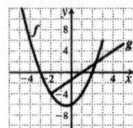

83. The problem states that Domain of $m = \{x \,|\, -1 < x < 5\}$.

Since $n(x) = 0$ when $2x - 3 = 0$, we have $n(x) = 0$ when

$x = \dfrac{3}{2}$. We conclude that Domain of m/n

$= \left\{x \,\Big|\, x \text{ is a real number and } -1 < x < 5 \text{ and } x \neq \dfrac{3}{2}\right\}$.

85. Answers may vary. $f(x) = \dfrac{1}{x+2}$, $g(x) = \dfrac{1}{x-5}$

87. *Graphing Calculator Exercise*

Chapter 2 Review

1. False; see page 83 in the text.

3. True; see page 98 in the text.

5. True; see page 110 in the text.

7. True; see page 109 in the text.

9. False; see page 122 in the text.

11. $\dfrac{x = 2y + 12}{-2 \,|\, 2 \cdot 8 + 12}$

$\qquad -2 \overset{?}{=} 28$

No

13. The first coordinate is negative and the second is positive,

so the point $(-3, 5)$ is in quadrant II.

15. We can use the coordinates of any two points on the line.

Let's use $(2, 75)$ and $(8, 120)$.

Rate of change $= \dfrac{\text{change in } y}{\text{change in } x} = \dfrac{75 - 120}{2 - 8} = \dfrac{-45}{-6} = 7.5$

The value of the apartment is increasing at a rate of

7500 per year.

17. $\text{Slope} = \dfrac{\text{difference in } y}{\text{difference in } x} = \dfrac{5-1}{4-(-3)} = \dfrac{4}{7}$

19. $\text{Slope} = \dfrac{-1-(-2)}{-5-(-1)} = \dfrac{1}{-4} = -\dfrac{1}{4}$

21. $g(x) = -5x - 11$

Slope is –5; y-intercept is $(0, -11)$

23. $C(t) = 11t + 1542$

11 signifies that the number of calories consumed each day increases by 11 per year, for years after 1971. 1542 signifies that the number of calories consumed each day in 1971 was 1542.

25. $-2x = 9$

$$x = -\dfrac{9}{2}$$

The graph of $x = -\dfrac{9}{2}$ is a vertical line. Since $-2x = 9$ is equivalent to $x = -\dfrac{9}{2}$, the slope is undefined.

27. Graph $f(x) = -3x + 2$.

Slope is –3 or $\dfrac{-3}{1}$; y-intercept is $(0, 2)$.

From the y-intercept, we go *down* 3 units and to the *right* 1 unit. This gives us $(1, -1)$. We can now draw the graph.

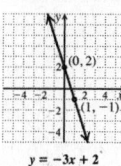

$y = -3x + 2$

29. Graph $y = 6$.

This is a horizontal line that crosses the y-axis at $(0, 6)$. If we find some ordered pairs, note that, for any x-value chosen, y must be 6.

x	y
3	6
0	6
6	6

31. Graph $8x + 32 = 0$.

Since y does not appear, we solve for x.

$$8x + 32 = 0$$
$$x = -4.$$

This is a vertical line that crosses the x-axis at $(-4, 0)$.

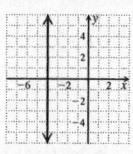

$8x + 32 = 0$

33. Graph $f(x) = \dfrac{1}{2}x - 3$.

Slope is $\dfrac{1}{2}$; y-intercept is $(0, -3)$.

From the y-intercept, we go *up* 1 unit and *right* 2 units. This gives us the point $(2, -2)$. We can now draw the graph.

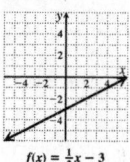

$f(x) = \frac{1}{2}x - 3$

35. $2 - x = 5 + 2x$

Graph $f(x) = 2 - x$ and $g(x) = 5 + 2x$ on the same grid.

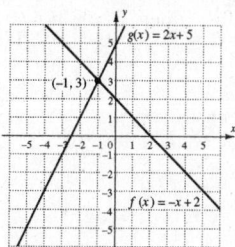

The lines appear to intersect at $(-1, 3)$ so the solution is apparently –1.

Check:

$$
\begin{array}{c|c}
\multicolumn{2}{c}{2 - x = 5 + 2x} \\
\hline
2 - (-1) & 5 + 2(-1) \\
2 + 1 & 5 - 2 \\
& \overset{?}{} \\
3 & = 3 \qquad \text{TRUE}
\end{array}
$$

The solution is –1.

37. First solve for y and determine the slope of each line.

$$y + 5 = -x$$
$$y = -x - 5$$

The slope of $y + 5 = -x$ is –1.

$$x - y = 2$$
$$y = x - 2$$

The slope of $x - y = 2$ is 1.

The product of their slopes is $(-1)(1)$, or –1; the lines are perpendicular.

39. Use the slope-intercept equation $f(x) = mx + b$, with $m = \dfrac{2}{9}$ and $b = -4$.

$$f(x) = mx + b$$
$$f(x) = \dfrac{2}{9}x - 4$$

41. First find the slope of the line:

$$m = \dfrac{6-5}{-2-2} = -\dfrac{1}{4}$$

Use the point-slope equation with $m = -\dfrac{1}{4}$ and

$(2,\,5) = (x_1,\,y_1)$.

$$y - y_1 = m(x - x_1)$$
$$y - 5 = -\frac{1}{4}(x - 2)$$
$$y - 5 = -\frac{1}{4}x + \frac{1}{2}$$
$$y = -\frac{1}{4}x + \frac{11}{2}$$
$$f(x) = -\frac{1}{4}x + \frac{11}{2} \qquad \text{Using function notation}$$

43. First solve the equation for y and determine the slope of the given line.

$$3x - 5y = 9$$
$$y = \frac{3}{5}x - \frac{9}{5}$$

The slope of the given line is $\frac{3}{5}$.

The slope of a perpendicular line is given by the opposite of the reciprocal of $\frac{3}{5}$, $-\frac{5}{3}$.

We find the equation of the line with slope $-\frac{5}{3}$ containing the point (2, –5).

$$y - (-5) = -\frac{5}{3}(x - 2)$$
$$y + 5 = -\frac{5}{3}x + \frac{10}{3}$$
$$y = -\frac{5}{3}x - \frac{5}{3}$$

45. To estimate the cost per gallon in 2008, extend the graph and extrapolate. It appears that in 2008, the cost per gallon is about $2.40.

47.
$$2x - 7 = 0$$
$$x = \frac{7}{2}$$

The equation is linear. The graph is a vertical line.

49. $2x^3 - 7y = 5$

The equation is not linear, because it has an x^3-term.

51. a) Locate 2 on the horizontal axis and find the point on the graph for which 2 is the first coordinate. From this point, look to the vertical axis to find the corresponding y-coordinate, 3. Thus $f(2) = 3$.

b) The set of all x-values in the graph extends from –2 to 4, so the domain is $\{x | -2 \le x \le 4\}$.

c) To determine which member(s) of the domain are with 2, locate 2 on the vertical axis. From there, look left and right to the graph to find any points for which 2 is the second coordinate. One such point exists. Its first coordinate is –1. Thus $f(-1) = 2$.

d) The set of all y-values in the graph extends from 1 to 5, so the range is $\{y | 1 \le y \le 5\}$.

53. We can use the vertical-line test. Visualize moving a vertical line across the graph. No vertical line will intersect the graph more than once. Thus, the graph is a graph of a function.

55. $g(0) = 3 \cdot 0 - 6 = 0 - 6 = -6$

57. $g(a + 5) = 3(a + 5) - 6 = 3a + 15 - 6 = 3a + 9$

59. $\left(\dfrac{g}{h}\right)(-1) = \dfrac{g(-1)}{h(-1)} = \dfrac{3(-1) - 6}{(-1)^2 + 1} = \dfrac{-3 - 6}{1 + 1} = -\dfrac{9}{2}$

61. $g(x) = 3x - 6$

Since we can compute $3x - 6$ for any real number x, the domain is the set of all real numbers.

63. Domain of g = Domain of h = $\{x | x$ is a real number$\}$.

Since $g(x) = 0$ when $3x - 6 = 0$, we have $g(x) = 0$ when $x = 2$. We conclude that

Domain of $h / g = \{x | x$ is a real number and $x \ne 2\}$.

65. *Writing Exercise.* The slope of a line is the rise between two points on the line divided by the run between those points. For a vertical line, there is no run between any two points, and division by 0 is undefined; therefore, the slope is undefined. For a horizontal line, there is no rise between any two points, so the slope is 0/run, or 0.

67. Write both equations in slope-intercept form.

$$y = \frac{3}{4}x - 3 \qquad m = \frac{3}{4}$$
$$y = -\frac{a}{6}x - \frac{3}{2} \qquad m = -\frac{a}{6}$$

For parallel lines, the slopes are the same, or

$$\frac{3}{4} = -\frac{a}{6}$$
$$a = -\frac{9}{2}$$

69. a) Graph III indicates the beginning and end, walking at the same rate, with a rapid rate in between.

b) Graph IV indicates fast bike riding, followed by not as fast running, finishing with slower walking.

c) Graph I indicates fast rate of motorboat followed by constant rate of 0 for fishing, finishing with fast rate of motorboat.

d) Graph II indicates constant rate of 0 while waiting, followed by fast rate of train, and finishing with not as fast run.

Chapter 2 Test

1.
$$x + 4y = -20$$
$$\frac{12 + 4(-3)}{12 - 12} \bigg|\; \frac{-20}{}$$
$$0 \overset{?}{=} -20$$

Since $0 = -20$ is false, $(12, -3)$ is not a solution.

3. Rate of change $= \dfrac{\text{change in } y}{\text{change in } x} = \dfrac{150 - 100}{2007 - 2005} = \dfrac{50}{2} = 25$

The number of people on the National Do Not Call Registry is increasing at a rate of 25 million people per year.

5. Slope $= \dfrac{\text{change in } y}{\text{change in } x} = \dfrac{5.2 - 5.2}{-4.4 - (-3.1)} = \dfrac{0}{-4.4 + 3.1} = 0$

7. Convert to slope-intercept equation.
$$-5y - 2x = 7$$
$$-5y = 2x + 7$$
$$y = -\frac{2}{5}x - \frac{7}{5}$$

Slope is $-\dfrac{2}{5}$; y-intercept is $\left(0, -\dfrac{7}{5}\right)$.

9.
$$x - 5 = 11$$
$$x = 16$$

The graph of $x = 16$ is a vertical line. Since $x - 5 = 11$ is equivalent to $x = 16$, the slope of $x - 5 = 11$ is undefined.

11. Graph $f(x) = -3x + 4$.

Slope is -3 or $\dfrac{-3}{1}$; y-intercept is $(0, 4)$.

From the y-intercept, we go *down* 3 units and to the *right* 1 unit. This gives us $(1, 1)$. We can now draw the graph.

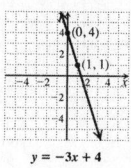

13. Convert to slope-intercept equation.
$$-2x + 5y = 20$$
$$y = \frac{2}{5}x + 4$$

Slope is $\dfrac{2}{5}$; y-intercept is $(0, 4)$.

From the y-intercept, we go *up* 2 units and *right* 5 units. This gives us the point $(5, 6)$. We can now draw the graph.

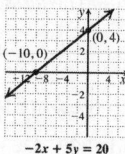

15. $x + 3 = 2x$

Graph $f(x) = x + 3$ and $g(x) = 2x$ on the same grid.

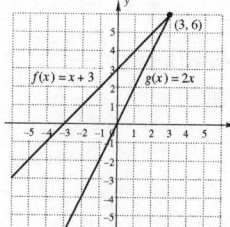

The lines appear to intersect at $(3, 6)$ so the solution is apparently 3.

Check: $x + 3 = 2x$
$$\frac{3 + 3}{} \bigg|\; 2 \cdot 3$$
$$6 \overset{?}{=} 6 \qquad \text{TRUE}$$

The solution is 3.

17. a) $\quad 8x - 7 = 0$
$$x = \frac{7}{8}$$

The equation is linear. (Its graph is a vertical line.)

b) $\quad 4x - 9y^2 = 12$

The equation is not linear, because it has a y^2-term.

c) $\quad 2x - 5y = 3$

The equation is linear.

19. Write both equations in slope-intercept form.
$$y = -2x + 5 \qquad 2y - x = 6$$
$$m = -2 \qquad\qquad y = \frac{1}{2}x + 3$$
$$m = \frac{1}{2}$$

The product of their slopes is $(-2)\left(\dfrac{1}{2}\right)$, or -1; the lines are perpendicular.

21. $y-(-4)=4[x-(-2)]$ or $y+4=4(x+2)$

23. $2x-5y=8$

$$y=\frac{2}{5}x-\frac{8}{5}$$

The slope is $\frac{2}{5}$.

$$y-2=\frac{2}{5}(x+3)$$
$$y-2=\frac{2}{5}x+\frac{6}{5}$$
$$y=\frac{2}{5}x+\frac{16}{5}$$

25. a)   Let $m=$ the number of miles, $C(m)$ represents the cost. We form the pairs (250, 100) and (300, 115).

$$\text{Slope}=\frac{115-100}{300-250}=\frac{15}{50}=\frac{3}{10}=0.3$$

$$y-y_1=m(x-x_1)$$
$$C-100=0.3(m-250)$$
$$C=0.3m+25$$
$$C(m)=0.3m+25$$

b) $C(500)=0.3(500)+25=\$175$

27. $h(-5)=2(-5)+1=-9$

29. Domain of g: $\{x|x \text{ is a real number and } x \neq 0\}$

31. $h(x)=2x+1=0$, if $x=-\frac{1}{2}$

Domain of g/h:

$$\left\{x\Big|x \text{ is a real number and } x \neq 0 \text{ and } x \neq -\frac{1}{2}\right\}$$

33. If the graph is to be parallel to the line $3x-2y=7$, then we must determine the slope:

$$3x-2y=7$$
$$-2y=-3x+7$$
$$y=\frac{3}{2}x-\frac{7}{2} \qquad m=\frac{3}{2}$$

The line must contain the two points $(r, 3)$ and $(7, s)$.

The slope of this line must be $\frac{3}{2}$.

$$\frac{3}{2}=\frac{s-3}{7-r}$$
$$3(7-r)=2(s-3)$$
$$21-3r=2s-6$$
$$21-3r+6=2s$$
$$27-3r=2s$$
$$2s=-3r+27$$
$$s=-\frac{3}{2}r+\frac{27}{2} \text{ or } s=\frac{27-3r}{2}$$

Chapter 3

Solving Systems of Linear Equations and Problem Solving

Exercise Set 3.1

1. False; see Example 4(b).

3. True; see page 150 in the text.

5. True; see Example 4(b).

7. False; see page 153 in the text.

9. We use alphabetical order for the variables. We replace x by 2 and y by 3.

$$\frac{2x-y=1}{\begin{array}{c|c} 2\cdot 2 - 3 & 1 \\ 4 - 3 & \end{array}}$$
$$\overset{?}{1=1} \quad \text{TRUE}$$

$$\frac{5x-3y=1}{\begin{array}{c|c} 5\cdot 2 - 3\cdot 3 & 1 \\ 10 - 9 & \end{array}}$$
$$\overset{?}{1=1} \quad \text{TRUE}$$

The pair (2, 3) makes both equations true, so is it a solution of the system.

11. We use alphabetical order for the variables. We replace x by –5 and y by 1.

$$\frac{x+5y=0}{\begin{array}{c|c} -5+5\cdot 1 & 0 \\ -5+5 & \end{array}}$$
$$\overset{?}{0=0} \quad \text{TRUE}$$

$$\frac{y=2x+9}{\begin{array}{c|c} 1 & 2(-5)+9 \\ & -10+9 \end{array}}$$
$$\overset{?}{1=-1} \quad \text{FALSE}$$

The pair (–5, 1) is not a solution of $y = 2x + 9$.

Therefore, it is not a solution of the system of equations.

13. We replace x by 0 and y by –5.

$$\frac{x-y=5}{\begin{array}{c|c} 0-(-5) & 5 \\ 0+5 & \end{array}}$$
$$\overset{?}{5=5} \quad \text{TRUE}$$

$$\frac{y=3x-5}{\begin{array}{c|c} -5 & 3\cdot 0 - 5 \\ & 0 - 5 \end{array}}$$
$$\overset{?}{-5=-5} \quad \text{TRUE}$$

The pair (0, –5) makes both equations true, so is it a solution of the system.

15. Observe that if we multiply both sides of the first equation by 2, we get the second equation. Thus, if we find that the given points makes the one equation true, we will also know that it makes the other equation true. We replace x by 3 and y by –1 in the first equation.

$$\frac{3x-4y=13}{\begin{array}{c|c} 3\cdot 3 - 4(-1) & 13 \\ 9+4 & \end{array}}$$
$$\overset{?}{13=13} \quad \text{TRUE}$$

The pair (3, –1) makes both equations true, so is it a solution of the system.

17. Graph both equations.

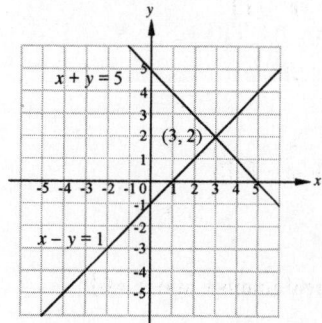

The solution (point of intersection) is apparently (3, 2).

Check:

$$\frac{x-y=1}{\begin{array}{c|c} 3-2 & 1 \end{array}}$$
$$\overset{?}{1=1} \quad \text{TRUE}$$

$$\frac{x+y=5}{\begin{array}{c|c} 3+2 & 5 \end{array}}$$
$$\overset{?}{5=5} \quad \text{TRUE}$$

The solution is (3, 2).

19. Graph the equations.

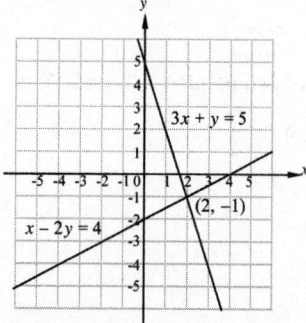

The solution (point of intersection) is apparently (2, –1).

Check:

$$\frac{3x+y=5}{\begin{array}{c|c} 3\cdot 2 + (-1) & 5 \\ 6-1 & \end{array}}$$
$$\overset{?}{5=5} \quad \text{TRUE}$$

$$\frac{x-2y=4}{\begin{array}{c|c} 2-2(-1) & 4 \\ 2+2 & \end{array}}$$
$$\overset{?}{4=4} \quad \text{TRUE}$$

The solution is (2, –1).

21. Graph both equations.

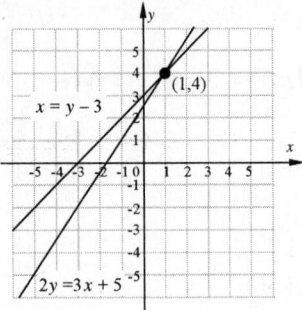

The solution (point of intersection) is apparently (1, 4).

Check:

$$\begin{array}{c|c} 2y=3x+5 \\ \hline 2\cdot 4 \;\big|\; 3\cdot 1+5 \\ 8 \;\big|\; 3+5 \\ ? \\ 8=8 \quad \text{TRUE} \end{array} \qquad \begin{array}{c|c} x=y-3 \\ \hline 1 \;\big|\; 4-3 \\ ? \\ 1=1 \quad \text{TRUE} \end{array}$$

The solution is (1, 4).

23. Graph both equations.

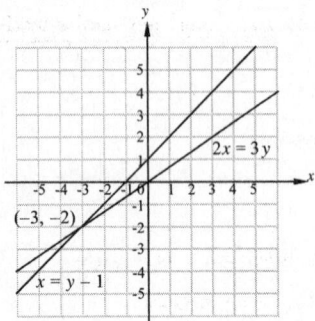

The solution (point of intersection) is apparently (−3, −2).

Check:

$$\begin{array}{c|c} x=y-1 \\ \hline -3 \;\big|\; -2-1 \\ ? \\ -3=-3 \quad \text{TRUE} \end{array} \qquad \begin{array}{c|c} 2x=3y \\ \hline 2(-3) \;\big|\; 3(-2) \\ ? \\ -6=-6 \quad \text{TRUE} \end{array}$$

The solution is (−3, −2).

25. Graph both equations.

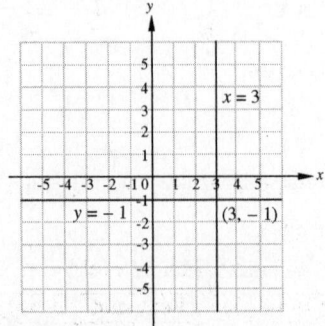

The ordered pairs (3, −1) checks in both equations. It is the solution.

27. Graph both equations.

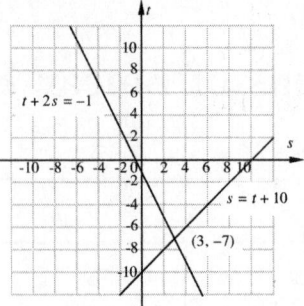

The solution (point of intersection) is apparently (3, −7).

Check:

$$\begin{array}{c|c} t+2s=-1 \\ \hline -7+2\cdot 3 \;\big|\; -1 \\ -7+6 \\ ? \\ -1=-1 \quad \text{TRUE} \end{array} \qquad \begin{array}{c|c} s=t+10 \\ \hline 3 \;\big|\; -7+10 \\ ? \\ 3=3 \quad \text{TRUE} \end{array}$$

The solution is (3, −7).

29. Graph both equations.

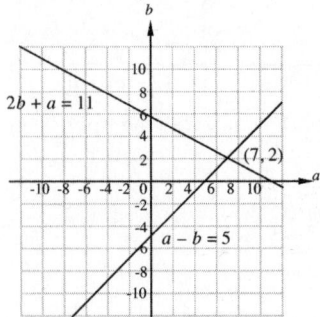

The solution (point of intersection) is apparently (7, 2).

Check:

$$\begin{array}{c|c} 2b+a=11 \\ \hline 2\cdot 2+7 \;\big|\; 11 \\ 4+7 \\ ? \\ 11=11 \quad \text{TRUE} \end{array} \qquad \begin{array}{c|c} a-b=5 \\ \hline 7-2 \;\big|\; 5 \\ ? \\ 5=5 \quad \text{TRUE} \end{array}$$

The solution is (7, 2).

31. Graph both equations.

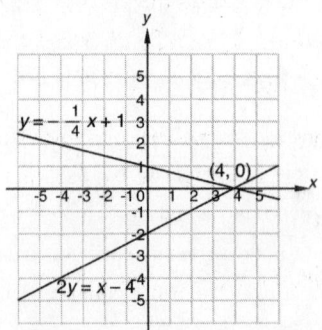

The solution (point of intersection) is apparently (4, 0).

Check:

$$\begin{array}{c|c} y=-\frac{1}{4}x+1 \\ \hline 0 & -\frac{1}{4}\cdot 4+1 \\ & -1+1 \\ \text{?} \\ 0=0 & \text{TRUE} \end{array} \qquad \begin{array}{c|c} 2y=x-4 \\ \hline 2\cdot 0 & 4-4 \\ \text{?} \\ 0=0 & \text{TRUE} \end{array}$$

The solution is $(4, 0)$.

33. Graph both equations.

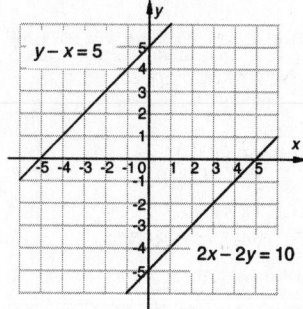

The lines are parallel. The system has no solution.

35. Graph both equations.

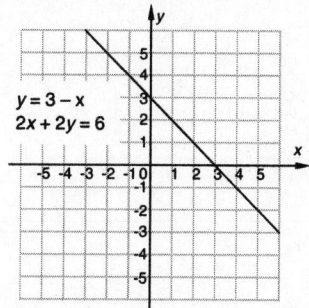

The graphs are the same. Any solution of one equation is a solution of the other. Each equation has infinitely many solutions. The solution set is the set of all pairs (x, y) for which $y=3-x$, or $\{(x, y)\,|\,y=3-x\}$. (In place of $y=3-x$ we could have used $2x+2y=6$ since the two equations are equivalent.)

37. A system of equations is consistent if it has at least one solution. Of the systems under consideration, only the one in Exercise 33 has no solution. Therefore, all except the system in Exercise 33 are consistent.

39. A system of two equations in two variables is dependent if it has infinitely many solutions. Only the system in Exercise 35 is dependent.

41. *Familiarize*. Let $x =$ the first number and $y =$ the second number.

Translate.

The sum of the numbers is 10.

$$x + y = 10$$

The first number is $\frac{2}{3}$ of the second number.

$$x = \frac{2}{3} \times y$$

We have a system of equations:

$$x+y=10,$$
$$x=\frac{2}{3}y.$$

43. *Familiarize*. Let $p =$ the number or personal e-mails and $b =$ the number if business e-mails.

Translate.

Personal e-mails and Business e-mails is 578.

$$p + b = 578$$

Business e-mails was 30 more than personal e-mails.

$$b = 30 + p$$

We have a system of equations:

$$p+b=578,$$
$$b=p+30.$$

45. *Familiarize*. Let $x =$ the measure of one angle and $y =$ the measure of the other angle.

Translate.

Two angles are supplementary.

Rewording: The sum of the measures is $180°$.

$$x+y = 180$$

One angle is 3 less than twice the other.

Rewording: One angle is twice the other angle minus $3°$.

$$x = 2y - 3$$

We have a system of equations:

$$x+y=180,$$
$$x=2y-3$$

47. Familiarize. Let $g =$ the number of field goals and $t =$ the number of foul shots made.

Translate. We organize the information in a table.

Kind of shot	Field goal	Foul shot	Total
Number scored	g	t	64
Points per score	2	1	
Points scored	$2g$	t	100

From the "Number scored" row of the table we get one equation:

$$g + t = 64$$

The "Points scored" row gives us another equation:

$$2g + t = 100$$

We have a system of equations:

$$g + t = 64,$$
$$2g + t = 100$$

49. *Familiarize*. Let $x =$ the number of hats sold and $y =$ the number of tee shirts sold.

***Translate*.** We organize the information in a table.

Souvenir	Hat	Tee shirt	Total
Number sold	x	y	45
Price	$14.50	$19.50	
Amount taken in	$14.50x$	$19.50y$	697.50

The "Number sold" row of the table gives us one equation:

$$x + y = 45.$$

The "Amount taken in" row gives us a second equation:

$$14.50x + 19.50y = 697.50.$$

We can multiply both sides of the second equation by 10 to clear the decimals.

$$x + y = 45,$$
$$145x + 195y = 6975.$$

51. *Familiarize*. Let $h =$ the number of vials of Humalog sold and $n =$ the number of vials of Lantus sold.

***Translate*.** We organize the information in a table.

Brand	Humalog	Lantus	Total
Number sold	h	n	50
Price	$83.29	$76.76	
Amount taken in	$83.29h$	$76.76n$	3981.66

The "Number sold" row of the table gives us one equation:

$$h + n = 50.$$

The "Amount taken in" row gives us a second equation:

$$83.29h + 76.76n = 3981.66.$$

We can multiply both sides of the second equation by 100 to clear the decimals.

$$h + n = 50,$$
$$8329h + 7676n = 398{,}166.$$

53. *Familiarize*. The lacrosse field is a rectangular with perimeter 340 yd. Let $l =$ the length, in yards, and $w =$ the width, in yards. Recall that for a rectangle with length l and width w, the perimeter P is given by

$$P = 2l + 2w.$$

***Translate*.** The formula for perimeter gives us one equation:

$$2l + 2w = 340.$$

The statement relating length and width gives us another equation:

$$l = w + 50.$$

We have a system of equations:

$$2l + 2w = 340,$$
$$l = w + 50.$$

55. *Writing Exercise*. Answers may vary.

The Fever made 37 field goals in a basketball game, some 3-pointers and some 2-pointers. Altogether the 37 baskets counted for 80 points. How many of each type of field goal was made?

57.
$$
\begin{aligned}
3x + 2(5x - 1) &= 6 & \\
3x + 10x - 2 &= 6 & \text{Removing parentheses} \\
13x - 2 &= 6 & \text{Collecting like terms} \\
13x &= 8 & \text{Adding 2 to both sides} \\
x &= \frac{8}{13} &
\end{aligned}
$$

The solution is $\dfrac{8}{13}$.

59.
$$
\begin{aligned}
9y &= 5 - (y + 6) & \\
9y &= 5 - y - 6 & \text{Removing parentheses} \\
9y &= -y - 1 & \text{Collecting like terms} \\
10y &= -1 & \text{Adding } y \text{ to both sides} \\
y &= -\frac{1}{10} &
\end{aligned}
$$

The solution is $-\dfrac{1}{10}$.

61.
$$
\begin{aligned}
3x - y &= 4 & \\
-y &= -3x + 4 & \text{Adding } -3x \text{ to both sides} \\
y &= 3x - 4 & \text{Multiplying both sides by } -1
\end{aligned}
$$

63. *Writing Exercise*. At the point where the lines representing TV and Radio intersect, there was no clear leader because those media had the same market share. Overall, the graph clearly indicates that Newspapers have maintained a higher advertising market share.

65. a) There are many correct answers. One can be found by expressing the sum and difference of the two numbers:

$$x + y = 6,$$
$$x - y = 4$$

b) There are many correct answers. For example, write an equation in two variables. Then write a second equation by multiplying the left side of the first equation by one nonzero constant and multiplying the right side by another nonzero constant.

$$x + y = 1,$$
$$2x + 2y = 3$$

c) There are many correct answers. One can be found by writing an equation in two variables and then writing a nonzero constant multiple of that equation:

$$x + y = 1,$$
$$2x + 2y = 2$$

67. Substitute 4 for x and -5 for y in the first equation:

$$A(4) - 6(-5) = 13$$
$$4A + 30 = 13$$
$$4A = -17$$
$$A = -\frac{17}{4}$$

Substitute 4 for x and -5 for y in the second equation:

$$4 - B(-5) = -8$$
$$4 + 5B = -8$$
$$5B = -12$$
$$B = -\frac{12}{5}$$

We have $A = -\frac{17}{4}$, $B = -\frac{12}{5}$.

69. *Familiarize*. Let $x =$ the number of years Dell has taught and $y =$ the number of years Juanita has taught. Two years ago, Dell and Juanita had taught $x - 2$ and $y - 2$ years, respectively.

Translate.

$$\underbrace{\text{Together, the number of years of service}}_{\downarrow} \quad \underset{\downarrow}{\text{is}} \quad \underset{\downarrow}{46.}$$
$$x + y \qquad\qquad = \quad 46$$

$$\underbrace{\text{Two years ago}}_{}$$
$$\underbrace{\text{Dell had taught 2.5 times as many years as Juanita.}}_{}$$
$$x - 2 = 2.5(y - 2)$$

We have a system of equations:

$$x + y = 46,$$
$$x - 2 = 2.5(y - 2)$$

71. *Familiarize*. Let $b =$ the number of ounces of baking soda and $v =$ the number of ounces of vinegar to be used. The amount of baking soda in the mixture will be four times the amount of vinegar.

Translate.

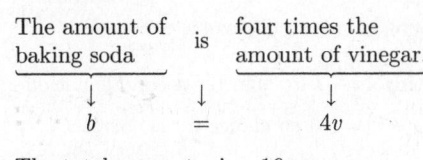

$$\underbrace{\text{The amount of baking soda}}_{\downarrow} \quad \underset{\downarrow}{\text{is}} \quad \underbrace{\text{four times the amount of vinegar.}}_{\downarrow}$$
$$b \qquad\qquad = \qquad\qquad 4v$$

$$\underline{\text{The total amount}} \quad \text{is} \quad \underline{16 \text{ oz.}}$$
$$\downarrow \qquad\qquad \downarrow \qquad \downarrow$$
$$b + v \qquad\qquad = \qquad 16$$

We have a system of equations.

$$b = 4v,$$
$$b + v = 16$$

73. From Exercise 44, graph both equations:

$$v + m = 16$$
$$m = 2v + 4$$

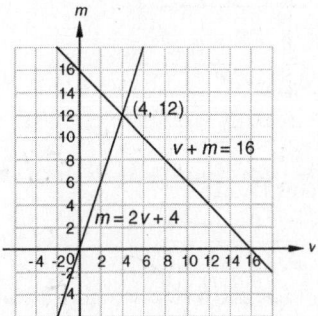

The lines intersect at $v = 4$, $m = 12$. The solution is 4 oz of vinegar and 12 oz of mineral oil.

75. Graph both equations.

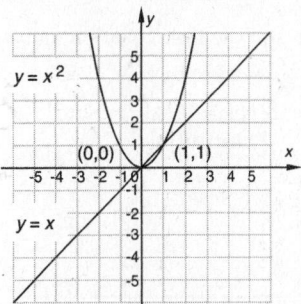

The solutions are apparently $(0, 0)$ and $(1, 1)$. Both pairs check.

77. $(0.07, -7.95)$

79. $(0.00, 1.25)$

Exercise Set 3.2

1. Adding the equations, we get $8x = 7$, so choice (d) is correct.

3. Multiplying the first equation by –5 gives us the system of equations in (a), so choice (a) is correct.

5. Substituting $4x - 7$ for y in the second equation gives us

$6x + 3(4x - 7) = 19$, so choice (c) is correct.

7. $y = 3 - 2x$,   (1)
  $3x + y = 5$ (2)

We substitute $3 - 2x$ for y into the second equation and solve for x.

$$
\begin{array}{ll}
3x + y = 5 & (2) \\
3x + (3 - 2x) = 5 & \text{Substituting} \\
x + 3 = 5 & \\
x = 2 &
\end{array}
$$

Next substitute 2 for x in either equation of the original system and solve for y.

$$
\begin{array}{ll}
y = 3 - 2x & (1) \\
y = 3 - 2 \cdot 2 & \text{Substitute} \\
y = 3 - 4 & \\
y = -1 &
\end{array}
$$

We check the ordered pair $(2, -1)$.

$$
\begin{array}{c|c}
y = 3 - 2x & 3x + y = 5 \\
\hline
-1 \mid 3 - 2 \cdot 2 & 3 \cdot 2 + (-1) \mid 5 \\
 3 - 4 & 6 - 1 \\
\end{array}
$$

$$
\begin{array}{cc}
\overset{?}{-1 = -1} \quad \text{TRUE} & \overset{?}{5 = 5} \quad \text{TRUE}
\end{array}
$$

Since $(2, -1)$ checks, it is the solution.

9. $3x + 5y = 3$,   (1)
  $x = 8 - 4y$ (2)

We substitute $8 - 4y$ for x in the first equation and solve for y.

$$
\begin{array}{ll}
3x + 5y = 3 & (1) \\
3(8 - 4y) + 5y = 3 & \text{Substituting} \\
24 - 12y + 5y = 3 & \\
24 - 7y = 3 & \\
-7y = -21 & \\
y = 3 &
\end{array}
$$

Next we substitute 3 for y in either equation of the original system and solve for x.

$$
\begin{array}{l}
x = 8 - 4y \quad (2) \\
x = 8 - 4 \cdot 3 = 8 - 12 = -4
\end{array}
$$

We check the ordered pair $(-4, 3)$.

$$
\begin{array}{c|c}
x = 8 - 4y & 3x + 5y = 3 \\
\hline
-4 \mid 8 - 4 \cdot 3 & 3(-4) + 5 \cdot 3 \mid 3 \\
 8 - 12 & -12 + 15 \\
\end{array}
$$

$$
\begin{array}{cc}
\overset{?}{-4 = -4} \quad \text{TRUE} & \overset{?}{3 = 3} \quad \text{TRUE}
\end{array}
$$

Since $(-4, 3)$ checks, it is the solution.

11. $3s - 4t = 14$,   (1)
  $5s + t = 8$ (2)

We solve the second equation for t.

$$
\begin{array}{ll}
5s + t = 8 & (2) \\
t = 8 - 5s & (3)
\end{array}
$$

We substitute $8 - 5s$ for t in the first equation and solve for s.

$$
\begin{array}{ll}
3s - 4t = 14 & (1) \\
3s - 4(8 - 5s) = 14 & \text{Substituting} \\
3s - 32 + 20s = 14 & \\
23s - 32 = 14 & \\
23s = 46 & \\
s = 2 &
\end{array}
$$

Next we substitute 2 for s in Equation (1), (2), or (3). It is easiest to use Equation (3) since it is already solved for t.

$$
t = 8 - 5 \cdot 2 = 8 - 10 = -2
$$

We check the ordered pair $(2, -2)$.

$$
\begin{array}{c|c}
3s - 4t = 14 & 5s + t = 8 \\
\hline
3 \cdot 2 - 4(-2) \mid 14 & 5 \cdot 2 + (-2) \mid 8 \\
 6 + 8 & 10 - 2 \\
\end{array}
$$

$$
\begin{array}{cc}
\overset{?}{14 = 14} \quad \text{TRUE} & \overset{?}{8 = 8} \quad \text{TRUE}
\end{array}
$$

Since $(2, -2)$ checks, it is the solution.

13. $4x - 2y = 6$,   (1)
  $2x - 3 = y$ (2)

We substitute $2x - 3$ for y in the first equation and solve for x.

$$
\begin{array}{ll}
4x - 2y = 6 & (1) \\
4x - 2(2x - 3) = 6 & \\
4x - 4x + 6 = 6 & \\
6 = 6 &
\end{array}
$$

We have an identity, or an equation that is always true. The equations are dependent and the solution set is infinite: $\{(x, y) \mid 2x - 3 = y\}$.

15. $-5s + t = 11$,   (1)
  $4s + 12t = 4$ (2)

We solve the first equation for t.

$$
\begin{array}{ll}
-5s + t = 11 & (1) \\
t = 5s + 11 & (3)
\end{array}
$$

We substitute $5s + 11$ for t in the second equation and solve for s.

$$
\begin{array}{ll}
4s + 12t = 4 & (2) \\
4s + 12(5s + 11) = 4 & \\
4s + 60s + 132 = 4 & \\
64s + 132 = 4 & \\
64s = -128 & \\
s = -2 &
\end{array}
$$

Next we substitute –2 for s in Equation (3).

$$
t = 5s + 11 = 5(-2) + 11 = -10 + 11 = 1
$$

We check the ordered pair $(-2, 1)$.

$$\begin{array}{c|c} -5s+t=11 \\ \hline -5(-2)+1 \mid 11 \\ 10+1 \mid \\ ? \\ 11=11 \quad \text{TRUE} \end{array} \qquad \begin{array}{c|c} 4s+12t=4 \\ \hline 4(-2)+12\cdot 1 \mid 4 \\ -8+12 \mid \\ ? \\ 4=4 \quad \text{TRUE} \end{array}$$

Since $(-2, 1)$ checks, it is the solution.

17. $2x+2y=2,$ (1)
$3x-y=1$ (2)

We solve the second equation for y.

$$\begin{aligned} 3x-y&=1 &&(2) \\ -y&=-3x+1 \\ y&=3x-1 &&(3) \end{aligned}$$

We substitute $3x-1$ for y in the first equation and solve for x.

$$\begin{aligned} 2x+2y&=2 &&(1) \\ 2x+2(3x-1)&=2 \\ 2x+6x-2&=2 \\ 8x-2&=2 \\ 8x&=4 \\ x&=\tfrac{1}{2} \end{aligned}$$

Next we substitute $\tfrac{1}{2}$ for x in Equation (3).

$$y=3x-1=3\cdot\tfrac{1}{2}-1=\tfrac{3}{2}-1=\tfrac{1}{2}$$

The ordered pair $\left(\tfrac{1}{2}, \tfrac{1}{2}\right)$ checks in both equations. It is the solution.

19. $2a+6b=4,$ (1)
$3a-b=6$ (2)

We solve the second equation for b.

$$\begin{aligned} 3a-b&=6 &&(2) \\ -b&=-3a+6 \\ b&=3a-6 &&(3) \end{aligned}$$

We substitute $3a-6$ for b in the first equation and solve for a.

$$\begin{aligned} 2a+6b&=4 &&(1) \\ 2a+6(3a-6)&=4 \\ 2a+18a-36&=4 \\ 20a-36&=4 \\ 20a&=40 \\ a&=2 \end{aligned}$$

We substitute 2 for a in Equation 3 and solve for y.

$$\begin{aligned} b&=3a-6 &&(3) \\ b&=3\cdot 2-6 \\ b&=6-6 \\ b&=0 \end{aligned}$$

The ordered pair $(2, 0)$ checks in both equations. It is the solution.

21. $2x-3=y$ (1)
$y-2x=1,$ (2)

We substitute $2x-3$ for y in the second equation and solve for x.

$$\begin{aligned} y-2x&=1 &&(2) \\ 2x-3-2x&=1 &&\text{Substituting} \\ -3&=1 &&\text{Collecting like terms} \end{aligned}$$

We have a contradiction, or an equation that is always false. Therefore, there is no solution.

23. $x+3y=7$ (1)
$\underline{-x+4y=7}$ (2)
$0+7y=14$ Adding
$7y=14$
$y=2$

Substitute 2 for y in one of the original equations and solve for x.

$$\begin{aligned} x+3y&=7 &&(1) \\ x+3\cdot 2&=7 &&\text{Substituting} \\ x+6&=7 \\ x&=1 \end{aligned}$$

Check:

$$\begin{array}{c|c} x+3y=7 \\ \hline 1+3\cdot 2 \mid 7 \\ 1+6 \mid \\ ? \\ 7=7 \quad \text{TRUE} \end{array} \qquad \begin{array}{c|c} -x+4y=7 \\ \hline -1+4\cdot 2 \mid 7 \\ -1+8 \mid \\ ? \\ 7=7 \quad \text{TRUE} \end{array}$$

Since $(1, 2)$ checks, it is the solution.

25. $x-2y=11$ (1)
$\underline{3x+2y=17}$ (2)
$4x=28$ Adding
$x=7$

Substitute 7 for x in Equation (1) and solve for y.

$$\begin{aligned} x-2y&=11 &&(1) \\ 7-2y&=11 &&\text{Substituting} \\ -2y&=4 \\ y&=-2 \end{aligned}$$

We obtain $(7, -2)$. This checks, so it is the solution.

27. $9x+3y=-3$ (1)
$\underline{2x-3y=-8}$ (2)
$11x+0=-11$ Adding
$11x=-11$
$x=-1$

Substitute -1 for x in Equation (1) and solve for y.

$$\begin{aligned} 9x+3y&=-3 \\ 9(-1)+3y&=-3 &&\text{Substituting} \\ -9+3y&=-3 \\ 3y&=6 \\ y&=2 \end{aligned}$$

We obtain $(-1, 2)$. This checks, so it is the solution.

29. $5x+3y=19,$ (1)
$x-6y=11$ (2)

We multiply Equation (1) by 2 to make two terms become opposites.

$$10x + 6y = 38 \qquad \text{Multiplying (1) by 2}$$
$$\underline{x - 6y = 11}$$
$$11x \qquad = 49$$
$$x = \frac{49}{11}$$

Substitute $\frac{49}{11}$ for x in Equation (1) and solve for y.

$$5x + 3y = 19$$
$$5 \cdot \frac{49}{11} + 3y = 19$$
$$3y = -\frac{36}{11}$$
$$y = -\frac{12}{11}$$

We obtain $\left(\frac{49}{11}, -\frac{12}{11}\right)$. This checks, so it is the solution.

31. $5r - 3s = 24, \quad$ (1)
$\quad\;\; 3r + 5s = 28 \quad$ (2)

We multiply twice to make two terms become additive inverses.

From (1): $25r - 15s = 120 \qquad$ Multiplying by 5
From (2): $\underline{9r + 15s = 84} \qquad$ Multiplying by 3
$\qquad\quad 34r + \;\; 0 \; = 204 \qquad$ Adding
$\qquad\qquad\qquad r = 6$

Substitute 6 for r in Equation (2) and solve for s.

$$3r + 5s = 28$$
$$3 \cdot 6 + 5s = 28 \qquad \text{Substituting}$$
$$18 + 5s = 28$$
$$5s = 10$$
$$s = 2$$

We obtain $(6, 2)$. This checks, so it is the solution.

33. $6s + 9t = 12, \quad$ (1)
$\quad\;\; 4s + 6t = 5 \quad$ (2)

We multiply twice to make two terms become opposites.

From (1): $12s + 18t = 24 \qquad$ Multiplying by 2
From (2): $\underline{-12s - 18t = -15} \qquad$ Multiplying by -3
$\qquad\qquad\quad 0 = 9$

We get a contradiction, or an equation that is always false. The system has no solution.

35. $\frac{1}{2}x - \frac{1}{6}y = 10, \quad$ (1)
$\quad\;\; \frac{2}{5}x + \frac{1}{2}y = 8 \quad$ (2)

We first multiply each equation by the LCM of the denominators to clear fractions.

$3x - y = 60, \quad$ (3) $\quad$ Multiplying (1) by 6
$4x + 5y = 80 \quad$ (4) $\quad$ Multiplying (2) by 10

We multiply Equation (1) by 5 and then add.

$15x - 5y = 300 \qquad$ Multiplying (3) by 5
$\underline{4x + 5y = 80} \qquad$ (4)
$19x \qquad = 380$
$\qquad x = 20$

Substitute 20 for x in one of the equations in which the

fractions were cleared and solve for y.

$$3x - y = 60 \qquad (3)$$
$$3 \cdot 20 - y = 60$$
$$60 - y = 60$$
$$-y = 0$$
$$y = 0$$

We obtain $(20, 0)$. This checks, so it is the solution.

37. $\frac{x}{2} + \frac{y}{3} = \frac{7}{6}, \quad$ (1)
$\quad\;\; \frac{2x}{3} + \frac{3y}{4} = \frac{5}{4} \quad$ (2)

We first multiply each equation by the LCM of the denominators to clear fractions.

$3x + 2y = 7 \quad$ (3) $\quad$ Multiplying (1) by 6
$8x + 9y = 15 \quad$ (4) $\quad$ Multiplying (2) by 12

We multiply twice to make two terms become opposites.

From (3): $27x + 18y = 63 \qquad$ Multiplying by 9
From (4): $\underline{-16x - 18y = -30} \qquad$ Multiplying by -2
$\qquad\quad 11x \qquad = 33 \qquad$ Adding
$\qquad\qquad\quad x = 3$

Substitute 3 for x in one of the equations in which the fractions were cleared and solve for y.

$$3x + 2y = 7 \qquad (3)$$
$$3 \cdot 3 + 2y = 7 \qquad \text{Substituting}$$
$$9 + 2y = 7$$
$$2y = -2$$
$$y = -1$$

We obtain $(3, -1)$. This checks, so it is the solution.

39. $12x - 6y = -15, \quad$ (1)
$\quad\;\; -4x + 2y = 5 \quad$ (2)

Observe that, if we multiply Equation (1) by $-\frac{1}{3}$, we obtain Equation (2). Thus, any pair that is a solution of Equation (1) is also a solution of Equation (2). The equations are dependent and the solution set is infinite: $\left\{(x,\ y)\,|\,-4x + 2y = 5\right\}$.

41. $0.3x + 0.2y = 0.3,$
$\quad\;\; 0.5x + 0.4y = 0.4$

We first multiply each equation by 10 to clear decimals.

$3x + 2y = 3, \quad$ (1)
$5x + 4y = 4 \quad$ (2)

We multiply Equation (1) by -2.

$-6x - 4y = -6 \qquad$ Multiplying (1) by -2
$\underline{5x + 4y = 4} \qquad$ (2)
$-x \qquad = -2$
$\qquad x = 2$

Substitute 2 for x in Equation (1) and solve for y.

$$3x + 2y = 3 \quad (1)$$
$$3 \cdot 2 + 2y = 3$$
$$6 + 2y = 3$$
$$2y = -3$$
$$y = -\frac{3}{2}$$

We obtain $\left(2, -\frac{3}{2}\right)$. This checks, so it is the solution.

43. $a - 2b = 16, \quad (1)$
$b + 3 = 3a \quad (2)$

We will use the substitution method. First solve Equation (1) for a.

$$a - 2b = 16$$
$$a = 2b + 16 \quad (3)$$

Now substitute $2b + 16$ for a in Equation (2) and solve for b.

$$b + 3 = 3a \qquad (2)$$
$$b + 3 = 3(2b + 16) \quad \text{Substituting}$$
$$b + 3 = 6b + 48$$
$$-45 = 5b$$
$$-9 = b$$

Substitute -9 for b in Equation (3).

$$a = 2(-9) + 16 = -2$$

We obtain $(-2, -9)$. This checks, so it is the solution.

45. $10x + y = 306, \quad (1)$
$10y + x = 90 \quad (2)$

We will use the substitution method. First solve Equation (1) for y.

$$10x + y = 306$$
$$y = -10x + 306 \quad (3)$$

Now substitute $-10x + 306$ for y in Equation (2) and solve for y.

$$10y + x = 90 \qquad (2)$$
$$10(-10x + 306) + x = 90 \quad \text{Substituting}$$
$$-100x + 3060 + x = 90$$
$$-99x + 3060 = 90$$
$$-99x = -2970$$
$$x = 30$$

Substitute 30 for x in Equation (3).

$$y = -10 \cdot 30 + 306 = 6$$

We obtain $(30, 6)$. This checks, so it is the solution.

47. $6x - 3y = 3, \quad (1)$
$4x - 2y = 2 \quad (2)$

Observe that, if we multiply Equation (1) by $\frac{3}{2}$, we obtain Equation (2). Thus, any pair that is a solution of Equation (1) is also a solution of Equation (2). The equations are dependent and the solution set infinite:
$\{(x, y) | 4x - 2y = 2\}$.

49. $3s - 7t = 5,$
$7t - 3s = 8$

First we rewrite the second equation with the variables in a different order. Then we use the elimination method.

$$3s - 7t = 5 \qquad (1)$$
$$\underline{-3s + 7t = 8} \qquad (2)$$
$$0 = 13$$

We get a contradiction, so the system has no solution.

51. $0.05x + 0.25y = 22,$
$0.15x + 0.05y = 24$

We first multiply each equation by 100 to clear decimals.

$$5x + 25y = 2200, \quad (1)$$
$$15x + 5y = 2400 \quad (2)$$

We multiply by -5 on both sides of the second equation and add.

$$5x + 25y = 2200 \qquad (1)$$
$$\underline{-75x - 25y = -12,000} \quad \text{Multiplying (2) by } -5$$
$$-70x \qquad = -9800 \quad \text{Adding}$$
$$x = \frac{-9800}{-70}$$
$$x = 140$$

Substitute 140 for x in one of the equations in which the decimals were cleared and solve for y.

$$5x + 25y = 2200 \quad (1)$$
$$5 \cdot 140 + 25y = 2200 \quad \text{Substituting}$$
$$700 + 25y = 2200$$
$$25y = 1500$$
$$y = 60$$

We obtain $(140, 60)$. This checks, so it is the solution.

53. $13a - 7b = 9, \quad (1)$
$2a - 8b = 6 \quad (2)$

We will use the elimination method. First we multiply the equations so that the b-terms can be eliminated.

$$\text{From (1): } 104a - 56b = 72 \qquad \text{Multiplying by 8}$$
$$\text{From (2): } \underline{-14a + 56b = -42} \quad \text{Multiplying by } -7$$
$$90a \qquad = 30 \qquad \text{Adding}$$
$$a = \frac{1}{3}$$

Substitute $\frac{1}{3}$ for a in one of the equations and solve for b.

$$2a - 8b = 6 \qquad (1)$$
$$2 \cdot \frac{1}{3} - 8b = 6 \quad \text{Substituting}$$
$$\frac{2}{3} - 8b = 6$$
$$-8b = \frac{16}{3}$$
$$b = -\frac{2}{3}$$

We obtain $\left(\frac{1}{3}, -\frac{2}{3}\right)$. This checks, so it is the solution.

55. *Writing Exercise.* Answers may vary. Form a linear expression in two variables and set it equal to two different constants.

57. *Familiarize.* Let w = the number of kilowatt hours of electricity used each month by the toaster oven. Then $4w$ = the number of kilowatt hours used by the convection oven. The sum of these two numbers is 15 kilowatt hours.

Translate.

kilowatt hours for toaster oven	plus	kilowatt hours for convection oven	is	15.
↓	↓	↓	↓	↓
w	$+$	$4w$	$=$	15

Carry out. We solve the equation.

$$w + 4w = 15$$
$$5w = 15$$
$$w = 3$$

If $w = 3$, then $4w = 12$.

Check. We have 3 kWh + 4(3) kWh = 15 kWh. The answer checks.

State. For the month, the toaster oven uses 3 kWh and the convection oven uses 12 kWh.

59. *Familiarize.* Let p = the original price of the house. The reduced price is $\frac{9}{10}p$ or \$94,500.

Translate.

reduced price	is	$\frac{9}{10}$	of	original price.
↓	↓	↓	↓	↓
$94,500$	$=$	$\frac{9}{10}$	$\cdot$	p

Carry out. We solve the equation.

$$94,500 = \frac{9}{10}p$$
$$945,000 = 9p$$
$$105,000 = p$$

Check. The reduced price is found by multiplying 105,000 by $\frac{9}{10}$ obtaining \$94,500. The answer checks.

State. The original price of the home was \$105,000.

61. *Familiarize.* Let l = the length of the first piece in inches. Then $2l$ = the length of the second piece and $\frac{1}{10}(2l)$ or $\frac{1}{5}l$ = the length of the third piece. The sum of these three lengths is 96 in.

Translate.

First length	plus	second length	plus	third length	is	96.
↓	↓	↓	↓	↓	↓	↓
l	$+$	$2l$	$+$	$\frac{1}{5}l$	$=$	96

Carry out. We solve the equation.

$$2l + l + \frac{1}{5}l = 96$$
$$5l + 10l + l = 480 \qquad \text{Clearing fractions}$$
$$16l = 480$$
$$l = 30$$

If $l = 30$, then $2l = 60$, and $\frac{1}{5}l = 6$. The pieces are 30 in., 60 in. and 6 in.

Check. We have $96 = 30 + 2(30) + \frac{1}{5}(30)$. The answer checks.

State. The pieces are 30 in., 60 in., and 6 in.

63. *Writing Exercise.* Answers may vary. See the box on page 163 of the text.

65. First write $f(x) = mx + b$ as $y = mx + b$. Then substitute 1 for x and 2 for y to get one equation and also substitute –3 for x and 4 for y to get a second equation:

$$2 = m \cdot 1 + b$$
$$4 = m(-3) + b$$

Solve the resulting system of equations.

$$2 = m + b$$
$$4 = -3m + b$$

Multiply the second equation by –1 and add.

$$\begin{aligned} 2 &= m + b \\ -4 &= 3m - b \\ \hline -2 &= 4m \\ -\frac{1}{2} &= m \end{aligned}$$

Substitute $-\frac{1}{2}$ for m in the first equation and solve for b.

$$2 = -\frac{1}{2} + b$$
$$\frac{5}{2} = b$$

Thus, $m = -\frac{1}{2}$ and $b = \frac{5}{2}$.

67. Substitute –4 for x and –3 for y in both equations and solve for a and b.

$$-4a - 3b = -26, \quad (1)$$
$$-4b + 3a = 7 \qquad (2)$$

$$\begin{aligned} -12a - 9b &= -78 \qquad \text{Multiplying (1) by 3} \\ 12a - 16b &= 28 \qquad \text{Multiplying (2) by 4} \\ \hline -25b &= -50 \\ b &= 2 \end{aligned}$$

Substitute 2 for b in Equation (2).

$-4\cdot 2+3a=7$

$3a=15$

$a=5$

Thus, $a=5$ and $b=2$.

69. $\dfrac{x+y}{2}-\dfrac{x-y}{5}=1,$

$\dfrac{x-y}{2}+\dfrac{x+y}{6}=-2$

After clearing fractions we have:

$3x+7y=10,$ (1)

$4x-2y=-12$ (2)

$\begin{array}{ll}6x+14y=20 & \text{Multiplying (1) by 2}\\ \underline{28x-14y=-84} & \text{Multiplying (2) by 7}\\ 34x\quad\quad=-64 \end{array}$

$x=-\dfrac{32}{17}$

Substitute $-\dfrac{32}{17}$ for x in Equation (1).

$3\left(-\dfrac{32}{17}\right)+7y=10$

$7y=\dfrac{266}{17}$

$y=\dfrac{38}{17}$

The solution is $\left(-\dfrac{32}{17},\ \dfrac{38}{17}\right)$.

71. $\dfrac{2}{x}+\dfrac{1}{y}=0,$ $2\cdot\dfrac{1}{x}+\dfrac{1}{y}=0,$.

or

$\dfrac{5}{x}+\dfrac{2}{y}=-5$ $5\cdot\dfrac{1}{x}+2\cdot\dfrac{1}{y}=-5$

Substitute u for $\dfrac{1}{x}$ and v for $\dfrac{1}{y}$.

$2u+v=0,$ (1)

$5u+2v=-5$ (2)

$\begin{array}{ll}-4u-2v=0 & \text{Multiplying (1) by }-2\\ \underline{5u+2v=-5} & \text{(2)}\\ u=-5 \end{array}$

Substitute -5 for u in Equation (1).

$2(-5)+v=0$

$-10+v=0$

$v=10$

If $u=-5$, then $\dfrac{1}{x}=-5$. Thus $x=-\dfrac{1}{5}$.

If $v=10$, then $\dfrac{1}{y}=10$. Thus $y=\dfrac{1}{10}$.

The solution is $\left(-\dfrac{1}{5},\ \dfrac{1}{10}\right)$.

73. *Writing Exercise.* One way is to show the student algebraically that the equations have different intercepts. Another is to solve the system using the elimination method and observe that a contradiction results.

Connecting the Concepts

1. $x=y,$ (1)

$x+y=2$ (2)

Substitute y for x in Equation (2) and solve for y.

$\begin{array}{ll}x+y=2 & \text{(2)}\\ y+y=2 & \text{Substituting}\\ 2y=2 \\ y=1 \end{array}$

Substitute 1 for y in Equation (1) and solve for x.

$x=y$ (1)

$x=1$

We obtain $(1,\ 1)$ as the solution.

3. $y=\dfrac{1}{2}x+1,$ (1)

$y=2x-5$ (2)

Substitute $2x-5$ for y in Equation (1) and solve for x.

$\begin{array}{ll}y=\dfrac{1}{2}x+1 & \text{(1)}\\ 2x-5=\dfrac{1}{2}x+1 & \text{Substituting}\\ 4x-10=x+2 & \text{Clearing fractions}\\ 3x=12 \\ x=4 \end{array}$

Substitute 4 for x in Equation (2) and solve for y.

$\begin{array}{l}y=2x-5 \quad (2)\\ y=2(4)-5=3 \end{array}$

We obtain $(4,\ 3)$ as the solution.

5. $x=5,$ (1)

$y=10$ (2)

We obtain $(5,\ 10)$ as the solution.

7. $2x-y=1,$ (1)

$2y-4x=3$ (2)

We multiply Equation (1) by 2.

$\begin{array}{ll}4x-2y=2 & \text{Multiplying (1) by 2}\\ \underline{-4x+2y=3} & \text{(2)}\\ 0=5 & \text{FALSE} \end{array}$

There is no solution.

9. $x+2y=3,$ (1)

$3x=4-y$ (2)

We solve Equation (2) for y.

$\begin{array}{ll}3x=4-y & \text{(2)}\\ y=4-3x \end{array}$

Substitute $4-3x$ for y in Equation (1) and solve for x.

$\begin{array}{ll}x+2y=3 & \text{(1)}\\ x+2(4-3x)=3 & \text{Substituting}\\ x+8-6x=3 \\ -5x=-5 \\ x=1 \end{array}$

Substitute 1 for x in Equation (2) and solve for y.

$$y = 4 - 3x \quad (1)$$
$$y = 4 - 3(1) = 1$$

We obtain $(1, 1)$ as the solution.

11. $\quad 10x + 20y = 40, \quad (1)$
$\quad\quad x - y = 7 \quad\quad\quad (2)$

We multiply Equation (2) by 20.

$$\begin{array}{ll} 10x + 20y = 40 & (1) \\ \underline{20x - 20y = 140} & \text{Multiplying (2) by 20} \\ 30x \quad\quad = 180 & \\ \quad\quad x = 6 & \end{array}$$

Substitute 6 for x in Equation (2) and solve for y.

$$\begin{array}{l} x - y = 7 \quad (2) \\ 6 - y = 7 \\ \quad -y = 1 \\ \quad\quad y = -1 \end{array}$$

We obtain $(6, -1)$ as the solution.

13. $\quad 2x - 5y = 1, \quad (1)$
$\quad\quad 3x + 2y = 11 \quad (2)$

We multiply Equation (1) by 2 and Equation (2) by 5.

$$\begin{array}{ll} 4x - 10y = 2 & \text{Multiplying (1) by 2} \\ \underline{15x + 10y = 55} & \text{Multiplying (2) by 5} \\ 19x \quad\quad = 57 & \\ \quad\quad x = 3 & \end{array}$$

Substitute 3 for x in Equation (2) and solve for y.

$$\begin{array}{l} 3x + 2y = 11 \quad (2) \\ 3(3) + 2y = 11 \\ 9 + 2y = 11 \\ 2y = 2 \\ y = 1 \end{array}$$

We obtain $(3, 1)$ as the solution.

15. $\quad \begin{array}{ll} 1.1x - 0.3y = 0.8 & (1) \\ \underline{2.3x + 0.3y = 2.6} & (2) \\ 3.4x \quad\quad = 3.4 & \text{Adding} \\ \quad\quad x = 1 & \end{array}$

Substitute 1 for x in Equation (1) and solve for y.

$$\begin{array}{ll} 1.1x - 0.3y = 0.8 & (1) \\ 1.1(1) - 0.3y = 0.8 & \text{Substituting} \\ -0.3y = -0.3 & \\ y = 1 & \end{array}$$

We obtain $(1, 1)$ as the solution.

17. $\quad x - 2y = 5, \quad\quad (1)$
$\quad\quad 3x - 15 = 6y \quad (2)$

We solve Equation (1) for x.

$$\begin{array}{l} x - 2y = 5 \quad (1) \\ x = 2y + 5 \end{array}$$

Substitute $2y + 5$ for x in Equation (2) and solve for y.

$$\begin{array}{ll} 3x - 15 = 6y & (2) \\ 3(2y + 5) - 15 = 6y & \text{Substituting} \\ 6y + 15 - 15 = 6y & \\ 0 = 0 & \end{array}$$

There are many solutions.

The solution set is $\{(x, y) | x - 2y = 5\}$.

19. $\quad 0.2x + 0.7y = 1.2,$
$\quad\quad 0.3x - 0.1y = 2.7$

We first multiply each equation by 10 to clear decimals.

$$\begin{array}{ll} 2x + 7y = 12, & (1) \\ 3x - y = 27 & (2) \end{array}$$

We multiply Equation (2) by 7.

$$\begin{array}{ll} 2x + 7y = 12 & (1) \\ \underline{21x - 7y = 189} & \text{Multiplying (2) by 7} \\ 23x \quad\quad = 201 & \\ \quad\quad x = \dfrac{201}{23} & \end{array}$$

Substitute $\dfrac{201}{23}$ for x in Equation (2) and solve for y.

$$\begin{array}{l} 3x - y = 27 \quad (2) \\ 3\left(\dfrac{201}{23}\right) - y = 27 \\ \dfrac{603}{23} - y = 27 \\ -y = \dfrac{18}{23} \\ y = -\dfrac{18}{23} \end{array}$$

We obtain $\left(\dfrac{201}{23}, -\dfrac{18}{23}\right)$ as the solution.

Exercise Set 3.3

1. The *Familiarize* and *Translate* steps were done in Exercise 41 of Exercise Set 3.1.

Carry out. We solve the system of equations.

$$\begin{array}{ll} x + y = 10, & (1) \\ x = \dfrac{2}{3}y & (2) \end{array}$$

where x is the first number and y is the second number. We use substitution.

$$\dfrac{2}{3}y + y = 10$$
$$\dfrac{5}{3}y = 10$$
$$y = 6$$

Now substitute 6 for y in Equation (2).

$$x = \dfrac{2}{3}(6) = 4$$

Check. The sum of the numbers is $4 + 6$, or 10 and $\dfrac{2}{3}$ times the second number, 6, is the first number 4. The answer checks.

State. The first number is 4, and the second number is 6.

3. The Familiarize and Translate steps were done in Exercise 43 of Exercise Set 3.1.

Carry out. We solve the system of equations.

$$p + b = 578, \quad (1)$$
$$b = p + 30 \quad (2)$$

where p and b represent the number of personal e-mails and business e-mails, respectively. We use substitution. Substitute $p + 30$ for b in Equation (1) and solve for p.

$$p + (p + 30) = 578$$
$$2p + 30 = 578$$
$$2p = 548$$
$$p = 274$$

Now substitute 274 for p in Equation (2).

$$b = 274 + 30 = 304$$

Check. The sum of the personal e-mails and the business e-mails is $274 + 304$, or 578. The answer checks.

State. There are 274 personal e-mails and 304 business e-mails.

5. The Familiarize and Translate steps were done in Exercise 45 of Exercise Set 3.1.

Carry out. We solve the system of equations

$$x + y = 180, \quad (1)$$
$$x = 2y - 3 \quad (2)$$

where x = the measure of one angle and y = the measure of the other angle. We use substitution.

Substitute $2y - 3$ for x in (1) and solve for y.

$$2y - 3 + y = 180$$
$$3y - 3 = 180$$
$$3y = 183$$
$$y = 61$$

Now substitute 61 for y in (2).

$$x = 2 \cdot 61 - 3 = 122 - 3 = 119$$

Check. The sum of the angle measures is $119° + 61°$, or $180°$, so the angles are supplementary. Also $2 \cdot 61° - 3° = 122° - 3° = 119°$. The answer checks.

State. The measures of the angles are $119°$ and $61°$.

7. The Familiarize and Translate steps were done in Exercise 47 of Exercise Set 3.1.

Carry out. We solve the system of equations

$$g + t = 64, \quad (1)$$
$$2g + t = 100 \quad (2)$$

where g = the number of field goals and t = the number of foul shots Chamberlain made. We use elimination.

$$-g - t = -64 \quad \text{Multiplying (1) by} -1$$
$$\underline{2g + t = 100}$$
$$g = 36$$

Substitute 36 for g in (1) and solve for t.

$$36 + t = 64$$
$$t = 28$$

Check. The total number of scores was $36 + 28$, or 64. The total number of points was

$2 \cdot 36 + 28 = 72 + 28 = 100$. The answer checks.

State. Chamberlain made 36 field goals and 28 foul shots.

9. The Familiarize and Translate steps were done in Exercise 49 of Exercise Set 3.1.

Carry out. We solve the system of equations.

$$x + y = 45, \quad (1)$$
$$145x + 195y = 6975 \quad (2)$$

where x is the number of hats and y is the number of tee shirts sold. We use elimination. Begin by multiplying Equation (1) by -145.

$$-145x - 145y = -6525 \quad \text{Multiplying (1) by} -145$$
$$\underline{145x + 195y = 6975} \quad (2)$$
$$50y = 450$$
$$y = 9$$

Substitute 9 for y in Equation (1) and solve for x.

$$x + 9 = 45$$
$$x = 36$$

Check. The number of hats and tee shirts sold is $9 + 36$, or 45. The amount taken in was

$\$14.50(36) + \$19.50(9) = \$522 + \$175.50 = \$697.50$. The answer checks.

State. 36 hats and 9 tee shirts were sold.

11. The Familiarize and Translate steps were done in Exercise 51 of Exercise Set 3.1.

Carry out. We solve the system of equations.

$$h + n = 50, \quad (1)$$
$$8329h + 7676n = 398{,}166 \quad (2)$$

where h = the number or vials of Humalog sold and n = the number of vials of Lantus sold. We use elimination.

$$-7676h - 7676n = -383{,}800 \quad \text{Multiplying (1) by} -7676$$
$$\underline{8329h + 7676n = 398{,}166} \quad (2)$$
$$653h \qquad\quad = 14{,}366$$
$$h = 22$$

Substitute 22 for h in Equation (1) and solve for n.

$$22 + n = 50$$
$$n = 28$$

Check. A total of $22 + 28$, or 50 vials, were sold. The amount collected was

$\$83.29(22) + \$76.76(28) = \$1832.38 + \$2149.28 = \$3981.66$. The answer checks.

State. 22 vials of Humalog and 28 vials of Lantus were sold.

13. The Familiarize and Translate steps were done in Exercise 53 of Exercise Set 3.1.

Carry out. We solve the system of equations.

$$2l + 2w = 340, \quad (1)$$
$$l = w + 50 \quad (2)$$

where l = the length, in yards, and w = the width, in yards of the lacrosse field. We use substitution. We substitute $w + 50$ for l in Equation (1) and solve for w.

$$2(w + 50) + 2w = 340$$
$$2w + 100 + 2w = 340$$
$$4w + 100 = 340$$
$$4w = 240$$
$$w = 60$$

Now substitute 60 for w in Equation (2).

$$l = 60 + 50 = 110$$

Check. The perimeter is $2 \cdot 110 + 2 \cdot 60 = 220 + 120 = 340$. The length, 110 yards, is 50 yards more than the width, 60 yards. The answer checks.

State. The length of the lacrosse field is 110 yards, and the width is 60 yards.

15. *Familiarize*. Let x = the number of reams of regular paper used and y = the number of reams of recycled paper.

Translate. We organize the information in a table.

	Regular	Recycled paper	Total
Number used	x	y	116
Price	$3.79	$5.49	
Total cost	$3.79x$	$5.49y$	582.44

We get one equation from the "Number used" row of the table: $x + y = 116$

The "Total cost" row yields a second equation. All costs are expressed in dollars.

$$3.79x + 5.49y = 582.44.$$

We have the problem translated to a system of equations.

$$x + y = 116 \quad (1)$$
$$3.79x + 5.49y = 582.44 \quad (2)$$

Carry out. We use the elimination method to solve the system of equations.

$$\begin{array}{ll} -379x - 379y = -43{,}964 & \text{Multiplying (1) by } -379 \\ \underline{379x + 549y = 58{,}244} & \text{Multiplying (2) by } 100 \\ 170y = 14{,}280 \\ y = 84 \end{array}$$

Substitute 84 for y in Equation (1) and solve for x.

$$x + 84 = 116$$
$$x = 32$$

Check. A total of 32 + 84, or 116 reams of paper were used. The total cost was $3.79(32) + $5.49(84)

$= \$121.28 + \$461.16 = \$582.44$. The answer checks.

State. 32 reams of regular paper and 84 reams of recycled paper were used.

17. *Familiarize*. Let x = the number of 13-watt bulbs and y = the number of 18-watt bulbs purchased.

Translate. We organize the information in a table.

	13-watt bulbs	18-watt bulbs	Total
Number purchased	x	y	200
Price	$5	$6	
Total cost	$5x$	$6y$	1140

We get our equation from the "Number purchased" row of the table:

$$x + y = 200$$

The "Total cost" row yields a second equation:

$$5x + 6y = 1140$$

We have translated to a system of equations:

$$x + y = 200 \quad (1)$$
$$5x + 6y = 1140 \quad (2)$$

Carry out. We use the elimination method to solve the system of equations.

$$\begin{array}{ll} -5x - 5y = -1000 & \text{Multiplying (1) by } -5 \\ \underline{5x + 6y = 1140} & (2) \\ y = 140 \end{array}$$

Substitute 140 for y in Equation (1) and solve for x.

$$x + 140 = 200$$
$$x = 60$$

Check. A total of 60 + 140, or 200 bulbs, were purchased. The total cost was

$\$5(60) + \$6(140) = \$300 + \$840 = \$1140$. The answer checks.

State. 60 13-watt bulbs and 140 18-watt bulbs were purchased.

19. *Familiarize*. Let a = the number of Apple cartridges purchased and h = the number of HP cartridges.

Translate.

	Apple	HP	Total
Number purchased	a	h	450
Price	$58.99	$64.99	
Total cost	$58.99a$	$64.99h$	27,625.50

We get one equation from the "Number purchased" row of the table:

$$a + h = 450$$

The "Total cost" row yields a second equation:

$$58.99a + 64.99h = 27{,}625.50$$

We have translated to a system of equations:
$$a + h = 450 \qquad (1)$$
$$58.99a + 64.99h = 27{,}625.50 \qquad (2)$$

Carry out. We use the elimination method to solve the system of equations.

$$
\begin{array}{ll}
-5899a - 5899h = -2{,}654{,}550 & \text{Multiplying (1) by } -5899 \\
\underline{5899a + 6499h = 2{,}762{,}550} & \text{Multiplying (2) by } 100 \\
\qquad\quad\; 600h = 108{,}000 & \\
\qquad\qquad\; h = 180 &
\end{array}
$$

Substitute 180 for h in Equation (1) and solve for a.
$$a + 180 = 450$$
$$a = 270$$

Check. A total of $270 + 180$, or 450 cartridges, were sold. The total cost was $\$58.99(270) + \$64.99(180)$
$= \$15{,}927.30 + \$11{,}698.20 = \$27{,}625.50$. The answer checks.

State. 270 Apple cartridges and 180 HP cartridges were purchased.

21. An immediate solution can be determined from the fact that \$12 is the average of equal parts of \$11 and \$13. So there is 14 lb of Mexican and 14 lb of Peruvian. Or we can solve using the usual method.

Familiarize. Let $m =$ the number of pounds of Mexican coffee and $p =$ the number of pounds of Peruvian coffee to be used in the mixture. The value of the mixture will be \$12(28), or \$336.

Translate. We organize the information in a table.

	Mexican	Peruvian	Mixture
Number of pounds	m	p	28
Price per pound	\$13	\$11	\$12
Value of coffee	$13m$	$11p$	336

The "Number of pounds" row of the table gives us one equation:
$$m + p = 28$$

The "Value of coffee" row yields a second equation:
$$13m + 11p = 336$$

We have translated to a system of equations:
$$m + p = 28 \qquad (1)$$
$$13m + 11p = 336 \qquad (2)$$

Carry out. We use the elimination method to solve the system of equations.

$$
\begin{array}{ll}
-11m - 11p = -308 & \text{Multiplying (1) by } -11 \\
\underline{13m + 11p = 336} & (2) \\
\quad 2m \qquad\quad = 28 & \\
\qquad m = 14 &
\end{array}
$$

Substitute 14 for m in Equation (1) and solve for p.
$$14 + p = 28$$
$$p = 14$$

Check. The total mixture contains 14 lb + 14 lb, or 28 lb. Its value is $\$13 \cdot 14 + \$11 \cdot 14$
$= \$182 + \$154 = \$336$. The answer checks.

State. 14 lb of Mexican coffee and 14 lb of Peruvian coffee should be used.

23. Familiarize. Let $x =$ the number of ounces of custom printed M&Ms and $y =$ the number of ounces of bulk M&Ms. The value of the mixture will be \$0.32(20), or \$6.40. Converting lb to oz: 20 lb = 20(16) = 320 oz.

Translate. We organize the information in a table.

	Custom printed	Bulk	Mixture
Number of ounces	x	y	320
Price per ounce	\$0.60	\$0.25	\$0.32
Value of M&Ms	$0.60x$	$0.25y$	102.40

The "Number of ounces" row of the table gives us one equation:
$$x + y = 320$$

The "Value of M&Ms" row yields a second equation:
$$0.60x + 0.25y = 102.40$$

After clearing decimals, we have the problem translated to a system of equations:
$$x + y = 320 \qquad (1)$$
$$60x + 25y = 10{,}240 \qquad (2)$$

Carry out. We use the elimination method to solve the system of equations.

$$
\begin{array}{ll}
-25x - 25y = -8000 & \text{Multiplying (1) by } -25 \\
\underline{60x + 25y = 10{,}240} & (2) \\
\quad 35x \qquad\quad = 2240 & \\
\qquad\; x = 64 &
\end{array}
$$

Substitute 64 for x in Equation (1) and solve for y.
$$64 + y = 320$$
$$y = 256$$

Check. The total mixture contains 64 oz + 256 oz, or 320 oz. Its value is $\$0.60 \cdot 64 + \$0.25 \cdot 256 = \$102.40$. The answer checks.

State. 64 oz of custom-printed M&Ms and 256 oz of bulk M&Ms should be used.

25. Familiarize. Let $x =$ the number of pounds of 50% chocolate candy and $y =$ the number of pounds of 10% chocolate candy. The amount of chocolate in the mixture is 25%(20 lb), or 0.25(20 lb) = 5 lb.

Translate. We organize the information in a table.

	50% chocolate	10% chocolate	Mixture
Number of pounds	x	y	20
Percent of chocolate	50%	10%	25%
Amount of chocolate	$0.50x$	$0.10y$	5

The "Number of pounds" row of the table gives us one equation:

$$x + y = 20$$

The last row of the table yields a second equation:

$$0.50x + 0.10y = 5$$

After clearing decimals, we have the problem translated to a system of equations:

$$x + y = 20 \quad (1)$$
$$5x + y = 50 \quad (2)$$

Carry out. We use the elimination method to solve the system of equations.

$$-x - y = -20 \quad \text{Multiplying (1) by } -1$$
$$\underline{5x + y = 50 \quad (2)}$$
$$4x \quad\quad = 30$$
$$x = 7.5$$

Substitute 7.5 for x in Equation (1) and solve for y.

$$7.5 + y = 20$$
$$y = 12.5$$

Check. The amount of the mixture is 7.5 lb + 12.5 lb, or 20 lb. The amount of chocolate is
$0.50(7.5) + 0.10(12.5) = 3.75$ lb $+ 1.25$ lb $= 5$ lb. The answer checks.

State. 7.5 lb of 50% chocolate and 12.5 lb of 10% chocolate should be used.

27. ***Familiarize.*** Let $x =$ the number of pounds of Deep Thought Granola and $y =$ the number of pounds of Oat Dream Granola to be used in the mixture. The amount of nuts and dried fruit in the mixture is 19%(20 lb), or $0.19(20 \text{ lb}) = 3.8$ lb.

Translate. We organize the information in a table.

	Deep Thought	Oat Dream	Mixture
Number of pounds	x	y	20
Percent of nuts and dried fruit	25%	10%	19%
Amount of nuts and dried fruit	$0.25x$	$0.10y$	3.8 lb

We get one equation from the "Number of pounds" row of

the table:

$$x + y = 20$$

The last row of the table yields a second equation:

$$0.25x + 0.1y = 3.8$$

After clearing decimals, we have the problem translated to a system of equations:

$$x + y = 20, \quad (1)$$
$$25x + 10y = 380 \quad (2)$$

Carry out. We use the elimination method to solve the system of equations.

$$-10x - 10y = -200 \quad \text{Multiplying (1) by } -10$$
$$\underline{25x + 10y = 380}$$
$$15x \quad\quad = 180$$
$$x = 12$$

Substitute 12 for x in (1) and solve for y.

$$12 + y = 20$$
$$y = 8$$

Check. The amount of the mixture is 12 lb + 8 lb, or 20 lb. The amount of nuts and dried fruit in the mixture is
$0.25(12 \text{ lb}) + 0.1(8 \text{ lb}) = 3 \text{ lb} + 0.8 \text{ lb} = 3.8 \text{ lb}$. The answer checks.

State. 12 lb of Deep Thought Granola and 8 lb of Oat Dream Granola should be mixed.

29. ***Familiarize.*** Let $x =$ the amount of the 6.5% loan and $y =$ the amount of the 7.2% loan. Recall that the formula for simple interest is

$$\text{Interest} = \text{Principal} \times \text{Rate} \times \text{Time}$$

Translate. We organize the information in a table.

	6.5% loan	7.2% loan	Total
Principal	x	y	12,000
Interest rate	6.5%	7.2%	
Time	1 yr	1 yr	
Interest	$0.065x$	$0.072y$	811.50

The "Principal" row of the table gives us one equation:

$$x + y = 12{,}000$$

The last row of the table yields a second equation:

$$0.065x + 0.072y = 811.50$$

After clearing decimals, we have the problem translated to a system of equations:

$$x + y = 12{,}000 \quad (1)$$
$$65x + 72y = 811{,}500 \quad (2)$$

Carry out. We use the elimination method to solve the system of equations.

$$-65x - 65y = -780{,}000 \quad \text{Multiplying (1) by } -65$$
$$\underline{65x + 72y = 811{,}500 \quad (2)}$$
$$7y = 31{,}500$$
$$y = 4500$$

Substitute 4500 for y in Equation (1) and solve for x.

$$x + 4500 = 12{,}000$$
$$x = 7500$$

Check. The loans total $7500 + $4500, or $12,000. The total interest is $0.065(\$7500) + 0.072(\$4500)$ $= \$487.50 + \$324 = \$811.50$. The answer checks.

State. The 6.5% loan was for $7500 and the 7.2% loan was for $4500.

31. Familiarize. Let $x =$ the number of liters of Steady State and $y =$ the number of liters of Even Flow in the mixture. The amount of alcohol in the mixture is $0.15(20\,\text{L}) = 3\,\text{L}$.

Translate. We organize the information in a table.

	18% solution	10% solution	Mixture
Number of liters	x	y	20
Percent of alcohol	18%	10%	15%
Amount of alcohol	$0.18x$	$0.10y$	3

We get one equation from the "Number of liters" row of the table:

$$x + y = 20$$

The last row of the table yields a second equation:

$$0.18x + 0.1y = 3$$

After clearing decimals we have the problem translated to a system of equations:

$$x + y = 20, \quad (1)$$
$$18x + 10y = 300 \quad (2)$$

Carry out. We use the elimination method to solve the system of equations.

$$-10x - 10y = -200 \quad \text{Multiplying (1) by } -10$$
$$\underline{18x + 10y = 300 \quad (2)}$$
$$8x \qquad = 100$$
$$x = 12.5$$

Substitute 12.5 for x in (1) and solve for y.

$$12.5 + y = 20$$
$$y = 7.5$$

Check. The total amount of the mixture is $12.5\,\text{L} + 7.5\,\text{L}$ or 20 L. The amount of alcohol in the mixture is $0.18(12.5\,\text{L}) + 0.1(7.5\,\text{L}) = 2.25\,\text{L} + 0.75\,\text{L} = 3\,\text{L}$. The answer checks.

State. 12.5 L of Steady State and 7.5 L of Even Flow should be used.

33. Familiarize. Let $x =$ the number of gallons of 87-octane gas and $y =$ the number of gallons of 95-octane gas in the mixture. The amount of octane in the mixture can be expressed as $93(10)$, or 930.

Translate. We organize the information in a table.

	87-octane	95-octane	Mixture
Number of gallons	x	y	10
Octane rating	87	95	93
Total octane	$87x$	$95y$	930

We get one equation from the "Number of gallons" row of the table :

$$x + y = 10$$

The last row of the table yields a second equation:

$$87x + 95y = 930$$

We have a system of equations:

$$x + y = 10 \quad (1)$$
$$87x + 95y = 930 \quad (2)$$

Carry out. We use the elimination method to solve the system of equations.

$$-87x - 87y = -870 \quad \text{Multiplying (1) by } -87$$
$$\underline{87x + 95y = 930 \quad (2)}$$
$$8y = 60$$
$$y = 7.5$$

Substitute 7.5 for y in Equation (1) and solve for x.

$$x + 7.5 = 10$$
$$x = 2.5$$

Check. The total amount of the mixture is 2.5 gal + 7.5 gal, or 10 gal. The amount of octane can be expressed as $87(2.5) + 95(7.5) = 217.5 + 712.5 = 930$. The answer checks.

State. 2.5 gal of 87-octane gas and 7.5 gal of 95-octane gas should be used.

35. Familiarize. From the bar graph we see that whole milk is 4% milk fat, milk for cream cheese is 8% milk fat, and cream is 30% milk fat. Let $x =$ the number of pounds of whole milk and $y =$ the number of pounds of cream to be used. The mixture contains 8%(200 lb), or $0.08(200\,\text{lb}) = 16\,\text{lb}$ of milk fat.

Translate. We organize the information in a table.

	Whole milk	Cream	Mixture
Number of pounds	x	y	200
Percent of milk fat	4%	30%	8%
Amount of milk fat	$0.04x$	$0.30y$	16 lb

We get one equation from the "Number of pounds" row of the table:

$$x + y = 200$$

The last row of the table yields a second equation:

$$0.04x + 0.3y = 16$$

After clearing decimals, we have the problem translated to a system of equations:

$$x + y = 200, \quad (1)$$
$$4x + 30y = 1600 \quad (2)$$

Carry out. We use the elimination method to solve the system of equations.

$$-4x - 4y = -800 \quad \text{Multiplying (1) by } -4$$
$$\underline{4x + 30y = 1600 \quad (2)}$$
$$26y = 800$$
$$y = \frac{400}{13}, \quad \text{or } 30\frac{10}{13}$$

Substitute $\frac{400}{13}$ for y in (1) and solve for x.

$$x + \frac{400}{13} = 200$$
$$x = \frac{2200}{13}, \quad \text{or } 169\frac{3}{13}$$

Check. The total amount of the mixture is

$\frac{2200}{13}$ lb $+ \frac{400}{13}$ lb $= \frac{2600}{13}$ lb $= 200$ lb . The amount

of milk fat in the mixture is $0.04\left(\frac{2200}{13} \text{ lb}\right) +$

$0.3\left(\frac{400}{13} \text{ lb}\right) = \frac{88}{13}$ lb $+ \frac{120}{13}$ lb $= \frac{208}{13}$ lb $= 16$ lb . The

answer checks.

State. $169\frac{3}{13}$ lb of whole milk and $30\frac{10}{13}$ lb of cream

should be mixed.

37. Familiarize. We first make a drawing.

Slow train,
d kilometers,   75 km/h   (t + 2) hr

Fast train,
d kilometers,   125 km/h   t hr

From the drawing we see that the distances are the same. Now complete the chart.

	Distance	Rate	Time	
	d	=	r	·	t
Slow train	d	75	t + 2	$\to d = 75(t+2)$
Fast train	d	125	t	$\to d = 125t$

Translate. Using $d = rt$ in each row of the table, we get a system of equations:

$$d = 75(t + 2),$$
$$d = 125t$$

Carry out. We solve the system of equations.

$$125t = 75(t + 2) \quad \text{Using substitution}$$
$$125t = 75t + 150$$
$$50t = 150$$
$$t = 3$$

Then $d = 125t = 125 \cdot 3 = 375$

Check. At 125 km/h, in 3 hr the fast train will travel $125 \cdot 3 = 375$ km. At 75 km/h, in $3 + 2$, or 5 hr the slow train will travel $75 \cdot 5 = 375$ km. The numbers check.

State. The trains will meet 375 km from the station.

39. Familiarize. We first make a drawing. Let $d =$ the distance and $r =$ the speed of the canoe in still water. Then when the canoe travels downstream, its speed is $r + 6$, and its speed upstream is $r - 6$. From the drawing we see that the distances are the same.

Downstream, 6 mph current
d mi, r + 6, 4 hr

Upstream, 6 mph current
d mi, r − 6, 10 hr

Organize the information in a table.

	Distance	Rate	Time
Downstream	d	r + 6	4
Upstream	d	r − 6	10

Translate. Using $d = rt$ in each row of the table, we get a system of equations:

$$d = 4(r + 6), \qquad d = 4r + 24,$$
$$\text{or}$$
$$d = 10(r - 6) \qquad d = 10r - 60$$

Carry out. Solve the system of equations.

$$4r + 24 = 10r - 60$$
$$24 = 6r - 60$$
$$84 = 6r$$
$$14 = r$$

Check. When $r = 14$, then $r + 6 = 14 + 6 = 20$, and the distance traveled in 4 hours is $4 \cdot 20 = 80$ km. Also $r - 6 = 14 - 6 = 8$, and the distance traveled in 10 hours is $10 \cdot 8 = 80$ km. The answer checks.

State. The speed of the canoe in still water is 14 km/h.

41. Familiarize. We make a drawing. Note that the plane's speed traveling toward London is $360 + 50$, or 410 mph, and the speed traveling toward New York City is $360 - 50$, or 310 mph. Also, when the plane is d mi from New York City, it is $3458 - d$ mi from London.

New York City   London
310 mph   t hours   t hours   410 mph

├────────── 3458 mi ──────────┤

├────── d ──────┼────── 3458 mi − d ──────┤

Organize the information in a table.

	Distance	Rate	Time
Toward NYC	d	310	t
Toward London	3458 − d	410	t

Translate. Using $d = rt$ in each row of the table, we get a system of equations:

$$d = 310t, \quad (1)$$
$$3458 - d = 410t \quad (2)$$

Carry out. We solve the system of equations.

$$3458 - 310t = 410t \quad \text{Using substitution}$$
$$3458 = 720t$$
$$4.8028 \approx t$$

Substitute 4.8028 for t in (1).

$$d \approx 310(4.8028) \approx 1489$$

Check. If the plane is 1489 mi from New York City, it can return to New York City, flying at 310 mph, in $1489 / 310 \approx 4.8 \, \text{hr}$. If the plane is $3458 - 1489$, or 1969 mi from London, it can fly to London, traveling at 410 mph, in $1969 / 410 \approx 4.8 \, \text{hr}$. Since the times are the same, the answer checks.

State. The point of no return is about 1489 mi from New York City.

43. *Familiarize.* Let $l =$ the length, in feet, and $w =$ the width, in feet. Recall that the formula for the perimeter P of a rectangle with length l and width w is $P = 2l + 2w$.

Translate.

The perimeter	is	860 ft.
↓	↓	↓
$2l + 2w$	$=$	860

The length	is	100 ft.	more than	the width.
↓	↓	↓	↓	↓
l	$=$	100	$+$	w

We have translated to a system of equations:

$$2l + 2w = 860, \quad (1)$$
$$l = 100 + w \quad (2)$$

Carry out. We use the substitution method to solve the system of equations.

Substitute $100 + w$ for l in (1) and solve for w.

$$2(100 + w) + 2w = 860$$
$$200 + 2w + 2w = 860$$
$$200 + 4w = 860$$
$$4w = 660$$
$$w = 165$$

Now substitute 165 for w in (2).

$$l = 100 + 165 = 265$$

Check. The perimeter is

$2 \cdot 265 \, \text{ft} + 2 \cdot 165 \, \text{ft} = 530 \, \text{ft} + 330 \, \text{ft} = 860 \, \text{ft}$. The length, 265 ft, is 100 ft more than the width, 165 ft. The answer checks.

State. The length is 265 ft, and the width is 165 ft.

45. *Familiarize.* Let $x =$ the number of Wii game machines and $y =$ the number of PlayStation consoles.

Translate.

Number of Wiis	is	three	times	number of PlayStations.
↓	↓	↓	↓	↓
x	$=$	3	$\times$	y

Total number sold	is	4.84 million.
↓	↓	↓
$x + y$	$=$	4.84

We have a system of equations:

$$x = 3y \quad (1)$$
$$x + y = 4.84 \quad (2)$$

Carry out. We use the substitution method to solve the system of equations.

Substitute $3y$ for x in Equation (2) and solve for y.

$$3y + y = 4.84$$
$$4y = 4.84$$
$$y = 1.21$$

Now substitute 1.21 for y in Equation (1).

$$x = 3(1.21) = 3.63$$

Check. The total number of machines sold is 3.63 million + 1.21 million = 4.84 million. The number of Wii game machines sold, 3.63 million is 3 times 1.21 million, the number of PlayStation consoles sold. The answer checks.

State. 3.63 million Wii game machines and 1.21 million PlayStation consoles were sold.

47. *Familiarize.* Let $x =$ the number of unlimited 1 DVD plans for \$8.99 and $y =$ the number of limit of 2 DVD per month plans for \$4.99.

Translate.

Total number of plans	is	250.
↓	↓	↓
$x + y$	$=$	250

Total value of plans	is	1975.50.
↓	↓	↓
$8.99x + 4.99y$	$=$	1975.50

After clearing decimals, we have a system of equations:

$$x + y = 250 \quad (1)$$
$$899x + 499y = 197{,}550 \quad (2)$$

Carry out. We use the elimination method to solve the system of equations.

$$\begin{array}{ll} -499x - 499y = -124{,}750 & \text{Multiplying (1) by } -499 \\ \underline{899x + 499y = 197{,}550} & (2) \\ 400x = 72{,}800 & \\ x = 182 & \end{array}$$

Substitute 182 for x in Equation (1) and solve for y.
$$182 + y = 250$$
$$y = 68$$

Check. The total number of plans is $182 + 68$, or 250. The total value of the plans is $8.99(182) + 4.99(68)$
$= 1636.18 + 339.32 = 1975.50$. The answer checks.

State. 182 of the $8.99 plans and 68 of the $4.99 plans were purchased.

49. Familiarize. The change from the $9.25 purchase is $20 – $9.25, or $10.75. Let $x =$ the number of quarters and $y =$ the number of fifty-cent pieces. The total value of the quarters, in dollars, is $0.25x$ and the total value of the fifty-cent pieces, in dollars, is $0.50y$.

Translate.

The total number of coins is 30.
$$x + y = 30$$

The total value of the coins is $10.75.
$$0.25x + 0.50y = 10.75$$

After clearing decimals we have the following system of equations:
$$x + y = 30, \quad (1)$$
$$25x + 50y = 1075 \quad (2)$$

Carry out. We use the elimination method to solve the system of equations.
$$\begin{array}{r} -25x - 25y = -750 \quad \text{Multiplying (1) by } -25 \\ 25x + 50y = 1075 \\ \hline 25y = 325 \\ y = 13 \end{array}$$

Substitute 13 for y in (1) and solve for x.
$$x + 13 = 30$$
$$x = 17$$

Check. The total number of coins is $17 + 13$, or 30. The total value of the coins is $\$0.25(17) + \$0.50(13)$
$= \$4.25 + \$6.50 = \$10.75$. The answer checks.

State. There were 17 quarters and 13 fifty-cent pieces.

51. Writing Exercise. Both can be considered mixture problems or total value problems. In each system the first equation pertains to a total amount and the second pertains to a total value.

53. $2x - 3y - z = 2(5) - 3(2) - 3$
$\qquad = 10 - 6 - 3 = 1$

55. $x + y + 2z = 1 + (-4) + 2(-5)$
$\qquad = 1 - 4 - 10 = -13$

57. $a - 2b - 3c = -2 - 2(3) - 3(-5)$
$\qquad = -2 - 6 + 15 = 7$

59. Writing Exercise. We might not have found l after first finding a. However, in order to check, it is necessary to find l, so the method of solving need not change.

61. Familiarize. Let $x =$ the number of reams of 0% post-consumer fiber paper purchased and $y =$ the number of reams of 30% post-consumer fiber paper.

Translate. We organize the information in a table.

	0% post-consumer	30% post-consumer	Total
Reams purchased	x	y	60
Percent of post-consumer fiber	0%	30%	20%
Total post-consumer fiber	$0 \cdot x$, or 0	$0.3y$	0.2(60), or 12

We get one equation from the "Reams purchased" row of the table:
$$x + y = 60$$

The last row of the table yields a second equation:
$$0x + 0.3y = 12, \text{ or } 0.3y = 12$$

After clearing the decimal we have the problem translated to a system of equations.
$$x + y = 60, \quad (1)$$
$$3y = 120 \quad (2)$$

Carry out. First we solve (2) for y.
$$3y = 120$$
$$y = 40$$

Now substitute 40 for y in (1) and solve for x.
$$x + 40 = 60$$
$$x = 20$$

Check. The total purchase is $20 + 40$, or 60 reams. The post-consumer fiber can be expressed as $0 \cdot 20 + 0.3(40) = 12$. The answer checks.

State. 20 reams of 0% post-consumer fiber paper and 40 reams of 30% post-consumer fiber paper would have to be purchased.

63. Familiarize. Let $x =$ the number of ounces of pure silver.

Translate. We organize the information in a table.

	Coin	Pure silver	New Mixture
Number of ounces	32	x	$32 + x$
Percent silver	90%	100%	92.5%
Amount of silver	0.9(32)	$1 \cdot x$	0.925(32 + x)

The last row gives the equation
$$0.9(32) + x = 0.925(32 + x).$$

Carry out. We solve the equation.
$$28.8 + x = 29.6 + 0.925x$$
$$0.075x = 0.8$$
$$x = \frac{32}{3} = 10\frac{2}{3}$$

Check. 90% of 32, or 28.8 + 100% of $10\frac{2}{3}$, or $10\frac{2}{3}$ is $39\frac{7}{15}$ and 92.5% of $\left(32 + 10\frac{2}{3}\right)$ is $0.925\left(42\frac{2}{3}\right)$, or $39\frac{7}{15}$. The answer checks.

State. $10\frac{2}{3}$ ounces of pure silver should be added.

65. Familiarize. Let x = the number of 3-volume sets and y = the number of single volumes. Note that $3x$ is the number of volumes in the set.

Translate.

Total number of volumes is 51.
$$3x + y = 51$$

Total value sales is 1641.
$$88x + 39y = 1641$$

We have a system of equations:
$$3x + y = 51 \qquad (1)$$
$$88x + 39y = 1641 \qquad (2)$$

Carry out. We use the elimination method to solve the system of equations.
$$\begin{array}{ll} -117x - 39y = -1989 & \text{Multiplying (1) by } -39 \\ 88x + 39y = 1641 & (2) \\ \hline -29x \qquad = -348 & \\ x = 12 & \end{array}$$

Although the problem asks for the number of 3-volume sets, we will also find y in order to check. Substitute 12 for x in Equation (1) and solve for y.
$$3(12) + y = 51$$
$$y = 15$$

Check. The total number of books is $3(12) + 15 = 36 + 15 = 51$. The total sales is $\$88(12) + \$39(15) = \$1056 + \$585 = \$1641$. The answer checks.

State. 12 3-volume sets were ordered.

67. Familiarize. Let x = the number of gallons of pure brown and y = the number of gallons of neutral stain that should be added to the original 0.5 gal. Note that a total of 1 gal of stain needs to be added to bring the amount of stain up to 1.5 gal. The original 0.5 gal of stain

contains 20%(0.5 gal), or 0.2(0.5 gal) = 0.1 gal of brown stain. The final solution contains 60%(1.5 gal), or 0.6(1.5 gal) = 0.9 gal of brown stain. This is composed of the original 0.1 gal and the x gal that are added.

Translate.

The amount of stain added was 1 gal.
$$x + y = 1$$

The amount of brown stain in the final solution is 0.9 gal.
$$0.1 + x = 0.9$$

We have a system of equations.
$$x + y = 1, \qquad (1)$$
$$0.1 + x = 0.9 \qquad (2)$$

Carry out. First we solve (2) for x.
$$0.1 + x = 0.9$$
$$x = 0.8$$

Then substitute 0.8 for x in (1) and solve for y.
$$0.8 + y = 1$$
$$y = 0.2$$

Check. Total amount of stain: $0.5 + 0.8 + 0.2 = 1.5$ gal

Total amount of brown stain: $0.1 + 0.8 = 0.9$ gal

Total amount of neutral stain:
$$0.8(0.5) + 0.2 = 0.4 + 0.2 = 0.6 \text{ gal} = 0.4(1.5 \text{ gal})$$
The answer checks.

State. 0.8 gal of pure brown and 0.2 gal of neutral stain should be added.

69. Familiarize. Let x and y represent the number of city miles and highway miles that were driven, respectively. Then in city driving, $\frac{x}{18}$ gallons of gasoline are used; in highway driving, $\frac{y}{24}$ gallons are used.

Translate. We organize the information in a table.

Type of driving	City	Highway	Total
Number of miles	x	y	465
Gallons of gasoline used	$\frac{x}{18}$	$\frac{y}{24}$	23

The first row of the table gives us one equation:
$$x + y = 465$$

The second row gives us another equation:
$$\frac{x}{18} + \frac{y}{24} = 23$$

After clearing fractions, we have the following system of equations:

$$x + y = 465, \quad (1)$$
$$24x + 18y = 9936 \quad (2)$$

Carry out. We solve the system of equations using the elimination method.

$$\begin{array}{ll} -18x - 18y = -8370 & \text{Multiplying (1) by } -18 \\ \underline{24x + 18y = 9936} & (2) \\ 6x \qquad\quad = 1566 \\ \quad x = 261 \end{array}$$

Now substitute 261 for x in Equation (1) and solve for y.

$$261 + y = 465$$
$$y = 204$$

Check. The total mileage is 261 + 204, or 465. In 216 city miles, $261/18$, or 14.5 gal of gasoline are used; in 204 highway miles, $204/24$, or 8.5 gal are used. Then a total of 14.5 + 8.5, or 23 gal of gasoline are used. The answer checks.

State. 261 miles were driven in the city, and 204 miles were driven on the highway.

71. The 1.5 gal mixture contains $0.1 + x$ gal of pure brown stain. (See Exercise 67.). Thus, the function

$$P(x) = \frac{0.1 + x}{1.5}$$ gives the percentage of brown in the

mixture as a decimal quantity. Using the Intersect feature, we confirm that when $x = 0.8$, then $P(x) = 0.6$ or 60%.

Exercise Set 3.4

1. The equation is equivalent to one in the form

$Ax + By + Cz = D$, so the statement is true.

3. False; see page 183.

5. True; see Example 6.

7. Substitute $(2, -1, -2)$ into the three equations, using alphabetical order.

$$\frac{x + y - 2z = 5}{2 + (-1) - 2(-2) \,\big|\, 5}$$
$$\qquad 2 - 1 + 4 \,\big|$$
$$\qquad\qquad ?$$
$$\qquad\qquad 5 = 5 \quad \text{TRUE}$$

$$\frac{2x - y - z = 7}{2 \cdot 2 - (-1) - (-2) \,\big|\, 7}$$
$$\qquad 4 + 1 + 2 \,\big|$$
$$\qquad\qquad ?$$
$$\qquad\qquad 7 = 7 \quad \text{TRUE}$$

$$\frac{-x - 2y - 3z = 6}{-2 - 2(-1) - 3(-2) \,\big|\, 6}$$
$$\qquad -2 + 2 + 6 \,\big|$$
$$\qquad\qquad ?$$
$$\qquad\qquad 6 = 6 \quad \text{TRUE}$$

Since the triple $(2, -1, -2)$ is true in all three equations, it is a solution of the system.

9. $\quad x - y - z = 0, \quad (1)$
$\quad\quad 2x - 3y + 2z = 7, \quad (2)$
$\quad\quad -x + 2y + z = 1 \quad (3)$

1., 2. The equations are already in standard form with no fractions or decimals.

3. Use Equations (1) and (2) to eliminate x.

$$\begin{array}{ll} -2x + 2y + 2z = 0 & \text{Multiplying (1) by } -2 \\ \underline{2x - 3y + 2z = 7} & (2) \\ \quad -y + 4z = 7 & (4) \text{ Adding} \end{array}$$

4. Use a different pair of equations and eliminate x.

$$\begin{array}{ll} x - y - z = 0 & (1) \\ \underline{-x + 2y + z = 1} & (3) \\ \quad y \qquad = 1 & \text{Adding} \end{array}$$

5. When we used Equation (1) and (3) to eliminate x, we also eliminated z and found $y = 1$. Substitute 1 for y in Equation (4) to find z.

$$\begin{array}{ll} -1 + 4z = 7 & \text{Substituting 1 for} y \text{ in (4)} \\ 4z = 8 \\ z = 2 \end{array}$$

6. Substitute in one of the original equations to find x.

$$x - 1 - 2 = 0$$
$$x = 3$$

We obtain $(3, 1, 2)$. This checks, so it is the solution.

11. $\quad x - y - z = 1, \quad (1)$
$\quad\quad 2x + y + 2z = 4, \quad (2)$
$\quad\quad x + y + 3z = 5 \quad (3)$

1., 2. The equations are already in standard form with no fractions or decimals.

3. Use Equations (1) and (2) to eliminate y.

$$\begin{array}{ll} x - y \ - z = 1 & (1) \\ \underline{2x + y + 2z = 4} & (2) \\ 3x \quad\ + z = 5 & (4) \text{ Adding} \end{array}$$

4. Use a different pair of equations and eliminate y.

$$\begin{array}{ll} x - y \ - z = 1 & (1) \\ \underline{x + y + 3z = 5} & (3) \\ 2x \quad\ + 2z = 6 & (5) \text{ Adding} \end{array}$$

5. Now solve the system of Equations (4) and (5).

$$3x + z = 5 \quad (4)$$
$$2x + 2z = 6 \quad (5)$$

$$\begin{array}{ll} -6x - 2z = -10 & \text{Multiplying (4) by } -2 \\ \underline{2x + 2z = 6} & (5) \\ -4x \qquad = -4 & \text{Adding} \\ \quad x = 1 \end{array}$$

$3 \cdot 1 + z = 5$ Substituting 1 for x in (4)
$\quad 3 + z = 5$
$\qquad z = 2$

6. Substitute in one of the original equations to find y.

$1 + y + 3 \cdot 2 = 5$ Substituting 1 for x and
$\qquad\qquad\qquad$ 2 for z in (3)
$\quad 1 + y + 6 = 5$
$\qquad y + 7 = 5$
$\qquad\quad y = -2$

We obtain $(1, -2, 2)$. This checks, so it is the solution.

13. $3x + 4y - 3z = 4,$ (1)
$\quad 5x - y + 2z = 3,$ (2)
$\quad\, x + 2y - z = -2$ (3)

1., 2. The equations are already in standard form with no fractions or decimals.

3., 4. We eliminate y from two different pairs of equations.

$\quad 3x + 4y - 3z = 4$ (1)
$20x - 4y + 8z = 12$ Multiplying (2) by 4
$\overline{23x \qquad + 5z = 16}$ (4)

$10x - 2y + 4z = 6$ Multiplying (2) by 2
$\quad\; x + 2y\; - z = -2$ (3)
$\overline{11x \qquad + 3z = 4}$ (5)

5. Now solve the system of Equations (4) and (5).

$23x + 5z = 16$ (4)
$11x + 3z = 4$ (5)

$-69x - 15z = -48$ Multiplying (4) by -3
$\;\; 55x + 15z = 20$ Multiplying (5) by 5
$\overline{-14x \qquad = -28}$ Adding
$\qquad x = 2$

$11 \cdot 2 + 3z = 4$ Substituting 2 for x in (5)
$\qquad 3z = -18$
$\qquad\; z = -6$

6. Substitute in one of the original equations to find y.

$3 \cdot 2 + 4y - 3(-6) = 4$ Substituting 2 for x and
$\qquad\qquad\qquad\qquad$ -6 for z in (1)
$\quad 6 + 4y + 18 = 4$
$\qquad\qquad 4y = -20$
$\qquad\qquad\; y = -5$

We obtain $(2, -5, -6)$. This checks, so it is the solution.

15. $\quad x + y + z = 0,$ (1)
$\; 2x + 3y + 2z = -3,$ (2)
$\, -x - 2y - z = 1$ (3)

1., 2. The equations are already in standard form with no fractions or decimals.

3., 4. We eliminate x from two different pairs of equations.

$-2x - 2y - 2z = 0$ Multiplying (1) by -2
$\;\; 2x + 3y + 2z = -3$ (2)
$\overline{\qquad y \qquad = -3}$

We eliminated not only x but also z and found that

$y = -3$.

5., 6. Substitute -3 for y in two of the original equations to produce a system of two equations in two variables. Then solve this system.

$\quad x - 3 + z = 0$ Substituting in (1)
$-x - 2(-3) - z = 1$ Substituting in (3)

Simplifying we have

$\quad x + z = 3$
$-x - z = -5$
$\overline{\qquad 0 = -2}$

We get a false equation, so there is no solution.

17. $\; 2x - 3y - z = -9,$ (1)
$\;\; 2x + 5y + z = 1,$ (2)
$\qquad x - y + z = 3$ (3)

1., 2. The equations are already in standard form with no fractions or decimals.

3., 4. We eliminate z from two different pairs of equations.

$\; 2x - 3y - z = -9$ (1)
$\; 2x + 5y + z = 1$ (2)
$\overline{4x + 2y \qquad = -8}$ (4)

$2x - 3y - z = -9$ (1)
$\; x\; - y + z = 3$ (2)
$\overline{3x - 4y \qquad = -6}$ (5)

5. Now solve the system of Equations (4) and (5).

$4x + 2y = -8$ (4)
$3x - 4y = -6$ (5)

$8x + 4y = -16$ Multiplying (4) by 2
$3x - 4y = -6$ (5)
$\overline{11x \qquad = -22}$ Adding
$\qquad x = -2$

$4(-2) + 2y = -8$ Substituting -2 for x in (4)
$\quad -8 + 2y = -8$
$\qquad\qquad y = 0$

6. Substitute in one of the original equations to find z.

$2(-2) + 5 \cdot 0 + z = 1$ Substituting -2 for x and
$\qquad\qquad\qquad\qquad$ 0 for y in (1)
$\qquad -4 + z = 1$
$\qquad\qquad z = 5$

We obtain $(-2, 0, 5)$. This checks, so it is the solution.

19. $\quad a + b + c = 5,$ (1)
$\; 2a + 3b - c = 2,$ (2)
$\; 2a + 3b - 2c = 4$ (3)

1., 2. The equations are already in standard form with no fractions or decimals.

3., 4. We eliminate a from two different pairs of equations.

$-2a - 2b - 2c = -10$ Multiplying (1) by -2
$\;\; 2a + 3b\; - c = 2$ (2)
$\overline{\qquad b - 3c = -8}$ (4)

$$\begin{array}{ll} -2a-3b+c=-2 & \text{Multiplying (1) by } -1 \\ \underline{2a+3b-2c=4} & \text{(3)} \\ \quad\quad -c=2 \\ \quad\quad\quad c=-2 \end{array}$$

We eliminate not only a, but also b and found $c=-2$.

5. Substitute -2 for c in Equation (4) to find b.

$$\begin{array}{ll} b-3(-2)=-8 & \text{Substituting } -2 \text{ for } c \text{ in (4)} \\ \quad b+6=-8 \\ \quad\quad b=-14 \end{array}$$

6. Substitute in one of the original equations to find a.

$$\begin{array}{ll} a-14-2=5 & \text{Substituting } -2 \text{ for } c \text{ and} \\ & \quad -14 \text{ for } b \text{ in (1)} \\ \quad a-16=5 \\ \quad\quad a=21 \end{array}$$

We obtain $(21, -14, -2)$. This checks, so it is the solution.

21. $\begin{array}{ll} -2x+8y+2z=4, & \text{(1)} \\ x+6y+3z=4, & \text{(2)} \\ 3x-2y+z=0 & \text{(3)} \end{array}$

1., 2. The equations are already in standard form with no fractions or decimals.

3., 4. We eliminate z from two different pairs of equations.

$$\begin{array}{ll} -2x+8y+2z=4 & \text{(1)} \\ \underline{-6x+4y-2z=0} & \text{Multiplying (3) by } -2 \\ -8x+12y\quad\quad=4 & \text{(4)} \end{array}$$

$$\begin{array}{ll} x+6y+3z=4 & \text{(2)} \\ \underline{-9x+6y-3z=0} & \text{Multiplying (3) by } -3 \\ -8x+12y\quad\quad=4 & \text{(5)} \end{array}$$

5.   Now solve the system of Equations (4) and (5).

$$\begin{array}{ll} -8x+12y=4 & \text{(4)} \\ -8x+12y=4 & \text{(5)} \end{array}$$

$$\begin{array}{ll} -8x+12y=4 & \text{(4)} \\ \underline{8x-12y=-4} & \text{Multiplying (5) by } -1 \\ \quad\quad 0=0 & \text{(6)} \end{array}$$

Equation (6) indicates that Equations (1), (2), and (3) are dependent. (Note that if Equation (1) is subtracted from Equation (2), the result is Equation (3).) We could also have concluded that the equations are dependent by observing that Equations (4) and (5) are identical.

23. $\begin{array}{ll} 2u-4v-w=8, & \text{(1)} \\ 3u+2v+w=6, & \text{(2)} \\ 5u-2v+3w=2 & \text{(3)} \end{array}$

1., 2. The equations are already in standard form with no fractions or decimals.

3., 4. We eliminate w from two different pairs of equations.

$$\begin{array}{ll} 2u-4v-w=8 & \text{(1)} \\ \underline{3u+2v+w=6} & \text{(2)} \\ 5u-2v\quad\quad=14 & \text{(4)} \end{array}$$

$$\begin{array}{ll} 6u-12v-3w=24 & \text{Multiplying (1) by 3} \\ \underline{5u-2v+3w=2} & \text{(3)} \\ 11u-14v\quad\quad=26 & \text{(5)} \end{array}$$

5. Now solve the system of Equations (4) and (5).

$$\begin{array}{ll} 5u-2v=14 & \text{(4)} \\ 11u-14v=26 & \text{(5)} \end{array}$$

$$\begin{array}{ll} -35u+14v=-98 & \text{Multiplying (4) by } -7 \\ \underline{11u-14v=26} & \text{(5)} \\ -24u\quad\quad=-72 \\ \quad\quad u=3 \end{array}$$

$$\begin{array}{ll} 5\cdot3-2v=14 & \text{Substituting 3 for } u \text{ in (4)} \\ 15-2v=14 \\ \quad -2v=-1 \\ \quad\quad v=\dfrac{1}{2} \end{array}$$

6. Substitute in one of the original equations to find v.

$$\begin{array}{ll} 2\cdot3-4\left(\dfrac{1}{2}\right)-2=8 & \text{Substituting 3 for } u \text{ and } \dfrac{1}{2} \\ & \quad \text{for } v \text{ in (1)} \\ \quad 6-2-w=8 \\ \quad\quad\quad w=-4 \end{array}$$

We obtain $\left(3, \dfrac{1}{2}, -4\right)$. This checks, so it is the solution.

25. $\begin{array}{l} r+\dfrac{3}{2}s+6t=2, \\ 2r-3s+3t=0.5, \\ r+s+t=1 \end{array}$

1. All equations are already in standard form.

2. Multiply the first equation by 2 to clear the fraction. Also, multiply the second equation by 10 to clear the decimal.

$$\begin{array}{ll} 2r+3s+12t=4, & \text{(1)} \\ 20r-30s+30t=5, & \text{(2)} \\ r+s+t=1 & \text{(3)} \end{array}$$

3., 4. We eliminate s from two different pairs of equations.

$$\begin{array}{ll} 20r+30s+120t=40 & \text{Multiplying (1) by 10} \\ \underline{20r-30s\ +30t=5} & \text{(2)} \\ 40r\quad\quad+150t=45 & \text{(4) Adding} \end{array}$$

$$\begin{array}{ll} 20r+30s+120t=40 & \text{Multiplying (1) by 10} \\ \underline{-30r-30s\ -30t=-30} & \text{Multiplying (3) by } -30 \\ -10r\quad\quad+90t=10 & \text{(5) Adding} \end{array}$$

5. Solve the system of Equations (4) and (5).

$$\begin{array}{ll} 40r+150t=45, & \text{(4)} \\ -10r+90t=10, & \text{(5)} \end{array}$$

$$\begin{array}{ll} 40r+150t=45 & \text{(4)} \\ \underline{-40r+360t=40} & \text{Multiplying (5) by 4} \\ \quad\quad 510t=85 \\ \quad\quad\quad t=\dfrac{85}{510} \\ \quad\quad\quad t=\dfrac{1}{6} \end{array}$$

$40r + 150\left(\dfrac{1}{6}\right) = 45$ Substituting $\dfrac{1}{6}$ for t in (4)

$\qquad 40r + 25 = 45$

$\qquad\qquad 40r = 20$

$\qquad\qquad\quad r = \dfrac{1}{2}$

6. Substitute in one of the original equations to find s.

$\dfrac{1}{2} + s + \dfrac{1}{6} = 1$ Substituting $\dfrac{1}{2}$ for r

$\qquad\qquad\qquad$ and $\dfrac{1}{6}$ for t in (3)

$\qquad s + \dfrac{2}{3} = 1$

$\qquad\qquad s = \dfrac{1}{3}$

We obtain $\left(\dfrac{1}{2},\ \dfrac{1}{3},\ \dfrac{1}{6}\right)$. This checks, so it is the solution.

27. $\quad 4a + 9b \quad\ \ = 8,\quad (1)$

$\qquad 8a \quad\ + 6c = -1,\quad (2)$

$\qquad\quad 6b + 6c = -1\quad (3)$

1., 2. The equations are already in standard form with no fractions or decimals.

3., 4. Note that there is no c in Equation (1). We will use Equations (2) and (3) to obtain another equation with no c-term.

$\qquad 8a \quad\ + 6c = -1 \quad (2)$

$\qquad\ \underline{\quad -6b - 6c = 1}\quad$ Multiplying (3) by -1

$\qquad 8a \quad\ - 6b = 0 \quad\ (4)$ Adding

5. Now solve the system of Equations (1) and (4).

$\qquad -8a - 18b = -16 \quad$ Multiplying (1) by -2

$\qquad\ \underline{\quad 8a - 6b = 0}$

$\qquad\qquad -24b = -16$

$\qquad\qquad\quad\ \ b = \dfrac{2}{3}$

$8a - 6\left(\dfrac{2}{3}\right) = 0$ Substituting $\dfrac{2}{3}$ for b in (4)

$\qquad 8a - 4 = 0$

$\qquad\quad 8a = 4$

$\qquad\quad\ a = \dfrac{1}{2}$

6. Substitute in Equation (2) or (3) to find c.

$8\left(\dfrac{1}{2}\right) + 6c = -1$ Substituting $\dfrac{1}{2}$ for a in (2)

$\qquad 4 + 6c = -1$

$\qquad\qquad 6c = -5$

$\qquad\qquad\ c = -\dfrac{5}{6}$

We obtain $\left(\dfrac{1}{2},\ \dfrac{2}{3}, -\dfrac{5}{6}\right)$. This checks, so it is the solution.

29. $\quad x + y + z = 57,\quad (1)$

$\qquad -2x + y \quad\ = 3,\quad (2)$

$\qquad\quad x \quad\ - z = 6\quad (3)$

1., 2. The equations are already in standard form with no fractions or decimals.

3., 4. Note that there is no z in Equation (2). We will use Equations (1) and (3) to obtain another equation with no

z-term.

$\qquad x + y + z = 57 \quad (1)$

$\qquad \underline{\ x \quad\ - z = 6}\quad\ (3)$

$\qquad 2x + y \quad\ = 63 \quad (4)$

5. Now solve the system of Equations (2) and (4).

$\qquad -2x + y = 3 \quad (2)$

$\qquad \underline{\ 2x + y = 63}\quad (4)$

$\qquad\qquad 2y = 66$

$\qquad\qquad\ y = 33$

$2x + 33 = 63$ Substituting 33 for y in (4)

$\qquad 2x = 30$

$\qquad\ x = 15$

6. Substitute in Equation (1) or (3) to find z.

$15 - z = 6$ Substituting 15 for x in (3)

$\qquad 9 = z$

We obtain $(15,\ 33,\ 9)$. This checks, so it is the solution.

31. $\quad a \quad\ - 3c = 6,\quad (1)$

$\qquad\quad b + 2c = 2,\quad (2)$

$\quad 7a - 3b - 5c = 14 \quad (3)$

1., 2. The equations are already in standard form with no fractions or decimals.

3., 4. Note that there is no b in Equation (1). We will use Equations (2) and (3) to obtain another equation with no b-term.

$\qquad\quad 3b + 6c = 6 \quad$ Multiplying (2) by 3

$\qquad \underline{7a - 3b - 5c = 14}\quad (3)$

$\qquad 7a \quad\quad + c = 20 \quad (4)$

5. Now solve the system of Equations (1) and (4).

$\qquad a - 3c = 6 \quad (1)$

$\qquad 7a + c = 20 \quad (4)$

$\qquad\quad a - 3c = 6 \quad (1)$

$\qquad \underline{21a + 3c = 60}\quad$ Multiplying (4) by 3

$\qquad 22a \quad\ = 66$

$\qquad\qquad a = 3$

$3 - 3c = 6$ Substituting in (1)

$\quad -3c = 3$

$\qquad c = -1$

6. Substitute in Equation (2) or (3) to find b.

$b + 2(-1) = 2$ Substituting in (2)

$\qquad b - 2 = 2$

$\qquad\quad b = 4$

We obtain $(3,\ 4,\ -1)$. This checks, so it is the solution.

33. $\quad x + y + z = 83,\quad (1)$

$\qquad\qquad y = 2x + 3,\quad (2)$

$\qquad\qquad z = 40 + x \quad (3)$

Observe, from Equations (2) and (3), that we can substitute $2 + 3x$ for y and $40 + x$ for z in Equation (1) and solve for x.

$$x+y+z=83$$
$$x+(2x+3)+(40+x)=83$$
$$4x+43=83$$
$$4x=40$$
$$x=10$$

Now substitute 10 for x in Equation (2).

$$y=2x+3=2\cdot10+3=20+3=23$$

Finally, substitute 10 for x in Equation (3).

$$z=40+x=40+10=50\,.$$

We obtain $(10, 23, 50)$. This checks, so it is the solution.

35.
$$x \quad\;\; +z=0, \quad (1)$$
$$x+y+2z=3, \quad (2)$$
$$y \;+z=2 \quad (3)$$

1., 2. The equations are already in standard form with no fractions or decimals.

3., 4. Note that there is no y in Equation (1). We use Equations (2) and (3) to obtain another equation with no y-term.

$$\begin{array}{ll} x+y+2z=3 & (2) \\ -y\;-z=-2 & \text{Multiplying (3) by}-1 \\ \hline x \quad\;\; +z=1 & (4)\ \text{Adding} \end{array}$$

5. Now solve the system of Equations (1) and (4).

$$\begin{array}{ll} x+z=0 & (1) \\ -x-z=-1 & \text{Multiplying (4) by}-1 \\ \hline 0=-1 & \text{Adding} \end{array}$$

We get a false equation, or contradiction. There is no solution.

37.
$$x+y+z=1, \quad (1)$$
$$-x+2y+z=2, \quad (2)$$
$$2x-y \quad\;\; =-1 \quad (3)$$

1., 2. The equations are already in standard form with no fractions or decimals.

3. Note that there is no z in Equation (3). We will use Equations (1) and (2) to eliminate z.

$$\begin{array}{ll} x+y+z=1 & (1) \\ x-2y-z=-2 & \text{Multiplying (2) by}-1 \\ \hline 2x-y \quad\;\; =-1 & \text{Adding} \end{array}$$

Equations (3) and (4) are identical, so Equations (1), (2), and (3) are dependent. (We have seen that if Equation (2) is multiplied by –1 and added to Equation (1), the result is Equation (3).)

39. *Writing Exercise.* This approach will work, since any of the variables may be selected to be eliminated. Sometimes the coefficients of x may be larger, which will make the calculations a bit more difficult.

41. Let x and y represent the numbers; $x=\frac{1}{2}y$

43. Let x represent the first number,

Let $x+1$ represent the second number,

Let $x+2$ represent the third number.

$$x+(x+1)+(x+2)=100$$

45. Let x, y, and z represent the numbers; $xy=5z$

47. *Writing Exercise.* No; the graph of each equation is a plane and all three planes can only have no points in common, exactly one point in common, or infinitely many points in common.

49.
$$\frac{x+2}{3}-\frac{y+4}{2}+\frac{z+1}{6}=0,$$
$$\frac{x-4}{3}+\frac{y+1}{4}-\frac{z-2}{2}=-1,$$
$$\frac{x+1}{2}+\frac{y}{2}+\frac{z-1}{4}=\frac{3}{4}$$

1., 2. We clear fractions and write each equation in standard form.

To clear fractions, we multiply both sides of each equation by the LCM of its denominators. The LCM's are 6, 12, and 4, respectively.

$$6\left(\frac{x+2}{3}-\frac{y+4}{2}+\frac{z+1}{6}\right)=6\cdot0$$
$$2(x+2)-3(y+4)+(z+1)=0$$
$$2x+4-3y-12+z+1=0$$
$$2x-3y+z=7$$

$$12\left(\frac{x-4}{3}+\frac{y+1}{4}-\frac{z-2}{2}\right)=12\cdot(-1)$$
$$4(x-4)+3(y+1)-6(z-2)=-12$$
$$4x-16+3y+3-6z+12=-12$$
$$4x+3y-6z=-11$$

$$4\left(\frac{x+1}{2}+\frac{y}{2}+\frac{z-1}{4}\right)=4\cdot\frac{3}{4}$$
$$2(x+1)+2(y)+(z-1)=3$$
$$2x+2+2y+z-1=3$$
$$2x+2y+z=2$$

The resulting system is

$$\begin{array}{ll} 2x-3y+z=7, & (1) \\ 4x+3y-6z=-11, & (2) \\ 2x+2y+z=2 & (3) \end{array}$$

3., 4. We eliminate z from two different pairs of equations.

$$\begin{array}{ll} 12x-18y+6z=42 & \text{Multiplying (1) by 6} \\ 4x+3y-6z=-11 & (2) \\ \hline 16x-15y \quad\;\; =31 & (4)\ \text{Adding} \end{array}$$

$$\begin{array}{ll} 2x-3y+z=7 & (1) \\ -2x-2y-z=-2 & \text{Multiplying (3) by}-1 \\ \hline -5y \quad\;\; =5 & (5)\ \text{Adding} \end{array}$$

5. Solve (5) for y:
$$-5y=5$$
$$y=-1$$

Substitute –1 for y in (4):

$$16x - 15(-1) = 31$$
$$16x + 15 = 31$$
$$16x = 16$$
$$x = 1$$

6. Substitute 1 for x and –1 for y in (1):

$$2 \cdot 1 - 3(-1) + z = 7$$
$$5 + z = 7$$
$$z = 2$$

We obtain $(1, -1, 2)$. This checks, so it is the solution.

51.
$$w + x + y + z = 2, \quad (1)$$
$$w + 2x + 2y + 4z = 1, \quad (2)$$
$$w - x + y + z = 6, \quad (3)$$
$$w - 3x - y + z = 2 \quad (4)$$

The equations are already in standard form with no fractions or decimals.

Start by eliminating w from three different pairs of equations.

$$
\begin{array}{ll}
w + x + y + z = 2 & (1) \\
-w - 2x - 2y - 4z = -1 & \text{Multiplying (2) by} -1 \\
\hline
-x - y - 3z = 1 & (5) \text{ Adding}
\end{array}
$$

$$
\begin{array}{ll}
w + x + y + z = 2 & (1) \\
-w + x - y - z = -6 & \text{Multiplying (3) by} -1 \\
\hline
2x = -4 & (6) \text{ Adding}
\end{array}
$$

$$
\begin{array}{ll}
w + x + y + z = 2 & (1) \\
-w + 3x + y - z = -2 & \text{Multiplying (4) by} -1 \\
\hline
4x + 2y = 0 & (7) \text{ Adding}
\end{array}
$$

We can solve (6) for x:

$$2x = -4$$
$$x = -2$$

Substitute –2 for x in (7):

$$4(-2) + 2y = 0$$
$$-8 + 2y = 0$$
$$2y = 8$$
$$y = 4$$

Substitute –2 for x and 4 for y in (5):

$$-(-2) - 4 - 3z = 1$$
$$-2 - 3z = 1$$
$$-3z = 3$$
$$z = -1$$

Substitute –2 for x, 4 for y, and –1 for z in (1):

$$w - 2 + 4 - 1 = 2$$
$$w + 1 = 2$$
$$w = 1$$

We obtain $(1, -2, 4, -1)$. This checks, so it is the solution.

53.
$$\frac{2}{x} - \frac{1}{y} - \frac{3}{z} = -1,$$
$$\frac{2}{x} - \frac{1}{y} + \frac{1}{z} = -9,$$
$$\frac{1}{x} + \frac{2}{y} - \frac{4}{z} = 17$$

Let u represent $\frac{1}{x}$, v represent $\frac{1}{y}$, and w represent $\frac{1}{z}$.

Substituting, we have

$$2u - v - 3w = -1, \quad (1)$$
$$2u - v + w = -9, \quad (2)$$
$$u + 2v - 4w = 17 \quad (3)$$

1., 2. The equations in u, v, and w are in standard form with no fractions or decimals.

3., 4. We eliminate v from two different pairs of equations.

$$
\begin{array}{ll}
2u - v - 3w = -1 & (1) \\
-2u + v - w = 9 & \text{Multiplying (2) by} -1 \\
\hline
-4w = 8 & (4) \text{ Adding}
\end{array}
$$

$$
\begin{array}{ll}
4u - 2v - 6w = -2 & \text{Multiplying (1) by 2} \\
u + 2v - 4w = 17 & (3) \\
\hline
5u - 10w = 15 & (5) \text{ Adding}
\end{array}
$$

5. We can solve (4) for w:

$$-4w = 8$$
$$w = -2$$

Substitute –2 for w in (5):

$$5u - 10(-2) = 15$$
$$5u + 20 = 15$$
$$5u = -5$$
$$u = -1$$

6. Substitute –1 for u and –2 for w in (1):

$$2(-1) - v - 3(-2) = -1$$
$$-v + 4 = -1$$
$$-v = -5$$
$$v = 5$$

Solve for x, y, and z. We substitute –1 for u, 5 for v and –2 for w.

$$u = \frac{1}{x} \qquad v = \frac{1}{y} \qquad w = \frac{1}{z}$$
$$-1 = \frac{1}{x} \qquad 5 = \frac{1}{y} \qquad -2 = \frac{1}{z}$$
$$x = -1 \qquad y = \frac{1}{5} \qquad z = -\frac{1}{2}$$

We obtain $\left(-1, \frac{1}{5}, -\frac{1}{2}\right)$. This checks, so it is the solution.

55.
$$5x - 6y + kz = -5, \quad (1)$$
$$x + 3y - 2z = 2, \quad (2)$$
$$2x - y + 4z = -1 \quad (3)$$

Eliminate y from two different pairs of equations.

$$
\begin{array}{ll}
5x - 6y + kz = -5 & (1) \\
2x + 6y - 4z = 4 & \text{Multiplying (2) by 2} \\
\hline
7x + (k-4)z = -1 & (4)
\end{array}
$$

$$x+3y\ -2z=2 \quad (2)$$
$$\underline{6x-3y+12z=-3} \quad \text{Multiplying (3) by 3}$$
$$7x\ \ +10z=-1 \quad (5)$$

Solve the system of Equations (4) and (5).

$$7x+(k-4)z=-1 \quad (4)$$
$$7x+10z=-1 \quad (5)$$

$$-7x-(k-4)z=1 \quad \text{Multiplying (4) by} -1$$
$$\underline{7x\ \ +10z=-1} \quad (5)$$
$$(-k+14)z=0 \quad (6)$$

The system is dependent for the value of k that makes Equation (6) true. This occurs when $-k+14$ is 0. We solve for k:

$$-k+14=0$$
$$14=k$$

57. $z=b-mx-ny$

Three solutions are $(1, 1, 2)$, $(3, 2, -6)$, and $\left(\dfrac{3}{2}, 1, 1\right)$.

We substitute for x, y, and z and then solve for b, m, and n.

$$2=b-m-n,$$
$$-6=b-3m-2n,$$
$$1=b-\tfrac{3}{2}m-n$$

1., 2. Write the equations in standard form. Also, clear the fraction in the last equation.

$$b\ -m\ -n=2, \quad (1)$$
$$b-3m-2n=-6, \quad (2)$$
$$2b-3m-2n=2 \quad (3)$$

3., 4. Eliminate b from two different pairs of equations.

$$b\ -m\ -n=2 \quad (1)$$
$$\underline{-b+3m+2n=6} \quad \text{Multiplying (2) by} -1$$
$$2m\ +n=8 \quad (4)\ \text{Adding}$$

$$-2b+2m+2n=-4 \quad \text{Multiplying (1) by} -2$$
$$\underline{2b-3m-2n=2} \quad (3)$$
$$-m\ \ =-2 \quad (5)\ \text{Adding}$$

5. We solve Equation (5) for m:

$$-m=-2$$
$$m=2$$

Substitute in Equation (4) and solve for n.

$$2\cdot2+n=8$$
$$4+n=8$$
$$n=4$$

6. Substitute in one of the original equations to find b.

$$b-2-4=2 \quad \text{Substituting 2 for } m \text{ and 4 for } n \text{ in (1)}$$
$$b-6=2$$
$$b=8$$

The solution is $(8, 2, 4)$, so the equation is

$$z=8-2x-4y\,.$$

59. *Writing Exercise.*

$$x+2y-z=1, \quad (1)$$
$$-x-2y+z=3, \quad (2)$$
$$2x+4y-2z=2 \quad (3)$$

Kadi determined that Equations (1) and (3) are dependent, since multiplying Equation (1) by 2 gives Equation (3). Ahmed found there is no solution by adding Equations (1) and (2) resulted in a contradiction.

Exercise Set 3.5

1. *Familiarize.* Let $x =$ the first number, $y =$ the second number, and $z =$ the third number.

Translate.

The sum of the three numbers is 85.
$$x+y+z = 85$$

The second is seven more than the first.
$$y = 7 + x$$

The third is two more than four times the second.
$$z = 2 + 4 \times y$$

We have a system of equations:

$$x+y+z=85, \quad \text{or} \quad x+y+z=85$$
$$y=7+x \qquad\quad -x+y\ \ =7$$
$$z=2+4y \qquad\quad -4y+z=2$$

Carry out. Solving the system we get $(8, 15, 62)$.

Check. The sum of the three numbers is $8 + 15 + 62$, or 85. The second number 15, is 7 more than the first number 8. The third number, 62 is 2 more than four times the first number, 8. The numbers check.

State. The numbers are 8, 15, and 62.

3. *Familiarize.* Let $x =$ the first number, $y =$ the second number, and $z =$ the third number.

Translate.

The sum of three numbers is 26.
$$x+y+z = 26$$

Twice the first minus the second is the third less 2.
$$2x - y = z - 2$$

The third is the second minus 3 times the first.
$$z = y - 3x$$

We now have a system of equations.

$$x+y+z=26, \quad \text{or} \quad x+y+z=26,$$
$$2x-y=z-2, \quad 2x-y-z=-2,$$
$$z=y-3x \quad 3x-y+z=0$$

Carry out. Solving the system we get (8, 21, –3).

Check. The sum of the numbers is $8+21-3$, or 26. Twice the first minus the second is $2\cdot 8-21$, or –5, which is 2 less than the third. The second minus three times the first is $21-3\cdot 8$, or –3, which is the third. The numbers check.

State. The numbers are 8, 21, and –3.

5. *Familiarize.* We first make a drawing.

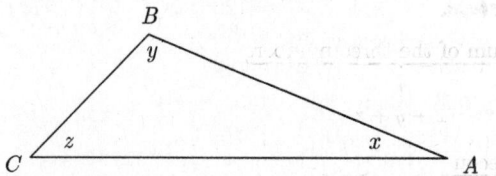

We let x, y, and z represent the measures of angles A, B, and C, respectively. The measures of the angles of a triangle add up to 180°.

Translate.

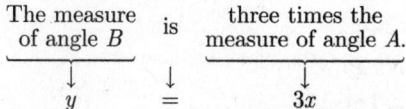

The sum of the measures is 180.

$$x+y+z = 180$$

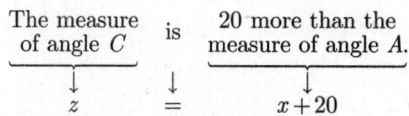

The measure of angle B is three times the measure of angle A.

$$y = 3x$$

The measure of angle C is 20 more than the measure of angle A.

$$z = x+20$$

We now have a system of equations.

$$x+y+z=180,$$
$$y=3x,$$
$$z=x+20$$

Carry out. Solving the system we get (32, 96, 52).

Check. The sum of the measures is $32°+96°+52°$, or 180°. Three times the measure of angle A is $3\cdot 32°$, or 96°, the measure of angle B. 20 more than the measure of angle A is $32°+20°$, or 52°, the measure of angle C. The numbers check.

State. The measures of angles A, B, and C are 32°, 96°, and 52°, respectively.

7. *Familiarize.* Let x, y and z represent the SAT critical reading score, mathematics score, and writing score, respectively.

Translate.

The average total score is 1511.

$$x+y+z = 1511$$

average math score is 13 more than reading score.

$$y = 13 + x$$

average writing score is reading score less 8.

$$z = x - 8$$

We have a system of equations:

$$x+y+z=1511, \quad \text{or} \quad x+y+z=1511$$
$$y=13+x \quad -x+y=13$$
$$z=x-8 \quad -x+z=-8$$

Carry out. Solving the system we get (502, 515, 494).

Check. The sum of the scores is $502 + 515 + 494$, or 1511. The mathematics score, 515, is 13 more than the reading score, 502. The writing score, 494 is 8 less than the reading score, 502. The answer checks.

State. The average score for reading was 502, for mathematics 515, and for writing 494.

9. *Familiarize.* Let x, y, and z represent the number of grams of fiber in 1 bran muffin, 1 banana, and a 1-cup serving of Wheaties, respectively.

Translate.

Two bran muffins, 1 banana, and a 1-cup serving of Wheaties contain 9 g of fiber, so we have

$$2x+y+z=9.$$

One bran muffin, 2 bananas, and a 1-cup serving of Wheaties contain 10.5 g of fiber, so we have

$$x+2y+z=10.5$$

Two bran muffins and a 1-cup serving of Wheaties contain 6 g of fiber, so we have

$$2x+z=6.$$

We now have a system of equations.

$$2x+y+z=9,$$
$$x+2y+z=10.5,$$
$$2x+z=6$$

Carry out. Solving the system, we get (1.5, 3, 3).

Check. Two bran muffins, 1 banana, and a 1-cup serving of Wheaties contain $2(1.5)+3+3$, or 9 g of fiber. One bran muffin, 2 bananas, and a 1-cup serving of Wheaties contain $1.5+2\cdot 3+3$, or 10.5 g of fiber. Two bran muffins and a 1-cup serving of Wheaties contain $2(1.5)+3$, or 6 g of fiber. The answer checks.

State. A bran muffin has 1.5 g of fiber, a banana has 3 g, and a 1-cup serving of Wheaties has 3 g.

11. Observe that the basic model plus tow package costs $30,815 and when a rear backup camera is added the price rises to $31,565. This tells us that the price of a rear backup camera is $31,565 – $30,815, or $750. Now observe that the basic model and rear backup camera costs $31,360 so the basic model costs $31,360 – $750, or $30,610. Finally we know the tow package is $30,815 – $30,610, or $205.

13. *Familiarize*. Let $x =$ the number of 12-oz cups, $y =$ the number of 16-oz cups, and $z =$ the number of 20-oz cups that Reba filled. Note that six 144-oz brewers contain $6 \cdot 144$, or 864 oz of coffee. Also, x 12-oz cups contain a total of $12x$ oz of coffee and bring in $1.65x$, y 16-oz cups contain $16y$ oz and bring in $1.85y$, and z 20-oz cups contain $20z$ oz and bring in $1.95z$.

Translate.

The total number of coffees served was 55.
$$x + y + z = 55$$

The total amount of coffee served was 864 oz.
$$12x + 16y + 20z = 864$$

The total amount collected was $99.65.
$$1.65x + 1.85y + 1.95z = 99.65$$

Now we have a system of equations.
$$x + y + z = 55,$$
$$12x + 16y + 20z = 864,$$
$$1.65x + 1.85y + 1.95z = 99.65$$

Carry out. Solving the system we get (17, 25, 13).

Check. The total number of coffees served was $17 + 25 + 13$, or 55. The total amount of coffee served was $12 \cdot 17 + 16 \cdot 25 + 20 \cdot 13 = 204 + 400 + 260 = 864$ oz. The total amount collected was $\$1.65(17) + \$1.85(25) + \$1.95(13) = \$28.05 + \$46.25 + \$25.35 = \$85.90$. The numbers check.

State. Reba filled 17 12-oz cups, 25 16-oz cups, and 13 20-oz cups.

15. *Familiarize*. Let $x =$ the amount of the loan at 8%, $y =$ the amount of the loan at 5%, and $z =$ the amount of the loan at 4%.

Translate.

Total of the three loans is $120,000.
$$x + y + z = 120,000$$

Total interest due is $5750.
$$0.08x + 0.05y + 0.04z = 5750$$

Mortgage interest is $1600 more than bank loan interest.
$$0.04z = 1600 + 0.08x$$

We have a system of equations:
$$x + y + z = 120,000$$
$$0.08x + 0.05y + 0.04z = 5750$$
$$-0.08x + 0.04z = 1600$$

Carry out. Solving the system we get (15,000, 35,000, 70,000).

Check. The total of the three loans is $15,000 + \$35,000 + \$70,000 = \$120,000$. The total interest due is $0.08(\$15,000) + 0.05(\$35,000) + 0.04(\$70,000) = \$1200 + \$1750 + \$2800 = \$5750$. The interest from the mortgage, $0.04(\$70,000)$, or $2800 is $1600 more than the bank loan interest $0.08(\$15,000)$, or $1200. The numbers check.

State. The bank loan was $15,000, the small-business loan was $35,000, and the mortgage loan was $70,000.

17. *Familiarize*. Let $x =$ the price of 1 g of gold, $y =$ the the price of 1 g of silver, and $z =$ the price of 1 g of copper.

Translate.

Cost of 100 g of red gold is $2265.40.
$$100(0.75x + 0.05y + 0.20z) = 2265.40$$

Cost of 100 g of yellow gold is $2287.75.
$$100(0.75x + 0.125y + 0.125z) = 2287.75$$

Cost of 100 g of white gold is $1312.50.
$$100(0.375x + 0.625y) = 1312.50$$

We have a system of equations:
$$75x + 5y + 20z = 2265.40$$
$$750x + 125y + 125z = 22,877.50$$
$$375x + 625y = 13,125.00$$

Carry out. Solving the system we get (30, 3, 0.02).

Check. The cost of 100 g of red gold is
$$100[0.75(\$30) + 0.05(\$3) + 0.20(\$0.02)]$$

$= \$2250 + \$15 + \$0.40 = \$2265.40$. The cost of 100 g of yellow gold is $100[0.75(\$30) + 0.125(\$3) + 0.125(\$0.02)]$ $= \$2250 + \$37.50 + \$0.25 = \$2287.75$. The cost of 100 g of white gold is $100[0.375(\$30) + 0.625(\$3)]$ $= \$1125 + \$187.50 = \$1312.50$. The numbers check.

State. The price of 1 g of gold is \$30, of silver is \$3 and of copper is \$0.02.

19. *Familiarize*. Let $r =$ the number of servings of roast beef, $p =$ the number of baked potatoes, and $b =$ the number of servings of broccoli. Then r servings of roast beef contain $300r$ Calories, $20r$ g of protein, and no vitamin C. In p baked potatoes there are $100p$ Calories, $5p$ g of protein, and $20p$ mg of vitamin C. And b servings of broccoli contain $50b$ Calories, $5b$ g of protein, and $100b$ mg of vitamin C. The patient requires 800 Calories, 55 g of protein, and 220 mg of vitamin C.

Translate. Write equations for the total number of calories, the total amount of protein, and the total amount of vitamin C.

$$\begin{aligned} 300r + 100p + 50b &= 800 \quad \text{(Calories)} \\ 20r + 5p + 5b &= 55 \quad \text{(protein)} \\ 20p + 100b &= 220 \quad \text{(vitamin C)} \end{aligned}$$

We now have a system of equations.

Carry out. Solving the system we get $(2, 1, 2)$.

Check. Two servings of roast beef provide 600 Calories, 40 g of protein, and no vitamin C. One baked potato provides 100 Calories, 5 g of protein, and 20 mg of vitamin C. And 2 servings of broccoli provide 100 Calories, 10 g of protein, and 200 mg of vitamin C. Together, then, they provide 800 Calories, 55 g of protein, and 220 mg of vitamin C. The values check.

State. The dietician should prepare 2 servings of roast beef, 1 baked potato, and 2 servings of broccoli.

21. *Familiarize*. Let x, y, and z be the number of tickets sold for the first mezzanine, main floor and second mezzanine, respectively.

Translate.

Total number of tickets   is   40.
$$x + y + z \quad = \quad 40$$

Number of tickets for first mezzanine and main floor   is   number of tickets for second mezzanine.
$$x + y \quad = \quad z$$

Total cost of tickets   is   \$1432.
$$52x + 38y + 28z \quad = \quad 1432$$

We have a system of equations:
$$\begin{aligned} x + y + z &= 40 \\ x + y &= z \\ 52x + 38y + 28z &= 1432 \end{aligned}$$

Carry out. Solving the system we get $(8, 12, 20)$.

Check. The total number of tickets is $8 + 12 + 20$, or 40. The sum of first mezzanine and main floor is $8 + 12$, is the number of second mezzanine, 20. The total cost is $\$52(8) + \$38(12) + \$28(20) = \$416 + \$456 + \$560 = \$1432$. The numbers check.

State. There were 8 first mezzanine tickets, 12 main floor tickets and 20 second mezzanine tickets sold.

23. *Familiarize*. Let x, y, and z represent the populations of Asia, Africa, and the rest of the world, respectively, in billions, in 2050.

Translate.

The total world population   will be   9.4 billion.
$$x + y + z \quad = \quad 9.4$$

Population of Asia   will be   3.5 billion   more than   Population of Africa
$$x \quad = \quad 3.5 \quad + \quad y$$

Population of the rest of the world   will be   two-fifths   of   population of Asia   less   0.3 billion.
$$z \quad = \quad \frac{2}{5} \quad \times \quad x \quad - \quad 0.3$$

We have a system of equations.
$$\begin{aligned} x + y + z &= 9.4 \\ x &= 3.5 + y \\ z &= \frac{2}{5}x - 0.3 \end{aligned}$$

Carry out. Solving the system we get $(5.5, 2.0, 1.9)$.

Check. The total population will be $5.5 + 2.0 + 1.9$, or 9.4 billion. The population of Asia, 5.5 billion, is 3.5 billion more than the population of Africa, 2.0 billion. Also, the rest of the world population, 1.9 billion is 0.3 billion less than $\frac{2}{5}$ of 5.5, or 2.2 billion. The numbers check.

State. In 2050, the population of Asia will be 5.5 billion, the population of Africa will be 2.0 billion and the population of the rest of the world will be 1.9 billion.

25. *Writing Exercise.* Problems like Exercises 13 and 14 have three unknowns whereas those in Section 3.3 have two unknowns.

27. $-2(2x - 3y) = -4x + 6y$

29. $-6(x - 2y) + (6x - 5y) = -6x + 12y + 6x - 5y = 7y$

31. $-(2a - b - 6c) = -2a + b + 6c$

33. $-2(3x - y + z) + 3(-2x + y - 2z)$
$= -6x + 2y - 2z - 6x + 3y - 6z$
$= -12x + 5y - 8z$

35. *Writing Exercise.* If the Knicks made 19 more 2-pointers than foul shots, and no foul shots were made, then there would have been 19 2-pointers. If there were 50 baskets and 19 were 2-pointers and 31 3-pointers yield a total of $2 \cdot 19 + 3 \cdot 31$, or 131 points, rather than 92 points. Thus, if no foul shots were made, there is no solution.

37. *Familiarize.* Let w, x, y, and z represent the monthly rates for an applicant, a spouse, the first child and the second child, respectively.

Translate.

$$\underbrace{\text{Rate for applicant and spouse}}_{\underset{\downarrow}{w + x}} \quad \underset{\downarrow}{\text{is}} \quad \underset{\downarrow}{\underbrace{\$174.}}$$
$$w + x \qquad = \qquad 174$$

$$\underbrace{\substack{\text{Rate for applicant,} \\ \text{spouse and one child}}}_{\underset{\downarrow}{w + x + y}} \quad \underset{\downarrow}{\text{is}} \quad \underset{\downarrow}{\underbrace{\$221.}}$$
$$w + x + y \qquad = \qquad 221$$

$$\underbrace{\substack{\text{Rate for applicant, spouse} \\ \text{and two children}}}_{\underset{\downarrow}{w + x + y + z}} \quad \underset{\downarrow}{\text{is}} \quad \underset{\downarrow}{\underbrace{\$263.}}$$
$$w + x + y + z \qquad = \qquad 263$$

$$\underbrace{\substack{\text{Rate for applicant,} \\ \text{and one child}}}_{\underset{\downarrow}{w + y}} \quad \underset{\downarrow}{\text{is}} \quad \underset{\downarrow}{\underbrace{\$134.}}$$
$$w + y \qquad = \qquad 134$$

We have a system of equations:
$$\begin{array}{ll} w + x = 174 & (1) \\ w + x + y = 221 & (2) \\ w + x + y + z = 263 & (3) \\ w + y = 134 & (4) \end{array}$$

Carry out. We solve the system of equations. First substitute 221 for $w + x + y$ in (3) and solve for z.
$$221 + z = 263$$
$$z = 42$$

Next substitute 174 for $w + x$ in (2) and solve for y.

$$174 + y = 221$$
$$y = 47$$

Substitute 47 for y in (4) and solve for w.
$$w + 47 = 134$$
$$w = 87$$

Finally, substitute 87 for w in (1) and solve for x.
$$87 + x = 174$$
$$x = 87$$

The solution is $(87, 87, 47, 42)$.

Check. The check is left to the student.

State. The separate monthly rates for an applicant, a spouse, the first child, and the second child are $87, $87, $47, and $42, respectively.

39. *Familiarize.* Let w, x, y, and z represent the ages of Tammy, Carmen, Dennis, and Mark respectively.

Translate.

Tammy's age is the sum of the ages of Carmen and Dennis, so we have
$$w = x + y.$$

Carmen's age is 2 more than the sum of the ages of Dennis and Mark, so we have
$$x = 2 + y + z.$$

Dennis's age is four times Mark's age, so we have
$$y = 4z.$$

The sum of all four ages is 42, so we have
$$w + x + y + z = 42.$$

Now we have a system of equations.
$$\begin{array}{ll} w = x + y, & (1) \\ x = 2 + y + z, & (2) \\ y = 4z, & (3) \\ w + x + y + z = 42 & (4) \end{array}$$

Carry out. We solve the system of equations. First we will express w, x, and y in terms of z and then solve for z. From (3) we know that $y = 4z$. Substitute $4z$ for y in (2):
$$x = 2 + 4z + z = 2 + 5z.$$

Substitute $2 + 5z$ for x and $4z$ for y in (1):
$$w = 2 + 5z + 4z = 2 + 9z.$$

Now substitute $2 + 9z$ for w, $2 + 5z$ for x, and $4z$ for y in (4) and solve for z.
$$2 + 9z + 2 + 5z + 4z + z = 42$$
$$19z + 4 = 42$$
$$19z = 38$$
$$z = 2$$

Then we have:
$$w = 2 + 9z = 2 + 9 \cdot 2 = 20,$$
$$x = 2 + 5z = 2 + 5 \cdot 2 = 12, \text{ and}$$
$$y = 4z = 4 \cdot 2 = 8$$

Although we were asked to find only Tammy's age, we found all of the ages so that we can check the result.

Check. The check is left to the student.

State. Tammy is 20 years old.

41. Let T, G, and H represent the number of tickets Tom, Gary, and Hal begin with, respectively. After Hal gives tickets to Tom and Gary, each has the following number of tickets:

$$\text{Tom}: \quad T + T, \text{ or } 2T,$$
$$\text{Gary}: \quad G + G, \text{ or } 2G,$$
$$\text{Hal}: \quad H - T - G.$$

After Tom gives tickets to Gary and Hal, each has the following number of tickets:

$$\text{Gary}: \quad 2G + 2G, \text{ or } 4G,$$
$$\text{Hal}: \quad (H - T - G) + (H - T - G), \text{ or }$$
$$\quad\quad 2(H - T - G),$$
$$\text{Tom}: \quad 2T - 2G - (H - T - G), \text{ or }$$
$$\quad\quad 3T - H - G$$

After Gary gives tickets to Hal and Tom, each has the following number of tickets:

$$\text{Hal}: \quad 2(H - T - G) + 2(H - T - G), \text{ or }$$
$$\quad\quad 4(H - T - G)$$
$$\text{Tom}: \quad (3T - H - G) + (3T - H - G), \text{ or }$$
$$\quad\quad 2(3T - H - G),$$
$$\text{Gary}: \quad 4G - 2(H - T - G) - (3T - H - G), \text{ or }$$
$$\quad\quad 7G - H - T.$$

Since Hal, Tom, and Gary each finish with 40 tickets, we write the following system of equations:

$$4(H - T - G) = 40,$$
$$2(3T - H - G) = 40,$$
$$7G - H - T = 40$$

Solving the system we find that $T = 35$, so Tom started with 35 tickets.

Exercise Set 3.6

1. matrix; see page 194 in the text.

3. entry; see page 196 in the text.

5. rows; see page 196 in the text.

7. $x + 2y = 11$,
   $3x - y = 5$

Write a matrix using only the constants.

$$\begin{bmatrix} 1 & 2 & | & 11 \\ 3 & -1 & | & 5 \end{bmatrix}$$

Multiply the first row by -3 and add it to the second row.

$$\begin{bmatrix} 1 & 2 & | & 11 \\ 0 & -7 & | & -28 \end{bmatrix} \quad \text{New Row 2} = -3(\text{Row 1}) + \text{Row 2}$$

Reinserting the variables, we have

$$x + 2y = 11, \quad (1)$$
$$-7y = -28 \quad (2)$$

Solve Equation (2) for y.

$$-7y = -28$$
$$y = 4$$

Substitute 4 for y in Equation (1) and solve for x.

$$x + 2(4) = 11$$
$$x + 8 = 11$$
$$x = 3$$

The solution is $(3, 4)$.

9. $3x + y = -1$,
   $6x + 5y = 13$

We first write a matrix using only the constants.

$$\begin{bmatrix} 3 & 1 & | & -1 \\ 6 & 5 & | & 13 \end{bmatrix}$$

Multiply the first row by -2 and add it to the second row.

$$\begin{bmatrix} 3 & 1 & | & -1 \\ 0 & 3 & | & 15 \end{bmatrix} \quad \text{New Row 2} = -2(\text{Row 1}) + \text{Row 2}$$

Reinserting the variables, we have

$$3x + y = -1, \quad (1)$$
$$3y = 15 \quad (2)$$

Solve Equation (2) for y.

$$3y = 15$$
$$y = 5$$

Substitute 5 for y in Equation (1) and solve for x.

$$3x + 5 = -1$$
$$3x = -6$$
$$x = -2$$

The solution is $(-2, 5)$.

11. $6x - 2y = 4$,
    $7x + y = 13$

Write a matrix using only the constants.

$$\begin{bmatrix} 6 & -2 & | & 4 \\ 7 & 1 & | & 13 \end{bmatrix}$$

Multiply the second row by 6 to make the first number in row 2 a multiple of 6.

$$\begin{bmatrix} 6 & -2 & | & 4 \\ 42 & 6 & | & 78 \end{bmatrix} \quad \text{New Row 2} = 6(\text{Row 2})$$

Now multiply the first row by -7 and add it to the second row.

$$\begin{bmatrix} 6 & -2 & | & 4 \\ 0 & 20 & | & 50 \end{bmatrix} \quad \text{New Row 2} = -7(\text{Row 1}) + \text{Row 2}$$

Reinserting the variables, we have

$$6x - 2y = 4, \quad (1)$$
$$20y = 50. \quad (2)$$

Solve Equation (2) for y.

$$20y = 50$$
$$y = \frac{5}{2}$$

Substitute $\frac{5}{2}$ for y in Equation (1) and solve for x.

$$6x - 2y = 4$$
$$6x - 2\left(\frac{5}{2}\right) = 4$$
$$6x - 5 = 4$$
$$6x = 9$$
$$x = \frac{3}{2}$$

The solution is $\left(\frac{3}{2}, \frac{5}{2}\right)$.

13. $3x + 2y + 2z = 3,$
$\quad x + 2y - z = 5,$
$\quad 2x - 4y + z = 0$

We first write a matrix using only the constants.

$$\begin{bmatrix} 3 & 2 & 2 & | & 3 \\ 1 & 2 & -1 & | & 5 \\ 2 & -4 & 1 & | & 0 \end{bmatrix}$$

First interchange rows 1 and 2 so that each number below the first number in the first row is a multiple of that number.

$$\begin{bmatrix} 1 & 2 & -1 & | & 5 \\ 3 & 2 & 2 & | & 3 \\ 2 & -4 & 1 & | & 0 \end{bmatrix}$$

Multiply row 1 by –3 and add it to row 2.

Multiply row 1 by –2 and add it to row 3.

$$\begin{bmatrix} 1 & 2 & -1 & | & 5 \\ 0 & -4 & 5 & | & -12 \\ 0 & -8 & 3 & | & -10 \end{bmatrix}$$

Multiply row 2 by –2 and add it to row 3.

$$\begin{bmatrix} 1 & 2 & -1 & | & 5 \\ 0 & -4 & 5 & | & -12 \\ 0 & 0 & -7 & | & 14 \end{bmatrix}$$

Reinserting the variables, we have

$$x + 2y - z = 5, \quad (1)$$
$$-4y + 5z = -12, \quad (2)$$
$$-7z = 14. \quad (3)$$

Solve (3) for z.

$$-7z = 14$$
$$z = -2$$

Substitute –2 for z in (2) and solve for y.

$$-4y + 5(-2) = -12$$
$$-4y - 10 = -12$$
$$-4y = -2$$
$$y = \frac{1}{2}$$

Substitute $\frac{1}{2}$ for y and –2 for z in (1) and solve for x.

$$x + 2 \cdot \frac{1}{2} - (-2) = 5$$
$$x + 1 + 2 = 5$$
$$x + 3 = 5$$
$$x = 2$$

The solution is $\left(2, \frac{1}{2}, -2\right)$.

15. $p - 2q - 3r = 3,$
$\quad 2p - q - 2r = 4,$
$\quad 4p + 5q + 6r = 4$

We first write a matrix using only the constants.

$$\begin{bmatrix} 1 & -2 & -3 & | & 3 \\ 2 & -1 & -2 & | & 4 \\ 4 & 5 & 6 & | & 4 \end{bmatrix}$$

$$\begin{bmatrix} 1 & -2 & -3 & | & 3 \\ 0 & 3 & 4 & | & -2 \\ 0 & 13 & 18 & | & -8 \end{bmatrix} \begin{matrix} \\ \text{New Row 2} = -2(\text{Row 1}) + \text{Row 2} \\ \text{New Row 3} = -4(\text{Row 1}) + \text{Row 3} \end{matrix}$$

$$\begin{bmatrix} 1 & -2 & -3 & | & 3 \\ 0 & 3 & 4 & | & -2 \\ 0 & 39 & 54 & | & -24 \end{bmatrix} \begin{matrix} \\ \\ \text{New Row 3} = 3(\text{Row 3}) \end{matrix}$$

$$\begin{bmatrix} 1 & -2 & -3 & | & 3 \\ 0 & 3 & 4 & | & -2 \\ 0 & 0 & 2 & | & 2 \end{bmatrix} \begin{matrix} \\ \\ \text{New Row 3} = -13(\text{Row 2}) + \text{Row 3} \end{matrix}$$

Reinserting the variables, we have

$$p - 2q - 3r = 3, \quad (1)$$
$$3q + 4r = -2, \quad (2)$$
$$2r = 2 \quad (3)$$

Solve (3) for r.

$$2r = 2$$
$$r = 1$$

Substitute 1 for r in (2) and solve for q.

$$3q + 4 \cdot 1 = -2$$
$$3q + 4 = -2$$
$$3q = -6$$
$$q = -2$$

Substitute –2 for q and 1 for r in (1) and solve for p.

$$p - 2(-2) - 3 \cdot 1 = 3$$
$$p + 4 - 3 = 3$$
$$p + 1 = 3$$
$$p = 2$$

The solution is $(2, -2, 1)$.

17. $3p + 2r = 11,$
$\quad q - 7r = 4,$
$\quad p - 6q = 1$

We first write a matrix using only the constants.

$$\begin{bmatrix} 3 & 0 & 2 & | & 11 \\ 0 & 1 & -7 & | & 4 \\ 1 & -6 & 0 & | & 1 \end{bmatrix}$$

$$\begin{bmatrix} 1 & -6 & 0 & | & 1 \\ 0 & 1 & -7 & | & 4 \\ 3 & 0 & 2 & | & 11 \end{bmatrix} \begin{matrix} \\ \text{Interchange} \\ \text{Row 1 and Row 3} \end{matrix}$$

$$\begin{bmatrix} 1 & -6 & 0 & | & 1 \\ 0 & 1 & -7 & | & 4 \\ 0 & 18 & 2 & | & 8 \end{bmatrix} \quad \text{New Row } 3 = -3(\text{Row } 1) + \text{Row } 3$$

$$\begin{bmatrix} 1 & -6 & 0 & | & 1 \\ 0 & 1 & -7 & | & 4 \\ 0 & 0 & 128 & | & -64 \end{bmatrix} \quad \text{New Row } 3 = -18(\text{Row } 2) + \text{Row } 3$$

Reinserting the variables, we have

$$\begin{aligned} p - 6q &= 1, & (1) \\ q - 7r &= 4, & (2) \\ 128r &= -64. & (3) \end{aligned}$$

Solve (3) for r.

$$128r = -64$$
$$r = -\frac{1}{2}$$

Substitute $-\frac{1}{2}$ for r in (2) and solve for q.

$$q - 7r = 4$$
$$q - 7\left(-\frac{1}{2}\right) = 4$$
$$q + \frac{7}{2} = 4$$
$$q = \frac{1}{2}$$

Substitute $\frac{1}{2}$ for q in (1) and solve for p.

$$p - 6 \cdot \frac{1}{2} = 1$$
$$p - 3 = 1$$
$$p = 4$$

The solution is $\left(4, \ \frac{1}{2}, \ -\frac{1}{2}\right)$.

19. We will rewrite the equations with the variables in alphabetical order:

$$\begin{aligned} -2w + 2x + 2y - 2z &= -10, \\ w + x + y + z &= -5, \\ 3w + x - y + 4z &= -2, \\ w + 3x - 2y + 2z &= -6 \end{aligned}$$

Write a matrix using only the constants.

$$\begin{bmatrix} -2 & 2 & 2 & -2 & | & -10 \\ 1 & 1 & 1 & 1 & | & -5 \\ 3 & 1 & -1 & 4 & | & -2 \\ 1 & 3 & -2 & 2 & | & -6 \end{bmatrix}$$

$$\begin{bmatrix} -1 & 1 & 1 & -1 & | & -5 \\ 1 & 1 & 1 & 1 & | & -5 \\ 3 & 1 & -1 & 4 & | & -2 \\ 1 & 3 & -2 & 2 & | & -6 \end{bmatrix} \quad \text{New Row } 1 = \frac{1}{2}(\text{Row } 1)$$

$$\begin{bmatrix} -1 & 1 & 1 & -1 & | & -5 \\ 0 & 2 & 2 & 0 & | & -10 \\ 0 & 4 & 2 & 1 & | & -17 \\ 0 & 4 & -1 & 1 & | & -11 \end{bmatrix} \quad \begin{array}{l} \text{New Row } 2 = \text{Row } 1 + \text{Row } 2 \\ \text{New Row } 3 = 3(\text{Row } 1) + \text{Row } 3 \\ \text{New Row } 4 = \text{Row } 1 + \text{Row } 4 \end{array}$$

$$\begin{bmatrix} -1 & 1 & 1 & -1 & | & -5 \\ 0 & 2 & 2 & 0 & | & -10 \\ 0 & 0 & -2 & 1 & | & 3 \\ 0 & 0 & -5 & 1 & | & 9 \end{bmatrix} \quad \begin{array}{l} \text{New Row } 3 = -2(\text{Row } 2) + \text{Row } 3 \\ \text{New Row } 4 = -2(\text{Row } 2) + \text{Row } 4 \end{array}$$

$$\begin{bmatrix} -1 & 1 & 1 & -1 & | & -5 \\ 0 & 2 & 2 & 0 & | & -10 \\ 0 & 0 & -2 & 1 & | & 3 \\ 0 & 0 & -10 & 2 & | & 18 \end{bmatrix} \quad \text{New Row } 4 = 2(\text{Row } 4)$$

$$\begin{bmatrix} -1 & 1 & 1 & -1 & | & -5 \\ 0 & 2 & 2 & 0 & | & -10 \\ 0 & 0 & -2 & 1 & | & 3 \\ 0 & 0 & 0 & -3 & | & 3 \end{bmatrix} \quad \text{New Row } 4 = -5(\text{Row } 3) + \text{Row } 4$$

Reinserting the variables, we have

$$\begin{aligned} -w + x + y - z &= -5, & (1) \\ 2x + 2y &= -10, & (2) \\ -2y + z &= 3, & (3) \\ -3z &= 3. & (4) \end{aligned}$$

Solve (4) for z.

$$-3z = 3$$
$$z = -1$$

Substitute -1 for z in (3) and solve for y.

$$-2y + (-1) = 3$$
$$-2y = 4$$
$$y = -2$$

Substitute -2 for y in (2) and solve for x.

$$2x + 2(-2) = -10$$
$$2x - 4 = -10$$
$$2x = -6$$
$$x = -3$$

Substitute -3 for x, -2 for y, and -1 for z in (1) and solve for w.

$$-w + (-3) + (-2) - (-1) = -5$$
$$-w - 3 - 2 + 1 = -5$$
$$-w - 4 = -5$$
$$-w = -1$$
$$w = 1$$

The solution is $(1, -3, -2, -1)$.

21. *Familiarize.* Let $d =$ the number of dimes and $n =$ the number of nickels. The value of d dimes is $\$0.10d$, and the value of n nickels is $\$0.05n$.

Translate.

$$\underline{\text{Total number of coins}} \quad \text{is} \quad 42.$$
$$d + n \qquad\qquad = \quad 42$$

$$\underline{\text{Total value of coins}} \quad \text{is} \quad \$3.$$
$$0.10d + 0.05n \qquad = \quad 3$$

After clearing decimals, we have this system.

$$\begin{aligned} d + n &= 42, \\ 10d + 5n &= 300 \end{aligned}$$

Carry out. Solve using matrices.

$$\begin{bmatrix} 1 & 1 & | & 42 \\ 10 & 5 & | & 300 \end{bmatrix}$$

$$\begin{bmatrix} 1 & 1 & | & 42 \\ 0 & -5 & | & -120 \end{bmatrix} \quad \text{New Row } 2 = -10(\text{Row } 1) + \text{Row } 2$$

Reinserting the variables, we have

$$d + n = 42, \quad (1)$$
$$-5n = -120 \quad (2)$$

Solve (2) for n.

$$-5n = -120$$
$$n = 24$$
$$d + 24 = 42 \quad \text{Back-substituting}$$
$$d = 18$$

Check. The sum of the two numbers is 42. The total value is $\$0.10(18) + \$0.05(24) = \$1.80 + \$1.20 = \$3$. The numbers check.

State. There are 18 dimes and 24 nickels.

23. **Familiarize.** Let $x =$ the number of pounds of dried fruit and $y =$ the number of pounds of macadamia nuts. We organize the information in a table.

	Dried fruit	Macadamia nuts	Mixture
Number of pounds	x	y	15
Price per pound	$5.80	$14.75	$9.38
Value of coffee	$5.80x$	$14.75y$	140.70

Translate.

The total number of pounds is 15.

$$x + y = 15$$

The total value of the mixture is $140.70.

$$5.80x + 14.75y = 140.70$$

After clearing decimals, we have this system:

$$x + y = 15,$$
$$580x + 1475y = 14,070$$

Carry out. Solve using matrices.

$$\begin{bmatrix} 1 & 1 & | & 15 \\ 580 & 1475 & | & 14{,}070 \end{bmatrix}$$

Multiply the first row by -580 and add it to the second row.

$$\begin{bmatrix} 1 & 1 & | & 15 \\ 0 & 895 & | & 5370 \end{bmatrix} \quad \text{New Row } 2 = -580(\text{Row } 1) + \text{Row } 2$$

Reinserting the variables, we have

$$x + y = 15$$
$$895y = 5370$$
$$y = 6$$

Back-substitute 6 for y in Equation (1) and solve for x.

$$x + 6 = 15$$
$$x = 9$$

Check. The sum of the numbers, $6 + 9 = 15$. The total value is $\$5.80(9) + \$14.75(6)$, or $\$52.20 + \$88.50$, or $\$140.70$. The numbers check.

State. 9 pounds of dried fruit and 6 pounds of macadamia nuts should be used.

25. **Familiarize.** We let x, y, and z represent the amounts invested at 7%, 8%, and 9%, respectively. Recall the formula for simple interest:

Interest = Principal × Rate × Time

Translate. We organize the information in a table.

	First Investment	Second Investment	Third Investment	Total
P	x	y	z	$2500
R	7%	8%	9%	
T	1 yr	1 yr	1 yr	
I	$0.07x$	$0.08y$	$0.09z$	$212

The first row gives us one equation:

$$x + y + z = 2500$$

The last row gives a second equation:

$$0.07x + 0.08y + 0.09z = 212$$

Amount invested at 9%	is	$1100	more than	amount invested at 8%.
↓	↓	↓	↓	↓
z	=	$1100	+	y

After clearing decimals, we have this system:

$$x + y + z = 2500,$$
$$7x + 8y + 9z = 21{,}200,$$
$$-y + z = 1100$$

Carry out. Solve using matrices.

$$\begin{bmatrix} 1 & 1 & 1 & | & 2500 \\ 7 & 8 & 9 & | & 21{,}200 \\ 0 & -1 & 1 & | & 1100 \end{bmatrix}$$

$$\begin{bmatrix} 1 & 1 & 1 & | & 2500 \\ 0 & 1 & 2 & | & 3700 \\ 0 & -1 & 1 & | & 1100 \end{bmatrix} \quad \text{New Row } 2 = -7(\text{Row } 1) + \text{Row } 2$$

$$\begin{bmatrix} 1 & 1 & 1 & | & 2500 \\ 0 & 1 & 2 & | & 3700 \\ 0 & 0 & 3 & | & 4800 \end{bmatrix} \quad \text{New Row } 3 = \text{Row } 2 + \text{Row } 3$$

Reinserting the variables, we have

$$x + y + z = 2500, \quad (1)$$
$$y + 2z = 3700, \quad (2)$$
$$3z = 4800 \quad (3)$$

Solve (3) for z.

$$3z = 4800$$
$$z = 1600$$

Back-substitute 1600 for z in (2) and solve for y.

$$y + 2 \cdot 1600 = 3700$$
$$y + 3200 = 3700$$
$$y = 500$$

Back-substitute 500 for y and 1600 for z in (1) and solve for x.

$$x + 500 + 1600 = 2500$$
$$x + 2100 = 2500$$
$$x = 400$$

Check. The total investment is $400 + $500 + $1600, or

$2500. The total interest is

$0.07(\$400) + 0.08(\$500) + 0.09(\$1600)$

$= \$28 + \$40 + \$144 = \$212$. The amount invested at 9%, $1600, is $1100 more than the amount invested at 8%, $500. The numbers check.

State. $400 is invested at 7%, $500 is invested at 8%, and $1600 is invested at 9%.

27. *Writing Exercise.* Row-equivalent operations yield a row containing all 0's. (This corresponds to an equation that is true for all values of the variables.)

29. $3(-1) - (-4)(5) = -3 - (-20) = -3 + 20 = 17$

31. $-2(5 \cdot 3 - 4 \cdot 6) - 3(2 \cdot 7 - 15) + 4(3 \cdot 8 - 5 \cdot 4)$
$= -2(15 - 24) - 3(14 - 15) + 4(24 - 20)$
$= -2(-9) - 3(-1) + 4(4)$
$= 18 + 3 + 16$
$= 21 + 16$
$= 37$

33. *Writing Exercise.* No; two systems of equations with different coefficients and constants can have the same solution. Because the coefficients and constants are different, corresponding entries in the matrices used to solve the systems are not all equal to each other.

35. *Familiarize.* Let w, x, y, and z represent the thousand's, hundred's, ten's, and one's digits, respectively.

Translate.

The sum of the digits is 10.
$$w + x + y + z = 10$$

Twice the sum of the thousand's and ten's digits is the sum of the hundred's and one's digits less one.
$$2(w + y) = x + z - 1$$

The ten's digit is twice the thousand's digit.
$$y = 2 \cdot w$$

The one's digit equals the sum of the thousand's and hundred's digits.
$$z = w + x$$

We have a system of equations which can be written as

$$\begin{aligned} w + x + y + z &= 10, \\ 2w - x + 2y - z &= -1, \\ -2w + y &= 0, \\ w + x - z &= 0. \end{aligned}$$

Carry out. We can use matrices to solve the system. We get $(1, 3, 2, 4)$.

Check. The sum of the digits is 10. Twice the sum of 1 and 2 is 6. This is one less than the sum of 3 and 4. The ten's digit, 2, is twice the thousand's digit, 1. The one's digit, 4, equals $1 + 3$. The numbers check.

State. The number is 1324.

Exercise Set 3.7

1. True; see page 198 in the text.

3. True; see page 198 in the text.

5. False; see page 199 in the text.

7. $\begin{vmatrix} 3 & 5 \\ 4 & 8 \end{vmatrix} = 3 \cdot 8 - 4 \cdot 5 = 24 - 20 = 4$

9. $\begin{vmatrix} 10 & 8 \\ -5 & -9 \end{vmatrix} = 10(-9) - 8(-5) = -90 + 40 = -50$

11. $\begin{vmatrix} 1 & 4 & 0 \\ 0 & -1 & 2 \\ 3 & -2 & 1 \end{vmatrix}$

$= 1 \begin{vmatrix} -1 & 2 \\ -2 & 1 \end{vmatrix} - 0 \begin{vmatrix} 4 & 0 \\ -2 & 1 \end{vmatrix} + 3 \begin{vmatrix} 4 & 0 \\ -1 & 2 \end{vmatrix}$
$= 1[-1 \cdot 1 - (-2) \cdot 2] - 0 + 3[4 \cdot 2 - (-1) \cdot 0]$
$= 1 \cdot 3 - 0 + 3 \cdot 8$
$= 3 - 0 + 24$
$= 27$

13. $\begin{vmatrix} -1 & -2 & -3 \\ 3 & 4 & 2 \\ 0 & 1 & 2 \end{vmatrix}$

$= -1 \begin{vmatrix} 4 & 2 \\ 1 & 2 \end{vmatrix} - 3 \begin{vmatrix} -2 & -3 \\ 1 & 2 \end{vmatrix} + 0 \begin{vmatrix} -2 & -3 \\ 4 & 2 \end{vmatrix}$
$= -1[4 \cdot 2 - 1 \cdot 2] - 3[-2 \cdot 2 - 1(-3)] + 0$
$= -1 \cdot 6 - 3 \cdot (-1) + 0$
$= -6 + 3 + 0$
$= -3$

15. $\begin{vmatrix} -4 & -2 & 3 \\ -3 & 1 & 2 \\ 3 & 4 & -2 \end{vmatrix}$

$= -4 \begin{vmatrix} 1 & 2 \\ 4 & -2 \end{vmatrix} - (-3) \begin{vmatrix} -2 & 3 \\ 4 & -2 \end{vmatrix} + 3 \begin{vmatrix} -2 & 3 \\ 1 & 2 \end{vmatrix}$
$= -4[1(-2) - 4 \cdot 2] + 3[-2(-2) - 4 \cdot 3] + 3[-2 \cdot 2 - 1 \cdot 3]$
$= -4(-10) + 3(-8) + 3(-7)$
$= 40 - 24 - 21 = -5$

17. $5x + 8y = 1,$
$3x + 7y = 5$

We compute D, D_x, and D_y.

$$D = \begin{vmatrix} 5 & 8 \\ 3 & 7 \end{vmatrix} = 35 - 24 = 11$$

$$D_x = \begin{vmatrix} 1 & 8 \\ 5 & 7 \end{vmatrix} = 7 - 40 = -33$$

$$D_y = \begin{vmatrix} 5 & 1 \\ 3 & 5 \end{vmatrix} = 25 - 3 = 22$$

Then,

$$x = \frac{D_x}{D} = \frac{-33}{11} = -3$$

and

$$y = \frac{D_y}{D} = \frac{22}{11} = 2.$$

The solution is $(-3, 2)$.

19. $5x - 4y = -3,$
$7x + 2y = 6$

We compute D, D_x, and D_y.

$$D = \begin{vmatrix} 5 & -4 \\ 7 & 2 \end{vmatrix} = 10 - (-28) = 38$$

$$D_x = \begin{vmatrix} -3 & -4 \\ 6 & 2 \end{vmatrix} = -6 - (-24) = 18$$

$$D_y = \begin{vmatrix} 5 & -3 \\ 7 & 6 \end{vmatrix} = 30 - (-21) = 51$$

Then,

$$x = \frac{D_x}{D} = \frac{18}{38} = \frac{9}{19}$$

and

$$y = \frac{D_y}{D} = \frac{51}{38}.$$

The solution is $\left(\frac{9}{19}, \frac{51}{38}\right)$.

21. $3x - y + 2z = 1,$
$x - y + 2z = 3,$
$-2x + 3y + z = 1$

We compute D, D_x, and D_y.

$$D = \begin{vmatrix} 3 & -1 & 2 \\ 1 & -1 & 2 \\ -2 & 3 & 1 \end{vmatrix}$$

$$= 3\begin{vmatrix} -1 & 2 \\ 3 & 1 \end{vmatrix} - 1\begin{vmatrix} -1 & 2 \\ 3 & 1 \end{vmatrix} - 2\begin{vmatrix} -1 & 2 \\ -1 & 2 \end{vmatrix}$$

$$= 3(-7) - 1(-7) - 2(0)$$
$$= -21 + 7 - 0$$
$$= -14$$

$$D_x = \begin{vmatrix} 1 & -1 & 2 \\ 3 & -1 & 2 \\ 1 & 3 & 1 \end{vmatrix}$$

$$= 1\begin{vmatrix} -1 & 2 \\ 3 & 1 \end{vmatrix} - 3\begin{vmatrix} -1 & 2 \\ 3 & 1 \end{vmatrix} + 1\begin{vmatrix} -1 & 2 \\ -1 & 2 \end{vmatrix}$$

$$= 1(-7) - 3(-7) + 1(0)$$
$$= -7 + 21 + 0$$
$$= 14$$

$$D_y = \begin{vmatrix} 3 & 1 & 2 \\ 1 & 3 & 2 \\ -2 & 1 & 1 \end{vmatrix}$$

$$= 3\begin{vmatrix} 3 & 2 \\ 1 & 1 \end{vmatrix} - 1\begin{vmatrix} 1 & 2 \\ 1 & 1 \end{vmatrix} - 2\begin{vmatrix} 1 & 2 \\ 3 & 2 \end{vmatrix}$$

$$= 3 \cdot 1 - 1(-1) - 2(-4)$$
$$= 3 + 1 + 8$$
$$= 12$$

Then,

$$x = \frac{D_x}{D} = \frac{14}{-14} = -1$$

and

$$y = \frac{D_y}{D} = \frac{12}{-14} = -\frac{6}{7}.$$

Substitute in the third equation to find z.

$$-2(-1) + 3\left(-\frac{6}{7}\right) + z = 1$$

$$2 - \frac{18}{7} + z = 1$$

$$-\frac{4}{7} + z = 1$$

$$z = \frac{11}{7}$$

The solution is $\left(-1, -\frac{6}{7}, \frac{11}{7}\right)$.

23. $2x - 3y + 5z = 27,$
$x + 2y - z = -4,$
$5x - y + 4z = 27$

We compute D, D_x, and D_y.

$$D = \begin{vmatrix} 2 & -3 & 5 \\ 1 & 2 & -1 \\ 5 & -1 & 4 \end{vmatrix}$$

$$= 2\begin{vmatrix} 2 & -1 \\ -1 & 4 \end{vmatrix} - 1\begin{vmatrix} -3 & 5 \\ -1 & 4 \end{vmatrix} + 5\begin{vmatrix} -3 & 5 \\ 2 & -1 \end{vmatrix}$$

$$= 2(7) - 1(-7) + 5(-7)$$
$$= 14 + 7 - 35 = -14$$

$$D_x = \begin{vmatrix} 27 & -3 & 5 \\ -4 & 2 & -1 \\ 27 & -1 & 4 \end{vmatrix}$$

$$= 27\begin{vmatrix} 2 & -1 \\ -1 & 4 \end{vmatrix} - (-4)\begin{vmatrix} -3 & 5 \\ -1 & 4 \end{vmatrix} + 27\begin{vmatrix} -3 & 5 \\ 2 & -1 \end{vmatrix}$$

$$= 27(7) + 4(-7) + 27(-7)$$
$$= 189 - 28 - 189$$
$$= -28$$

$$D_y = \begin{vmatrix} 2 & 27 & 5 \\ 1 & -4 & -1 \\ 5 & 27 & 4 \end{vmatrix}$$

$$= 2\begin{vmatrix} -4 & -1 \\ 27 & 4 \end{vmatrix} - 1\begin{vmatrix} 27 & 5 \\ 27 & 4 \end{vmatrix} + 5\begin{vmatrix} 27 & 5 \\ -4 & -1 \end{vmatrix}$$

$$= 2(11) - 1(-27) + 5(-7)$$
$$= 22 + 27 - 35$$
$$= 14$$

Then,

$$x = \frac{D_x}{D} = \frac{-28}{-14} = 2,$$

and

$$y = \frac{D_y}{D} = \frac{14}{-14} = -1.$$

We substitute in the second equation to find z.

$$2 + 2(-1) - z = -4$$
$$2 - 2 - z = -4$$
$$-z = -4$$
$$z = 4$$

The solution is $(2, -1, 4)$.

25. $r - 2s + 3t = 6,$
$2r - s - t = -3,$
$r + s + t = 6$

We compute D, D_r, and D_s.

$$D = \begin{vmatrix} 1 & -2 & 3 \\ 2 & -1 & -1 \\ 1 & 1 & 1 \end{vmatrix}$$

$$= 1\begin{vmatrix} -1 & -1 \\ 1 & 1 \end{vmatrix} - 2\begin{vmatrix} -2 & 3 \\ 1 & 1 \end{vmatrix} + 1\begin{vmatrix} -2 & 3 \\ -1 & -1 \end{vmatrix}$$

$$= 1(0) - 2(-5) + 1(5)$$
$$= 0 + 10 + 5$$
$$= 15$$

$$D_r = \begin{vmatrix} 6 & -2 & 3 \\ -3 & -1 & -1 \\ 6 & 1 & 1 \end{vmatrix}$$

$$= 6\begin{vmatrix} -1 & -1 \\ 1 & 1 \end{vmatrix} - (-3)\begin{vmatrix} -2 & 3 \\ 1 & 1 \end{vmatrix} + 6\begin{vmatrix} -2 & 3 \\ -1 & -1 \end{vmatrix}$$

$$= 6(0) + 3(-5) + 6(5)$$
$$= 0 - 15 + 30$$
$$= 15$$

$$D_s = \begin{vmatrix} 1 & 6 & 3 \\ 2 & -3 & -1 \\ 1 & 6 & 1 \end{vmatrix}$$

$$= 1\begin{vmatrix} -3 & -1 \\ 6 & 1 \end{vmatrix} - 2\begin{vmatrix} 6 & 3 \\ 6 & 1 \end{vmatrix} + 1\begin{vmatrix} 6 & 3 \\ -3 & -1 \end{vmatrix}$$

$$= 1(3) - 2(-12) + 1(3)$$
$$= 3 + 24 + 3$$
$$= 30$$

Then,

$$r = \frac{D_r}{D} = \frac{15}{15} = 1,$$

and

$$s = \frac{D_s}{D} = \frac{30}{15} = 2.$$

Substitute in the third equation to find t.

$$1 + 2 + t = 6$$
$$3 + t = 6$$
$$t = 3$$

The solution is $(1, 2, 3)$.

27. *Writing Exercise.* Answers may vary. One pattern for Cramer's rule is that the system is dependent if the denominator is 0 and all the numerators are also 0.

29. $f(90) = 80(90) + 2500 = 7200 + 2500 = 9700$

31. $(g - f)(10) = g(10) - f(10)$
$$= 150(10) - [80(10) + 2500]$$
$$= 1500 - [800 + 2500]$$
$$= 1500 - 3300$$
$$= -1800$$

33.
$$f(x) = g(x)$$
$$80x + 2500 = 150x$$
$$2500 = 70x$$
$$\frac{250}{7} = x$$

35. *Writing Exercise.* If $a_1 x + b_1 y = c_1$ and $a_2 x + b_2 y = c_2$ are dependent, then one equation is a multiple of the other. That is, $a_1 = k a_2$ and $b_1 = k b_2$ for some constant k. Then

$$\begin{vmatrix} a_1 & b_1 \\ a_2 & b_2 \end{vmatrix} = \begin{vmatrix} k a_2 & k b_2 \\ a_2 & b_2 \end{vmatrix}$$
$$= k a_2 (b_2) - a_2 (k b_2)$$
$$= 0$$

37. $\begin{vmatrix} y & -2 \\ 4 & 3 \end{vmatrix} = 44$

$$y \cdot 3 - 4(-2) = 44 \quad \text{Evaluating the determinant}$$
$$3y + 8 = 44$$
$$3y = 36$$
$$y = 12$$

39. $\begin{vmatrix} m+1 & -2 \\ m-2 & 1 \end{vmatrix} = 27$

$$(m+1)(1) - (m-2)(-2) = 27 \quad \text{Evaluating}$$
$$ \text{the determinant}$$
$$m + 1 + 2m - 4 = 27$$
$$3m = 30$$
$$m = 10$$

Exercise Set 3.8

1. b; see page 203 in the text.

3. h; see page 204 in the text.

5. e; see page 203 in the text.

7. c; see page 205 in the text.

9. $C(x) = 35x + 200,000 \quad R(x) = 55x$

a) $\begin{aligned} P(x) &= R(x) - C(x) \\ &= 55x - (35x + 200,000) \\ &= 55x - 35x - 200,000 \\ &= 20x - 200,000 \end{aligned}$

b) Solve the system

$R(x) = 55x,$
$C(x) = 35x + 200,000.$

Since both $R(x)$ and $C(x)$ are in dollars and they are equal at the break-even point, we can rewrite the system:

$d = 55x, \qquad\qquad (1)$
$d = 35x + 200,000 \quad (2)$

We solve using substitution.

$55x = 35x + 200,000 \quad$ Substituting $55x$ for
$\qquad\qquad\qquad\qquad\qquad d$ in (2)
$20x = 200,000$
$\quad x = 10,000$

Thus, 10,000 units must be produced and sold in order to break even.

The revenue will be $R(10,000) = 55 \cdot 10,000 = 550,000$.

The break-even point is (10,000 units, $550,000).

11. $C(x) = 15x + 3100 \quad R(x) = 40x$

a) $\begin{aligned} P(x) &= R(x) - C(x) \\ &= 40x - (15x + 3100) \\ &= 40x - 15x - 3100 \\ &= 25x - 3100 \end{aligned}$

b) Solve the system

$R(x) = 40x,$
$C(x) = 15x + 3100.$

Since both $R(x)$ and $C(x)$ are in dollars and they are equal at the break-even point, we can rewrite the system:

$d = 40x, \qquad\qquad (1)$
$d = 15x + 3100 \quad (2)$

We solve using substitution.

$40x = 15x + 3100 \quad$ Substituting $40x$ for
$\qquad\qquad\qquad\qquad\qquad d$ in (2)
$25x = 3100$
$\quad x = 124$

Thus, 124 units must be produced and sold in order to break even.

The revenue will be $R(124) = 40 \cdot 124 = 4960$.

The break-even point is (124 units, $4960).

13. $C(x) = 40x + 22,500 \quad R(x) = 85x$

a) $\begin{aligned} P(x) &= R(x) - C(x) \\ &= 85x - (40x + 22,500) \\ &= 85x - 40x - 22,500 \\ &= 45x - 22,500 \end{aligned}$

b) Solve the system

$R(x) = 85x,$
$C(x) = 40x + 22,500.$

Since both $R(x)$ and $C(x)$ are in dollars and they are equal at the break-even point, we can rewrite the system:

$d = 85x, \qquad\qquad (1)$
$d = 40x + 22,500 \quad (2)$

We solve using substitution.

$85x = 40x + 22,500 \quad$ Substituting $85x$ for
$\qquad\qquad\qquad\qquad\qquad d$ in (2)
$45x = 22,500$
$\quad x = 500$

Thus, 500 units must be produced and sold in order to break even.

The revenue will be $R(500) = 85 \cdot 500 = 42,500$.

The break-even point is (500 units, $42,500).

15. $C(x) = 24x + 50,000 \quad R(x) = 40x$

a) $\begin{aligned} P(x) &= R(x) - C(x) \\ &= 40x - (24x + 50,000) \\ &= 40x - 24x - 50,000 \\ &= 16x - 50,000 \end{aligned}$

b) Solve the system

$R(x) = 40x,$
$C(x) = 24x + 50,000.$

Since both $R(x)$ and $C(x)$ are in dollars and they are equal at the break-even point, we can rewrite the system:

$d = 40x, \qquad\qquad (1)$
$d = 24x + 50,000 \quad (2)$

We solve using substitution.

$40x = 24x + 50,000 \quad$ Substituting $40x$ for
$\qquad\qquad\qquad\qquad\qquad d$ in (2)
$16x = 50,000$
$\quad x = 3125$

Thus, 3125 units must be produced and sold in order to break even.

The revenue will be $R(3125) = 40 \cdot 3125 = \$125,000$.

The break-even point is (3125 units, $125,000).

17. $C(x) = 75x + 100,000 \quad R(x) = 125x$

a) $\begin{aligned} P(x) &= R(x) - C(x) \\ &= 125x - (75x + 100,000) \\ &= 125x - 75x - 100,000 \\ &= 50x - 100,000 \end{aligned}$

b) Solve the system

$R(x) = 125x,$
$C(x) = 75x + 100,000.$

Since $R(x) = C(x)$ at the break-even point, we can rewrite the system:

$R(x) = 125x, \qquad\qquad (1)$
$C(x) = 75x + 100,000 \quad (2)$

We solve using substitution.

$125x = 75x + 100,000$ Substituting $125x$
$50x = 100,000$ for $R(x)$ in (2)
$x = 2000$

To break even 2000 units must be produced and sold.

The revenue will be $R(2000) = 125 \cdot 2000 = 250,000$. The break-even point is (2000 units, \$250,000).

19. $D(p) = 2000 - 15p$,
$S(p) = 740 + 6p$

Rewrite the system:

$q = 2000 - 15p$,   (1)
$q = 740 + 6p$ (2)

Substitute $2000 - 15p$ for q in (2) and solve.

$2000 - 15p = 740 + 6p$
$1260 = 21p$
$60 = p$

The equilibrium price is \$60 per unit.

To find the equilibrium quantity we substitute \$60 into either $D(p)$ or $S(p)$.

$D(60) = 2000 - 15(60) = 2000 - 900 = 1100$

The equilibrium quantity is 1100 units.

The equilibrium point is (\$60, 1100).

21. $D(p) = 760 - 13p$,
$S(p) = 430 + 2p$

Rewrite the system:

$q = 760 - 13p$,   (1)
$q = 430 + 2p$ (2)

Substitute $760 - 13p$ for q in (2) and solve.

$760 - 13p = 430 + 2p$
$330 = 15p$
$22 = p$

The equilibrium price is \$22 per unit.

To find the equilibrium quantity we substitute \$22 into either $D(p)$ or $S(p)$.

$S(22) = 430 + 2(22) = 430 + 44 = 474$

The equilibrium quantity is 474 units.

The equilibrium point is (\$22, 474).

23. $D(p) = 7500 - 25p$,
$S(p) = 6000 + 5p$

Rewrite the system:

$q = 7500 - 25p$,   (1)
$q = 6000 + 5p$ (2)

Substitute $7500 - 25p$ for q in (2) and solve.

$7500 - 25p = 6000 + 5p$
$1500 = 30p$
$50 = p$

The equilibrium price is \$50 per unit.

To find the equilibrium quantity we substitute \$50 into either $D(p)$ or $S(p)$.

$D(50) = 7500 - 25(50) = 7500 - 1250 = 6250$

The equilibrium quantity is 6250 units.

The equilibrium point is (\$50, 6250).

25. $D(p) = 1600 - 53p$,
$S(p) = 320 + 75p$

Rewrite the system:

$q = 1600 - 53p$,   (1)
$q = 320 + 75p$ (2)

Substitute $1600 - 53p$ for q in (2) and solve.

$1600 - 53p = 320 + 75p$
$1280 = 128p$
$10 = p$

The equilibrium price is \$10 per unit.

To find the equilibrium quantity we substitute \$10 into either $D(p)$ or $S(p)$.

$S(10) = 320 + 75(10) = 320 + 750 = 1070$

The equilibrium quantity is 1070 units.

The equilibrium point is (\$10, 1070).

27. a) $C(x) =$ Fixed costs + Variable costs
$C(x) = 45,000 + 40x$,

where x is the number of MP3/cell phones.

b)   Each MP3/cell phone sells for \$130. The total revenue is 130 times the number of MP3/cell phone sold. We assume that all MP3/cell phones produced are sold.

$R(x) = 130x$

c) $P(x) = R(x) - C(x)$
$P(x) = 130x - (45,000 + 40x)$
$= 130x - 45,000 - 40x$
$= 90x - 45,000$

d) $P(3000) = 90(3000) - 45,000$
$= 270,000 - 45,000$
$= \$225,000$

The company will realize a profit of \$225,000 when 3000 MP3/cell phones are produced and sold.

$P(400) = 90(400) - 45,000$
$= 36,000 - 45,000$
$= -\$9000$

The company will realize a \$9000 loss when 400 MP3/cell phones are produced and sold.

e)   Solve the system
$R(x) = 130x$,
$C(x) = 45,000 + 40x$.

Since both $R(x)$ and $C(x)$ are in dollars and they are equal at the break-even point, we can rewrite the system:

$d = 130x$,   (1)
$d = 45,000 + 40x$ (2)

We solve using substitution.

$130x = 45,000 + 40x$ Substituting $130x$ for d
            in (2)
$90x = 45,000$
$x = 500$

The firm will break even if it produces and sells 500 MP3/cell phones and takes in a total of $R(500) = 130 \cdot 500 = \$65,000$ in revenue. Thus, the break-even point is (500 MP3/cell phones, \$65,000).

29. a) $C(x) = \text{Fixed costs} + \text{Variable costs}$
        $C(x) = 10,000 + 30x,$

where x is the number of pet car seats produced.

b) Each pet car seat sells for \$80. The total revenue is 80 times the number of seats sold. We assume that all seats produced are sold.

   $R(x) = 80x$

c) $P(x) = R(x) - C(x)$
   $P(x) = 80x - (10,000 + 30x)$
   $= 80x - 10,000 - 30x$
   $= 50x - 10,000$

d) $P(2000) = 50(2000) - 10,000$
   $= 100,000 - 10,000$
   $= 90,000$

The company will realize a profit of \$90,000 when 2000 seats are produced and sold.

   $P(50) = 50(50) - 10,000$
   $= 2500 - 10,000$
   $= -7500$

The company will realize a loss of \$7500 when 50 seats are produced and sold.

e) Solve the system
   $R(x) = 80x,$
   $C(x) = 10,000 + 30x.$

Since both $R(x)$ and $C(x)$ are in dollars and they are equal at the break-even point, we can rewrite the system:

   $d = 80x,$ (1)
   $d = 10,000 + 30x$ (2)

We solve using substitution.

   $80x = 10,000 + 30x$ Substituting $80x$ for d
            in (2)
   $50x = 10,000$
   $x = 200$

The firm will break even if it produces and sells 200 seats and takes in a total of $R(200) = 80 \cdot 200 = \$16,000$ in revenue. Thus, the break-even point is (200 seats, \$16,000).

31. *Writing Exercise.* Since $P(x) = R(x) - C(x)$, we can also say $R(x) = P(x) + C(x)$. Thus, the slope of the line representing revenue is the sum of the slopes of the other two lines, which represent cost and profit.

33. $4x - 3 = 21$
        $4x = 24$
        $x = 6$

35. $3x - 5 = 12x + 6$
        $-11 = 9x$
        $-\dfrac{11}{9} = x$

37. $3 - (x + 2) = 7$
        $3 - x - 2 = 7$
        $1 - x = 7$
        $-x = 6$
        $x = -6$

39. *Writing Exercise.* No; there will be variable costs in the production of the birdbaths, so Rosie will need more than $\$300 \cdot 10$, or \$3000, in revenue in order to break even.

41. The supply function contains the points (\$2, 100) and (\$8, 500). We find its equation:

$$m = \frac{500 - 100}{8 - 2} = \frac{400}{6} = \frac{200}{3}$$
$$y - y_1 = m(x - x_1) \qquad \text{Point-slope form}$$
$$y - 100 = \frac{200}{3}(x - 2)$$
$$y - 100 = \frac{200}{3}x - \frac{400}{3}$$
$$y = \frac{200}{3}x - \frac{100}{3}$$

We can equivalently express supply S as a function of price p:

$$S(p) = \frac{200}{3}p - \frac{100}{3}$$

The demand function contains the points (\$1, 500) and (\$9, 100). We find its equation:

$$m = \frac{100 - 500}{9 - 1} = \frac{-400}{8} = -50$$
$$y - y_1 = m(x - x_1)$$
$$y - 500 = -50(x - 1)$$
$$y - 500 = -50x + 50$$
$$y = -50x + 550$$

We can equivalently express demand D as a function of price p:

$$D(p) = -50p + 550$$

We have a system of equations

$$S(p) = \frac{200}{3}p - \frac{100}{3},$$
$$D(p) = -50p + 550.$$

Rewrite the system:

$$q = \frac{200}{3}p - \frac{100}{3}, \quad (1)$$
$$q = -50p + 550 \quad (2)$$

Substitute $\frac{200}{3}p - \frac{100}{3}$ for q in (2) and solve.

$$\frac{200}{3}p - \frac{100}{3} = -50p + 550$$
$$200p - 100 = -150p + 1650 \quad \text{Multiplying by 3}$$
$$350p - 100 = 1650 \qquad \text{to clear fractions}$$
$$350p = 1750$$
$$p = 5$$

The equilibrium price is $5 per unit.

To find the equilibrium quantity, we substitute $5 into either $S(p)$ or $D(p)$.

$$D(5) = -50(5) + 550 = -250 + 550 = 300$$

The equilibrium quantity is 300 yo-yo's.

The equilibrium point is ($5, 300 yo-yo's).

43. a) Use a graphing calculator to find the first coordinate of the point of intersection of $y_1 = -14.97x + 987.35$ and $y_2 = 98.55x - 5.13$, to the nearest hundredth. It is 8.74, so the price per unit that should be charged is $8.74.

b) Use a graphing calculator to find the first coordinate of the point of intersection of $y_1 = 87,985 + 5.15x$ and $y_2 = 8.74x$. It is about 24,508.4, so 24,509 units must be sold in order to break even.

Chapter 3 Review

1. substitution; see Section 3.2.

3. graphical; see Sections 3.1 and 3.2.

5. inconsistent; see Section 3.1.

7. parallel; see Section 3.1.

9. determinant; see Section 3.7.

11. Graph the equations.

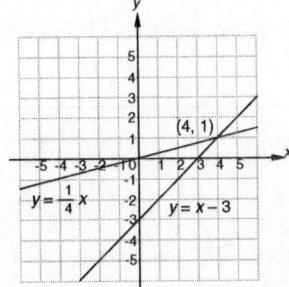

The solution (point of intersection) is (4, 1).

13. $5x - 2y = 4, \quad (1)$
$\quad\;\; x = y - 2 \quad (2)$

We substitute $y - 2$ for x in Equation (1) and solve for y.

$$5x - 2y = 4 \qquad (1)$$
$$5(y - 2) - 2y = 4 \qquad \text{Substituting}$$
$$5y - 10 - 2y = 4$$
$$3y - 10 = 4$$
$$3y = 14$$
$$y = \frac{14}{3}$$

Next we substitute $\frac{14}{3}$ for y in either equation of the original system and solve for x.

$$x = y - 2 \qquad (1)$$
$$x = \frac{14}{3} - 2 = \frac{8}{3}$$

Since $\left(\frac{8}{3}, \frac{14}{3}\right)$ checks, it is the solution.

15. $x - 3y = -2, \quad (1)$
$\quad\;\; 7y - 4x = 6 \quad (2)$

We solve Equation (1) for x.

$$x - 3y = -2$$
$$x = 3y - 2 \qquad (3)$$

We substitute $3y - 2$ for x in the second equation and solve for y.

$$7y - 4x = 6 \qquad (2)$$
$$7y - 4(3y - 2) = 6 \qquad \text{Substituting}$$
$$7y - 12y + 8 = 6$$
$$-5y + 8 = 6$$
$$-5y = -2$$
$$y = \frac{2}{5}$$

We substitute $\frac{2}{5}$ for y in Equation (3) and solve for x.

$$x = 3y - 2 \qquad (3)$$
$$x = 3 \cdot \frac{2}{5} - 2 = \frac{6}{5} - 2 = -\frac{4}{5}$$

Since $\left(-\frac{4}{5}, \frac{2}{5}\right)$ checks, it is the solution.

17. $4x - 7y = 18, \quad (1)$
$\quad\;\; 9x + 14y = 40 \quad (2)$

We multiply Equation (1) by 2.

$$\begin{array}{ll} 8x - 14y = 36 & \text{Multiplying (1) by 2} \\ \underline{9x + 14y = 40} & (2) \\ 17x = 76 & \text{Adding} \\ x = \frac{76}{17} \end{array}$$

Substitute $\frac{76}{17}$ for x in Equation (1) and solve for y.

$$4x - 7y = 18$$
$$4\left(\frac{76}{17}\right) - 7y = 18 \qquad \text{Substituting}$$
$$\frac{304}{17} - 7y = 18$$
$$-7y = \frac{2}{17}$$
$$y = -\frac{2}{119}$$

We obtain $\left(\frac{76}{17}, -\frac{2}{119}\right)$. This checks, so it is the solution.

19. $1.5x - 3 = -2y,$ (1)
$3x + 4y = 6$ (2)

Rewriting the equations in standard form.

$1.5x + 2y = 3,$ (1)
$3x + 4y = 6$ (2)

Observe that, if we multiply Equation (1) by 2, we obtain Equation (2). Thus, any pair that is a solution of Equation (1) is also a solution of Equation (2). The equations are dependent and the solution set is infinite:

$\{(x,\ y)\,|\,3x + 4y = 6\}.$

21. *Familiarize*. Let t = the number of hours for the passenger train, and $t + 1$ = the number of hours for the freight train.

Now complete the chart.

$$d \quad = \quad r \quad \cdot \quad t$$

	Distance	Rate	Time	
Passenger train	d	55	t	$\rightarrow d = 55t$
Freight train	d	44	$t + 1$	$\rightarrow d = 44(t+1)$

Translate. Using $d = rt$ in each row of the table, we get a system of equations:

$d = 55t,$
$d = 44(t + 1)$

Carry out. We solve the system of equations.

$55t = 44(t + 1)$　Using substitution
$55t = 44t + 44$
$11t = 44$
$t = 4$

Check. At 55 mph, in 4 hr the passenger train will travel $55 \cdot 4 = 220$ mi. At 44 mph, in $4 + 1$, or 5 hr the freight train will travel $44 \cdot 5 = 220$ mi. The numbers check.

State. The passenger train overtakes the freight train after 4 hours.

23. $x + 4y + 3z = 2,$ (1)
$2x + y + z = 10,$ (2)
$-x + y + 2z = 8$ (3)

1., 2. The equations are already in standard form with no fractions or decimals.

3. Use Equations (1) and (3) to eliminate x.

$\begin{array}{ll} x + 4y + 3z = 2 & (1) \\ \underline{-x + y + 2z = 8} & (3) \\ 5y + 5z = 10 & (4)\ \text{Adding} \end{array}$

4. Use a different pair of equations and eliminate x.

$\begin{array}{ll} 2x + y + z = 10 & (2) \\ \underline{-2x + 2y + 4z = 16} & \text{Multiplying (3) by 2} \\ 3y + 5z = 26 & (5)\ \text{Adding} \end{array}$

5. Now solve the system of Equations (4) and (5).

$5y + 5z = 10$ (4)
$3y + 5z = 26$ (5)

$\begin{array}{ll} 5y + 5z = 10 & (4) \\ \underline{-3y - 5z = -26} & \text{Multiplying (5) by } -1 \\ 2y = -16 & \text{Adding} \\ y = -8 \end{array}$

$\begin{array}{ll} 5(-8) + 5z = 10 & \text{Substituting } -8 \text{ for } y \text{ in (4)} \\ -40 + 5z = 10 \\ 5z = 50 \\ z = 10 \end{array}$

6. Substitute in one of the original equations to find x.

$\begin{array}{ll} x + 4(-8) + 3(10) = 2 & \text{Substituting } -8 \text{ for } y \text{ and} \\ & \quad 10 \text{ for } z \text{ in (1)} \\ x - 2 = 2 \\ x = 4 \end{array}$

We obtain $(4, -8, 10)$. This checks, so it is the solution.

25. $2x - 5y - 2z = -4,$ (1)
$7x + 2y - 5z = -6,$ (2)
$-2x + 3y + 2z = 4$ (3)

1., 2. The equations are already in standard form with no fractions or decimals.

3. Use Equations (1) and (2) to eliminate x.

$\begin{array}{ll} 14x - 35y - 14z = -28 & \text{Multiplying (1) by 7} \\ \underline{-14x - 4y + 10z = 12} & \text{Multiplying (2) by } -2 \\ -39y - 4z = -16 & (4)\ \text{Adding} \end{array}$

4. Use Equations (1) and (3) to eliminate x.

$\begin{array}{ll} 2x - 5y - 2z = -4 & (1) \\ \underline{-2x + 3y + 2z = 4} & (3) \\ -2y = 0 & (5)\ \text{Adding} \\ y = 0 \end{array}$

5. When we used Equation (1) and (3) to eliminate x, we also eliminated z and found $y = 0$. Substitute 0 for y in Equation (4) to find z.

$\begin{array}{ll} -39 \cdot 0 - 4z = -16 & \text{Substituting 0 for } y \text{ in (4)} \\ -4z = -16 \\ z = 4 \end{array}$

6. Substitute in one of the original equations to find x.

$2x - 5 \cdot 0 - 2 \cdot 4 = -4$
$2x = 4$
$x = 2$

We obtain $(2, 0, 4)$. This checks, so it is the solution.

27. *Familiarize*. We let x, y, and z represent the measures of angles A, B, and C, respectively. The measures of the angles of a triangle add up to $180°$.

Translate.

$$\underline{\text{The sum of the measures}} \quad \text{is} \quad 180.$$
$$\downarrow \qquad\qquad\qquad\quad \downarrow \quad\ \downarrow$$
$$x + y + z \qquad\qquad\quad = \quad 180$$

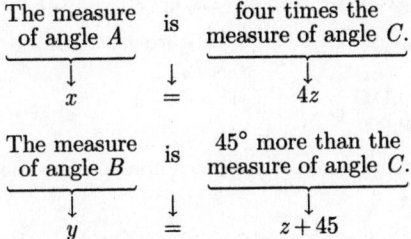

The measure of angle A is four times the measure of angle C.
$$x = 4z$$

The measure of angle B is $45°$ more than the measure of angle C.
$$y = z + 45$$

We now have a system of equations.
$$\begin{aligned} x+y+z &= 180, \\ x &= 4z, \\ y &= z+45 \end{aligned}$$

Carry out. Solving the system we get $(90, 67.5, 22.5)$.

Check. The sum of the measures is $90° + 67.5° + 22.5°$, or $180°$. Four times the measure of angle C is $4 \cdot 22.5°$, or $90°$, the measure of angle A. 45 more than the measure of angle C is $45° + 22.5°$, or $67.5°$, the measure of angle B. The numbers check.

State. The measures of angles A, B, and C are $90°$, $67.5°$, and $22.5°$, respectively.

29. Familiarize. Let x, y, and z represent the average number of cries for a man, woman and a one-year-old child, respectively.

Translate.

The sum for each month is 56.7, so we have
$$x+y+z = 56.7.$$
The number of cries for a woman is 3.9 more than the man, so we have
$$y = 3.9 + x.$$
The number of cries for the child is 43.3 more than the sum for the man and woman, so we have
$$z = 43.3 + x + y.$$
We now have a system of equations.
$$\begin{aligned} x+y+z &= 56.7, \\ y &= 3.9 + x, \\ z &= 43.3 + x + y \end{aligned}$$

Carry out. Solving the system, we get $(1.4, 5.3, 50)$.

Check. The sum of the average number of cry times for a man, a woman, and a one-year-old child is $1.4 + 5.3 + 50$, or 56.7. A woman cries $3.9 + 1.4$, or 5.3 times a month. A one-year-old child cries $43.3 + 1.4 + 5.3$, or 50 times a month. The answer checks.

State. The monthy average number of cries for a man is 1.4, for a woman is 5.3, and a one-year-old child is 50.

31.
$$\begin{aligned} 3x - y + z &= -1, \\ 2x + 3y + z &= 4, \\ 5x + 4y + 2z &= 5 \end{aligned}$$

We first write a matrix using only the constants.

$$\begin{bmatrix} 3 & -1 & 1 & | & -1 \\ 2 & 3 & 1 & | & 4 \\ 5 & 4 & 2 & | & 5 \end{bmatrix}$$

$$\begin{bmatrix} 3 & -1 & 1 & | & -1 \\ 0 & 11 & 1 & | & 14 \\ 0 & 17 & 1 & | & 20 \end{bmatrix} \quad \begin{array}{l} \text{New Row 2} = -2(\text{Row 1}) + 3(\text{Row 2}) \\ \text{New Row 3} = -5(\text{Row 1}) + 3(\text{Row 3}) \end{array}$$

$$\begin{bmatrix} 3 & -1 & 1 & | & -1 \\ 0 & 11 & 1 & | & 14 \\ 0 & 187 & 11 & | & 220 \end{bmatrix} \quad \text{New Row 3} = 11(\text{Row 3})$$

$$\begin{bmatrix} 3 & -1 & 1 & | & -1 \\ 0 & 11 & 1 & | & 14 \\ 0 & 0 & -6 & | & -18 \end{bmatrix} \quad \text{New Row 3} = -17(\text{Row 2}) + \text{Row 3}$$

Reinserting the variables, we have
$$\begin{aligned} 3x - y + z &= -1, \quad (1) \\ 11y + z &= 14, \quad (2) \\ -6z &= -18 \quad (3) \end{aligned}$$

Solve (3) for z.
$$\begin{aligned} -6z &= -18 \\ z &= 3 \end{aligned}$$

Substitute 3 for z in (2) and solve for y.
$$\begin{aligned} 11y + 3 &= 14 \\ 11y &= 11 \\ y &= 1 \end{aligned}$$

Substitute 1 for y and 3 for z in (1) and solve for x.
$$\begin{aligned} 3x - 1 + 3 &= -1 \\ 3x + 2 &= -1 \\ 3x &= -3 \\ x &= -1 \end{aligned}$$

The solution is $(-1, 1, 3)$.

33.
$$\begin{vmatrix} 2 & 3 & 0 \\ 1 & 4 & -2 \\ 2 & -1 & 5 \end{vmatrix}$$
$$= 2 \begin{vmatrix} 4 & -2 \\ -1 & 5 \end{vmatrix} - 1 \begin{vmatrix} 3 & 0 \\ -1 & 5 \end{vmatrix} + 2 \begin{vmatrix} 3 & 0 \\ 4 & -2 \end{vmatrix}$$
$$= 2[4 \cdot 5 - (-2)(-1)] - 1[3 \cdot 5 - 0(-1)] + 2[3(-2) - 0 \cdot 4]$$
$$= 2(18) - 1(15) + 2(-6)$$
$$= 36 - 15 - 12 = 9$$

35.
$$\begin{aligned} 2x + y + z &= -2, \\ 2x - y + 3z &= 6, \\ 3x - 5y + 4z &= 7 \end{aligned}$$

We compute D, D_x, and D_y.

$$D = \begin{vmatrix} 2 & 1 & 1 \\ 2 & -1 & 3 \\ 3 & -5 & 4 \end{vmatrix}$$
$$= 2 \begin{vmatrix} -1 & 3 \\ -5 & 4 \end{vmatrix} - 2 \begin{vmatrix} 1 & 1 \\ -5 & 4 \end{vmatrix} + 3 \begin{vmatrix} 1 & 1 \\ -1 & 3 \end{vmatrix}$$
$$= 2(11) - 2(9) + 3(4) = 22 - 18 + 12 = 16$$

$$D_x = \begin{vmatrix} -2 & 1 & 1 \\ 6 & -1 & 3 \\ 7 & -5 & 4 \end{vmatrix}$$

$$= -2\begin{vmatrix} -1 & 3 \\ -5 & 4 \end{vmatrix} - 6\begin{vmatrix} 1 & 1 \\ -5 & 4 \end{vmatrix} + 7\begin{vmatrix} 1 & 1 \\ -1 & 3 \end{vmatrix}$$

$$= -2(11) - 6(9) + 7(4)$$

$$= -22 - 54 + 28 = -48$$

$$D_y = \begin{vmatrix} 2 & -2 & 1 \\ 2 & 6 & 3 \\ 3 & 7 & 4 \end{vmatrix}$$

$$= 2\begin{vmatrix} 6 & 3 \\ 7 & 4 \end{vmatrix} - 2\begin{vmatrix} -2 & 1 \\ 7 & 4 \end{vmatrix} + 3\begin{vmatrix} -2 & 1 \\ 6 & 3 \end{vmatrix}$$

$$= 2(3) - 2(-15) + 3(-12)$$

$$= 6 + 30 - 36 = 0$$

Then,

$$x = \frac{D_x}{D} = \frac{-48}{16} = -3,$$

and

$$y = \frac{D_y}{D} = \frac{0}{16} = 0.$$

We substitute in the second equation to find z.

$$2(-3) + 0 + z = -2$$
$$-6 + z = -2$$
$$z = 4$$

The solution is $(-3, 0, 4)$.

37. a) $C(x) = 4.75x + 54,000$, where x is the number of pints of honey.

b)   Each pint of honey sells for \$9.25. The total revenue is 9.25 times the number of pints of honey. We assume all the honey is sold. $R(x) = 9.25x$

c) $P(x) = R(x) - C(x)$
$$= 9.25x - (4.75x + 54,000)$$
$$= 9.25x - 4.75x - 54,000$$
$$= 4.5x - 54,000$$

d) $P(5000) = 4.5(5000) - 54,000$
$$= 22,500 - 54,000$$
$$= -31,500$$

Danae will realize a \$31,500 loss when 5000 pints of honey are produced and sold.

$$P(15,000) = 4.5(15,000) - 54,000$$
$$= 67,500 - 54,000$$
$$= 13,500$$

Danae will realize a profit of \$13,500 when 15,000 pints of honey are produced and sold.

e)   Solve the system
$$R(x) = 9.25x,$$
$$C(x) = 4.75x + 54,000.$$

Since both $R(x)$ and $C(x)$ are in dollars and they are equal at the break-even point, we can rewrite the system:

$$d = 9.25x, \qquad (1)$$
$$d = 4.75x + 54,000 \quad (2)$$

We solve using substitution.

$$9.25x = 4.75x + 54,000 \quad \text{Substituting } 9.25x \text{ for}$$
$$\qquad\qquad\qquad\qquad\qquad\qquad d \text{ in (2)}$$
$$4.5x = 54,000$$
$$x = 12,000$$

Thus, 12,000 units must be produced and sold in order to break even.

The revenue will be $R(12,000) = 9.25 \cdot 12,000 = \$111,000$.

The break-even point is (12,000 pints, \$111,000).

39. *Writing Exercise.* A system of equations can be both dependent and inconsistent if it is equivalent to a system with fewer equations that has no solution. An example is a system of three equations in three unknowns in which two of the equations represent the same plane, and the third represents a parallel plane.

41. *Familiarize.* Let $x =$ the number of 2-count packs of pencils and $y =$ the number of 12-count packs of pencils.
Translate.

The total number of pencils is 138, so we have

$$2x + 12y = 138.$$

The purchase price is \$157.26, so we have

$$5.99x + 7.49y = 157.26.$$

After clearing decimals, we have the problem translated to a system of equations:

$$2x + 12y = 138 \qquad (1)$$
$$599x + 749y = 15,726 \quad (2)$$

Carry out. We use the elimination method to solve the system of equations.

$$\begin{array}{ll} -1198x - 7188y = -82,662 & \text{Multiplying (1) by } -599 \\ \underline{1198x + 1498y = 31,452} & \text{Multiplying (2) by 2} \\ \qquad\quad -5690y = -51,210 & \\ \qquad\qquad\quad y = 9 & \end{array}$$

Substitute 9 for y in Equation (1) and solve for x.

$$2x + 12 \cdot 9 = 138$$
$$2x + 108 = 138$$
$$2x = 30$$
$$x = 15$$

Check. The total number of pencils is

$12(15) + 12(9) = 30 + 108,$ or 138 pencils. The purchase price is \$5.99(15) + \$7.49(9) = \$89.85 + \$67.41 = \$157.26.
The answer checks.

State. Wiese Accounting purchased 15 packs of Round Stic Grip and 9 packs of Matic Grip pencils.

43. $f(x) = ax^2 + bx + c$

The three points are $(-2, 3)$, $(1, 1)$ and $(0, 3)$.

We substitute for x and $f(x)$ and solve for a, b, and c.

$$3 = 4a - 2b + c$$
$$1 = a + b + c$$
$$3 = c$$

1., 2. We put the equations in standard form. There are no fractions or decimals.

$$4a - 2b + c = 3, \quad (1)$$
$$a + b + c = 1, \quad (2)$$
$$c = 3 \quad (3)$$

3., 4. Note that there is no a or b in Equation (3). We will use Equations (1) and (2) to obtain another equation with no b-term.

$$4a - 2b + c = 3 \quad (1)$$
$$\underline{2a + 2b + 2c = 2} \quad \text{Multiplying (2) by 2}$$
$$6a + 3c = 5 \quad (4)$$

5. Now substitute 3 for c in (4) to solve for a.

$$6a + 3 \cdot 3 = 5 \qquad \text{Substituting 3 for } c \text{ in (4)}$$
$$6a = -4$$
$$a = -\frac{2}{3}$$

6. Substitute in Equation (2) to find b.

$$-\frac{2}{3} + b + 3 = 1 \qquad \text{Substituting}$$
$$b = -\frac{4}{3}$$

We obtain $\left(-\frac{2}{3}, -\frac{4}{3}, 3\right)$. So, the function is

$$f(x) = -\frac{2}{3}x^2 - \frac{4}{3}x + 3 \,.$$

Chapter 3 Test

1. Graph the equations.

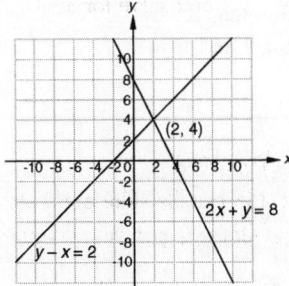

3. $2x - 4y = -6 \qquad (1)$
$x = 2y - 3 \qquad (2)$

We substitute $2y - 3$ for x in the first equation and solve for y.

$$2x - 4y = -6$$
$$2(2y - 3) - 4y = -6$$
$$4y - 6 - 4y = -6$$
$$-6 = -6$$

We have an identity, or an equation that is always true. The equations are dependent and the solution set is infinite: $\left\{(x, y) | 2y - 3 = x\right\}$.

5. $4y + 2x = 18,$
$3x + 6y = 26$

Rewrite the equations in standard form.

$$2x + 4y = 18, \quad (1)$$
$$3x + 6y = 26 \quad (2)$$

We multiply Equation (1) by -3 and Equation (2) by 2.

$$-6x - 12y = -54 \qquad \text{Multiplying (1) by } -3$$
$$\underline{6x + 12y = 52} \qquad \text{Multiplying (2) by 2}$$
$$0 = -2 \qquad \text{Adding}$$

We have a contradiction, or an equation that is always false. Therefore, there is no solution.

7. *Familiarize.* Let $w =$ the width of the basketball court and let $l =$ the length of the basketball court. Recall that the perimeter of a rectangle is given by the formula

$$P = 2w + 2l.$$

Translate.

The perimeter of the court is 288 ft, so we have

$$2w + 2l = 288.$$

The length if 44 longer than the width, so we have

$$l = 44 + w.$$

We now have a system of equations.

$$2w + 2l = 288. \quad (1)$$
$$l = 44 + w \quad (2)$$

Carry out. Substitute $44 + w$ for l in the first equation and solve for w.

$$2w + 2l = 288$$
$$2w + 2(44 + w) = 288$$
$$2w + 88 + 2w = 288$$
$$4w + 88 = 288$$
$$4w = 200$$
$$w = 50$$

Finally substitute 50 for w in Equation (2) and solve for l.

$$l = 44 + w$$
$$l = 44 + 50 = 94$$

Check. The perimeter of the court is

$2(94 \text{ ft}) + 2(50 \text{ ft}) = 188 \text{ ft} + 100 \text{ ft} = 288 \text{ ft}$. The length, 94 ft, is 44 ft more than the width, or $44 + 50 = 94$.

State. The basketball court is 94 ft long and 50 ft wide.

9. *Familiarize.* Let $d =$ the distance. Let $t =$ the time it takes the car to travel at 65 mph. Then $t + 1$ is the time it takes the car to travel at 55 mph. The distances are the same.

Organize the information in a table.

	Distance	Rate	Time
Truck	d	55	$t+1$
Car	d	65	t

Translate. Using $d = rt$ in each row of the table, we get a system of equations:

$$d = 55(t+1)$$
$$d = 65t$$

Carry out. Solve the system of equations.

$$55(t+1) = 65t \quad \text{Substituting}$$
$$55t + 55 = 65t$$
$$55 = 10t$$
$$5.5 = t$$

Check. At 55 mph in $5.5 + 1$, or 6.5 the truck travels $55(6.5)$, or 357.5 miles. At 65 mph, in 5.5 hr, the car travels $65(5.5)$, or 357.5 miles. The answer checks.

State. The car will catch up to the truck in 5.5 hr.

11.
$$6x + 2y - 4z = 15, \quad (1)$$
$$-3x - 4y + 2z = -6, \quad (2)$$
$$4x - 6y + 3z = 8 \quad (3)$$

1., 2. The equations are already in standard form with no fractions or decimals.

3., 4. We eliminate x from two different pairs of equations.

$$\begin{array}{ll} 6x + 2y - 4z = 15 & (1) \\ \underline{-6x - 8y + 4z = -12} & \text{Multiplying (2) by 2} \\ -6y = 3 & (4) \end{array}$$

$$y = -\frac{1}{2}$$

$$\begin{array}{ll} -12x - 16y + 8z = -24 & \text{Multiplying (2) by 4} \\ \underline{12x - 18y + 9z = 24} & \text{Multiplying (3) by 3} \\ -34y + 17z = 0 & (5) \end{array}$$

5. Now substitute $-\frac{1}{2}$ for y in (5) to solve for z.

$$-34\left(-\frac{1}{2}\right) + 17z = 0 \quad \text{Substituting } -\frac{1}{2} \text{ for } y \text{ in (5)}$$
$$17 + 17z = 0$$
$$z = -1$$

6. Substitute in Equation (1) to find x.

$$6x + 2\left(-\frac{1}{2}\right) - 4(-1) = 15 \quad \text{Substituting}$$
$$6x - 1 + 4 = 15$$
$$6x = 12$$
$$x = 2$$

We obtain $\left(2, -\frac{1}{2}, -1\right)$. This checks, so it is the solution.

13.
$$\begin{array}{ll} 3x + 3z = 0, & (1) \\ 2x + 2y = 2, & (2) \\ 3y + 3z = 3 & (3) \end{array}$$

1., 2. The equations are already in standard form with no fractions or decimals.

3. , 4. Note that there is no z in Equation (2). We will use Equations (1) and (3) to obtain another equation with no z-term.

$$\begin{array}{ll} 3x + 3z = 0 & (1) \\ \underline{-3y - 3z = -3} & \text{Multiplying (3) by } -1 \\ 3x - 3y = -3 & (4) \end{array}$$

5. Now solve the system of Equations (2) and (4).

$$2x + 2y = 2 \quad (2)$$
$$3x - 3y = -3 \quad (4)$$

$$\begin{array}{ll} 6x + 6y = 6 & \text{Multiplying (2) by 3} \\ \underline{-6x + 6y = 6} & \text{Multiplying (4) by } -2 \\ 12y = 12 \\ y = 1 \end{array}$$

$$2x + 2 \cdot 1 = 2 \quad \text{Substituting in (2)}$$
$$2x = 0$$
$$x = 0$$

6. Substitute in Equation (1) or (3) to find z.

$$3 \cdot 0 + 3z = 0 \quad \text{Substituting in (1)}$$
$$3z = 0$$
$$z = 0$$

We obtain $(0, 1, 0)$. This checks, so it is the solution.

15.
$$x + 3y - 3z = 12,$$
$$3x - y + 4z = 0,$$
$$-x + 2y - z = 1$$

We first write a matrix using only the constants.

$$\begin{bmatrix} 1 & 3 & -3 & | & 12 \\ 3 & -1 & 4 & | & 0 \\ -1 & 2 & -1 & | & 1 \end{bmatrix}$$

$$\begin{bmatrix} 1 & 3 & -3 & | & 12 \\ 0 & -10 & 13 & | & -36 \\ 0 & 5 & -4 & | & 13 \end{bmatrix} \begin{array}{l} \text{New Row 2} = -3(\text{Row 1}) + \text{Row 2} \\ \text{New Row 3} = (\text{Row 1}) + \text{Row 3} \end{array}$$

$$\begin{bmatrix} 1 & 3 & -3 & | & 12 \\ 0 & 5 & -4 & | & 13 \\ 0 & -10 & 13 & | & -36 \end{bmatrix} \begin{array}{l} \text{Interchange} \\ \text{Row 2 and Row 3} \end{array}$$

$$\begin{bmatrix} 1 & 3 & -3 & | & 12 \\ 0 & 5 & -4 & | & 13 \\ 0 & 0 & 5 & | & -10 \end{bmatrix} \text{New Row 3} = 2(\text{Row 2}) + \text{Row 3}$$

Reinserting the variables, we have

$$\begin{array}{ll} x + 3y - 3z = 12, & (1) \\ 5y - 4z = 13, & (2) \\ 5z = -10. & (3) \end{array}$$

Solve (3) for z.

$$5z = -10$$
$$z = -2$$

Substitute -2 for z in (2) and solve for y.

$$5y - 4z = 13$$
$$5y - 4(-2) = 13$$
$$5y = 5$$
$$y = 1$$

Substitute 1 for y in (1) and solve for x.

$$x + 3 \cdot 1 - 3(-2) = 12$$
$$x + 3 + 6 = 12$$
$$x = 3$$

The solution is $(3, 1, -2)$.

17.
$$\begin{vmatrix} 3 & 4 & 2 \\ -2 & -5 & 4 \\ 0 & 5 & -3 \end{vmatrix}$$

$$= 3 \begin{vmatrix} -5 & 4 \\ 5 & -3 \end{vmatrix} - (-2) \begin{vmatrix} 4 & 2 \\ 5 & -3 \end{vmatrix} + 0 \begin{vmatrix} 4 & 2 \\ -5 & 4 \end{vmatrix}$$

$$= 3[(-5)(-3) - 5 \cdot 4] + 2[4(-3) - 5 \cdot 2]$$
$$\quad + 0[4 \cdot 4 - 2(-5)]$$
$$= 3(-5) + 2(-22) + 0$$
$$= -15 - 44 = -59$$

19. *Familiarize*. Let x, y, and z represent the number of hours for the electrician, carpenter and plumber, respectively.

Translate.

The total number of hours worked is 21.5, so we have

$$x + y + z = 21.5.$$

The total earnings is \$673, so we have

$$30x + 28.50y + 34z = 673.$$

The plumber worked 2 more hours than the carpenter, so we have

$$z = 2 + y.$$

We have a system of equations:

$$x + y + z = 21.5,$$
$$30x + 28.50y + 34z = 673,$$
$$z = 2 + y$$

Carry out. Solving the system we get $(3.5, 8, 10)$.

Check. The total number of hours is $3.5 + 8 + 10$, or 21.5 h. The total amount earned is

$\$30(3.5) + \$28.50(8) + \$34(10) = \$105 + \$228 + \$340$, or $\$673$. The plumber worked 10 h, which is 2 h more than the 8 h worked by the carpenter. The numbers check.

State. The electrician worked 3.5 h, the carpenter worked 8 h and the plumber worked 10 h.

21. a) $C(x) = 25x + 44,000$, where x is the number of hammocks produced.

b) $R(x) = 80x$ We assume all hammocks are sold.

c) $P(x) = R(x) - C(x)$
$$= 80x - (25x + 44,000)$$
$$= 80x - 25x - 44,000$$
$$= 55x - 44,000$$

d) $P(300) = 55(300) - 44,000 = 16,500 - 44,000$
$$= -27,500$$

The company will realize a \$27,500 loss when 300 hammocks are produced and sold.

$$P(900) = 55(900) - 44,000 = 49,500 - 44,000$$
$$= 5500$$

The company will realize a profit of \$5500 when 900 hammocks are produced and sold.

e) Solve the system
$$R(x) = 80x,$$
$$C(x) = 25x + 44,000.$$

Since both $R(x)$ and $C(x)$ are in dollars and they are equal at the break-even point, we can rewrite the system:

$$d = 80x, \qquad (1)$$
$$d = 25x + 44,000 \quad (2)$$

We solve using substitution.

$$80x = 25x + 44,000 \quad \text{Substituting } 80x \text{ for}$$
$$\qquad\qquad\qquad\qquad\qquad d \text{ in (2)}$$
$$55x = 44,000$$
$$x = 800$$

Thus, 800 hammocks must be produced and sold in order to break even.

The revenue will be $R(800) = 80 \cdot 800 = \$64,000$.

The break-even point is (800 hammocks, \$64,000).

23. *Familiarize*. Let $x =$ the number of pounds of Kona coffee

Translate. We organize the information in a table.

	Kona	Mexican	Mixture
Number of pounds	x	40	$x + 40$
Percent Kona	100%	0%	30%
Amount of Kona	$1.00x$	0	$0.30(x + 40)$

The last row of the table gives us one equation:

$$1.00x + 0 = 0.3(x + 40).$$

Carry out. After clearing decimals, we solve the equation.

$$100x + 0 = 30(x + 40)$$
$$100x = 30x + 1200$$
$$70x = 1200$$
$$x = \frac{120}{7}$$

Check. 30% of $\left(\frac{120}{7} + 40\right)$ lb is $0.3\left(\frac{400}{7}\right)$, or $\frac{120}{7}$ lb. of Kona coffee. The answer checks.

State. At least $\frac{120}{7}$ lb of Kona coffee is added to the 40 lb of Mexican coffee to market the mixture as Kona Blend.

Chapter 4

Inequalities and Problem Solving

Exercise Set 4.1

1. If we add $3x$ to both sides of the equation $5x + 7 = 6 - 3x$, we get the equation $8x + 7 = 6$, so these are equivalent equations.

3. If we add 7 to both sides of the inequality $x - 7 > -2$, we get the inequality $x > 5$ so these are equivalent inequalities.

5. The solution set of $-4t \leq 12$ is $\{t \mid t \geq -3\}$ and the solution set of $t \leq -3$ is $\{t \mid t \leq -3\}$. The solution sets are not the same, so the inequalities are not equivalent.

7. The expressions are equivalent by the distributive law.

9. The solution set of $-\frac{1}{2}x < 7$ is $\{x \mid x > -14\}$ and the solution set of $x > 14$ is $\{x \mid x > 14\}$. The solution sets are not the same, so the inequalities are not equivalent.

11. $x - 4 \geq 1$
    a) −4: We substitute and get $-4 - 4 \geq 1$, or $-8 \geq 1$, a false sentence. Therefore, −4 is not a solution.

    b) 4: We substitute and get $4 - 4 \geq 1$, or $0 \geq 1$, a false sentence. Therefore, 4 is not a solution.

    c) 5: We substitute and get $5 - 4 \geq 1$, or $1 \geq 1$, a true sentence. Therefore, 5 is a solution.

    d) 8: We substitute and get $8 - 4 \geq 1$, or $4 \geq 1$, a true sentence. Therefore, 8 is a solution.

13. $2y + 3 < 6 - y$
    a) 0: We substitute and get $2(0) + 3 < 6 - 0$, or $3 < 6$, a true sentence. Therefore, 0 is a solution.

    b) 1: We substitute and get $2(1) + 3 < 6 - 1$, or $5 < 5$, a false sentence. Therefore, 1 is not a solution.

    c) −1: We substitute and get $2(-1) + 3 < 6 - (-1)$, or $1 < 7$, a true sentence. Therefore, −1 is a solution.

    d) 4: We substitute and get $2(4) + 3 < 6 - 4$, or $11 < 2$, a false sentence. Therefore, 4 is not a solution.

15. $y < 6$

Graph: The solutions consist of all real numbers less than 6, so we shade all numbers to the left of 6 and use a parenthesis at 6 to indicate that it is not a solution.

Set builder notation: $\{y \mid y < 6\}$

Interval notation: $(-\infty, 6)$

17. $x \geq -4$

Graph: We shade all numbers to the right of −4 and use a bracket at −4 to indicate that it is also a solution.

Set builder notation: $\{x \mid x \geq -4\}$

Interval notation: $[-4, \infty)$

19. $t > -3$

Graph: We shade all numbers to the right of −3 and use a parenthesis at −3 to indicate that it is not a solution.

Set builder notation: $\{t \mid t > -3\}$

Interval notation: $(-3, \infty)$

21. $x \leq -7$

Graph: We shade all numbers to the left of −7 and use a bracket at −7 to indicate that it is also a solution.

Set builder notation: $\{x \mid x \leq -7\}$

Interval notation: $(-\infty, -7]$

23.
$$x + 2 > 1$$
$$x + 2 + (-2) > 1 + (-2) \qquad \text{Adding } -2$$
$$x > -1$$
The solution set is $\{x \mid x > -1\}$, or $(-1, \infty)$.

25.
$$t - 6 \leq 4$$
$$t - 6 + 6 \leq 4 + 6 \qquad \text{Adding } 6$$
$$t \leq 10$$
The solution set is $\{t \mid t \leq 10\}$, or $(-\infty, 10]$.

27.
$$x - 12 \geq -11$$
$$x - 12 + 12 \geq -11 + 12 \qquad \text{Adding 12}$$
$$x \geq 1$$
The solution set is $\{x \mid x \geq 1\}$, or $[1, \infty)$.

29.
$$9t < -81$$
$$\frac{1}{9} \cdot 9t < \frac{1}{9}(-81) \qquad \text{Multiplying by } \frac{1}{9}$$
$$t < -9$$
The solution set is $\{t \mid t < -9\}$, or $(-\infty, -9)$.

31.
$$-0.3x > -15$$
$$-\frac{1}{0.3}(-0.3x) < -\frac{1}{0.3}(-15) \qquad \text{Multiplying by } -\frac{1}{0.3} \text{ and reversing the inequality symbol}$$
$$x < 50$$
The solution set is $\{x \mid x < 50\}$, or $(-\infty, 50)$.

33.
$$-9x \geq 8.1$$
$$-\frac{1}{9}(-9x) \leq -\frac{1}{9}(8.1) \qquad \text{Multiplying by } -\frac{1}{9} \text{ and reversing the inequality symbol}$$
$$x \leq -0.9$$
The solution set is $\{x \mid x \leq -0.9\}$, or $(-\infty, -0.9]$.

35.
$$\frac{3}{4}y \geq -\frac{5}{8}$$
$$\frac{4}{3}\left(\frac{3}{4}y\right) \geq \frac{4}{3}\left(-\frac{5}{8}\right) \qquad \text{Multiplying by } \frac{4}{3}$$
$$y \geq -\frac{5}{6}$$
The solution set is $\left\{y \mid y \geq -\frac{5}{6}\right\}$, or $\left[-\frac{5}{6}, \infty\right)$.

37.
$$3x + 1 < 7$$
$$3x < 6 \qquad \text{Adding } -1$$
$$x < 2 \qquad \text{Dividing by 3}$$
The solution set is $\{x \mid x < 2\}$, or $(-\infty, 2)$.

39.
$$3 - x \geq 12$$
$$-x \geq 9 \qquad \text{Adding } -3$$
$$x \leq -9 \qquad \text{Dividing by } -1 \text{ and reversing the inequality symbol}$$
The solution set is $\{x \mid x \leq -9\}$, or $(-\infty, -9]$.

41.
$$\frac{2x + 7}{5} < -9$$
$$5 \cdot \frac{2x + 7}{5} < 5(-9) \qquad \text{Multiplying by 5}$$
$$2x + 7 < -45$$
$$2x < -52 \qquad \text{Adding } -7$$
$$x < -26 \qquad \text{Dividing by 2}$$
The solution set is $\{x \mid x < -26\}$, or $(-\infty, -26)$.

43.
$$\frac{3t - 7}{-4} \leq 5$$
$$-4 \cdot \frac{3t - 7}{-4} \geq -4 \cdot 5 \qquad \text{Multiplying by } -4 \text{ and reversing the inequality symbol}$$
$$3t - 7 \geq -20$$
$$3t \geq -13 \qquad \text{Adding 7}$$
$$t \geq -\frac{13}{3} \qquad \text{Dividing by 3}$$
The solution set is $\left\{t \mid t \geq -\frac{13}{3}\right\}$, or $\left[-\frac{13}{3}, \infty\right)$.

45.
$$\frac{9 - x}{-2} \geq -6$$
$$9 - x \leq 12 \qquad \text{Multiplying by } -2 \text{ and reversing the inequality symbol}$$
$$-x \leq 3 \qquad \text{Adding } -9$$
$$x \geq -3 \qquad \text{Multiplying by } -1 \text{ and reversing the inequality symbol}$$
The solution set is $\{x \mid x \geq -3\}$, or $[-3, \infty)$.

47. $f(x) = 7 - 3x$, $g(x) = 2x - 3$
$$f(x) \leq g(x)$$
$$7 - 3x \leq 2x - 3$$
$$7 - 5x \leq -3 \qquad \text{Adding } -2x$$
$$-5x \leq -10 \qquad \text{Adding } -7$$
$$x \geq 2 \qquad \text{Multiplying by } -\frac{1}{5} \text{ and reversing the inequality symbol}$$
The solution set is $\{x \mid x \geq 2\}$, or $[2, \infty)$.

49. $f(x) = 2x - 7$, $g(x) = 5x - 9$
$$f(x) < g(x)$$
$$2x - 7 < 5x - 9$$
$$-3x - 7 < -9 \qquad \text{Adding } -5x$$
$$-3x < -2 \qquad \text{Adding 7}$$
$$x > \frac{2}{3} \qquad \text{Dividing by } -3$$
The solution set is $\left\{x \mid x > \frac{2}{3}\right\}$, or $\left(\frac{2}{3}, \infty\right)$.

51. $y_1 = \frac{3}{8} + 2x$, $y_2 = 3x - \frac{1}{8}$

$$y_2 \geq y_1$$
$$3x - \frac{1}{8} \geq \frac{3}{8} + 2x$$
$$x - \frac{1}{8} \geq \frac{3}{8} \qquad \text{Adding } -2x$$
$$x \geq \frac{1}{2} \qquad \text{Adding } \frac{1}{8}$$

The solution set is $\left\{ x \mid x \geq \frac{1}{2} \right\}$, or $\left[\frac{1}{2}, \infty \right)$.

53.
$$3 - 8y \geq 9 - 4y$$
$$-4y + 3 \geq 9$$
$$-4y \geq 6$$
$$y \leq -\frac{3}{2}$$

The solution set is $\left\{ y \mid y \leq -\frac{3}{2} \right\}$, or $\left(-\infty, -\frac{3}{2} \right]$.

55.
$$5(t - 3) + 4t < 2(7 + 2t)$$
$$5t - 15 + 4t < 14 + 4t$$
$$9t - 15 < 14 + 4t$$
$$5t - 15 < 14$$
$$5t < 29$$
$$t < \frac{29}{5}$$

The solution set is $\left\{ t \mid t < \frac{29}{5} \right\}$, or $\left(-\infty, \frac{29}{5} \right)$.

57.
$$5 \left[3m - (m + 4) \right] > -2(m - 4)$$
$$5(3m - m - 4) > -2(m - 4)$$
$$5(2m - 4) > -2(m - 4)$$
$$10m - 20 > -2m + 8$$
$$12m - 20 > 8$$
$$12m > 28$$
$$m > \frac{28}{12}$$
$$m > \frac{7}{3}$$

The solution set is $\left\{ m \mid m > \frac{7}{3} \right\}$, or $\left(\frac{7}{3}, \infty \right)$.

59.
$$19 - (2x + 3) \leq 2(x + 3) + x$$
$$19 - 2x - 3 \leq 2x + 6 + x$$
$$16 - 2x \leq 3x + 6$$
$$16 - 5x \leq 6$$
$$-5x \leq -10$$
$$x \geq 2$$

The solution set is $\{ x \mid x \geq 2 \}$, or $[2, \infty)$.

61. $\frac{1}{4}(8y + 4) - 17 < -\frac{1}{2}(4y - 8)$

$$2y + 1 - 17 < -2y + 4$$
$$2y - 16 < -2y + 4$$
$$4y - 16 < 4$$
$$4y < 20$$
$$y < 5$$

The solution set is $\{ y \mid y < 5 \}$, or $(-\infty, 5)$.

63.
$$2[8 - 4(3 - x)] - 2 \geq 8[2(4x - 3) + 7] - 50$$
$$2[8 - 12 + 4x] - 2 \geq 8[8x - 6 + 7] - 50$$
$$2[-4 + 4x] - 2 \geq 8[8x + 1] - 50$$
$$-8 + 8x - 2 \geq 64x + 8 - 50$$
$$8x - 10 \geq 64x - 42$$
$$-56x - 10 \geq -42$$
$$-56x \geq -32$$
$$x \leq \frac{32}{56}$$
$$x \leq \frac{4}{7}$$

The solution set is $\left\{ x \mid x \leq \frac{4}{7} \right\}$, or $\left(-\infty, \frac{4}{7} \right]$.

65. $f(x) = \sqrt{x - 10}$

$$x - 10 \geq 0 \qquad x - 10 \text{ must be nonnegative.}$$
$$x \geq 10 \qquad \text{Adding 10}$$

When $x \geq 10$, the expression $x - 10$ is nonnegative. Thus the domain of f is $\{ x \mid x \geq 10 \}$, or $[10, \infty)$.

67. $f(x) = \sqrt{3 - x}$

$$3 - x \geq 0 \qquad 3 - x \text{ must be nonnegative.}$$
$$-x \geq -3 \qquad \text{Adding } -3$$
$$x \leq 3 \qquad \text{Multiplying by } -1 \text{ and reversing the inequality symbol}$$

When $x \leq 3$, the expression $3 - x$ is nonnegative. Thus the domain of f is $\{ x \mid x \leq 3 \}$, or $(-\infty, 3]$.

69. $f(x) = \sqrt{2x + 7}$

$$2x + 7 \geq 0 \qquad 2x + 7 \text{ must be nonnegative.}$$
$$2x \geq -7 \qquad \text{Adding } -7$$
$$x \geq -\frac{7}{2} \qquad \text{Dividing by 2}$$

When $x \geq -\frac{7}{2}$, the expression $2x + 7$ is nonnegative.

Thus the domain of f is $\left\{ x \mid x \geq -\frac{7}{2}, \right\}$, or $\left[-\frac{7}{2}, \infty \right)$.

71. $f(x) = \sqrt{8 - 2x}$

$$8 - 2x \geq 0 \qquad 8 - 2x \text{ must be nonnegative.}$$
$$-2x \geq -8 \qquad \text{Adding } -8$$
$$x \leq 4 \qquad \text{Dividing by } -2 \text{ and reversing the inequality symbol}$$

When $x \leq 4$, the expression $8 - 2x$ is nonnegative.

Thus the domain of f is $\{ x \mid x \leq 4 \}$, or $(-\infty, 4]$.

73. *Familiarize.* Let $n =$ the number of hours. Then the total fee using the hourly plan is $120n$.

Translate. We write an inequality stating that the hourly plan costs less than the flat fee.

$$120n < 900$$

Carry out.

$$120n < 900$$
$$n < \frac{15}{2}, \text{ or } 7\frac{1}{2}$$

Check. We can do a partial check by substituting a value for n less than $\frac{15}{2}$. When $n = 7$, the hourly plan costs $120(7)$, or $840, so the hourly plan is less than the flat fee of $900. When $n = 8$, the hourly plan costs $120(8)$, or $960, so the hourly plan is more expensive than the flat fee of $900.

State. The hourly rate is less expensive for lengths of time less than $7\frac{1}{2}$ hours.

75. Familiarize. Let n = the number of correct answers. Then the points earned are $2n$, and the points deducted are $\frac{1}{2}$ of the rest of the questions, $80 - n$, or $\frac{1}{2}(80 - n)$.

Translate. We write an inequality stating the score is at least 100.

$$2n - \frac{1}{2}(80 - n) \geq 100.$$

Carry out.

$$2n - \frac{1}{2}(80 - n) \geq 100$$
$$2n - 40 + \frac{1}{2}n \geq 100$$
$$\frac{5}{2}n - 40 \geq 100$$
$$\frac{5}{2}n \geq 140$$
$$n \geq 56$$

Check. When $n = 56$, the score earned is

$2(56) - \frac{1}{2}(80 - 56)$, or $112 - \frac{1}{2}(24)$, or $112 - 12$, or 100.

When $n = 58$, the score earned is $2(58) - \frac{1}{2}(80 - 58)$, or

$116 - \frac{1}{2}(22)$, or $116 - 11$, or 105. Since the score is exactly 100 when 56 questions are answered correctly and more than 100 when 58 questions are correct, we have performed a partial check.

State. At least 56 questions are correct for a score of at least 100.

77. Familiarize. Let m = the number of peak local minutes used. Then the charge for the minutes used is $0.01m$ and the total monthly charge is $13.40 + 0.01m$.

Translate. We write an inequality stating that the monthly charge is at most $41.40.

$$13.40 + 0.01m \geq 41.40$$

Carry out.

$$13.40 + 0.01m \geq 41.40$$
$$0.01m \geq 28$$
$$m \geq 2800$$

Check. We can do a partial check by substituting a value for m less than 2800. When $m = 2790$, the monthly charge is $13.40 + 0.01(2790) \approx 41.30$. This is less than the maximum charge of $41.40.

State. A customer must speak on the phone for 2800 local peak minutes or more if the maximum charge is to apply.

79. Familiarize. Let t = the number of out-of-network transactions. Then the Local plan will cost $5 + 3t$ per month and the Anywhere plan will cost $15 + 1.75t$ per month.

Translate. We write an inequality stating that the Anywhere plan costs less than the Local plan.

$$5 + 3t > 15 + 1.75t$$

Carry out.

$$5 + 3t > 15 + 1.75t$$
$$1.25t > 10$$
$$t > 8$$

Check. We do a partial check by substituting a value for t less than 8 and a value for t greater than 8. When $t = 7$, the Local plan costs $5 + 3(7)$, or $26 and the Anywhere plan costs $15 + 1.75(7)$, or $27.25. So the Local plan is less expensive. When $t = 9$, the Local plan costs $5 + 3(9)$, or $32, and the Anywhere plan costs $15 + 1.75(9)$, or $30.75. So the Anywhere plan is less expensive.

State. The Anywhere plan costs less for more than 8 transactions per month.

81. Familiarize. We list the given information in a table.

Plan A: Monthly Income	Plan B: Monthly Income
$400 salary	$610
8% of sales	5% of sales
Total: 400+8% of sales	Total: 610+5% of sales

Suppose Toni had gross sales of $5000 one month. Then under plan A she would earn

$400 + 0.08(\$5000)$, or $800.

Under plan B she would earn

$610 + 0.05(\$5000)$, or $860.

This shows that, for gross sales of $5000, plan B is better. If Toni had gross sales of $10,000 one month, then under plan A she would earn

$400 + 0.08(\$10,000)$, or $1200.

Under plan B she would earn

$610 + 0.05(\$10,000)$, or $1110.

This shows that, for gross sales of $10,000, plan A is better. To determine all values for which plan A is better we solve an inequality.

Translate.

Income from plan A	is greater than	Income from plan B.
↓	↓	↓
$400 + 0.08s$	$>$	$610 + 0.05s$

Carry out.

$$400 + 0.08s > 610 + 0.05s$$
$$400 + 0.03s > 610$$
$$0.03s > 210$$
$$s > 7000$$

Check. For $s = \$7000$, the income from plan A is

$$\$400 + 0.08(\$7000), \text{ or } \$960$$

and the income from plan B is

$$\$610 + 0.05(\$7000), \text{ or } \$960.$$

This shows that for sales of $7000 Toni's income is the same from each plan. In the Familiarize step we show that, for a value less than $7000, plan B is better and, for a value greater than $7000, plan A is better. Since we cannot check all possible values, we stop here.

State. Toni should select plan A for gross sales greater than $7000.

83. *Familiarize.* Let m = the medical bill. Then the "Green Badge" medical insurance plan will cost $2000 + 0.30($m$ - 2000) and the "Blue Seal" plan will cost $2500 + 0.20($m$ - 2500).

Translate. We write an inequality stating that the "Blue Seal" plan costs less than the "Green Badge" plan.

$$2000 + 0.30(m - 2000) > 2500 + 0.20(m - 2500)$$

Carry out.

$$2000 + 0.30(m - 2000) > 2500 + 0.20(m - 2500)$$
$$2000 + 0.3m - 600 > 2500 + 0.2m - 500$$
$$1400 + 0.3m > 2000 + 0.2m$$
$$0.1m > 600$$
$$m > 6000$$

Check. When $m = 5000$, the "Green Badge" plan cost is $2000 + 0.30($5000 - 2000), or $2000 + 0.30($3000$), or $2900 and the "Blue Seal" plan cost is $2500 + 0.20($5000 - 2500), or $2500 + 0.20($2500$), or $3000. So the "Green Badge" plan is less expensive. When $m = 7000$, the "Green Badge" plan cost is $2000 + 0.30($7000 - 2000), or $2000 + 0.30($5000$), or $3500 and the "Blue Seal" plan is $2500 + 0.20($7000 - 2500), or $2500 + 0.20($4500$), or

$3400. So the "Blue Seal" plan is less expensive.

State. For medical bills of more than $6000, the "Blue Seal" plan is less expensive.

85. *Familiarize.* Find the values of t for which $C(t) < 1750$.

Translate. $-40.5t + 2159 < 1750$

Carry out.

$$-40.5t + 2159 < 1750$$
$$-40.5t < -409$$
$$t > \frac{818}{81}, \text{ or } 10\frac{8}{81}$$

Check. $C\left(\frac{818}{81}\right) = 1750.$

When $t = 10$, $C(10) = -40.5(10) + 2159$, or 1754.

When $t = 11$, $C(11) = -40.5(11) + 2159$, or 1713.5.

State. The domestic production will drop below 1750 million barrels later than 10 years after 2000, or the years after 2010.

87. a) *Familiarize.* Find the values of d for which $F(d) > 25$.

Translate.

$$\left(\frac{4.95}{d} - 4.50\right) \times 100 > 25$$

Carry out.

$$\left(\frac{4.95}{d} - 4.50\right) \times 100 > 25$$
$$\frac{495}{d} - 450 > 25$$
$$\frac{495}{d} > 475$$
$$495 > 475d$$
$$\frac{495}{475} > d$$
$$\frac{99}{95} > d, \text{ or } d < 1.04$$

Check. When $d = 1$,

$$F(1) = \left(\frac{4.95}{1} - 4.50\right) \times 100, \text{ or } 45 \text{ percent.}$$

When $d = 1.05$,

$$F(1.05) = \left(\frac{4.95}{1.05} - 4.50\right) \times 100, \text{ or } 21 \text{ percent.}$$

State. A man is considered obese for body density less than $\frac{99}{95}$ kg/L, or about 1.04 kg/L.

b) *Familiarize.* Find the values of d for which $F(d) > 32$.

Translate.

$$\left(\frac{4.95}{d} - 4.50\right) \times 100 > 32$$

Carry out.

$$\left(\frac{4.95}{d} - 4.50\right) \times 100 > 32$$

$$\frac{495}{d} - 450 > 32$$

$$\frac{495}{d} > 482$$

$$495 > 482d$$

$$\frac{495}{482} > d, \text{ or } d < 1.03$$

Check. Our check from part (a) leads to the result that $F(d) > 32$ when $d < 1.03$.

State. A woman is considered obese for body density less than $\frac{495}{482}$ kg/L, or about 1.03 kg/L.

89. a) **Familiarize.** Find the values of x for which $R(x) < C(x)$.

   Translate.

   $$48x < 90,000 + 25x$$

   Carry out.

   $$23x < 90,000$$
   $$x < 3913\tfrac{1}{23}$$

   Check. $R\left(3913\tfrac{1}{23}\right) = \$187,826.09 = C\left(3913\tfrac{1}{23}\right).$

   Calculate $R(x)$ and $C(x)$ for some x greater than $3913\tfrac{1}{23}$ and for some x less than $3913\tfrac{1}{23}$.
   Suppose $x = 4000$:
   $$R(x) = 48(4000) = 192,000 \text{ and}$$
   $$C(x) = 90,000 + 25(4000) = 190,000.$$

   In this case $R(x) > C(x)$.
   Suppose $x = 3900$:
   $$R(x) = 48(3900) = 187,200 \text{ and}$$
   $$C(x) = 90,000 + 25(3900) = 187,500.$$

   In this case $R(x) < C(x)$.

   Then for $x < 3913\tfrac{1}{23}$, $R(x) < C(x)$.

   State. We will state the result in terms of integers, since the company cannot sell a fraction of a lamp. For 3913 or fewer lamps the company loses money.

   b) Our check in part a) shows that for $x > 3913\tfrac{1}{23}$, $R(x) > C(x)$ and the company makes a profit. Again, we will state the result in terms of an integer. For more than 3913 lamps the company makes money.

91. *Writing Exercise.* In solving the equation our solution is one value of x that makes the statement true. In solving either inequality, our solution is an infinite set of numbers bounded by one value of x. Solving the equation, we have $x = 5$. Solving the inequality $x + 3 > 8$, we have $x > 5$, so 5 is a lower bound of the solution set. Solving the

inequality $x + 3 < 8$, we have $x < 5$, so 5 is an upper bound of the solution set.

93. $f(x) = \frac{5}{x}$, the domain is

   $\{x \mid x \text{ is a real number and } x \neq 0\}$.

95. $f(x) = \frac{x-2}{2x+1}$, $2x + 1 \neq 0$, so $x \neq -\frac{1}{2}$. the domain is

   $\left\{x \middle| x \text{ is a real number and } x \neq -\frac{1}{2}\right\}$.

97. $f(x) = \frac{x+10}{8}$ The domain is $\{x \mid x \text{ is a real number}\}$.

99. *Writing Exercise.* Since the percentage cannot exceed 100%, the function $P(t)$ will make sense only for the years where $P(t) < 100\%$, or about 11 years.

101. $$3ax + 2x \geq 5ax - 4$$
   $$2x - 2ax \geq -4$$
   $$2x(1 - a) \geq -4$$
   $$x(1 - a) \geq -2$$
   $$x \leq -\frac{2}{1-a}, \text{ or } \frac{2}{a-1}$$

   We reversed the inequality symbol when we divided because when $a > 1$, then $1 - a < 0$.

   The solution set is $\left\{x \middle| x \leq \frac{2}{a-1}\right\}$.

103. $$a(by - 2) \geq b(2y + 5)$$
   $$aby - 2a \geq 2by + 5b$$
   $$aby - 2by \geq 2a + 5b$$
   $$y(ab - 2b) \geq 2a + 5b$$
   $$y \geq \frac{2a+5b}{ab-2b}, \text{ or } \frac{2a+5b}{b(a-2)}$$

   The inequality symbol remained unchanged when we divided because when $a > 2$ and $b > 0$, then

   $$ab - 2b > 0.$$

   The solution set is $\left\{y \middle| y \geq \frac{2a+5b}{b(a-2)}\right\}$.

105. $$c(2 - 5x) + dx > m(4 + 2x)$$
   $$2c - 5cx + dx > 4m + 2mx$$
   $$-5cx + dx - 2mx > 4m - 2c$$
   $$x(-5c + d - 2m) > 4m - 2c$$
   $$x[d - (5c + 2m)] > 4m - 2c$$
   $$x > \frac{4m-2c}{d-(5c+2m)}$$

   The inequality symbol remained unchanged when we divided because when $5c + 2m < d$, then

   $$d - (5c + 2m) > 0.$$

   The solution set is $\left\{x \middle| x > \frac{4m-2c}{d-(5c+2m)}\right\}$.

107. False. If $a = 2$, $b = 3$, $c = 4$, and $d = 5$, then $2 < 3$ and $4 < 5$ but $2 - 4 = 3 - 5$.

109. *Writing Exercise.* Not equivalent

0 is a solution of $x < 3$ but not of $x + \frac{1}{x} < 3 + \frac{1}{x}$.

111. $x + 5 \leq 5 + x$
$\quad\quad 5 \leq 5 \quad\quad$ Subtracting x

We get an inequality that is true for all real numbers x. Thus the solution set is all real numbers.

113. $0^2 = 0$, $x^2 > 0$ for $x \neq 0$

The solution is $\{x \mid x$ is a real number *and* $x \neq 0\}$.

115. *Graphing Calculator Exercise*

Exercise Set 4.2

1. h

3. f

5. e

7. b

9. c

11. $\{2, 4, 16\} \cap \{4, 16, 256\}$

The numbers 4 and 16 are common to both sets, so the intersection is $\{4, 16\}$.

13. $\{0, 5, 10, 15\} \cup \{5, 15, 20\}$

The numbers in either or both sets are 0, 5, 10, 15, and 20, so the union is $\{0, 5, 10, 15, 20\}$.

15. $\{a, b, c, d, e, f\} \cap \{b, d, f\}$

The letters b, d, and f are common to both sets, so the intersection is $\{b, d, f\}$.

17. $\{x, y, z\} \cup \{u, v, x, y, z\}$

The letters in either or both sets are u, v, x, y, and z, so the union is $\{u, v, x, y, z\}$.

19. $\{3, 6, 9, 12\} \cap \{5, 10, 15\}$

There are no numbers common to both sets, so the solution set has no members. It is $\varnothing$.

21. $\{1, 3, 5\} \cup \varnothing$

The numbers in either or both sets are 1, 3, and 5, so the union is $\{1, 3, 5\}$.

23. $1 < x < 3$

This inequality is an abbreviation for the conjunction $1 < x$ *and* $x < 3$. The graph is the intersection of two separate solution sets:

$\{x \mid 1 < x\} \cap \{x \mid x < 3\} = \{x \mid 1 < x < 3\}$.

Interval notation: $(1, 3)$

25. $-6 \leq y \leq 0$

This inequality is an abbreviation for the conjunction $-6 \leq y$ and $y \leq 0$.

Interval notation: $[-6, 0]$

27. $x \leq -1$ *or* $x > 4$

The graph of this disjunction is the union of the graphs of the individual solution sets $\{x \mid x \leq -1\}$ and $\{x \mid x > 4\}$.

Interval notation: $(-\infty, -1] \cup (4, \infty)$

29. $x \leq -2$ *or* $x > 1$

Interval notation: $(-\infty, -2] \cup (1, \infty)$

31. $-4 \leq -x < 2$
$\quad\;\; 4 \geq x > -2 \quad$ Multiplying by -1 and
$\quad\quad\quad\quad\quad\quad$ reversing the inequality symbols
$\;\; -2 < x \leq 4 \quad$ Rewriting

Interval notation: $(-2, 4]$

33. $x > -2$ *and* $x < 4$

This conjunction can be abbreviated as $-2 < x < 4$.

Interval notation: $(-2, 4)$

35. $5 > a$ *or* $a > 7$

Interval notation: $(-\infty, 5) \cup (7, \infty)$

37. $x \geq 5 \ \ or \ \ -x \geq 4$

Multiplying the second inequality by –1 and reversing the inequality symbols, we get $x \geq 5$ or $x \leq -4$.

Interval notation: $(-\infty, -4] \cup [5, \infty)$

39. $7 > y \ \ and \ \ y \geq -3$

This conjunction can be abbreviated as $-3 \leq y < 7$.

Interval notation: $[-3, 7)$

41. $-x < 7 \ and -x \geq 0$

Multiplying the inequalities by –1 and reversing the inequality symbols, we get $x > -7 \ and \ x \leq 0$.

Interval notation: $(-7, 0]$

43. $t < 2 \ \ or \ \ t < 5$

Observe that every number that is less than 2 is also less than 5. Then $t < 2 \ \ or \ \ t < 5$ is equivalent to $t < 5$ and the graph of this disjunction is the set $\{t \mid t < 5\}$.

Interval notation: $(-\infty, 5)$

44. $t > 4 \ or \ t > -1$
$$\{t \mid t > 4\} \cup \{t \mid t > -1\} = \{t \mid t > -1\}$$

Interval notation: $(-1, \infty)$

45. $-3 \leq x + 2 < 9$
$$-3 - 2 \leq x < 9 - 2$$
$$-5 \leq x < 7$$

The solution set is $\{x \mid -5 \leq x < 7\}$ or $[-5, 7)$.

47. $0 < t - 4 \ \ and \ \ t - 1 \leq 7$
$$4 < t \quad\quad and \quad\quad t \leq 8$$

We can abbreviate the answer as $4 < t \leq 8$.

The solution set is $\{t \mid 4 < t \leq 8\}$, or $(4, 8]$.

49. $-7 \leq 2a - 3 \ \ and \ \ 3a + 1 < 7$
$$-4 \leq 2a \quad\quad and \quad\quad 3a < 6$$
$$-2 \leq a \quad\quad and \quad\quad a < 2$$

We can abbreviate the answer as $-2 \leq a < 2$. The solution set is $\{a \mid -2 \leq a < 2\}$, or $[-2, 2)$.

51. $x + 3 \leq -1 \ or \ x + 3 > -2$

Observe that any real number is either less than or equal to –1 or greater than or equal to –2. Then the solution set is $\{x \mid x$ is a real number$\}$, or $(-\infty, \infty)$.

53. $-10 \leq 3x - 1 \leq 5$
$$-9 \leq 3x \leq 6$$
$$-3 \leq x \leq 2$$

The solution set is $\{x \mid -3 \leq x \leq 2\}$, or $[-3, 2]$.

55. $5 > \dfrac{x - 3}{4} > 1$
$$20 > x - 3 > 4 \quad \text{Multiplying by 4}$$
$$23 > x > 7, \ or$$
$$7 < x < 23$$

The solution set is $\{x \mid 7 < x < 23\}$, or $(7, 23)$.

57. $-2 \leq \dfrac{x + 2}{-5} \leq 6$
$$10 \geq x + 2 \geq -30 \quad\quad \text{Multiplying by } -5$$
$$8 \geq x \geq -32, \ or$$
$$-32 \leq x \leq 8$$

The solution set is $\{x \mid -32 \leq x \leq 8\}$, or $[-32, 8]$.

59. $2 \leq 3x - 1 \leq 8$
$$3 \leq 3x \leq 9$$
$$1 \leq x \leq 3$$

The solution set is $\{x \mid 1 \leq x \leq 3\}$, or $[1, 3]$.

61. $-21 \leq -2x - 7 < 0$
$$-14 \leq -2x < 7$$
$$7 \geq x > -\frac{7}{2}, \ or$$
$$-\frac{7}{2} < x \leq 7$$

The solution set is $\left\{x \mid -\dfrac{7}{2} < x \leq 7\right\}$, or $\left(-\dfrac{7}{2}, 7\right]$.

63. $5t + 3 < 3 \quad\ or \quad\ 5t + 3 > 8$
$$5t < 0 \quad\ or \quad\quad 5t > 5$$
$$t < 0 \quad\ or \quad\quad\ t > 1$$

The solution set is $\{t \mid t < 0 \ or \ t > 1\}$, or $(-\infty, 0) \cup (1, \infty)$.

65. $6 > 2a - 1$ *or* $-4 \leq -3a + 2$
        $7 > 2a$ *or* $-6 \leq -3a$
        $\dfrac{7}{2} > a$ *or* $2 \geq a$

The solution set is $\left\{a \mid \dfrac{7}{2} > a\right\} \cup \left\{a \mid 2 \geq a\right\} = \left\{a \mid \dfrac{7}{2} > a\right\}$, or

$\left\{a \mid a < \dfrac{7}{2}\right\}$, or $\left(-\infty, \dfrac{7}{2}\right)$.

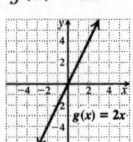

67. $a + 3 < -2$ *and* $3a - 4 < 8$
        $a < -5$ *and* $3a < 12$
        $a < -5$ *and* $a < 4$

The solution set is $\{a \mid a < -5\} \cap \{a \mid a < 4\} = \{a \mid a < -5\}$,

or $(-\infty, -5)$.

69. $3x + 2 < 2$ *and* $3 - x < 1$
        $3x < 0$ *and* $-x < -2$
        $x < 0$ *and* $x > 2$

The solution set is $\{x \mid x < 0\} \cap \{x \mid x > 2\} = \varnothing$.

71. $2t - 7 \leq 5$ *or* $5 - 2t > 3$
        $2t \leq 12$ *or* $-2t > -2$
        $t \leq 6$ *or* $t < 1$

The solution set is $\{t \mid t \leq 6\} \cup \{t \mid t < 1\} = \{t \mid t \leq 6\}$, or

$(-\infty, 6]$.

73. $f(x) = \dfrac{9}{x + 6}$

$f(x)$ cannot be computed when the denominator is 0.

Since $x + 6 = 0$ is equivalent to $x = -6$, we have

Domain of $f = \{x \mid x \text{ is a real number } and \ x \neq -6\}$

$= (-\infty, -6) \cup (-6, \infty)$.

75. $f(x) = \dfrac{1}{x}$

$f(x)$ cannot be computed when the denominator is 0.

We have Domain of f

$= \{x \mid x \text{ is a real number and } x \neq 0\} = (-\infty, 0) \cup (0, \infty)$.

77. $f(x) = \dfrac{x + 3}{2x - 8}$

$f(x)$ cannot be computed when the denominator is 0.

Since $2x - 8 = 0$ is equivalent to $x = 4$, we have

Domain of $f = \{x \mid x \text{ is a real number } and \ x \neq 4\}$, or

$(-\infty, 4) \cup (4, \infty)$.

79. *Writing Exercise.* By definition, the notation $2 < x < 5$

indicates that $2 < x \ and \ x < 5$. A solution of the

disjunction $2 < x \ or \ x < 5$ must be in at least one of

these sets but not necessarily in both, so the disjunction

cannot be written as $2 < x < 5$.

81. $g(x) = 2x$

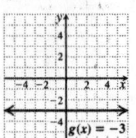

83. $g(x) = -3$

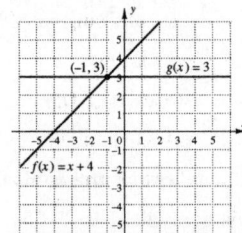

85. Graph both equations.

The point of intersection is $(-1, 3)$.

$\dfrac{y = x + 4}{3 \mid -1 + 4}$ $\dfrac{y = 3}{3 \mid 3}$
$\quad ?$ $\quad ?$
$3 = 3$ TRUE $3 = 3$ TRUE

The solution is -1.

87. *Writing Exercise.* When $[a, \ b] \cup [c, \ d] = [a, \ d]$, we know

that $a < b$, $c < d$, and $b \geq c$.

89. From the graph we observe that the values of x for which

$2x - 5 > -7 \ and \ 2x - 5 < 7$ are $\{x \mid -1 < x < 6\}$, or

$(-1, \ 6)$.

91. Solve $18{,}000 < S(t) < 21{,}000$, or

$18{,}000 < 500t + 16{,}500 < 21{,}000$
$\quad\quad 1500 < 500t < 4500$
$\quad\quad\quad\quad 3 < t < 9$

Thus, from 3 through 9 years after 2000, or between 2003

and 2009.

93. Solve $32 < f(x) < 46$, or $32 < 2(x + 10) < 46$.

$\quad\quad 32 < 2(x + 10) < 46$
$\quad\quad 32 < 2x + 20 < 46$
$\quad\quad 12 < 2x < 26$
$\quad\quad\quad 6 < x < 13$

For U.S. dress sizes between 6 and 13, dress sizes in Italy

will be between 32 and 46.

95. Solve $25 \le F(d) \le 31$.

$$25 \le (4.95/d - 4.50) \times 100 \le 31$$
$$25 \le \frac{495}{d} - 450 \le 31$$
$$475 \le \frac{495}{d} \le 481$$
$$\frac{1}{475} \ge \frac{d}{495} \ge \frac{1}{481}$$
$$\frac{495}{475} \ge d \ge \frac{495}{481},$$
$$\text{or } 1.03 \le d \le 1.04$$

Acceptable body densities are between 1.03 kg/L and 1.04 kg/L.

97. Let $c =$ the number of crossings in six months. Then at the $6 per crossing rate, the total cost of c crossings is $6c$. A six-month pass costs $50 and additional $2 per crossing toll brings the total cost of c crossings to $50 + $2c$. An unlimited crossing pass costs $300.

We write an inequality that states that the cost of c crossings using six-month passes is less than the cost using $6 per crossing toll and is less than the cost of using the unlimited-trip pass. Solve:

$$50 + 2c < 6c \quad and \quad 50 + 2c < 300.$$

We get $12.5 < c \ and \ c < 125$.

For more than 12 crossings but less than 125 crossings in six months, the reduced-fare pass is more economical.

99.
$$\begin{array}{lll} 4m - 8 > 6m + 5 & or & 5m - 8 < -2 \\ -13 > 2m & or & 5m < 6 \\ -\frac{13}{2} > m & or & m < \frac{6}{5} \end{array}$$
$$\left\{ m \middle| m < \frac{6}{5} \right\}, \text{ or } \left(-\infty, \ \frac{6}{5} \right)$$

101.
$$\begin{aligned} 3x &< 4 - 5x < 5 + 3x \\ 0 &< 4 - 8x < 5 \\ -4 &< -8x < 1 \\ \frac{1}{2} &> x > -\frac{1}{8} \end{aligned}$$
The solution set is $\left\{ x \middle| -\frac{1}{8} < x < \frac{1}{2} \right\}$, or $\left(-\frac{1}{8}, \ \frac{1}{2} \right)$.

103. Let $a = b = c = 2$. Then $a \le c$ and $c \le b$, but $b \not> a$. The given statement is false.

105. If $-a < c$, then $-1(-a) > -1 \cdot c$, or $a > -c$. Then if $a > -c$ and $-c > b$, we have $a > -c > b$, so $a > b$ and the given statement is true.

107. $f(x) = \dfrac{\sqrt{3 - 4x}}{x + 7}$

$3 - 4x \ge 0$ is equivalent to $x \le \frac{3}{4}$ and $x + 7 = 0$ is equivalent to $x = -7$. Then we have Domain of $f = \left\{ x \middle| x \le \frac{3}{4} \ and \ x \ne -7 \right\}$, or $(-\infty, -7) \cup \left(-7, \ \frac{3}{4} \right]$.

109. Observe that the graph of y_2 lies on or above the graph of y_1 and below the graph of y_3 for x in the interval $[-3, \ 4)$.

111. *Graphing Calculator Exercise*

Exercise Set 4.3

1. $|x| \ge 0$, so the statement is true.

3. True; see page 242 in the text.

5. True; see page 244 in the text.

7. False; see page 245 in the text.

9. g

11. d

13. a

15. $|x| = 10$

$x = -10 \ or \ x = 10$ Using the absolute value principle
The solution set is $\{-10, 10\}$.

17. $|x| = -1$

The absolute value of a number is always nonnegative. Therefore, the solution set is $\varnothing$.

19. $|p| = 0$

The only number whose absolute value is 0 is 0. The solution set is $\{0\}$.

21. $|2x - 3| = 4$

$$\begin{array}{lll} 2x - 3 = -4 & or & 2x - 3 = 4 \\ 2x = -1 & or & 2x = 7 \\ x = -\frac{1}{2} & or & x = \frac{7}{2} \end{array}$$
     Absolute-value principle

The solution set is $\left\{ -\frac{1}{2}, \ \frac{7}{2} \right\}$.

23. $|3x+5|=-8$

Absolute value is always nonnegative, so the equation has no solution. The solution set is $\varnothing$.

25. $|x-2|=6$

$$x-2=-6 \quad or \quad x-2=6 \quad \text{Absolute-value}$$
$$\qquad\qquad\qquad\qquad\qquad\qquad \text{principle}$$
$$x=-4 \quad or \qquad x=8$$

The solution set is $\{-4,\ 8\}$.

27. $|x-7|=1$

$$x-7=-1 \quad or \quad x-7=1$$
$$x=6 \qquad or \qquad x=8$$

The solution set is $\{6,\ 8\}$.

29. $|t|+1.1=6.6$

$$|t|=5.5 \qquad \text{Adding } -1.1$$
$$t=-5.5 \quad or \quad t=5.5$$

The solution set is $\{-5.5,\ 5.5\}$.

31. $|5x|-3=37$

$$|5x|=40 \qquad \text{Adding } 3$$
$$5x=-40 \quad or \quad 5x=40$$
$$x=-8 \quad or \qquad x=8$$

The solution set is $\{-8,\ 8\}$.

33. $7|q|+2=9$

$$7|q|=7 \qquad \text{Adding } -2$$
$$|q|=1 \qquad \text{Multiplying by } \tfrac{1}{7}$$
$$q=-1 \ or \ q=1$$

The solution set is $\{-1,\ 1\}$.

35. $\left|\dfrac{2x-1}{3}\right|=4$

$$\frac{2x-1}{3}=-4 \quad or \quad \frac{2x-1}{3}=4$$
$$2x-1=-12 \quad or \quad 2x-1=12$$
$$2x=-11 \quad or \qquad 2x=13$$
$$x=-\frac{11}{2} \quad or \qquad x=\frac{13}{2}$$

The solution set is $\left\{-\dfrac{11}{2},\ \dfrac{13}{2}\right\}$.

37. $|5-m|+9=16$

$$|5-m|=7 \qquad \text{Adding } -9$$
$$5-m=-7 \quad or \quad 5-m=7$$
$$-m=-12 \quad or \quad -m=2$$
$$m=12 \quad or \qquad m=-2$$

The solution set is $\{-2,\ 12\}$.

39. $5-2|3x-4|=-5$

$$-2|3x-4|=-10$$
$$|3x-4|=5$$
$$3x-4=-5 \quad or \quad 3x-4=5$$
$$3x=-1 \quad or \qquad 3x=9$$
$$x=-\frac{1}{3} \quad or \qquad x=3$$

The solution set is $\left\{-\dfrac{1}{3},\ 3\right\}$.

41. $|2x+6|=8$

$$2x+6=-8 \quad or \quad 2x+6=8$$
$$2x=-14 \quad or \qquad 2x=2$$
$$x=-7 \quad or \qquad x=1$$

The solution set is $\{-7,\ 1\}$.

43. $|x|-3=5.7$

$$|x|=8.7$$
$$x=-8.7 \ or \ x=8.7$$

The solution set is $\{-8.7,\ 8.7\}$.

45. $\left|\dfrac{1-2x}{5}\right|=2$

$$\frac{1-2x}{5}=-2 \quad or \quad \frac{1-2x}{5}=2$$
$$1-2x=-10 \quad or \quad 1-2x=10$$
$$-2x=-11 \quad or \qquad -2x=9$$
$$x=\frac{11}{2} \quad or \qquad x=-\frac{9}{2}$$

The solution set is $\left\{-\dfrac{9}{2},\ \dfrac{11}{2}\right\}$.

47. $|x-7|=|2x+1|$

$$x-7=2x+1 \quad or \quad x-7=-(2x+1)$$
$$-x=8 \qquad or \quad x-7=-2x-1$$
$$x=-8 \qquad or \qquad 3x=6$$
$$\qquad\qquad\qquad\qquad x=2$$

The solution set is $\{-8,\ 2\}$.

49. $|x+4|=|x-3|$

$$x+4=x-3 \quad or \quad x+4=-(x-3)$$
$$4=-3 \qquad or \quad x+4=-x+3$$
$$\text{False} \qquad\qquad 2x=-1$$
$$\qquad\qquad\qquad\qquad x=-\frac{1}{2}$$

The solution set is $\left\{-\dfrac{1}{2}\right\}$.

51. $|3a-1|=|2a+4|$

$$3a-1=2a+4 \quad or \quad 3a-1=-(2a+4)$$
$$a-1=4 \qquad or \quad 3a-1=-2a-4$$
$$a=5 \qquad or \quad 5a-1=-4$$
$$\qquad\qquad\qquad\qquad 5a=-3$$
$$\qquad\qquad\qquad\qquad a=-\frac{3}{5}$$

The solution set is $\left\{-\dfrac{3}{5},\ 5\right\}$.

53. Since $|n-3|$ and $|3-n|$ are equivalent expressions, the solution set of $|n-3|=|3-n|$ is the set of all real numbers.

55. $|7-4a|=|4a+5|$

$$7-4a=4a+5 \quad or \quad 7-4a=-(4a+5)$$
$$-8a=-2 \quad or \quad 7-4a=-4a-5$$
$$a=\frac{1}{4} \quad or \quad 7=-5$$
$$\text{False}$$

The solution set is $\left\{\frac{1}{4}\right\}$.

57. $|a|\leq 3$

$-3\leq a\leq 3$ Part (b)

The solution set is $\{a|-3\leq a\leq 3\}$, or $[-3, 3]$.

59. $|t|>0$

$t<0$ or $0<t$ Part (c)

The solution set is $\{t|t<0 \ or \ t>0\}$, or $\{t|t\neq 0\}$, or $(-\infty, 0)\cup(0, \infty)$.

61. $|x-1|<4$

$-4<x-1<4$ Part (b)

$-3<x<5$

The solution set is $\{x|-3<x<5\}$, or $(-3, 5)$.

63. $|n+2|\leq 6$

$-6\leq n+2\leq 6$ Part (b)

$-8\leq n\leq 4$

The solution set is $\{n|-8\leq n\leq 4\}$, or $[-8, 4]$.

65. $|x-3|+2>7$

$|x-3|>5$ Adding -2

$x-3<-5 \quad or \quad 5<x-3$ Part(c)

$x<-2 \quad or \quad 8<x$

The solution set is $\{x|x<-2 \ or \ x>8\}$, or $(-\infty,-2)\cup(8, \infty)$.

67. $|2y-9|>-5$

Since absolute value is never negative, any value of $2y-9$, and hence any value of y, will satisfy the inequality. The solution set is the set of all real numbers, or $(-\infty, \infty)$.

69. $|3a+4|+2\geq 8$

$|3a+4|\geq 6$ Adding -2

$3a+4\leq -6 \quad or \quad 6\leq 3a+4$ Part (c)

$3a\leq -10 \quad or \quad 2\leq 3a$

$a\leq -\frac{10}{3} \quad or \quad \frac{2}{3}\leq a$

The solution set is $\left\{a\middle|a\leq -\frac{10}{3} \ or \ a\geq \frac{2}{3}\right\}$, or $\left(-\infty,-\frac{10}{3}\right]\cup\left[\frac{2}{3}, \infty\right)$.

71. $|y-3|<12$

$-12<y-3<12$ Part (b)

$-9<y<15$ Adding 3

The solution set is $\{y|-9<y<15\}$, or $(-9, 15)$.

73. $9-|x+4|\leq 5$

$-|x+4|\leq -4$

$|x+4|\geq 4$

$x+4\leq -4 \quad or \quad 4\leq x+4$ Part (c)

$x\leq -8 \quad or \quad 0\leq x$

The solution set is $\{x|x\leq -8 \ or \ x\geq 0\}$, or $(-\infty,-8]\cup[0, \infty)$.

75. $6+|3-2x|>10$

$|3-2x|>4$

$3-2x<-4 \quad or \quad 4<3-2x$

$-2x<-7 \quad or \quad 1<-2x$

$x>\frac{7}{2} \quad or \quad -\frac{1}{2}>x$

The solution set is $\left\{x\middle|x<-\frac{1}{2} \ or \ x>\frac{7}{2}\right\}$, or $\left(-\infty,-\frac{1}{2}\right)\cup\left(\frac{7}{2}, \infty\right)$.

77. $|5-4x|<-6$

Absolute value is always nonnegative, so the inequality has no solution. The solution set is $\varnothing$.

79. $\left|\dfrac{1+3x}{5}\right|>\dfrac{7}{8}$

$$\frac{1+3x}{5}<-\frac{7}{8} \quad or \quad \frac{7}{8}<\frac{1+3x}{5}$$
$$1+3x<-\frac{35}{8} \quad or \quad \frac{35}{8}<1+3x$$
$$3x<-\frac{43}{8} \quad or \quad \frac{27}{8}<3x$$
$$x<-\frac{43}{24} \quad or \quad \frac{9}{8}<x$$

The solution set is $\left\{x\middle|x<-\frac{43}{24} \ or \ x>\frac{9}{8}\right\}$, or

$\left(-\infty, -\frac{43}{24}\right) \cup \left(\frac{9}{8}, \infty\right)$.

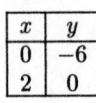

81. $|m+3|+8 \leq 14$

$\qquad |m+3| \leq 6 \quad$ Adding -8

$\qquad -6 \leq m+3 \leq 6$

$\qquad -9 \leq m \leq 3$

The solution set is $\{m \,|\, -9 \leq m \leq 3\}$, or $[-9, 3]$.

83. $25-2|a+3| > 19$

$\qquad -2|a+3| > -6$

$\qquad |a+3| < 3 \quad$ Multiplying by $-\frac{1}{2}$

$\qquad -3 < a+3 < 3 \quad$ Part(b)

$\qquad -6 < a < 0$

The solution set is $\{a \,|\, -6 < a < 0\}$, or $(-6, 0)$.

85. $|2x-3| \leq 4$

$\qquad -4 \leq 2x-3 \leq 4 \quad$ Part (b)

$\qquad -1 \leq 2x \leq 7 \quad$ Adding 3

$\qquad -\frac{1}{2} \leq x \leq \frac{7}{2} \quad$ Multiplying by $\frac{1}{2}$

The solution set is $\left\{x \,\middle|\, -\frac{1}{2} \leq x \leq \frac{7}{2}\right\}$, or $\left[-\frac{1}{2}, \frac{7}{2}\right]$.

87. $5+|3x-4| \geq 16$

$\qquad |3x-4| \geq 11$

$\qquad 3x-4 \leq -11 \quad or \quad 11 \leq 3x-4 \quad$ Part (c)

$\qquad 3x \leq -7 \quad or \quad 15 \leq 3x$

$\qquad x \leq -\frac{7}{3} \quad or \quad 5 \leq x$

The solution set is $\left\{x \,\middle|\, x \leq -\frac{7}{3} \ or \ x \geq 5\right\}$, or

$\left(-\infty, -\frac{7}{3}\right] \cup [5, \infty)$.

89. $7+|2x-1| < 16$

$\qquad |2x-1| < 9$

$\qquad -9 < 2x-1 < 9 \quad$ Part (b)

$\qquad -8 < 2x < 10$

$\qquad -4 < x < 5$

The solution set is $\{x \,|\, -4 < x < 5\}$, or $(-4, 5)$.

91. *Writing Exercise.* The solutions of $|x| < 5$ are those numbers whose distance from zero is less than 5. Since

the distance of -7 from 0 is not less than 5, then -7 is not a solution of the inequality.

93. $3x-y = 6$

x	y
0	-6
2	0

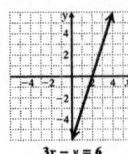

95. $x = -2$

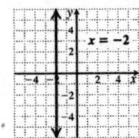

97. Solve using elimination.

$\qquad x-3y = 8 \quad (1)$

$\qquad \underline{2x+3y = 4} \quad (2)$

$\qquad 3x \quad\ = 12$

$\qquad x = 4$

Substituting 4 for x in Equation (1).

$\qquad 4-3y = 8$

$\qquad -3y = 4$

$\qquad y = -\frac{4}{3}$

The solution is $\left(4, -\frac{4}{3}\right)$.

99. Solve using substitution.

$\qquad y = 1-5x \quad (1)$

$\qquad 2x-y = 4 \quad\quad (2)$

Substitute $1-5x$ for y into Equation (2).

$\qquad 2x-(1-5x) = 4$

$\qquad 2x-1+5x = 4$

$\qquad 7x = 5$

$\qquad x = \frac{5}{7}$

When $x = \frac{5}{7}$, then $y = 1-5x = 1-5\left(\frac{5}{7}\right) = -\frac{18}{7}$.

The solution is $\left(\frac{5}{7}, -\frac{18}{7}\right)$.

101. *Writing Exercise.* Graph $y = g(x)$ and $y = c$ on the same axes. The solution set consists of x-values for which $(x, g(x))$ is below the horizontal line $y = c$.

103. From the definition of absolute value, $|3t-5| = 3t-5$ only when $3t-5 \geq 0$. Solve $3t-5 \geq 0$.

$\qquad 3t-5 \geq 0$

$\qquad 3t \geq 5$

$\qquad t \geq \frac{5}{3}$

The solution set is $\left\{t \,\middle|\, t \geq \frac{5}{3}\right\}$, or $\left[\frac{5}{3}, \infty\right)$.

105. $|x+2| > x$

The inequality is true for all $x < 0$ (because absolute value must be nonnegative). The solution set in this case is $\{x\,|\,x < 0\}$. If $x = 0$, we have $|0+2| > 0$, which is true. The solution set in this case is $\{0\}$. If $x > 0$, we have the following:

$$x+2 < -x \quad or \quad x < x+2$$
$$2x < -2 \quad or \quad 0 < 2$$
$$x < -1$$

Although $x > 0$ *and* $x < -1$ yields no solution, $x > 0$ *and* $2 > 0$ (true for all x) yields the solution set $\{x\,|\,x > 0\}$ in this case. The solution set for the inequality is $\{x\,|\,x < 0\} \cup \{0\} \cup \{x\,|\,x > 0\}$, or $\{x\,|\,x \text{ is a real number}\}$, or $(-\infty,\ \infty)$.

107. $|5t-3| = 2t+4$

From the definition of absolute value, we know that $2t+4 \geq 0$, or $t \geq -2$. So we have

$t \geq -2$ and

$$5t-3 = -(2t+4) \quad or \quad 5t-3 = 2t+4$$
$$5t-3 = -2t-4 \quad or \quad 3t = 7$$
$$7t = -1 \quad or \quad t = \frac{7}{3}$$
$$t = -\frac{1}{7} \quad or \quad t = \frac{7}{3}$$

Since $-\frac{1}{7} \geq -2$ *and* $\frac{7}{3} \geq -2$, the solution set is $\left\{-\frac{1}{7},\ \frac{7}{3}\right\}$.

109. Using part (b), we find that $-3 < x < 3$ is equivalent to $|x| < 3$.

111. $x \leq -6$ or $6 \leq x$
$|x| \geq 6$ Using part (c)

113.
$$x < -8 \quad or \quad 2 < x$$
$$x+3 < -5 \quad or \quad 5 < x+3 \quad \text{Adding 3}$$
$$|x+3| > 5 \qquad\qquad\qquad \text{Using part (c)}$$

115. The distance from x to 7 is $|x-7|$ or $|7-x|$, so we have $|x-7| < 2$, or $|7-x| < 2$.

117. The length of the segment from –1 to 7 is $|-1-7| = |-8| = 8$ units. The midpoint of the segment is $\frac{-1+7}{2} = \frac{6}{2} = 3$. Thus, the interval extends 8/2, or 4, units on each side of 3. An inequality for which the closed interval is the solution set is then $|x-3| \leq 4$.

119. The length of the segment from –7 to –1 is $|-7-(-1)| = |-6| = 6$ units. The midpoint of the segment is $\frac{-7+(-1)}{2} = \frac{-8}{2} = -4$. Thus, the interval

extends 6/2, or 3, units on each side of –4. An inequality for which the open interval is the solution set is $|x-(-4)| < 3$, or $|x+4| < 3$.

121. $|d-60\text{ft}| \leq 10\text{ft}$
$$-10\text{ft} \leq d-60\text{ft} \leq 10\text{ft}$$
$$50\text{ft} \leq d \leq 70\text{ft}$$

When the bungee jumper is 50 ft above the river, she is $150-50$, or 100 ft, from the bridge. When she is 70 ft above the river, she is $150-70$, or 80 ft, from the bridge. Thus, at any given time, the bungee jumper is between 80 ft and 100 ft from the bridge.

123. Graph $g(x) = 4$ on the same axes as $f(x) = |2x-6|$.

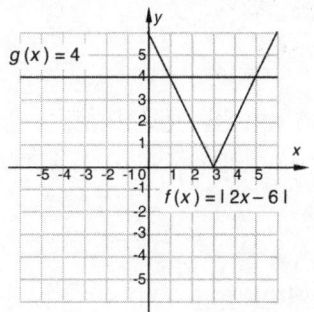

The solution set consists of the x-values for which $(x, f(x))$ is on or below the horizontal line $g(x) = 4$. These x-values comprise the interval $[1,\ 5]$.

125. *Writing Exercise.* No portion of the graph of the function $y = \text{abs}(x-3)$, the V-shaped graph, can lie below the x-axis, because absolute value is always nonnegative.

Connecting the Concepts

1. $2x+3 = 7$
$$2x = 4$$
$$x = 2$$
The solution is 2.

3. $3(t-5) = 4-(t+1)$
$$3t-15 = 4-t-1$$
$$3t-15 = -t+3$$
$$4t-15 = 3$$
$$4t = 18$$
$$t = \frac{18}{4}, \text{ or } \frac{9}{2}$$
The solution is $\frac{9}{2}$.

5. $-x \leq 6$
$$x \geq -6 \qquad \text{Reversing the inequality symbol}$$
The solution is $\{x\,|\,x \geq -6\}$, or $[-6,\ \infty)$.

7. $2(3n+6)-n=4-3(n+1)$

     $6n+12-n=4-3n-3$

        $5n+12=-3n+1$

            $8n=-11$

              $n=-\dfrac{11}{8}$

The solution is $-\dfrac{11}{8}$.

9. $2+|3x|=10$

      $|3x|=8$

   $3x=-8 \quad or \quad 3x=8$

     $x=-\dfrac{8}{3} \quad or \quad x=\dfrac{8}{3}$

The solution is $\left\{-\dfrac{8}{3}, \dfrac{8}{3}\right\}$.

11. $\dfrac{1}{2}x-7=\dfrac{3}{4}+\dfrac{1}{4}x$

      $\dfrac{1}{4}x=\dfrac{31}{4}$

       $x=31$

The solution is 31.

13. $|2x+5|+1\geq 13$

     $|2x+5|\geq 12$

   $2x+5\leq -12 \quad or \quad 12\leq 2x+5$

      $2x\leq -17 \quad or \quad 7\leq 2x$

        $x\leq -\dfrac{17}{2} \quad or \quad \dfrac{7}{2}\leq x$

The solution is $\left\{x\Big| x\leq -\dfrac{17}{2} \text{ or } x\geq \dfrac{7}{2}\right\}$, or

$\left(-\infty,-\dfrac{17}{2}\right]\cup\left[\dfrac{7}{2}, \infty\right)$.

15. $|m+6|-8<10$

     $|m+6|<18$

  $-18<m+6<18$

   $-24<m<12$

The solution is $\{m|-24<m<12\}$, or $(-24, 12)$.

17. $4-|7-t|\leq 1$

     $-|7-t|\leq -3$

      $|7-t|\geq 3$

   $7-t\leq -3 \quad or \quad 3\leq 7-t$

     $-t\leq -10 \quad or \quad -4\leq -t$

       $t\geq 10 \quad or \quad 4\geq t$

The solution is $\{t|t\leq 4 \text{ or } t\geq 10\}$, or $(-\infty, 4]\cup[10, \infty)$.

19. $8-5|a+6|>3$

    $-5|a+6|>-5$

      $|a+6|<1$

   $-1<a+6<1$

    $-7<a<-5$

The solution is $\{a|-7<a<-5\}$, or $(-7, -5)$.

Exercise Set 4.4

1. e; see pages 251 and 256 in the text.

3. d; see pages 255-257 in the text.

5. b; see page 257 in the text.

7. We replace x with -2 and y with 3.

$$\dfrac{2x-3y<-4}{2(-2)-3 \mid -4}$$

       $-7\overset{?}{<}-4 \quad$ FALSE

Since $-7<-4$ is false, $(-2, 3)$ is not a solution.

9. We replace x with 5 and y with 8.

$$\dfrac{3y-5x\leq 0}{3\cdot 8-5\cdot 5 \mid 0}$$

     $24-25 \mid$

        $-1\overset{?}{\leq}0 \quad$ TRUE

Since $-1\leq 0$ is true, $(5, 8)$ is a solution.

11. Graph: $y\geq \dfrac{1}{2}x$

We first graph the line $y=\dfrac{1}{2}x$. We draw the line solid since the inequality symbol is $\geq$. To determine which half-plane to shade, test a point not on the line, $(0, 1)$:

$$\dfrac{y\geq \frac{1}{2}x}{1 \mid \frac{1}{2}\cdot 0}$$

  $1\overset{?}{\geq}0 \quad$ TRUE

Since $1\geq 0$ is true, $(0, 1)$ is a solution as are all of the points in the half-plane containing $(0, 1)$. We shade that half-plane and obtain the graph.

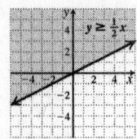

13. Graph: $y>x-3$.

First graph the line $y=x-3$. Draw it dashed since the inequality symbol is $>$. Test the point $(0, 0)$ to determine if it is a solution.

$$\dfrac{y>x-3}{0 \mid 0-3}$$

  $0\overset{?}{>}-3 \quad$ TRUE

Since $0>-3$ is true, we shade the half-plane that

contains $(0, 0)$ and obtain the graph.

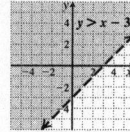

15. Graph: $y \le x + 2$.

First graph the line $y = x + 2$. Draw it solid since the inequality symbol is $\le$. Test the point $(0, 0)$ to determine if it is a solution.

$$\frac{y \le x+2}{0 \mid 0+2}$$

$$0 \overset{?}{\le} 2 \quad \text{TRUE}$$

Since $0 \le 2$ is true, we shade the half-plane that contains $(0, 0)$ and obtain the graph.

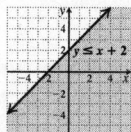

17. Graph: $x - y \le 4$

First graph the line $x - y = 4$. Draw a solid line since the inequality symbol is $\le$. Test the point $(0, 0)$ to determine if it is a solution.

$$\frac{x - y \le 4}{0 - 0 \mid 4}$$

$$0 \overset{?}{\le} 4 \quad \text{TRUE}$$

Since $0 \le 4$ is true, we shade the half-plane that contains $(0, 0)$ and obtain the graph.

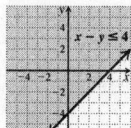

19. Graph: $2x + 3y < 6$

First graph $2x + 3y = 6$. Draw the line dashed since the inequality symbol is $<$. Test the point $(0, 0)$ to determine if it is a solution.

$$\frac{2x + 3y < 6}{2 \cdot 0 + 3 \cdot 0 \mid 6}$$

$$0 \overset{?}{<} 6 \quad \text{TRUE}$$

Since $0 < 6$ is true, we shade the half-plane containing $(0, 0)$ and obtain the graph.

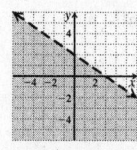

$2x + 3y < 6$

21. Graph: $2y - x \le 4$

We first graph $2y - x = 4$. Draw the line solid since the inequality symbol is $\le$. Test the point $(0, 0)$ to determine if it is a solution.

$$\frac{2y - x \le 4}{2 \cdot 0 - 0 \mid 4}$$

$$0 \overset{?}{\le} 4 \quad \text{TRUE}$$

Since $0 \le 4$ is true, we shade the half-plane containing $(0, 0)$ and obtain the graph.

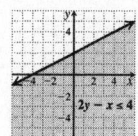

23. Graph: $2x - 2y \ge 8 + 2y$
$$2x - 4y \ge 8$$

First graph $2x - 4y = 8$. Draw the line solid since the inequality symbol is $\ge$. Test the point $(0, 0)$ to determine if it is a solution.

$$\frac{2x - 4y \ge 8}{2 \cdot 0 - 4 \cdot 0 \mid 8}$$

$$0 \overset{?}{\ge} 8 \quad \text{FALSE}$$

Since $0 \ge 8$ is false, we shade the half-plane that does not contain $(0, 0)$ and obtain the graph.

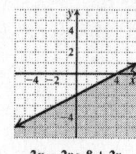

$2x - 2y \ge 8 + 2y$

25. Graph: $x > -2$.

We first graph $x = -2$. We draw the line dashed since the inequality symbol is $>$. Test the point $(0, 0)$ to determine if it is a solution.

$$\frac{x > -2}{0 \mid -2}$$

$$0 \overset{?}{>} -2 \quad \text{TRUE}$$

Since $0 > -2$ is true, we shade the half-plane that contains $(0, 0)$ and obtain the graph.

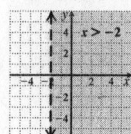

27. Graph: $y \le 6$.

We first graph $y = 6$. We draw the line solid since the inequality symbol is $\le$. Test the point $(0, 0)$ to determine if it is a solution.

$$\frac{y \le 6}{0 \mid 6}$$

$$0 \overset{?}{\le} 6 \quad \text{TRUE}$$

Since $0 \le 6$ is true, we shade the half-plane that contains $(0, 0)$ and obtain the graph.

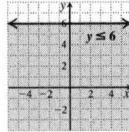

29. Graph: $-2 < y < 7$

This is a system of inequalities:

$$-2 < y,$$
$$y < 7$$

The graph of $-2 < y$ is the half-plane above the line $-2 = y$; the graph of $y < 7$ is the half-plane below the line $y = 7$. We shade the intersection of these graphs.

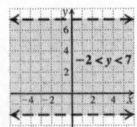

31. Graph: $-5 \le x < 4$.

This is a system of inequalities:

$$-5 \le x$$
$$x < 4$$

Graph $-5 \le x$ and $x < 4$. Then shade the intersection of these graphs.

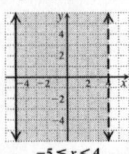

33. Graph: $0 \le y \le 3$

This is a system of inequalities:

$$0 \le y,$$
$$y \le 3$$

Graph $0 \le y$ and $y \le 3$.

Then we shade the intersection of these graphs.

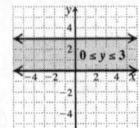

35. Graph: $y > x$,
$$y < -x + 3.$$

We graph the lines $y = x$ and $y = -x + 3$, using dashed lines. We indicate the region for each inequality by the arrows at the ends of the lines. Note where the regions

overlap and shade the region of solutions.

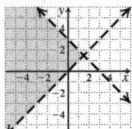

37. Graph: $y \le x$,
$$y \le 2x - 5.$$

We graph the lines $y = x$ and $y = 2x - 5$, using solid lines. Indicate the region for each inequality by arrows and shade the region where they overlap

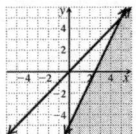

39. Graph: $y \le -3$,
$$x \ge -1$$

Graph $y = -3$ and $x = -1$ using solid lines. Indicate the region for each inequality by arrows, and shade the region where they overlap.

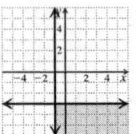

41. Graph: $x > -4$,
$$y < -2x + 3$$

Graph the lines $x = -4$ and $y = -2x + 3$, using dashed lines. Indicate the region for each inequality by arrows, and shade the region where they overlap.

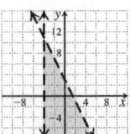

43. Graph: $y \le 5$,
$$y \ge -x + 4$$

Graph the lines $y = 5$ and $y = -x + 4$, using solid lines. Indicate the region for each inequality by arrows, and shade the region where they overlap.

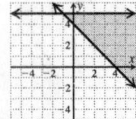

45. Graph: $x + y \le 6$,
$$x - y \le 4$$

Graph the lines $x + y = 6$ and $x - y = 4$, using solid lines. Indicate the region for each inequality by arrows, and shade the region where they overlap.

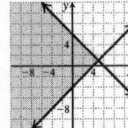

47. Graph: $y + 3x > 0$,
$\quad\quad\quad y + 3x < 2$

Graph the lines $y + 3x = 0$ and $y + 3x = 2$, using dashed lines. Indicate the region for each inequality by arrows, and shade the region where they overlap.

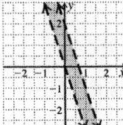

49. Graph: $y \leq 2x - 3$, $\quad$ (1)
$\quad\quad\quad y \geq -2x + 1$, $\quad$ (2)
$\quad\quad\quad x \leq 5$ $\quad\quad\quad\quad$ (3)

Graph the lines $y = 2x - 3$, $y = -2x + 1$, and $x = 5$ using solid lines. Indicate the region for each inequality by arrows, and shade the region where they overlap.

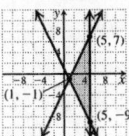

To find the vertex we solve three different systems of related equations.

From (1) and (2) we have $y = 2x - 3$,
$\quad\quad\quad\quad\quad\quad\quad\quad\quad\quad\quad\quad y = -2x + 1$.

Solving, we obtain the vertex $(1, -1)$.

From (1) and (3) we have $y = 2x - 3$,
$\quad\quad\quad\quad\quad\quad\quad\quad\quad\quad\quad\quad x = 5$.

Solving, we obtain the vertex $(5, 7)$.

From (2) and (3) we have $y = -2x + 1$,
$\quad\quad\quad\quad\quad\quad\quad\quad\quad\quad\quad\quad x = 5$.

Solving, we obtain the vertex $(5, -9)$.

51. Graph: $x + 2y \leq 12$, $\quad$ (1)
$\quad\quad\quad 2x + y \leq 12$ $\quad\quad$ (2)
$\quad\quad\quad x \geq 0$, $\quad\quad\quad\quad\quad$ (3)
$\quad\quad\quad y \geq 0$ $\quad\quad\quad\quad\quad$ (4)

Graph the lines $x + 2y = 12$, $2x + y = 12$, $x = 0$, and $y = 0$ using solid lines. Indicate the region for each inequality by arrows, and shade the region where they overlap.

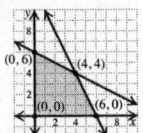

To find the vertices we solve four different systems of equations.

From (1) and (2) we have $x + 2y = 12$,
$\quad\quad\quad\quad\quad\quad\quad\quad\quad\quad\quad\quad 2x + y = 12$.

Solving, we obtain the vertex $(4, 4)$.

From (1) and (3) we have $x + 2y = 12$,
$\quad\quad\quad\quad\quad\quad\quad\quad\quad\quad\quad\quad x = 0$.

Solving, we obtain the vertex $(0, 6)$.

From (2) and (4) we have $2x + y = 12$,
$\quad\quad\quad\quad\quad\quad\quad\quad\quad\quad\quad\quad y = 0$.

Solving, we obtain the vertex $(6, 0)$.

From (3) and (4) we have $x = 0$,
$\quad\quad\quad\quad\quad\quad\quad\quad\quad\quad\quad\quad y = 0$.

Solving, we obtain the vertex $(0, 0)$.

53. Graph: $8x + 5y \leq 40$, $\quad$ (1)
$\quad\quad\quad x + 2y \leq 8$ $\quad\quad\quad$ (2)
$\quad\quad\quad x \geq 0$, $\quad\quad\quad\quad\quad$ (3)
$\quad\quad\quad y \geq 0$ $\quad\quad\quad\quad\quad$ (4)

Graph the lines $8x + 5y = 40$, $x + 2y = 8$, $x = 0$, and $y = 0$ using solid lines. Indicate the region for each inequality by arrows, and shade the region where they overlap.

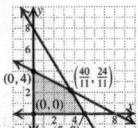

To find the vertices we solve four different systems of equations.

From (1) and (2) we have $8x + 5y = 40$,
$\quad\quad\quad\quad\quad\quad\quad\quad\quad\quad\quad\quad x + 2y = 8$.

Solving, we obtain the vertex $\left(\frac{40}{11}, \frac{24}{11}\right)$.

From (1) and (4) we have $8x + 5y = 40$,
$\quad\quad\quad\quad\quad\quad\quad\quad\quad\quad\quad\quad y = 0$.

Solving, we obtain the vertex $(5, 0)$.

From (2) and (3) we have $x + 2y = 8$,
$\quad\quad\quad\quad\quad\quad\quad\quad\quad\quad\quad\quad x = 0$.

Solving, we obtain the vertex $(0, 4)$.

From (3) and (4) we have $x = 0$,
$\quad\quad\quad\quad\quad\quad\quad\quad\quad\quad\quad\quad y = 0$.

Solving, we obtain the vertex $(0, 0)$.

55. Graph: $y - x \geq 2$, $\quad\quad$ (1)
$\quad\quad\quad y - x \leq 4$, $\quad\quad$ (2)
$\quad\quad\quad 2 \leq x \leq 5$ $\quad\quad$ (3)

Think of (3) as two inequalities:
$\quad\quad 2 \leq x$, $\quad$ (4)
$\quad\quad x \leq 5$ $\quad\quad$ (5)

Graph the lines $y - x = 2$, $y - x = 4$, $x = 2$, and $x = 5$, using solid lines. Indicate the region for each inequality by arrows, and shade the region where they overlap.

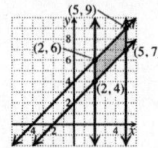

To find the vertices we solve four different systems of equations.

From (1) and (4) we have $y - x = 2$,

$$x = 2.$$

Solving, we obtain the vertex $(2, 4)$.

From (1) and (5) we have $y - x = 2$,

$$x = 5.$$

Solving, we obtain the vertex $(5, 7)$.

From (2) and (4) we have $y - x = 4$,

$$x = 2.$$

Solving, we obtain the vertex $(2, 6)$.

From (2) and (5) we have $y - x = 4$,

$$x = 5.$$

Solving, we obtain the vertex $(5, 9)$.

57. *Writing Exercise.* The boundary line is drawn dashed for the symbols $<$ and $>$ to show that the line is a border to the solution, but not part of the solution. The boundary line is drawn solid for the symbols $\leq$ and $\geq$ to show that both the line and the half-plane are the solution.

59. *Familiarize.* The formula for interest I, with principal P, rate r, and time t is $I = Prt$.

Translate. Substitute $25.35 for I, $1560 for P and $\frac{1}{2}$ for t in the formula.

$$I = Prt$$
$$\$25.35 = \$1560 \cdot r \cdot \frac{1}{2}$$

Carry out. We solve the equation.

$$25.35 = 1560 \cdot r \cdot \frac{1}{2}$$
$$25.35 = 780r \qquad \text{Multiplying}$$
$$0.0325 = r \qquad \text{Dividing by 780 on both sides}$$

Check. Interest is $\$15.60(0.0325)\left(\frac{1}{2}\right)$ or $25.35. These numbers check.

State. The rate of interest is 0.0325, or 3.25%.

61. *Familiarize.* Let $x =$ the amount invested at 5% and $y =$ the amount invested at 3%.

Translate. The interest from the first investment is $0.05x$ and the interest from the second investment is $0.03y$.

The sum of the investments is

$$x + y = 10,000.$$

The sum of the interests is

$$0.05x + 0.03y = 428.$$

Carry out. We solve the system of equations.

$$x + y = 10,000, \qquad (1)$$
$$0.05x + 0.03y = 428. \qquad (2)$$

We solve equation (1) for y, and substitute into equation (2), and solve for x.

$$0.05x + 0.03(10,000 - x) = 428$$
$$0.05x + 300 - 0.03x = 428$$
$$0.02x + 300 = 428$$
$$0.02x = 128$$
$$x = 6400$$

$$y = 10,000 - x = 10,000 - 6400 = 3600$$

Check. The interest of the first investment is $0.05(\$6400)$, or $320 and the interest of the second investment is $0.03(\$3600)$, or $108. The total interest is $320 + $108, or $428. These numbers check.

State. $6400 is invested at 5% and $3600 is invested at 3%.

63. *Familiarize.* Let $x =$ the number of student tickets sold and $y =$ the number of adult tickets sold. We arrange the information in a table.

	Student	Adult	Total
Price	$1	$3	
Number sold	x	y	170
Money taken in	$1x$	$3y$	386

Translate. The last two rows of the table give us two equations. The total number of tickets sold was 170, so we have

$$x + y = 170.$$

The total amount of money collected was $386, so we have

$$1x + 3y = 386.$$

Carry out. Solve the system using the elimination method.

$$x + y = 170, \qquad (1)$$
$$x + 3y = 386. \qquad (2)$$

$$\begin{array}{r} -x - y = -170 \qquad \text{Multiplying (1) by } -1 \\ \underline{x + 3y = 386} \\ 2y = 216 \\ y = 108 \end{array}$$

We go back to Equation (1) and substitute 108 for y.

$$x + y = 170$$
$$x + 108 = 170$$
$$x = 62$$

Check. The number of tickets sold was 108+62, or 170. The money collected was $1(62) + $3(108), or $62+$324, or \$386. These numbers check.

State. 62 student tickets and 108 adult tickets were sold.

65. *Writing Exercise.* The intersection of the individual graphs is a single point. An example is the following system:

$$y \le -x,$$
$$x \ge 0,$$
$$y \ge 0$$

The solution set is $(0, 0)$.

67. Graph: $x + y > 8,$
$$x + y \le -2$$

Graph the line $x + y = 8$ using a dashed line and graph $x + y = -2$, using a solid line. Indicate the region for each inequality by arrows. The regions do not overlap (the solution set is $\varnothing$), so we do not shade any portion of the graph.

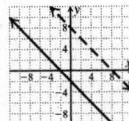

69. Graph: $x - 2y \le 0,$
$$-2x + y \le 2,$$
$$x \le 2,$$
$$y \le 2,$$
$$x + y \le 4$$

Graph the five inequalities above, and shade the region where they overlap.

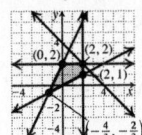

71. Both the width and the height must be positive, so we have

$$w > 0,$$
$$h > 0.$$

To be checked as luggage, the sum of the width, height, and length cannot exceed 62 in., so we have

$$w + h + 30 \le 62, \text{or}$$
$$w + h \le 32.$$

The girth is represented by $2w + 2h$ and the length is 30 in. In order to meet postal regulations the sum of the girth and the length cannot exceed 130 in., so we have:

$$2w + 2h + 30 < 130, \text{or}$$
$$2w + 2h \le 100, \text{or}$$
$$w + h \le 50$$

Thus, have a system of inequalities:

$$w > 0,$$
$$h > 0$$
$$w + h \le 32,$$
$$w + h \le 50$$

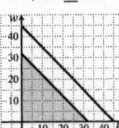

73. We graph the following inequalities:

$$q \ge 700$$
$$v \ge 400$$
$$q + v \ge 1150$$
$$q \le 800$$
$$v \le 800$$

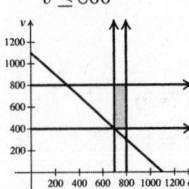

75. Graph: $35c + 75a > 1000,$
$$c \ge 0,$$
$$a \ge 0$$

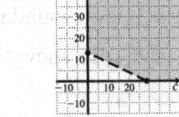

77. $h < 2w$
$$w \le 1.5h$$
$$h \le 3200$$
$$h \ge 0$$
$$w \ge 0$$

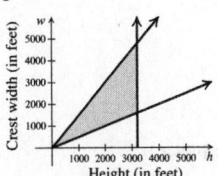

79. a) $3x + 6y > 2$

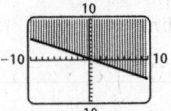

b) $x - 5y \le 10$

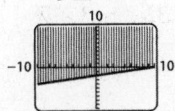

c) $13x - 25y + 10 \le 0$

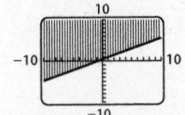

d) $2x + 5y > 0$

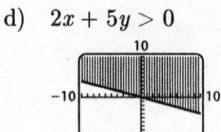

Connecting the Concepts

1. $x + 2 = 7$
    $x = 5$

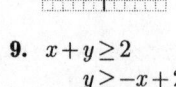

3. $x + 2 \leq 7$
    $x \leq 5$

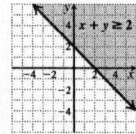

5. $6 - 2x \geq 8$
    $-2x \geq 2$
    $x \leq -1$

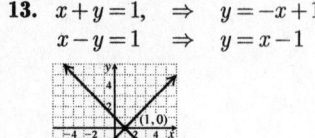

7. $x + y = 2$
    $y = -x + 2$

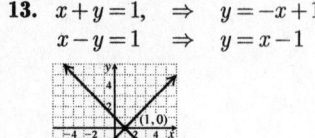

9. $x + y \geq 2$
    $y \geq -x + 2$

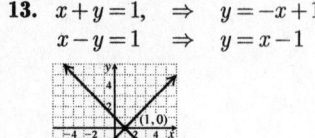

11. $x = 4$

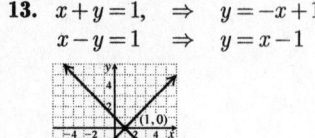

13. $x + y = 1, \quad \Rightarrow \quad y = -x + 1,$
     $x - y = 1 \quad \Rightarrow \quad y = x - 1$

15. $2x + y < 6$
     $y < -2x + 6$

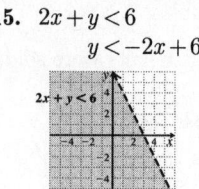

17. $4x = 3y$
     $y = \dfrac{4}{3}x$

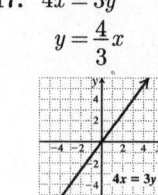

19. $x - y \leq 3, \quad \Rightarrow \quad y \geq x - 3,$
     $y \geq 2x, \quad \Rightarrow \quad y \geq 2x,$
     $2y - x \leq 2 \quad \Rightarrow \quad y \leq \dfrac{1}{2}x + 1$

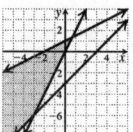

Exercise Set 4.5

1. Objective; see page 263 in the text.

3. Corner; see page 264 in the text.

5. Vertices; see page 264 in the text.

7. Find the maximum and minimum values of
    $$F = 2x + 14y \,,$$
    subject to
    $$5x + 3y \leq 34, \quad (1)$$
    $$3x + 5y \leq 30, \quad (2)$$
    $$x \geq 0, \quad (3)$$
    $$y \geq 0. \quad (4)$$
    Graph the system of inequalities and find the coordinates of the vertices.

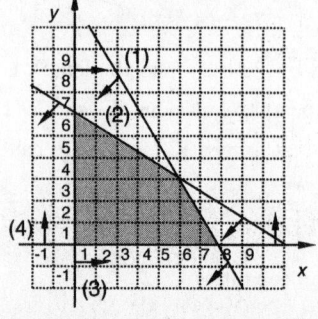

To find one vertex we solve the system

$x = 0,$
$y = 0.$

This vertex is $(0,0)$.

To find a second vertex we solve the system

$5x + 3y = 34,$
$y = 0.$

This vertex is $\left(\frac{34}{5}, \, 0\right)$.

To find a third vertex we solve the system

$5x + 3y = 34,$
$3x + 5y = 30.$

This vertex is $(5, \, 3)$.

To find the fourth vertex we solve the system

$3x + 5y = 30,$
$x = 0.$

This vertex is $(0, \, 6)$.

Now find the value of F at each of these points.

Vertex $(x, \ y)$	$F = 2x + 14y$	
$(0, \ 0)$	$2\cdot 0 + 14\cdot 0 = 0 + 0 = 0$	⟵Minimum
$\left(\frac{34}{5}, \, 0\right)$	$2\cdot\frac{34}{5} + 14\cdot 0 = \frac{68}{5} + 0 = 13\frac{3}{5}$	
$(5, \ 3)$	$2\cdot 5 + 14\cdot 3 = 10 + 42 = 52$	
$(0, \ 6)$	$2\cdot 0 + 14\cdot 6 = 0 + 84 = 84$	⟵Maximum

The maximum value of F is 84 when $x = 0$ and $y = 6$.

The minimum value of F is 0 when $x = 0$ and $y = 0$.

9. Find the maximum and minimum values of

$$P = 8x - y + 20\,,$$

subject to

$6x + 8y \leq 48,$ (1)
$0 \leq y \leq 4,$ (2)
$0 \leq x \leq 7.$ (3)

Think of (2) as $0 \leq y,$ (4)
             $y \leq 4.$ (5)

Think of (3) as $0 \leq x,$ (6)
             $x \leq 7.$ (7)

Graph the system of inequalities.

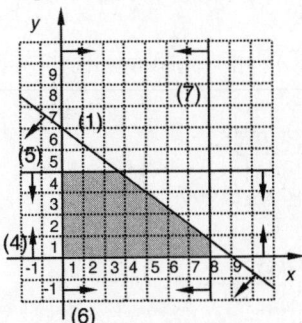

To determine the coordinates of the vertices, we solve the following systems:

$x = 0,$ $x = 7,$ $6x + 8y = 48,$
$y = 0;$ $y = 0;$ $x = 7;$

$6x + 8y = 48,$ $x = 0,$
$y = 4;$ $y = 4$

The vertices are $(0, \ 0)$, $(7, 0)$, $\left(7, \ \frac{3}{4}\right)$, $\left(\frac{8}{3}, \ 4\right)$, and $(0, \ 4)$, respectively. Compute the value of P at each of these points.

Vertex $(x, \ y)$	$P = 8x - y + 20$	
$(0, \ 0)$	$8\cdot 0 - 0 + 20$ $= 0 - 0 + 20 = 20$	
$(7, \ 0)$	$8\cdot 7 - 0 + 20$ $= 56 - 0 + 20 = 76$	⟵ Maximum
$\left(7, \ \frac{3}{4}\right)$	$8\cdot 7 - \frac{3}{4} + 20$ $= 56 - \frac{3}{4} + 20 = 75\frac{1}{4}$	
$\left(\frac{8}{3}, \ 4\right)$	$8\cdot\frac{8}{3} - 4 + 20$ $= \frac{64}{3} - 4 + 20 = 37\frac{1}{3}$	
$(0,4)$	$8\cdot 0 - 4 + 20$ $= 0 - 4 + 20 = 16$	⟵ Minimum

The maximum is 76 when $x = 7$ and $y = 0$. The minimum is 16 when $x = 0$ and $y = 4$.

11. Find the maximum and minimum values of

$$F = 2y - 3x\,,$$

subject to

$y \leq 2x + 1,$ (1)
$y \geq -2x + 3,$ (2)
$x \leq 3$ (3)

Graph the system of inequalities and find the coordinates of the vertices.

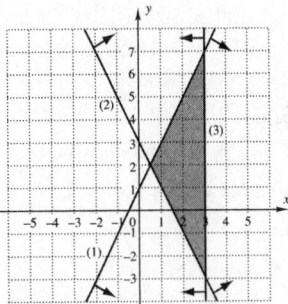

To determine the coordinates of the vertices, we solve the following systems:

$y = 2x + 1,$ $y = 2x + 1,$ $y = -2x + 3,$
$y = -2x + 3;$ $x = 3;$ $x = 3$

The solutions of the systems are $\left(\frac{1}{2}, \ 2\right)$, $(3, \ 7)$, and $(3, -3)$, respectively. Now find the value of F at each of these points.

Vertex (x, y)	$F = 2y - 3x$	
$\left(\frac{1}{2}, 2\right)$	$2 \cdot 2 - 3 \cdot \frac{1}{2} = \frac{5}{2}$	
$(3, 7)$	$2 \cdot 7 - 3 \cdot 3 = 5$	← Maximum
$(3, -3)$	$2(-3) - 3 \cdot 3 = -15$	← Minimum

The maximum value is 5 when $x = 3$ and $y = 7$. The minimum value is -15 when $x = 3$ and $y = -3$.

13. *Familiarize.* Let $x =$ the number of gumbo orders and $y =$ the number of sandwiches sold each day.

Translate. The profit P is given by

$$P = \$1.65x + \$1.05y.$$

We wish to maximize P subject to these constraints:

$$10 \le x \le 40$$
$$30 \le y \le 70$$
$$x + y \le 90.$$

Carry out. We graph the system of inequalities, determine the vertices and evaluate P at each vertex.

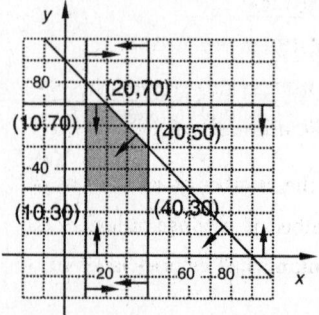

Vertex	$P = \$1.65x + \$1.05y$
$(10, 30)$	$\$1.65(10) + \$1.05(30) = \$48$
$(40, 30)$	$\$1.65(40) + \$1.05(30) = \$97.50$
$(40, 50)$	$\$1.65(40) + \$1.05(50) = \$118.50$
$(20, 70)$	$\$1.65(20) + \$1.05(70) = \$106.50$
$(10, 70)$	$\$1.65(10) + \$1.05(70) = \$90$

The largest profit in the table is $118.50 obtained when 40 orders of gumbo and 50 sandwiches are sold.

Check. Go over the algebra and arithmetic.

State. The maximum profit occurs when 40 orders of gumbo and 50 sandwiches are sold.

15. *Familiarize.* Let $x =$ the number of 4 photo pages, and $y =$ the number of 6 photo pages.

Translate. The number of photos N is given by

$$N = 4x + 6y.$$

We wish to maximize N subject to these constraints.

$$x + y \le 20$$
$$3x + 5y \le 90$$
$$x \ge 0$$
$$y > 0.$$

Carry out. We graph the system of inequalities,

determine the vertices, and evaluate N at each vertex.

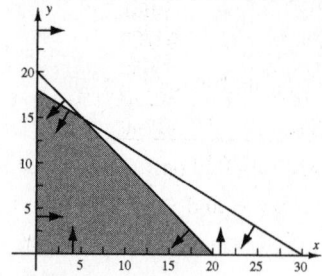

Vertex	$N = 4x + 6y$
$(0, 0)$	$4 \cdot 0 + 6 \cdot 0 = 0$
$(0, 18)$	$4 \cdot 0 + 6 \cdot 18 = 108$
$(20, 0)$	$4 \cdot 20 + 6 \cdot 0 = 80$
$(5, 15)$	$4 \cdot 5 + 6 \cdot 15 = 110$

The greatest number of photos is 110, obtained when 5 pages of 4-photos and 15 pages of 6-photos are used.

Check. Go over the algebra and arithmetic.

State. The maximum number of photos is achieved by using 5 pages or 4-photos and 15 pages of 6-photos.

17. In order to earn the most interest Rosa should invest the entire $40,000. She should also invest as much as possible in the type of investment that has the higher interest rate. Thus, she should invest $22,000 in corporate bonds and the remaining $18,000 in municipal bonds. The maximum income is

$$0.08(\$22,000) + 0.075(\$18,000) = \$3110.$$

We can also solve this problem as follows.

Let $x =$ the amount invested in corporate bonds and $y =$ the amount invested in municipal bonds. Find the maximum value of

$$I = 0.08x + 0.075y$$

subject to

$$x + y \le \$40,000,$$
$$\$6000 \le x \le \$22,000$$
$$0 \le y \le \$30,000.$$

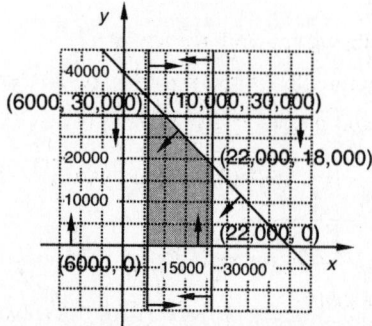

Vertex	$I = 0.08x + 0.075y$
($6000, $0)	$480
($6000, $30,000)	$2730
($10,000, $30,000)	$3050
($22,000, $18,000)	$3110
($22,000, $0)	$1760

The maximum income of $3110 occurs when $22,000 is invested in corporate bonds and $18,000 is invested in municipal bonds.

19. *Familiarize*. Let $x =$ the number of short-answer questions and $y =$ the number of essay questions answered.

Translate. The score S is given by

$S = 10x + 15y$.

We wish to maximize S subject to these constraints:

$x + y \leq 16$
$3x + 6y \leq 60$
$x \geq 0$
$y \geq 0$.

Carry out. We graph the system of inequalities, determine the vertices, and evaluate S at each vertex.

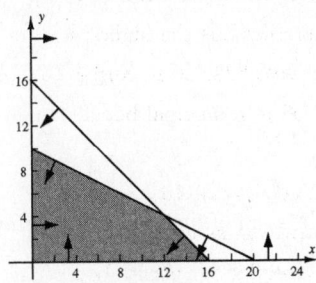

Vertex	$S = 10x + 15y$
(0, 0)	$10 \cdot 0 + 15 \cdot 0 = 0$
(0, 10)	$10 \cdot 0 + 15 \cdot 10 = 150$
(16, 0)	$10 \cdot 16 + 15 \cdot 0 = 160$
(12, 4)	$10 \cdot 12 + 15 \cdot 4 = 180$

The greatest score in the table is 180, obtained when 12 short-answer questions and 4 essay questions are answered.

Check. Go over the algebra and arithmetic.

State. The maximum score is 180 points when 12 short-answer questions and 4 essay questions are answered.

21. *Familiarize*. Let $x =$ the Merlot acreage and $y =$ the Cabernet acreage.

Translate. The profit P is given by

$P = \$400x + \$300y$.

We wish to maximize P subject to these constraints:

$x + y \leq 240,$
$2x + y \leq 320,$
$x \geq 0,$
$y \geq 0.$

Carry out. We graph the system of inequalities, determine the vertices, and evaluate P at each vertex.

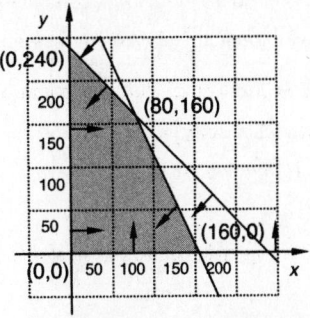

Vertex	$P = \$400x + \$300y$
(0, 0)	$0
(0, 240)	$72,000
(80, 160)	$80,000
(160, 0)	$64,000

Check. Go over the algebra and arithmetic.

State. The maximum profit occurs by planting 80 acres of Merlot grapes and 160 acres of Cabernet grapes.

23. *Familiarize*. Let $x =$ the number of servings of goat cheese and $y =$ the number of servings of hazelnuts.

Translate. The total number of calories is given by

$C = 264x + 628y$.

We wish to minimize C subject to these constraints.

$15 \leq x + 5y$
$x + 5y \leq 45$
$1500 \leq 500x + 100y$
$500x + 100y \leq 2500$

Carry out. We graph the system of inequalities, determine the vertices and evaluate C at each vertex.

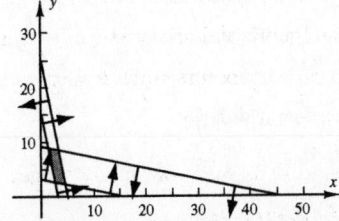

Vertex	$C = 264x + 628y$
(2.5, 2.5)	$264 \cdot 2.5 + 628 \cdot 2.5 = 2230$
(1.25, 8.75)	$264 \cdot 1.25 + 628 \cdot 8.75 = 5825$
$\left(\dfrac{55}{12}, \dfrac{25}{12}\right)$	$264 \cdot \dfrac{55}{12} + 628 \cdot \dfrac{25}{12} = 2518.3 = \dfrac{30,220}{12}$
$\left(\dfrac{10}{3}, \dfrac{25}{3}\right)$	$264 \cdot \dfrac{10}{3} + 628 \cdot \dfrac{25}{3} = 6113.3 = \dfrac{18,340}{3}$

The least number of calories in the table is 2230, obtained with 2.5 servings of each.

Check. Go over the algebra and arithmetic.

State. The minimum calories consumed is 2230 with 2.5 servings of each.

25. *Writing Exercise*. Answers may vary. Four sections that might be considered are sections on graphing linear equations, graphing linear inequalities, solving systems of equations, and evaluating algebraic expressions.

27. For $3x^3 - 5x^2 - 8x + 7$, when $x = -1$,
$3(-1)^3 - 5(-1)^2 - 8(-1) + 7$
$= -3 - 5 + 8 + 7 = 7$

29. $3(2t - 7) + 5(3t + 1)$
$= 6t - 21 + 15t + 5$
$= 21t - 16$

31. $(8t + 6) - (7t + 6)$
$= 8t + 6 - 7t - 6$
$= t$

33. *Writing Exercise*. In regard to Exercise 17, see the solution for that exercise. In Exercise 18, it is logical to invest the maximum amount possible in the investment that generates the most income.

35. **Familiarize**. Let x represent the number of T3 planes and y represent the number of S5 planes. Organize the information in a table.

Plane	Number of planes	Passengers		
		First	Tourist	Economy
T3	x	$40x$	$40x$	$120x$
S5	y	$80y$	$30y$	$40y$

Plane	Cost per mile
T3	$30x$
S5	$25y$

Translate. Suppose C is the total cost per mile. Then $C = 30x + 25y$. We wish to minimize C subject to these facts (constraints) about x and y.
$$40x + 80y \geq 2000,$$
$$40x + 30y \geq 1500,$$
$$120x + 40y \geq 2400,$$
$$x \geq 0,$$
$$y \geq 0$$

Carry out. Graph the system of inequalities, determine the vertices, and evaluate C at each vertex.

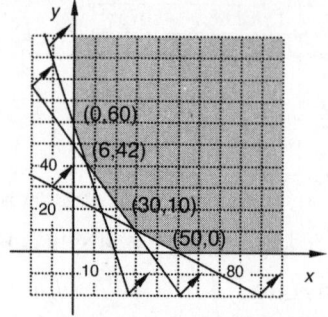

Vertex	$C = 30x + 25y$
(0, 60)	$30(0) + 25(60) = 1500$
(6, 42)	$30(6) + 25(42) = 1230$
(30, 10)	$30(30) + 25(10) = 1150$
(50, 0)	$30(50) + 25(0) = 1500$

Check. Go over the algebra and arithmetic.

State. In order to minimize the operating cost, 30 T3 planes and 10 S5 planes should be used.

37. **Familiarize**. Let $x =$ the number of chairs and $y =$ the number of sofas produced.

Translate. Find the maximum value of
$$I = \$80x + \$1200y$$
subject to
$$20x + 100y \leq 1900,$$
$$x + 50y \leq 500,$$
$$2x + 20y \leq 240,$$
$$x \geq 0,$$
$$y \geq 0.$$

Carry out. Graph the system of inequalities, determine the vertices, and evaluate I at each vertex.

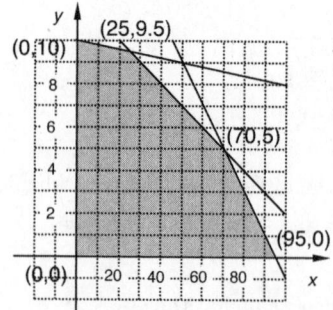

Vertex	$I = \$80x + \$1200y$
(0, 0)	$0
(0, 10)	$12,000
(25, 9.5)	$13,400
(70, 5)	$11,600
(95, 0)	$7600

Check. Go over the algebra and arithmetic.

State. The maximum income of $13,400 occurs when 25 chairs and 9.5 sofas are made. A more practical answer is that the maximum income of $12,800 is achieved when 25 chairs and 9 sofas are made.

Chapter 4 Review

1. True; see page 226 in the text.

3. True; see page 245 in the text.

5. True; see page 235 in the text.

7. True; see page 245 in the text.

9. False; see page 256 in the text.

11. Graph: $x \le -1$.

Set builder notation: $\{x \mid x \le -1\}$

Interval notation: $(-\infty, -1]$

13. $4y > -15$

Graph: $y > -\dfrac{15}{4}$.

Set builder notation: $\left\{y \middle| y > -\dfrac{15}{4}\right\}$

Interval notation: $\left(-\dfrac{15}{4}, \infty\right)$

15. $-6x - 5 < 4$
$\quad\quad -6x < 9$

Graph: $x > -\dfrac{3}{2}$

Set builder notation: $\left\{x \middle| x > -\dfrac{3}{2}\right\}$

Interval notation: $\left(-\dfrac{3}{2}, \infty\right)$

17. $0.3y - 7 < 2.6y + 15$
$\quad\quad -22 < 2.3y$
$\quad -\dfrac{22}{2.3} < y$
$\quad -\dfrac{220}{23} < y$

Graph: $-\dfrac{220}{23} < y$

Set builder notation: $\left\{y \middle| y > -\dfrac{220}{23}\right\}$

Interval notation: $\left(-\dfrac{220}{23}, \infty\right)$

19. $f(x) \le g(x)$
$\quad 3x + 2 \le 10 - x$
$\quad\quad\quad 4x \le 8$
$\quad\quad\quad\ x \le 2$

$\{x \mid x \le 2\}$ or $(-\infty, 2]$

21. Let $x =$ the amount invested at 3% and $9000 - x =$ the amount invested at 3.5%. The interest from the first investment is $0.03x$ and the interest from the second investment is $0.035(9000 - x)$. We solve the inequality.

$$0.03x + 0.035(9000 - x) \ge 300$$
$$0.03x + 315 - 0.035x \ge 300$$
$$-0.005x + 315 \ge 300$$
$$-0.005x \ge -15$$
$$x \le 3000$$

Clay should invest at most $3000 at 3%.

23. $\{a, b, c, d\} \cup \{a, c, e, f, g\} = \{a, b, c, d, e, f, g\}$

25. Graph: $x \le 3$ *or* $x > -5$

$(-\infty, \infty)$

27. $-15 < -4x - 5 < 0$
$\quad -10 < -4x < 5$
$\quad \dfrac{5}{2} > x > -\dfrac{5}{4}$, or $-\dfrac{5}{4} < x < \dfrac{5}{2}$

$\left\{x \middle| -\dfrac{5}{2} < x < \dfrac{5}{2}\right\}$

$\left(-\dfrac{5}{4}, \dfrac{5}{2}\right)$

29. $2x + 5 < -17 \quad or \quad -4x + 10 \le 34$
$\quad\ 2x < -22 \quad or \quad\quad -4x \le 24$
$\quad\quad\ x < -11 \quad or \quad\quad\quad x \ge -6$

$\{x \mid x < -11 \text{ or } x \ge -6\}$ or $(-\infty, -11) \cup [-6, \infty)$

31. $\quad f(x) < -5 \quad or \quad\ f(x) > 5$
$\quad 3 - 5x < -5 \quad or \quad 3 - 5x > 5$
$\quad\quad -5x < -8 \quad or \quad\quad -5x > 2$
$\quad\quad\quad\ x > \dfrac{8}{5} \quad or \quad\quad\quad x < -\dfrac{2}{5}$

$\left\{x \middle| x < -\dfrac{2}{5} \text{ or } x > \dfrac{8}{5}\right\}$ or $\left(-\infty, -\dfrac{2}{5}\right) \cup \left(\dfrac{8}{5}, \infty\right)$

33. $f(x) = \sqrt{5x - 10}$
$\quad 5x - 10 \ge 0$
$\quad\quad\ 5x \ge 10$
$\quad\quad\quad x \ge 2$

The domain of f is $[2, \infty)$.

35. $|x| = 11$

$x = -11 \quad or \quad x = 11$

$\{-11, 11\}$

37. $|x - 8| = 3$

$x - 8 = -3 \quad or \quad x - 8 = 3$
$x = 5 \quad\quad or \quad\quad x = 11$

$\{5, 11\}$

39. $|3x - 4| \geq 15$

$3x - 4 \leq -15 \quad or \quad 15 \leq 3x - 4$
$3x \leq -11 \quad or \quad 19 \leq 3x$
$x \leq -\dfrac{11}{3} \quad or \quad \dfrac{19}{3} \leq x$

$\left\{x \middle| x \leq -\dfrac{11}{3} \ or \ x \geq \dfrac{19}{3}\right\}$ or $\left(-\infty, -\dfrac{11}{3}\right] \cup \left[\dfrac{19}{3}, \infty\right)$

41. $|5n + 6| = -11$

Absolute value is never negative.

The solution is $\varnothing$.

43. $2|x - 5| - 7 > 3$

$2|x - 5| > 10$
$|x - 5| > 5$
$x - 5 < -5 \quad or \quad 5 < x - 5$
$x < 0 \quad\quad or \quad\quad 10 < x$

$\{x | x < 0 \ or \ x > 10\}$ or $(-\infty, 0) \cup (10, \infty)$

45. $|8x - 3| < 0$

Absolute value is never negative.

The solution is $\varnothing$.

47. Graph $x + 3y > -1$,
$\quad\quad\quad\quad x + 3y < 4$

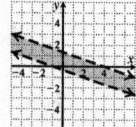

The lines are parallel, there are no vertices.

49. For $F = 3x + y + 4$, subject to

$y \leq 2x + 1$,
$x \leq 7$,
$y \geq 3$.

Vertices	$F = 3x + y + 4$
(7, 3)	$3 \cdot 7 + 3 + 4 = 28$
(1, 3)	$3 \cdot 1 + 3 + 4 = 10$
(7, 15)	$3 \cdot 7 + 15 + 4 = 40$

The maximum value of F is 40 at $x = 7$, $y = 15$.

The minimum value of F is 10 at $x = 1$, $y = 3$.

51. *Writing Exercise.* The equation $|X| = p$ has two solutions when p is positive because X can be either p or $-p$. The same equation has no solution when p is negative because no number has a negative absolute value.

53. $|2x + 5| \leq |x + 3|$

$2x + 5 \leq x + 3 \quad or \quad 2x + 5 \geq -(x + 3)$
$x \leq -2 \quad\quad or \quad 2x + 5 \geq -x - 3$
$\quad\quad\quad\quad\quad\quad\quad\quad 3x \geq -8$
$\quad\quad\quad\quad\quad\quad\quad\quad x \geq -\dfrac{8}{3}$

$\left\{x \middle| -\dfrac{8}{3} \leq x \leq -2\right\}$

55. $|d - 2.5| \leq 0.003$

Chapter 4 Test

1. $x - 3 < 8$
$\quad\quad x < 11$
$\{x \mid x < 11\}$ or $(-\infty, 11)$

3. $-4y - 3 \geq 5$
$\quad\quad -4y \geq 8$
$\quad\quad\quad y \leq -2$
$\{y \mid y \leq -2\}$ or $(-\infty, -2]$

5. $3(7 - x) < 2x + 5$
$\quad 21 - 3x < 2x + 5$
$\quad\quad 21 < 5x + 5$
$\quad\quad 16 < 5x$
$\quad\quad \dfrac{16}{5} < x$

$\left\{x \middle| x > \dfrac{16}{5}\right\}$ or $\left(\dfrac{16}{5}, \infty\right)$

7. $f(x) > g(x)$
$\quad -5x - 1 > -9x + 3$
$\quad\quad 4x - 1 > 3$
$\quad\quad\quad 4x > 4$
$\quad\quad\quad\quad x > 1$
$\{x | x > 1\}$ or $(1, \infty)$

9. Let $x =$ the number of additional hours. The cost is $\$80 + \$60x$ and $\$200$ is budgeted. We solve the inequality.

$$80 + 60x \le 200$$
$$60x \le 120$$
$$x \le 2$$

The time of service is $x + \frac{1}{2}$ hr or $2\frac{1}{2}$ hours or less.

11. $\{a, e, i, o, u\} \cup \{a, b, c, d, e\} = \{a, b, c, d, e, i, o, u\}$

13. $f(x) = \dfrac{x}{x-7}$
$$x - 7 = 0$$
$$x = 7$$
The domain of f is $(-\infty, 7) \cup (7, \infty)$.

15. $3x - 2 < 7 \quad or \quad x - 2 > 4$
$\quad\;\; 3x < 9 \quad or \quad\quad x > 6$
$\quad\;\;\; x < 3$

The solution set is $\{x | x < 3 \; or \; x > 6\}$ or $(-\infty, 3) \cup (6, \infty)$.

$$\xleftarrow{\qquad} \underset{0}{\;} \; \underset{3}{)} \; \underset{6}{(} \xrightarrow{\qquad}$$

17. $\quad 1 \le 3 - 2x \le 9$
$\quad -2 \le -2x \le 6$
$\quad\;\; 1 \ge x \ge -3$

The solution set is $\{x | -3 \le x \le 1\}$ or $[-3, 1].$.

$$\xleftarrow{\qquad} \underset{-3}{[} \; \underset{0}{\;} \; \underset{1}{]} \xrightarrow{\qquad}$$

19. $|a| > 5$

$a < -5 \; or \; 5 < a$

$\{a \mid a < -5 \; or \; a > 5\}$ or $(-\infty, -5) \cup (5, \infty)$

$$\xleftarrow{\qquad} \underset{-5}{)} \; \underset{0}{\;} \; \underset{5}{(} \xrightarrow{\qquad}$$

21. $|-5t - 3| \ge 10$

$-5t - 3 \le -10 \quad or \quad 10 \le -5t - 3$
$\quad\;\; -5t \le -7 \quad or \quad 13 \le -5t$
$\quad\quad\;\; t \ge \dfrac{7}{5} \quad or \quad -\dfrac{13}{5} \ge t$

$\left\{ t \left| t \le -\dfrac{13}{5} \; or \; t \ge \dfrac{7}{5} \right. \right\}$ or $\left(-\infty, -\dfrac{13}{5}\right] \cup \left[\dfrac{7}{5}, \infty\right)$

$$\xleftarrow{\qquad} \underset{-\frac{13}{5}}{]} \; \underset{0}{\;} \; \underset{\frac{7}{5}}{[} \xrightarrow{\qquad}$$

23. $\quad g(x) < -3 \quad or \quad\;\; g(x) > 3$
$\quad 4 - 2x < -3 \quad or \quad 4 - 2x > 3$
$\quad\;\; -2x < -7 \quad or \quad\;\; -2x > -1$
$\quad\quad\;\; x > \dfrac{7}{2} \quad or \quad\quad\;\; x < \dfrac{1}{2}$

$\left\{ x \left| x < \dfrac{1}{2} \; or \; x > \dfrac{7}{2} \right. \right\}$ or $\left(-\infty, \dfrac{1}{2}\right) \cup \left(\dfrac{7}{2}, \infty\right)$

$$\xleftarrow{\qquad} \underset{0\;\frac{1}{2}}{)} \quad\quad \underset{\frac{7}{2}}{(} \xrightarrow{\qquad}$$

25. $y \le 2x + 1$

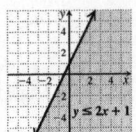

27. $2y - x \ge -7,$
$2y + 3x \le 15,$
$y \le 0,$
$x \le 0$

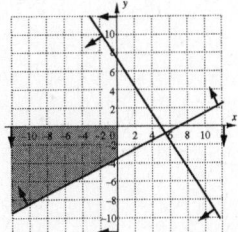

Vertices: $(0, 0)$ and $\left(0, -\dfrac{7}{2}\right)$

29. Let $x =$ the number of manicures and $y =$ the number of haircuts. The profit is $P = 12x + 18y$. We maximize the profit subject to the constraints:

$$30x + 50y \le 5(6)(60)$$
$$x + y \le 50$$
$$x \ge 0$$
$$y \ge 0$$

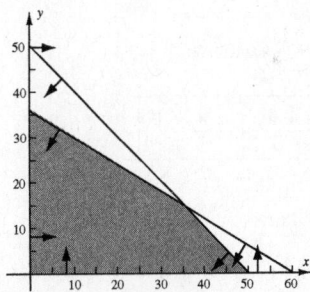

Vertices	$P = 12x + 18y$
$(0, 0)$	$12 \cdot 0 + 18 \cdot 0 = 0$
$(50, 0)$	$12 \cdot 50 + 18 \cdot 0 = 600$
$(0, 36)$	$12 \cdot 0 + 18 \cdot 36 = 648$
$(35, 15)$	$12 \cdot 35 + 18 \cdot 15 = 690$

The maximum profit is $\$690$ when there are 35 manicures and 15 haircuts.

31. $7x < 8 - 3x < 6 + 7x$
$\quad\;\; 0 < 8 - 10x < 6$
$\quad -8 < -10x < -2$
$\quad\;\; \dfrac{4}{5} > x > \dfrac{1}{5}$

$\left\{ x \left| \dfrac{1}{5} < x < \dfrac{4}{5} \right. \right\}$ or $\left(\dfrac{1}{5}, \dfrac{4}{5}\right)$

Chapter 5

Polynomials and Polynomial Functions

Exercise Set 5.1

1. g

3. a

5. b

7. j

9. f

11. $-5x^6 + x^4 + 7x^3 - 2x - 10$

a) Number of terms: 5

b)

Term	$-5x^6$	x^4	$7x^3$	$-2x$	-10
Degree	6	4	3	1	0

c) Degree of polynomial: 6

d) Leading term: $-5x^6$

e) Leading coefficient: -5

13. $7a^4 + a^3b^2 - 5a^2b + 3$

a) Number of terms: 4

b)

Term	$7a^4$	a^3b^2	$-5a^2b$	3
Degree	4	5	3	0

c) Degree of polynomial: 5

d) Leading term: a^3b^2

e) Leading coefficient: 1

15. $8y^2 + y^5 - 9 - 2y + 3y^4$

Term	$8y^2$	y^5	-9	$-2y$	$3y^4$
Degree	2	5	0	1	4
Degree of polynomial			5		

17. $3p^4 - 5pq + 2p^3q^3 + 8pq^2 - 7$

Term	$3p^4$	$-5pq$	$2p^3q^3$	$8pq^2$	-7
Degree	4	2	6	3	0
Degree of polynomial			6		

19. $-15t^4 + 2t^3 + 5t^2 - 8t + 4; \ -15t^4; \ -15$

21. $-x^6 + 6x^5 + 7x^2 + 3x - 5; \ -x^6; \ -1$

23. $-9 + 4x + 5x^3 - x^6$

25. $8y + 5xy^3 + 2x^2y - x^3$

27. $g(x) = x - 5x^2 + 4$

$g(3) = 3 - 5(3)^2 + 4 = -38$

29. $f(x) = -3x^4 + 5x^3 + 6x - 2$

$f(x) = -3(-1)^4 + 5(-1)^3 + 6(-1) - 2$

$\qquad = -3 - 5 - 6 - 2$

$\qquad = -16$

31. $F(x) = 2x^2 - 6x - 9$

$F(2) = 2 \cdot 2^2 - 6 \cdot 2 - 9 = 8 - 12 - 9 = -13$

$F(5) = 2 \cdot 5^2 - 6 \cdot 5 - 9 = 50 - 30 - 9 = 11$

33. $Q(y) = -8y^3 + 7y^2 - 4y - 9$

$Q(-3) = -8(-3)^3 + 7(-3)^2 - 4(-3) - 9$

$\qquad = 216 + 63 + 12 - 9 = 282$

$Q(0) = -8 \cdot 0^3 + 7 \cdot 0^2 - 4 \cdot 0 - 9$

$\qquad = 0 + 0 + 0 - 9 = -9$

35. We evaluate the function for $t = 1.32$.

$s(t) = 16t^2$

$s(1.32) = 16(1.32)^2 \approx 28$

He dropped about 28 ft.

37. $N(p) = p^3 - 3p^2 + 2p$

$N(20) = 20^3 - 3 \cdot 20^2 + 2 \cdot 20$

$\qquad = 8000 - 1200 + 40$

$\qquad = 6840$

A president, vice president, and treasurer can be elected in 6840 ways.

39. Evaluate the polynomial function for $v = 180$.

$h(v) = \dfrac{0.354}{8250} v^3$

$h(180) = \dfrac{0.354}{8250} \cdot (180)^3 \approx 250$

The race car needs about 250 horsepower.

41. Locate 10 on the horizontal axis. From there, move vertically to the function and then horizontally to the vertical axis. This locates a value of about 20 W.

43. To approximate $P(20)$, locate 20 on the horizontal axis. From there, move vertically to the function and then horizontally to the vertical axis. This locates the value of about 150. Thus, $P(20) \approx 150$.

45. Using this function, we find $N(3)$.

$$N(x) = \frac{1}{3}x^3 + \frac{1}{2}x^2 + \frac{1}{6}x$$
$$N(3) = \frac{1}{3}\cdot 3^3 + \frac{1}{2}\cdot 3^2 + \frac{1}{6}\cdot 3$$
$$= \frac{1}{3}\cdot 27 + \frac{1}{2}\cdot 9 + \frac{1}{6}\cdot 3$$
$$= 9 + \frac{9}{2} + \frac{1}{2} = 14$$

From the figure we see that the bottom layer has 9 spheres, the second layer has 4, and the third layer has 1. Thus there are 9 + 4 + 1, or 14 spheres.

Using either the function or the figure, we find that $N(3) = 14$.

To calculate the number of oranges in a pyramid with 5 layers, we evaluate the function for $x = 5$.

$$N(5) = \frac{1}{3}\cdot 5^3 + \frac{1}{2}\cdot 5^2 + \frac{1}{6}\cdot 5$$
$$= \frac{1}{3}\cdot 125 + \frac{1}{2}\cdot 25 + \frac{1}{6}\cdot 5$$
$$= \frac{125}{3} + \frac{25}{2} + \frac{5}{6}$$
$$= \frac{250}{6} + \frac{75}{6} + \frac{5}{6} = \frac{330}{6}$$
$$= 55 \text{ oranges}$$

47. Locate 2 on the horizontal axis. From there, move vertically to the graph and then horizontally to the $C(t)$-axis. This locates a value of about 2.3. Thus, about 2.3 mcg/mL of Gentamicin is in the bloodstream 2 hr after injection.

49. Locate 2 on the horizontal axis. From there, move vertically to the graph and then horizontally to the $C(t)$-axis. This locates a value of about 2.3. Thus, $C(2) \approx 2.3$.

51. We evaluate the polynomial for $h = 6.3$ and $r = 1.2$:

$$2\pi rh + 2\pi r^2 = 2\pi(1.2)(6.3) + 2\pi(1.2)^2 \approx 56.5$$

The surface area is about 56.5 in^2.

53. Evaluate the polynomial function for $x = 75$:

$$R(x) = 280x - 0.4x^2$$
$$R(75) = 280\cdot 75 - 0.4(75)^2$$
$$= 21,000 - 0.4(5625)$$
$$= 21,000 - 2250 = 18,750$$

The total revenue is $18,750.

55. Evaluate the polynomial function for $x = 75$:

$$C(x) = 5000 + 0.6x^2$$
$$C(75) = 5000 + 0.6(75)^2$$
$$= 5000 + 0.6(5625)$$
$$= 5000 + 3375$$
$$= 8375$$

The total cost is $8375.

57. $8x + 2 - 5x + 3x^3 - 4x - 1$
$$= 3x^3 + (8 - 5 - 4)x + 2 - 1$$
$$= 3x^3 - x + 1$$

59. $3a^2b + 4b^2 - 9a^2b - 7b^2$
$$= (3 - 9)a^2b + (4 - 7)b^2$$
$$= -6a^2b - 3b^2$$

61. $9x^2 - 3xy + 12y^2 + x^2 - y^2 + 5xy + 4y^2$
$$= (9 + 1)x^2 + (-3 + 5)xy + (12 - 1 + 4)y^2$$
$$= 10x^2 + 2xy + 15y^2$$

63. $\left(5t^4 - 2t^3 + t\right) + \left(-t^4 - t^3 + 6t^2\right)$
$$= (5 - 1)t^4 + (-2 - 1)t^3 + 6t^2 + t$$
$$= 4t^4 - 3t^3 + 6t^2 + t$$

65. $(x^2 + 2x - 3xy - 7) + (-3x^2 - x + 2y^2 + 6)$
$$= (1 - 3)x^2 + (2 - 1)x - 3xy + 2y^2 + (-7 + 6)$$
$$= -2x^2 + x - 3xy + 2y^2 - 1$$

67. $(8x^2y - 3xy^2 + 4xy) + (-2x^2y - xy^2 + xy)$
$$= (8 - 2)x^2y + (-3 - 1)xy^2 + (4 + 1)xy$$
$$= 6x^2y - 4xy^2 + 5xy$$

69. $(2r^2 + 12r - 11) + (6r^2 - 2r + 4) + (r^2 - r - 2)$
$$= (2 + 6 + 1)r^2 + (12 - 2 - 1)r + (-11 + 4 - 2)$$
$$= 9r^2 + 9r - 9$$

71. $\left(\frac{1}{8}xy - \frac{3}{5}x^3y^2 + 4.3y^3\right) + \left(-\frac{1}{3}xy - \frac{3}{4}x^3y^2 - 2.9y^3\right)$
$$= \left(\frac{1}{8} - \frac{1}{3}\right)xy + \left(-\frac{3}{5} - \frac{3}{4}\right)x^3y^2 + (4.3 - 2.9)y^3$$
$$= \left(\frac{3}{24} - \frac{8}{24}\right)xy + \left(-\frac{12}{20} - \frac{15}{20}\right)x^3y^2 + 1.4y^3$$
$$= -\frac{5}{24}xy - \frac{27}{20}x^3y^2 + 1.4y^3$$

73. $3t^4 + 8t^2 - 7t - 1$

a) $-\left(3t^4 + 8t^2 - 7t - 1\right)$ Writing the opposite of P as $-P$

b) $-3t^4 - 8t^2 + 7t + 1$ Changing the time of every term

75. $-12y^5 + 4ay^4 - 7by^2$

a) $-(-12y^5 + 4ay^4 - 7by^2)$

b) $12y^5 - 4ay^4 + 7by^2$

77. $(4x - 6) - (-3x + 2)$
$$= (4x - 6) + (3x - 2)$$
$$= 7x - 8$$

79. $(-3x^2 + 2x + 9) - (x^2 + 5x - 4)$
$$= (-3x^2 + 2x + 9) + (-x^2 - 5x + 4)$$
$$= -4x^2 - 3x + 13$$

81. $(8a - 3b + c) - (2a + 3b - 4c)$
$= (8a - 3b + c) + (-2a - 3b + 4c)$
$= 6a - 6b + 5c$

83. $(6a^2 + 5ab - 4b^2) - (8a^2 - 7ab + 3b^2)$
$= (6a^2 + 5ab - 4b^2) + (-8a^2 + 7ab - 3b^2)$
$= -2a^2 + 12ab - 7b^2$

85. $(6ab - 4a^2b + 6ab^2) - (3ab^2 - 10ab - 12a^2b)$
$= (6ab - 4a^2b + 6ab^2) + (-3ab^2 + 10ab + 12a^2b)$
$= 8a^2b + 16ab + 3ab^2$

86. $17xy + 5x^2y^2 - 7y^3$

87. $\left(\frac{5}{8}x^4 - \frac{1}{4}x^2 - \frac{1}{2}\right) - \left(-\frac{3}{8}x^4 + \frac{3}{4}x^2 + \frac{1}{2}\right)$
$= \left(\frac{5}{8}x^4 - \frac{1}{4}x^2 - \frac{1}{2}\right) + \left(\frac{3}{8}x^4 - \frac{3}{4}x^2 - \frac{1}{2}\right)$
$= x^4 - x^2 - 1$

89. $(6t^2 + 7) - (2t^2 + 3) + (t^2 + t)$
$= (6t^2 + 7) + (-2t^2 - 3) + t^2 + t$
$= (6 - 2 + 1)t^2 + t + 7 - 3$
$= 5t^2 + t + 4$

91. $(8r^2 - 6r) - (2r - 6) + (5r^2 - 7)$
$= (8r^2 - 6r) + (-2r + 6) + (5r^2 - 7)$
$= (8 + 5)r^2 + (-6 - 2)r + (6 - 7)$
$= 13r^2 - 8r - 1$

93. $(x^2 - 4x + 7) + (3x^2 - 9) - (x^2 - 4x + 7)$

Note that $x^2 - 4x + 7$ and $-(x^2 - 4x + 7)$ are opposites so their sum is 0. Then the result is $3x^2 - 9$.

95. $P(x) = R(x) - C(x)$
$P(x) = (280x - 0.4x^2) - (5000 + 0.6x^2)$
$P(x) = (280x - 0.4x^2) + (-5000 - 0.6x^2)$
$P(x) = 280x - x^2 - 5000$
$P(70) = 280 \cdot 70 - 70^2 - 5000$
$ = 19,600 - 4900 - 5000 = 9700$

The profit is $9700.

97. *Writing Exercise.* No; if the coefficients of at least one pair of like terms are opposites, then the sum is a monomial. For example, $(2x + 3) + (-2x + 1) = 4$, a monomial.

99. $x^5 \cdot x^3 = x^{5+3} = x^8$

101. $(a^2b^3)(a^4b) = a^{2+4}b^{3+1} = a^6b^4$

103. $(5y^3)^2 = 5^2y^{3\cdot2} = 25y^6$

105. *Writing Exercise.*
$P(x) = 0.0157x^3 + 0.1163x^2 - 1.3396x + 3.7063$
$P(20) = 0.0157(20)^3 + 0.1163(20)^2 - 1.3396(20) + 3.7063$
$ = 125.6 + 46.52 - 26.792 + 3.7063$
$ = 149.0343$
$P(10) = 0.0157(10)^3 + 0.1163(10)^2 - 1.3396(10) + 3.7063$
$ = 15.7 + 11.63 - 13.396 + 3.7063$
$ = 17.6403$
$\dfrac{P(20) - P(10)}{20 - 10} = \dfrac{149.0343 - 17.6403}{20 - 10} = 13.1394$

This number gives the slope of the line from the point $(10,\ P(10))$ to the point $(20,\ P(20))$. In this application it gives the average rate of change in power when the wind speed increases from 10 mph to 20 mph.

107. $2[P(x)] = 2(13x^5 - 22x^4 - 36x^3 + 40x^2 - 16x + 75)$
$ = 26x^5 - 44x^4 - 72x^3 + 80x^2 - 32x + 150$

Use columns to add:

$$26x^5 - 44x^4 - 72x^3 + 80x^2 - 32x + 150$$
$$\underline{42x^5 - 37x^4 + 50x^3 - 28x^2 + 34x + 100}$$
$$68x^5 - 81x^4 - 22x^3 + 52x^2 + 2x + 250$$

109. $2[Q(x)] = 2(42x^5 - 37x^4 + 50x^3 - 28x^2 + 34x + 100)$
$ = 84x^5 - 74x^4 + 100x^3 - 56x^2 + 68x + 200$
$3[P(x)] = 3(13x^5 - 22x^4 - 36x^3 + 40x^2 - 16x + 75)$
$ = 39x^5 - 66x^4 - 108x^3 + 120x^2 - 48x + 225$

Use columns to subtract, adding the opposite of $3[P(x)]$:

$$84x^5 - 74x^4 + 100x^3 - 56x^2 + 68x + 200$$
$$\underline{-39x^5 + 66x^4 + 108x^3 - 120x^2 + 48x - 225}$$
$$45x^5 - 8x^4 + 208x^3 - 176x^2 + 116x - 25$$

111. First we find the number of truffles in the display.

$N(x) = \frac{1}{6}x^3 + \frac{1}{2}x^2 + \frac{1}{3}x$
$N(5) = \frac{1}{6} \cdot 5^3 + \frac{1}{2} \cdot 5^2 + \frac{1}{3} \cdot 5$
$ = \frac{1}{6} \cdot 125 + \frac{1}{2} \cdot 25 + \frac{5}{3}$
$ = \frac{125}{6} + \frac{25}{2} + \frac{5}{3}$
$ = \frac{125}{6} + \frac{75}{6} + \frac{10}{6}$
$ = \frac{210}{6} = 35$

There are 35 truffles in the display. Now find the volume of one truffle. Each truffle's diameter is 3 cm, so the radius is $\frac{3}{2}$, or 1.5 cm.

$V(r) = \frac{4}{3}\pi r^3$

$V(1.5) \approx \frac{4}{3}(3.14)(1.5)^3 \approx 14.13$ cm^3

Finally, multiply the number of truffles and the volume of a truffle to find the total volume of chocolate.

$35(14.13 \text{ cm}^3) = 494.55 \text{ cm}^3$

The display contains about 494.55 cm^3 of chocolate.

113. The area of the base is $x \cdot x$, or x^2.

The area of each side is $x \cdot (x-2)$.

The total area of all four sides is $4x(x-2)$.

The surface area of this box can be expressed as a polynomial function.

$$S(x) = x^2 + 4x(x-2)$$
$$= x^2 + 4x^2 - 8x$$
$$= 5x^2 - 8x$$

115. $(2x^{2a} + 4x^a + 3) + (6x^{2a} + 3x^a + 4)$
$$= (2+6)x^{2a} + (4+3)x^a + (3+4)$$
$$= 8x^{2a} + 7x^a + 7$$

117. $(2x^{5b} + 4x^{4b} + 3x^{3b} + 8) - (x^{5b} + 2x^{3b} + 6x^{2b} + 9x^b + 8)$
$$= (2-1)x^{5b} + 4x^{4b} + (3-2)x^{3b} - 6x^{2b} - 9x^b + (8-8)$$
$$= x^{5b} + 4x^{4b} + x^{3b} - 6x^{2b} - 9x^b$$

119. *Graphing Calculator Exercise*

Exercise Set 5.2

1. False; the coefficient of $3x^5$ is 3.

3. True; see Example 2.

5. False; FOIL is intended to be used to multiply two binomials.

7. True; see page 293 in the text.

9. $3x^4 \cdot 5x = (3 \cdot 5)(x^4 \cdot x) = 15x^5$

11. $6a^2(-8ab^2) = 6(-8)(a^2 \cdot a)b^2 = -48a^3b^2$

13. $(-4x^3y^2)(-9x^2y^4) = (-4)(-9)(x^3 \cdot x^2)(y^2 \cdot y^4)$
$$= 36x^5y^6$$

15. $7x(3-x) = 7x \cdot 3 - 7x \cdot x$
$$= 21x - 7x^2$$

17. $5cd(4c^2d - 5cd^2)$
$$= 5cd \cdot 4c^2d - 5cd \cdot 5cd^2$$
$$= 20c^3d^2 - 25c^2d^3$$

19. $(x+3)(x+5)$
$$= x^2 + 5x + 3x + 15 \quad \text{FOIL}$$
$$= x^2 + 8x + 15$$

21. $(2a+3)(4a-1)$
$$= 8a^2 - 2a + 12a - 3 \quad \text{FOIL}$$
$$= 8a^2 + 10a - 3$$

23. $(x+2)(x^2 - 3x + 1)$
$$= x(x^2 - 3x + 1) + 2(x^2 - 3x + 1)$$
$$= x^3 - 3x^2 + x + 2x^2 - 6x + 2$$
$$= x^3 - x^2 - 5x + 2$$

25. $(t-5)(t^2 + 2t - 3)$
$$= t(t^2 + 2t - 3) - 5(t^2 + 2t - 3)$$
$$= t^3 + 2t^2 - 3t - 5t^2 - 10t + 15$$
$$= t^3 - 3t^2 - 13t + 15$$

27.

$$
\begin{array}{ll}
\quad\quad a^2 + \ a - 1 & \\
\quad\quad a^2 + 4a - 5 & \\
\hline
\quad -5a^2 - 5a + 5 & \text{Multiplying by } -5 \\
4a^3 + \ 4a^2 - 4a & \text{Multiplying by } 4a \\
a^4 + \ a^3 - \ a^2 & \text{Multiplying by } a^2 \\
\hline
a^4 + 5a^3 - \ 2a^2 - 9a + 5 & \text{Adding}
\end{array}
$$

29. $(x+3)(x^2 - 3x + 9)$
$$= (x+3)(x^2) + (x+3)(-3x) + (x+3)(9)$$
$$= x^3 + 3x^2 - 3x^2 - 9x + 9x + 27$$
$$= x^3 + 27$$

31. $(a-b)(a^2 + ab + b^2)$
$$= (a-b)(a^2) + (a-b)(ab) + (a-b)(b^2)$$
$$= a^3 - a^2b + a^2b - ab^2 + ab^2 - b^3$$
$$= a^3 - b^3$$

33. $(t-3)(t+2)$
$$= t^2 + 2t - 3t - 6$$
$$= t^2 - t - 6$$

35. $(5x+2y)(4x+y)$
$$= 20x^2 + 5xy + 8xy + 2y^2$$
$$= 20x^2 + 13xy + 2y^2$$

37. $\left(t - \frac{1}{3}\right)\left(t - \frac{1}{4}\right)$
$$= t^2 - \frac{1}{4}t - \frac{1}{3}t + \frac{1}{12} \quad \text{FOIL}$$
$$= t^2 - \frac{3}{12}t - \frac{4}{12}t + \frac{1}{12}$$
$$= t^2 - \frac{7}{12}t + \frac{1}{12}$$

39. $(1.2t + 3s)(2.5t - 5s)$
$$= 3t^2 - 6st + 7.5st - 15s^2 \quad \text{FOIL}$$
$$= 3t^2 + 1.5st - 15s^2$$

41. $(r+3)(r+2)(r-1)$
$= (r^2 + 2r + 3r + 6)(r-1)$ FOIL
$= (r^2 + 5r + 6)(r-1)$
$= (r^2 + 5r + 6) \cdot r + (r^2 + 5r + 6)(-1)$
$= r^3 + 5r^2 + 6r - r^2 - 5r - 6$
$= r^3 + 4r^2 + r - 6$

43. $(x+5)^2$
$= x^2 + 2 \cdot x \cdot 5 + 5^2$ $(A+B)^2 = A^2 + 2AB + B^2$
$= x^2 + 10x + 25$

45. $(2y-7)^2$
$= (2y)^2 - 2 \cdot 2y \cdot 7 + 7^2$ $(A-B)^2 = A^2 - 2AB + B^2$
$= 4y^2 - 28y + 49$

47. $(5c-2d)^2$
$= (5c)^2 - 2 \cdot 5c \cdot 2d + (2d)^2$
     $(A-B)^2 = A^2 - 2AB + B^2$
$= 25c^2 - 20cd + 4d^2$

49. $\left(3a^3 - 10b^2\right)^2$
$= \left(3a^3\right)^2 - 2 \cdot 3a^3 \cdot 10b^2 + \left(10b^2\right)^2$
$= 9a^6 - 60a^3b^2 + 100b^4$

51. $(x^3y^4 + 5)^2$
$= (x^3y^4)^2 + 2 \cdot x^3y^4 \cdot 5 + 5^2$
     $(A+B)^2 = A^2 + 2AB + B^2$
$= x^6y^8 + 10x^3y^4 + 25$

53. $P(x) \cdot Q(x) = (3x^2 - 5)(4x^2 - 7x + 1)$
$= (3x^2 - 5)(4x^2) + (3x^2 - 5)(-7x) + (3x^2 - 5)(1)$
$= 12x^4 - 20x^2 - 21x^3 + 35x + 3x^2 - 5$
$= 12x^4 - 21x^3 - 17x^2 + 35x - 5$

55. $P(x) \cdot P(x)$
$= (5x-2)(5x-2)$
$= (5x)^2 - 2 \cdot 5x \cdot 2 + 2^2$
     $(A-B)^2 = A^2 - 2AB + B^2$
$= 25x^2 - 20x + 4$

57. $[F(x)]^2$
$= \left(2x - \frac{1}{3}\right)^2$
$= (2x)^2 - 2 \cdot 2x \cdot \frac{1}{3} + \left(\frac{1}{3}\right)^2$
     $(A-B)^2 = A^2 - 2AB + B^2$
$= 4x^2 - \frac{4}{3}x + \frac{1}{9}$

59. $(c+7)(c-7)$
$= c^2 - 7^2$ $(A+B)(A-B) = A^2 - B^2$
$= c^2 - 49$

61. $(1-4x)(1+4x)$
$= 1^2 - (4x)^2$ $(A+B)(A-B) = A^2 - B^2$
$= 1 - 16x^2$

63. $\left(3m - \frac{1}{2}n\right)\left(3m + \frac{1}{2}n\right)$
$= (3m)^2 - \left(\frac{1}{2}n\right)^2$ $(A+B)(A-B) = A^2 - B^2$
$= 9m^2 - \frac{1}{4}n^2$

65. $(x^3 + yz)(x^3 - yz)$
$= \left(x^3\right)^2 - (yz)^2$ $(A+B)(A-B) = A^2 - B^2$
$= x^6 - y^2z^2$

67. $\left(-mn + 3m^2\right)\left(mn + 3m^2\right)$
$= \left(3m^2 - mn\right)\left(3m^2 + mn\right)$
$= \left(3m^2\right)^2 - (mn)^2$ $(A+B)(A-B) = A^2 - B^2$
$= 9m^4 - m^2n^2$, or $-m^2n^2 + 9m^4$

69. $(x+7)^2 - (x+3)(x-3)$
$= x^2 + 2 \cdot x \cdot 7 + 7^2 - (x^2 - 3^2)$
$= x^2 + 14x + 49 - (x^2 - 9)$
$= x^2 + 14x + 49 - x^2 + 9$
$= 14x + 58$

71. $(2m-n)(2m+n) - (m-2n)^2$
$= \left[(2m)^2 - n^2\right] - \left[m^2 - 2 \cdot m \cdot 2n + (2n)^2\right]$
$= 4m^2 - n^2 - (m^2 - 4mn + 4n^2)$
$= 4m^2 - n^2 - m^2 + 4mn - 4n^2$
$= 3m^2 + 4mn - 5n^2$

73. $(a+b+1)(a+b-1)$
$= [(a+b)+1][(a+b)-1]$
$= (a+b)^2 - 1^2$
$= a^2 + 2ab + b^2 - 1$

75. $(2x+3y+4)(2x+3y-4)$
$= [(2x+3y)+4][(2x+3y)-4]$
$= (2x+3y)^2 - 4^2$
$= 4x^2 + 12xy + 9y^2 - 16$

77. $A = P(1+r)^2$
$A = P(1 + 2r + r^2)$
$A = P + 2Pr + Pr^2$

79. a)   Replace x with $t-1$.
$f(t-1) = (t-1)^2 + 5$
$= t^2 - 2t + 1 + 5$
$= t^2 - 2t + 6$

b) $f(a+h)-f(a)$

$$=[(a+h)^2+5]-(a^2+5)$$
$$=a^2+2ah+h^2+5-a^2-5$$
$$=2ah+h^2$$

c) $f(a)-f(a-h)$

$$=(a^2+5)-[(a-h)^2+5]$$
$$=a^2+5-(a^2-2ah+h^2+5)$$
$$=a^2+5-a^2+2ah-h^2-5$$
$$=2ah-h^2$$

81. a) $f(a)+f(-a)$

$$=(a^2+a)+\left[(-a)^2+(-a)\right]$$
$$=a^2+a+a^2-a$$
$$=2a^2$$

b) $f(a+h)$

$$=(a+h)^2+(a+h)$$
$$=a^2+2ah+h^2+a+h$$

c) $f(a+h)-f(a)$

$$=(a+h)^2+(a+h)-(a^2+a)$$
$$=a^2+2ah+h^2+a+h-a^2-a$$
$$=2ah+h^2+h$$

83. *Writing Exercise.* Choose $x+5$ and $x-5$ since x^2-25 is of the form x^2-5^2.

85. $5x+15y-5=5(x+3y-1)$

87. $16t-64=16(t-4)$

89. $ax+bx-cx=x(a+b-c)$

91. *Writing Exercise.* $95\cdot105=(100-5)(100+5)=100^2-5^2$
$$=10,000-25=9975$$

93. $\left(x^2+y^n\right)\left(x^2-y^n\right)=\left(x^2\right)^2-\left(y^n\right)^2=x^4-y^{2n}$

95. $x^2y^3(5x^n+4y^n)=x^2y^3\cdot5x^n+x^2y^3\cdot4y^n$
$$=5x^{n+2}y^3+4x^2y^{n+3}$$

97. $(x^n-4)(x^{2n}+3x^n-2)$
$$=x^n(x^{2n}+3x^n-2)-4(x^{2n}+3x^n-2)$$
$$=x^{3n}+3x^{2n}-2x^n-4x^{2n}-12x^n+8$$
$$=x^{3n}-x^{2n}-14x^n+8$$

99. $(a-b+c-d)(a+b+c+d)$
$$=[(a+c)-(b+d)][(a+c)+(b+d)]$$
$$=(a+c)^2-(b+d)^2$$
$$=(a^2+2ac+c^2)-(b^2+2bd+d^2)$$
$$=a^2+2ac+c^2-b^2-2bd-d^2$$

101. $(x^2-3x+5)(x^2+3x+5)$
$$=[(x^2+5)-3x][(x^2+5)+3x]$$
$$=(x^2+5)^2-(3x)^2$$
$$=x^4+10x^2+25-9x^2$$
$$=x^4+x^2+25$$

103. $(x-1)(x^2+x+1)(x^3+1)$
$$=(x^3+x^2+x-x^2-x-1)(x^3+1)$$
$$=(x^3-1)(x^3+1)$$
$$=x^6-1$$

105. $\left(x^{a-b}\right)^{a+b}=x^{(a-b)(a+b)}=x^{a^2-b^2}$

107. $(x-a)(x-b)(x-c)\cdots(x-z)$
$$=(x-a)(x-b)\cdots(x-x)(x-y)(x-z)$$
$$=(x-a)(x-b)\cdots0\cdot(x-y)(x-z)$$
$$=0$$

109. $\dfrac{g(a+h)-g(a)}{h}$
$$=\dfrac{(a+h)^2-9-(a^2-9)}{h}$$
$$=\dfrac{a^2+2ah+h^2-9-a^2+9}{h}$$
$$=\dfrac{2ah+h^2}{h}=\dfrac{h(2a+h)}{h}$$
$$=2a+h$$

111. $(A-B)^2=A^2-2AB+B^2$

113. One method is as follows. For each equation, let y_1 represent the left-hand side and y_2 represent the right-hand side, and let $y_3=y_2-y_1$. Then use a graphing calculator to view the graph of y_3 and/or a table of values for y_3. If $y_3=0$, the equation is an identity. If $y_3\neq0$, the equation is not an identity.

a) Not an identity

b) Identity

c) Identity

d) Not an identity

e) Not an identity

Exercise Set 5.3

1. True; see page 300 in the text.

3. True; see page 300 in the text.

5. True; see page 301 in the text.

7. True; $-(a-b)=-a+b=b-a;\ -1(a-b)=-a+b=b-a.$

9. $10x^2+35$
$=5\cdot2x^2+5\cdot7$
$=5(2x^2+7)$

11. $2y^2-18y$
$=2y\cdot y-2y\cdot9$
$=2y(y-9)$

13. $5t^3-15t+5$
$=5\cdot t^3-5\cdot3t+5\cdot1$
$=5(t^3-3t+1)$

15. $a^6+2a^4-a^3$
$=a^3\cdot a^3+a^3\cdot2a-a^3\cdot1$
$=a^3(a^3+2a-1)$

17. $12x^4-30x^3+42x$
$=6x\cdot2x^3-6x\cdot5x^2+6x\cdot7$
$=6x(2x^3-5x^2+7)$

19. $6a^2b-2ab-9b$
$=b\cdot6a^2-b\cdot2a-b\cdot9$
$=b(6a^2-2a-9)$

21. $15m^4n+30m^5n^2+25m^3n^3$
$=5m^3n\cdot3m+5m^3n\cdot6m^2n+5m^3n\cdot5n^2$
$=5m^3n(3m+6m^2n+5n^2)$

23. $9x^3y^6z^2-12x^4y^4z^4+15x^2y^5z^3$
$=3x^2y^4z^2\cdot3xy^2-3x^2y^4z^2\cdot4x^2z^2+3x^2y^4z^2\cdot5yz$
$=3x^2y^4z^2(3xy^2-4x^2z^2+5yz)$

25. $-5x-40=-5(x+8)$

27. $-16t^2+96=-16(t^2-6)$

29. $-2x^2+12x+40=-2(x^2-6x-20)$

31. $5-10y=-5(-1+2y),$ or $-5(2y-1)$

33. $8d^2-12cd=-4d(-2d+3c),$ or $-4d(3c-2d)$

35. $-m^3+8=-1(m^3-8)$

37. $-p^3-2p^2-5p+2=-1(p^3+2p^2+5p-2)$

39. $a(b-5)+c(b-5)=(b-5)(a+c)$

41. $(x+7)(x-1)+(x+7)(x-2)$
$=(x+7)(x-1+x-2)$
$=(x+7)(2x-3)$

43. $a^2(x-y)+5(y-x)$
$=a^2(x-y)+5(-1)(x-y)$ Factoring out -1
$\qquad\qquad\qquad\qquad$ to reverse the second subtraction
$=a^2(x-y)-5(x-y)$ $\qquad$ Simplifying
$=(x-y)(a^2-5)$

45. $xy+xz+wy+wz$
$=x(y+z)+w(y+z)$
$=(y+z)(x+w)$

47. y^3-y^2+3y-3
$=y^2(y-1)+3(y-1)$
$=(y-1)(y^2+3)$

49. $t^3+6t^2-2t-12$
$=t^2(t+6)-2(t+6)$
$=(t+6)(t^2-2)$

51. $12a^4-21a^3-9a^2$
$=3a^2\cdot4a^2-3a^2\cdot7a-3a^2\cdot3$
$=3a^2(4a^2-7a-3)$

53. $y^8-1-y^7+y=y^8-y^7+y-1$
$=y^7(y-1)+1(y-1)$
$=(y-1)(y^7+1)$

55. $2xy+3x-x^2y-6=2xy-x^2y-6+3x$
$=xy(2-x)-3(2-x)$
$=(2-x)(xy-3)$

57. a) $h(t)=-16t^2+72t$
$h(t)=-8t(2t-9)$

    b) Using $h(t)=-16t^2+72t$:

    $h(1)=-16\cdot1^2+72\cdot1=-16\cdot1+72$
    $\quad=-16+72=56$ ft

    Using $h(t)=-8t(2t-9)$:

    $h(1)=-8(1)(2\cdot1-9)=-8(1)(-7)=56$ ft

    The expressions have the same value for $t=1$, so the factorization is probably correct.

59. $2\pi rh+\pi r^2=\pi r(2h+r)$

61. $P(t)=t^2-5t$
$P(t)=t(t-5)$

63. $R(x)=280x-0.4x^2$
$R(x)=0.4x(700-x)$

65. $P(n) = \frac{1}{2}n^2 - \frac{3}{2}n$

$P(n) = \frac{1}{2}(n^2 - 3n)$

67. $N(x) = \frac{1}{6}x^3 + \frac{1}{2}x^2 + \frac{1}{3}x$

$N(x) = \frac{1}{6}(x^3 + 3x^2 + 2x)$ Factoring out $\frac{1}{6}$

69. *Writing Exercise.* The prime factorization of a polynomial is the factorization into a product of relatively prime polynomial factors. Similarly, the prime factorization of a number is the factorization into a product of its prime factors.

71. Graph $f(x) = -\frac{1}{2}x + 3$.

Slope is $-\frac{1}{2}$; y-intercept is $(0, 3)$

From the y-intercept we go down 1 unit and to the right 2 units. This gives us $(2, 2)$. We can now draw the graph.

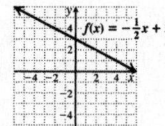

73. Graph $y - 1 = 2(x + 3)$.

From the point-slope equation, $m = 2$, and a point is $(-3, 1)$.

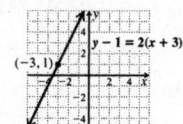

75. Graph $6x = 3$.

Since y does not appear, we solve for x.

$6x = 3$

$x = \frac{1}{2}$

This is a vertical line that crosses the x-axis at $\left(\frac{1}{2}, 0\right)$.

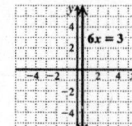

77. *Writing Exercise.* When coefficients and/or exponents are large, a polynomial is more easily evaluated after it has been factored.

79. $x^5y^4 + \quad = x^3y(\quad + xy^5)$

The term that goes in the first blank is the product of x^3y and xy^5, or x^4y^6.

The term that goes in the second blank is the expression

that is multiplied with x^3y to obtain x^5y^4, or x^2y^3.

Thus, we have $x^5y^4 + x^4y^6 = x^3y(x^2y^3 + xy^5)$.

81. $rx^2 - rx + 5r + sx^2 - sx + 5s$

$= r(x^2 - x + 5) + s(x^2 - x + 5)$

$= (x^2 - x + 5)(r + s)$

83. $a^4x^4 + a^4x^2 + 5a^4 + a^2x^4 + a^2x^2 + 5a^2 + 5x^4 + 5x^2 + 25$

$= a^4(x^4 + x^2 + 5) + a^2(x^4 + x^2 + 5) + 5(x^4 + x^2 + 5)$

$= (x^4 + x^2 + 5)(a^4 + a^2 + 5)$

85. $x^{-6} + x^{-9} + x^{-3}$

$= x^{-9} \cdot x^3 + x^{-9} \cdot 1 + x^{-9} \cdot x^6$

$= x^{-9}(x^3 + 1 + x^6)$

87. $x^{1/3} - 5x^{1/2} + 3x^{3/4}$

$= x^{4/12} - 5x^{6/12} + 3x^{9/12}$

$= x^{4/12}(1 - 5x^{2/12} + 3x^{5/12})$

$= x^{1/3}(1 - 5x^{1/6} + 3x^{5/12})$

89. $x^{-5/2} + x^{-3/2}$

$= x^{-5/2} \cdot 1 + x^{-5/2} \cdot x$

$= x^{-5/2}(1 + x)$

91. $x^{-4/5} - x^{-7/5} + x^{-1/3}$

$= x^{-7/5} \cdot x^{3/5} - x^{-7/5} \cdot 1 + x^{-7/5} \cdot x^{16/15}$

$= x^{-7/5}(x^{3/5} - 1 + x^{16/15})$

93. $3a^{n+1} + 6a^n - 15a^{n+2}$

$= 3a^n \cdot a + 3a^n \cdot 2 - 3a^n(5a^2)$

$= 3a^n(a + 2 - 5a^2)$

95. $7y^{2a+b} - 5y^{a+b} + 3y^{a+2b}$

$= y^{a+b} \cdot 7y^a - y^{a+b}(5) + y^{a+b} \cdot 3y^b$

$= y^{a+b}(7y^a - 5 + 3y^b)$

97. One method is to let $y_1 = \left(x^2 - 3x + 2\right)^4$ and let $y_2 = x^8 + 81x^4 + 16$. Then use a table to show that $y_1 = y_2$ for all values of x.

Connecting the Concepts

1. $\left(4t^3 - 2t + 6\right) + \left(8t^2 - 11t - 7\right)$

$= 4t^3 + 8t^2 - 2t - 11t + 6 - 7$

$= 4t^3 + 8t^2 - 13t - 1$

3. $4x^2y\left(3xy - 2x^3 + 6y^2\right) = 12x^3y^2 - 8x^5y + 24x^2y^3$

5. $\left(8n^2 + 5n - 2\right) - \left(-n^2 + 6n - 2\right)$

$= 8n^2 + 5n - 2 + n^2 - 6n + 2$

$= 9n^2 - n$

7. $\left(2x^2 - 5xy + y^2\right) + \left(-x^2 + 5xy - 9y^2\right)$
$= 2x^2 - 5xy + y^2 - x^2 + 5xy - 9y^2$
$= x^2 - 8y^2$

9. $(x+1)(x+7) = x^2 + 7x + x + 7 = x^2 + 8x + 7$

11. $p^3 - p^2 + 7p - 7$
$= p^2(p-1) + 7(p-1)$
$= (p-1)\left(p^2 + 7\right)$

13. $(3m-10)^2 = (3m-10)(3m-10)$
$= 9m^2 - 30m - 30m + 100$
$= 9m^2 - 60m + 100$

15. $(a+2)\left(a^2 - a - 6\right)$
$= a^3 - a^2 - 6a + 2a^2 - 2a - 12$
$= a^3 + a^2 - 8a - 12$

17. $(c+9)(c-9) = c^2 - 81$

19. $8x^2y^3z + 12x^3y^2 - 16x^2yz^3 = 4x^2y\left(2y^2z + 3xy - 4z^3\right)$

Exercise Set 5.4

1. True; see page 299 in the text.

3. False; the factors of a prime number c are c and 1.

5. False; whenever the product of a pair of factors is negative, the factors have different signs.

7. True; see Example 8.

9. $x^2 + 5x + 4$

We look for two numbers whose product is 4 and whose sum is 5. Since 4 and 5 are both positive, we need only consider positive factors. The only positive pair is 1 and 4. They are the numbers we need. The factorization is $(x+1)(x+4)$.

11. $y^2 - 12y + 27$

Since the constant term is positive and the coefficient of the middle term is negative, we look for a factorization of 27 in which both factors are negative. Their sum must be -12.

Pair of Factors	Sum of Factors
$-1, -27$	-28
$-3, -9$	-12

The numbers we need are -3 and -9. The factorization is $(y-3)(y-9)$.

13. $t^2 - 2t - 8$

Since the constant term is negative, we look for a factorization of -8 in which one factor is positive and one factor is negative. Their sum must be -2, so the negative factor must have the larger absolute value. Thus we consider only pairs of factors in which the negative factor has the larger absolute value.

Pair of Factors	Sum of Factors
$-8, 1$	-7
$-4, 2$	-2

The numbers we need are -4 and 2. The factorization is $(t-4)(t+2)$.

15. $a^2 + a - 2$

Since the constant term is negative, we look for a factorization of -2 in which one factor is positive and one factor is negative. Their sum must be 1, so the positive factor must have the larger absolute value. The only pair is 2 and -1. The factorization is $(a+2)(a-1)$.

17. $2x^2 + 6x - 108$
$= 2\left(x^2 + 3x - 54\right)$ Removing the common factor

We now factor $x^2 + 3x - 54$. Since the constant term is negative, we look for a factorization of -54 in which one factor is positive and one factor is negative. We consider only pairs of factors in which the positive factor has the larger absolute value, since the sum of the factors, 3, is positive.

Pair of Factors	Sum of Factors
$-1, 54$	53
$-2, 27$	25
$-3, 18$	15
$-6, 9$	3

The numbers we need are -6 and 9.
$$x^2 + 3x - 54 = (x-6)(x+9)$$
We must not forget to include the common factor 2.
$$2x^2 + 6x - 108 = 2(x-6)(x+9)$$

19. $14a + a^2 + 45 = a^2 + 14a + 45$

Since the constant term and the middle term are both positive, we look for a factorization of 45 in which both factors are positive. Their sum must be 14.

Pair of Factors	Sum of Factors
45, 1	46
15, 3	18
9, 5	14

The numbers we need are 9 and 5. The factorization is $(a+9)(a+5)$.

21. $p^3 - p^2 - 72p$

$= p\left(p^2 - p - 72\right)$ Removing the common factor

We now factor $p^2 - p - 72$. Since the constant term is negative, we look for a factorization of -72 in which one factor is positive and one factor is negative. We consider only pairs of factors in which the negative factor has the larger absolute value, since the sum of the factors, -1, is negative.

Pair of Factors	Sum of Factors
-72, 1	-71
-36, 2	-34
-24, 3	-21
-18, 4	-14
-12, 6	-6
-9, 8	-1

The numbers we need are -9 and 8.

$$p^2 - p - 72 = (p - 9)(p + 8)$$

We must not forget to include the common factor p.

$$p^3 - p^2 - 72p = p(p - 9)(p + 8)$$

23. $a^2 - 11a + 28$

Since the constant term is positive and the coefficient of the middle term is negative, we look for a factorization of 28 in which both factors are negative. Their sum must be -11.

Pair of Factors	Sum of Factors
-1, -28	-29
-2, -14	-16
-4, -7	-11

The numbers we need are -4 and -7. The factorization is $(a - 4)(a - 7)$.

25. $x + x^2 - 6 = x^2 + x - 6$

Since the constant term is negative, we look for a factorization of -6 in which one factor is positive and one factor is negative. We consider only pairs of factors in which the positive factor has the larger absolute value, since the sum of the factors, 1, is positive.

Pair of Factors	Sum of Factors
6, -1	5
3, -2	1

The numbers we need are 3 and -2. The factorization is $(x + 3)(x - 2)$.

27. $5y^2 + 40y + 35$

$= 5\left(y^2 + 8y + 7\right)$ Removing the common factor

We now factor $y^2 + 8y + 7$. We look for two numbers whose product is 7 and whose sum is 8. Since 7 and 8 are both positive, we need consider only positive factors. The only possible pair is 1 and 7. They are the numbers we need.

$$y^2 + 8y + 7 = (y + 1)(y + 7)$$

We must not forget to include the common factor 5.

$$5y^2 + 40y + 35 = 5(y + 1)(y + 7)$$

29. $32 + 4y - y^2 = -y^2 + 4y + 32 = -(y^2 - 4y - 32)$

We now factor $y^2 - 4y - 32$. Since the constant term is negative, we look for a factorization of -32 in which one factor is positive and one factor is negative. We consider only pairs of factors in which the negative factor has the larger absolute value, since the sum of the factors, -4, is negative.

Pair of Factors	Sum of Factors
-32, 1	-31
-16, 2	-14
-8, 4	-4

The numbers we need are -8 and 4. Thus, $y^2 - 4y - 32 = (y - 8)(y + 4)$. We must not forget to include the factor that was factored out earlier:

$$32 + 4y - y^2 = -(y - 8)(y + 4),$$

or $(-y + 8)(y + 4)$, or $(8 - y)(4 + y)$

31. $56x + x^2 - x^3$

There is a common factor, x. We also factor out -1 in order to make the leading coefficient positive.

$$56x + x^2 - x^3 = -x(-56 - x + x^2)$$
$$= -x(x^2 - x - 56)$$

Now we factor $x^2 - x - 56$. Since the constant term is negative, we look for a factorization of -56 in which one factor is positive and one factor is negative. We consider only pairs of factors in which the negative factor has the larger absolute value, since the sum of the factors, -1, is negative.

Pair of Factors	Sum of Factors
-56, 1	-55
-28, 2	-26
-14, 4	-10
-8, 7	-1

The numbers we need are -8 and 7. Thus, $x^2 - x - 56 = (x - 8)(x + 7)$. We must not forget to include the common factor:

$$56x + x^2 - x^3 = -x(x - 8)(x + 7),$$

or $x(-x + 8)(x + 7)$, or $x(8 - x)(7 + x)$

33. $y^4 + 5y^3 - 84y^2$

$= y^2(y^2 + 5y - 84)$ Removing the common factor

We now factor $y^2 + 5y - 84$. We look for pairs of factors of -84, one positive and one negative, such that the positive factor has the larger absolute value and the sum of the factors is 5.

Pair of Factors	Sum of Factors
84, −1	83
42, −2	40
28, −3	25
21, −4	17
14, −6	8
12, −7	5

The numbers we need are 12 and −7. Then

$$y^2 + 5y - 84 = (y + 12)(y - 7).$$

We must not forget to include the common factor:

$$y^4 + 5y^3 - 84y^2 = y^2(y + 12)(y - 7)$$

35. $x^2 - 3x + 5$

There are no factors of 5 whose sum is −3. This trinomial is not factorable into binomials with integer coefficients. The polynomial is prime.

37. $x^2 + 12xy + 27y^2$

We look for numbers p and q such that $x^2 + 12xy + 27y^2 = (x + py)(x + qy)$. Our thinking is much the same as if we were factoring $x^2 + 12x + 27$. Since the constant term is positive and the coefficient of the middle term is positive, we look for a factorization of 27 in which both factors are positive. Their sum must be 12.

Pair of Factors	Sum of Factors
1, 27	28
3, 9	12

The numbers we need are 3 and 9. The factorization is $(x + 3y)(x + 9y)$.

39. $x^2 - 14xy + 49y^2$

We look for numbers p and q such that $x^2 - 14xy + 49y^2 = (x + py)(x + qy)$. Our thinking is much the same as if we were factoring $x^2 - 14x + 49$. We look for factors of 49 whose sum is −14. Since the constant term is positive and the coefficient of the middle term is negative, both factors must be negative.

Pair of Factors	Sum of Factors
−49, −1	−50
−7, −7	−14

The numbers we need are −7 and −7. The factorization is $(x - 7y)(x - 7y)$, or $(x - 7y)^2$.

41. $n^5 - 80n^4 + 79n^3$

$= n^3(n^2 - 80n + 79)$ Removing the common factor

Now we factor $n^2 - 80n + 79$. We look for a pair of factors of 79 whose sum is −80. The numbers we need are −1 and −79. Then $n^2 - 80n + 79 = (n - 79)(n - 1)$. We must not forget to include the common factor.

$$n^5 - 80n^4 + 79n^3 = n^3(n - 79)(n - 1)$$

43. $x^6 + 2x^5 - 63x^4$

$= x^4(x^2 + 2x - 63)$ Removing the common factor

We now factor $x^2 + 2x - 63$. We look for a pair of factors of -63 whose sum is 2. The numbers we need are 9 and −7. Then $x^2 + 2x - 63 = (x + 9)(x - 7)$. We must not forget to include the common factor:

$$x^6 + 2x^5 - 63x^4 = x^4(x + 9)(x - 7)$$

45. $3x^2 - 4x - 4$

We will use the FOIL method.

1. There is no common factor (other than 1 or −1.)

2. Factor the first term, $3x^2$. The factors are $3x$, x. We have this possibility:
    $$(3x + \quad)(x + \quad).$$

3. Factor the last term, −4. The possibilities are $4(-1)$, $-4 \cdot 1$, and $2(-2)$.

4. We need factors for which the sum of the products (the "outer" and "inner" parts of FOIL) is the middle term, $-4x$. Try some possibilities and check by multiplying.
    $$(3x + 1)(x - 4) = 3x^2 - 3x - 4$$
    We try again.
    $$(3x + 2)(x - 2) = 3x^2 - 4x - 4$$

The factorization is $(3x + 2)(x - 2)$.

47. $6t^2 + t - 15$

We will use the grouping method.

1. Factor the trinomial $6t^2 + t - 15$. Multiply the leading coefficient, 6, and the constant, −15.
    $$6(-15) = -90$$

2. Try to factor −90 so the sum of the factors is 1. We need only consider pairs of factors in which the positive factor has the larger absolute value, since their sum is positive.

Pair of Factors	Sum of Factors
90, −1	89
45, −2	43
30, −3	27
18, −5	13
15, −6	9
10, −9	1

3. Split the middle term, t, using the results of step (2).

$$t = 10t - 9t$$

4. Factor by grouping.

$$6t^2 + t - 15 = 6t^2 + 10t - 9t - 15$$
$$= 2t(3t+5) - 3(3t+5)$$
$$= (3t+5)(2t-3)$$

49. $6p^2 - 20p + 16$

We will use the FOIL method.

1. Factor out the common factor 2.

$$2(3p^2 - 10p + 8)$$

2. Factor the first term $3p^2$. The factors are $3p$, p. We have this possibility:

$$(3p+\ )(p+\ ).$$

3. Factor the last term, 8. The possibilities are $8 \cdot 1$, $-8(-1)$, $4 \cdot 2$, and $-4(-2)$..

4. Look for factors such that the sum of the products is the middle term, $-10p$. Trial and error leads us to the correct factorization.

$$3p^2 - 10p + 8 = (3p-4)(p-2).$$

We must include the common factor.

$$6p^2 - 20p + 16 = 2(3p-4)(p-2)$$

51. $9a^2 + 18a + 8$

We will use the grouping method.

1. There is no common factor (other than 1 or −1).

2. Multiply the leading coefficient, 9, and the constant, 8: $9(8) = 72$

3. Try to factor 72 so the sum of the factors is 18. We need only consider pairs of positive factors since 72 and 18 are both positive.

Pair of Factors	Sum of Factors
72, 1	73
36, 2	38
24, 3	27
18, 4	22
12, 6	18
9, 8	17

4. Split $18a$ using the results of step (3): $18a = 12a + 6a$

5. Factor by grouping:

$$9a^2 + 18a + 8 = 9a^2 + 12a + 6a + 8$$
$$= 3a(3a+4) + 2(3a+4)$$
$$= (3a+4)(3a+2)$$

53. $8y^2 + 30y^3 - 6y = 30y^3 + 8y^2 - 6y$

We will use the FOIL method.

1. Factor out the common factor $2y$.

$$2y(15y^2 + 4y - 3)$$

2. Now we factor the trinomial $15y^2 + 4y - 3$.

Factor the first term, $15y^2$. The factors are $15y$, y and $5y$, $3y$. We have these possibilities:

$$(15y+\ )(y+\ ) \text{ and } (5y+\ )(3y+\ ).$$

3. Factor the last term, −3. The possibilities are $(1)(-3)$ and $(-1)3$ as well as $(-3)(1)$ and $3(-1)$.

4. Look for factors such that the sum of the products is the middle term, $4y$. Trial and error leads us to the correct factorization.

$$15y^2 + 4y - 3 = (5y+3)(3y-1)$$

We must include the common factor to get a factorization of the original trinomial.

$$8y^2 + 30y^3 - 6y = 2y(5y+3)(3y-1)$$

55. $18x^2 - 24 - 6x = 18x^2 - 6x - 24$

We will use the grouping method.

1. Factor out the common factor, 6:

$$6(3x^2 - x - 4)$$

2. Now we factor the trinomial $3x^2 - x - 4$. Multiply the leading coefficient, 3, and the constant, −4:

$$3(-4) = -12$$

3. Factor −12 so the sum of the factors is −1. We need only consider pairs of factors in which the negative factor has the larger absolute value, since their sum is negative.

Pair of Factors	Sum of Factors
−12, 1	−11
−6, 2	−4
−4, 3	−1

4. Split $-x$ using the results of step (3):

$$-x = -4x + 3x$$

5. Factor by grouping:

$$3x^2 - x - 4 = 3x^2 - 4x + 3x - 4$$
$$= x(3x-4) + (3x-4)$$
$$= (3x-4)(x+1)$$

We must include the common factor to get a

factorization of the original trinomial:

$$18x^2 - 24 - 6x = 6(3x - 4)(x + 1)$$

57. $t^8 + 5t^7 - 14t^6$

$= t^6(t^2 + 5t - 14)$ Removing the common factor

We now factor $t^2 + 5t - 14$. We look for a pair of factors of -14 whose sum is 5. The numbers we need are 7 and -2. Then we have $t^2 + 5t - 14 = (t + 7)(t - 2)$. We must not forget to include the common factor:

$$t^8 + 5t^7 - 14t^6 = t^6(t + 7)(t - 2)$$

59. $70x^4 - 68x^3 + 16x^2$

We will use the grouping method.

1. Factor out the common factor, $2x^2$:

    $2x^2(35x^2 - 34x + 8)$

2. Now we factor the trinomial $35x^2 - 34x + 8$. Multiply the leading coefficient, 35, and the constant, 8: $35 \cdot 8 = 280$

3. Factor 280 so the sum of the factors is -34. We need only consider pairs of negative factors since the sum is negative.

Pair of Factors	Sum of Factors
$-280, \ -1$	-281
$-140, \ -2$	-142
$-70, \ -4$	-74
$-56, \ -5$	-61
$-40, \ -7$	-47
$-35, \ -8$	-43
$-28, \ -10$	-38
$-20, \ -14$	-34

4. Split $-34x$ using the results of step (3):

    $-34x = -20x - 14x$

5. Factor by grouping:

    $35x^2 - 34x + 8 = 35x^2 - 20x - 14x + 8$
    $= 5x(7x - 4) - 2(7x - 4)$
    $= (7x - 4)(5x - 2)$

    We must include the common factor to get a factorization of the original trinomial:

    $70x^4 - 68x^3 + 16x^2 = 2x^2(7x - 4)(5x - 2)$

61. $18y^2 - 9y - 20$

We will use the FOIL method.

1. There is no common factor (other than 1 or -1).

2. Factor the first term, $18y^2$. The possibilities are

    $(18y + \ )(y + \ )$, $(9y + \ )(2y + \ )$ and $(6y + \ )(3y + \ )$.

3. Factor the last term, -20. The possibilities are

    $-20 \cdot 1$, $20(-1)$, $-10 \cdot 2$, $10(-2)$, $-5 \cdot 4$, $5(-4)$.

4. We need factors for which the sum of the products is the middle term, $-9y$. Trial and error leads us to the correct factorization.

    $18y^2 - 9y - 20 = (3y - 4)(6y + 5)$.

63. $16x^2 + 24x + 5$

We will use the grouping method.

1. There is no common factor (other than 1 or -1).

2. Multiply the leading coefficient and constant:

    $16(5) = 80$.

3. Factor 80 so the sum of the factors is 24. We need only consider pairs of positive factors since 80 and 24 are both positive.

Pair of Factors	Sum of Factors
80, 1	81
40, 2	42
20, 4	24
16, 5	21
10, 8	18

4. Split $24x$ using the results of step (3).

    $24x = 20x + 4x$

5. Factor by grouping.

    $16x^2 + 24x + 5 = 16x^2 + 20x + 4x + 5$
    $= 4x(4x + 5) + 1(4x + 5)$
    $= (4x + 5)(4x + 1)$

65. $5x^2 + 24x + 16$

We will use the FOIL method.

1. There is no common factor (other than 1 or -1).

2. Factor the first term, $5x^2$. The factors are $5x, x$. We have this possibility:

    $(5x + \ )(x + \ )$.

3. Factor the last term, 16. We consider only positive factors since both the middle term and the last term are positive. The possibilities are $16 \cdot 1$, $8 \cdot 2$, and $4 \cdot 4$.

4. We need factors for which the sum of products is the middle term, $24x$. Trial and error leads us to the correct factorization.

    $5x^2 + 24x + 16 = (5x + 4)(x + 4)$

67. $-8t^2 - 8t + 30$

We will use the grouping method.

1. Factor out -2: $-2(4t^2 + 4t - 15)$

2. Now we factor the trinomial $4t^2 + 4t - 15$. Multiply the leading coefficient and the constant:

    $4(-15) = -60$

3. Factor –60 so the sum of the factors is 4. The desired factorization is $10(-6)$.

4. Split $4t$ using the results of step (3): $4t = 10t - 6t$

5. Factor by grouping:
$$4t^2 + 4t - 15 = 4t^2 + 10t - 6t - 15$$
$$= 2t(2t + 5) - 3(2t + 5)$$
$$= (2t + 5)(2t - 3)$$

We must include the common factor to get a factorization of the original trinomial:
$$-8t^2 - 8t + 30 = -2(2t + 5)(2t - 3)$$

69. $18xy^3 + 3xy^2 - 10xy$

We will use the FOIL method.

1. Factor out the common factor, xy.
$$xy(18y^2 + 3y - 10)$$

2. We now factor the trinomial $18y^2 + 3y - 10$.
Factor the first term, $18y^2$. The possibilities are $(18y +)(y +)$, $(9y +)(2y +)$, and $(6y +)(3y +)$.

3. Factor the last term, -10. The possibilities are $-10 \cdot 1$, $-5 \cdot 2$, $10(-1)$ and $5(-2)$.

4. We need factors for which the sum of the products is the middle term, $3y$. Trial and error leads us to the correct factorization.
$$18y^2 + 3y - 10 = (6y + 5)(3y - 2)$$

We must include the common factor to get a factorization of the original trinomial:
$$18xy^3 + 3xy^2 - 10xy = xy(6y + 5)(3y - 2)$$

71. $24x^2 - 2 - 47x = 24x^2 - 47x - 2$

We will use the grouping method.

1. There is no common factor (other than 1 or –1).

2. Multiply the leading coefficient and the constant: $24(-2) = -48$

3. Factor –48 so the sum of the factors is –47. The desired factorization is $-48 \cdot 1$.

4. Split $-47x$ using the results of step (3):
$$-47x = -48x + x$$

5. Factor by grouping:
$$24x^2 - 47x - 2 = 24x^2 - 48x + x - 2$$
$$= 24x(x - 2) + (x - 2)$$
$$= (x - 2)(24x + 1)$$

73. $63x^3 + 111x^2 + 36x$

We will use the FOIL method.

1. Factor out the common factor, $3x$.
$$3x(21x^2 + 37x + 12)$$

2. Now we will factor the trinomial $21x^2 + 37x + 12$.
Factor the first term, $21x^2$. The factors are $21x$, x and $7x$, $3x$. We have these possibilities:
$(21x +)(x +)$ and $(7x +)(3x +)$.

3. Factor the last term, 12. The possibilities are $12 \cdot 1$, $(-12)(-1)$, $6 \cdot 2$, $(-6)(-2)$, $4 \cdot 3$, and $(-4)(-3)$ as well as $1 \cdot 12$, $(-1)(-12)$, $2 \cdot 6$, $(-2)(-6)$, $3 \cdot 4$, and $(-3)(-4)$.

4. Look for factors such that the sum of the products is the middle term, $37x$. Trial and error leads us to the correct factorization:
$$(7x + 3)(3x + 4)$$

We must include the common factor to get a factorization of the original trinomial:
$$63x^3 + 111x^2 + 36x = 3x(7x + 3)(3x + 4)$$

75. $48x^4 + 4x^3 - 30x^2$

We will use the grouping method.

1. We factor out the common factor, $2x^2$:
$$2x^2(24x^2 + 2x - 15)$$

2. We now factor $24x^2 + 2x - 15$. Multiply the leading coefficient and the constant:
$$24(-15) = -360$$

3. Factor –360 so the sum of the factors is 2. The desired factorization is $-18 \cdot 20$.

4. Split $2x$ using the results of step (3):
$$2x = -18x + 20x$$

5. Factor by grouping:
$$24x^2 + 2x - 15 = 24x^2 - 18x + 20x - 15$$
$$= 6x(4x - 3) + 5(4x - 3)$$
$$= (4x - 3)(6x + 5)$$

We must not forget to include the common factor:
$$48x^4 + 4x^3 - 30x^2 = 2x^2(4x - 3)(6x + 5)$$

77. $12a^2 - 17ab + 6b^2$

We will use the FOIL method. (Our thinking is much the same as if we were factoring $12a^2 - 17a + 6$.)

1. There is no common factor (other than 1 or –1).

2. Factor the first term, $12a^2$. The factors are $12a$, a and $6a$, $2a$ and $4a$, $3a$. We have these possibilities: $(12a +)(a +)$ and $(6a +)(2a +)$ and $(4a +)(3a +)$.

3. Factor the last term, $6b^2$. The possibilities are $6b \cdot b$, $(-6b)(-b)$, $3b \cdot 2b$, and $(-3b)(-2b)$ as well as $b \cdot 6b$,

$(-b)(-6b)$, $2b \cdot 3b$, and $(-2b)(-3b)$.

4. Look for factors such that the sum of the products is the middle term, $-17ab$. Trial and error leads us to the correct factorization:

$$(4a - 3b)(3a - 2b)$$

79. $2x^2 + xy - 6y^2$

We will use the grouping method.

1. There is no common factor (other than 1 or –1).

2. Multiply the coefficients of the first and last terms:

$$2(-6) = -12$$

3. Factor –12 so the sum of the factors is 1. The desired factorization is $4(-3)$.

4. Split xy using the results of step (3):

$$xy = 4xy - 3xy$$

5. Factor by grouping:

$$2x^2 + xy - 6y^2 = 2x^2 + 4xy - 3xy - 6y^2$$
$$= 2x(x + 2y) - 3y(x + 2y)$$
$$= (x + 2y)(2x - 3y)$$

81. $6x^2 - 29xy + 28y^2$

We will use the FOIL method.

1. There is no common factor (other than 1 or –1).

2. Factor the first term, $6x^2$. The factors are $6x$, x and $3x$, $2x$. We have these possibilities: $(6x + \ )(x + \ )$ and $(3x + \ )(2x + \ )$.

3. Factor the last term, $28y^2$. The possibilities are $28y \cdot y$, $(-28y)(-y)$, $14y \cdot 2y$, $(-14y)(-2y)$, $7y \cdot 4y$, and $(-7y)(-4y)$ as well as $y \cdot 28y$, $(-y)(-28y)$, $2y \cdot 14y$, $(-2y)(-14y)$, $4y \cdot 7y$, and $(-4y)(-7y)$.

4. Look for factors such that the sum of the products is the middle term, $-29xy$. Trial and error leads us to the correct factorization: $(3x - 4y)(2x - 7y)$

83. $9x^2 - 30xy + 25y^2$

We will use the grouping method.

1. There is no common factor (other than 1 or –1).

2. Multiply the coefficients of the first and last terms:

$$9(25) = 225$$

3. Factor 225 so the sum of the factors is –30. The desired factorization is $-15(-15)$.

4. Split $-30xy$ using the results of step (3):

$$-30xy = -15xy - 15xy$$

5. Factor by grouping:

$$9x^2 - 30xy + 25y^2 = 9x^2 - 15xy - 15xy + 25y^2$$
$$= 3x(3x - 5y) - 5y(3x - 5y)$$
$$= (3x - 5y)(3x - 5y), \text{ or } (3x - 5y)^2$$

85. $9x^2y^2 + 5xy - 4$

Let $u = xy$ and $u^2 = x^2y^2$. Factor $9u^2 + 5u - 4$. We will use the FOIL method.

1. There is no common factor (other than 1 or –1).

2. Factor the first term, $9u^2$. The factors are $9u$, u and $3u$, $3u$. We have these possibilities: $(9u + \ )(u + \ )$ and $(3u + \ )(3u + \ )$.

3. Factor the last term, –4. The possibilities are $-4 \cdot 1$, $-2 \cdot 2$, and $-1 \cdot 4$.

4. We need factors for which the sum of the products is the middle term, $5u$. Trial and error leads us to the factorization: $(9u - 4)(u + 1)$. Replace u by xy. We have $9x^2y^2 + 5xy - 4 = (9xy - 4)(xy + 1)$.

87. *Writing Exercise.* Note that the middle term and the last term of $x^2 + 5x + 200$ are both positive. Then the factorization would contain two positive numbers whose product is 200 and whose sum is 5. Without performing any trials, we can see that any pair of positive factors whose product is 200 will have a sum that is greater than 5.

89. $(5a)^2 = (5a)(5a) = 25a^2$

91. $(x + 3)^2 = x^2 + 2 \cdot 3x + 3^2 = x^2 + 6x + 9$

93. $(y + 1)(y - 1) = y^2 - 1^2 = y^2 - 1$

95. *Writing Exercise.* Answers will vary. Either the FOIL method or the grouping method could be described.

97. $60x^8y^6 + 35x^4y^3 + 5$
$= 5(12x^8y^6 + 7x^4y^3 + 1)$ Removing the common factor
To factor the trinomial $12x^8y^6 + 7x^4y^3 + 1$, first note that $(x^4y^3)^2 = x^8y^6$, so the trinomial is of the form $12u^2 + 7u + 1$. Trial and error leads us to the factorization:

$$12x^8y^6 + 7x^4y^3 + 1 = (4x^4y^3 + 1)(3x^4y^3 + 1)$$

Then the factorization of the original trinomial is

$$5(4x^4y^3 + 1)(3x^4y^3 + 1).$$

99. $y^2 - \dfrac{8}{49} + \dfrac{2}{7}y = y^2 + \dfrac{2}{7}y - \dfrac{8}{49}$

We look for factors of $-\dfrac{8}{49}$ whose sum is $\dfrac{2}{7}$. The factors are $\dfrac{4}{7}$ and $-\dfrac{2}{7}$. The factorization is $\left(y + \dfrac{4}{7}\right)\left(y - \dfrac{2}{7}\right)$.

101. $20a^3b^6 - 3a^2b^4 - 2ab^2$

Factor the common factor, ab^2.

$$20a^3b^6 - 3a^2b^4 - 2ab^2 = ab^2\left(20a^2b^4 - 3ab^2 - 2\right)$$

Substitute u for ab^2 (and u^2 for $\left(ab^2\right)^2 = a^2b^4$). We factor $20u^2 - 3u - 2$. Trial and error leads us to the factorization: $20u^2 - 3u - 2 = (4u + 1)(5u - 2)$. Replace u with ab^2: $20a^2b^4 - 3ab^2 - 2 = \left(4ab^2 + 1\right)\left(5ab^2 - 2\right)$. Include the common factor:

$$20a^3b^6 - 3a^2b^4 - 2ab^2 = ab^2\left(4ab^2 + 1\right)\left(5ab^2 - 2\right)$$

103. $x^{2a} + 5x^a - 24$

Substitute u for x^a (and u^2 for x^{2a}). We factor $u^2 + 5u - 24$. We look for factors of –24 whose sum is 5. The factors are 8 and –3. We have $u^2 + 5u - 24 = (u + 8)(u - 3)$. Replace u with x^a:

$x^{2a} + 5x^a - 24 = (x^a + 8)(x^a - 3)$.

105. $2ar^2 + 4asr + as^2 - asr$
$= 2ar^2 + 3asr + as^2$
$= a(2r^2 + 3sr + s^2)$
$= a(2r + s)(r + s)$

107. $(x + 3)^2 - 2(x + 3) - 35$

Substitute u for $x + 3$ (and u^2 for $(x + 3)^2$). We factor $u^2 - 2u - 35$. Look for factors of –35 whose sum is –2. The factors are –7 and 5. We have $u^2 - 2u - 35 = (u - 7)(u + 5)$. Replace u with $x + 3$:

$(x + 3)^2 - 2(x + 3) - 35 = [(x + 3) - 7][(x + 3) + 5]$, or

$(x - 4)(x + 8)$

109. $x^2 + mx + 75$

All such m are the sums of the factors of 75.

Pair of Factors	Sum of Factors
75, 1	76
−75, −1	−76
25, 3	28
−25, −3	−28
15, 5	20
−15, −5	−20

m can be 76, –76, 28, –28, 20, or –20.

111. Since $ax^2 + bx + c = (mx + r)(nx + s)$, from FOIL we know that $a = mn$, $c = rs$, and $b = ms + rn$. If $P = ms$ and $Q = rn$, then $b = P + Q$. Since $ac = mnrs = msrn$, we have $ac = PQ$.

113. *Graphing Calculator Exercise*

115. *Writing Exercise.* If it were true that $x^2 + 3x - 2 = (x - 2)(x + 1)$, then the graph would coincide with the x-axis. Since it does not, we know that $x^2 + 3x - 2 \neq (x - 2)(x + 1)$.

Exercise Set 5.5

1. $x^2 - 100 = (x)^2 - 10^2$

This is a difference of squares.

3. $36x^2 - 12x + 1 = (6x)^2 - 2 \cdot 6x \cdot 1 + 1^2$

This is a perfect-square trinomial.

5. $4r^2$ and 9 are squares but $8r \neq 2 \cdot 2r \cdot 3$ and $8r \neq -2 \cdot 2r \cdot 3$, so this trinomial is classified as none of these.

7. $4x^2 + 8x + 10 = 2(2x^2 + 4x + 5)$ and $2x^2 + 4x + 5$ cannot be factored, so this is a polynomial having a common factor.

9. $4t^2 + 9s^2 + 12st = 4t^2 + 12st + 9s^2$
$$= (2t)^2 + 2 \cdot 2t \cdot 3s + (3s)^2$$

This is a perfect-square trinomial.

11. $x^2 + 20x + 100 = (x + 10)^2$

Find the square terms and write the square roots with a plus sign between them.

13. $t^2 - 2t + 1 = (t - 1)^2$

Find the square terms and write the square roots with a minus sign between them.

15. $4a^2 - 24a + 36$
$= 4\left(a^2 - 6a + 9\right)$ Factoring out the common factor
$= 4(a - 3)^2$ Factoring the perfect-square trinomial

17. $y^2 + 36 + 12y$
$= y^2 + 12y + 36$ Changing order
$= (y + 6)^2$ Factoring the perfect-square trinomial

19. $-18y^2 + y^3 + 81y$

$= y^3 - 18y^2 + 81y$ Changing order

$= y(y^2 - 18y + 81)$ Factoring out the
                              common factor

$= y(y - 9)^2$

21. $2x^2 - 40x + 200$

$= 2(x^2 - 20x + 100)$ Factoring out the common factor

$= 2(x - 10)^2$ Factoring the perfect–square
                              trinomial

23. $1 - 8d + 16d^2$

$= (1 - 4d)^2$ Factoring the perfect-square trinomial

25. $-y^3 - 8y^2 - 16y$

$= -y(y^2 + 8y + 16)$

$= -y(y + 4)^2$

27. $0.25x^2 + 0.30x + 0.09 = (0.5x + 0.3)^2$

                            Find the square terms and write
                            the square roots with a plus
                            sign between them.

29. $p^2 - 2pq + q^2 = (p - q)^2$

31. $25a^2 + 30ab + 9b^2 = (5a + 3b)^2$

33. $5a^2 + 10ab + 5b^2$

$= 5(a^2 + 2ab + b^2)$

$= 5(a + b)^2$

35. $x^2 - 25 = x^2 - 5^2 = (x + 5)(x - 5)$

37. $m^2 - 64 = m^2 - 8^2 = (m + 8)(m - 8)$

39. $4a^2 - 81 = (2a)^2 - 9^2 = (2a + 9)(2a - 9)$

41. $12c^2 - 12d^2$

$= 12(c^2 - d^2)$

$= 12(c + d)(c - d)$

43. $7xy^4 - 7xz^4$

$= 7x(y^4 - z^4)$

$= 7x[(y^2)^2 - (z^2)^2]$

$= 7x(y^2 + z^2)(y^2 - z^2)$

$= 7x(y^2 + z^2)(y + z)(y - z)$

45. $4a^3 - 49a = a(4a^2 - 49)$

$= a[(2a)^2 - 7^2]$

$= a(2a + 7)(2a - 7)$

47. $3x^8 - 3y^8$

$= 3(x^8 - y^8)$

$= 3[(x^4)^2 - (y^4)^2]$

$= 3(x^4 + y^4)(x^4 - y^4)$

$= 3(x^4 + y^4)[(x^2)^2 - (y^2)^2]$

$= 3(x^4 + y^4)(x^2 + y^2)(x^2 - y^2)$

$= 3(x^4 + y^4)(x^2 + y^2)(x + y)(x - y)$

49. $p^2q^2 - 100 = (pq + 10)(pq - 10)$

51. $9a^4 - 25a^2b^4 = a^2(9a^2 - 25b^4)$

$= a^2[(3a)^2 - (5b^2)^2]$

$= a^2(3a + 5b^2)(3a - 5b^2)$

53. $y^2 - \dfrac{1}{4} = \left(y + \dfrac{1}{2}\right)\left(y - \dfrac{1}{2}\right)$

55. $\dfrac{1}{100} - x^2 = \left(\dfrac{1}{10} + x\right)\left(\dfrac{1}{10} - x\right)$

57. $(a + b)^2 - 36 = (a + b + 6)(a + b - 6)$

59. $x^2 - 6x + 9 - y^2$

$= (x^2 - 6x + 9) - y^2$ Grouping as a difference of squares

$= (x - 3)^2 - y^2$

$= (x - 3 + y)(x - 3 - y)$

61. $t^3 + 8t^2 - t - 8$

$= t^2(t + 8) - (t + 8)$ Factoring by

$= (t + 8)(t^2 - 1)$ grouping

$= (t + 8)(t + 1)(t - 1)$ Factoring the difference
                                  of squares

63. $r^3 - 3r^2 - 9r + 27$

$= r^2(r - 3) - 9(r - 3)$ Factoring by

$= (r - 3)(r^2 - 9)$ grouping

$= (r - 3)(r + 3)(r - 3),$ Factoring the difference
                                  of squares

or $(r - 3)^2(r + 3)$

65. $m^2 - 2mn + n^2 - 25$

$= (m^2 - 2mn + n^2) - 25$ Grouping as a difference
                                     of squares

$= (m - n)^2 - 5^2$

$= (m - n + 5)(m - n - 5)$

67. $81 - (x + y)^2 = [9 + (x + y)][9 - (x + y)]$

$= (9 + x + y)(9 - x - y)$

69. $r^2 - 2r + 1 - 4s^2$

$= (r^2 - 2r + 1) - 4s^2$ Grouping as a difference
                                  of squares

$= (r - 1)^2 - (2s)^2$

$= (r - 1 + 2s)(r - 1 - 2s)$

71. $16 - a^2 - 2ab - b^2$

$= 16 - (a^2 + 2ab + b^2)$ Grouping as a difference
                                  of squares

$= 4^2 - (a + b)^2$

$= [4 + (a + b)][4 - (a + b)]$

$= (4 + a + b)(4 - a - b)$

73. $x^3 + 5x^2 - 4x - 20$

$= x^2(x + 5) - 4(x + 5)$

$= (x + 5)(x^2 - 4)$

$= (x + 5)(x + 2)(x - 2)$

75. $a^3 - ab^2 - 2a^2 + 2b^2$

$= a(a^2 - b^2) - 2(a^2 - b^2)$ Factoring by

$= (a^2 - b^2)(a - 2)$ grouping

$= (a + b)(a - b)(a - 2)$ Factoring the
                                 difference of squares

77. *Writing Exercise.* A difference of squares only has two terms, each a perfect square, and one is negative.

79. $\left(2x^2 y^4\right)^3 = 2^3 \left(x^2\right)^3 \left(y^4\right)^3 = 8x^6 y^{12}$

81. $\left(-10x^{10}\right)^3 = (-10)^3 \left(x^{10}\right)^3 = -1000x^{30}$

83. $(x + 1)(x + 1)(x + 1) = \left(x^2 + x + x + 1\right)(x + 1)$ FOIL

$= \left(x^2 + 2x + 1\right)(x + 1)$

$= \left(x^2 + 2x + 1\right) \cdot x + \left(x^2 + 2x + 1\right) \cdot 1$

$= x^3 + 2x^2 + x + x^2 + 2x + 1$

$= x^3 + 3x^2 + 3x + 1$

85. $(p + q)^3 = (p + q)(p + q)^2$

$= (p + q)\left(p^2 + 2pq + q^2\right)$

$= p\left(p^2 + 2pq + q^2\right) + q\left(p^2 + 2pq + q^2\right)$

$= p^3 + 2p^2 q + pq^2 + p^2 q + 2pq^2 + q^3$

$= p^3 + 3p^2 q + 3pq^2 + q^3$

87. *Writing Exercise.* Although FOIL can be used to factor perfect-square trinomials or differences of squares, once recognized, both can be quickly factored.

89. $-\dfrac{8}{27} r^2 - \dfrac{10}{9} rs - \dfrac{1}{6} s^2 + \dfrac{2}{3} rs$

$= -\dfrac{8}{27} r^2 - \dfrac{4}{9} rs - \dfrac{1}{6} s^2$

$= -\dfrac{1}{54}\left(16r^2 + 24rs + 9s^2\right)$

$= -\dfrac{1}{54}(4r + 3s)^2$

91. $0.09x^8 + 0.48x^4 + 0.64 = \left(0.3x^4 + 0.8\right)^2$, or

$\dfrac{1}{100}\left(3x^4 + 8\right)^2$

93. $r^2 - 8r - 25 - s^2 - 10s + 16$

$= \left(r^2 - 8r + 16\right) - \left(s^2 + 10s + 25\right)$

$= (r - 4)^2 - (s + 5)^2$

$= [(r - 4) + (s + 5)][(r - 4) - (s + 5)]$

$= (r - 4 + s + 5)(r - 4 - s - 5)$

$= (r + s + 1)(r - s - 9)$

95. $x^{4a} - y^{2b} = \left(x^{2a}\right)^2 - \left(y^b\right)^2 = \left(x^{2a} + y^b\right)\left(x^{2a} - y^b\right)$

97. $25y^{2a} - \left(x^{2b} - 2x^b + 1\right)$

$= (5y^a)^2 - \left(x^b - 1\right)^2$

$= \left[5y^a + \left(x^b - 1\right)\right]\left[5y^a - \left(x^b - 1\right)\right]$

$= \left(5y^a + x^b - 1\right)\left(5y^a - x^b + 1\right)$

99. $3(x + 1)^2 + 12(x + 1) + 12 = 3\left[(x + 1)^2 + 4(x + 1) + 4\right]$

$= 3(x + 1 + 2)^2$, or $3(x + 3)^2$

101. $s^2 - 4st + 4t^2 + 4s - 8t + 4$

$= (s - 2t)^2 + 4(s - 2t) + 4$

$= (s - 2t + 2)^2$

103. $9x^{2n} - 6x^n + 1 = (3x^n)^2 - 6x^n + 1$

$= (3x^n - 1)^2$

105. If $P(x) = x^2$, then

$P(a + h) - P(a)$

$= (a + h)^2 - a^2$

$= [(a + h) + a][(a + h) - a]$

$= (2a + h)h, \text{ or } h(2a + h)$

107. a) $\pi R^2 h - \pi r^2 h = \pi h(R^2 - r^2)$

                      $= \pi h(R + r)(R - r)$

    b)   Note that 4 m = 400 cm.

        $\pi R^2 h - \pi r^2 h$

        $= \pi(50)^2 (400) - \pi(10)^2 (400)$

        $= 1,000,000\pi - 40,000\pi$

        $= 960,000\pi$ cm^3 (or 0.96π m^3)

        $\approx 3,014,400$ cm^3 Using 3.14 for π

        $\pi h(R + r)(R - r)$

        $= \pi(400)(50 + 10)(50 - 10)$

        $= \pi(400)(60)(40)$

        $= 960,000\pi$ cm^3 (or $0.96\ \pi$ m^3)

        $\approx 3,014,400$ cm^3 Using 3.14 for π

    If we use the π key on a calculator, the result is approximately 3,015,929 cm^3.

109. *Graphing Calculator Exercise*

Exercise Set 5.6

1. $x^3 - 1 = (x)^3 - 1^3$

This is a difference of two cubes.

3. $9x^4 - 25 = (3x^2)^2 - 5^2$

This is a difference of two squares.

5. $1000t^3 + 1 = (10t)^3 + 1^3$

This is a sum of two cubes.

7. $25x^2 + 8x$ has a common factor of x so it is not prime, but it does not fall into any of the other categories. It is classified as none of these.

9. $s^{21} - t^{15} = (s^7)^3 - (t^5)^3$

This is a difference of two cubes .

11. $x^3 - 64 = x^3 - 4^3$

$\qquad = (x - 4)(x^2 + 4x + 16)$

$\qquad A^3 - B^3 = (A - B)(A^2 + AB + B^2)$

13. $z^3 + 1 = z^3 + 1^3$

$\qquad = (z + 1)(z^2 - z + 1)$

$\qquad A^3 + B^3 = (A + B)(A^2 - AB + B^2)$

15. $t^3 - 1000 = t^3 - 10^3$

$\qquad = (t - 10)(t^2 + 10t + 100)$

$\qquad A^3 - B^3 = (A - B)(A^2 + AB + B^2)$

17. $27x^3 + 1 = (3x)^3 + 1^3$

$\qquad = (3x + 1)(9x^2 - 3x + 1)$

$\qquad A^3 + B^3 = (A + B)(A^2 - AB + B^2)$

19. $64 - 125x^3 = 4^3 - (5x)^3 = (4 - 5x)(16 + 20x + 25x^2)$

21. $x^3 - y^3 = (x - y)(x^2 + xy + y^2)$

23. $a^3 + \dfrac{1}{8} = a^3 + \left(\dfrac{1}{2}\right)^3 = \left(a + \dfrac{1}{2}\right)\left(a^2 - \dfrac{1}{2}a + \dfrac{1}{4}\right)$

25. $8t^3 - 8 = 8(t^3 - 1) = 8(t^3 - 1^3) = 8(t - 1)(t^2 + t + 1)$

27. $54x^3 + 2 = x(27x^3 + 1) = 2[(3x)^3 + 1^3]$

$\qquad = 2(3x + 1)(9x^2 - 3x + 1)$

29. $rs^4 + 64rs = rs(s^3 + 64)$

$\qquad = rs(s^3 + 4^3)$

$\qquad = rs(s + 4)(s^2 - 4s + 16)$

31. $5x^3 - 40z^3 = 5(x^3 - 8z^3)$

$\qquad = 5[x^3 - (2z)^3]$

$\qquad = 5(x - 2z)(x^2 + 2xz + 4z^2)$

33. $y^3 - \dfrac{1}{1000} = y^3 - \left(\dfrac{1}{10}\right)^3 = \left(y - \dfrac{1}{10}\right)\left(y^2 + \dfrac{1}{10}y + \dfrac{1}{100}\right)$

35. $x^3 + 0.001 = x^3 + (0.1)^3 = (x + 0.1)(x^2 - 0.1x + 0.01)$

37. $64x^6 - 8t^6 = 8(8x^6 - t^6)$

$\qquad = 8[(2x^2)^3 - (t^2)^3]$

$\qquad = 8(2x^2 - t^2)(4x^4 + 2x^2t^2 + t^4)$

39. $54y^4 - 128y = 2y(27y^3 - 64)$

$\qquad = 2y[(3y)^3 - 4^3]$

$\qquad = 2y(3y - 4)(9y^2 + 12y + 16)$

41. $z^6 - 1$

$\qquad = (z^3)^2 - 1^2$ $\qquad$ Writing as a difference of squares

$\qquad = (z^3 + 1)(z^3 - 1)$ $\qquad$ Factoring a difference of squares

$\qquad = (z + 1)(z^2 - z + 1)(z - 1)(z^2 + z + 1)$

$\qquad\qquad\qquad\qquad$ Factoring a sum and

$\qquad\qquad\qquad\qquad$ a difference of cubes

43. $t^6 + 64y^6 = (t^2)^3 + (4y^2)^3 = (t^2 + 4y^2)(t^4 - 4t^2y^2 + 16y^4)$

45. $x^{12} - y^3z^{12} = (x^4)^3 - (yz^4)^3$

$\qquad = (x^4 - yz^4)(x^8 + x^4yz^4 + y^2z^8)$

47. *Writing Exercise.*

$(x + y)^3 = (x + y)(x + y)(x + y)$

$\qquad = (x^2 + 2xy + y^2)(x + y)$

$\qquad = x^3 + 3x^2y + 3xy^2 + y^3$

$\qquad \neq x^3 + y^3$

49. *Familiarize.* Let x, $x + 1$, and $x + 2$ represent the lengths of the three sides of the triangle. The perimeter is the sum of the three sides.

Translate. The perimeter of the triangle is 108 cm, so we have one equation:

$\qquad x + (x + 1) + (x + 2) = 108.$

Carry out. We solve the equation.

$\qquad x + x + 1 + x + 2 = 108$

$\qquad\qquad\qquad 3x + 3 = 108$

$\qquad\qquad\qquad\quad 3x = 105$

$\qquad\qquad\qquad\quad\ x = 35$

When $x = 35$, $x + 1 = 36$, and $x + 2 = 37$.

Check. If the three sides are 35 cm, 36 cm and 37 cm, then the perimeter is 35 + 36 + 37, or 108 cm. Our answer checks.

State. The lengths of the sides are 35 cm, 36 cm, and 37 cm.

51. *Familiarize*. Let x, y, and z be the number of rolls of dimes, nickels and quarters, respectively.

Translate. The number of rolls of coins is 10, so we have one equation:

$$x + y + z = 10 \, .$$

The total value is $77, so we have a second equation:

$$50(0.10)x + 40(0.05)y + 40(0.25)z = 77 \, .$$

The number of rolls of quarters is twice the number of rolls of dimes, so we have a third equation:

$$z = 2x \, .$$

We now have a system of equations.

$$\begin{aligned} x + y + z &= 10 \\ 5x + 2y + 10z &= 77 \\ z &= 2x \end{aligned}$$

Carry out. Solving the system, we get $(3, 1, 6)$.

Check. The number of rolls is $3 + 1 + 6$, or 10. The total value is $50(0.10)(3) + 40(0.05)(1) + 40(0.25)(6)$

$= 15 + 2 + 60$, or $77. The number of rolls of quarters, 6, is twice the number of rolls of dimes. The answer checks.

State. Jenna has 3 rolls of dimes, 1 roll of nickels and 6 rolls of quarters.

53. *Familiarize*. Let x = the number of nests found by Kathy and $x + 8$ = the number of nests found by Ken.

Translate. The total number of nests found is 100, so we have an equation:

$$x + (x + 8) = 100.$$

Carry out. We solve the equation.

$$\begin{aligned} x + x + 8 &= 100 \\ 2x + 8 &= 100 \\ 2x &= 92 \\ x &= 46 \end{aligned}$$

$$x + 8 = 54$$

Check. The number of nests found by Ken, 54, is 8 more than 46, the number of nests found by Kathy. The total number of nests is $46 + 54$, or 100.

State. Kathy found 46 nests and Ken found 54 nests.

55. *Writing Exercise*. The model shows a cube with volume a^3 from which a portion whose volume is b^3 has been removed. This leaves a remaining volume which can be expressed as $a^2(a - b) + ab(a - b) + b^2(a - b)$, or $(a - b)(a^2 + ab + b^2)$. Thus, $a^3 - b^3 = (a - b)(a^2 + ab + b^2)$.

57. $x^{6a} - y^{3b} = (x^{2a})^3 - (y^b)^3$
$$= (x^{2a} - y^b)(x^{4a} + x^{2a}y^b + y^{2b})$$

58. $2x^{3a} + 16y^{3b} = 2(x^{3a} + 8y^{3b})$
$$= 2\left[(x^a)^3 + (2y^b)^3\right]$$
$$= 2(x^a + 2y^b)(x^{2a} - 2x^a y^b + 4y^{2b})$$

59. $(x + 5)^3 + (x - 5)^3$ Sum of cubes
$$= [(x + 5) + (x - 5)][(x + 5)^2 - (x + 5)(x - 5) + (x - 5)^2]$$
$$= 2x[(x^2 + 10x + 25) - (x^2 - 25) + (x^2 - 10x + 25)]$$
$$= 2x(x^2 + 10x + 25 - x^2 + 25 + x^2 - 10x + 25)$$
$$= 2x(x^2 + 75)$$

61. $5x^3 y^6 - \frac{5}{8}$
$$= 5\left(x^3 y^6 - \frac{1}{8}\right)$$
$$= 5\left(xy^2 - \frac{1}{2}\right)\left(x^2 y^4 + \frac{1}{2}xy^2 + \frac{1}{4}\right)$$

63. $x^{6a} - (x^{2a} + 1)^3$
$$= \left[x^{2a} - (x^{2a} + 1)\right]\left[x^{4a} + x^{2a}(x^{2a} + 1) + (x^{2a} + 1)^2\right]$$
$$= (x^{2a} - x^{2a} - 1)(x^{4a} + x^{4a} + x^{2a} + x^{4a} + 2x^{2a} + 1)$$
$$= -(3x^{4a} + 3x^{2a} + 1)$$

65. $t^4 - 8t^3 - t + 8$
$$= t^3(t - 8) - (t - 8)$$
$$= (t - 8)(t^3 - 1)$$
$$= (t - 8)(t - 1)(t^2 + t + 1)$$

67. If $Q(x) = x^6$, then
$$Q(a + h) - Q(a)$$
$$= (a + h)^6 - a^6$$
$$= \left[(a + h)^3 + a^3\right]\left[(a + h)^3 - a^3\right]$$
$$= [(a + h) + a] \cdot \left[(a + h)^2 - (a + h)a + a^2\right]$$
$$\cdot [(a + h) - a] \cdot \left[(a + h)^2 + (a + h)a + a^2\right]$$
$$= (2a + h) \cdot (a^2 + 2ah + h^2 - a^2 - ah + a^2) \cdot (h)$$
$$\cdot (a^2 + 2ah + h^2 + a^2 + ah + a^2)$$
$$= h(2a + h)(a^2 + ah + h^2)(3a^2 + 3ah + h^2)$$

69. *Graphing Calculator Exercise*

Exercise Set 5.7

1. b

3. f

5. c

7. a

9. $x^2 - 3x - 4$
  $= (x-4)(x+1)$ FOIL or grouping method

11. $2x^3 - 5x^2 - 2x + 5$
  $= x^2(2x-5) - 1(2x-5)$ Factoring by grouping
  $= (2x-5)(x^2-1)$ Difference of squares
  $= (2x-5)(x+1)(x-1)$

13. $24a^2 - 4a - 8$
  $= 4(6a^2 - a - 2)$
  $= 4(3a-2)(2a+1)$ FOIL or grouping method

15. $x^2 - 81$
  $= x^2 - 9^2$ Difference of squares
  $= (x+9)(x-9)$

17. $9m^4 - 900$
  $= 9(m^4 - 100)$
  $= 9\left[(m^2)^2 - 10^2\right]$ Difference of squares
  $= 9(m^2+10)(m^2-10)$

19. $2x^3 + 12x^2 + 16x$
  $= 2x(x^2 + 6x + 8)$
  $= 2x(x+4)(x+2)$ Trial and error

21. $a^2 + 25 + 10a$
  $= a^2 + 10a + 25$ Perfect-square trinomial
  $= (a+5)^2$

23. $2y^2 - 11y + 12$
  $= (2y-3)(y-4)$ FOIL or grouping method

25. $3x^2 + 15x - 252$
  $= 3(x^2 + 5x - 84)$
  $= 3(x+12)(x-7)$ FOIL or grouping method

27. $25x^2 - 9y^2$
  $= (5x)^2 - (3y)^2$ Difference of squares
  $= (5x+3y)(5x-3y)$

29. $t^6 + 1$
  $= \left(t^2\right)^3 + 1^3$ Sum of cubes
  $= (t^2+1)(t^4 - t^2 + 1)$

31. $x^2 + 6x - y^2 + 9$
  $= x^2 + 6x + 9 - y^2$
  $= (x+3)^2 - y^2$ Difference of squares
  $= \left[(x+3)+y\right]\left[(x+3)-y\right]$
  $= (x+y+3)(x-y+3)$

33. $128a^3 + 250b^3$
  $= 2(64a^3 + 125b^3)$ Sum of cubes
  $= 2(4a+5b)(16a^2 - 20ab + 25b^2)$

35. $7x^3 - 14x^2 - 105x$
  $= 7x(x^2 - 2x - 15)$
  $= 7x(x-5)(x+3)$ Trial and error

37. $-9t^2 + 16t^4$
  $= t^2(-9 + 16t^2)$
  $= t^2(16t^2 - 9)$ Difference of squares
  $= t^2(4t+3)(4t-3)$

39. $8m^3 + m^6 - 20$
  $= (m^3)^2 + 8m^3 - 20$
  $= (m^3-2)(m^3+10)$ Trial and error

41. $ac + cd - ab - bd$
  $= c(a+d) - b(a+d)$ Factoring by grouping
  $= (a+d)(c-b)$

43. $4c^2 - 4cd + d^2$ Perfect-square trinomial
  $= (2c-d)^2$

45. $40x^2 + 3xy - y^2$
  $= (5x+y)(8x-y)$ FOIL or grouping method

47. $4a - 5a^2 + 10 + 2a^3$
  $= 2a^3 - 5a^2 + 4a - 10$ Factoring by grouping
  $= a^2(2a-5) + 2(2a-5)$
  $= (2a-5)(a^2+2)$

49. $2x^3 + 6x^2 - 8x - 24$
  $= 2(x^3 + 3x^2 - 4x - 12)$
  $= 2\left[x^2(x+3) - 4(x+3)\right]$ Factoring by grouping
  $= 2(x+3)(x^2-4)$ Difference of squares
  $= 2(x+3)(x+2)(x-2)$

51. $54a^3 - 16b^3$
  $= 2(27a^3 - 8b^3)$
  $= 2\left[(3a)^3 - (2b)^3\right]$ Difference of cubes
  $= 2(3a-2b)(9a^2 + 6ab + 4b^2)$

53. $36y^2 - 35 + 12y$
  $= 36y^2 + 12y - 35$
  $= (6y-5)(6y+7)$ FOIL or grouping method

55. $4m^4 - 64n^4$
  $= 4(m^4 - 16n^4)$ Difference of squares
  $= 4(m^2 + 4n^2)(m^2 - 4n^2)$ Difference of squares
  $= 4(m^2 + 4n^2)(m+2n)(m-2n)$

57. $a^5b - 16ab^5$
  $= ab(a^4 - 16b^4)$
  $= ab\left[(a^2)^2 - (4b^2)^2\right]$ Difference of squares
  $= ab(a^2 + 4b^2)(a^2 - 4b^2)$
  $= ab(a^2 + 4b^2)\left[a^2 - (2b)^2\right]$ Difference of squares
  $= ab(a^2 + 4b^2)(a+2b)(a-2b)$

59. $34t^3 - 6t = 2t(17t^2 - 3)$

61. $(a-3)(a+7) + (a-3)(a-1)$
$= (a-3)(a+7+a-1)$
$= (a-3)(2a+6)$
$= (a-3)(2)(a+3)$
$= 2(a-3)(a+3)$

63. $7a^4 - 14a^3 + 21a^2 - 7a$
$= 7a(a^3 - 2a^2 + 3a - 1)$ Removing a common factor

65. $42ab + 27a^2b^2 + 8$
$= 27a^2b^2 + 42ab + 8$
$= (9ab+2)(3ab+4)$ FOIL or grouping method

67. $-10t^3 + 15t = -5t(2t^2 - 3)$

69. $-6x^4 + 8x^3 - 12x = -2x(3x^3 - 4x^2 + 6)$

71. $p - 64p^4$
$= p(1 - 64p^3)$ Sum of cubes
$= p(1 - 4p)(1 + 4p + 16p^2)$

73. $a^2 - b^2 - 6b - 9$
$= a^2 - (b^2 + 6b + 9)$ Factoring out -1
$= a^2 - (b+3)^2$ Difference of squares
$= [a + (b+3)][a - (b+3)]$
$= (a+b+3)(a-b-3)$

75. *Writing Exercise.* Both are correct. The factorizations are
equivalent:
$(a-b)(x-y) = -1(b-a)(-1)(y-x)$
$= (-1)(-1)(b-a)(y-x)$
$= (b-a)(y-x)$

77. $x + 2 = 0$
$x = -2$

79. $4x = 0$
$x = 0$

81. $f(x) = \dfrac{2x}{3x-2}$
$3x - 2 = 0$
$3x = 2$
$x = \dfrac{2}{3}$

The domain is $\left\{x \middle| x \text{ is a real number and } x \neq \dfrac{2}{3}\right\}$.

83. *Writing Exercise.* One way is to find the product of a
sum of two cubes and a difference of the same two cubes:
$(a^3 + b^3)(a^3 - b^3) = a^6 - b^6$

85. $28a^3 - 25a^2bc + 3ab^2c^2$
$= a(28a^2 - 25abc + 3b^2c^2)$
$= a(7a - bc)(4a - 3bc)$

87. $(x-p)^2 - p^2$
$= (x - p + p)(x - p - p)$
$= x(x - 2p)$

89. $(y-1)^4 - (y-1)^2$
$= (y-1)^2[(y-1)^2 - 1]$
$= (y-1)^2[(y-1)+1][(y-1)-1]$
$= (y-1)^2(y)(y-2), \text{ or } y(y-1)^2(y-2)$

91. $4x^2 + 4xy + y^2 - r^2 + 6rs - 9s^2$
$= (4x^2 + 4xy + y^2) - (r^2 - 6rs + 9s^2)$ Grouping
$= (2x+y)^2 - (r-3s)^2$ Difference of squares
$= [(2x+y) + (r-3s)][(2x+y) - (r-3s)]$
$= (2x + y + r - 3s)(2x + y - r + 3s)$

93. $\dfrac{x^{27}}{1000} - 1$
$= \left(\dfrac{x^9}{10}\right)^3 - 1^3$
$= \left(\dfrac{x^9}{10} - 1\right)\left(\dfrac{x^{18}}{100} + \dfrac{x^9}{10} + 1\right)$

95. $3(x+1)^2 - 9(x+1) - 12$

Substitute u for $x + 1$ (and u^2 for $(x+1)^2$.)
$3u^2 - 9u - 12 = 3(u^2 - 3u - 4)$
$= 3(u-4)(u+1)$

Now replace u by $x + 1$.
$3(x+1-4)(x+1+1) = 3(x-3)(x+2)$

97. $3(a+2)^2 + 30(a+2) + 75$

Substitute u for $a + 2$ (and u^2 for $(a+2)^2$.)
$3u^2 + 30u + 75 = 3(u^2 + 10u + 25)$
$= 3(u+5)^2$

Now replace u by $a + 2$.
$3(a+2+5)^2 = 3(a+7)^2$

99. $27x^{6s} + 64y^{3t}$
$= (3x^{2s})^3 + (4y^t)^3$ Sum of cubes
$= (3x^{2s} + 4y^t)(9x^{4s} - 12x^{2s}y^t + 16y^{2t})$

101. $a^{2w+1} + 2a^{w+1} + a$
$= a(a^{2w} + 2a^w + 1)$
$= a(a^w + 1)^2$

Connecting the Concepts

---

1. $t^2 - 2t + 1 = (t-1)^2$ Perfect-square trinomial

3. $x^3 - 64x$
$= x(x^2 - 64)$
$= x(x^2 - 8^2)$ Difference of squares
$= x(x+8)(x-8)$

5. $5t^3 + 500t = 5t(t^2 + 100)$

7. $4x^3 + 100x + 40x^2$
$= 4x^3 + 40x^2 + 100x$
$= 4x(x^2 + 10x + 25)$ Perfect square trinomial
$= 4x(x+5)^2$

9. $12y^3 + y^2 - 6y$
$= y(12y^2 + y - 6)$ Trial and error
$= y(4y+3)(3y-2)$

11. $7t^3 + 7$
$= 7(t^3 + 1)$ Sum of cubes
$= 7(t+1)(t^2 - t + 1)$

13. $x^3 + 3x^2 - x - 3$
$= x^2(x+3) - 1(x+3)$ Factoring by grouping
$= (x+3)(x^2 - 1)$ Difference of squares
$= (x+3)(x+1)(x-1)$

15. $0.25 - y^2 = (0.5+y)(0.5-y)$ Difference of squares

17. $x^4 + 4 - 5x^2$
$= x^4 - 5x^2 + 4$
$= (x^2 - 4)(x^2 - 1)$ Difference of squares
$= (x+2)(x-2)(x+1)(x-1)$

19. $1 - 64t^6$
$= 1 - (8t^3)^2$ Difference of squares
$= (1 + 8t^3)(1 - 8t^3)$ Sum and difference of cubes
$= [1 + (2t)^3][1 - (2t)^3]$
$= (1+2t)(1-2t+4t^2)(1-2t)(1+2t+4t^2)$

21. $2x^5 + 6x^4 + 3x^2 + 9x$
$= x(2x^4 + 6x^3 + 3x + 9)$ Factoring by grouping
$= x[2x^3(x+3) + 3(x+3)]$
$= x(x+3)(2x^3 + 3)$

23. $50a^2b^2 - 32c^4$
$= 2(25a^2b^2 - 16c^4)$ Difference of squares
$= 2[(5ab)^2 - (4c^2)^2]$
$= 2(5ab + 4c^2)(5ab - 4c^2)$

25. $2x^2 - 12xy - 32y^2$
$= 2(x^2 - 6xy - 16y^2)$
$= 2(x - 8y)(x + 2y)$

27. $m^2 - n^2 + 12n - 36$
$= m^2 - (n^2 - 12n + 36)$ Perfect-square trinomial
$= m^2 - (n-6)^2$ Difference of squares
$= [m + (n-6)][m - (n-6)]$
$= (m+n-6)(m-n+6)$

29. $p^2 + 121q^2 - 22pq$
$= p^2 - 22pq + 121q^2$ Perfect-square trinomial
$= (p - 11q)^2$

Exercise Set 5.8

1. True; see page 330 in the text.

3. False; see Example 2(b).

5. False; hypotenuse is defined only for a right triangle. See page 336 in the text.

7. $(x-2)(x-5) = 0$
$x - 2 = 0$ or $x - 5 = 0$ Principle of zero products
$x = 2$ or $x = 5$

The solutions are 2 and 5. The solution set is $\{2, 5\}$.

9. $x^2 + 8x + 7 = 0$
$(x+7)(x+1) = 0$ Factoring
$x + 7 = 0$ or $x + 1 = 0$ Principle of zero products
$x = -7$ or $x = -1$

The solutions are -7 and -1. The solution set is $\{-7, -1\}$.

11. $9t(2t+1) = 0$
$9t = 0$ or $2t + 1 = 0$ Principle of zero products
$t = 0$ or $2t = -1$
$t = 0$ or $t = -\dfrac{1}{2}$

The solutions are 0 and $-\dfrac{1}{2}$. The solution set is $\left\{0, -\dfrac{1}{2}\right\}$.

13. $15t^2 - 12t = 0$
$3t(5t - 4) = 0$ Factoring
$3t = 0$ or $5t - 4 = 0$ Principle of zero products
$t = 0$ or $5t = 4$
$t = 0$ or $t = \dfrac{4}{5}$

The solutions are 0 and $\dfrac{4}{5}$. The solution set is $\left\{0, \dfrac{4}{5}\right\}$.

15. $(2t+5)(t-7) = 0$
$2t + 5 = 0$ or $t - 7 = 0$ Principle of zero products
$2t = -5$ or $t = 7$
$t = -\dfrac{5}{2}$ or $t = 7$

The solutions are $-\frac{5}{2}$ and 7. The solution set is $\left\{-\frac{5}{2}, 7\right\}$.

17. $x^2 - 3x - 18 = 0$
$(x+3)(x-6) = 0$ Factoring
$x + 3 = 0$ or $x - 6 = 0$ Principle of zero products
$\quad x = -3$ or $x = 6$

The solutions are –3 and 6. The solution set is $\{-3, 6\}$.

19. $t^2 - 10t = 0$
$t(t - 10) = 0$ Factoring
$t = 0$ or $t - 10 = 0$ Principle of zero products
$t = 0$ or $t = 10$

The solutions are 0 and 10. The solution set is $\{0, 10\}$.

21. $(3x - 1)(4x - 5) = 0$
$3x - 1 = 0$ or $4x - 5 = 0$ Principle of zero products
$\quad 3x = 1$ or $4x = 5$
$\quad\;\; x = \frac{1}{3}$ or $x = \frac{5}{4}$

The solutions are $\frac{1}{3}$ and $\frac{5}{4}$. The solution set is $\left\{\frac{1}{3}, \frac{5}{4}\right\}$.

23. $\qquad 4a^2 = 10a$
$4a^2 - 10a = 0$ Getting 0 on one side
$2a(2a - 5) = 0$ Factoring
$2a = 0$ or $2a - 5 = 0$ Principle of zero products
$\;\; a = 0$ or $2a = 5$
$\;\; a = 0$ or $a = \frac{5}{2}$

The solutions are 0 and $\frac{5}{2}$. The solution set is $\left\{0, \frac{5}{2}\right\}$.

25. $t^2 - 6t - 16 = 0$
$(t - 8)(t + 2) = 0$ Factoring
$t - 8 = 0$ or $t + 2 = 0$ Principle of zero products
$\;\; t = 8$ or $t = -2$

The solutions are 8 and –2. The solution set is $\{-2, 8\}$.

27. $\qquad t^2 - 3t = 28$
$t^2 - 3t - 28 = 0$ Getting 0 on one side
$(t - 7)(t + 4) = 0$ Factoring
$t - 7 = 0$ or $t + 4 = 0$ Principle of zero products
$\;\; t = 7$ or $t = -4$

The solutions are 7 and –4. The solution set is $\{-4, 7\}$.

29. $\qquad r^2 + 16 = 8r$
$r^2 - 8r + 16 = 0$ Getting 0 on one side
$(r - 4)(r - 4) = 0$ Factoring
$r - 4 = 0$ or $r - 4 = 0$ Principle of zero products
$\;\; r = 4$ or $r = 4$

There is only one solution, 4. The solution set is $\{4\}$.

31. $a^2 + 20a + 100 = 0$
$(a + 10)(a + 10) = 0$ Factoring
$a + 10 = 0$ or $a + 10 = 0$ Principle of zero products
$\quad a = -10$ or $a = -10$

The solution is –10. The solution set is $\{-10\}$.

33. $8y + y^2 + 15 = 0$
$y^2 + 8y + 15 = 0$ Changing order
$(y + 5)(y + 3) = 0$ Factoring
$y + 5 = 0$ or $y + 3 = 0$ Principle of zero products
$\quad y = -5$ or $y = -3$

The solutions are –5 and –3. The solution set is $\{-5, -3\}$.

35. $n^2 - 81 = 0$

Observe that we can write this equation as $n^2 = 81$. Then the solutions are the numbers which, when squared are 81. These are the square roots of 81, –9 or 9. The solution set is $\{-9, 9\}$.

We could also use the principle of zero products to solve this equation, as shown below.

$$n^2 - 81 = 0$$
$$(n + 9)(n - 9) = 0$$
$$n + 9 = 0 \quad or \quad n - 9 = 0$$
$$n = -9 \quad or \quad n = 9$$

The solutions are –9 and 9. The solution set is $\{-9, 9\}$.

37. $\qquad x^3 - 2x^2 = 63x$
$x^3 - 2x^2 - 63x = 0$ Getting 0 on one side
$x(x^2 - 2x - 63) = 0$
$x(x - 9)(x + 7) = 0$
$x = 0$ or $x - 9 = 0$ or $x + 7 = 0$ Principle of zero products
$x = 0$ or $x = 9$ or $x = -7$

The solutions are 0, 9, and –7. The solution set is $\{-7, 0, 9\}$.

39. $t^2 = 25$

Using the reasoning in Exercise 35, we see that the solution set is composed of the square roots of 25, $\{-5, 5\}$. We could also do this exercise as follows:

$$t^2 = 25$$
$$t^2 - 25 = 0$$
$$(t + 5)(t - 5) = 0$$
$$t + 5 = 0 \quad or \quad t - 5 = 0$$
$$t = -5 \quad or \quad t = 5$$

The solutions are –5 and 5. The solution set is $\{-5, 5\}$.

41. $(a-4)(a+4)=20$
$a^2-16=20$
$a^2-36=0$
$(a+6)(a-6)=0$
$a+6=0 \quad or \quad a-6=0$
$a=-6 \quad or \quad\quad a=6$

The solutions are –6 and 6. The solution set is $\{-6,\ 6\}$.

43. $-9x^2+15x-4=0$
$9x^2-15x+4=0 \quad$ Multiplying by -1
$(3x-4)(3x-1)=0$
$3x-4=0 \quad or \quad 3x-1=0$
$3x=4 \quad or \quad\quad 3x=1$
$x=\dfrac{4}{3} \quad or \quad\quad x=\dfrac{1}{3}$

The solutions are $\dfrac{4}{3}$ and $\dfrac{1}{3}$. The solution set is $\left\{\dfrac{1}{3},\ \dfrac{4}{3}\right\}$.

45. $-8y^3-10y^2-3y=0$
$-y(8y^2+10y+3)=0$
$-y(2y+1)(4y+3)=0$
$-y=0 \quad or \quad 2y+1=0 \quad or \quad 4y+3=0$
$y=0 \quad or \quad\quad 2y=-1 \quad or \quad\quad 4y=-3$
$y=0 \quad or \quad\quad y=-\dfrac{1}{2} \quad or \quad\quad y=-\dfrac{3}{4}$

The solutions are 0, $-\dfrac{1}{2}$, and $-\dfrac{3}{4}$. The solution set is
$\left\{-\dfrac{3}{4},-\dfrac{1}{2},\ 0\right\}$.

47. $(z+4)(z-2)=-5$
$z^2+2z-8=-5 \quad$ Multiplying
$z^2+2z-3=0$
$(z+3)(z-1)=0$
$z+3=0 \quad or \quad z-1=0$
$z=-3 \quad or \quad\quad z=1$

The solutions are –3 and 1. The solution set is $\{-3,\ 1\}$.

49. $x(5+12x)=28$
$5x+12x^2=28 \quad$ Multiplying
$5x+12x^2-28=0$
$12x^2+5x-28=0 \quad$ Rearranging
$(4x+7)(3x-4)=0$
$4x+7=0 \quad or \quad 3x-4=0$
$4x=-7 \quad or \quad\quad 3x=4$
$x=-\dfrac{7}{4} \quad or \quad\quad x=\dfrac{4}{3}$

The solutions are $-\dfrac{7}{4}$ and $\dfrac{4}{3}$. The solution set is
$\left\{-\dfrac{7}{4},\ \dfrac{4}{3}\right\}$.

51. $a^2-\dfrac{1}{100}=0$
$\left(a+\dfrac{1}{10}\right)\left(a-\dfrac{1}{10}\right)=0$
$a+\dfrac{1}{10}=0 \quad or \quad a-\dfrac{1}{10}=0$
$a=-\dfrac{1}{10} \quad or \quad\quad a=\dfrac{1}{10}$

The solutions are $-\dfrac{1}{10}$ and $\dfrac{1}{10}$. The solution set is
$\left\{-\dfrac{1}{10},\ \dfrac{1}{10}\right\}$.

53. $t^4-26t^2+25=0$
$(t^2-1)(t^2-25)=0$
$(t+1)(t-1)(t+5)(t-5)=0$
$t+1=0 \quad or \quad t-1=0 \quad or \quad t+5=0 \quad or \quad t-5=0$
$t=-1 \quad or \quad\quad t=1 \quad or \quad\quad t=-5 \quad or \quad\quad t=5$

The solutions are –1, 1, –5, and 5. The solution set is
$\{-5,-1,\ 1,\ 5\}$.

55. We set $f(a)$ equal to 8.
$a^2+12a+40=8$
$a^2+12a+32=0$
$(a+8)(a+4)=0$
$a+8=0 \quad or \quad a+4=0$
$a=-8 \quad or \quad\quad a=-4$

The values of a for which $f(a)=8$ are –8 and –4.

57. We set $g(a)$ equal to 12.
$2a^2+5a=12$
$2a^2+5a-12=0$
$(2a-3)(a+4)=0$
$2a-3=0 \quad or \quad a+4=0$
$2a=3 \quad or \quad\quad a=-4$
$a=\dfrac{3}{2} \quad or \quad\quad a=-4$

The values of a for which $g(a)=12$ are $\dfrac{3}{2}$ and –4.

59. We set $h(a)$ equal to -27.
$12a+a^2=-27$
$a^2+12a+27=0$
$(a+3)(a+9)=0$
$a+3=0 \quad or \quad a+9=0$
$a=-3 \quad or \quad\quad a=-9$

The values of a for which $h(a)=-27$ are –3 and –9.

61.
$$f(x) = g(x)$$
$$12x^2 - 15x = 8x - 5$$
$$12x^2 - 23x + 5 = 0$$
$$(4x - 1)(3x - 5) = 0$$
$$4x - 1 = 0 \quad or \quad 3x - 5 = 0$$
$$4x = 1 \quad or \quad 3x = 5$$
$$x = \frac{1}{4} \quad or \quad x = \frac{5}{3}$$

The values of x for which $f(x) = g(x)$ are $\frac{1}{4}$ and $\frac{5}{3}$.

63.
$$f(x) = g(x)$$
$$2x^3 - 5x = 10x - 7x^2$$
$$2x^3 + 7x^2 - 15x = 0$$
$$x(2x^2 + 7x - 15) = 0$$
$$x(2x - 3)(x + 5) = 0$$
$$x = 0 \quad or \quad 2x - 3 = 0 \quad or \quad x + 5 = 0$$
$$x = 0 \quad or \quad 2x = 3 \quad or \quad x = -5$$
$$x = 0 \quad or \quad x = \frac{3}{2} \quad or \quad x = -5$$

The values of x for which $f(x) = g(x)$ are 0, $\frac{3}{2}$, and -5.

65. $f(x) = \dfrac{3}{x^2 - 3x - 4}$

$f(x)$ cannot be calculated for any x-value for which the denominator, $x^2 - 3x - 4$, is 0. To find the excluded values, we solve:

$$x^2 - 3x - 4 = 0$$
$$(x - 4)(x + 1) = 0$$
$$x - 4 = 0 \quad or \quad x + 1 = 0$$
$$x = 4 \quad or \quad x = -1$$

The domain of f is
$\{x \mid x \text{ is a real number } and \ x \neq 4 \ and \ x \neq -1\}$, or
$(-\infty, -1) \cup (-1, \ 4) \cup (4, \ \infty)$.

67. $f(x) = \dfrac{x}{6x^2 - 54}$

$f(x)$ cannot be calculated for any x-value for which the denominator, $6x^2 - 54$, is 0. To find the excluded values, we solve:

$$6x^2 - 54 = 0$$
$$6(x^2 - 9) = 0$$
$$6(x + 3)(x - 3) = 0$$
$$x + 3 = 0 \quad or \quad x - 3 = 0$$
$$x = -3 \quad or \quad x = 3$$

The domain of f is
$\{x \mid x \text{ is a real number } and \ x \neq -3 \ and \ x \neq 3\}$, or
$(-\infty, -3) \cup (-3, \ 3) \cup (3, \ \infty)$.

69. $f(x) = \dfrac{x - 5}{9x - 18x^2}$

$f(x)$ cannot be calculated for any x-value for which the

denominator, $9x - 18x^2$, is 0. To find the excluded values, we solve:

$$9x - 18x^2 = 0$$
$$9x(1 - 2x) = 0$$
$$9x = 0 \quad or \quad 1 - 2x = 0$$
$$x = 0 \quad or \quad -2x = -1$$
$$x = 0 \quad or \quad x = \frac{1}{2}$$

The domain of f is
$\left\{x \mid x \text{ is a real number } and \ x \neq 0 \ and \ x \neq \frac{1}{2}\right\}$, or
$(-\infty, \ 0) \cup \left(0, \ \frac{1}{2}\right) \cup \left(\frac{1}{2}, \ \infty\right)$.

71. $f(x) = \dfrac{7}{5x^3 - 35x^2 + 50x}$

$f(x)$ cannot be calculated for any x-value for which the denominator, $5x^3 - 35x^2 + 50x$, is 0. To find the excluded values, we solve:

$$5x^3 - 35x^2 + 50x = 0$$
$$5x(x^2 - 7x + 10) = 0$$
$$5x(x - 2)(x - 5) = 0$$
$$5x = 0 \quad or \quad x - 2 = 0 \quad or \quad x - 5 = 0$$
$$x = 0 \quad or \quad x = 2 \quad or \quad x = 5$$

The domain of f is
$\{x \mid x \text{ is a real number } and \ x \neq 0 \ and \ x \neq 2 \ and \ x \neq 5\}$,
or $(-\infty, \ 0) \cup (0, \ 2) \cup (2, \ 5) \cup (5, \ \infty)$.

73. *Familiarize.* We let w represent the width and $w + 3$ represent the length. We make a drawing and label it.

Recall that the formula for the area of a rectangle is
$$A = \text{length} \times \text{width}.$$

Translate.

$$\underline{\text{Area}} \quad \text{is} \quad \underline{180 \text{ in}^2}.$$
$$\downarrow \qquad\quad \downarrow \qquad\quad \downarrow$$
$$w(w + 3) \quad = \quad 180$$

Carry out. We solve the equation.
$$w(w + 3) = 180$$
$$w^2 + 3w = 180$$
$$w^2 + 3w - 180 = 0$$
$$(w + 15)(w - 12) = 0$$
$$w + 15 = 0 \quad or \quad w - 12 = 0$$
$$w = -15 \quad or \quad w = 12$$

Check. The number -15 is not a solution, because width cannot be negative. If the width is 12 in. and the length is 3 in. more, or 15 in., then the area is $12 \cdot 15$, or 180 in^2. This is a solution.

State. The length is 15 in. and the width is 12 in.

75. ***Familiarize***. We make a drawing and label it. We let x represent the length of a side of the original square, in meters.

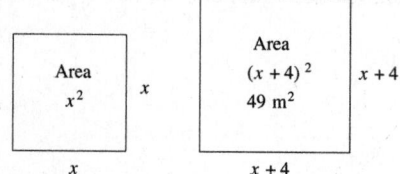

Translate.

$$\underbrace{\text{Area of new square}}_{(x+4)^2} \quad \underset{=}{\text{is}} \quad \underbrace{49 \text{ m}^2}_{49}$$

Carry out. We solve the equation:

$$(x+4)^2 = 49$$
$$x^2 + 8x + 16 = 49$$
$$x^2 + 8x - 33 = 0$$
$$(x-3)(x+11) = 0$$
$$x - 3 = 0 \quad or \quad x + 11 = 0$$
$$x = 3 \quad or \quad x = -11$$

Check. We check only 3 since the length of a side cannot be negative. If we increase the length by 4, the new length is $3 + 4$, or 7 m. Then the new area is $7 \cdot 7$, or 49 m^2. We have a solution.

State. The length of a side of the original square is 3 m.

77. ***Familiarize***. We make a drawing and label it with both known and unknown information. We let x represent the width of the frame.

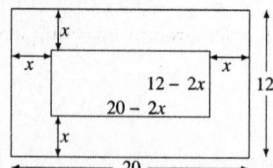

The length and width of the picture that shows are represented by $20 - 2x$ and $12 - 2x$. The area of the picture that shows is 84 cm^2.

Translate. Using the formula for the area of a rectangle, $A = l \cdot w$, we have

$$84 = (20 - 2x)(12 - 2x).$$

Carry out. We solve the equation:

$$84 = 240 - 64x + 4x^2$$
$$84 = 4(60 - 16x + x^2)$$
$$21 = 60 - 16x + x^2 \qquad \text{Dividing by 4}$$
$$0 = x^2 - 16x + 39$$
$$0 = (x-3)(x-13)$$
$$x - 3 = 0 \quad or \quad x - 13 = 0$$
$$x = 3 \quad or \quad x = 13$$

Check. We see that 13 is not a solution because when $x = 13$, $20 - 2x = -6$ and $12 - 2x = -14$, and the length and width of the frame cannot be negative. We check 3. When $x = 3$, $20 - 2x = 14$ and $12 - 2x = 6$ and $14 \cdot 6 = 84$. The area is 84. The value checks.

State. The width of the frame is 3 cm.

79. ***Familiarize***. Let $x =$ the length on each side. The length of the table cloth is $2x + 60$, and the width is $2x + 40$. The area of a rectangle is $A = \text{length} \times \text{width}$.

Translate.

$$\underbrace{\text{Area of table cloth}}_{(2x+60)(2x+40)} \quad \underset{=}{\text{is}} \quad \underbrace{\text{Twice the area of the table.}}_{2(60 \cdot 40)}$$

Carry out. We solve the equation.

$$(2x+60)(2x+40) = 2(60 \cdot 40)$$
$$4x^2 + 200x + 2400 = 4800$$
$$4x^2 + 200x - 2400 = 0$$
$$x^2 + 50x - 600 = 0$$
$$(x+60)(x-10) = 0$$
$$x + 60 = 0 \quad or \quad x - 10 = 0$$
$$x = -60 \quad or \quad x = 10$$

Check. We check only 10 since the length cannot be negative. The length of the table cloth is $2(10) + 60$, or 80 in. and the width is $2(10) + 40$, or 60 in. The area is $80 \cdot 60$, or 4800 in^2 which is twice 2400 in^2, the area of the table. The answer checks.

State. On each side, the table cloth hangs down 10 in.

81. ***Familiarize***. Let x represent the first integer, $x + 2$ the second, and $x + 4$ the third.

Translate.

$$\underbrace{\text{Square of the third}}_{(x+4)^2} \quad \underset{=}{\text{is}} \quad \underset{76}{76} \quad \underset{+}{\text{more than}} \quad \underbrace{\text{square of the second.}}_{(x+2)^2}$$

Carry out. We solve the equation:

$$(x+4)^2 = 76 + (x+2)^2$$
$$x^2 + 8x + 16 = 76 + x^2 + 4x + 4$$
$$x^2 + 8x + 16 = x^2 + 4x + 80$$
$$4x = 64$$
$$x = 16$$

Check. We check the integers 16, 18, and 20. The square of 20, or 400, is 76 more than 324, the square of 18. The answer checks.

State. The integers are 16, 18, and 20.

83. **Familiarize**. Using the labels on the drawing in the text, we let x represent the height of the triangle and $x + 20$ represent the base. Recall that the formula for the area of the triangle with base b and height h is $\frac{1}{2}bh$.

Translate.

$$\underline{\text{Area}} \quad \text{is} \quad \underline{750 \text{ in}^2}.$$
$$\downarrow \qquad \downarrow \qquad \downarrow$$
$$\frac{1}{2}x(x+20) \quad = \quad 750$$

Carry out. We solve the equation.

$$\frac{1}{2}x(x+20) = 750$$
$$x(x+20) = 1500$$
$$x^2 + 20x = 1500$$
$$x^2 + 20x - 1500 = 0$$
$$(x+50)(x-30) = 0$$
$$x + 50 = 0 \quad or \quad x - 30 = 0$$
$$x = -50 \quad or \qquad x = 30$$

Check. We check only 30 since the height cannot be negative. If the height is 30 in., then the base is $30 + 20$, or 50 in., and the area is $\frac{1}{2}(30)(50)$, or 750 in^2. The answer checks.

State. The height is 30 in. and the base is 50 in.

85. **Familiarize**. We make a drawing. Let $x =$ one side and $x + 5$ is the other side of the triangle.

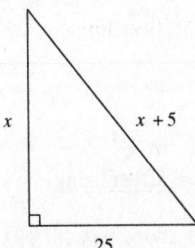

Translate. We use the Pythagorean theorem.

$$25^2 + x^2 = (x+5)^2$$

Carry out. We solve the equation:

$$625 + x^2 = x^2 + 10x + 25$$
$$600 = 10x$$
$$60 = x$$

Check. If $x = 60$, then $x + 5 = 65$;

$$25^2 + 60^2 = 625 + 3600 = 4225 = 65^2,$$ so the answer checks.

State. One side is 60 ft and the other side is 65 ft.

87. **Familiarize**. We make a drawing. Let $h =$ the height the ladder reaches on the wall. Then the length of the ladder is $h + 1$.

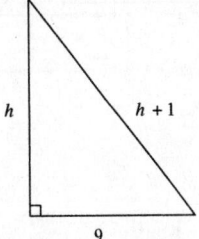

Translate. We use the Pythagorean theorem.

$$9^2 + h^2 = (h+1)^2$$

Carry out. We solve the equation.

$$81 + h^2 = h^2 + 2h + 1$$
$$80 = 2h$$
$$40 = h$$

Check. If $h = 40$, then $h + 1 = 41$;

$$9^2 + 40^2 = 81 + 1600 = 1681 = 41^2,$$ so the answer checks.

State. The ladder is 41 ft long.

89. **Familiarize**. Let w represent the width and $w + 25$ represent the length, in meters. Make a drawing.

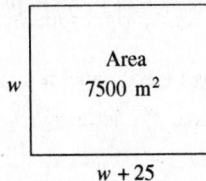

Recall that the formula for the area of a rectangle is $A = \text{length} \times \text{width}$.

Translate.

$$\underline{\text{Area}} \quad \text{is} \quad \underline{7500 \text{ m}^2}.$$
$$\downarrow \qquad \downarrow \qquad \downarrow$$
$$w(w+25) \quad = \quad 7500$$

Carry out. We solve the equation.

$$w(w+25) = 7500$$
$$w^2 + 25w = 7500$$
$$w^2 + 25w - 7500 = 0$$
$$(w+100)(w-75) = 0$$
$$w + 100 = 0 \qquad or \quad w - 75 = 0$$
$$w = -100 \quad or \qquad w = 75$$

Check. The number –100 is not a solution because width cannot be negative. If the width is 75 m and the length is 25 m more, or 100 m, then the area will be $75 \cdot 100$, or 7500 m^2. This is a solution.

State. The dimensions will be 100 m by 75 m.

91. Familiarize. The firm breaks even when the cost and the revenue are the same. We use the functions given in the text.

Translate.

$$\underset{\downarrow}{\underline{\text{Cost}}} \quad \text{equals} \quad \underset{\downarrow}{\underline{\text{revenue.}}}$$
$$x^2 - 2x + 10 \quad = \quad 2x^2 + x$$

Carry out. We solve the equation.

$$x^2 - 2x + 10 = 2x^2 + x$$
$$x^2 + 3x - 10 = 0$$
$$(x+5)(x-2) = 0$$
$$x+5=0 \quad or \quad x-2=0$$
$$x=-5 \quad or \quad x=2$$

Check. We check only 2 since the number of cabinet sets cannot be negative. If 2 cabinet sets are produced, the cost is $C(2) = (2)^2 - 2(2) + 10 = \$10$ thousand. If 2 cabinet sets are sold, the revenue is $R(2) = 2(2)^2 + 2 = 8 + 2 = \$10$ thousand. The answer checks.

State. The firm breaks even when 2 sets of cabinets are produced and sold.

93. Familiarize. We will use the formula in Example 6,

$$h(t) = -15t^2 + 75t + 10$$

Translate. We need to find the value of t for which $h(t) = 100$.

$$-15t^2 + 75t + 10 = 100$$

Carry out. We solve the equation.

$$-15t^2 + 75t + 10 = 100$$
$$15t^2 - 75t + 90 = 0$$
$$15(t^2 - 5t + 6) = 0$$
$$15(t-2)(t-3) = 0$$
$$t-2=0 \quad or \quad t-3=0$$
$$t=2 \quad or \quad t=3$$

Check. We have

$$h(2) = -15(2)^2 + 75(2) + 10 = 100$$
$$h(3) = -15(3)^2 + 75(3) + 10 = 100$$

The solutions of the equation are 2 and 3. We reject 3 since it indicates when the height of the tee shirt was 100 ft on the way down.

State. The tee shirt was airborne for 2 sec before it was caught.

95. Familiarize. We will use the formula

$$h(t) = -16t^2 + 64t + 80 .$$

Translate. We find the value of t for which $h(t) = 0$. We have

$$-16t^2 + 64t + 80 = 0 .$$

Carry out. We solve the equation.

$$-16t^2 + 64t + 80 = 0$$
$$t^2 - 4t - 5 = 0 \qquad \text{Diving by } -16$$
$$(t+1)(t-5) = 0$$
$$t+1=0 \quad or \quad t-5=0$$
$$t=-1 \quad or \quad t=5$$

Check. Time cannot be negative in this application.

$$h(5) = -16(5)^2 + 64(5) + 80 = 0 \quad \text{The answer checks.}$$

State. The cardboard will reach the ground 5 sec after it is launched.

97. Familiarize. We will use the formula

$$a(t) = \frac{1}{2}t^2 - \frac{19}{10}t + 8 .$$

Translate. We find the value of t for which $a(t) = 11$. We have

$$\frac{1}{2}t^2 - \frac{19}{10}t + 8 = 11 .$$

Carry out. We solve the equation.

$$\frac{1}{2}t^2 - \frac{19}{10}t + 8 = 11$$
$$\frac{1}{2}t^2 - \frac{19}{10}t - 3 = 0$$
$$5t^2 - 19t - 30 = 0 \qquad \text{Multiplying by 10}$$
$$(5t+6)(t-5) = 0$$
$$5t+6=0 \quad or \quad t-5=0$$
$$t=-\frac{6}{5} \quad or \quad t=5$$

Check. We check only 5 since time cannot be negative in this application. $a(5) = \frac{1}{2}(5)^2 - \frac{19}{10}(5) + 8$

$$= \frac{25}{2} - \frac{19}{2} + 8 = 11 \quad \text{The answer checks.}$$

State. In the year 2000 + 5, or 2005, US companies spend about \$11 billion in online advertising.

99. Writing Exercise. Find the first coordinate(s) of the x-intercept(s) of $y = p(x)$. They are the solutions of $p(x) = 0$.

101. The reciprocal of $\frac{2}{3}$ is $\frac{3}{2}$.

103. $\dfrac{5}{12}\cdot\left(-\dfrac{45}{8}\right)=-\dfrac{75}{32}$

105. $\dfrac{6-4\cdot8}{-2+3\cdot5}=\dfrac{6-32}{-2+15}=\dfrac{-26}{13}=-2$

107. $\dfrac{240}{280}=\dfrac{6\cdot40}{7\cdot40}=\dfrac{6}{7}$

109. *Writing Exercise.*
$$x=-3 \quad or \quad x=5$$
$$x+3=0 \quad or \quad x-5=0$$
$$(x+3)(x-5)=0$$
$$x^2-2x-15=0$$

111. $(8x+11)(12x^2-5x-2)=0$
$$(8x+11)(3x-2)(4x+1)=0$$
$$8x+11=0 \quad or \quad 3x-2=0 \quad or \quad 4x+1=0$$
$$8x=-11 \quad or \quad 3x=2 \quad or \quad 4x=-1$$
$$x=-\frac{11}{8} \quad or \quad x=\frac{2}{3} \quad or \quad x=-\frac{1}{4}$$

The solution set is $\left\{-\dfrac{11}{8},-\dfrac{1}{4},\dfrac{2}{3}\right\}$.

113.

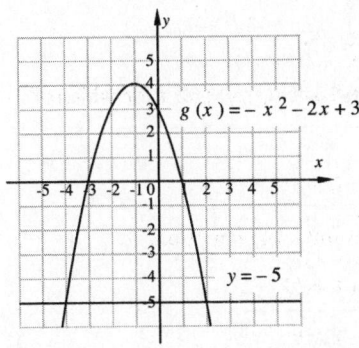

$g(x)=-x^2-2x+3$

$y=-5$

The solutions of $-x^2-2x+3=0$ are the first coordinates of the x-intercepts. From the graph we see that these are -3 and 1. The solution set is $\{-3,1\}$.

To solve $-x^2-2x+3\geq-5$ we find the x-values for which $g(x)\geq-5$. From the graph we see that these are the values in the interval $[-4,2]$. The solution set can also be expressed as $\{x\,|-4\leq x\leq2\}$.

115. Answers may vary. A polynomial function of lowest degree that meets the given criteria is of the form $f(x)=ax^3+bx^2+cx+d$. Substituting, we have
$$a\cdot2^3+b\cdot2^2+c\cdot2+d=0,$$
$$a(-1)^3+b(-1)^2+c(-1)+d=0,$$
$$a\cdot3^3+b\cdot3^2+c\cdot3+d=0,$$
$$a\cdot0^3+b\cdot0^2+c\cdot0+d=30, \text{ or}$$
$$8a+4b+2c+d=0,$$
$$-a+b-c+d=0,$$
$$27a+9b+3c+d=0,$$
$$d=30.$$

Solving the system of equations we get $(5,-20,5,30)$, so the corresponding function is $f(x)=5x^3-20x^2+5x+30$.

117. *Familiarize.* Using the labels on the drawing in the text, we let x represent the width of the piece of tin and $2x$ represent the length. Then the width and length of the base of the box are represented by $x-4$ and $2x-4$, respectively. Recall that the formula for the volume of a rectangular solid with length l, width w, and height h is $l\cdot w\cdot h$.

Translate.

$$\underset{\downarrow}{\underline{\text{The volume}}} \quad \underset{}{\text{is}} \quad \underset{\downarrow}{\underline{480\text{ cm}^3}}.$$
$$(2x-4)(x-4)(2) \;\;=\;\; 480$$

Carry out. We solve the equation:
$$(2x-4)(x-4)(2)=480$$
$$(2x-4)(x-4)=240 \quad \text{Dividing by 2}$$
$$2x^2-12x+16=240$$
$$2x^2-12x-224=0$$
$$x^2-6x-112=0 \quad \text{Dividing by 2}$$
$$(x+8)(x-14)=0$$
$$x+8=0 \quad or \quad x-14=0$$
$$x=-8 \quad or \quad x=14$$

Check. We check only 14 since the width cannot be negative. If the width of the piece of tin is 14 cm, then its length is $2\cdot14$, or 28 cm, and the dimensions of the base of the box are $14-4$, or 10 cm by $28-4$, or 24 cm. The volume of the box is $24\cdot10\cdot2$, or 480 cm^3. The answer checks.

State. The dimensions of the piece of tin are 14 cm by 28 cm.

119. Graph $y_1=11.12(x+1)^2$ and $y_2=15.4x^2$ in a window that shows the point of intersection of the graphs. The window $[0,10,0,500]$, Xscl = 1, Yscl = 100 is one good choice. Then find the first coordinate of the point of intersection. It is approximately 5.7, so it will take the camera about 5.7 sec to catch up to the skydiver.

121. *Graphing Calculator Exercise*

123. $\{6.90\}$

125. $\{3.48\}$

Connecting the Concepts

1. $x^2 + 5x + 6 = (x+2)(x+3)$

3.
$$x^2 + 6 = 5x$$
$$x^2 - 5x + 6 = 0$$
$$(x-2)(x-3) = 0$$
$$x-2 = 0 \quad or \quad x-3 = 0$$
$$x = 2 \quad or \quad x = 3$$

The solution set is $\{2, 3\}$.

5. $(3x^2 - x) - (x^2 - 5) = 3x^2 - x - x^2 + 5$
$$= 2x^2 - x + 5$$

7. $(a+1)(a-1) = a^2 - 1$

9.
$$x^2 = 19x$$
$$x^2 - 19x = 0$$
$$x(x-19) = 0$$
$$x = 0 \quad or \quad x - 19 = 0$$
$$x = 0 \quad or \quad x = 19$$

The solution set is $\{0, 19\}$.

11. $3x^2 - 14 + x^2 + 3x = 4x^2 + 3x - 14$

13. $t^3 + 1 = (t+1)(t^2 - t + 1)$

15. $(5n-6)^2$
$$= (5n)^2 - 2 \cdot 5n \cdot 6 + 6^2$$
$$= 25n^2 - 60n + 36$$

17. $2x^2 + 8x - 42 = 2(x^2 + 4x - 21)$
$$= 2(x+7)(x-3)$$

19. $(x^2 - 5x - 7) - (4x^2 - 5x + 2)$
$$= x^2 - 5x - 7 - 4x^2 + 5x - 2$$
$$= -3x^2 - 9$$

Chapter 5 Review

1. g

3. a

5. e

7. h

9. i

11.

Term	$2xy^6$	$-7x^8y^3$	$2x^3$	9
Degree	7	11	3	0

Degree of polynomial: 11

13. $-3x^2 + 2x^3 + 8x^6y - 7x^8y^3$

15. $P(x) = x^2 + 10x$

$P(a+h) - P(a) = (a+h)^2 + 10(a+h) - (a^2 + 10a)$
$$= a^2 + 2ah + h^2 + 10a + 10h - a^2 - 10a$$
$$= 2ah + h^2 + 10h$$

17. $4x^2y - 3xy^2 - 5x^2y + xy^2$
$$= (4-5)x^2y + (-3+1)xy^2$$
$$= -x^2y - 2xy^2$$

19. $(4n^3 + 2n^2 - 12n + 7) + (-6n^3 + 9n + 4 + n)$
$$= (4-6)n^3 + 2n^2 + (-12+9+1)n + 7 + 4$$
$$= -2n^3 + 2n^2 - 2n + 11$$

21. $(8x-5) - (-6x+2)$
$$= 8x - 5 + 6x - 2$$
$$= 14x - 7$$

23. $(8x^2 - 4xy + y^2) - (2x^2 - 3y^2 - 9y)$
$$= 8x^2 - 4xy + y^2 - 2x^2 + 3y^2 + 9y$$
$$= 6x^2 - 4xy + 4y^2 + 9y$$

25.
$$
\begin{array}{r}
x^4 \quad - 2x^2 + 3 \\
x^4 + x^2 \quad - 1 \\
\hline
- x^4 + 2x^2 - 3 \\
x^6 \quad - 2x^4 + 3x^2 \\
x^8 - 2x^6 + 3x^4 \\
\hline
x^8 - x^6 \quad + 5x^2 - 3
\end{array}
$$

27. $(7t+1)(7t-1) = 49t^2 - 1$ Difference of squares

29. $(x+3)(2x-1)$
$$= 2x^2 - x + 6x - 3 \qquad \text{FOIL}$$
$$= 2x^2 + 5x - 3$$

31. $(3t-5)^2 - (2t+3)^2$
$$= 9t^2 - 30t + 25 - (4t^2 + 12t + 9)$$
$$= 5t^2 - 42t + 16$$

33. $7x^2 + 6x = x(7x+6)$

35. $15x^4 - 18x^3 + 21x^2 - 3x = 3x(5x^3 - 6x^2 + 7x - 1)$

37. $3m^2 - 10m - 8$

We will use the FOIL method.

1. There is no common factor (other than 1 or –1).

2. Factor the first term, $3m^2$. The factors are $3m$, m.
    We have this possibility:
    $$(3m + \quad)(m + \quad).$$

3. Factor the last term, -8. The possibilities are $8(-1)$, $-8 \cdot 1$, $4(-2)$, and $-4 \cdot 2$.

4. We need factors for which the sum of the products (the "outer" and "inner" parts of FOIL) is the middle term, $-10m$. Try some possibilities and check by multiplying. Trial and error leads us to

$$(3m+2)(m-4) = 3m^2 - 10m - 8$$

The factorization is $(3m+2)(m-4)$.

39. $4y^2 - 16$
$$= 4\left(y^2 - 4\right) \qquad \text{Difference of squares}$$
$$= 4(y+2)(y-2)$$

41. $ax + 2bx - ay - 2by$
$$= x(a+2b) - y(a+2b) \qquad \text{Grouping}$$
$$= (a+2b)(x-y)$$

43. $a^4 - 81 = a^4 - 3^4 \qquad \text{Difference of squares}$
$$= \left(a^2 + 3^2\right)\left(a^2 - 3^2\right) \qquad \text{Difference of squares}$$
$$= \left(a^2 + 9\right)(a+3)(a-3)$$

45. $27x^3 + 8 = (3x)^3 + 2^3 \qquad \text{Sum of cubes}$
$$= (3x+2)\left(9x^2 - 6x + 4\right)$$

47. $490t^2 - 640t^4$
$$= 10t^2\left(49 - 64t^2\right) \qquad \text{Difference of squares}$$
$$= 10t^2(7+8t)(7-8t)$$

49. $0.01x^4 - 1.44y^6$
$$= \left(0.1x^2\right)^2 - \left(1.2y^3\right)^2 \qquad \text{Difference of squares}$$
$$= \left(0.1x^2 + 1.2y^3\right)\left(0.1x^2 - 1.2y^3\right)$$

51. $6t^2 + 17pt + 5p^2$

We will use the grouping method.

1. There is no common factor (other than 1 or -1).

2. Multiply the leading coefficient, 6, and the last coefficient, 5: $6(5) = 30$

3. Try to factor 30 so the sum of the factors is 17. We need only consider pairs of positive factors since 30 and 17 are both positive.

Pair of Factors	Sum of Factors
30, 1	31
15, 2	17
10, 3	13
5, 6	11

4. Split $17pt$ using the results of step (3):

$$17pt = 15pt + 2pt$$

5. Factor by grouping:

$$6t^2 + 17pt + 5p^2 = 6t^2 + 15pt + 2pt + 5p^2$$
$$= 3t(2t+5p) + p(2t+5p)$$
$$= (2t+5p)(3t+p)$$

53. $a^2 - 2ab + b^2 - 4t^2$
$$= \left(a^2 - 2ab + b^2\right) - 4t^2 \qquad \text{Grouping as a difference}$$
$$= (a-b)^2 - 4t^2 \qquad\qquad\quad \text{of squares}$$
$$= (a-b+2t)(a-b-2t)$$

55. $\qquad 6b^2 + 6 = 13b$
$$6b^2 - 13b + 6 = 0$$
$$(3b-2)(2b-3) = 0$$
$$3b - 2 = 0 \quad or \quad 2b - 3 = 0$$
$$b = \frac{2}{3} \quad or \qquad b = \frac{3}{2}$$

The solution set is $\left\{\dfrac{2}{3}, \dfrac{3}{2}\right\}$.

57. $\qquad 3r^2 = 12$
$$3r^2 - 12 = 0$$
$$3\left(r^2 - 4\right) = 0$$
$$3(r+2)(r-2) = 0$$
$$r + 2 = 0 \quad or \quad r - 2 = 0$$
$$r = -2 \quad or \qquad r = 2$$

The solution set is $\{-2, 2\}$.

59. $(y-1)(y-4) = 10$
$$y^2 - 5y + 4 = 10$$
$$y^2 - 5y - 6 = 0$$
$$(y+1)(y-6) = 0$$
$$y + 1 = 0 \quad or \quad y - 6 = 0$$
$$y = -1 \quad or \qquad y = 6$$

The solution set is $\{-1, 6\}$.

61. $f(x) = \dfrac{x-5}{x^2 - x - 56}$

$f(x)$ cannot be calculated for any x-value for which the denominator, $x^2 - x - 56$, is 0. To find the excluded values, we solve:

$$x^2 - x - 56 = 0$$
$$(x-8)(x+7) = 0$$
$$x - 8 = 0 \quad or \quad x + 7 = 0$$
$$x = 8 \quad or \qquad x = -7$$

The domain of f is

$\{x \mid x \text{ is a real number } and \ x \neq -7 \ and \ x \neq 8\}$, or

$(-\infty, -7) \cup (-7, 8) \cup (8, \infty)$.

63. *Familiarize*. Let x represent the first integer, $x+2$ the second, and $x+4$ the third.

Translate.

The sum of the squares of the numbers is 155, so we have

$$x^2 + (x+2)^2 + (x+4)^2 = 155$$

Carry out. We solve the equation:

$$x^2 + (x+2)^2 + (x+4)^2 = 155$$
$$x^2 + x^2 + 4x + 4 + x^2 + 8x + 16 = 155$$
$$3x^2 + 12x + 20 = 155$$
$$3x^2 + 12x - 135 = 0$$
$$3(x^2 + 4x - 45) = 0$$
$$3(x+9)(x-5) = 0$$
$$x + 9 = 0 \quad or \quad x - 5 = 0$$
$$x = -9 \quad or \quad x = 5$$

Check. We check the integers –9, –7, –5, and 5, 7, 9. The sum of the squares of the first set of numbers is

$(-9)^2 + (-7)^2 + (-5)^2 = 81 + 49 + 25 = 155$. The sum of the squares of the second set of numbers is

$5^2 + 7^2 + 9^2 = 25 + 49 + 81 = 155$. The answer checks.

State. The numbers are –9, –7, –5, or 5, 7, 9.

65. *Familiarize*. Let x = second leg and $x + 2$ is the hypotenuse of the triangle. The first leg is 8 ft.

Translate. We use the Pythagorean theorem.

$$8^2 + x^2 = (x+2)^2$$

Carry out. We solve the equation:

$$8^2 + x^2 = (x+2)^2$$
$$64 + x^2 = x^2 + 4x + 4$$
$$60 = 4x$$
$$15 = x$$

Check. If $x = 15$, then $x + 2 = 17$;

$8^2 + 15^2 = 64 + 225 = 289 = 17^2$, so the answer checks.

State. The lengths of the other sides are 15 ft and 17 ft.

67. *Writing Exercise*. When multiplying polynomials, we begin with a product and carry out the multiplication to write a sum of terms. When factoring a polynomial, we write an equivalent expression that is a product.

69. $128x^6 - 2y^6$

$$= 2(64x^6 - y^6) \qquad 64 = 2^6$$
$$= 2[(2x)^6 - y^6] \qquad \text{Difference of squares}$$
$$= 2[(2x)^3 - y^3][(2x)^3 + y^3] \qquad \text{Difference/sum of cubes}$$
$$= 2(2x - y)(4x^2 + 2xy + y^2)(2x + y)(4x^2 - 2xy + y^2)$$

71. $[a - (b-1)][(b-1)^2 + a(b-1) + a^2]$

$$= [a - (b-1)][a^2 + a(b-1) + (b-1)^2] \quad \text{Difference of cubes}$$
$$= a^3 - (b-1)^3$$
$$= a^3 - (b^3 - 3b^2 + 3b - 1)$$
$$= a^3 - b^3 + 3b^2 - 3b + 1$$

73.
$$(x+1)^3 = x^2(x+1)$$
$$x^3 + 3x^2 + 3x + 1 = x^3 + x^2$$
$$2x^2 + 3x + 1 = 0$$
$$(x+1)(2x+1) = 0$$
$$x + 1 = 0 \quad or \quad 2x + 1 = 0$$
$$x = -1 \quad or \quad x = -\frac{1}{2}$$

The solution set is $\left\{-1, -\frac{1}{2}\right\}$.

Chapter 5 Test

1.

Term	$8xy^3$	$-14x^2y$	$5x^5y^4$	$-9x^4y$
Degree	4	3	9	5

Degree of polynomial: 9

3. Leading term: $-5a^3$

5. $P(x) = x^2 - 3x$

$$P(a+h) - P(a) = (a+h)^2 - 3(a+h) - (a^2 - 3a)$$
$$= a^2 + 2ah + h^2 - 3a - 3h - a^2 + 3a$$
$$= 2ah + h^2 - 3h$$

7. $(-4y^3 + 6y^2 - y) + (3y^3 - 9y - 7)$

$$= (-4 + 3)y^3 + 6y^2 + (-1 - 9)y - 7$$
$$= -y^3 + 6y^2 - 10y - 7$$

9. $(8a - 4b) - (3a + 4b)$

$$= 8a - 4b - 3a - 4b$$
$$= 5a - 8b$$

11. $(-4x^2y^3)(-16xy^5)$

$$= (-4)(-16)x^{2+1}y^{3+5}$$
$$= 64x^3y^8$$

13. $(x-y)(x^2 - xy - y^2)$

$$= x(x^2 - xy - y^2) - y(x^2 - xy - y^2)$$
$$= x^3 - x^2y - xy^2 - x^2y + xy^2 + y^3$$
$$= x^3 - 2x^2y + y^3$$

15. $(5a^3 + 9)^2$

$$= (5a^3)^2 + 2 \cdot 5a^3 \cdot 9 + 9^2 \qquad (A+B)^2 = A^2 + 2AB + B^2$$
$$= 25a^6 + 90a^3 + 81$$

17. $45x^2 + 5x^4 = 5x^2\left(9 + x^2\right)$

19. $p^2 - 12p - 28$

Since the constant term is negative, we look for a factorization of -28 in which one factor is positive and one factor is negative. Their sum must be -12, so the negative factor must have the larger absolute value. Thus we consider only pairs of factors in which the negative factor has the larger absolute value.

Pair of Factors	Sum of Factors
$-28,\ 1$	-27
$-14,\ 2$	-12
$-7,\ 4$	-3

The numbers we need are -14 and 2. The factorization is $(p - 14)(p + 2)$.

21. $9y^2 - 25$ Difference of squares
$= (3y + 5)(3y - 5)$

23. $9x^2 + 25 - 30x$
$= 9x^2 - 30x + 25$ Perfect-square trinomial
$= (3x - 5)(3x - 5)$
$= (3x - 5)^2$

25. $y^2 + 8y + 16 - 100t^2$
$= \left(y^2 + 8y + 16\right) - 100t^2$ Grouping as a difference
$= (y + 4)^2 - 100t^2$ of squares
$= (y + 4 + 10t)(y + 4 - 10t)$

27. $24x^2 - 46x + 10$
$= 2\left(12x^2 - 23x + 5\right)$
$= 2(4x - 1)(3x - 5)$

29. $4y^4x + 36yx^2 + 8y^2x^3 + 4xy$
$= 4xy \cdot y^3 + 4xy \cdot 9x + 4xy \cdot 2x^2y + 4xy \cdot 1$
$= 4xy\left(y^3 + 9x + 2x^2y + 1\right)$

31. $\qquad 5y^2 = 125$
$\qquad 5y^2 - 125 = 0$
$\qquad 5\left(y^2 - 25\right) = 0$
$\qquad 5(y + 5)(y - 5) = 0$
$y + 5 = 0 \quad or \quad y - 5 = 0$
$\quad y = -5 \quad or \qquad y = 5$
The solution set is $\{-5,\ 5\}$.

33. $\quad 12r^2 + 6r = 0$
$\quad 6r(2r + 1) = 0$
$6r = 0 \quad or \quad 2r + 1 = 0$
$\ r = 0 \quad or \qquad r = -\dfrac{1}{2}$

The solution set is $\left\{-\dfrac{1}{2},\ 0\right\}$.

35. $f(x) = \dfrac{8 - x}{x^2 + 2x + 1}$

$f(x)$ cannot be calculated for any x-value for which the denominator, $x^2 + 2x + 1$, is 0. To find the excluded values, we solve:

$$x^2 + 2x + 1 = 0$$
$$(x + 1)(x + 1) = 0$$
$$x + 1 = 0 \quad or \quad x + 1 = 0$$
$$x = -1 \quad or \qquad x = -1$$

The domain of f is $\{x \mid x$ is a real number $and\ x \neq -1\}$, or $(-\infty, -1) \cup (-1,\ \infty)$.

37. *Familiarize*. Let $x =$ one side and $2x + 4$ is the other side of the triangle. The hypotenuse is 26 cm.

Translate. We use the Pythagorean theorem.

$$x^2 + (2x + 4)^2 = 26^2$$

Carry out. We solve the equation:

$$x^2 + (2x + 4)^2 = 26^2$$
$$x^2 + 4x^2 + 16x + 16 = 676$$
$$5x^2 + 16x + 16 = 676$$
$$5x^2 + 16x - 660 = 0$$
$$(5x + 66)(x - 10) = 0$$
$$5x + 66 = 0 \quad or \quad x - 10 = 0$$
$$x = -13.2 \quad or \qquad x = 10$$

Check. Since the length of a leg cannot be negative, -13.2 is not a solution. If $x = 10$, then $2x + 4 = 24$; $10^2 + 24^2 = 100 + 576 = 676 = 26^2$, so the answer checks.

State. The lengths of the legs are 10 cm and 24 cm.

39. a)
$$
\begin{array}{r}
x^2 + x + 1 \\
x^3 - x^2 + 1 \\
\hline
x^2 + x + 1 \qquad \text{Multiplying by 1}\\
- x^4 - x^3 - x^2 \qquad \text{Multiplying by } -x^2\\
x^5 + x^4 + x^3 \qquad\qquad \text{Multiplying by } x^3\\
\hline
x^5 \qquad\qquad\quad + x + 1 \qquad \text{Adding}
\end{array}
$$

b) From part (a), we see that since $\left(x^2 + x + 1\right)\left(x^3 - x^2 + 1\right) = x^5 + x + 1$, then $x^5 + x + 1 = \left(x^2 + x + 1\right)\left(x^3 - x^2 + 1\right)$.

Chapter 6

Rational Expressions, Equations, and Functions

Exercise Set 6.1

1. Since $x - 5 = 0$ when $x = 5$, choice (e) is correct.

3. Since $x - 2 = 0$ when $x = 2$, choice (g) is correct.

5. Since $(x - 2)(x - 5) = 0$ when $x = 2$ or $x = 5$, choice (i) is correct.

7. Since $(x - 2)(x + 5) = 0$ when $x = 2$ or $x = -5$, choice (a) is correct.

9. Since $x + 3 = 0$ when $x = -3$, choice (d) is correct.

11. $f(x) = \dfrac{2x^2 - x - 5}{x - 1}$

   a) $f(0) = \dfrac{2 \cdot 0^2 - 0 - 5}{0 - 1} = \dfrac{-5}{-1} = 5$

   b) $f(-1) = \dfrac{2 \cdot (-1)^2 - (-1) - 5}{(-1) - 1} = \dfrac{2 + 1 - 5}{-2} = 1$

   c) $f(3) = \dfrac{2 \cdot 3^2 - 3 - 5}{3 - 1} = \dfrac{18 - 3 - 5}{2} = 5$

13. $r(t) = \dfrac{t^2 - 8t - 9}{t^2 - 4}$

   a) $r(0) = \dfrac{0^2 - 8 \cdot 0 - 9}{0^2 - 4} = \dfrac{-9}{-4} = \dfrac{9}{4}$

   b) $r(2) = \dfrac{2^2 - 8 \cdot 2 - 9}{2^2 - 4} = \dfrac{4 - 16 - 9}{4 - 4} = \dfrac{-21}{0}$ does not exist

   c) $r(-1) = \dfrac{(-1)^2 - 8(-1) - 9}{(-1)^2 - 4} = \dfrac{1 + 8 - 9}{1 - 4} = \dfrac{0}{-3} = 0$

15. $H(t) = \dfrac{t^2 + t}{2t + 1}$

    $H(5) = \dfrac{5^2 + 5}{2 \cdot 5 + 1} = \dfrac{25 + 5}{10 + 1} = \dfrac{30}{11}$ hr, or $2\dfrac{8}{11}$ hr

17. $\dfrac{9x}{9x} \cdot \dfrac{x + 2}{x - 5} = \dfrac{9x(x + 2)}{9x(x - 5)}$

19. $\dfrac{t - 2}{t + 3} \cdot \dfrac{-1}{-1} = \dfrac{(t - 2)(-1)}{(t + 3)(-1)}$

21. $\dfrac{8t^4}{40t}$

    $= \dfrac{8t \cdot t^3}{8t \cdot 5}$ Factoring; the greatest common factor is $8t$.

    $= \dfrac{8t}{8t} \cdot \dfrac{t^3}{5}$ Factoring the rational expression

    $= 1 \cdot \dfrac{t^3}{5}$ $\dfrac{8t}{8t} = 1$

    $= \dfrac{t^3}{5}$ Removing a factor equal to 1

23. $\dfrac{24x^3 y}{30x^5 y^8}$

    $= \dfrac{6x^3 y \cdot 4}{6x^3 y \cdot 5x^2 y^7}$ Factoring the numerator and the denominator

    $= \dfrac{6x^3 y}{6x^3 y} \cdot \dfrac{4}{5x^2 y^7}$ Factoring the rational expression

    $= \dfrac{4}{5x^2 y^7}$ Removing a factor equal to 1

25. $\dfrac{2a - 10}{2} = \dfrac{2(a - 5)}{2 \cdot 1} = \dfrac{2}{2} \cdot \dfrac{a - 5}{1} = a - 5$

27. $\dfrac{15}{25y - 30} = \dfrac{5 \cdot 3}{5(5y - 6)} = \dfrac{5}{5} \cdot \dfrac{3}{5y - 6} = \dfrac{3}{5y - 6}$

29. $\dfrac{3x - 12}{3x + 15} = \dfrac{3(x - 4)}{3(x + 5)} = \dfrac{3}{3} \cdot \dfrac{x - 4}{x + 5} = \dfrac{x - 4}{x + 5}$

31. $f(x) = \dfrac{5x + 30}{x^2 + 6x}$

    $= \dfrac{5(x + 6)}{x(x + 6)}$ Note that $x \neq 0$ and $x \neq -6$

    $= \dfrac{5}{x} \cdot \dfrac{x + 6}{x + 6}$

    $= \dfrac{5}{x}$

    Thus, $f(x) = \dfrac{5}{x}$, $x \neq 0, -6$.

33. $g(x) = \dfrac{x^2 - 9}{5x + 15}$

    $= \dfrac{(x + 3)(x - 3)}{5(x + 3)}$ Note that $x \neq -3$.

    $= \dfrac{x + 3}{x + 3} \cdot \dfrac{x - 3}{5}$

    $= \dfrac{x - 3}{5}$

    Thus, $g(x) = \dfrac{x - 3}{5}$, $x \neq -3$.

35. $h(x) = \dfrac{2-x}{7x-14}$

$\quad = \dfrac{-1(x-2)}{7(x-2)} \quad$ Note that $x \neq 2$

$\quad = \dfrac{-1}{7} \cdot \dfrac{x-2}{x-2}$

$\quad = -\dfrac{1}{7}$

Thus, $h(x) = -\dfrac{1}{7},\ x \neq 2$.

37. $f(t) = \dfrac{t^2-16}{t^2-8t+16}$

$\quad = \dfrac{(t+4)(t-4)}{(t-4)(t-4)} \quad$ Note that $t \neq 4$.

$\quad = \dfrac{t+4}{t-4} \cdot \dfrac{t-4}{t-4}$

$\quad = \dfrac{t+4}{t-4}$

Thus, $f(t) = \dfrac{t+4}{t-4},\ t \neq 4$.

39. $g(t) = \dfrac{21-7t}{3t-9}$

$\quad = \dfrac{-7(t-3)}{3(t-3)} \quad$ Note that $t \neq 3$.

$\quad = \dfrac{-7}{3} \cdot \dfrac{t-3}{t-3}$

$\quad = -\dfrac{7}{3}$

Thus, $g(t) = -\dfrac{7}{3},\ t \neq 3$.

41. $h(t) = \dfrac{t^2+5t+4}{t^2-8t-9}$

$\quad = \dfrac{(t+1)(t+4)}{(t+1)(t-9)} \quad$ Note that $t \neq -1$ and $t \neq 9$.

$\quad = \dfrac{t+1}{t+1} \cdot \dfrac{t+4}{t-9}$

$\quad = \dfrac{t+4}{t-9}$

Thus, $h(t) = \dfrac{t+4}{t-9},\ t \neq -1,\ 9$.

43. $f(x) = \dfrac{9x^2-4}{3x-2}$

$\quad = \dfrac{(3x+2)(3x-2)}{3x-2} \quad$ Note that $x \neq \dfrac{2}{3}$.

$\quad = \dfrac{3x+2}{1} \cdot \dfrac{3x-2}{3x-2}$

$\quad = 3x+2$

Thus, $f(x) = 3x+2,\ x \neq \dfrac{2}{3}$.

45. $g(t) = \dfrac{16-t^2}{t^2-8t+16}$

$\quad = \dfrac{(4+t)(4-t)}{(t-4)(t-4)} \quad$ Note that $t \neq 4$.

$\quad = \dfrac{(4+t)(4-t)}{-1(4-t)(t-4)}$

$\quad = \dfrac{4-t}{4-t} \cdot \dfrac{4+t}{-1(t-4)}$

$\quad = \dfrac{4+t}{4-t}$

Thus, $g(t) = \dfrac{4+t}{4-t},\ t \neq 4$. (We could also write this

as $g(t) = \dfrac{-t-4}{t-4},\ t \neq 4$.)

47. $\dfrac{3y^3}{5z} \cdot \dfrac{10z^4}{7y^6}$

$\quad = \dfrac{3y^3 \cdot 10z^4}{5z \cdot 7y^6} \quad$ Multiplying the numerators and also the denominators

$\quad = \dfrac{3 \cdot y^3 \cdot 2 \cdot 5 \cdot z \cdot z^3}{5 \cdot z \cdot 7 \cdot y^3 \cdot y^3} \quad$ Factoring the numerator and denominator

$\quad = \dfrac{3 \cdot y^3 \cdot 2 \cdot 5 \cdot z \cdot z^3}{5 \cdot z \cdot 7 \cdot y^3 \cdot y^3} \quad$ Removing a factor equal to 1

$\quad = \dfrac{6z^3}{7y^3}$

49. $\dfrac{8x-16}{5x} \cdot \dfrac{x^3}{5x-10} = \dfrac{(8x-16)(x^3)}{5x(5x-10)}$

$\quad = \dfrac{8(x-2)(x)(x^2)}{5 \cdot x \cdot 5(x-2)}$

$\quad = \dfrac{8(x-2)(x)(x^2)}{5 \cdot x \cdot 5(x-2)}$

$\quad = \dfrac{8x^2}{25}$

51. $\dfrac{y^2-9}{y^2} \cdot \dfrac{y^2-3y}{y^2-y-6} = \dfrac{(y^2-9)(y^2-3y)}{y^2(y^2-y-6)}$

$\quad = \dfrac{(y+3)(y-3)(y)(y-3)}{y \cdot y(y-3)(y+2)}$

$\quad = \dfrac{(y+3)(y-3)(y)(y-3)}{y \cdot y(y-3)(y+2)}$

$\quad = \dfrac{(y+3)(y-3)}{y(y+2)}$

53. $\dfrac{7a-14}{4-a^2} \cdot \dfrac{5a^2+6a+1}{35a+7}$

$= \dfrac{(7a-14)(5a^2+6a+1)}{(4-a^2)(35a+7)}$

$= \dfrac{7(a-2)(5a+1)(a+1)}{(2+a)(2-a)(7)(5a+1)}$

$= \dfrac{7(-1)(2-a)(5a+1)(a+1)}{(2+a)(2-a)(7)(5a+1)}$

$= \dfrac{\cancel{7}(-1)\cancel{(2-a)}\,\cancel{(5a+1)}\,(a+1)}{(2+a)\cancel{(2-a)}\,\cancel{(7)}\,\cancel{(5a+1)}}$

$= \dfrac{-1(a+1)}{2+a}$

$= \dfrac{-a-1}{2+a}, \text{ or } -\dfrac{a+1}{2+a}$

55. $\dfrac{t^3-4t}{t-t^4} \cdot \dfrac{t^4-t}{4t-t^3}$

$= \dfrac{t^3-4t}{t-t^4} \cdot \dfrac{-1(t-t^4)}{-1(t^3-4t)}$

$= \dfrac{\cancel{(t^3-4t)}\,\cancel{(-1)}\,\cancel{(t-t^4)}}{\cancel{(t-t^4)}\,\cancel{(-1)}\,\cancel{(t^3-4t)}}$

$= 1$

57. $\dfrac{c^3+8}{c^5-4c^3} \cdot \dfrac{c^6-4c^5+4c^4}{c^2-2c+4}$

$= \dfrac{(c^3+8)(c^6-4c^5+4c^4)}{(c^5-4c^3)(c^2-2c+4)}$

$= \dfrac{(c+2)(c^2-2c+4)(c^4)(c-2)(c-2)}{c^3(c+2)(c-2)(c^2-2c+4)}$

$= \dfrac{c^3(c+2)(c^2-2c+4)(c-2)}{c^3(c+2)(c^2-2c+4)(c-2)} \cdot \dfrac{c(c-2)}{1}$

$= c(c-2)$

59. $\dfrac{a^3-b^3}{3a^2+9ab+6b^2} \cdot \dfrac{a^2+2ab+b^2}{a^2-b^2}$

$= \dfrac{(a^3-b^3)(a^2+2ab+b^2)}{(3a^2+9ab+6b^2)(a^2-b^2)}$

$= \dfrac{(a-b)(a^2+ab+b^2)(a+b)(a+b)}{3(a+b)(a+2b)(a+b)(a-b)}$

$= \dfrac{\cancel{(a-b)}\,(a^2+ab+b^2)\,\cancel{(a+b)}\,\cancel{(a+b)}}{3\,\cancel{(a+b)}\,(a+2b)\,\cancel{(a+b)}\,\cancel{(a-b)}}$

$= \dfrac{a^2+ab+b^2}{3(a+2b)}$

61. $\dfrac{12a^3}{5b^2} \div \dfrac{4a^2}{15b}$

$= \dfrac{12a^3}{5b^2} \cdot \dfrac{15b}{4a^2}$ Multiplying by the reciprocal

                    of the divisor

$= \dfrac{12a^3(15b)}{5b^2(4a^2)}$

$= \dfrac{3\cdot4\cdot a^2\cdot5\cdot3\cdot b}{5\cdot b\cdot b\cdot4\cdot a^2}$

$= \dfrac{3\cdot\cancel{4}\cdot\cancel{a^2}\cdot\cancel{5}\cdot3\cdot\cancel{b}}{\cancel{5}\cdot\cancel{b}\cdot b\cdot\cancel{4}\cdot\cancel{a^2}}$

$= \dfrac{9a}{b}$

63. $\dfrac{5x+20}{x^6} \div \dfrac{x+4}{x^2} = \dfrac{5x+20}{x^6} \cdot \dfrac{x^2}{x+4}$

$= \dfrac{(5x+20)(x^2)}{x^6(x+4)}$

$= \dfrac{5\cancel{(x+4)}\,\cancel{(x^2)}}{\cancel{x^2}(x^4)\cancel{(x+4)}}$

$= \dfrac{5}{x^4}$

65. $\dfrac{25x^2-4}{x^2-9} \div \dfrac{2-5x}{x+3} = \dfrac{25x^2-4}{x^2-9} \cdot \dfrac{x+3}{2-5x}$

$= \dfrac{(25x^2-4)(x+3)}{(x^2-9)(2-5x)}$

$= \dfrac{(5x+2)(5x-2)(x+3)}{(x+3)(x-3)(-1)(5x-2)}$

$= \dfrac{(5x+2)\cancel{(5x-2)}\,\cancel{(x+3)}}{\cancel{(x+3)}\,(x-3)(-1)\cancel{(5x-2)}}$

$= \dfrac{5x+2}{-x+3}, \text{ or } -\dfrac{5x+2}{x-3}$

67. $\dfrac{5y-5x}{15y^3} \div \dfrac{x^2-y^2}{3x+3y} = \dfrac{5y-5x}{15y^3} \cdot \dfrac{3x+3y}{x^2-y^2}$

$= \dfrac{(5y-5x)(3x+3y)}{15y^3(x^2-y^2)}$

$= \dfrac{5(y-x)(3)(x+y)}{5\cdot3\cdot y^3(x+y)(x-y)}$

$= \dfrac{5(-1)(x-y)(3)(x+y)}{5\cdot3\cdot y^3(x+y)(x-y)}$

$= \dfrac{\cancel{5}(-1)\cancel{(x-y)}\,\cancel{(3)}\,\cancel{(x+y)}}{\cancel{5}\cdot\cancel{3}\cdot y^3\cancel{(x+y)}\,\cancel{(x-y)}}$

$= \dfrac{-1}{y^3}, \text{ or } -\dfrac{1}{y^3}$

69. $\dfrac{y^2-36}{y^2-8y+16} \div \dfrac{3y-18}{y^2-y-12}$

$= \dfrac{y^2-36}{y^2-8y+16} \cdot \dfrac{y^2-y-12}{3y-18}$

$= \dfrac{(y+6)(y-6)(y-4)(y+3)}{(y-4)(y-4)3(y-6)}$

$= \dfrac{(y+6)\cancel{(y-6)}\,\cancel{(y-4)}\,(y+3)}{\cancel{(y-4)}(y-4)(3)\cancel{(y-6)}}$

$= \dfrac{(y+6)(y+3)}{3(y-4)}$

71. $\dfrac{x^3-64}{x^3+64} \div \dfrac{x^2-16}{x^2-4x+16}$

$= \dfrac{x^3-64}{x^3+64} \cdot \dfrac{x^2-4x+16}{x^2-16}$

$= \dfrac{(x^3-64)(x^2-4x+16)}{(x^3+64)(x^2-16)}$

$= \dfrac{(x-4)(x^2+4x+16)(x^2-4x+16)}{(x+4)(x^2-4x+16)(x+4)(x-4)}$

$= \dfrac{(x-4)(x^2-4x+16)}{(x-4)(x^2-4x+16)} \cdot \dfrac{x^2+4x+16}{(x+4)(x+4)}$

$= \dfrac{x^2+4x+16}{(x+4)(x+4)}, \text{ or } \dfrac{x^2+4x+16}{(x+4)^2}$

73. $f(t) = \dfrac{t^2-100}{5t+20} \cdot \dfrac{t+4}{t-10}$

The denominators of the rational expressions are $5t+20$ and $t-10$, so the domain of f cannot contain -4 or 10.

$f(t) = \dfrac{t^2-100}{5t+20} \cdot \dfrac{t+4}{t-10}$

$= \dfrac{(t^2-100)(t+4)}{(5t+20)(t-10)}$

$= \dfrac{(t+10)(t-10)(t+4)}{5(t+4)(t-10)}$

$= \dfrac{(t+10)\cancel{(t-10)}\,\cancel{(t+4)}}{5\cancel{(t+4)}\,\cancel{(t-10)}}$

$= \dfrac{t+10}{5}, \quad t \neq -4, \, 10$

75. $g(x) = \dfrac{x^2-2x-35}{2x^3-3x^2} \cdot \dfrac{4x^3-9x}{7x-49}$

First we find the values of x for which the denominator $2x^3-3x^2 = 0$.

$2x^3-3x^2 = 0$

$x^2(2x-3) = 0$

$x \cdot x \cdot (2x-3) = 0$

$x = 0 \quad or \quad x = 0 \quad or \quad 2x-3 = 0$

$x = 0 \quad or \quad x = 0 \quad or \quad x = \dfrac{3}{2}$

We see that the domain cannot contain 0 or $\dfrac{3}{2}$. Since

$7x-49 = 0$ when $x = 7$, the number 7 must also be

excluded from the domain.

$g(x) = \dfrac{x^2-2x-35}{2x^3-3x^2} \cdot \dfrac{4x^3-9x}{7x-49}$

$= \dfrac{(x^2-2x-35)(4x^3-9x)}{(2x^3-3x^2)(7x-49)}$

$= \dfrac{(x-7)(x+5)(x)(2x+3)(2x-3)}{x \cdot x \cdot (2x-3)(7)(x-7)}$

$= \dfrac{\cancel{(x-7)}(x+5)\cancel{(x)}(2x+3)\cancel{(2x-3)}}{\cancel{x} \cdot x \cdot \cancel{(2x-3)}(7)\cancel{(x-7)}}$

$= \dfrac{(x+5)(2x+3)}{7x}, \quad x \neq 0, \, \dfrac{3}{2}, \, 7$

77. $f(x) = \dfrac{x^2-4}{x^3} \div \dfrac{x^5-2x^4}{x+4}$

$x^3 = 0$ when $x = 0$ and $x+4 = 0$ when $x = -4$, so the domain cannot contain 0 or -4. Also, division by $(x^5-2x^4)/(x+4)$ is defined only when this expression is nonzero. We find the values of x for which x^5-2x^4 is zero.

$x^5-2x^4 = 0$

$x^4(x-2) = 0$

The solutions of this equation are 0 and 2 so, in addition to 0 and -4, we must also exclude 2 from the domain.

$f(x) = \dfrac{x^2-4}{x^3} \div \dfrac{x^5-2x^4}{x+4}$

$= \dfrac{x^2-4}{x^3} \cdot \dfrac{x+4}{x^5-2x^4}$

$= \dfrac{(x^2-4)(x+4)}{x^3(x^5-2x^4)}$

$= \dfrac{(x+2)(x-2)(x+4)}{x^3 \cdot x^4(x-2)}$

$= \dfrac{(x+2)\cancel{(x-2)}(x+4)}{x^7\cancel{(x-2)}}$

$= \dfrac{(x+2)(x+4)}{x^7}, \quad x \neq -4, \, 0, \, 2$

79. $h(n) = \dfrac{n^3+3n}{n^2-9} \div \dfrac{n^2+5n-14}{n^2+4n-21}$

First we find the values of n for which the denominators of the rational expressions are zero.

$n^2-9 = 0$

$(n+3)(n-3) = 0$

$n+3 = 0 \quad or \quad n-3 = 0$

$n = -3 \quad or \quad n = 3$

$n^2+4n-21 = 0$

$(n+7)(n-3) = 0$

$n+7 = 0 \quad or \quad n-3 = 0$

$n = -7 \quad or \quad n = 3$

We see that the domain cannot contain -7, -3, or 3. Now we find the values of n for which

$(n^2+5n-14)/(n^2+4n-21)$ is zero.

$$n^2+5n-14=0$$
$$(n+7)(n-2)=0$$
$$n+7=0 \quad or \quad n-2=0$$
$$n=-7 \quad or \quad n=2$$

We must also exclude 2 from the domain.

$$h(n)=\frac{n^3+3n}{n^2-9}\div\frac{n^2+5n-14}{n^2+4n-21}$$
$$=\frac{n^3+3n}{n^2-9}\cdot\frac{n^2+4n-21}{n^2+5n-14}$$
$$=\frac{(n^3+3n)(n^2+4n-21)}{(n^2-9)(n^2+5n-14)}$$
$$=\frac{n(n^2+3)(n+7)(n-3)}{(n+3)(n-3)(n+7)(n-2)}$$
$$=\frac{n(n^2+3)\cancel{(n+7)}\cancel{(n-3)}}{(n+3)\cancel{(n-3)}\cancel{(n+7)}(n-2)}$$
$$=\frac{n(n^2+3)}{(n+3)(n-2)},\quad n\neq -7,\ -3,\ 2,\ 3$$

81. $\dfrac{4x^2-9y^2}{8x^3-27y^3}\div\dfrac{4x+6y}{3x-9y}\cdot\dfrac{4x^2+6xy+9y^2}{4x^2-8xy+3y^2}$

$$=\frac{4x^2-9y^2}{8x^3-27y^3}\cdot\frac{3x-9y}{4x+6y}\cdot\frac{4x^2+6xy+9y^2}{4x^2-8xy+3y^2}$$
$$=\frac{(4x^2-9y^2)(3x-9y)(4x^2+6xy+9y^2)}{(8x^3-27y^3)(4x+6y)(4x^2-8xy+3y^2)}$$
$$=\frac{(2x+3y)(2x-3y)(3)(x-3y)(4x^2+6xy+9y^2)}{(2x-3y)(4x^2+6xy+9y^2)(2)(2x+3y)(2x-y)(2x-3y)}$$
$$=\frac{(2x+3y)(2x-3y)(4x^2+6xy+9y^2)}{(2x+3y)(2x-3y)(4x^2+6xy+9y^2)}\cdot\frac{3(x-3y)}{2(2x-y)(2x-3y)}$$
$$=\frac{3(x-3y)}{2(2x-y)(2x-3y)}$$

83. $\dfrac{a^3-ab^2}{2a^2+3ab+b^2}\cdot\dfrac{4a^2-b^2}{a^2-2ab+b^2}\div\dfrac{a^2+a}{a-1}$

$$=\frac{a^3-ab^2}{2a^2+3ab+b^2}\cdot\frac{4a^2-b^2}{a^2-2ab+b^2}\cdot\frac{a-1}{a^2+a}$$
$$=\frac{(a^3-ab^2)(4a^2-b^2)(a-1)}{(2a^2+3ab+b^2)(a^2-2ab+b^2)(a^2+a)}$$
$$=\frac{a(a+b)(a-b)(2a+b)(2a-b)(a-1)}{(2a+b)(a+b)(a-b)(a-b)(a)(a+1)}$$
$$=\frac{a(a+b)(a-b)(2a+b)}{a(a+b)(a-b)(2a+b)}\cdot\frac{(2a-b)(a-1)}{(a-b)(a+1)}$$
$$=\frac{(2a-b)(a-1)}{(a-b)(a+1)}$$

85. *Writing Exercise.* Nancy's misconception is that x is a factor of the numerator. $\left(\dfrac{x+2}{x}=3 \text{ only for } x=1.\right)$

87. $\dfrac{7}{12}-\dfrac{2}{15}=\dfrac{7}{12}\cdot\dfrac{5}{5}-\dfrac{2}{15}\cdot\dfrac{4}{4}$
$$=\frac{35}{60}-\frac{8}{60}$$
$$=\frac{27}{60}$$
$$=\frac{9}{20}$$

89. $\dfrac{1}{5}\cdot\dfrac{3}{4}-\dfrac{7}{10}\cdot\dfrac{3}{5}=\dfrac{3}{20}-\dfrac{21}{50}$
$$=\frac{3}{20}\cdot\frac{5}{5}-\frac{21}{50}\cdot\frac{2}{2}$$
$$=\frac{15}{100}-\frac{42}{100}$$
$$=-\frac{27}{100}$$

91. $\left(5x^2-6x+1\right)-\left(x^2-6x+3\right)$
$$=5x^2-6x+1-x^2+6x-3$$
$$=4x^2-2$$

93. $(y+1)(y-2)-(y+3)(y-5)$
$$=y^2-y-2-\left(y^2-2y-15\right)$$
$$=y^2-y-2-y^2+2y+15$$
$$=y+13$$

95. *Writing Exercise.* The graph of $g(x)=\dfrac{5x^2}{x}$ is identical to the graph of $f(x)=5x$ except that it does not contain the point $(0, 0)$.

97. $m=\dfrac{f(a+h)-f(a)}{(a+h)-a}$
$$=\frac{(a+h)^2+5-(a^2+5)}{a+h-a}$$
$$=\frac{a^2+2ah+h^2+5-a^2-5}{h}$$
$$=\frac{2ah+h^2}{h}=\frac{h(2a+h)}{h}$$
$$=2a+h$$

99. To find the domain of f we set
$$x-3=0$$
$$x=3.$$

The domain of $f=\{x\,|\,x$ is a real number and $x\neq 3\}$.

Simplify: $f(x)=\dfrac{x^2-9}{x-3}=\dfrac{(x+3)(x-3)}{(x-3)\cdot 1}$
$$=\frac{x-3}{x-3}\cdot\frac{x+3}{1}=x+3$$

Graph $f(x)=x+3$ using the domain found above.

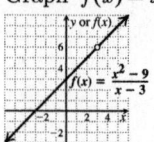

101. $\dfrac{d^2-d}{d^2-6d+8}\cdot\dfrac{d-2}{d^2+5d}\div\left(\dfrac{5d^2}{d^2-9d+20}\right)^2$

$=\dfrac{d^2-d}{d^2-6d+8}\cdot\dfrac{d-2}{d^2+5d}\cdot\dfrac{(d^2-9d+20)^2}{(5d^2)^2}$

$=\dfrac{d(d-1)(d-2)[(d-4)(d-5)]^2}{(d-2)(d-4)(d)(d+5)(25d^4)}$

$=\dfrac{\cancel{d}(d-1)\cancel{(d-2)}\cancel{(d-4)}(d-4)(d-5)^2}{\cancel{(d-2)}\cancel{(d-4)}\cancel{(d)}(d+5)(25d^4)}$

$=\dfrac{(d-1)(d-4)(d-5)^2}{25d^4(d+5)}$

103. $\dfrac{m^2-t^2}{m^2+t^2+m+t+2mt}$

$=\dfrac{m^2-t^2}{(m^2+2mt+t^2)+(m+t)}$

$=\dfrac{(m+t)(m-t)}{(m+t)^2+(m+t)}$

$=\dfrac{(m+t)(m-t)}{(m+t)[(m+t)+1]}$

$=\dfrac{\cancel{(m+t)}(m-t)}{\cancel{(m+t)}(m+t+1)}$

$=\dfrac{m-t}{m+t+1}$

105. $\dfrac{x^3+x^2-y^3-y^2}{x^2-2xy+y^2}$

$=\dfrac{(x^3-y^3)+(x^2-y^2)}{x^2-2xy+y^2}$

$=\dfrac{(x-y)(x^2+xy+y^2)+(x+y)(x-y)}{(x-y)^2}$

$=\dfrac{\cancel{(x-y)}(x^2+xy+y^2+x+y)}{(x-y)\cancel{(x-y)}}$

$=\dfrac{x^2+xy+y^2+x+y}{x-y}$

107. $\dfrac{x^5-x^3+x^2-1-(x^3-1)(x+1)^2}{(x^2-1)^2}$

$=\dfrac{x^5-x^3+(x^2-1)-\left[(x^3-1)(x+1)^2\right]}{(x^2-1)^2}$

$=\dfrac{x^3(x^2-1)+(x^2-1)-\left[(x-1)(x^2+x+1)(x+1)(x+1)\right]}{(x^2-1)^2}$

$=\dfrac{x^3(x^2-1)+(x^2-1)-\left[(x^2-1)(x+1)(x^2+x+1)\right]}{(x^2-1)^2}$

$=\dfrac{(x^2-1)\left[x^3+1-(x^3+x^2+x+x^2+x+1)\right]}{(x^2-1)(x^2-1)}$

$=\dfrac{\cancel{(x^2-1)}(-2x^2-2x)}{(x^2-1)\cancel{(x^2-1)}}$

$=\dfrac{-2x^2-2x}{x^2-1}$

$=\dfrac{-2x(x+1)}{(x+1)(x-1)}$

$=\dfrac{-2x\cancel{(x+1)}}{\cancel{(x+1)}(x-1)}$

$=\dfrac{-2x}{x-1}$, or $-\dfrac{2x}{x-1}$

109. a) $(f\cdot g)(x)=\dfrac{4}{x^2-1}\cdot\dfrac{4x^2+8x+4}{x^3-1}$

$=\dfrac{4(4x^2+8x+4)}{(x^2-1)(x^3-1)}$

$=\dfrac{4\cdot4\,(x+1)(x+1)}{\cancel{(x+1)}(x-1)(x-1)(x^2+x+1)}$

$=\dfrac{4\cdot4(x+1)(x+1)}{(x+1)(x-1)(x-1)(x^2+x+1)}$

$=\dfrac{16(x+1)}{(x-1)^2(x^2+x+1)}$

(Note that $x\neq-1$ is an additional restriction, since -1 is not in the domain of f.)

b) $(f/g)(x)=\dfrac{4}{x^2-1}\div\dfrac{4x^2+8x+4}{x^3-1}$

$=\dfrac{4}{x^2-1}\cdot\dfrac{x^3-1}{4x^2+8x+4}$

$=\dfrac{4(x^3-1)}{(x^2-1)(4x^2+8x+4)}$

$=\dfrac{4(x-1)(x^2+x+1)}{(x+1)(x-1)(4)(x+1)(x+1)}$

$=\dfrac{\cancel{4}\,(x-1)(x^2+x+1)}{(x+1)\cancel{(x-1)}\,\cancel{(4)}(x+1)(x+1)}$

$=\dfrac{x^2+x+1}{(x+1)^3}$

(Note that $x\neq1$ is an additional restriction, since 1 is not in the domain of either f or g.)

c) $(g/f)(x) = \dfrac{1}{(f/g)(x)}$

$= \dfrac{(x+1)^3}{x^2+x+1}$ (See part (b) above)

(Note that $x \ne -1$ and $x \ne 1$ are restrictions, since -1 is not in the domain of f and 1 is not in the domain of either f or g.)

111. *Graphing Calculator Exercise*

113. *Writing Exercise.* Look at a table of values for y_1 and y_2 and observe that, for $x = 0$, the entry in the Y1-column is ERROR but the entry in the Y2-column is 1.5.

Exercise Set 6.2

1. True; see page 363 in the text.

3. False; see page 363 in the text.

5. False; see Example 4.

7. False; see page 364 in the text.

9. $\dfrac{4}{3a} + \dfrac{11}{3a} = \dfrac{15}{3a}$ Adding the numerators. The denominator is unchanged.

$= \dfrac{3 \cdot 5}{3 \cdot a}$

$= \dfrac{\cancel{3} \cdot 5}{\cancel{3} \cdot a}$

$= \dfrac{5}{a}$

11. $\dfrac{5}{3m^2n^2} - \dfrac{4}{3m^2n^2} = \dfrac{1}{3m^2n^2}$

13. $\dfrac{x-3y}{x+y} + \dfrac{x+5y}{x+y} = \dfrac{2x+2y}{x+y} = \dfrac{2(x+y)}{x+y} = \dfrac{2\cancel{(x+y)}}{1\cancel{(x+y)}} = 2$

15. $\dfrac{3t+2}{t-4} - \dfrac{t-2}{t-4} = \dfrac{3t+2-(t-2)}{t-4} = \dfrac{3t+2-t+2}{t-4} = \dfrac{2t+4}{t-4}$

17. $\dfrac{5-7x}{x^2-3x-10} + \dfrac{8x-3}{x^2-3x-10} = \dfrac{x+2}{x^2-3x-10}$

$= \dfrac{x+2}{(x+2)(x-5)}$

$= \dfrac{\cancel{x+2}}{\cancel{(x+2)}(x-5)}$

$= \dfrac{1}{x-5}$

19. $\dfrac{a-2}{a^2-25} - \dfrac{2a-7}{a^2-25} = \dfrac{a-2-(2a-7)}{a^2-25}$

$= \dfrac{a-2-2a+7}{a^2-25}$

$= \dfrac{-a+5}{a^2-25}$

$= \dfrac{-1(a-5)}{(a+5)(a-5)}$

$= \dfrac{-1\cancel{(a-5)}}{(a+5)\cancel{(a-5)}}$

$= \dfrac{-1}{a+5}$

21. $8x^2 = 2^3 x^2$, $12x^5 = 2^2 \cdot 3x^5$

LCM $= 2^3 \cdot 3x^5 = 24x^5$

23. $x^2 - 9 = (x+3)(x-3)$, $x^2 - 6x + 9 = (x-3)^2$

LCM $= (x+3)(x-3)^2$

25. $f(x) = \dfrac{2x+1}{x^2+6x+5} + \dfrac{x-2}{x^2+6x+5}$ Note that $x \ne -5, -1$

$= \dfrac{3x-1}{x^2+6x+5}$, $x \ne -5, -1$

27. $f(x) = \dfrac{x-4}{x^2-1} - \dfrac{2x+1}{x^2-1}$ Note that $x \ne -1, 1$

$= \dfrac{x-4-(2x+1)}{x^2-1}$

$= \dfrac{x-4-2x-1}{x^2-1}$

$= \dfrac{-x-5}{x^2-1}$, $x \ne -1, 1$

29. $\dfrac{2}{15x^2} + \dfrac{3}{5x}$ LCD is $15x^2$

$= \dfrac{2}{15x^2} + \dfrac{3}{5x} \cdot \dfrac{3x}{3x}$

$= \dfrac{2}{15x^2} + \dfrac{9x}{15x^2}$

$= \dfrac{9x+2}{15x^2}$

31. $\dfrac{y+1}{y-2} - \dfrac{y-1}{2y-4} = \dfrac{y+1}{y-2} - \dfrac{y-1}{2(y-2)}$ LCD is $2(y-2)$

$= \dfrac{y+1}{y-2} \cdot \dfrac{2}{2} - \dfrac{y-1}{2(y-2)}$

$= \dfrac{2(y+1)-(y-1)}{2(y-2)}$

$= \dfrac{2y+2-y+1}{2(y-2)}$

$= \dfrac{y+3}{2(y-2)}$

33. $\dfrac{4xy}{x^2-y^2}+\dfrac{x-y}{x+y}$

$=\dfrac{4xy}{(x+y)(x-y)}+\dfrac{x-y}{x+y}$ LCD is $(x+y)(x-y)$.

$=\dfrac{4xy}{(x+y)(x-y)}+\dfrac{x-y}{x+y}\cdot\dfrac{x-y}{x-y}$

$=\dfrac{4xy+x^2-2xy+y^2}{(x+y)(x-y)}$

$=\dfrac{x^2+2xy+y^2}{(x+y)(x-y)}=\dfrac{(x+y)(x+y)}{(x+y)(x-y)}$

$=\dfrac{\cancel{(x+y)}(x+y)}{\cancel{(x+y)}(x-y)}=\dfrac{x+y}{x-y}$

35. $\dfrac{8}{2x^2-7x+5}+\dfrac{3x+2}{2x^2-x-10}$

$=\dfrac{8}{(2x-5)(x-1)}+\dfrac{3x+2}{(2x-5)(x+2)}$

     LCD is $(2x-5)(x-1)(x+2)$.

$=\dfrac{8}{(2x-5)(x-1)}\cdot\dfrac{x+2}{x+2}+\dfrac{3x+2}{(2x-5)(x+2)}\cdot\dfrac{x-1}{x-1}$

$=\dfrac{8x+16+3x^2-x-2}{(2x-5)(x-1)(x+2)}$

$=\dfrac{3x^2+7x+14}{(2x-5)(x-1)(x+2)}$

37. $\dfrac{5ab}{a^2-b^2}-\dfrac{a-b}{a+b}$

$=\dfrac{5ab}{(a+b)(a-b)}-\dfrac{a-b}{a+b}$ LCD is $(a+b)(a-b)$.

$=\dfrac{5ab}{(a+b)(a-b)}-\dfrac{a-b}{a+b}\cdot\dfrac{a-b}{a-b}$

$=\dfrac{5ab-\left(a^2-2ab+b^2\right)}{(a+b)(a-b)}$

$=\dfrac{5ab-a^2+2ab-b^2}{(a+b)(a-b)}$

$=\dfrac{-a^2+7ab-b^2}{(a+b)(a-b)}$

39. $\dfrac{x}{x^2+9x+20}-\dfrac{4}{x^2+7x+12}$

$=\dfrac{x}{(x+5)(x+4)}-\dfrac{4}{(x+3)(x+4)}$

     [LCD is $(x+5)(x+4)(x+3)$.]

$=\dfrac{x}{(x+5)(x+4)}\cdot\dfrac{x+3}{x+3}-\dfrac{4}{(x+3)(x+4)}\cdot\dfrac{x+5}{x+5}$

$=\dfrac{x^2+3x-(4x+20)}{(x+5)(x+4)(x+3)}$

$=\dfrac{x^2+3x-4x-20}{(x+5)(x+4)(x+3)}$

$=\dfrac{x^2-x-20}{(x+5)(x+4)(x+3)}$

$=\dfrac{(x-5)(x+4)}{(x+5)(x+4)(x+3)}$

$=\dfrac{(x-5)\cancel{(x+4)}}{(x+5)\cancel{(x+4)}(x+3)}$

$=\dfrac{x-5}{(x+5)(x+3)}$

41. $\dfrac{3}{t}-\dfrac{6}{-t}=\dfrac{3}{t}-\dfrac{-1}{-1}\cdot\dfrac{6}{-t}=\dfrac{3}{t}+\dfrac{6}{t}=\dfrac{9}{t}$

43. $\dfrac{s^2}{r-s}+\dfrac{r^2}{s-r}=\dfrac{s^2}{r-s}+\dfrac{-1}{-1}\cdot\dfrac{r^2}{s-r}$

$=\dfrac{s^2}{r-s}+\dfrac{-r^2}{r-s}$

$=\dfrac{s^2-r^2}{r-s}$

$=\dfrac{(s-r)(s+r)}{-1(s-r)}$

$=\dfrac{\cancel{(s-r)}(s+r)}{-1\cancel{(s-r)}}$

$=-(s+r)$

45. $\dfrac{a+2}{a-4}+\dfrac{a-2}{a+3}$ LCD is $(a-4)(a+3)$.

$=\dfrac{a+2}{a-4}\cdot\dfrac{a+3}{a+3}+\dfrac{a-2}{a+3}\cdot\dfrac{a-4}{a-4}$

$=\dfrac{(a^2+5a+6)+(a^2-6a+8)}{(a-4)(a+3)}$

$=\dfrac{2a^2-a+14}{(a-4)(a+3)}$

47. $4+\dfrac{x-3}{x+1}=\dfrac{4}{1}+\dfrac{x-3}{x+1}$ [LCD is $x+1$.]

$=\dfrac{4}{1}\cdot\dfrac{x+1}{x+1}+\dfrac{x-3}{x+1}$

$=\dfrac{(4x+4)+(x-3)}{x+1}$

$=\dfrac{5x+1}{x+1}$

49. $\dfrac{x+6}{5x+10}-\dfrac{x-2}{4x+8}$

$=\dfrac{x+6}{5(x+2)}-\dfrac{x-2}{4(x+2)}$ [LCD is $5\cdot4(x+2)$.]

$=\dfrac{x+6}{5(x+2)}\cdot\dfrac{4}{4}-\dfrac{x-2}{4(x+2)}\cdot\dfrac{5}{5}$

$=\dfrac{4(x+6)-5(x-2)}{5\cdot4(x+2)}$

$=\dfrac{4x+24-5x+10}{5\cdot4(x+2)}$

$=\dfrac{-x+34}{5\cdot4(x+2)}$, or $\dfrac{-x+34}{20(x+2)}$

51. $\dfrac{4}{x+1}+\dfrac{x+2}{x^2-1}+\dfrac{3}{x-1}$

$=\dfrac{4}{x+1}+\dfrac{x+2}{(x+1)(x-1)}+\dfrac{3}{x-1}$ [LCD is $(x+1)(x-1)$.]

$=\dfrac{4}{x+1}\cdot\dfrac{x-1}{x-1}+\dfrac{x+2}{(x+1)(x-1)}+\dfrac{3}{x-1}\cdot\dfrac{x+1}{x+1}$

$=\dfrac{4x-4+x+2+3x+3}{(x+1)(x-1)}$

$=\dfrac{8x+1}{(x+1)(x-1)}$

53. $\dfrac{y-4}{y^2-25}-\dfrac{9-2y}{25-y^2}=\dfrac{y-4}{y^2-25}-\dfrac{-1}{-1}\cdot\dfrac{9-2y}{25-y^2}$

$$=\dfrac{y-4}{y^2-25}-\dfrac{2y-9}{y^2-25}$$

$$=\dfrac{y-4-(2y-9)}{y^2-25}$$

$$=\dfrac{y-4-2y+9}{y^2-25}$$

$$=\dfrac{-y+5}{y^2-25}$$

$$=\dfrac{-1(y-5)}{(y+5)(y-5)}$$

$$=\dfrac{-1\cancel{(y-5)}}{(y+5)\cancel{(y-5)}}$$

$$=\dfrac{-1}{y+5},\text{ or }-\dfrac{1}{y+5}$$

55. $\dfrac{y^2-5}{y^4-81}+\dfrac{4}{81-y^4}$

$$=\dfrac{y^2-5}{y^4-81}+\dfrac{-1}{-1}\cdot\dfrac{4}{81-y^4}$$

$$=\dfrac{y^2-5}{y^4-81}+\dfrac{-4}{y^4-81}$$

$$=\dfrac{y^2-5+(-4)}{y^4-81}$$

$$=\dfrac{y^2-9}{y^4-81}=\dfrac{y^2-9}{(y^2+9)(y^2-9)}$$

$$=\dfrac{y^2-9}{y^2-9}\cdot\dfrac{1}{y^2+9}$$

$$=\dfrac{1}{y^2+9}$$

57. $\dfrac{r-6s}{r^3-s^3}-\dfrac{5s}{s^3-r^3}=\dfrac{r-6s}{r^3-s^2}-\dfrac{-1}{-1}\cdot\dfrac{5s}{s^3-r^3}$

$$=\dfrac{r-6s}{r^3-s^3}-\dfrac{-5s}{r^3-s^3}$$

$$=\dfrac{r-6s-(-5s)}{r^3-s^3}$$

$$=\dfrac{r-s}{(r-s)(r^2+rs+s^2)}$$

$$=\dfrac{\cancel{r-s}}{\cancel{(r-s)}\,(r^2+rs+s^2)}$$

$$=\dfrac{1}{r^2+rs+s^2}$$

59. $\dfrac{3y}{y^2-7y+10}-\dfrac{2y}{y^2-8y+15}$

$$=\dfrac{3y}{(y-5)(y-2)}-\dfrac{2y}{(y-5)(y-3)}$$
   [LCD is $(y-5)(y-2)(y-3)$.]

$$=\dfrac{3y}{(y-5)(y-2)}\cdot\dfrac{y-3}{y-3}-\dfrac{2y}{(y-5)(y-3)}\cdot\dfrac{y-2}{y-2}$$

$$=\dfrac{3y^2-9y-(2y^2-4y)}{(y-5)(y-2)(y-3)}$$

$$=\dfrac{3y^2-9y-2y^2+4y}{(y-5)(y-2)(y-3)}$$

$$=\dfrac{y^2-5y}{(y-5)(y-2)(y-3)}=\dfrac{y(y-5)}{(y-5)(y-2)(y-3)}$$

$$=\dfrac{y\cancel{(y-5)}}{\cancel{(y-5)}(y-2)(y-3)}=\dfrac{y}{(y-2)(y-3)}$$

61. $\dfrac{2x+1}{x-y}+\dfrac{5x^2-5xy}{x^2-2xy+y^2}=\dfrac{2x+1}{x-y}+\dfrac{5x(x-y)}{(x-y)(x-y)}$

$$=\dfrac{2x+1}{x-y}+\dfrac{5x\cancel{(x-y)}}{\cancel{(x-y)}(x-y)}$$

$$=\dfrac{2x+1}{x-y}+\dfrac{5x}{x-y}$$

$$=\dfrac{7x+1}{x-y}$$

63. $\dfrac{2y-6}{y^2-9}-\dfrac{y}{y-1}+\dfrac{y^2+2}{y^2+2y-3}$

$$=\dfrac{2(y-3)}{(y+3)(y-3)}-\dfrac{y}{y-1}+\dfrac{y^2+2}{(y+3)(y-1)}$$

$$=\dfrac{2}{y+3}-\dfrac{y}{y-1}+\dfrac{y^2+2}{(y+3)(y-1)}$$
   [LCD is $(y+3)(y-1)$.]

$$=\dfrac{2}{y+3}\cdot\dfrac{y-1}{y-1}-\dfrac{y}{y-1}\cdot\dfrac{y+3}{y+3}+\dfrac{y^2+2}{(y+3)(y-1)}$$

$$=\dfrac{2y-2-y^2-3y+y^2+2}{(y+3)(y-1)}$$

$$=\dfrac{-y}{(y+3)(y-1)},\text{ or }-\dfrac{y}{(y+3)(y-1)}$$

65. $\dfrac{5y}{1-4y^2}-\dfrac{2y}{2y+1}+\dfrac{5y}{4y^2-1}$

Observe that $\dfrac{5y}{1-4y^2}$ and $\dfrac{5y}{4y^2-1}$ are opposites, so their

sum is 0. Then the result is the remaining expression,

$$-\dfrac{2y}{2y+1}.$$

67. $f(x) = 2 + \dfrac{x}{x-3} - \dfrac{18}{x^2-9}$

$= \dfrac{2}{1} + \dfrac{x}{x-3} - \dfrac{18}{(x+3)(x-3)}$

Note that $x \neq -3,\ 3$.

LCD is $(x+3)(x-3)$.

$= \dfrac{2}{1} \cdot \dfrac{(x+3)(x-3)}{(x+3)(x-3)} + \dfrac{x}{x-3} \cdot \dfrac{x+3}{x+3} - \dfrac{18}{(x+3)(x-3)}$

$= \dfrac{2(x+3)(x-3) + x(x+3) - 18}{(x+3)(x-3)}$

$= \dfrac{2x^2 - 18 + x^2 + 3x - 18}{(x+3)(x-3)}$

$= \dfrac{3x^2 + 3x - 36}{(x+3)(x-3)}$

$= \dfrac{3(x+4)(x-3)}{(x+3)(x-3)}$

$= \dfrac{3(x+4)}{x+3},\ x \neq -3,\ 3$

69. $f(x) = \dfrac{3x-1}{x^2+2x-3} - \dfrac{x+4}{x^2-16}$

$= \dfrac{3x-1}{(x+3)(x-1)} - \dfrac{x+4}{(x+4)(x-4)}$

Note that $x \neq -3,\ 1, -4,\ 4$.

$= \dfrac{3x-1}{(x+3)(x-1)} - \dfrac{1}{x-4}$ Removing a factor

equal to 1; LCD is

$(x+3)(x-1)(x-4)$.

$= \dfrac{3x-1}{(x+3)(x-1)} \cdot \dfrac{x-4}{x-4} - \dfrac{1}{x-4} \cdot \dfrac{(x+3)(x-1)}{(x+3)(x-1)}$

$= \dfrac{(3x-1)(x-4) - (x+3)(x-1)}{(x+3)(x-1)(x-4)}$

$= \dfrac{3x^2 - 13x + 4 - (x^2 + 2x - 3)}{(x+3)(x-1)(x-4)}$

$= \dfrac{3x^2 - 13x + 4 - x^2 - 2x + 3}{(x+3)(x-1)(x-4)}$

$= \dfrac{2x^2 - 15x + 7}{(x+3)(x-1)(x-4)}$

$= \dfrac{(2x-1)(x-7)}{(x+3)(x-1)(x-4)},\ x \neq -4, -3,\ 1,\ 4$

71. $f(x) = \dfrac{1}{x^2+5x+6} - \dfrac{2}{x^2+3x+2} - \dfrac{1}{x^2+5x+6}$

$= \dfrac{1}{(x+3)(x+2)} - \dfrac{2}{(x+2)(x+1)} - \dfrac{1}{(x+3)(x+2)}$

Note that $x \neq -3,\ -2,\ -1$.

Observe that the sum of the first and third terms is 0, so the result is the remaining term.

$f(x) = -\dfrac{2}{(x+2)(x+1)},\ \text{or}\ \dfrac{-2}{(x+2)(x+1)},$

$x \neq -3,\ -2, -1$

73. *Writing Exercise.* No;

$\dfrac{3-x}{x-5} = \dfrac{-x+3}{x-5} = \dfrac{-1(-x+3)}{-1(x-5)} = \dfrac{x-3}{-x+5} = \dfrac{x-3}{5-x}.$

75. $2x^{-1} = \dfrac{2}{x}$

77. $ab(a+b)^{-2} = \dfrac{ab}{(a+b)^2}$

79. $9x^3\left(\dfrac{1}{x^2} - \dfrac{2}{3x^3}\right) = \dfrac{9x^3}{x^2} - \dfrac{18x^3}{3x^3} = 9x - 6$

81. *Writing Exercise.* If the product of the denominators were used as the common denominator, the resulting numerator would be $y^3 - 5y^2 - 2y - 24$. This can be factored as $(y-6)(y^2 + y + 4)$, allowing $\dfrac{y-6}{y-6}$ to be removed from the result, but this factorization is not readily apparent.

83. We find the least common multiple of 14 (2 weeks = 14 days), 20, and 30.

$14 = 2 \cdot 7$
$20 = 2 \cdot 2 \cdot 5$
$30 = 2 \cdot 3 \cdot 5$
$\text{LCM} = 2 \cdot 2 \cdot 3 \cdot 5 \cdot 7 = 420$

It will be 420 days until Jinney can refill all three prescriptions on the same day.

85. The smallest number of parts possible is the least common multiple of 6 and 4.

$6 = 2 \cdot 3$
$4 = 2 \cdot 2$
$\text{LCM} = 2 \cdot 3 \cdot 2,\ \text{or}\ 12$

A measure should be divided into 12 parts.

87. $x^8 - x^4 = x^4(x^2+1)(x+1)(x-1)$
$x^5 - x^2 = x^2(x-1)(x^2+x+1)$
$x^5 - x^3 = x^3(x+1)(x-1)$
$x^5 + x^2 = x^2(x+1)(x^2-x+1)$

The LCM is

$x^4(x^2+1)(x+1)(x-1)(x^2+x+1)(x^2-x+1).$

89. The LCM is $8a^4b^7$.

One expression is $2a^3b^7$.

Then the other expression must contain 8, a^4, and one of the following:

no factor of b, b, b^2, b^3, b^4, b^5, b^6, or b^7.

Thus, all the possibilities for the other expression are

$8a^4$, $8a^4b$, $8a^4b^2$, $8a^4b^3$, $8a^4b^4$, $8a^4b^5$, $8a^4b^6$, $8a^4b^7$.

91.
$$(f+g)(x) = \frac{x^3}{x^2-4} + \frac{x^2}{x^2+3x-10}$$
$$= \frac{x^3}{(x+2)(x-2)} + \frac{x^2}{(x+5)(x-2)}$$
$$= \frac{x^3(x+5)+x^2(x+2)}{(x+2)(x-2)(x+5)}$$
$$= \frac{x^4+5x^3+x^3+2x^2}{(x+2)(x-2)(x+5)}$$
$$= \frac{x^4+6x^3+2x^2}{(x+2)(x-2)(x+5)}$$

93.
$$(f \cdot g)(x) = \frac{x^3}{x^2-4} \cdot \frac{x^2}{x^2+3x-10}$$
$$= \frac{x^5}{(x^2-4)(x^2+3x-10)}$$

95.
$$x^{-2}+2x^{-1} = \frac{1}{x^2}+\frac{2}{x} \qquad \text{LCD is } x^2$$
$$= \frac{1}{x^2}+\frac{2}{x}\cdot\frac{x}{x}$$
$$= \frac{1}{x^2}+\frac{2x}{x^2}$$
$$= \frac{2x+1}{x^2}$$

97.
$$5(x-3)^{-1}+4(x+3)^{-1}-2(x+3)^{-2}$$
$$= \frac{5}{x-3}+\frac{4}{x+3}-\frac{2}{(x+3)^2}$$
$$\qquad\qquad [\text{LCD is } (x-3)(x+3)^2.]$$
$$= \frac{5(x+3)^2+4(x-3)(x+3)-2(x-3)}{(x-3)(x+3)^2}$$
$$= \frac{5x^2+30x+45+4x^2-36-2x+6}{(x-3)(x+3)^2}$$
$$= \frac{9x^2+28x+15}{(x-3)(x+3)^2}$$

99.
$$\frac{x+4}{6x^2-20x}\left(\frac{x}{x^2-x-20}+\frac{2}{x+4}\right)$$
$$= \frac{x+4}{2x(3x-10)}\left(\frac{x}{(x-5)(x+4)}+\frac{2}{x+4}\right)$$
$$= \frac{x+4}{2x(3x-10)}\left(\frac{x+2(x-5)}{(x-5)(x+4)}\right)$$
$$= \frac{x+4}{2x(3x-10)}\left(\frac{x+2x-10}{(x-5)(x+4)}\right)$$
$$= \frac{(x+4)(3x-10)}{2x(3x-10)(x-5)(x+4)}$$
$$= \frac{(x+4)(3x-10)(1)}{2x(3x-10)(x-5)(x+4)}$$
$$= \frac{1}{2x(x-5)}$$

101.
$$\frac{8t^5}{2t^2-10t+12} \div \left(\frac{2t}{t^2-8t+15} - \frac{3t}{t^2-7t+10}\right)$$
$$= \frac{8t^5}{2t^2-10t+12} \div \left(\frac{2t}{(t-5)(t-3)} - \frac{3t}{(t-5)(t-2)}\right)$$
$$= \frac{8t^5}{2t^2-10t+12} \div \left(\frac{2t(t-2)-3t(t-3)}{(t-5)(t-3)(t-2)}\right)$$
$$= \frac{8t^5}{2t^2-10t+12} \div \left(\frac{2t^2-4t-3t^2+9t}{(t-5)(t-3)(t-2)}\right)$$
$$= \frac{8t^5}{2t^2-10t+12} \div \frac{-t^2+5t}{(t-5)(t-3)(t-2)}$$
$$= \frac{8t^5}{2(t-3)(t-2)} \cdot \frac{(t-5)(t-3)(t-2)}{-t(t-5)}$$
$$= \frac{2\cdot4\cdot t\cdot t^4\,(t-5)\,(t-3)\,(t-2)}{2\,(t-3)\,(t-2)(-1)(t)(t-5)}$$
$$= -4t^4$$

103. *Graphing Calculator Exercise*

Exercise Set 6.3

1. (b); see page 374 of the text.

3. (f); see page 374 of the text.

5. (d); see page 375 of the text, Example 3

7.
$$\frac{\dfrac{x+5}{x-3}}{\dfrac{x-2}{x+1}} = \frac{x+5}{x-3} \div \frac{x-2}{x+1}$$
$$= \frac{x+5}{x-3}\cdot\frac{x+1}{x-2}$$
$$= \frac{(x+1)(x+5)}{(x-3)(x-2)}$$

9.
$$\frac{\dfrac{3}{x}+\dfrac{2}{x^3}}{\dfrac{5}{x}-\dfrac{3}{x^2}} = \frac{\dfrac{3}{x}+\dfrac{2}{x^3}}{\dfrac{5}{x}-\dfrac{3}{x^2}}\cdot\frac{x^3}{x^3} \qquad \text{Multiplying by 1 using the LCD}$$
$$= \frac{\dfrac{3}{x}\cdot x^3+\dfrac{2}{x^3}\cdot x^3}{\dfrac{5}{x}\cdot x^3-\dfrac{3}{x^2}\cdot x^3} \qquad \text{Multiplying the numerators and the denominators}$$
$$= \frac{3x^2+2}{5x^2-3x}$$
$$= \frac{3x^2+2}{x(5x-3)}$$

11.
$$\frac{\dfrac{3}{m-n}}{\dfrac{5}{m+n}} = \frac{3}{m-n} \div \frac{5}{m+n}$$
$$= \frac{3}{m-n}\cdot\frac{m+n}{5}$$
$$= \frac{3(m+n)}{5(m-n)}$$

13. $\dfrac{\frac{6}{r}-\frac{1}{s}}{\frac{2}{r}+\frac{3}{s}}=\dfrac{\frac{6}{r}-\frac{1}{s}}{\frac{2}{r}+\frac{3}{s}}\cdot\dfrac{rs}{rs}$ Multiplying by 1 using the LCD

$=\dfrac{\frac{6}{r}\cdot rs-\frac{1}{s}\cdot rs}{\frac{2}{r}\cdot rs+\frac{3}{s}\cdot rs}$

$=\dfrac{6s-r}{2s+3r}$

15. $\dfrac{\frac{3}{z^2}+\frac{2}{yz}}{\frac{4}{zy^2}-\frac{1}{y}}=\dfrac{\frac{3}{z^2}+\frac{2}{yz}}{\frac{4}{zy^2}-\frac{1}{y}}\cdot\dfrac{y^2z^2}{y^2z^2}$ Multiplying by 1 using the LCD

$=\dfrac{\frac{3}{z^2}\cdot y^2z^2+\frac{2}{yz}\cdot y^2z^2}{\frac{4}{zy^2}\cdot y^2z^2-\frac{1}{y}\cdot y^2z^2}$

$=\dfrac{3y^2+2yz}{4z-yz^2},\ \text{or}\ \dfrac{y(3y+2z)}{z(4-yz)}$

17. $\dfrac{\frac{a^2-b^2}{ab}}{\frac{a-b}{b}}=\dfrac{a^2-b^2}{ab}\div\dfrac{a-b}{b}$

$=\dfrac{a^2-b^2}{ab}\cdot\dfrac{b}{a-b}$

$=\dfrac{(a+b)(a-b)(b)}{a\cdot b\cdot(a-b)}$

$=\dfrac{(a+b)\cancel{(a-b)}\cancel{(b)}}{a\cdot\cancel{b}\cdot\cancel{(a-b)}}$

$=\dfrac{a+b}{a}$

19. $\dfrac{1-\frac{2}{3x}}{x-\frac{4}{9x}}=\dfrac{1-\frac{2}{3x}}{x-\frac{4}{9x}}\cdot\dfrac{9x}{9x}$ Multiplying by 1, using the LCD

$=\dfrac{1\cdot 9x-\frac{2}{3x}\cdot 9x}{x\cdot 9x-\frac{4}{9x}\cdot 9x}$

$=\dfrac{9x-6}{9x^2-4}$

$=\dfrac{3(3x-2)}{(3x+2)(3x-2)}$

$=\dfrac{3\cancel{(3x-2)}}{(3x+2)\cancel{(3x-2)}}$

$=\dfrac{3}{3x+2}$

21. $\dfrac{y^{-1}-x^{-1}}{\frac{x^2-y^2}{xy}}=\dfrac{\frac{1}{y}-\frac{1}{x}}{\frac{x^2-y^2}{xy}}$ Rewriting with positive exponents

$=\dfrac{\frac{1}{y}-\frac{1}{x}}{\frac{x^2-y^2}{xy}}\cdot\dfrac{xy}{xy}$ Multiplying by 1, using the LCD

$=\dfrac{\frac{1}{y}\cdot xy-\frac{1}{x}\cdot xy}{\frac{x^2-y^2}{xy}\cdot xy}$

$=\dfrac{x-y}{x^2-y^2}=\dfrac{x-y}{(x+y)(x-y)}$

$=\dfrac{\cancel{x-y}}{(x+y)\cancel{(x-y)}}$

$=\dfrac{1}{x+y}$

23. $\dfrac{\frac{1}{x+h}-\frac{1}{x}}{h}=\dfrac{\frac{1}{x+h}\cdot\frac{x}{x}-\frac{1}{x}\cdot\frac{x+h}{x+h}}{h}$ Adding in the numerator

$=\dfrac{\frac{x-x-h}{x(x+h)}}{h}=\dfrac{\frac{-h}{x(x+h)}}{h}$

$=\dfrac{-h}{x(x+h)}\cdot\dfrac{1}{h}$ Multiplying by the reciprocal of the divisor

$=\dfrac{-1\cdot\cancel{h}\cdot 1}{x(x+h)(\cancel{h})}$ $(-h=-1\cdot h)$

$=-\dfrac{1}{x(x+h)}$

25. $\dfrac{\frac{a^2-4}{a^2+3a+2}}{\frac{a^2-5a-6}{a^2-6a-7}}$

$=\dfrac{a^2-4}{a^2+3a+2}\cdot\dfrac{a^2-6a-7}{a^2-5a-6}$ Multiplying by the reciprocal of the divisor

$=\dfrac{(a+2)(a-2)}{(a+2)(a+1)}\cdot\dfrac{(a+1)(a-7)}{(a+1)(a-6)}$

$=\dfrac{(a+2)(a-2)(a+1)(a-7)}{(a+2)(a+1)(a+1)(a-6)}$

$=\dfrac{\cancel{(a+2)}(a-2)\cancel{(a+1)}(a-7)}{\cancel{(a+2)}\cancel{(a+1)}(a+1)(a-6)}$

$=\dfrac{(a-2)(a-7)}{(a+1)(a-6)}$

27. $\dfrac{\dfrac{x}{x^2+3x-4}-\dfrac{1}{x^2+3x-4}}{\dfrac{x}{x^2+6x+8}+\dfrac{3}{x^2+6x+8}}$

$=\dfrac{\dfrac{x-1}{x^2+3x-4}}{\dfrac{x+3}{x^2+6x+8}}$ Adding in the numerator
and the denominator

$=\dfrac{x-1}{x^2+3x-4}\cdot\dfrac{x^2+6x+8}{x+3}$

$=\dfrac{(x-1)(x+4)(x+2)}{(x+4)(x-1)(x+3)}$

$=\dfrac{\cancel{(x-1)}\,\cancel{(x+4)}(x+2)}{\cancel{(x+4)}\,\cancel{(x-1)}(x+3)}=\dfrac{x+2}{x+3}$

29. $\dfrac{\dfrac{1}{y}+2}{\dfrac{1}{y}-3}=\dfrac{\dfrac{1}{y}+2}{\dfrac{1}{y}-3}\cdot\dfrac{y}{y}$ Multiplying by 1, using
the LCD

$=\dfrac{\left(\dfrac{1}{y}+2\right)y}{\left(\dfrac{1}{y}-3\right)y}$

$=\dfrac{\dfrac{1}{y}\cdot y+2\cdot y}{\dfrac{1}{y}\cdot y-3\cdot y}$

$=\dfrac{1+2y}{1-3y}$

31. $\dfrac{y+y^{-2}}{y-y^{-2}}=\dfrac{y+\dfrac{1}{y^2}}{y-\dfrac{1}{y^2}}$ Rewriting with positive exponents

$=\dfrac{y+\dfrac{1}{y^2}}{y-\dfrac{1}{y^2}}\cdot\dfrac{y^2}{y^2}$ Multiplying by 1, using the LCD

$=\dfrac{y\cdot y^2+\dfrac{1}{y^2}\cdot y^2}{y\cdot y^2-\dfrac{1}{y^2}\cdot y^2}$

$=\dfrac{y^3+1}{y^3-1}$

(Although the numerator and denominator can be factored, doing so does not lead to further simplification.)

33. $\dfrac{\dfrac{1}{x-2}+\dfrac{3}{x-1}}{\dfrac{2}{x-1}+\dfrac{5}{x-2}}$

$=\dfrac{\dfrac{1}{x-2}+\dfrac{3}{x-1}}{\dfrac{2}{x-1}+\dfrac{5}{x-2}}\cdot\dfrac{(x-2)(x-1)}{(x-2)(x-1)}$ Multiplying by 1, using the LCD

$=\dfrac{\dfrac{1}{x-2}\cdot(x-2)(x-1)+\dfrac{3}{x-1}\cdot(x-2)(x-1)}{\dfrac{2}{x-1}\cdot(x-2)(x-1)+\dfrac{5}{x-2}\cdot(x-2)(x-1)}$

$=\dfrac{x-1+3(x-2)}{2(x-2)+5(x-1)}$

$=\dfrac{x-1+3x-6}{2x-4+5x-5}$

$=\dfrac{4x-7}{7x-9}$

35. $\dfrac{a(a+3)^{-1}-2(a-1)^{-1}}{a(a+3)^{-1}-(a-1)^{-1}}$

$=\dfrac{\dfrac{a}{a+3}-\dfrac{2}{a-1}}{\dfrac{a}{a+3}-\dfrac{1}{a-1}}$

$=\dfrac{\dfrac{a}{a+3}-\dfrac{2}{a-1}}{\dfrac{a}{a+3}-\dfrac{1}{a-1}}\cdot\dfrac{(a+3)(a-1)}{(a+3)(a-1)}$ Multiplying by 1, using the LCD

$=\dfrac{\dfrac{a}{a+3}\cdot(a+3)(a-1)-\dfrac{2}{a-1}\cdot(a+3)(a-1)}{\dfrac{a}{a+3}\cdot(a+3)(a-1)-\dfrac{1}{a-1}\cdot(a+3)(a-1)}$

$=\dfrac{a(a-1)-2(a+3)}{a(a-1)-(a+3)}$

$=\dfrac{a^2-a-2a-6}{a^2-a-a-3}=\dfrac{a^2-3a-6}{a^2-2a-3}$

(Although the denominator can be factored, doing so does not lead to further simplification.)

37. $\dfrac{\dfrac{2}{a^2-1}+\dfrac{1}{a+1}}{\dfrac{3}{a^2-1}+\dfrac{2}{a-1}}$

$=\dfrac{\dfrac{2}{(a+1)(a-1)}+\dfrac{1}{a+1}}{\dfrac{3}{(a+1)(a-1)}+\dfrac{2}{a-1}}$

$=\dfrac{\dfrac{2}{(a+1)(a-1)}+\dfrac{1}{a+1}}{\dfrac{3}{(a+1)(a-1)}+\dfrac{2}{a-1}}\cdot\dfrac{(a+1)(a-1)}{(a+1)(a-1)}$ Multiplying by 1, using the LCD

$=\dfrac{\dfrac{2}{(a+1)(a-1)}\cdot(a+1)(a-1)+\dfrac{1}{a+1}\cdot(a+1)(a-1)}{\dfrac{3}{(a+1)(a-1)}\cdot(a+1)(a-1)+\dfrac{2}{a-1}\cdot(a+1)(a-1)}$

$=\dfrac{2+a-1}{3+2(a+1)}=\dfrac{a+1}{3+2a+2}=\dfrac{a+1}{2a+5}$

39.
$$\dfrac{\dfrac{5}{x^2-4}-\dfrac{3}{x-2}}{\dfrac{4}{x^2-4}-\dfrac{2}{x+2}}$$

$$=\dfrac{\dfrac{5}{(x+2)(x-2)}-\dfrac{3}{x-2}}{\dfrac{4}{(x+2)(x-2)}-\dfrac{2}{x+2}}$$

$$=\dfrac{\dfrac{5}{(x+2)(x-2)}-\dfrac{3}{x-2}}{\dfrac{4}{(x+2)(x-2)}-\dfrac{2}{x+2}}\cdot\dfrac{(x+2)(x-2)}{(x+2)(x-2)}\qquad\text{Multiplying by 1, using the LCD}$$

$$=\dfrac{\dfrac{5}{(x+2)(x-2)}\cdot(x+2)(x-2)-\dfrac{3}{x-2}\cdot(x+2)(x-2)}{\dfrac{4}{(x+2)(x-2)}\cdot(x+2)(x-2)-\dfrac{2}{x+2}\cdot(x+2)(x-2)}$$

$$=\dfrac{5-3(x+2)}{4-2(x-2)}=\dfrac{5-3x-6}{4-2x+4}=\dfrac{-1-3x}{8-2x},\text{ or }\dfrac{3x+1}{2x-8}$$

41.
$$\dfrac{\dfrac{y^3}{y^2-4}+\dfrac{125}{4-y^2}}{\dfrac{y}{y^2-4}+\dfrac{5}{4-y^2}}$$

$$=\dfrac{\dfrac{y^3}{y^2-4}+\dfrac{-1}{-1}\cdot\dfrac{125}{4-y^2}}{\dfrac{y}{y^2-4}+\dfrac{-1}{-1}\cdot\dfrac{5}{4-y^2}}$$

$$=\dfrac{\dfrac{y^3}{y^2-4}-\dfrac{125}{y^2-4}}{\dfrac{y}{y^2-4}-\dfrac{5}{y^2-4}}$$

$$=\dfrac{\dfrac{y^3-125}{y^2-4}}{\dfrac{y-5}{y^2-4}}\qquad\begin{array}{l}\text{Adding the numerator}\\\text{and the denominator}\end{array}$$

$$=\dfrac{y^3-125}{y^2-4}\cdot\dfrac{y^2-4}{y-5}\qquad\begin{array}{l}\text{Multiplying by the}\\\text{reciprocal of the divisor}\end{array}$$

$$=\dfrac{(y-5)(y^2+5y+25)(y^2-4)}{(y^2-4)(y-5)}$$

$$=y^2+5y+25$$

43.
$$\dfrac{\dfrac{y^2}{y^2-25}-\dfrac{y}{y-5}}{\dfrac{y}{y^2-25}-\dfrac{1}{y+5}}$$

$$=\dfrac{\dfrac{y^2}{(y+5)(y-5)}-\dfrac{y}{y-5}}{\dfrac{y}{(y+5)(y-5)}-\dfrac{1}{y+5}}\cdot\dfrac{(y+5)(y-5)}{(y+5)(y-5)}\qquad\begin{array}{l}\text{Multiplying by 1,}\\\text{using the LCD}\end{array}$$

$$=\dfrac{\dfrac{y^2}{(y+5)(y-5)}\cdot(y+5)(y-5)-\dfrac{y}{y-5}\cdot(y+5)(y-5)}{\dfrac{y}{(y+5)(y-5)}\cdot(y+5)(y-5)-\dfrac{1}{y+5}\cdot(y+5)(y-5)}$$

$$=\dfrac{y^2-y(y+5)}{y-(y-5)}$$

$$=\dfrac{y^2-y^2-5y}{y-y+5}=\dfrac{-5y}{5}=-y$$

45.
$$\dfrac{\dfrac{a}{a+2}+\dfrac{5}{a}}{\dfrac{a}{2a+4}+\dfrac{1}{3a}}$$

$$=\dfrac{\dfrac{a}{a+2}+\dfrac{5}{a}}{\dfrac{a}{2(a+2)}+\dfrac{1}{3a}}$$

$$=\dfrac{\dfrac{a}{a+2}+\dfrac{5}{a}}{\dfrac{a}{2(a+2)}+\dfrac{1}{3a}}\cdot\dfrac{6a(a+2)}{6a(a+2)}\qquad\begin{array}{l}\text{Multiplying by 1,}\\\text{using the LCD}\end{array}$$

$$=\dfrac{\dfrac{a}{a+2}\cdot6a(a+2)+\dfrac{5}{a}\cdot6a(a+2)}{\dfrac{a}{2(a+2)}\cdot6a(a+2)+\dfrac{1}{3a}\cdot6a(a+2)}$$

$$=\dfrac{6a^2+30(a+2)}{3a^2+2(a+2)}$$

$$=\dfrac{6a^2+30a+60}{3a^2+2a+4}$$

(Although the numerator can be factored, doing so does not lead to further simplification.)

47. $\dfrac{\dfrac{1}{x^2-3x+2}+\dfrac{1}{x^2-4}}{\dfrac{1}{x^2+4x+4}+\dfrac{1}{x^2-4}}$

$=\dfrac{\dfrac{1}{(x-1)(x-2)}+\dfrac{1}{(x+2)(x-2)}}{\dfrac{1}{(x+2)(x+2)}+\dfrac{1}{(x+2)(x-2)}}$

$=\dfrac{\dfrac{1}{(x-1)(x-2)}+\dfrac{1}{(x+2)(x-2)}}{\dfrac{1}{(x+2)(x+2)}+\dfrac{1}{(x+2)(x-2)}}$

$\quad\cdot\dfrac{(x-1)(x-2)(x+2)(x+2)}{(x-1)(x-2)(x+2)(x+2)}$ Multiplying by 1, using the LCD

$=\dfrac{(x+2)(x+2)+(x-1)(x+2)}{(x-1)(x-2)+(x-1)(x+2)}$

$=\dfrac{x^2+4x+4+x^2+x-2}{x^2-3x+2+x^2+x-2}$

$=\dfrac{2x^2+5x+2}{2x^2-2x}$

$=\dfrac{(2x+1)(x+2)}{2x(x-1)}$

49. $\dfrac{\dfrac{3}{a^2-4a+3}+\dfrac{3}{a^2-5a+6}}{\dfrac{3}{a^2-3a+2}+\dfrac{3}{a^2+3a-10}}$

$=\dfrac{\dfrac{3}{(a-1)(a-3)}+\dfrac{3}{(a-2)(a-3)}}{\dfrac{3}{(a-1)(a-2)}+\dfrac{3}{(a+5)(a-2)}}$

$=\dfrac{\dfrac{3}{(a-1)(a-3)}+\dfrac{3}{(a-2)(a-3)}}{\dfrac{3}{(a-1)(a-2)}+\dfrac{3}{(a+5)(a-2)}}$

$\quad\cdot\dfrac{(a-1)(a-3)(a-2)(a+5)}{(a-1)(a-3)(a-2)(a+5)}$ Multiplying by 1, using the LCD

$=\dfrac{3(a-2)(a+5)+3(a-1)(a+5)}{3(a-3)(a+5)+3(a-1)(a-3)}$

$=\dfrac{3[(a-2)(a+5)+(a-1)(a+5)]}{3[(a-3)(a+5)+(a-1)(a-3)]}$

$=\dfrac{a^2+3a-10+a^2+4a-5}{a^2+2a-15+a^2-4a+3}$

$=\dfrac{2a^2+7a-15}{2a^2-2a-12}$

$=\dfrac{(2a-3)(a+5)}{2(a^2-a-6)}$

$=\dfrac{(2a-3)(a+5)}{2(a-3)(a+2)}$

51. $\dfrac{\dfrac{y}{y^2-4}-\dfrac{2y}{y^2+y-6}}{\dfrac{2y}{y^2+y-6}-\dfrac{y}{y^2-4}}$

Observe that $\dfrac{y}{y^2-4}-\dfrac{2y}{y^2+y-6}=-\left(\dfrac{2y}{y^2+y-6}-\dfrac{y}{y^2-4}\right).$

Then, the numerator and denominator are opposites and thus their quotient is –1.

53. $\dfrac{\dfrac{3}{x^2+2x-3}-\dfrac{1}{x^2-3x-10}}{\dfrac{3}{x^2-6x+5}-\dfrac{1}{x^2+5x+6}}$

$=\dfrac{\dfrac{3}{(x+3)(x-1)}-\dfrac{1}{(x-5)(x+2)}}{\dfrac{3}{(x-5)(x-1)}-\dfrac{1}{(x+3)(x+2)}}$

$=\dfrac{\dfrac{3}{(x+3)(x-1)}-\dfrac{1}{(x-5)(x+2)}}{\dfrac{3}{(x-5)(x-1)}-\dfrac{1}{(x+3)(x+2)}}$

$\quad\cdot\dfrac{(x+3)(x-1)(x-5)(x+2)}{(x+3)(x-1)(x-5)(x+2)}$ Multiplying by 1, using the LCD

$=\dfrac{3(x-5)(x+2)-(x+3)(x-1)}{3(x+3)(x+2)-(x-1)(x-5)}$

$=\dfrac{3(x^2-3x-10)-(x^2+2x-3)}{3(x^2+5x+6)-(x^2-6x+5)}$

$=\dfrac{3x^2-9x-30-x^2-2x+3}{3x^2+15x+18-x^2+6x-5}$

$=\dfrac{2x^2-11x-27}{2x^2+21x+13}$

55. *Writing Exercise.* Michael is treating the *terms* b^{-1} and c^{-1} as *factors*. To convince him that his result is incorrect, let $a=1$, $b=2$, and $c=3$. Then

$\dfrac{a+b^{-1}}{a+c^{-1}}=\dfrac{1+2^{-1}}{1+3^{-1}}=\dfrac{1+\dfrac{1}{2}}{1+\dfrac{1}{3}}=\dfrac{\dfrac{3}{2}}{\dfrac{4}{3}}=\dfrac{3}{2}\cdot\dfrac{3}{4}=\dfrac{9}{8},$ but

$\dfrac{a+c}{a+b}=\dfrac{1+3}{1+2}=\dfrac{4}{3}.$

57. $2(y+3)-5(y-1)=10y$

$\quad\quad 2y+6-5y+5=10y$

$\quad\quad\quad\quad -3y+11=10y$

$\quad\quad\quad\quad\quad\quad 11=13y$

$\quad\quad\quad\quad\quad\dfrac{11}{13}=y$

The solution is $\dfrac{11}{13}$.

59. $x^2=25$

$\quad x=\pm\sqrt{25}$

$\quad x=\pm 5$

The solutions are –5 and 5.

61.
$$\frac{1}{3}x - \frac{1}{4} = \frac{1}{6} - \frac{1}{2}x$$

$$12\left(\frac{1}{3}x - \frac{1}{4}\right) = 12\left(\frac{1}{6} - \frac{1}{2}x\right) \quad \text{Clearing fractions}$$

$$4x - 3 = 2 - 6x$$

$$10x - 3 = 2$$

$$10x = 5$$

$$x = \frac{1}{2}$$

63. *Writing Exercise.* Write $\frac{a}{b} \div \frac{c}{d}$ as a complex rational

expression, $\dfrac{\frac{a}{b}}{\frac{c}{d}}$. Simplifying using Method 1, we have:

$$\frac{\frac{a}{b}}{\frac{c}{d}} = \frac{\frac{a}{b}}{\frac{c}{d}} \cdot \frac{bd}{bd} = \frac{ad}{cb}.$$

Note that $\dfrac{ad}{cb} = \dfrac{a}{b} \cdot \dfrac{d}{c}$. Thus, when dividing $\dfrac{a}{b}$ by $\dfrac{d}{c}$, we

can "invert and multiply."

65.
$$\frac{5x^{-2} + 10x^{-1}y^{-1} + 5y^{-2}}{3x^{-2} - 3y^{-2}}$$

$$= \frac{\frac{5}{x^2} + \frac{10}{xy} + \frac{5}{y^2}}{\frac{3}{x^2} - \frac{3}{y^2}}$$

$$= \frac{\frac{5}{x^2} + \frac{10}{xy} + \frac{5}{y^2}}{\frac{3}{x^2} - \frac{3}{y^2}} \cdot \frac{x^2 y^2}{x^2 y^2}$$

$$= \frac{5y^2 + 10xy + 5x^2}{3y^2 - 3x^2}$$

$$= \frac{5(y^2 + 2xy + x^2)}{3(y^2 - x^2)}$$

$$= \frac{5(y+x)(y+x)}{3(y+x)(y-x)}$$

$$= \frac{5\cancel{(y+x)}(y+x)}{3\cancel{(y+x)}(y-x)}$$

$$= \frac{5(y+x)}{3(y-x)}$$

67. Substitute $\dfrac{c}{4}$ for both v_1 and v_2.

$$\frac{\frac{c}{4} + \frac{c}{4}}{1 + \frac{c}{4} \cdot \frac{c}{4}} = \frac{\frac{2c}{4}}{1 + \frac{c^2}{16}} = \frac{\frac{c}{2}}{1 + \frac{c^2}{16} \cdot \frac{1}{c^2}}$$

$$= \frac{\frac{c}{2}}{1 + \frac{1}{16}} = \frac{\frac{c}{2}}{\frac{17}{16}} = \frac{c}{2} \cdot \frac{16}{17}$$

$$= \frac{8c}{17}$$

The observed speed is $\dfrac{8c}{17}$, or $\dfrac{8}{17}$ the speed of light.

69. $f(x) = \dfrac{3}{x}$, $f(x+h) = \dfrac{3}{x+h}$

$$\frac{f(x+h) - f(x)}{h} = \frac{\frac{3}{x+h} - \frac{3}{x}}{h}$$

$$= \frac{\frac{3x - 3(x+h)}{x(x+h)}}{h}$$

$$= \frac{3x - 3(x+h)}{x(x+h)} \cdot \frac{1}{h}$$

$$= \frac{3x - 3x - 3h}{xh(x+h)}$$

$$= \frac{-3h}{xh(x+h)}$$

$$= \frac{-3\cancel{h}}{x\cancel{h}(x+h)}$$

$$= \frac{-3}{x(x+h)}$$

71. $f(x) = \dfrac{2x}{1+x}$, $f(x+h) = \dfrac{2(x+h)}{1+x+h}$

$$\frac{f(x+h) - f(x)}{h}$$

$$= \frac{\frac{2(x+h)}{1+x+h} - \frac{2x}{1+x}}{h}$$

$$= \frac{\frac{2(x+h)(1+x) - 2x(1+x+h)}{(1+x+h)(1+x)}}{h}$$

$$= \frac{2(x+h)(1+x) - 2x(1+x+h)}{(1+x+h)(1+x)} \cdot \frac{1}{h}$$

$$= \frac{2x + 2x^2 + 2h + 2hx - 2x - 2x^2 - 2xh}{(1+x+h)(1+x)h}$$

$$= \frac{2 \cdot \cancel{h}}{(1+x+h)(1+x)\cancel{h}}$$

$$= \frac{2}{(1+x+h)(1+x)}$$

73. Division by zero occurs in $\dfrac{1}{x^2 - 1}$ when $x = 1$ or $x = -1$.

Division by zero occurs in $\dfrac{1}{x^2 - 16}$ when $x = 4$ or

$x = -4$. To avoid division in the complex fraction we solve:

$$\frac{1}{9} - \frac{1}{x^2 - 16} = 0$$
$$x^2 - 16 - 9 = 0 \qquad \text{Multiplying by } 9(x^2 - 16)$$
$$x^2 - 25 = 0$$
$$(x + 5)(x - 5) = 0$$
$$x + 5 = 0 \quad or \quad x - 5 = 0$$
$$x = -5 \quad or \qquad x = 5.$$

The domain of $G = \{x \mid x$ is a real number $and \ x \neq 1$ and $x \neq -1$ and $x \neq 4$ and $x \neq -4$ and $x \neq -5$ and $x \neq 5\}$.

75.
$$f(x) = \frac{2}{2 + x}$$
$$f(a) = \frac{2}{2 + a}$$
$$f(f(a)) = \frac{2}{2 + \dfrac{2}{2 + a}} \qquad \text{Note that } a \neq -2, -3.$$
$$= \frac{2}{2 + \dfrac{2}{2 + a}} \cdot \frac{2 + a}{2 + a}$$
$$= \frac{2(2 + a)}{2(2 + a) + \dfrac{2}{2 + a} \cdot 2 + a}$$
$$= \frac{4 + 2a}{4 + 2a + 2}$$
$$= \frac{4 + 2a}{6 + 2a} = \frac{2(2 + a)}{2(3 + a)}$$
$$= \frac{\cancel{2}(2 + a)}{\cancel{2}(3 + a)} = \frac{2 + a}{3 + a}, \quad a \neq -2, -3$$

77.
$$\left[\frac{\dfrac{x + 3}{x - 3} + 1}{\dfrac{x + 3}{x - 3} - 1}\right]^4 = \left[\frac{\dfrac{x + 3}{x - 3} + 1}{\dfrac{x + 3}{x - 3} - 1} \cdot \frac{x - 3}{x - 3}\right]^4$$
$$= \left[\frac{x + 3 + x - 3}{x + 3 - x + 3}\right]^4$$
$$= \left(\frac{2x}{6}\right)^4 = \left(\frac{x}{3}\right)^4 = \frac{x^4}{81}$$

Division by zero occurs in both the numerator and the denominator of the original fraction when $x = 3$. To avoid division by zero in the complex fraction we solve:

$$\frac{x + 3}{x - 3} - 1 = 0$$
$$\frac{x + 3}{x - 3} = 1$$
$$x + 3 = x - 3$$
$$3 = -3$$

The equation has no solution, so the denominator of the complex fraction cannot be zero. Thus, the domain of $f = \{x \mid x$ is a real number $and \ x \neq 3\}$.

79.
$$\frac{30,000 \cdot \dfrac{0.075}{12}}{\left(1 + \dfrac{0.075}{12}\right)^{120} - 1} = \frac{30,000(0.00625)}{(1 + 0.00625)^{120} - 1}$$
$$= \frac{187.5}{(1.00625)^{120} - 1}$$
$$\approx \frac{187.5}{2.112064637 - 1}$$
$$\approx \frac{187.5}{1.112064637}$$
$$\approx 168.61$$

Alexis' monthly investment is \$168.61.

Exercise Set 6.4

1. Equation; see page 381 in the text.

3. Expression; see page 381 in the text.

5. Equation; see page 381 in the text.

7. Equation; see page 381 in the text.

9. Expression; see page 381 in the text.

11. Note that there is no value of t that makes a denominator 0.

$$\frac{t}{10} + \frac{t}{15} = 1, \quad \text{LCD is 30}$$
$$30\left(\frac{t}{10} + \frac{t}{15}\right) = 30 \cdot 1$$
$$3t + 2t = 30$$
$$5t = 30$$
$$t = 6$$

Check: $\dfrac{t}{10} + \dfrac{t}{15} = 1$

$$\begin{array}{c|c} \dfrac{6}{10} + \dfrac{6}{15} & 1 \\[2mm] \dfrac{18}{30} + \dfrac{12}{30} & \\ & \overset{?}{} \\ 1 = 1 & \text{TRUE} \end{array}$$

The solution is 6.

13. Because $\dfrac{1}{x}$ is undefined when x is 0, at the outset we state the restriction that $x \neq 0$.

$$\frac{3}{4} - \frac{1}{x} = \frac{7}{8}, \quad \text{LCD is } 8x$$
$$8x\left(\frac{3}{4} - \frac{1}{x}\right) = 8x\left(\frac{7}{8}\right)$$
$$8x \cdot \frac{3}{4} - 8x \cdot \frac{1}{x} = 8x \cdot \frac{7}{8}$$
$$6x - 8 = 7x$$
$$-8 = x$$

Check: $\dfrac{3}{4} - \dfrac{1}{x} = \dfrac{7}{8}$

$$\dfrac{3}{4} - \dfrac{1}{-8} \,\bigg|\, \dfrac{7}{8}$$

$$\dfrac{6}{8} + \dfrac{1}{8} \,\bigg|$$

$$\overset{?}{\dfrac{7}{8}} = \dfrac{7}{8} \qquad \text{TRUE}$$

The solution is -8.

15. Note that there is no value of a that makes a denominator 0.

$$\dfrac{a+1}{3} + \dfrac{a-4}{5} = \dfrac{2a}{9}, \quad \text{LCD is } 45$$

$$45\left(\dfrac{a+1}{3} + \dfrac{a-4}{5}\right) = 45 \cdot \dfrac{2a}{9}$$

$$45 \cdot \dfrac{a+1}{3} + 45 \cdot \dfrac{a-4}{5} = 45 \cdot \dfrac{2a}{9}$$

$$15(a+1) + 9(a-4) = 5 \cdot 2a$$

$$15a + 15 + 9a - 36 = 10a$$

$$24a - 21 = 10a$$

$$-21 = -14a$$

$$\dfrac{3}{2} = a$$

This number checks. The solution is $\dfrac{3}{2}$.

17. Because $\dfrac{1}{3t}$ and $\dfrac{1}{t}$ are undefined when t is 0, at the outset we state the restriction that $t \neq 0$.

$$\dfrac{1}{3t} + \dfrac{1}{t} = \dfrac{1}{2}, \quad \text{LCD is } 6t$$

$$6t\left(\dfrac{1}{3t} + \dfrac{1}{t}\right) = 6t\left(\dfrac{1}{2}\right)$$

$$6t \cdot \dfrac{1}{3t} + 6t \cdot \dfrac{1}{t} = 6t \cdot \dfrac{1}{2}$$

$$2 + 6 = 3t$$

$$8 = 3t$$

$$\dfrac{8}{3} = t$$

Check: $\dfrac{1}{3t} + \dfrac{1}{t} = \dfrac{1}{2}$

$$\dfrac{1}{3 \cdot \frac{8}{3}} + \dfrac{1}{\frac{8}{3}} \,\bigg|\, \dfrac{1}{2}$$

$$\dfrac{1}{8} + \dfrac{3}{8} \,\bigg|$$

$$\overset{?}{\dfrac{1}{2}} = \dfrac{1}{2} \qquad \text{TRUE}$$

The solution is $\dfrac{8}{3}$.

19. To ensure that no denominator is 0, at the outset we state the restriction that $x \neq 1$.

$$\dfrac{3}{x-1} + \dfrac{3}{10} = \dfrac{5}{2x-2}$$

$$\dfrac{3}{x-1} + \dfrac{3}{10} = \dfrac{5}{2(x-1)},$$

$$\qquad\qquad \text{LCD is } 10(x-1)$$

$$10(x-1)\left(\dfrac{3}{x-1} + \dfrac{3}{10}\right) = 10(x-1) \cdot \dfrac{5}{2(x-1)}$$

$$10(x-1) \cdot \dfrac{3}{x-1} + 10(x-1) \cdot \dfrac{3}{10} = 10(x-1) \cdot \dfrac{5}{2(x-1)}$$

$$30 + 3(x-1) = 5 \cdot 5$$

$$30 + 3x - 3 = 25$$

$$3x + 27 = 25$$

$$3x = -2$$

$$x = -\dfrac{2}{3}$$

Check: $\dfrac{3}{x-1} + \dfrac{3}{10} = \dfrac{5}{2x-2}$

$$\dfrac{3}{-\frac{2}{3}-1} + \dfrac{3}{10} \,\bigg|\, \dfrac{5}{2\left(-\frac{2}{3}\right)-2}$$

$$\dfrac{3}{-\frac{5}{3}} + \dfrac{3}{10} \,\bigg|\, \dfrac{5}{-\frac{4}{3}-2}$$

$$-\dfrac{9}{5} + \dfrac{3}{10} \,\bigg|\, \dfrac{5}{1} \cdot \dfrac{-3}{10}$$

$$-\dfrac{18}{10} + \dfrac{3}{10} \,\bigg|$$

$$\overset{?}{-\dfrac{15}{10}} = -\dfrac{15}{10} \qquad \text{TRUE}$$

The solution is $-\dfrac{2}{3}$.

21. $\dfrac{2}{6} + \dfrac{1}{2x} = \dfrac{1}{3}$

Because $\dfrac{1}{2x}$ is undefined when x is 0, at the outset we state the restriction that $x \neq 0$. Observe that $\dfrac{2}{6}$ is equivalent to $\dfrac{1}{3}$. This means that $\dfrac{1}{2x}$ must be 0 in order for the equation to be true. But there is no value of x for which $\dfrac{1}{2x} = 0$, so the equation has no solution.

23. $y + \dfrac{4}{y} = -5$

Because $\dfrac{4}{y}$ is undefined when y is 0, we note at the outset that $y \neq 0$. Then we multiply both sides by the LCD, y.

$$y\left(y + \dfrac{4}{y}\right) = y(-5)$$

$$y \cdot y + y \cdot \dfrac{4}{y} = -5y$$

$$y^2 + 4 = -5y$$

$$y^2 + 5y + 4 = 0$$

$$(y+1)(y+4) = 0$$

$$y+1=0 \quad or \quad y+4=0$$
$$y=-1 \quad or \quad y=-4$$

Both values check. The solutions are –1 and –4.

25. Because $\dfrac{12}{x}$ is undefined when x is 0, at the outset we

state the restriction that $x \neq 0$.

$$x - \frac{12}{x} = 4, \quad \text{LCD is } x$$

$$x\left(x - \frac{12}{x}\right) = x \cdot 4$$

$$x \cdot x - x \cdot \frac{12}{x} = x \cdot 4$$

$$x^2 - 12 = 4x$$

$$x^2 - 4x - 12 = 0$$

$$(x-6)(x+2) = 0$$

$$x = 6 \quad or \quad x = -2$$

Both numbers check. The solutions are –2 and 6.

27. Because $\dfrac{1}{y}$ is undefined when y is 0, at the outset we

state the restriction that $y \neq 0$.

$$\frac{9}{10} = \frac{1}{y}, \quad \text{LCD is } 10y$$

$$10y\left(\frac{9}{10}\right) = 10y \cdot \frac{1}{y}$$

$$9y = 10$$

$$y = \frac{10}{9}$$

This number checks. The solution is $\dfrac{10}{9}$.

29. $\dfrac{t-1}{t-3} = \dfrac{2}{t-3}$

To ensure that neither denominator is 0, we note at the outset that $t \neq 3$. Then we multiply both sides by the LCD, $t-3$.

$$(t-3) \cdot \frac{t-1}{t-3} = (t-3) \cdot \frac{2}{t-3}$$

$$t - 1 = 2$$

$$t = 3$$

Recall that, because of the restriction above, 3 cannot be a solution. A check confirms this.

Check: $\dfrac{t-1}{t-3} = \dfrac{2}{t-3}$

$$\frac{3-1}{3-3} \bigg| \frac{2}{3-3}$$

$$\overset{?}{}$$

$$\frac{2}{0} = \frac{2}{0} \quad \text{UNDEFINED}$$

The equation has no solution.

31. $\dfrac{x}{x-5} = \dfrac{25}{x^2 - 5x}$

$$\frac{x}{x-5} = \frac{25}{x(x-5)}$$

To ensure that neither denominator is 0, we note at the outset that $x \neq 0$ and $x \neq 5$. Then we multiply both

sides by the LCD, $x(x-5)$.

$$x(x-5) \cdot \frac{x}{x-5} = x(x-5) \cdot \frac{25}{x(x-5)}$$

$$x^2 = 25$$

$$x^2 - 25 = 0$$

$$(x+5)(x-5) = 0$$

$$x = -5 \quad or \quad x = 5$$

Recall that, because of the restrictions above, 5 cannot be a solution. The number –5 checks and is the solution.

33. $\dfrac{5}{4t} = \dfrac{7}{5t-2}$

To ensure that neither denominator is 0, we note at the outset that $t \neq 0$ and $t \neq \dfrac{2}{5}$. Then we multiply both sides by the LCD, $4t(5t-2)$.

$$4t(5t-2) \cdot \frac{5}{4t} = 4t(5t-2) \cdot \frac{7}{5t-2}$$

$$5(5t-2) = 4t \cdot 7$$

$$25t - 10 = 28t$$

$$-10 = 3t$$

$$-\frac{10}{3} = t$$

This value checks. The solution is $-\dfrac{10}{3}$.

35. $\dfrac{x^2+4}{x-1} = \dfrac{5}{x-1}$

To ensure that neither denominator is 0, we note at the outset that $x \neq 1$. Then we multiply both sides by the LCD, $x-1$.

$$(x-1) \cdot \frac{x^2+4}{x-1} = (x-1) \cdot \frac{5}{x-1}$$

$$x^2 + 4 = 5$$

$$x^2 - 1 = 0$$

$$(x+1)(x-1) = 0$$

$$x+1 = 0 \quad or \quad x-1 = 0$$

$$x = -1 \quad or \quad x = 1$$

Recall that, because of the restriction above, 1 cannot be a solution. The number –1 checks and is the solution. We might also observe that since the denominators are the same, the numerators must be the same. Solving $x^2 + 4 = 5$, we get $x = -1$ or $x = 1$ as shown above. Again, because of the restriction $x \neq 1$, only –1 is a solution of the equation.

37. $\dfrac{6}{a+1} = \dfrac{a}{a-1}$

To ensure that neither denominator is 0, we note at the outset that $a \neq -1$ and $a \neq 1$. Then we multiply both sides by the LCD, $(a+1)(a-1)$.

$$(a+1)(a-1)\cdot\frac{6}{a+1}=(a+1)(a-1)\cdot\frac{a}{a-1}$$
$$6(a-1)=a(a+1)$$
$$6a-6=a^2+a$$
$$0=a^2-5a+6$$
$$0=(a-2)(a-3)$$
$$a-2=0 \quad or \quad a-3=0$$
$$a=2 \quad or \quad a=3$$

Both values check. The solutions are 2 and 3.

39. $\dfrac{60}{t-5}-\dfrac{18}{t}=\dfrac{40}{t}$

To ensure that none of the denominators is 0, we note at the outset that $t\neq 5$ and $t\neq 0$. Then we multiply both sides by the LCD, $t(t-5)$.

$$t(t-5)\left(\frac{60}{t-5}-\frac{18}{t}\right)=t(t-5)\cdot\frac{40}{t}$$
$$60t-18(t-5)=40(t-5)$$
$$60t-18t+90=40t-200$$
$$2t=-290$$
$$t=-145$$

This value checks. The solution is –145.

41. $\dfrac{4}{y^2+y-12}=\dfrac{1}{y+4}-\dfrac{2}{y-3}$

$$\frac{4}{(y+4)(y-3)}=\frac{1}{y+4}-\frac{2}{y-3}$$

To ensure that none of the denominators is 0, we note at the outset that $y\neq -4$ and $y\neq 3$. Then we multiply both sides by the LCD, $(y+4)(y-3)$.

$$(y+4)(y-3)\cdot\frac{4}{(y+4)(y-3)}=(y+4)(y-3)\left(\frac{1}{y+4}-\frac{2}{y-3}\right)$$
$$4=y-3-2(y+4)$$
$$4=y-3-2y-8$$
$$4=-y-11$$
$$15=-y$$
$$-15=y$$

This value checks. The solution is –15.

43. $\dfrac{3}{x-3}+\dfrac{5}{x+2}=\dfrac{5x}{x^2-x-6}$

$$\frac{3}{x-3}+\frac{5}{x+2}=\frac{5x}{(x-3)(x+2)}$$

To ensure that none of the denominators is 0, we note at the outset that $x\neq 3$ and $x\neq -2$. Then we multiply both sides by the LCD, $(x-3)(x+2)$.

$$(x-3)(x+2)\left(\frac{3}{x-3}+\frac{5}{x+2}\right)=(x-3)(x+2)\cdot\frac{5x}{(x-3)(x+2)}$$
$$3(x+2)+5(x-3)=5x$$
$$3x+6+5x-15=5x$$
$$8x-9=5x$$
$$-9=-3x$$
$$3=x$$

Recall that, because of the restriction above, 3 cannot be a solution. Thus, the equation has no solution.

45. $\dfrac{3}{x}+\dfrac{x}{x+2}=\dfrac{4}{x^2+2x}$

$$\frac{3}{x}+\frac{x}{x+2}=\frac{4}{x(x+2)}$$

To ensure that none of the denominators is 0, we note at the outset that $x\neq 0$ and $x\neq -2$. Then we multiply both sides by the LCD, $x(x+2)$.

$$x(x+2)\left(\frac{3}{x}+\frac{x}{x+2}\right)=x(x+2)\cdot\frac{4}{x(x+2)}$$
$$3(x+2)+x\cdot x=4$$
$$3x+6+x^2=4$$
$$x^2+3x+2=0$$
$$(x+1)(x+2)=0$$
$$x+1=0 \quad or \quad x+2=0$$
$$x=-1 \quad or \quad x=-2$$

Recall that, because of the restrictions above, –2 cannot be a solution. The number –1 checks. The solution is –1.

47. $\dfrac{2}{t-4}+\dfrac{1}{t}=\dfrac{t}{4-t}$

$$\frac{2}{t-4}+\frac{1}{t}=\frac{-t}{t-4}$$

To ensure that none of the denominators is 0, we note at the outset that $t\neq 0$ and $t\neq 4$. Then we multiply both sides by the LCD, $t(t-4)$.

$$t(t-4)\left(\frac{2}{t-4}+\frac{1}{t}\right)=t(t-4)\cdot\frac{-t}{t-4}$$
$$t\cdot 2+t-4=t(-t)$$
$$2t+t-4=-t^2$$
$$t^2+3t-4=0$$
$$(t+4)(t-1)=0$$
$$t+4=0 \quad or \quad t-1=0$$
$$t=-4 \quad or \quad t=1$$

Both values check. The solutions are –4 and 1.

49. $\dfrac{5}{x+2}-\dfrac{3}{x-2}=\dfrac{2x}{4-x^2}$

$$\frac{5}{x+2}-\frac{3}{x-2}=\frac{2x}{(2+x)(2-x)}$$
$$\frac{5}{x+2}+\frac{3}{2-x}=\frac{2x}{(2+x)(2-x)} \qquad \left(-\frac{3}{x-2}=\frac{3}{2-x}\right)$$

First note that $x\neq -2$ and $x\neq 2$. Then multiply both sides by the LCD, $(2+x)(2-x)$.

$$(2+x)(2-x)\left(\frac{5}{x+2}+\frac{3}{2-x}\right)=(2+x)(2-x)\cdot\frac{2x}{(2+x)(2-x)}$$
$$5(2-x)+3(2+x)=2x$$
$$10-5x+6+3x=2x$$
$$16-2x=2x$$
$$16=4x$$
$$4=x$$

This value checks. The solution is 4.

51.
$$\frac{1}{x^2+2x+1}=\frac{x-1}{3x+3}+\frac{x+2}{5x+5}$$
$$\frac{1}{(x+1)(x+1)}=\frac{x-1}{3(x+1)}+\frac{x+2}{5(x+1)}$$

Note that $x\neq-1$. Then multiply both sides by the LCD, $15(x+1)^2$.

$$15(x+1)^2\cdot\frac{1}{(x+1)(x+1)}=15(x+1)^2\left(\frac{x-1}{3(x+1)}+\frac{x+2}{5(x+1)}\right)$$
$$15=5(x+1)(x-1)+3(x+1)(x+2)$$
$$15=5x^2-5+3x^2+9x+6$$
$$15=8x^2+9x+1$$
$$0=8x^2+9x-14$$
$$0=(x+2)(8x-7)$$
$$x=-2\ \ or\ \ x=\frac{7}{8}$$

Both values check. The solutions are -2 and $\frac{7}{8}$.

53.
$$\frac{3-2y}{y+1}-\frac{10}{y^2-1}=\frac{2y+3}{1-y}$$
$$\frac{3-2y}{y+1}-\frac{10}{(y+1)(y-1)}=\frac{2y+3}{1-y}$$
$$\frac{3-2y}{y+1}+\frac{10}{(y+1)(1-y)}=\frac{2y+3}{1-y}$$

Note that $y\neq-1$ and $y\neq1$.

$$(y+1)(1-y)\left(\frac{3-2y}{y+1}+\frac{10}{(y+1)(1-y)}\right)=(y+1)(1-y)\cdot\frac{2y+3}{1-y}$$
$$(1-y)(3-2y)+10=(y+1)(2y+3)$$
$$3-5y+2y^2+10=2y^2+5y+3$$
$$10=10y$$
$$1=y$$

Recall that because of the restriction above, 1 is not a solution. Thus, the equation has no solution.

55. We find all values of a for which $2a-\frac{15}{a}=7$. First note that $a\neq0$. Then multiply both sides by the LCD, a.

$$a\left(2a-\frac{15}{a}\right)=a\cdot7$$
$$a\cdot2a-a\cdot\frac{15}{a}=7a$$
$$2a^2-15=7a$$
$$2a^2-7a-15=0$$
$$(2a+3)(a-5)=0$$
$$a=-\frac{3}{2}\ \ or\ \ a=5$$

Both values check. The solutions are $-\frac{3}{2}$ and 5.

57. We find all values of a for which $\frac{a-5}{a+1}=\frac{3}{5}$. First note that $a\neq-1$. Then multiply both sides by the LCD, $5(a+1)$.

$$5(a+1)\cdot\frac{a-5}{a+1}=5(a+1)\cdot\frac{3}{5}$$
$$5(a-5)=3(a+1)$$
$$5a-25=3a+3$$
$$2a=28$$
$$a=14$$

This value checks. The solution is 14.

59. We find all values of a for which $\frac{12}{a}-\frac{12}{2a}=8$. First note that $a\neq0$. Then multiply both sides by the LCD, $2a$.

$$2a\left(\frac{12}{a}-\frac{12}{2a}\right)=2a\cdot8$$
$$2a\cdot\frac{12}{a}-2a\cdot\frac{12}{2a}=16a$$
$$24-12=16a$$
$$12=16a$$
$$\frac{3}{4}=a$$

This value checks. The solution is $\frac{3}{4}$.

61.
$$f(a)=g(a)$$
$$\frac{3a-1}{a^2-7a+10}=\frac{a-1}{a^2-4}+\frac{2a+1}{a^2-3a-10}$$
$$\frac{3a-1}{(a-2)(a-5)}=\frac{a-1}{(a+2)(a-2)}+\frac{2a+1}{(a-5)(a+2)}$$

First note that $a\neq2$, $a\neq5$, and $a\neq-2$. Then multiply both sides by the LCD, $(a-2)(a-5)(a+2)$.

$$(a-2)(a-5)(a+2)\cdot\frac{3a-1}{(a-2)(a-5)}$$
$$=(a-2)(a-5)(a+2)\left(\frac{a-1}{(a+2)(a-2)}+\frac{2a+1}{(a-5)(a+2)}\right)$$
$$(a+2)(3a-1)=(a-5)(a-1)+(a-2)(2a+1)$$
$$3a^2+5a-2=a^2-6a+5+2a^2-3a-2$$
$$3a^2+5a-2=3a^2-9a+3$$
$$5a-2=-9a+3$$
$$14a-2=3$$
$$14a=5$$
$$a=\frac{5}{14}$$

This number checks. Then $f(a)=g(a)$ for $a=\frac{5}{14}$.

63.
$$f(a)=g(a)$$
$$\frac{2}{a^2-8a+7}=\frac{3}{a^2-2a-3}-\frac{1}{a^2-1}$$
$$\frac{2}{(a-1)(a-7)}=\frac{3}{(a+1)(a-3)}-\frac{1}{(a+1)(a-1)}$$

First note that $a\neq1$, $a\neq7$, $a\neq-1$, and $a\neq3$. Then multiply both sides by the LCD, $(a-1)(a-7)(a+1)(a-3)$.

$$(a-1)(a-7)(a+1)(a-3)\cdot\frac{2}{(a-1)(a-7)}$$
$$=(a-1)(a-7)(a+1)(a-3)\left(\frac{3}{(a+1)(a-3)}-\frac{1}{(a+1)(a-1)}\right)$$

$$2(a+1)(a-3) = 3(a-1)(a-7) - (a-7)(a-3)$$
$$2(a^2 - 2a - 3) = 3(a^2 - 8a + 7) - (a^2 - 10a + 21)$$
$$2a^2 - 4a - 6 = 3a^2 - 24a + 21 - a^2 + 10a - 21$$
$$2a^2 - 4a - 6 = 2a^2 - 14a$$
$$-4a - 6 = -14a$$
$$-6 = -10a$$
$$\frac{3}{5} = a$$

This number checks. Then $f(a) = g(a)$ for $a = \frac{3}{5}$.

65. *Writing Exercise.* Note that the domain of f is all real numbers while the domain of g is all real numbers except 2. Then we see that the graph on the right represents f, and the graph of the left represents g.

67. *Familiarize.* Let $t =$ the time to travel 200 mi. We use the formula Time = Distance/Rate.

Translate. The distance is 200 mi, the rate is 135 mph + speed of the tailwind, 15 mph.

$$t = \frac{200}{135 + 15}$$

Carry out. We solve the equation.

$$t = \frac{200}{135 + 15} = \frac{4}{3}, \text{ or } 1\frac{1}{3}$$

Check. We recheck the calculation. The number checks.

State. It will take $1\frac{1}{3}$ hr to go 200 mi.

69. *Familiarize.* Let $x =$ the amount of plastic sorted by Brenton, in pounds. Then $2x =$ the amount of plastic sorted by Kylie.

Translate.

From together they sorted 123 pounds of plastic, we have

$$x + 2x = 123$$

Carry out. Solve the equation.

$$x + 2x = 123$$
$$3x = 123$$
$$x = 41$$

Then $2x = 82$

Check. Since $41 + 82 = 123$, the solution checks.

State. Brenton sorted 41 lb and Kylie sorted 82 lb.

71. *Familiarize.* Let x represent the width of the mat, in inches. Recall that area $A = lw$.

Translate.

From the picture in the text, we know that length of the photo is $13 - 2x$ and the width of the photo is $10 - 2x$.

So the area is
$$A = lw$$
$$70 = (13 - 2x)(10 - 2x)$$

Carry out. Solve the equation.

$$70 = (13 - 2x)(10 - 2x)$$
$$70 = 130 - 46x + 4x^2$$
$$0 = 4x^2 - 46x + 60$$
$$0 = 2(2x^2 - 23x + 30)$$
$$0 = 2(2x - 3)(x - 10)$$
$$x = 1.5 \text{ or } x = 10$$

Check. Since the width of the mat cannot equal the width of the picture, 10 is not a solution. We check 1.5. For the length, $13 - 2(1.5) = 10$. And the width, $10 - 2(1.5) = 7$. Length times width is $10 \cdot 7 = 70$. The number checks.

State. The width of the mat is 1.5 in.

73. *Writing Exercise.* False; let $a = 1$, $b = 2$, and $c = 0$. Then $1 \cdot 0 = 2 \cdot 0$, but $1 \neq 2$.

75.
$$f(a) = g(a)$$

$$\frac{a - \frac{2}{3}}{a + \frac{1}{2}} = \frac{a + \frac{2}{3}}{a - \frac{3}{2}}$$

$$\frac{a - \frac{2}{3}}{a + \frac{1}{2}} \cdot \frac{6}{6} = \frac{a + \frac{2}{3}}{a - \frac{3}{2}} \cdot \frac{6}{6}$$

$$\frac{6a - \frac{2}{3} \cdot 6}{6a + \frac{1}{2} \cdot 6} = \frac{6a + \frac{2}{3} \cdot 6}{6a - \frac{3}{2} \cdot 6}$$

$$\frac{6a - 4}{6a + 3} = \frac{6a + 4}{6a - 9}$$

$$\frac{6a - 4}{3(2a + 1)} = \frac{6a + 4}{3(2a - 3)}$$

To ensure that neither denominator is 0, we note at the outset that $a \neq -\frac{1}{2}$ and $a \neq \frac{3}{2}$. Then we multiply both sides by the LCD, $3(2a + 1)(2a - 3)$.

$$3(2a + 1)(2a - 3) \cdot \frac{6a - 4}{3(2a + 1)} = 3(2a + 1)(2a - 3) \cdot \frac{6a + 4}{3(2a - 3)}$$
$$(2a - 3)(6a - 4) = (2a + 1)(6a + 4)$$
$$12a^2 - 26a + 12 = 12a^2 + 14a + 4$$
$$-26a + 12 = 14a + 4$$
$$-40a + 12 = 4$$
$$-40a = -8$$
$$a = \frac{1}{5}$$

This number checks. For $a = \frac{1}{5}$, $f(a) = g(a)$.

77. $\dfrac{a+3}{a+2} - \dfrac{a+4}{a+3} = \dfrac{a+5}{a+4} - \dfrac{a+6}{a+5}$

Note that $a \neq -2$ and $a \neq -3$ and $a \neq -4$ and $a \neq -5$.

$$(a+2)(a+3)(a+4)(a+5)\left(\dfrac{a+3}{a+2} - \dfrac{a+4}{a+3}\right)$$

$$= (a+2)(a+3)(a+4)(a+5)\left(\dfrac{a+5}{a+4} - \dfrac{a+6}{a+5}\right)$$

$$(a+3)(a+4)(a+5)(a+3) - (a+2)(a+4)(a+5)(a+4)$$
$$= (a+2)(a+3)(a+5)(a+5) - (a+2)(a+3)(a+4)(a+6)$$

$$a^4 + 15a^3 + 83a^2 + 201a + 180 - (a^4 + 15a^3 + 82a^2 + 192a + 160)$$
$$= a^4 + 15a^3 + 81a^2 + 185a + 150 - (a^4 + 15a^3 + 80a^2 + 180a + 144)$$

$$a^2 + 9a + 20 = a^2 + 5a + 6$$
$$4a = -14$$
$$a = -\dfrac{7}{2}$$

This value checks. When $a = -\dfrac{7}{2}$, $f(a) = g(a)$.

79. Set $f(a)$ equal to $g(a)$ and solve for a.

$$\dfrac{0.793}{a} + 18.15 = \dfrac{6.034}{a} - 43.17$$

Note that $a \neq 0$. Then multiply on both sides by the LCD, a.

$$a\left(\dfrac{0.793}{a} + 18.15\right) = a\left(\dfrac{6.034}{a} - 43.17\right)$$
$$0.793 + 18.15a = 6.034 - 43.17a$$
$$6.132a = 5.241$$
$$a \approx 0.0854697$$

This value checks. When $a \approx 0.0854697$, $f(a) = g(a)$.

81. $\dfrac{x^2 + 6x - 16}{x - 2} = x + 8, \quad x \neq 2$

$$\dfrac{(x+8)(x-2)}{x-2} = x + 8$$
$$\dfrac{(x+8)(x-2)}{x-2} = x + 8$$
$$x + 8 = x + 8$$
$$8 = 8$$

Since $8 = 8$ is true for all values of x, the original equation is true for any possible replacements of the variable. It is an identity.

83. *Graphing Calculator Exercise*

85. Let $y_1 = \dfrac{x^2 - 4}{x - 2}$ and observe that for $x = 2$ the entry in the Y1-column of the table is "ERROR."

Connecting the Concepts

1. $\dfrac{5x^2 - 10x}{5x^2 + 5x} = \dfrac{5x(x-2)}{5x(x+1)}$ Factoring

$$= \dfrac{5x}{5x} \cdot \dfrac{x-2}{x+1}$$

$$= \dfrac{x-2}{x+1}$$ Removing a factor of 1

3. $\dfrac{t}{2} + \dfrac{t}{3} = 5, \quad$ LCD is 6

$$6\left(\dfrac{t}{2} + \dfrac{t}{3}\right) = 6 \cdot 5$$
$$3t + 2t = 30$$
$$5t = 30$$
$$t = 6$$

The solution is 6.

5. $\dfrac{\dfrac{1}{z} + 1}{\dfrac{1}{z^2} - 1} = \dfrac{\dfrac{1}{z} + 1}{\dfrac{1}{z^2} - 1} \cdot \dfrac{z^2}{z^2}$

$$= \dfrac{\dfrac{1}{z} \cdot z^2 + 1 \cdot z^2}{\dfrac{1}{z^2} \cdot z^2 - 1 \cdot z^2}$$

$$= \dfrac{z + z^2}{1 - z^2}$$

$$= \dfrac{z(1 + z)}{(1 + z)(1 - z)}$$

$$= \dfrac{z}{1 - z}$$

7. $\dfrac{5}{x + 3} = \dfrac{3}{x + 2}, \quad$ LCD is $(x+2)(x+3)$

Note that $x \neq -3$ and $x \neq -2$.

$$(x+2)(x+3)\dfrac{5}{x+3} = (x+2)(x+3)\dfrac{3}{x+2}$$
$$5(x+2) = 3(x+3)$$
$$5x + 10 = 3x + 9$$
$$2x = -1$$
$$x = -\dfrac{1}{2}$$

The solution is $-\dfrac{1}{2}$.

9. $\dfrac{27a^2}{8} \div \dfrac{12}{5a} = \dfrac{27a^2}{8} \cdot \dfrac{5a}{12}$

$$= \dfrac{3 \cdot 9a^2 \cdot 5a}{8 \cdot 4 \cdot 3} = \dfrac{\cancel{3} \cdot 9a^2 \cdot 5a}{8 \cdot 4 \cdot \cancel{3}}$$

$$= \dfrac{45a^3}{32}$$

11. $\dfrac{2n-1}{n-2}-\dfrac{n-3}{n+1}=\dfrac{2n-1}{n-2}\cdot\dfrac{n+1}{n+1}-\dfrac{n-3}{n+1}\cdot\dfrac{n-2}{n-2}$

$\qquad\qquad =\dfrac{2n^2+n-1}{(n-2)(n+1)}-\dfrac{n^2-5n+6}{(n-2)(n+1)}$

$\qquad\qquad =\dfrac{2n^2+n-1-\left(n^2-5n+6\right)}{(n-2)(n+1)}$

$\qquad\qquad =\dfrac{n^2+6n-7}{(n-2)(n+1)},\ \text{or}\ \dfrac{(n+7)(n-1)}{(n-2)(n+1)}$

13. $\dfrac{8t+8}{2t^2+t-1}\cdot\dfrac{t^2-1}{t^2-2t+1}$

$\qquad =\dfrac{8(t+1)(t+1)(t-1)}{(2t-1)(t+1)(t-1)(t-1)}$

$\qquad =\dfrac{8(t+1)\,\cancel{(t+1)}\,\cancel{(t-1)}}{(2t-1)\,\cancel{(t+1)}\,(t-1)\,\cancel{(t-1)}}$

$\qquad =\dfrac{8(t+1)}{(2t-1)(t-1)}$

15. $\dfrac{5}{t}=\dfrac{4}{3},\quad$ LCD is $3t$

$\qquad\qquad$ Note that $t\neq 0$.

$3t\cdot\dfrac{5}{t}=3t\cdot\dfrac{4}{3}$

$\qquad 15=4t$

$\qquad \dfrac{15}{4}=t$

The solution is $\dfrac{15}{4}$.

17. $\qquad\dfrac{1}{6x}=\dfrac{x}{x+1},\quad$ LCD is $6x(x+1)$

$\qquad\qquad\qquad$ Note that $x\neq -1$ and $x\neq 0$.

$6x(x+1)\cdot\dfrac{1}{6x}=6x(x+1)\cdot\dfrac{x}{x+1}$

$\qquad\quad x+1=6x^2$

$\qquad\qquad 0=6x^2-x-1$

$\qquad\qquad 0=(3x+1)(2x-1)$

$\qquad 3x+1=0\quad or\quad 2x-1=0$

$\qquad\quad x=-\dfrac{1}{3}\quad or\qquad x=\dfrac{1}{2}$

The solutions are $-\dfrac{1}{3}$ and $\dfrac{1}{2}$.

19. $\dfrac{4}{x^2-6x-16}+\dfrac{x}{x^2-x-6}$

$\qquad =\dfrac{4}{(x-8)(x+2)}+\dfrac{x}{(x-3)(x+2)}$

$\qquad =\dfrac{4}{(x-8)(x+2)}\cdot\dfrac{x-3}{x-3}+\dfrac{x}{(x-3)(x+2)}\cdot\dfrac{x-8}{x-8}$

$\qquad =\dfrac{4x-12+x^2-8x}{(x-8)(x-3)(x+2)}$

$\qquad =\dfrac{x^2-4x-12}{(x-8)(x-3)(x+2)}$

$\qquad =\dfrac{(x-6)(x+2)}{(x-8)(x-3)(x+2)}$

$\qquad =\dfrac{x-6}{(x-8)(x-3)}\cdot\dfrac{x+2}{x+2}$

$\qquad =\dfrac{x-6}{(x-8)(x-3)}$

Exercise Set 6.5

1. 1 cake in 2 hours $=\dfrac{1\ \text{cake}}{2\ \text{hr}}=\dfrac{1}{2}$ cake per hour

3. Sandy: $\dfrac{1\ \text{cake}}{2\ \text{hr}}=\dfrac{1}{2}$ cake per hour

Eric: $\dfrac{1\ \text{cake}}{3\ \text{hr}}=\dfrac{1}{3}$ cake per hour

Together: $\dfrac{1}{2}+\dfrac{1}{3}=\dfrac{3}{6}+\dfrac{2}{6}=\dfrac{5}{6}$ cake per hour

5. 1 lawn in 3 hours $=\dfrac{1\ \text{lawn}}{3\ \text{hr}}=\dfrac{1}{3}$ lawn per hour

7. *Familiarize.* Let $x=$ the number.

Translate.

The reciprocal of 3	plus	the reciprocal of 6	is	the reciprocal of the number.
$\downarrow$	$\downarrow$	$\downarrow$	$\downarrow$	$\downarrow$
$\dfrac{1}{3}$	$+$	$\dfrac{1}{6}$	$=$	$\dfrac{1}{x}$

Carry out. We solve the equation.

$\qquad\dfrac{1}{3}+\dfrac{1}{6}=\dfrac{1}{x},\quad$ LCD is $6x$

$6x\left(\dfrac{1}{3}+\dfrac{1}{6}\right)=6x\cdot\dfrac{1}{x}$

$\qquad 2x+x=6$

$\qquad\quad 3x=6$

$\qquad\quad x=2$

Check. $\dfrac{1}{3}+\dfrac{1}{6}=\dfrac{2}{6}+\dfrac{1}{6}=\dfrac{3}{6}=\dfrac{1}{2}$. This is the reciprocal of 2, so the result checks.

State. The number is 2.

9. *Familiarize.* We let $x=$ the number.

Translate.

A number	plus	6	times	its reciprocal	is	-5.
$\downarrow$	$\downarrow$	$\downarrow$	$\downarrow$	$\downarrow$	$\downarrow$	$\downarrow$
x	$+$	6	$\cdot$	$\dfrac{1}{x}$	$=$	-5

Carry out. We solve the equation.

$\qquad x+\dfrac{6}{x}=-5,\quad$ LCD is x

$\qquad x\left(x+\dfrac{6}{x}\right)=x(-5)$

$\qquad\quad x^2+6=-5x$

$\qquad x^2+5x+6=0$

$(x+3)(x+2)=0$

$\qquad\qquad x=-3\ \ or\ \ x=-2$

Check. The possible solutions are -3 and -2.

We check -3 in the conditions of the problem.

Number: -3

6 times the reciprocal
of the number: $6\left(-\dfrac{1}{3}\right) = -2$

Sum of the number and $-3 + (-2) = -5$
6 times its reciprocal:

The number -3 checks.

Now we check -2:

Number: -2

6 times the reciprocal
of the number: $6\left(-\dfrac{1}{2}\right) = -3$

Sum of the number and $-2 + (-3) = -5$
6 times its reciprocal:

The number -2 also checks.

State. The number is -3 or -2.

11. ***Familiarize.*** Let $x =$ the first integer. Then $x + 1 =$ the second, and their product $= x(x + 1)$.

Translate.

$$\underbrace{\text{Reciprocal of the product}}_{\dfrac{1}{x(x+1)}} \quad \underset{=}{\text{is}} \quad \underset{\dfrac{1}{90}}{\dfrac{1}{90}}.$$

Carry out. We solve the equation.

$$\frac{1}{x(x+1)} = \frac{1}{90}, \quad \text{LCD is } 90x(x+1)$$

$$90x(x+1)\left(\frac{1}{x(x+1)}\right) = 90x(x+1)\cdot\frac{1}{90}$$

$$90 = x(x+1)$$
$$90 = x^2 + x$$
$$0 = x^2 + x - 90$$
$$0 = (x+10)(x-9)$$
$$x = -10 \quad or \quad x = 9$$

Check. The possible solutions are -10 and 9.

If the first is -10, then the second is -9. $\dfrac{1}{-10(-9)} = \dfrac{1}{90}$.

If the first is 9, then the second is 10. $\dfrac{1}{9(10)} = \dfrac{1}{90}$.

Both values check.

State. The integers are -10 and -9, or 9 and 10.

13. ***Familiarize.*** Let t represent the number of hours it takes Bryan and Caroline to refinish the floors working together.

Translate. Bryan takes 8 hr and Caroline takes 6 hr to complete the job, so we have

$$\frac{t}{8} + \frac{t}{6} = 1$$

Carry out. We solve the equation. Multiply on both sides by the LCD, 24.

$$24\left(\frac{t}{8} + \frac{t}{6}\right) = 24 \cdot 1$$
$$3t + 4t = 24$$
$$7t = 24$$
$$t = \frac{24}{7}, \text{ or } 3\frac{3}{7}$$

Check. If Bryan does the job alone in 8 hr, then in $3\frac{3}{7}$ hr he does $\frac{24/7}{8}$, or $\frac{3}{7}$ of the job. If Caroline does the job alone in 6 hr, then in $3\frac{3}{7}$ hr she does $\frac{24/7}{6}$, or $\frac{4}{7}$ of the job. Together, they do $\frac{3}{7} + \frac{4}{7}$, or 1 entire job. The result checks.

State. It would take Bryan and Caroline $3\frac{3}{7}$ hr to finish the job working together.

15. ***Familiarize.*** The pool can be filled in 12 hours with only the pipe and in 30 hours with only the hose. Then in 1 hour, the pipe fills $\frac{1}{12}$ of the pool, and the hose fills $\frac{1}{30}$ of the pool. Using both the pipe and the hose, $\frac{1}{12} + \frac{1}{30}$ of the pool can be filled in 1 hour.

Suppose that it takes t hours to fill the pool using both the pipe and hose.

Translate. We need to find t such that

$$t\left(\frac{1}{12}\right) + t\left(\frac{1}{30}\right) = 1, \text{ or } \frac{t}{12} + \frac{t}{30} = 1,$$

where 1 represents the entire job.

Carry out. We solve the equation. Multiply on both sides by the LCD, 60.

$$60\left(\frac{t}{12} + \frac{t}{30}\right) = 60 \cdot 1$$
$$5t + 2t = 60$$
$$7t = 60$$
$$t = \frac{60}{7}$$

Check. The possible solution is $\frac{60}{7}$ hours. If the pipe is used $\frac{60}{7}$ hours, it fills $\frac{1}{12} \cdot \frac{60}{7}$, or $\frac{5}{7}$ of the pool. If the hose is used $\frac{60}{7}$ hours, it fills $\frac{1}{30} \cdot \frac{60}{7}$, or $\frac{2}{7}$ of the pool. Using both, $\frac{5}{7} + \frac{2}{7}$ of the pool, or all of it, will be filled in $\frac{60}{7}$ hours.

State. Using both the pipe and the hose, it will take $\frac{60}{7}$ hours, or $8\frac{4}{7}$ hours, to fill the pool.

17. ***Familiarize.*** In 1 minute the Wayne pump does $\frac{1}{42}$ of the job and the Craftsman pump does $\frac{1}{35}$ of the job.

Working together, they do $\frac{1}{42}+\frac{1}{35}$ of the job in 1 minute. Suppose it takes t minutes to do the job working together.

Translate. We find t such that

$$t\left(\frac{1}{42}\right)+t\left(\frac{1}{35}\right)=1,\ \text{or}\ \frac{t}{42}+\frac{t}{35}=1.$$

Carry out. We solve the equation. We multiply both sides by the LCD, 210.

$$210\left(\frac{t}{42}+\frac{t}{35}\right)=210\cdot 1$$
$$5t+6t=210$$
$$11t=210$$
$$t=\frac{210}{11}$$

Check. In $\frac{210}{11}$ min the Wayne pump does

$\frac{210}{11}\cdot\frac{1}{42}$, or $\frac{5}{11}$ of the job and the Craftsman pump does

$\frac{210}{11}\cdot\frac{1}{35}$, or $\frac{6}{11}$ of the job. Together they do $\frac{5}{11}+\frac{6}{11}$, or

1 entire job. The answer checks.

State. The two pumps can pump out the basement in

$\frac{210}{11}$ min, or $19\frac{1}{11}$ min, working together.

19. Familiarize. Let t represent the time, in minutes, that it takes the K5400 to print the brochures working alone. Then $2t$ represents the time it takes the H470 to do the job, working alone. In 1 minute the K5400 does $\frac{1}{t}$ of the job and the H470 does $\frac{1}{2t}$ of the job.

Translate. Working together, they can do the entire job in 45 min, so we want to find t such that

$$45\left(\frac{1}{t}\right)+45\left(\frac{1}{2t}\right)=1,\ \text{or}\ \frac{45}{t}+\frac{45}{2t}=1.$$

Carry out. We solve the equation. We multiply both sides by the LCD, $2t$.

$$2t\left(\frac{45}{t}+\frac{45}{2t}\right)=2t\cdot 1$$
$$90+45=2t$$
$$135=2t$$
$$\frac{135}{2}=t,\ \text{or}\ 67\frac{1}{2}$$

Check. If the K5400 can do the job in $\frac{135}{2}$ min, then in

45 min it does $45\cdot\frac{1}{135/2}$, or $\frac{2}{3}$ of the job. If it takes the

H470 $2\cdot\frac{135}{2}$, or 135 min, to do the job, then in 45 min it

does $45\cdot\frac{1}{135}$, or $\frac{1}{3}$ of the job. Working together, the two

machines do $\frac{2}{3}+\frac{1}{3}$, or 1 entire job, in 45 min.

State. Working alone, it takes the K5400 $67\frac{1}{2}$ min and the H470 135 min to print the brochure.

21. Familiarize. Let t represent the number of minutes it takes the Airgle machine to purify the air working alone. Then $t-15$ represents the time it takes the Austin machine to purify the air, working alone. In 1 minute the Airgle does $\frac{1}{t}$ of the job and the Austin does $\frac{1}{t-15}$ of the job.

Translate. Working together, the two machines can purify the air in 10 min to find t such that

$$10\left(\frac{1}{t}\right)+10\left(\frac{1}{t-15}\right)=1,\ \text{or}\ \frac{10}{t}+\frac{10}{t-15}=1.$$

Carry out. We solve the equation. First we multiply both sides by the LCD, $t(t-15)$.

$$t(t-15)\left(\frac{10}{t}+\frac{10}{t-15}\right)=t(t-15)\cdot 1$$
$$10(t-15)+10t=t(t-15)$$
$$10t-150+10t=t^2-15t$$
$$0=t^2-35t+150$$
$$0=(t-5)(t-30)$$
$$t=5\ \ or\ \ t=30$$

Check. If $t=5$, then $t-15=5-15=-10$. Since negative time has no meaning in this application, 5 cannot be a solution. If $t=30$, then $t-15=30-15=15$.

In 10 min the Airgle machine does $10\cdot\frac{1}{30}$, or $\frac{1}{3}$ of the

job. In 10 min the Austin does $10\cdot\frac{1}{15}$, or $\frac{2}{3}$ of the job.

Together they do $\frac{1}{3}+\frac{2}{3}$, or 1 entire job. The answer

checks.

State. Working alone, the Airgle machine can purify the air in 30 min and the Austin machine can purify the air in 15 min.

23. Familiarize. Let t represent the number of hours it takes Elliot to deliver the papers alone. Then $3t$ represents the number of hours it takes Sara to deliver the papers alone.

Translate. In 1 hr Elliot and Sara will do one entire job, so we have

$$1\left(\frac{1}{t}\right)+1\left(\frac{1}{3t}\right)=1,\ \text{or}\ \frac{1}{t}+\frac{1}{3t}=1$$

Carry out. We solve the equation. Multiply on both sides by the LCD, $3t$.

$$3t\left(\frac{1}{t}+\frac{1}{3t}\right)=3t\cdot 1$$
$$3+1=3t$$
$$4=3t$$
$$\frac{4}{3}=t$$

Check. If Elliot does the job alone in $\frac{4}{3}$ hr, then in 1 hr he does $\frac{1}{4/3}$, or $\frac{3}{4}$ of the job. If Sara does the job alone in $3\cdot\frac{4}{3}$, or 4 hr, then in 1 hr she does $\frac{1}{4}$ of the job. Together, they do $\frac{3}{4}+\frac{1}{4}$, or 1 entire job, in 1 hr. The result checks.

State. It would take Elliot $\frac{4}{3}$ hours and it would take Sara 4 hours to deliver the papers alone.

25. Familiarize. Let t represent the number of hours it takes Lia to paint the floor working alone. Then $t+3$ represents the time it takes Zeno to paint the floor working alone.

Translate. In 2 hr Lia and Zeno will do one entire job, so we have

$$2\left(\frac{1}{t}\right)+2\left(\frac{1}{t+3}\right)=1,\ \text{or}\ \frac{2}{t}+\frac{2}{t+3}=1$$

Carry out. We solve the equation. Multiply on both sides by the LCD, $t(t+3)$.

$$t(t+3)\left(\frac{2}{t}+\frac{2}{t+3}\right)=t(t+3)\cdot 1$$
$$2(t+3)+2t=t(t+3)$$
$$2t+6+2t=t^2+3t$$
$$0=t^2-t-6$$
$$0=(t-3)(t+2)$$
$$t=3\ \text{or}\ t=-2$$

Check. If Lia does the job alone in 3 hr, then in 2 hr she does $\frac{2}{3}$ of the job. If Zeno does the job alone in $3+3$, or 6 hr, then in 2 hr he does $\frac{2}{6}$, or $\frac{1}{3}$ of the job. Together, they do $\frac{2}{3}+\frac{1}{3}$, or 1 entire job, in 1 hr. The result checks.

State. It would take Lia 3 hours and it would take Zeno 6 hours to paint a floor alone.

27. Familiarize. Let t represent the number of minutes it takes Chris to do the job working alone. Then $t+120$ represents the time it takes Kim to do the job working alone.

We will convert hours to minutes:

$$2\ \text{hr}=2\cdot 60\ \text{min}=120\ \text{min}$$
$$2\ \text{hr}\ 55\ \text{min}=120\ \text{min}+55\ \text{min}=175\ \text{min}$$

Translate. In 175 min Chris and Kim will do one entire job, so we have

$$175\left(\frac{1}{t}\right)+175\left(\frac{1}{t+120}\right)=1,\ \text{or}\ \frac{175}{t}+\frac{175}{t+120}=1$$

Carry out. We solve the equation. Multiply on both sides by the LCD, $t(t+120)$.

$$t(t+120)\left(\frac{175}{t}+\frac{175}{t+120}\right)=t(t+120)\cdot 1$$
$$175(t+120)+175t=t(t+120)$$
$$175t+21{,}000+175t=t^2+120t$$
$$0=t^2-230t-21{,}000$$
$$0=(t-300)(t+70)$$
$$t=300\ \text{or}\ t=-70$$

Check. Since negative time has no meaning in this problem -70 is not a solution of the original problem. If Chris does the job alone in 300 min, then in 175 min he does $\frac{175}{300}=\frac{7}{12}$ of the job. If Kim does the job alone in $300+120$, or 420 min, then in 175 min she does $\frac{175}{420}=\frac{5}{12}$ of the job. Together, they do $\frac{7}{12}+\frac{5}{12}$, or 1 entire job, in 175 min. The result checks.

State. It would take Chris 300 min, or 5 hours to do the job alone.

29. Familiarize. We first make a drawing. Let $r=$ the kayak's speed in still water in mph. Then $r-3=$ the speed upstream and $r+3=$ the speed downstream.

$$\text{Upstream}\qquad 4\ \text{miles}\qquad r-3\ \text{mph}$$
$$\text{10 miles}\qquad r+3\ \text{mph}\qquad \text{Downstream}$$

We organize the information in a table. The time is the same both upstream and downstream so we use t for each time.

	Distance	Speed	Time
Upstream	4	$r-3$	t
Downstream	10	$r+3$	t

Translate. Using the formula Time = Distance/Rate in each row of the table and the fact that the times are the same, we can write an equation.

$$\frac{4}{r-3}=\frac{10}{r+3}$$

Carry out. We solve the equation.

$$\frac{4}{r-3}=\frac{10}{r+3},\quad \text{LCD is}\ (r-3)(r+3)$$
$$(r-3)(r+3)\cdot\frac{4}{r-3}=(r-3)(r+3)\cdot\frac{10}{r+3}$$
$$4(r+3)=10(r-3)$$
$$4r+12=10r-30$$
$$42=6r$$
$$7=r$$

Check. If $r=7$ mph, then $r-3$ is 4 mph and $r+3$ is 10 mph. The time upstream is $\frac{4}{4}$, or 1 hour. The time downstream is $\frac{10}{10}$, or 1 hour. Since the times are the same, the answer checks.

State. The speed of the kayak in still water is 7 mph.

31. *Familiarize*. We first make a drawing. Let r = Roslyn's speed on a nonmoving sidewalk in ft/sec. Then her speed moving forward on the moving sidewalk is $r + 1.8$, and her speed in the opposite direction is $r - 1.8$.

Forward $r + 1.8$ 105 ft

           Opposite

     51 ft $r - 1.8$ direction

We organize the information in a table. The time is the same both forward and in the opposite direction, so we use t for each time.

	Distance	Speed	Time
Forward	105	$r + 1.8$	t
Opposite direction	51	$r - 1.8$	t

Translate. Using the formula Time = Distance/Rate in each row of the table and the fact that the times are the same, we can write an equation.

$$\frac{105}{r+1.8} = \frac{51}{r-1.8}$$

Carry out. We solve the equation.

$$\frac{105}{r+1.8} = \frac{51}{r-1.8}$$
$$\text{LCD is } (r+1.8)(r-1.8)$$
$$(r+1.8)(r-1.8)\frac{105}{r+1.8} = (r+1.8)(r-1.8)\frac{51}{r-1.8}$$
$$105(r-1.8) = 51(r+1.8)$$
$$105r - 189 = 51r + 91.8$$
$$54r = 280.8$$
$$r = 5.2$$

Check. If Roslyn's speed on a nonmoving sidewalk is 5.2 ft/sec, then her speed moving forward on the moving sidewalk is 5.2 + 1.8, or 7 ft/sec, and her speed moving in the opposite direction on the sidewalk is 5.2 – 1.8, or 3.4 ft/sec. Moving 105 ft at 7 ft/sec takes $\frac{105}{7} = 15$ sec.

Moving 51 ft at 3.4 ft/sec takes $\frac{51}{3.4} = 15$ sec. Since the times are the same, the answer checks.

State. Roslyn's would be walking 5.2 ft/sec on a nonmoving sidewalk.

33. *Familiarize*. Let r = the speed of the passenger train in mph. Then $r - 14$ = the speed of the freight train in mph. We organize the information in a table. The time is the same for both trains so we use t for each time.

	Distance	Speed	Time
Passenger train	400	r	t
Freight train	330	$r - 14$	t

Translate. Using the formula Time = Distance/Rate in each row of the table and the fact that the times are the same, we can write an equation.

$$\frac{400}{r} = \frac{330}{r-14}$$

Carry out. We solve the equation.

$$\frac{400}{r} = \frac{330}{r-14}, \quad \text{LCD is } r(r-14)$$
$$r(r-14)\cdot\frac{400}{r} = r(r-14)\cdot\frac{330}{r-14}$$
$$400(r-14) = 330r$$
$$400r - 5600 = 330r$$
$$-5600 = -70r$$
$$80 = r$$

Check. If the passenger train's speed is 80 mph, then the freight train's speed is $80 - 14$, or 66 mph. Traveling 400 mi at 80 mph takes $\frac{400}{80} = 5$ hr. Traveling 330 mi at 66 mph takes $\frac{330}{66} = 5$ hr. Since the times are the same, the answer checks.

State. The speed of the passenger train is 80 mph; the speed of the freight train is 66 mph.

35. Note that 38 mi is 7 mi less than 45 mi and that the local bus travels 7 mph slower than the express. Then the express travels 45 mi in one hr, or 45 mph, and the local bus travels 38 mi in one hr, or 38 mph.

37. *Familiarize*. We let r = the speed of the river. Then $15 + r$ = LeBron's speed downstream in km/h and $15 - r$ = his speed upstream in km/h. The times are the same. Let t represent the time. We organize the information in a table.

	Distance	Speed	Time
Upstream	140	$15 + r$	t
Downstream	35	$15 - r$	t

Translate. Using the formula Time = Distance/Rate in each row of the table and the fact that the times are the same, we can write an equation.

$$\frac{140}{15+r} = \frac{35}{15-r}$$

Carry out. We solve the equation.

$$\frac{140}{15+r} = \frac{35}{15-r}, \quad \text{LCD is } (15+r)(15-r)$$

$$(15+r)(15-r) \cdot \frac{140}{15+r} = (15+r)(15-r) \cdot \frac{35}{15-r}$$

$$140(15-r) = 35(15+r)$$

$$2100 - 140r = 525 + 35r$$

$$1575 = 175r$$

$$9 = r$$

Check. If $r = 9$, then the speed downstream is $15 + 9$, or 24 km/h and the speed upstream is $15 - 9$, or 6 km/h. The time for the trip downstream is $\frac{140}{24}$, or $5\frac{5}{6}$ hours. The time for the trip upstream is $\frac{35}{6}$, or $5\frac{5}{6}$ hours. The times are the same. The values check.

State. The speed of the river is 9 km/h.

39. Familiarize. Let $c =$ the speed of the current, in km/h. Then $7 + c =$ the speed downriver and $7 - c =$ the speed upriver. We organize the information in a table.

	Distance	Speed	Time
Downriver	45	$7+c$	t_1
Upriver	45	$7-c$	t_2

Translate. Using the formula Time = Distance/Rate we see that $t_1 = \frac{45}{7+c}$ and $t_2 = \frac{45}{7-c}$. The total time upriver and back is 14 hr, so $t_1 + t_2 = 14$, or

$$\frac{45}{7+c} + \frac{45}{7-c} = 14.$$

Carry out. We solve the equation. Multiply both sides by the LCD, $(7+c)(7-c)$.

$$(7+c)(7-c)\left(\frac{45}{7+c} + \frac{45}{7-c}\right) = (7+c)(7-c)14$$

$$45(7-c) + 45(7+c) = 14(49 - c^2)$$

$$315 - 45c + 315 + 45c = 686 - 14c^2$$

$$14c^2 - 56 = 0$$

$$14(c+2)(c-2) = 0$$

$$c+2 = 0 \quad or \quad c-2 = 0$$

$$c = -2 \quad or \quad c = 2$$

Check. Since speed cannot be negative in this problem, -2 cannot be a solution of the original problem. If the speed of the current is 2 km/h, the barge travels upriver at $7 - 2$, or 5 km/h. At this rate it takes $\frac{45}{5}$, or 9 hr, to travel 45 km. The barge travels downriver at $7 + 2$, or 9 km/h. At this rate it takes $\frac{45}{9}$, or 5 hr, to travel 45 km. The total travel time is $9 + 5$, or 14 hr. The answer checks.

State. The speed of the current is 2 km/h.

41. Familiarize. Let $w =$ the wind speed, in mph. Then speed into the wind is $460 - w$, and the speed with the wind is $460 + w$. We organize the information in a table.

	Distance	Speed	Time
Into the wind	525	$460-w$	t_1
With the wind	525	$460+w$	t_2

Translate. Using the formula Time = Distance/Rate, we see that $t_1 = \frac{525}{460+w}$ and $t_2 = \frac{525}{460-w}$. The total time into the wind and back is 2.3 hr, so $t_1 + t_2 = 2.3$, or

$$\frac{525}{460+w} + \frac{525}{460-w} = 2.3.$$

Carry out. We solve the equation. Multiply both sides by the LCD, $(460+w)(460-w)$.

$$(460+w)(460-w)\left(\frac{525}{460+w} + \frac{525}{460-w}\right) = (460+w)(460-w)2.3$$

$$525(460-w) + 525(460+w) = 2.3(211{,}600 - w^2)$$

$$241{,}500 - 525w + 241{,}500 + 525w = 486{,}680 - 2.3w^2$$

$$483{,}000 = 486{,}680 - 2.3w^2$$

$$2.3w^2 - 3680 = 0$$

$$2.3(w^2 - 1600) = 0$$

$$2.3(w+40)(w-40) = 0$$

$$w = -40 \quad or \quad w = 40$$

Check. We check only 40 since the wind speed cannot be negative. If the wind speed is 40 mph, then the plane's speed into the wind is $460 - 40$, or 420 mph, and the speed with the wind is $460 + 40$, or 500 mph. Flying 525 mi into the wind takes $\frac{525}{420} = 1.25$ hr. Flying 525 mi with the wind takes $\frac{525}{500} = 1.05$ hr. The total time is $1.25 + 1.05$, or 2.3 hr. The answer checks.

State. The wind speed is 40 mph.

43. Familiarize. Let $r =$ the speed at which the train actually traveled in mph, and let $t =$ the actual travel time in hours. We organize the information in a table.

	Distance	Speed	Time
Actual speed	120	r	t
Faster speed	120	$r+10$	$t-2$

Translate. From the first row of the table we have $120 = rt$, and from the second row we have $120 = (r+10)(t-2)$. Solving the first equation for t, we have $t = \frac{120}{r}$. Substituting for t in the second equation, we have

$$120 = (r+10)\left(\frac{120}{r} - 2\right).$$

Carry out. We solve the equation.

$$120 = (r+10)\left(\frac{120}{r} - 2\right)$$
$$120 = 120 - 2r + \frac{1200}{r} - 20$$
$$20 = -2r + \frac{1200}{r}$$
$$r \cdot 20 = r\left(-2r + \frac{1200}{r}\right)$$
$$20r = -2r^2 + 1200$$
$$2r^2 + 20r - 1200 = 0$$
$$2(r^2 + 10r - 600) = 0$$
$$2(r+30)(r-20) = 0$$
$$r = -30 \ or \ r = 20$$

Check. Since speed cannot be negative in this problem, -30 cannot be a solution of the original problem. If the speed is 20 mph, it takes $\frac{120}{20}$, or 6 hr, to travel 120 mi. If the speed is 10 mph faster, or 30 mph, it takes $\frac{120}{30}$, or 4 hr, to travel 120 mi. Since 4 hr is 2 hr less time than 6 hr, the answer checks.

State. The speed was 20 mph.

45. *Writing Exercise*. Yes; if the steamrollers working together take more than half as long as the slower steamroller would working alone, then they do more than one entire job. That is, in half the time it takes the slower steamroller to do the job alone, the faster steamroller can do more than half of the job alone, and together they do more than $\frac{1}{2} + \frac{1}{2}$, or 1 entire job in that time.

47. $\dfrac{42x^8y^9}{7x^2y} = \dfrac{42}{7}x^{8-2}y^{9-1} = 6x^6y^8$

49. $\dfrac{4x^2y}{-xy^2} = \dfrac{4}{-1}x^{2-1}y^{1-2} = -4xy^{-1}$, or $\dfrac{-4x}{y}$

51.
$$\begin{array}{r} 4x^3 - 3x^2 \quad\ -7 \\ \underline{-\left(4x^3 - 8x^2 + 4x\right)} \\ 5x^2 - 4x - 7 \end{array}$$

53. *Writing Exercise*. Answers may vary.
Beth can paint a storage shed in 6 hr. It takes Leanne 12 hr to complete the same job. How long will it take them to paint the shed working together?

55. *Familiarize*. If the drainage gate is closed, $\frac{1}{9}$ of the bog is filled in 1 hr. If the bog is not being filled, $\frac{1}{11}$ of the bog is drained in 1 hr. If the bog is being filled with the drainage gate left open, $\frac{1}{9} - \frac{1}{11}$ of the bog is filled in 1 hr. Let $t = $ the time it takes to fill the bog with the drainage

gate left open.

Translate. We want to find t such that
$$t\left(\frac{1}{9} - \frac{1}{11}\right) = 1 \text{, or } \frac{t}{9} - \frac{t}{11} = 1.$$

Carry out. We solve the equation. First we multiply by the LCD, 99.
$$99\left(\frac{t}{9} - \frac{t}{11}\right) = 99 \cdot 1$$
$$11t - 9t = 99$$
$$2t = 99$$
$$t = \frac{99}{2}$$

Check. In $\frac{99}{2}$ hr, we have $\frac{99}{2}\left(\frac{1}{9} - \frac{1}{11}\right) = \frac{11}{2} - \frac{9}{2} = \frac{2}{2} = 1$ full bog.

State. It will take $\frac{99}{2}$, or $49\frac{1}{2}$ hr, to fill the bog.

57. Sean's speed downstream is $7 + 3$, or 10 mph. Using Time = Distance/Rate, we find that the time it will take Sean to kayak 5 mi downstream is 5/10, or 1/2 hr, or 30 min.

59. *Familiarize*. Let $p = $ the number of people per hour moved by the 60 cm-wide escalator. Then $2p = $ the number of people per hour moved by the 100 cm-wide escalator. We convert 1575 people per 14 minutes to people per hour:
$$\frac{1575 \text{ people}}{14 \text{ min}} \cdot \frac{60 \text{ min}}{1 \text{ hr}} = 6750 \text{ people}/\text{hr}$$

Translate. We use the information that together the escalators move 6750 people per hour to write an equation.
$$p + 2p = 6750$$

Carry out. We solve the equation.
$$p + 2p = 6750$$
$$3p = 6750$$
$$p = 2250$$

Check. If the 60 cm-wide escalator moves 2250 people per hour, then the 100 cm-wide escalator moves $2 \cdot 2250$, or 4500 people per hour. Together, they move $2250 + 4500$, or 6750 people per hour. The answer checks.

State. The 60 cm-wide escalator moves 2250 people per hour.

61. *Familiarize*. Let $d = $ the distance, in miles, the paddleboat can cruise upriver before it is time to turn around. The boat's speed upriver is $12 - 5$, or 7 mph, and its speed downriver is $12 + 5$, or 17 mph. We organize the information in a table.

	Distance	Speed	Time
Upriver	d	7	t_1
Downriver	d	17	t_2

Translate. Using the formula Time = Distance/Rate we see that $t_1 = \dfrac{d}{7}$ and $t_2 = \dfrac{d}{17}$. The time upriver and back is 3 hr, so $t_1 + t_2 = 3$, or $\dfrac{d}{7} + \dfrac{d}{17} = 3$.

Carry out. We solve the equation.

$$7 \cdot 17\left(\frac{d}{7} + \frac{d}{17}\right) = 7 \cdot 17 \cdot 3$$
$$17d + 7d = 357$$
$$24d = 357$$
$$d = \frac{119}{8}$$

Check. Traveling $\dfrac{119}{8}$ mi upriver at a speed of 7 mph takes $\dfrac{119/8}{7} = \dfrac{17}{8}$ hr. Traveling $\dfrac{119}{8}$ mi downriver at a speed of 17 mph takes $\dfrac{119/8}{17} = \dfrac{7}{8}$ hr. The total time is $\dfrac{17}{8} + \dfrac{7}{8} = \dfrac{24}{8} = 3$ hr. The answer checks.

State. The pilot can go $\dfrac{119}{8}$, or $14\dfrac{7}{8}$ mi upriver before it is time to turn around.

63. *Familiarize.* Let t represent the time it takes the printers to print 500 pages working together.
Translate. The faster machine can print 500 pages in 40 min, and it takes the slower printer 50 min to do the same job. Then we have

$$\frac{t}{40} + \frac{t}{50} = 1.$$

Carry out. We solve the equation.

$$\frac{t}{40} + \frac{t}{50} = 1, \quad \text{LCD is } 200$$
$$200\left(\frac{t}{40} + \frac{t}{50}\right) = 200 \cdot 1$$
$$5t + 4t = 200 \cdot 1$$
$$9t = 200$$
$$t = \frac{200}{9}$$

In $\dfrac{200}{9}$ min, the faster printer does $\dfrac{200/9}{40}$, or $\dfrac{200}{9} \cdot \dfrac{1}{40}$, or $\dfrac{5}{9}$ of the job. Then starting at page 1, it would print $\dfrac{5}{9} \cdot 500$, or $277\dfrac{7}{9}$ pages. Thus, in $\dfrac{200}{9}$ min, the two machines will meet on page 278.

Check. We can check to see that the slower machine is also printing page 278 after $\dfrac{200}{9}$ min. In $\dfrac{200}{9}$ min, the slower machine does $\dfrac{200/9}{50}$, or $\dfrac{200}{9} \cdot \dfrac{1}{50}$, or $\dfrac{4}{9}$ of the job. Then it would print $\dfrac{4}{9} \cdot 500$, or $222\dfrac{2}{9}$ pages.

Working backward from page 500, this machine would be on page $500 - 222\dfrac{2}{9}$, or $277\dfrac{2}{9}$. Thus, both machines are printing page 278 after $\dfrac{200}{9}$ min. The answer checks.

State. The two machines will meet on page 278.

65. *Familiarize.* Express the position of the hands in terms of minute units on the face of the clock. At 10:30 the hour hand is at $\dfrac{10.5}{12}$ hr $\times \dfrac{60 \text{ min}}{1 \text{ hr}}$, or 52.5 minutes, and the minute hand is at 30 minutes. The rate of the minute hand is 12 times the rate of the hour hand. (When the minute hand moves 60 minutes, the hour hand moves 5 minutes.) Let $t =$ the number of minutes after 10:30 that the hands will first be perpendicular. After t minutes the minute hand has moved t units, and the hour hand has moved $\dfrac{t}{12}$ units. The position of the hour hand will be 15 units "ahead" of the position of the minute hand when they are first perpendicular.

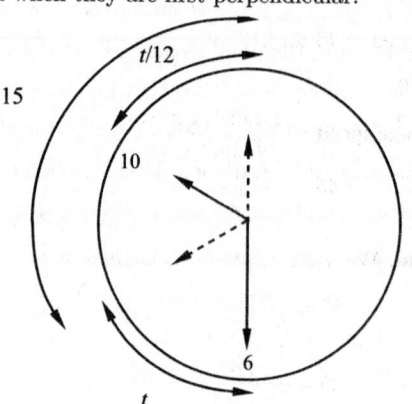

Translate.

Position of hour hand after t min	is	position of minute hand after t min	plus	15 min.
$\downarrow$	$\downarrow$	$\downarrow$	$\downarrow$	$\downarrow$
$52.5 + \dfrac{t}{12}$	$=$	$30 + t$	$+$	15

Carry out. We solve the equation.

$$52.5 + \frac{t}{12} = 30 + t + 15$$
$$52.5 + \frac{t}{12} = 45 + t, \quad \text{LCM is } 12$$
$$12\left(52.5 + \frac{t}{12}\right) = 12(45 + t)$$
$$630 + t = 540 + 12t$$
$$90 = 11t$$
$$\frac{90}{11} = t, \text{ or } 8\frac{2}{11} = t$$

Check. At $\dfrac{90}{11}$ min after 10:30, the position of the hour hand is at $52.5 + \dfrac{90/11}{12}$, or $53\dfrac{2}{11}$ min. The minute hand is at $30 + \dfrac{90}{11}$, or $38\dfrac{2}{11}$ min. The hour hand is 15 minutes ahead of the minute hand so the hands are perpendicular. The answer checks.

State. After 10:30 the hands of a clock will first be perpendicular in $8\dfrac{2}{11}$ min. The time is $10:38\dfrac{2}{11}$, or $21\dfrac{9}{11}$ min before 11:00.

67. *Familiarize*. Let $r =$ the speed in mph Garry would have to travel for the last half of the trip in order to average a speed of 45 mph for the entire trip. We organize the information in a table.

	Distance	Speed	Time
First half	50	40	t_1
Last half	50	r	t_2

The total distance is $50 + 50$, or 100 mi.

The total time is $t_1 + t_2$, or $\dfrac{50}{40} + \dfrac{50}{r}$, or $\dfrac{5}{4} + \dfrac{50}{r}$. The average speed is 45 mph.

Translate.

$$\text{Average speed} = \dfrac{\text{Total distance}}{\text{Total time}}$$
$$45 = \dfrac{100}{\dfrac{5}{4} + \dfrac{50}{r}}$$

Carry out. We solve the equation.

$$45 = \dfrac{100}{\dfrac{5}{4} + \dfrac{50}{r}}$$
$$45 = \dfrac{100}{\dfrac{5r + 200}{4r}}$$
$$45 = 100 \cdot \dfrac{4r}{5r + 200}$$
$$45 = \dfrac{400r}{5r + 200}$$
$$(5r + 200)(45) = (5r + 200) \cdot \dfrac{400r}{5r + 200}$$
$$225r + 9000 = 400r$$
$$9000 = 175r$$
$$\dfrac{360}{7} = r$$

Check. Traveling 50 mi at 40 mph takes $\dfrac{50}{40}$, or $\dfrac{5}{4}$ hr. Traveling 50 mi at $\dfrac{360}{7}$ mph takes $\dfrac{50}{360/7}$, or $\dfrac{35}{36}$ hr. Then the total time is $\dfrac{5}{4} + \dfrac{35}{36} = \dfrac{80}{36} = \dfrac{20}{9}$ hr. The average speed when traveling 100 mi for $\dfrac{20}{9}$ hr is $\dfrac{100}{20/9} = 45$ mph. The answer checks.

State. Garry would have to travel at a speed of $\dfrac{360}{7}$, or

$51\dfrac{3}{7}$ mph for the last half of the trip so that the average speed for the entire trip would be 45 mph.

Exercise Set 6.6

1. The divisor is $x - 3$.

3. The quotient is $x + 2$.

5. The degree of the divisor, $x - 3$, is 1.

7. $\dfrac{36x^6 + 18x^5 - 27x^2}{9x^2} = \dfrac{36x^6}{9x^2} + \dfrac{18x^5}{9x^2} - \dfrac{27x^2}{9x^2}$
$\qquad = 4x^4 + 2x^3 - 3$

9. $\dfrac{21a^3 + 7a^2 - 3a - 14}{-7a}$
$= \dfrac{21a^3}{-7a} + \dfrac{7a^2}{-7a} - \dfrac{3a}{-7a} - \dfrac{14}{-7a}$
$= -3a^2 - a + \dfrac{3}{7} + \dfrac{2}{a}$

11. $\dfrac{16y^4z^2 - 8y^6z^4 + 12y^8z^3}{-4y^4z}$
$= \dfrac{16y^4z^2}{-4y^4z} - \dfrac{8y^6z^4}{-4y^4z} + \dfrac{12y^8z^3}{-4y^4z}$
$= -4z + 2y^2z^3 - 3y^4z^2$

13. $\dfrac{16y^3 - 9y^2 - 8y}{2y^2}$
$= \dfrac{16y^3}{2y^2} - \dfrac{9y^2}{2y^2} - \dfrac{8y}{2y^2}$
$= 8y - \dfrac{9}{2} - \dfrac{4}{y}$

15. $\dfrac{15x^7 - 21x^4 - 3x^2}{-3x^2}$
$= \dfrac{15x^7}{-3x^2} + \dfrac{-21x^4}{-3x^2} + \dfrac{-3x^2}{-3x^2}$
$= -5x^5 + 7x^2 + 1$

17. $(a^2b - a^3b^3 - a^5b^5) \div (a^2b)$
$= \dfrac{a^2b}{a^2b} - \dfrac{a^3b^3}{a^2b} - \dfrac{a^5b^5}{a^2b}$
$= 1 - ab^2 - a^3b^4$

19. $(x^2 + 10x + 21) \div (x + 7)$
$= \dfrac{(x + 7)(x + 3)}{x + 7}$
$= \dfrac{\cancel{(x + 7)}(x + 3)}{\cancel{x + 7}}$
$= x + 3$

The answer is $x + 3$.

21.
$$\begin{array}{r} y-5 \\ y-5\overline{\smash{\big)}\,y^2-10y-25} \\ \underline{y^2-5y} \\ -5y-25 \\ \underline{-5y+25} \\ -50 \end{array}$$
$(y^2-10y)-(y^2-5y)=-5y$

$(-5y-25)-(-5y+25)=-50$

The answer is $y-5$, R -50, or $y-5+\dfrac{-50}{y-5}$.

23.
$$\begin{array}{r} x-5 \\ x-4\overline{\smash{\big)}\,x^2-9x+21} \\ \underline{x^2-4x} \\ -5x+21 \\ \underline{-5x+20} \\ 1 \end{array}$$

The answer is $x-5$, R 1, or $x-5+\dfrac{1}{x-4}$.

25. $(y^2-25)\div(y+5)=\dfrac{y^2-25}{y+5}$

$=\dfrac{(y+5)(y-5)}{y+5}$

$=\dfrac{\cancel{(y+5)}(y-5)}{\cancel{y+5}}$

$=y-5$

We could also find this quotient as follows.

$$\begin{array}{r} y-5 \\ y+5\overline{\smash{\big)}\,y^2+0y-25} \\ \underline{y^2+5y} \\ -5y-25 \\ \underline{-5y-25} \\ 0 \end{array}$$
Writing in the missing term

The answer is $y-5$.

27.
$$\begin{array}{r} y^2-2y-1 \\ y-2\overline{\smash{\big)}\,y^3-4y^2+3y-6} \\ \underline{y^3-2y^2} \\ -2y^2+3y \\ \underline{-2y^2+4y} \\ -y-6 \\ \underline{-y+2} \\ -8 \end{array}$$

The answer is y^2-2y-1, R -8, or $y^2-2y-1+\dfrac{-8}{y-2}$.

29.
$$\begin{array}{r} 2x^2-x+1 \\ x+2\overline{\smash{\big)}\,2x^3+3x^2-x-3} \\ \underline{2x^3+4x^2} \\ -x^2-x \\ \underline{-x^2-2x} \\ x-3 \\ \underline{x+2} \\ -5 \end{array}$$

The answer is $2x^2-x+1$, R -5, or

$2x^2-x+1+\dfrac{-5}{x+2}$.

31.
$$\begin{array}{r} a^2-4a+6 \\ a+4\overline{\smash{\big)}\,a^3+0a^2-10a+24} \\ \underline{a^3+4a^2} \\ -4a^2-10a \\ \underline{-4a^2-16a} \\ 6a+24 \\ \underline{6a+24} \\ 0 \end{array}$$

The answer is a^2-4a+6.

33.
$$\begin{array}{r} 2y^2+2y-1 \\ 5y-2\overline{\smash{\big)}\,10y^3+6y^2-9y+10} \\ \underline{10y^3-4y^2} \\ 10y^2-9y \\ \underline{10y^2-4y} \\ -5y+10 \\ \underline{-5y+2} \\ 8 \end{array}$$

The answer is $2y^2+2y-1$, R 8, or $2y^2+2y-1+\dfrac{8}{5y-2}$.

35.
$$\begin{array}{r} 3x^2+x+1 \\ x^2-3\overline{\smash{\big)}\,3x^4+x^3-8x^2-3x-3} \\ \underline{3x^4-9x^2} \\ x^3+x^2-3x \\ \underline{x^3-3x} \\ x^2-3 \\ \underline{x^2-3} \\ 0 \end{array}$$

The answer is $3x^2+x+1$.

37.
$$\begin{array}{r}
2x^2 - x - 9 \\
x^2+2\overline{\smash{\big)}\,2x^4 - x^3 - 5x^2 + x - 6} \\
\underline{2x^4 \phantom{{}-x^3} + 4x^2} \\
-x^3 - 9x^2 + x \\
\underline{-x^3 \phantom{{}-9x^2} - 2x} \\
-9x^2 + 3x - 6 \\
\underline{-9x^2 \phantom{{}+3x} - 18} \\
3x + 12
\end{array}$$

The answer is $2x^2 - x - 9$, R $3x + 12$, or

$2x^2 - x - 9 + \dfrac{3x+12}{x^2+2}$.

39. $F(x) = \dfrac{f(x)}{g(x)} = \dfrac{6x^2 - 11x - 10}{3x+2}$

$$\begin{array}{r}
2x - 5 \\
3x+2\overline{\smash{\big)}\,6x^2 - 11x - 10} \\
\underline{6x^2 + 4x} \\
-15x - 10 \\
\underline{-15x - 10} \\
0
\end{array}$$

Since $g(x)$ is 0 for $x = -\dfrac{2}{3}$, we have

$F(x) = 2x - 5$, provided $x \neq -\dfrac{2}{3}$.

41. $F(x) = \dfrac{f(x)}{g(x)} = \dfrac{8x^3 - 27}{2x - 3}$

$$\begin{array}{r}
4x^2 + 6x + 9 \\
2x-3\overline{\smash{\big)}\,8x^3 - 27} \\
\underline{8x^3 - 12x^2} \\
12x^2 + 0x \\
\underline{12x^2 - 18x} \\
18x - 27 \\
\underline{18x - 27} \\
0
\end{array}$$

Since $g(x)$ is 0 for $x = \dfrac{3}{2}$, we have

$F(x) = 4x^2 + 6x + 9$, provided $x \neq \dfrac{3}{2}$.

43. $F(x) = \dfrac{f(x)}{g(x)} = \dfrac{x^4 - 24x^2 - 25}{x^2 - 25}$

$$\begin{array}{r}
x^2 \phantom{{}-24x^2} + 1 \\
x^2-25\overline{\smash{\big)}\,5x^4 - 24x^2 - 25} \\
\underline{x^4 - 25x^2} \\
x^2 - 25 \\
\underline{x^2 - 25} \\
0
\end{array}$$

Since $g(x)$ is 0 for $x = -5$ or $x = 5$, we have

$F(x) = x^2 + 1$, provided $x \neq -5$ and $x \neq 5$.

45. We rewrite $f(x)$ in descending order.

$F(x) = \dfrac{f(x)}{g(x)} = \dfrac{2x^5 - 3x^4 - 2x^3 + 8x^2 - 5}{x^2 - 1}$

$$\begin{array}{r}
2x^3 - 3x^2 + 5 \\
x^2-1\overline{\smash{\big)}\,2x^5 - 3x^4 - 2x^3 + 8x^2 - 5} \\
\underline{2x^5 \phantom{{}-3x^4} - 2x^3} \\
-3x^4 \phantom{{}-2x^3} + 8x^2 \\
\underline{-3x^4 \phantom{{}-2x^3} + 3x^2} \\
5x^2 - 5 \\
\underline{5x^2 - 5} \\
0
\end{array}$$

Since $g(x)$ is 0 for $x = -1$ or $x = 1$, we have

$F(x) = 2x^3 - 3x^2 + 5$, provided $x \neq -1$ and $x \neq 1$.

47. *Writing Exercise.* Factor the numerator of $F(x)$ and simplify by removing a factor equal to 1.

49. Graph $3x - y = 9$
$$y = 3x - 9$$

This graph is a line with y-intercept $(0, -9)$ and slope 3.

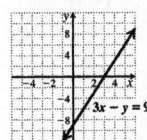

51. Graph $y < \dfrac{5}{2}x$

The graph is a dotted line with y-intercept $(0, 0)$ and shading above the line.

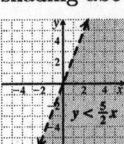

53. Graph $y = -\dfrac{3}{4}x + 1$

This graph is a line with slope $-\dfrac{3}{4}$ and y-intercept $(0, 1)$.

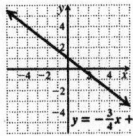

55. *Writing Exercise.* Answers may vary. One such polynomial is formed by writing $(x + c)^4 + 2$ in the form $a_4 x^4 + a_3 x^3 + a_2 x^2 + a_1 x + a_0$.

57.
$$\begin{array}{r}
a^2 + ab \\
a^2+3ab+2b^2\overline{\smash{\big)}\,a^4 + 4a^3 b + 5a^2 b^2 + 2ab^3} \\
\underline{a^4 + 3a^3 b + 2a^2 b^2} \\
a^3 b + 3a^2 b^2 + 2ab^3 \\
\underline{a^3 b + 3a^2 b^2 + 2ab^3} \\
0
\end{array}$$

The answer is $a^2 + ab$.

59.

$$a+b \overline{)\begin{array}{l} a^6 - a^5b + a^4b^2 - a^3b^3 + a^2b^4 - ab^5 + b^6 \\ a^7 \qquad\qquad\qquad\qquad\qquad\qquad + b^7 \end{array}}$$

$$\begin{array}{r} a^7 + a^6b \\ \hline -a^6b \\ -a^6b - a^5b^2 \\ \hline a^5b^2 \\ a^5b^2 + a^4b^3 \\ \hline -a^4b^3 \\ -a^4b^3 - a^3b^4 \\ \hline a^3b^4 \\ a^3b^4 + a^2b^5 \\ \hline -a^2b^5 \\ -a^2b^5 - ab^6 \\ \hline ab^6 + b^7 \\ ab^6 + b^7 \\ \hline 0 \end{array}$$

The answer is $a^6 - a^5b + a^4b^2 - a^3b^3 + a^2b^4 - ab^5 + b^6$.

61.

$$x+2 \overline{)\begin{array}{l} x - 5 \\ x^2 - 3x + 2k \end{array}}$$

$$\begin{array}{r} x^2 + 2x \\ \hline -5x + 2k \\ -5x \quad -10 \\ \hline 2k + 10 \end{array}$$

The remainder is 7. Thus, we solve the following equation for k.

$$2k + 10 = 7$$
$$2k = -3$$
$$k = -\frac{3}{2}$$

63. *Writing Exercise.* Answers may vary. One method would be to show that $(x^2 - 1)\left(x + 9 + \dfrac{x+4}{x^2-1}\right) \neq x^3 + 9x^2 - 6$.

Another method would be to show that

$(x^3 + 9x^2 - 6) \div (x^2 - 1) \neq x + 9 + \dfrac{x+4}{x^2-1}$ for some specific

value of x. For example, for $x = 0$,

$(x^3 + 9x - 6) \div (x^2 - 1) = -6 \div (-1) = 6$, but

$x + 9 + \dfrac{x+4}{x^2-1} = 9 - 4 = 5$.

65. *Graphing Calculator Exercise*

Exercise Set 6.7

1. True; see page 407 in the text.

3. True

5. True; see page 405 in the text.

7. $\left(x^3 - 4x^2 - 2x + 5\right) \div (x - 1)$

$$\begin{array}{r} \underline{1|} \quad 1 \quad -4 \quad -2 \quad 5 \\ \quad\quad 1 \quad -3 \quad -5 \\ \hline \quad 1 \quad -3 \quad -5 \quad |\, 0 \end{array}$$

The answer is $x^2 - 3x - 5$.

9. $\left(a^2 + 8a + 11\right) \div (a + 3)$

$= \left(a^2 + 8a + 11\right) \div [a - (-3)]$

$$\begin{array}{r} \underline{-3|} \quad 1 \quad 8 \quad 11 \\ \quad\quad -3 \quad -15 \\ \hline \quad 1 \quad 5 \quad |\, -4 \end{array}$$

The answer is $a + 5$, R -4, or $a + 5 + \dfrac{-4}{a+3}$.

11. $\left(2x^3 - x^2 - 7x + 14\right) \div (x + 2)$

$= \left(2x^3 - x^2 - 7x + 14\right) \div [x - (-2)]$

$$\begin{array}{r} \underline{-2|} \quad 2 \quad -1 \quad -7 \quad 14 \\ \quad\quad -4 \quad 10 \quad -6 \\ \hline \quad 2 \quad -5 \quad 3 \quad |\, 8 \end{array}$$

The answer is $2x^2 - 5x + 3$, R 8, or $2x^2 - 5x + 3 + \dfrac{8}{x+2}$.

13. $\left(a^3 - 10a + 12\right) \div (a - 2)$

$= \left(a^3 + 0a^2 - 10a + 12\right) \div (a - 2)$

$$\begin{array}{r} \underline{2|} \quad 1 \quad 0 \quad -10 \quad 12 \\ \quad\quad 2 \quad 4 \quad -12 \\ \hline \quad 1 \quad 2 \quad -6 \quad |\, 0 \end{array}$$

The answer is $a^2 + 2a - 6$.

15. $\left(3y^3 - 7y^2 - 20\right) \div (y - 3)$

$= \left(3y^3 - 7y^2 + 0y - 20\right) \div (y - 3)$

$$\begin{array}{r} \underline{3|} \quad 3 \quad -7 \quad 0 \quad -20 \\ \quad\quad 9 \quad 6 \quad 18 \\ \hline \quad 3 \quad 2 \quad 6 \quad |\, -2 \end{array}$$

The answer is $3y^2 + 2y + 6$, R -2, or $3y^2 + 2y + 6 + \dfrac{-2}{y-3}$.

17. $\left(x^5 - 32\right) \div (x - 2)$

$= \left(x^5 + 0x^4 + 0x^3 + 0x^2 + 0x - 32\right) \div (x - 2)$

$$\begin{array}{r} \underline{2|} \quad 1 \quad 0 \quad 0 \quad 0 \quad 0 \quad -32 \\ \quad\quad 2 \quad 4 \quad 8 \quad 16 \quad 32 \\ \hline \quad 1 \quad 2 \quad 4 \quad 8 \quad 16 \quad |\, 0 \end{array}$$

The answer is $x^4 + 2x^3 + 4x^2 + 8x + 16$.

19. $\left(3x^3+1-x+7x^2\right)\div\left(x+\dfrac{1}{3}\right)$

$=\left(3x^3+7x^2-x+1\right)\div\left[x-\left(-\dfrac{1}{3}\right)\right]$

$$\begin{array}{r|rrrr} -\frac{1}{3} & 3 & 7 & -1 & 1 \\ & & -1 & -2 & 1 \\ \hline & 3 & 6 & -3 & \underline{|\,2} \end{array}$$

The answer is $3x^2+6x-3$, R 2, or $3x^2+6x-3+\dfrac{2}{x+\frac{1}{3}}$

21.
$$\begin{array}{r|rrrrr} -3 & 5 & 12 & 0 & 28 & 9 \\ & & -15 & 9 & -27 & -3 \\ \hline & 5 & -3 & 9 & 1 & \underline{|\,6} \end{array}$$

The remainder tells us that $f(-3)=6$.

23.
$$\begin{array}{r|rrrrr} -3 & 2 & -1 & -7 & 1 & 2 \\ & & -6 & 21 & -42 & 123 \\ \hline & 2 & -7 & 14 & -41 & \underline{|\,125} \end{array}$$

The remainder tells us that $P(-3)=125$.

25.
$$\begin{array}{r|rrrrr} 4 & 1 & -6 & 11 & -17 & 20 \\ & & 4 & -8 & 12 & -20 \\ \hline & 1 & -2 & 3 & -5 & \underline{|\,0} \end{array}$$

The remainder tells us that $f(4)=0$.

27. *Writing Exercise.* When performing synthetic division we reverse the sign of the constant in the divisor and as a result we add rather than subtract as we do in long division.

29. $ac=b$

$c=\dfrac{b}{a}$

31. $pq-rq=st$

$q(p-r)=st$

$q=\dfrac{st}{p-r}$

33. $ab-cd=3b+d$

$ab-3b=cd+d$

$b(a-3)=cd+d$

$b=\dfrac{cd+d}{a-3}$

35. *Writing Exercise.* If $p(3)=0$, it follows by the principle of zero products that $Q(3)=0$. If $Q(3)=0$, it does not follow that $p(3)=0$. For example, let $Q(x)=x^2-x-6$. Then $p(x)=x+2$ is a factor of $Q(x)$ and $Q(3)=0$, but $p(3)\ne0$.

37. a) The degree of the remainder must be less than the degree of the divisor. Thus, the degree of the remainder must be 0, so R must be a constant.

b) $P(x)=(x-r)\cdot Q(x)+R$
$P(r)=(r-r)\cdot Q(r)+R=0\cdot Q(r)+R=R$

39.
$$\begin{array}{r|rrrr} -3 & 4 & 16 & -3 & -45 \\ & & -12 & -12 & 45 \\ \hline & 4 & 4 & -15 & \underline{|\,0} \end{array}$$

The remainder tells us that $f(-3)=0$.

$f(x)=(x+3)(4x^2+4x-15)=(x+3)(2x+5)(2x-3)$

Solve $f(x)=0$:

$(x+3)(2x+5)(2x-3)=0$

$x+3=0\quad or\quad 2x+5=0\quad or\quad 2x-3=0$

$x=-3\ or\qquad x=-\dfrac{5}{2}\ or\qquad x=\dfrac{3}{2}$

The solutions are -3 , $-\dfrac{5}{2}$, and $\dfrac{3}{2}$.

41. *Graphing Calculator Exercise*

43. $f(x)=4x^3+16x^2-3x-45$

$=x(4x^2+16x-3)-45$

$=x(x(4x+16)-3)-45$

$f(-3)=-3(-3(4(-3)+16)-3)-45$

$=-3(-3(-12+16)-3)-45$

$=-3(-3\cdot4-3)-45$

$=-3(-12-3)-45$

$=-3(-15)-45$

$=45-45$

$=0$

Exercise Set 6.8

1. (d) LCD

3. (e) Product

5. (a) Directly

7. As the number of painters increases, the time required to scrape the house decreases, so we have inverse variation.

9. As the number of laps increases, the time required to swim them increases, so we have direct variation.

11. As the number of volunteers increases, the time required to wrap the toys decreases, so we have inverse variation.

13. $f=\dfrac{L}{d}$

$df=L\qquad$ Multiplying by d

$d=\dfrac{L}{f}\qquad$ Dividing by f

15.
$$s = \frac{(v_1 + v_2)t}{2}$$
$$2s = (v_1 + v_2)t \quad \text{Multiplying by 2}$$
$$\frac{2s}{t} = v_1 + v_2 \quad \text{Dividing by } t$$
$$\frac{2s}{t} - v_2 = v_1$$

This result can also be expressed as $v_1 = \frac{2s - tv_2}{t}$.

17.
$$\frac{t}{a} + \frac{t}{b} = 1$$
$$ab\left(\frac{t}{a} + \frac{t}{b}\right) = ab \cdot 1 \quad \text{Multiplying by the LCD}$$
$$ab \cdot \frac{t}{a} + ab \cdot \frac{t}{b} = ab$$
$$bt + at = ab$$
$$at = ab - bt$$
$$at = b(a - t) \quad \text{Factoring}$$
$$\frac{at}{a - t} = b$$

19.
$$R = \frac{gs}{g + s}$$
$$(g + s) \cdot R = (g + s) \cdot \frac{gs}{g + s} \quad \text{Multiplying by the LCD}$$
$$Rg + Rs = gs$$
$$Rs = gs - Rg$$
$$Rs = g(s - R) \quad \text{Factoring out } g$$
$$\frac{Rs}{s - R} = g \quad \text{Multiplying by } \frac{1}{s - R}$$

21.
$$I = \frac{nE}{R + nr}$$
$$I(R + nr) = \frac{nE}{R + nr} \cdot (R + nr) \quad \begin{array}{l}\text{Multiplying} \\ \text{by the LCD}\end{array}$$
$$IR + Inr = nE$$
$$IR = nE - Inr$$
$$IR = n(E - Ir)$$
$$\frac{IR}{E - Ir} = n$$

23.
$$\frac{1}{p} + \frac{1}{q} = \frac{1}{f}$$
$$pqf\left(\frac{1}{p} + \frac{1}{q}\right) = pqf \cdot \frac{1}{f} \quad \begin{array}{l}\text{Multiplying by} \\ \text{the LCD}\end{array}$$
$$qf + pf = pq$$
$$pf = pq - qf$$
$$pf = q(p - f)$$
$$\frac{pf}{p - f} = q$$

25.
$$S = \frac{H}{m(t_1 - t_2)}$$
$$(t_1 - t_2)S = \frac{H}{m} \quad \text{Multiplying by } t_1 - t_2$$
$$t_1 - t_2 = \frac{H}{Sm} \quad \text{Dividing by } S$$
$$t_1 = \frac{H}{Sm} + t_2, \text{ or } \frac{H + Smt_2}{Sm}$$

27.
$$\frac{E}{e} = \frac{R + r}{r}$$
$$er \cdot \frac{E}{e} = er \cdot \frac{R + r}{r} \quad \text{Multiplying by the LCD}$$
$$Er = e(R + r)$$
$$Er = eR + er$$
$$Er - er = eR$$
$$r(E - e) = eR$$
$$r = \frac{eR}{E - e}$$

29.
$$S = \frac{a}{1 - r}$$
$$(1 - r)S = a \quad \text{Multiplying by the LCD, } 1 - r$$
$$1 - r = \frac{a}{S} \quad \text{Dividing by } S$$
$$1 - \frac{a}{S} = r \quad \text{Adding } r \text{ and } -\frac{a}{S}$$

This result can also be expressed as $r = \frac{S - a}{S}$.

31.
$$c = \frac{f}{(a + b)c}$$
$$\frac{a + b}{c} \cdot c = \frac{a + b}{c} \cdot \frac{f}{(a + b)c}$$
$$a + b = \frac{f}{c^2}$$

33.
$$P = \frac{A}{1 + r}$$
$$P(1 + r) = \frac{A}{1 + r} \cdot (1 + r)$$
$$P(1 + r) = A$$
$$1 + r = \frac{A}{P}$$
$$r = \frac{A}{P} - 1, \text{ or } \frac{A - P}{P}$$

35.
$$v = \frac{d_2 - d_1}{t_2 - t_1}$$
$$(t_2 - t_1)v = (t_2 - t_1) \cdot \frac{d_2 - d_1}{t_2 - t_1}$$
$$(t_2 - t_1)v = d_2 - d_1$$
$$t_2 - t_1 = \frac{d_2 - d_1}{v}$$
$$-t_1 = -t_2 + \frac{d_2 - d_1}{v}$$
$$t_1 = t_2 - \frac{d_2 - d_1}{v}, \text{ or } \frac{t_2 v - d_2 + d_1}{v}$$

37.
$$\frac{1}{t} = \frac{1}{a} + \frac{1}{b}$$
$$tab \cdot \frac{1}{t} = tab\left(\frac{1}{a} + \frac{1}{b}\right)$$
$$ab = tb + ta$$
$$ab = t(b + a)$$
$$\frac{ab}{b + a} = t$$

39.
$$A = \frac{2Tt + Qq}{2T + Q}$$
$$(2T + Q) \cdot A = (2T + Q) \cdot \frac{2Tt + Qq}{2T + Q}$$
$$2AT + AQ = 2Tt + Qq$$
$$AQ - Qq = 2Tt - 2AT \quad \text{Adding} -2AT \text{ and} -Qq$$
$$Q(A - q) = 2Tt - 2AT$$
$$Q = \frac{2Tt - 2AT}{A - q}$$

41.
$$p = \frac{-98.42 + 4.15c - 0.082w}{w}$$
$$pw = -98.42 + 4.15c - 0.082w$$
$$pw + 0.082w = -98.42 + 4.15c$$
$$w(p + 0.082) = -98.42 + 4.15c$$
$$w = \frac{-98.42 + 4.15c}{p + 0.082}$$

43.
$y = kx$
$30 = k \cdot 5 \quad$ Substituting
$6 = k$

The variation constant is 6.

The equation of variation is $y = 6x$.

45.
$y = kx$
$3.4 = k \cdot 2 \quad$ Substituting
$1.7 = k$

The variation constant is 1.7.

The equation of variation is $y = 1.7x$.

47.
$y = kx$
$2 = k \cdot \frac{1}{5} \quad$ Substituting
$10 = k \quad$ Multiplying by 5

The variation constant is 10.

The equation of variation is $y = 10x$.

49.
$y = \frac{k}{x}$
$5 = \frac{k}{20} \quad$ Substituting
$100 = k$

The variation constant is 100.

The equation of variation is $y = \frac{100}{x}$.

51.
$y = \frac{k}{x}$
$11 = \frac{k}{4} \quad$ Substituting
$44 = k$

The variation constant is 44.

The equation of variation is $y = \frac{44}{x}$.

53.
$y = \frac{k}{x}$
$27 = \frac{k}{\frac{1}{3}} \quad$ Substituting
$9 = k$

The variation constant is 9.

The equation of variation is $y = \frac{9}{x}$.

55. *Familiarize*. Because of the phrase "d ... varies directly as ... m," we express the distance as a function of the mass. Thus we have $d(m) = km$. We know that $d(3) = 20$.

***Translate*.** We find the variation constant and then find the equation of variation.

$$d(m) = km$$
$$d(3) = k \cdot 3 \quad \text{Replacing } m \text{ with 3}$$
$$20 = k \cdot 3 \quad \text{Replacing } d(3) \text{ with 20}$$
$$\frac{20}{3} = k \quad \text{Variation constant}$$

The equation of variation is $d(m) = \frac{20}{3}m$.

***Carry out*.** We compute $d(5)$.

$$d(m) = \frac{20}{3}m$$
$$d(5) = \frac{20}{3} \cdot 5 \quad \text{Replacing } m \text{ with 5}$$
$$= \frac{100}{3}, \text{ or } 33\frac{1}{3}$$

***Check*.** Reexamine the calculations.

***State*.** The distance is $33\frac{1}{3}$ cm.

57. *Familiarize*. Because T varies inversely as P, we write $T(P) = k/P$. We know that $T(7) = 5$.

***Translate*.** We find the variation constant and the equation of variation.

$$T(P) = \frac{k}{P}$$
$$T(7) = \frac{k}{7} \quad \text{Replacing } P \text{ with 7}$$
$$5 = \frac{k}{7} \quad \text{Replacing } T(P) \text{ with 5}$$
$$35 = k \quad \text{Variation constant}$$
$$T(P) = \frac{35}{P} \quad \text{Equation of variation}$$

***Carry out*.** We find $T(10)$.

$$T(10) = \frac{35}{10} = 3.5$$

***Check*.** Reexamine the calculations.

***State*.** It would take 3.5 hr for 10 volunteers to complete the job.

59. *Familiarize.* Because W varies directly as S, we write $W(S) = kS$. We know that $W(150) = 16.8$.

Translate. We find the variation constant and the equation of variation.

$$W(S) = kS$$
$$W(150) = k \cdot 150 \qquad \text{Replacing } S \text{ with } 150$$
$$16.8 = k \cdot 150 \qquad \text{Replacing } W(150) \text{ with } 16.8$$
$$0.112 = k \qquad \text{Variation constant}$$
$$W(S) = 0.112S \qquad \text{Equation of variation}$$

Carry out. We find $W(500)$.

$$W(500) = 0.112 \cdot 500 = 56$$

Check. Reexamine the calculations.

State. 56 in. of water will replace 500 in. of snow.

61. Since we have direct variation and $48 = \frac{1}{2} \cdot 96$, then the result is $\frac{1}{2} \cdot 64$ kg, or 32 kg. We could also do this problem as follows.

Familiarize. Because W varies directly as the total mass, we write $W(m) = km$. We know that $W(96) = 64$.

Translate.

$$W(m) = km$$
$$W(96) = k \cdot 96 \qquad \text{Replacing } m \text{ with } 96$$
$$64 = k \cdot 96 \qquad \text{Replacing } W(96) \text{ with } 64$$
$$\frac{2}{3} = k \qquad \text{Variation constant}$$
$$W(m) = \frac{2}{3}m \qquad \text{Equation of variation}$$

Carry out. Find $W(48)$.

$$W(m) = \frac{2}{3}m$$
$$W(48) = \frac{2}{3} \cdot 48 = 32$$

Check. Reexamine the calculations.

State. There are 32 kg of water in a 64 kg person.

63. *Familiarize.* Because the frequency, f varies inversely as length L, we write $f(L) = k/L$. We know that $f(33) = 260$.

Translate. We find the variation constant and the equation of variation.

$$f(L) = \frac{k}{L}$$
$$f(33) = \frac{k}{33} \qquad \text{Replacing } L \text{ with } 33$$
$$260 = \frac{k}{33} \qquad \text{Replacing } f(33) \text{ with } 260$$
$$8580 = k \qquad \text{Variation constant}$$
$$f(L) = \frac{8580}{L} \qquad \text{Equation of variation}$$

Carry out. We find $f(30)$.

$$f(30) = \frac{8580}{30} = 286$$

Check. Reexamine the calculations.

State. If the string was shortened to 30 cm the new frequency would be 286 Hz.

65. *Familiarize.* Because of the phrase "t varies inversely as ... u," we write $t(u) = k/u$. We know that $t(4) = 75$.

Translate. We find the variation constant and then we find the equation of variation.

$$t(u) = \frac{k}{u}$$
$$t(4) = \frac{k}{4} \qquad \text{Replacing } u \text{ with } 4$$
$$75 = \frac{k}{4} \qquad \text{Replacing } t(4) \text{ with } 75$$
$$300 = k \qquad \text{Variation constant}$$
$$t(u) = \frac{300}{u} \qquad \text{Equation of variation}$$

Carry out. We find $t(14)$.

$$t(14) = \frac{300}{14} \approx 21$$

Check. Reexamine the calculations. Note that, as expected, as the UV rating increases, the time it takes to burn goes down.

State. It will take about 21 min to burn when the UV rating is 14.

67. *Familiarize.* The amount A of carbon monoxide released, in tons, varies directly as the population P. We write A as a function of P: $A(P) = kP$. We know that $A(2.6) = 0.94$.

Translate.

$$A(P) = kP$$
$$A(2.6) = k \cdot 2.6 \qquad \text{Replacing } P \text{ with } 2.6$$
$$0.94 = k \cdot 2.6 \qquad \text{Replacing } A(2.6) \text{ with } 0.94$$
$$\frac{94}{260} = k \qquad \text{Variation constant}$$
$$A(P) = \frac{94}{260}P \qquad \text{Equation of variation}$$

Carry out. Find $A(305,000,000)$.

$$A(P) = \frac{94}{260}P$$
$$A(305,000,000) = \frac{94}{260}(305,000,000)$$
$$\approx 110,000,000$$

Check. Reexamine the calculations. Answers may vary slightly due to rounding differences.

State. About 110,000,000 tons of carbon monoxide were released nationally.

69. $y = kx^2$

$50 = k(10)^2$ Substituting

$50 = k \cdot 100$

$\dfrac{1}{2} = k$ Variation constant

The equation of variation is $y = \dfrac{1}{2}x^2$.

71. $y = \dfrac{k}{x^2}$

$50 = \dfrac{k}{(10)^2}$ Substituting

$50 = \dfrac{k}{100}$

$5000 = k$ Variation constant

The equation of variation is $y = \dfrac{5000}{x^2}$.

73. $y = kxz$

$105 = k \cdot 14 \cdot 5$ Substituting

$105 = k \cdot 70$

$1.5 = k$ Variation constant

The equation of variation is $y = 1.5xz$.

75. $y = k \cdot \dfrac{wx^2}{z}$

$49 = k \cdot \dfrac{3 \cdot 7^2}{12}$ Substituting

$4 = k$ Variation constant

The equation of variation is $y = \dfrac{4wx^2}{z}$.

77. *Familiarize.* Because the stopping distance d, in feet varies directly as the square of the speed r, in mph, we write $d = kr^2$. We know that $d = 138$ when $r = 60$.

Translate. Find k and the equation of variation.

$d = kr^2$

$138 = k(60)^2$

$\dfrac{23}{600} = k$

$d = \dfrac{23}{600}r^2$ Equation of variation

Carry out. We find the value of d when r is 40.

$d = \dfrac{23}{600}(40)^2 \approx 61.3$ ft

Check. Reexamine the calculations.

State. It would take a car going 40 mph about 61.3 ft to stop.

79. *Familiarize.* Because V varies directly as T and inversely as P, we write $V = kT/P$. We know that $V = 231$ when $T = 300$ and $P = 20$.

Translate. Find k and the equation of variation.

$V = \dfrac{kT}{P}$

$231 = \dfrac{k \cdot 300}{20}$

$\dfrac{20}{300} \cdot 231 = k$

$15.4 = k$

$V = \dfrac{15.4T}{P}$ Equation of variation

Carry out. Substitute 320 for T, 16 for P and find V.

$V = \dfrac{15.4(320)}{16} = 308$

Check. Reexamine the calculations.

State. The volume is 308 cm^3 when $T = 320$ K and $P = 16$ lb/cm^2.

81. *Familiarize.* The drag W varies jointly as the surface area A and velocity v, so we write $W = kAv$. We know that $W = 222$ when $A = 37.8$ and $v = 40$.

Translate. Find k.

$W = kAv$

$222 = k(37.8)(40)$

$\dfrac{222}{37.8(40)} = k$

$\dfrac{37}{252} = k$

$W = \dfrac{37}{252}Av$ Equation of variation

Carry out. Substitute 51 for A and 430 for W and solve for v.

$430 = \dfrac{37}{252} \cdot 51 \cdot v$

$57.42 \text{ mph} \approx v$

(If we had used the rounded value 0.1468 for k, the resulting speed would have been approximately 57.43 mph.)

Check. Reexamine the calculations.

State. The car must travel about 57.42 mph.

83. *Writing Exercise.* Yes; let $y(x) = kx$. Then $y(2x) = k(2x) = k \cdot 2x = 2 \cdot kx = 2 \cdot y(x)$. Thus, doubling x causes y to be doubled.

85. $f(x) = 4x - 7$

$f(a) + h = 4a - 7 + h$

87. $f(x) = \dfrac{x - 5}{2x + 1}$

$2x + 1 = 0$

$x = -\dfrac{1}{2}$

The domain is $\left\{ x \middle| x \text{ is a real number } and \ x \neq -\dfrac{1}{2} \right\}$ or $\left(-\infty, -\dfrac{1}{2} \right) \cup \left(-\dfrac{1}{2}, \infty \right)$.

89. $f(x) = \sqrt{2x+8}$

$$2x + 8 \geq 0$$
$$2x \geq -8$$
$$x \geq -4$$

The domain is $\{x \mid x$ is a real number *and* $x \geq -4\}$ or $[-4, \infty)$.

91. *Writing Exercise.* Let C represent the number of complaints, and let E represent the number of employees. We then have

$$C(E) = \frac{k}{E},$$

where k is a positive constant. This can be graphed as shown.

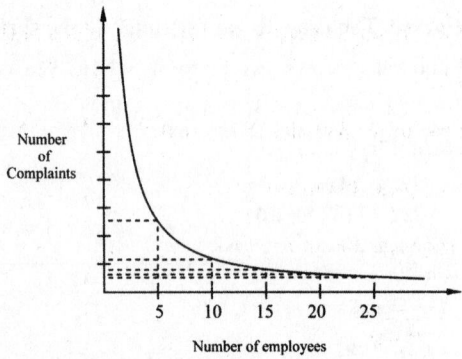

Note that regardless of the scale used on the vertical axis $C(10) - C(5) > C(25) - C(20)$. Thus, a greater reduction in the number of complaints occurs when the firm expands from 5 to 10 employees.

93. Use the result of Example 2.

$$h = \frac{2R^2 g}{V^2} - R$$

We have $V = 6.5$ mi/sec, $R = 3960$ mi, and $g = 32.2$ ft/sec^2. We must convert 32.2 ft/sec^2 to mi/sec^2 so all units of length are the same.

$$32.2 \frac{\text{ft}}{\text{sec}^2} \cdot \frac{1\,\text{mi}}{5280\,\text{ft}} \approx 0.0060984 \frac{\text{mi}}{\text{sec}^2}$$

Now we substitute and compute.

$$h = \frac{2(3960)^2(00060984)}{(65)^2} - 3960$$
$$h \approx 567$$

The satellite is about 567 mi from the surface of Earth.

95. $c = \dfrac{a}{a+12} \cdot d$

$$c = \frac{2a}{2a+12} \cdot d \qquad \text{Doubling } a$$
$$= \frac{2a}{2(a+6)} \cdot d$$
$$= \frac{a}{a+6} \cdot d \qquad \text{Simplifying}$$

The ratio of the larger dose to the smaller dose is

$$\frac{\dfrac{a}{a+6} \cdot d}{\dfrac{a}{a+12} \cdot d} = \frac{\dfrac{ad}{a+6}}{\dfrac{ad}{a+12}}$$
$$= \frac{ad}{a+6} \cdot \frac{a+12}{ad}$$
$$= \frac{ad(a+12)}{(a+6)\,ad}$$
$$= \frac{a+12}{a+6}$$

The amount by which the dosage increases is

$$\frac{a}{a+6} \cdot d - \frac{a}{a+12} \cdot d$$
$$= \frac{ad}{a+6} - \frac{ad}{a+12}$$
$$= \frac{ad}{a+6} \cdot \frac{a+12}{a+12} - \frac{ad}{a+12} \cdot \frac{a+6}{a+6}$$
$$= \frac{ad(a+12) - ad(a+6)}{(a+6)(a+12)}$$
$$= \frac{a^2 d + 12ad - a^2 d - 6ad}{(a+6)(a+12)}$$
$$= \frac{6ad}{(a+6)(a+12)}$$

Then the percent by which the dosage increases is

$$\frac{\dfrac{6ad}{(a+6)(a+12)}}{\dfrac{a}{a+12} \cdot d} = \frac{\dfrac{6ad}{(a+6)(a+12)}}{\dfrac{ad}{a+12}}$$
$$= \frac{6ad}{(a+6)(a+12)} \cdot \frac{a+12}{ad}$$
$$= \frac{6 \cdot ad \cdot (a+12)}{(a+6)(a+12) \cdot ad}$$
$$= \frac{6}{a+6}$$

This is a decimal representation for the percent of increase. To give the result in percent notation we multiply by 100 and use a percent symbol. We have $\dfrac{6}{a+6} \cdot 100\%$, or $\dfrac{600}{a+6}\%$.

97.
$$a = \frac{\dfrac{d_4 - d_3}{t_4 - t_3} - \dfrac{d_2 - d_1}{t_2 - t_1}}{t_4 - t_2}$$

$$a(t_4 - t_2) = \frac{d_4 - d_3}{t_4 - t_3} - \frac{d_2 - d_1}{t_2 - t_1} \qquad \begin{array}{l}\text{Multiplying}\\ \text{by } t_4 - t_2\end{array}$$

$$a(t_4-t_2)(t_4-t_3)(t_2-t_1) = (d_4-d_3)(t_2-t_1) - (d_2-d_1)(t_4-t_3)$$

$$\text{Multiplying by } (t_4 - t_3)(t_2 - t_1)$$

$$a(t_4 - t_2)(t_4 - t_3)(t_2 - t_1) - (d_4 - d_3)(t_2 - t_1)$$
$$= -(d_2 - d_1)(t_4 - t_3)$$

$$(t_2 - t_1)[a(t_4 - t_2)(t_4 - t_3) - (d_4 - d_3)]$$
$$= -(d_2 - d_1)(t_4 - t_3)$$

$$t_2 - t_1 = \frac{-(d_2 - d_1)(t_4 - t_3)}{a(t_4 - t_2)(t_4 - t_3) - (d_4 - d_3)}$$

$$t_2 + \frac{(d_2 - d_1)(t_4 - t_3)}{a(t_4 - t_2)(t_4 - t_3) + d_3 - d_4} = t_1$$

99. Let w = the wattage of the bulb. Then we have $I = \dfrac{kw}{d^2}$.

Now substitute $2w$ for w and $2d$ for d.

$$I = \frac{k(2w)}{(2d)^2} = \frac{2kw}{4d^2} = \frac{kw}{2d^2} = \frac{1}{2} \cdot \frac{kw}{d^2}$$

We see that the intensity is halved.

101. *Familiarize.* We write $T = kml^2 f^2$. We know that $T = 100$ when $m = 5$, $l = 2$, and $f = 80$.

Translate. Find k.

$$T = kml^2 f^2$$
$$100 = k(5)(2)^2(80)^2$$
$$0.00078125 = k$$
$$T = 0.00078125 ml^2 f^2$$

Carry out. Substitute 72 for T, 5 for m, and 80 for f and solve for l.

$$72 = 000078125(5)(l^2)(80)^2$$
$$2.88 = l^2$$
$$1.697 \approx l$$

Check. Recheck the calculations.

State. The string should be about 1.697 m long.

103. *Familiarize.* Because d varies inversely as s, we write $d(s) = k/s$. We know that $d(0.56) = 50$.

Translate.

$$d(s) = \frac{k}{s}$$
$$d(0.56) = \frac{k}{0.56} \qquad \text{Replacing } s \text{ with } 0.56$$
$$50 = \frac{k}{0.56} \qquad \text{Replacing } d(0.56) \text{ with } 50$$
$$28 = k$$
$$d(s) = \frac{28}{s} \qquad \text{Equation of variation}$$

Carry out. Find $d(0.40)$.

$$d(0.40) = \frac{28}{0.40} = 70$$

Check. Reexamine the calculations. Also observe that, as expected, when d decreases, then s increases.

State. The equation of variation is $d(s) = \dfrac{28}{s}$. The distance is 70 yd.

Chapter 6 Review

1. True; when $x = -2$ or $x = 2$, the denominator is zero, so these values are not included in the domain of f.

3. False; $3 - x$ can be written as $-1(x - 3)$, which is the LCM.

5. False; see pages 389-392 in the text.

7. True; see page 392 in the text.

9. True; see page 404 in the text.

11. $f(t) = \dfrac{t^2 - 3t + 2}{t^2 - 9}$

a) $f(0) = \dfrac{0^2 - 3 \cdot 0 + 2}{0^2 - 9} = -\dfrac{2}{9}$

b) $f(-1) = \dfrac{(-1)^2 - 3(-1) + 2}{(-1)^2 - 9} = \dfrac{1 + 3 + 2}{1 - 9} = \dfrac{6}{-8} = -\dfrac{3}{4}$

c) $f(1) = \dfrac{1^2 - 3 \cdot 1 + 2}{1^2 - 9} = \dfrac{1 - 3 + 2}{1 - 9} = 0$

13. $x^2 + 8x - 20 = (x + 10)(x - 2)$
$x^2 + 7x - 30 = (x + 10)(x - 3)$
$\text{LCM} = (x + 10)(x - 2)(x - 3)$

15. $\dfrac{12a^2 b^3}{5c^3 d^2} \cdot \dfrac{25c^9 d^4}{9a^7 b} = \dfrac{3 \cdot 4 \cdot 5 \cdot 5a^2 b^3 c^9 d^4}{5 \cdot 3 \cdot 3a^7 bc^3 d^2}$

$= \dfrac{20}{3} a^{2-7} b^{3-1} c^{9-3} d^{4-2}$

$= \dfrac{20}{3} a^{-5} b^2 c^6 d^2$, or $\dfrac{20b^2 c^6 d^2}{3a^5}$

17. $\dfrac{x^3 - 8}{x^2 - 25} \cdot \dfrac{x^2 + 10x + 25}{x^2 + 2x + 4}$

$= \dfrac{(x - 2)(x^2 + 2x + 4)}{(x + 5)(x - 5)} \cdot \dfrac{(x + 5)(x + 5)}{x^2 + 2x + 4}$

$= \dfrac{(x - 2)(x^2 + 2x + 4)(x + 5)(x + 5)}{(x + 5)(x - 5)(x^2 + 2x + 4)}$

$= \dfrac{(x + 5)(x^2 + 2x + 4)}{(x + 5)(x^2 + 2x + 4)} \cdot \dfrac{(x - 2)(x + 5)}{x - 5}$

$= \dfrac{(x - 2)(x + 5)}{x - 5}$

19. $\dfrac{x}{x^2+5x+6}-\dfrac{2}{x^2+3x+2}$

$=\dfrac{x}{(x+2)(x+3)}-\dfrac{2}{(x+1)(x+2)}$

[LCD is $(x+1)(x+2)(x+3)$.]

$=\dfrac{x}{(x+2)(x+3)}\cdot\dfrac{x+1}{x+1}-\dfrac{2}{(x+1)(x+2)}\cdot\dfrac{x+3}{x+3}$

$=\dfrac{x^2+x-(2x+6)}{(x+1)(x+2)(x+3)}$

$=\dfrac{x^2+x-2x-6}{(x+1)(x+2)(x+3)}$

$=\dfrac{x^2-x-6}{(x+1)(x+2)(x+3)}$

$=\dfrac{(x-3)(x+2)}{(x+1)(x+2)(x+3)}$

$=\dfrac{(x-3)\cancel{(x+2)}}{(x+1)\cancel{(x+2)}(x+3)}$

$=\dfrac{x-3}{(x+1)(x+3)}$

21. $\dfrac{5a^2}{a-b}+\dfrac{5b^2}{b-a}=\dfrac{5a^2}{a-b}+\dfrac{-1}{-1}\cdot\dfrac{5b^2}{b-a}$

$=\dfrac{5a^2}{a-b}+\dfrac{-5b^2}{a-b}$

$=\dfrac{5a^2-5b^2}{a-b}=\dfrac{5(a^2-b^2)}{a-b}$

$=\dfrac{5(a+b)(a-b)}{a-b}$

$=\dfrac{5(a+b)\cancel{(a-b)}}{1\cdot\cancel{(a-b)}}=5(a+b)$

23. $f(x)=\dfrac{4x-2}{x^2-5x+4}-\dfrac{3x+2}{x^2-5x+4}$ Note that $x\neq 1,\ 4$

$=\dfrac{x-4}{(x-1)(x-4)}$

$=\dfrac{1}{x-1},\ x\neq 1,\ 4$

25. $f(x)=\dfrac{9x^2-1}{x^2-9}\div\dfrac{3x+1}{x+3}$

$x^2-9=(x+3)(x-3)$ is zero when $x=-3$ or $x=3$;

$x+3$ is zero when $x=-3$. Also, $3x+1$ is zero when

$x=-\dfrac{1}{3}$. Thus, the domain cannot contain -3, $-\dfrac{1}{3}$, or 3.

$f(x)=\dfrac{9x^2-1}{x^2-9}\div\dfrac{3x+1}{x+3}$

$=\dfrac{9x^2-1}{x^2-9}\cdot\dfrac{x+3}{3x+1}$

$=\dfrac{\cancel{(3x+1)}(3x-1)\cancel{(x+3)}}{\cancel{(x+3)}(x-3)\cancel{(3x+1)}}$

$=\dfrac{3x-1}{x-3},\ x\neq -3,\ -\dfrac{1}{3},\ 3$

27. $\dfrac{\dfrac{3}{a}+\dfrac{3}{b}}{\dfrac{6}{a^3}+\dfrac{6}{b^3}}=\dfrac{\dfrac{3}{a}+\dfrac{3}{b}}{\dfrac{6}{a^3}+\dfrac{6}{b^3}}\cdot\dfrac{a^3b^3}{a^3b^3}$ Multiplying by 1, using the LCD

$=\dfrac{\dfrac{3}{a}\cdot a^3b^3+\dfrac{3}{b}\cdot a^3b^3}{\dfrac{6}{a^3}\cdot a^3b^3+\dfrac{6}{b^3}\cdot a^3b^3}$

$=\dfrac{3a^2b^3+3a^3b^2}{6b^3+6a^3}=\dfrac{3a^2b^2(a+b)}{6(b+a)(b^2-ab+a^2)}$

$=\dfrac{a^2b^2}{2(b^2-ab+a^2)}$

29. $\dfrac{\dfrac{5}{x^2-9}-\dfrac{3}{x+3}}{\dfrac{4}{x^2+6x+9}+\dfrac{2}{x-3}}$

$=\dfrac{\dfrac{5}{(x+3)(x-3)}-\dfrac{3}{x+3}}{\dfrac{4}{(x+3)(x+3)}+\dfrac{2}{x-3}}$

$=\dfrac{\dfrac{5}{(x+3)(x-3)}-\dfrac{3}{x+3}}{\dfrac{4}{(x+3)(x+3)}+\dfrac{2}{x-3}}\cdot\dfrac{(x+3)^2(x-3)}{(x+3)^2(x-3)}$ Multiplying by 1, using the LCD

$=\dfrac{\dfrac{5}{(x+3)(x-3)}\cdot(x+3)^2(x-3)-\dfrac{3}{x+3}\cdot(x+3)^2(x-3)}{\dfrac{4}{(x+3)(x+3)}\cdot(x+3)^2(x-3)+\dfrac{2}{x-3}\cdot(x+3)^2(x-3)}$

$=\dfrac{5(x+3)-3(x-3)(x+3)}{4(x-3)+2(x+3)^2}=\dfrac{5x+15-3x^2+27}{4x-12+2x^2+12x+18}$

$=\dfrac{-3x^2+5x+42}{2x^2+16x+6}=\dfrac{(14-3x)(x+3)}{2x^2+16x+6}$

31. $\dfrac{5}{3x+2}=\dfrac{3}{2x}$

To ensure that neither denominator is 0, we note at the

outset that $x\neq-\dfrac{2}{3}$ and $x\neq 0$. Then we multiply both

sides by the LCD, $2x(3x+2)$.

$2x(3x+2)\cdot\dfrac{5}{3x+2}=2x(3x+2)\cdot\dfrac{3}{2x}$

$2x(5)=(3x+2)(3)$

$10x=9x+6$

$x=6$

The solution is 6.

33. $\dfrac{x+6}{x^2+x-6}+\dfrac{x}{x^2+4x+3}=\dfrac{x+2}{x^2-x-2}$

$\dfrac{x+6}{(x+3)(x-2)}+\dfrac{x}{(x+1)(x+3)}=\dfrac{x+2}{(x+1)(x-2)}$

To ensure that none of the denominators is 0, we note at

the outset that $x\neq-3$, $x\neq-1$ and $x\neq 2$. Then we

multiply both sides by the LCD, $(x+1)(x-2)(x+3)$.

$(x+1)(x-2)(x+3)\cdot\left(\dfrac{x+6}{(x+3)(x-2)}+\dfrac{x}{(x+1)(x+3)}\right)$

$=(x+1)(x-2)(x+3)\cdot\dfrac{x+2}{(x+1)(x-2)}$

$$(x+1)(x+6)+x(x-2)=(x+2)(x+3)$$
$$x^2+7x+6+x^2-2x=x^2+5x+6$$
$$x^2=0$$
$$x=0$$

The solution is 0.

35. We find all values of a for which $\dfrac{2}{a-1}+\dfrac{2}{a+2}=1$. First

note that $a \neq 1$ and $a \neq -2$. Then multiply both sides by

the LCD, $(a-1)(a+2)$.

$$(a-1)(a+2)\left(\frac{2}{a-1}+\frac{2}{a+2}\right)=(a-1)(a+2)\cdot 1$$
$$2(a+2)+2(a-1)=(a-1)(a+2)$$
$$2a+4+2a-2=a^2+a-2$$
$$0=a^2-3a-4$$
$$0=(a+1)(a-4)$$
$$a=-1 \ or \ a=4$$

Both values check. The solutions are –1 and 4.

37. *Familiarize*. Let t represent the number of seconds it

takes the Core 2 Duo processor to process a data file

working alone. Then $t-15$ represents the time it takes

the Core 2 Quad processor to process a data file, working

alone. In 1 second the Core 2 Duo does $\dfrac{1}{t}$ of the job and

the Core 2 Quad does $\dfrac{1}{t-15}$ of the job.

Translate. Working together, the two processors can

process the file in 18 sec to find t such that

$$18\left(\frac{1}{t}\right)+18\left(\frac{1}{t-15}\right)=1, \ or \ \frac{18}{t}+\frac{18}{t-15}=1. \ .$$

Carry out. We solve the equation. First we multiply

both sides by the LCD, $t(t-15)$.

$$t(t-15)\left(\frac{18}{t}+\frac{18}{t-15}\right)=t(t-15)\cdot 1$$
$$18(t-15)+18t=t(t-15)$$
$$18t-270+18t=t^2-15t$$
$$0=t^2-51t+270$$
$$0=(t-6)(t-45)$$
$$t=6 \ or \ t=45$$

Check. If $t=6$, then $t-15=6-15=-9$. Since

negative time has no meaning in this application, 6

cannot be a solution. If $t=45$, then $t-15=45-15=30$.

In 18 sec the Core 2 Duo does $18\cdot\dfrac{1}{45}$, or $\dfrac{2}{5}$ of the job. In

18 sec the Core 2 Quad does $18\cdot\dfrac{1}{30}$, or $\dfrac{3}{5}$ of the job.

Together they do $\dfrac{2}{5}+\dfrac{3}{5}$, or 1 entire job. The answer

checks.

State. Working alone, the Core 2 Duo can process a file

in 45 sec and the Core 2 Quad can process a file in 30 sec.

39. *Familiarize*. Let $r=$ the speed of the motorcycle in

mph. Then $r+8=$ the speed of the car in mph. We

organize the information in a table. The time is the same

for both so we use t for each time.

	Distance	Speed	Time
Motorcycle	93	r	t
Car	105	$r+8$	t

Translate. Using the formula Time = Distance/Rate in

each row of the table and the fact that the times are the

same, we can write an equation.

$$\frac{93}{r}=\frac{105}{r+8}$$

Carry out. We solve the equation.

$$\frac{93}{r}=\frac{105}{r+8}, \quad \text{LCD is } r(r+8)$$
$$r(r+8)\cdot\frac{93}{r}=r(r+8)\cdot\frac{105}{r+8}$$
$$93(r+8)=105r$$
$$93r+744=105r$$
$$744=12r$$
$$62=r$$

Check. If the motorcycle's speed is 62 mph, then the

car's speed is 62 + 8, or 70 mph. Traveling 93 mi at

62 mph takes $\dfrac{93}{62}=1.5\,\text{hr}$. Traveling 105 mi at 70 mph

takes $\dfrac{105}{70}=1.5\,\text{hr}$. Since the times are the same, the

answer checks.

State. The speed of the motorcycle is 62 mph; the speed

of the car is 70 mph.

41. $\dfrac{y^3+8}{y+2}=\dfrac{(y+2)\left(y^2-2y+4\right)}{y+2}=y^2-2y+4$

43. $\left(x^3+3x^2+2x-6\right)\div(x-3)$

$$\begin{array}{r|rrrr} 3 & 1 & 3 & 2 & -6 \\ & & 3 & 18 & 60 \\ \hline & 1 & 6 & 20 & \mid \ 54 \end{array}$$

$x^2+6x+20$, R 54, or $x^2+6x+20+\dfrac{54}{x-3}$

45. $$I=\frac{2V}{R+2r}$$
$$(R+2r)I=(R+2r)\cdot\frac{2V}{R+2r}$$
$$IR+2rI=2V$$
$$2rI=2V-IR$$
$$r=\frac{2V-IR}{2I}, \ or \ \frac{V}{I}-\frac{R}{2}$$

47. $$\frac{1}{ac}=\frac{2}{ab}-\frac{3}{bc}$$
$$abc\cdot\frac{1}{ac}=abc\cdot\frac{2}{ab}-abc\cdot\frac{3}{bc}$$
$$b=2c-3a$$
$$b+3a=2c$$
$$\frac{b+3a}{2}=c$$

49. *Familiarize.* Because of the phrase "*t* varies inversely as

...*u*," we write $t(u) = k/u$. We know that $t(6) = 10$.

Translate. We find the variation constant and then we

find the equation of variation.

$$t(u) = \frac{k}{u}$$

$$t(6) = \frac{k}{6} \qquad \text{Replacing } u \text{ with } 6$$

$$10 = \frac{k}{6} \qquad \text{Replacing } t(6) \text{ with } 10$$

$$60 = k \qquad \text{Variation constant}$$

$$t(u) = \frac{60}{u} \qquad \text{Equation of variation}$$

Carry out. We find $t(4)$.

$$t(4) = \frac{60}{4} = 15$$

Check. Reexamine the calculations. Note that, as

expected, as the UV rating increases, the time it takes to

burn goes down.

State. It will take 15 min to burn when the UV rating is

4.

51. *Familiarize.* Because the time *t*, in seconds varies

inversely as the current *I*, in amperes, we write $t = \frac{k}{I^2}$.

We know that $t = 3.4$ when $I = 0.089$.

Translate. Find *k* and the equation of variation.

$$t = \frac{k}{I^2}$$

$$3.4 = \frac{k}{(0.089)^2}$$

$$0.0269314 = k$$

$$t = \frac{0.0269314}{I^2} \qquad \text{Equation of variation}$$

Carry out. We find the value of *t* when *I* is 0.096.

$$t = \frac{0.0269314}{(0.096)^2} \approx 2.9 \text{ sec}$$

Check. Reexamine the calculations.

State. A 0.096-amp current would be deadly after about

2.9 sec.

53. *Writing Exercise.* A rational *expression* is a quotient of

two polynomials. Expressions can be simplified,

multiplied, or added, but they cannot be solved for a

variable. A rational *equation* is an equation containing

rational expressions. In a rational equation, we often can

solve for a variable.

55. There is more than one approach to solving this equation.

To ensure that none of the denominators is 0, we note at

the outset that $x \neq -5$ and $x \neq 5$.

$$\frac{\dfrac{x}{x^2-25} + \dfrac{2}{x-5}}{\dfrac{3}{x-5} - \dfrac{4}{x^2-10x+25}} = 1$$

$$\frac{\dfrac{x}{(x+5)(x-5)} + \dfrac{2}{x-5}}{\dfrac{3}{x-5} - \dfrac{4}{(x-5)(x-5)}} = 1$$

$$\frac{\dfrac{x}{(x+5)(x-5)} + \dfrac{2}{x-5}}{\dfrac{3}{x-5} - \dfrac{4}{(x-5)(x-5)}} \cdot \frac{(x-5)^2(x+5)}{(x-5)^2(x+5)} = 1$$

$$\frac{x(x-5) + 2(x-5)(x+5)}{3(x-5)(x+5) - 4(x+5)} = 1$$

$$\frac{x^2 - 5x + 2x^2 - 50}{3x^2 - 75 - 4x - 20} = 1$$

$$\frac{3x^2 - 5x - 50}{3x^2 - 4x - 95} = 1$$

$$3x^2 - 5x - 50 = 3x^2 - 4x - 95$$

$$45 = x$$

The solution is 45.

Chapter 6 Test

---

1. $\dfrac{t+1}{t+3} \cdot \dfrac{5t+15}{4t^2-4} = \dfrac{t+1}{t+3} \cdot \dfrac{5(t+3)}{4(t+1)(t-1)}$

$\qquad = \dfrac{(t+1) \cdot 5(t+3)}{(t+3) \cdot 4(t+1)(t-1)}$

$\qquad = \dfrac{(t+1)(t+3)}{(t+1)(t+3)} \cdot \dfrac{5}{4(t-1)}$

$\qquad = \dfrac{5}{4(t-1)}$

3. $\dfrac{25x}{x+5} + \dfrac{x^3}{x+5} = \dfrac{25x+x^3}{x+5} = \dfrac{x(25+x^2)}{x+5}$

5. $\dfrac{4ab}{a^2-b^2} + \dfrac{a^2+b^2}{a+b} = \dfrac{4ab}{(a+b)(a-b)} + \dfrac{a^2+b^2}{a+b}$

$\qquad = \dfrac{4ab}{(a+b)(a-b)} + \dfrac{a^2+b^2}{a+b} \cdot \dfrac{a-b}{a-b}$

$\qquad = \dfrac{4ab + (a^2+b^2)(a-b)}{(a+b)(a-b)}$

$\qquad = \dfrac{4ab + a^3 - a^2b + ab^2 - b^3}{(a+b)(a-b)}$

$\qquad = \dfrac{a^3 - a^2b + 4ab + ab^2 - b^3}{(a+b)(a-b)}$

7. $\dfrac{4}{x+3} - \dfrac{x}{x-2} + \dfrac{x^2+4}{x^2+x-6}$

$= \dfrac{4}{x+3} - \dfrac{x}{x-2} + \dfrac{x^2+4}{(x+3)(x-2)}$

Note that $x \neq -3,\ 2$.

$= \dfrac{4}{x+3} \cdot \dfrac{x-2}{x-2} - \dfrac{x}{x-2} \cdot \dfrac{x+3}{x+3} + \dfrac{x^2+4}{(x+3)(x-2)}$

$= \dfrac{4(x-2) - x(x+3) + x^2 + 4}{(x+3)(x-2)}$

$= \dfrac{4x - 8 - x^2 - 3x + x^2 + 4}{(x+3)(x-2)}$

$= \dfrac{x-4}{(x+3)(x-2)},\quad x \neq -3,\ 2$

9. $\dfrac{\frac{2}{a} + \frac{3}{b}}{\frac{5}{ab} + \frac{1}{a^2}} = \dfrac{\frac{2}{a} + \frac{3}{b}}{\frac{5}{ab} + \frac{1}{a^2}} \cdot \dfrac{a^2 b}{a^2 b}$ Multiplying by 1, using the LCD

$= \dfrac{\frac{2}{a} \cdot a^2 b + \frac{3}{b} \cdot a^2 b}{\frac{5}{ab} \cdot a^2 b + \frac{1}{a^2} \cdot a^2 b}$

$= \dfrac{2ab + 3a^2}{5a + b}$

$= \dfrac{a(2b + 3a)}{5a + b}$

11. $\dfrac{\dfrac{2}{x+3} - \dfrac{1}{x^2 - 3x + 2}}{\dfrac{3}{x-2} + \dfrac{4}{x^2 + 2x - 3}}$

$= \dfrac{\dfrac{2}{x+3} - \dfrac{1}{(x-1)(x-2)}}{\dfrac{3}{x-2} + \dfrac{4}{(x+3)(x-1)}}$

$= \dfrac{\dfrac{2}{x+3} - \dfrac{1}{(x-1)(x-2)}}{\dfrac{3}{x-2} + \dfrac{4}{(x+3)(x-1)}} \cdot \dfrac{(x-1)(x-2)(x+3)}{(x-1)(x-2)(x+3)}$ Multiplying by 1, using the LCD

$= \dfrac{\dfrac{2}{x+3} \cdot (x-1)(x-2)(x+3) - \dfrac{1}{(x-1)(x-2)} \cdot (x-1)(x-2)(x+3)}{\dfrac{3}{x-2} \cdot (x-1)(x-2)(x+3) + \dfrac{4}{(x+3)(x-1)} \cdot (x-1)(x-2)(x+3)}$

$= \dfrac{2(x-1)(x-2) - 1(x+3)}{3(x-1)(x+3) + 4(x-2)} = \dfrac{2x^2 - 6x + 4 - x - 3}{3x^2 + 6x - 9 + 4x - 8}$

$= \dfrac{2x^2 - 7x + 1}{3x^2 + 10x - 17}$

13. $\dfrac{t+11}{t^2 - t - 12} + \dfrac{1}{t-4} = \dfrac{4}{t+3}$

$\dfrac{t+11}{(t-4)(t+3)} + \dfrac{1}{t-4} = \dfrac{4}{t+3}$

To ensure that none of the denominators is 0, we note at the outset that $t \neq 4$ and $t \neq -3$. Then we multiply both sides by the LCD, $(t-4)(t+3)$.

$(t-4)(t+3) \cdot \left(\dfrac{t+11}{(t-4)(t+3)} + \dfrac{1}{t-4} \right) = (t-4)(t+3) \cdot \dfrac{4}{t+3}$

$t + 11 + t + 3 = 4(t-4)$

$2t + 14 = 4t - 16$

$30 = 2t$

$15 = t$

The solution is 15.

15. $f(x) = \dfrac{x+5}{x-1}$

$f(0) = \dfrac{0+5}{0-1} = -5$

$f(-3) = \dfrac{-3+5}{-3-1} = \dfrac{2}{-4} = -\dfrac{1}{2}$

17. *Familiarize*. Let t represent the number of hours it takes Ella and Sari to install a countertop, working together.

Translate. Ella takes 5 hr and Sari takes 4 hr to complete the job, so we have

$\dfrac{t}{5} + \dfrac{t}{4} = 1$

Carry out. We solve the equation. Multiply on both sides by the LCD, 20.

$20 \left(\dfrac{t}{5} + \dfrac{t}{4} \right) = 20 \cdot 1$

$4t + 5t = 20$

$9t = 20$

$t = \dfrac{20}{9},\ \text{or}\ 2\dfrac{2}{9}$

Check. If Ella does the job alone in 5 hr, then in $2\dfrac{2}{9}$ hr she does $\dfrac{20/9}{5}$, or $\dfrac{4}{9}$ of the job. If Sari does the job alone in 4 hr, then in $2\dfrac{2}{9}$ hr she does $\dfrac{20/9}{4}$, or $\dfrac{5}{9}$ of the job. Together, they do $\dfrac{4}{9} + \dfrac{5}{9}$, or 1 entire job. The result checks.

State. It would take Ella and Sari $2\dfrac{2}{9}$ hr to finish the job working together.

19.
$$\begin{array}{r} y - 14 \\ y - 6 \overline{)\, y^2 - 20y + 64} \\ \underline{y^2 - 6y} \\ -14y + 64 \\ \underline{-14y + 84} \\ -20 \end{array}$$

The answer is $y - 14$, R -20, or $y - 14 + \dfrac{-20}{y-6}$.

21. $\left(x^3 + 5x^2 + 4x - 7\right) \div (x - 2)$

$$
\begin{array}{r|rrrr}
2\!\!\!| & 1 & 5 & 4 & -7 \\
  &   & 2 & 14 & 36 \\
\hline
  & 1 & 7 & 18 & \;|\; 29
\end{array}
$$

$x^2 + 7x + 18$, R 29, or $x^2 + 7x + 18 + \dfrac{29}{x-2}$

23.
$$R = \frac{gs}{g+s}$$
$$(g+s)R = (g+s) \cdot \frac{gs}{g+s}$$
$$gR + sR = gs$$
$$gr = gs - sR$$
$$gR = s(g - R)$$
$$\frac{gR}{g-R} = s$$

25. *Familiarize*. Let $w =$ the speed of the wind, in mph. Then $12 + w =$ the speed with the wind and $12 - w =$ the speed against the wind. We organize the information in a table.

	Distance	Speed	Time
With	14	$12+w$	t
Against	8	$12-w$	t

***Translate*.** Using the formula Time = Distance/Rate in each row of the table and the fact that the times are the same, we can write an equation.

$$\frac{14}{12+w} = \frac{8}{12-w}$$

***Carry out*.** We solve the equation. Multiply both sides by the LCD, $(12+w)(12-w)$.

$$(12+w)(12-w) \cdot \frac{14}{12+w} = (12+w)(12-w) \cdot \frac{8}{12-w}$$
$$14(12-w) = 8(12+w)$$
$$168 - 14w = 96 + 8w$$
$$72 = 22w$$
$$\frac{36}{11} = r, \text{ or } r = 3\frac{3}{11}$$

***Check*.** If $w = 3\frac{3}{11}$, then the speed with the wind is $12 + 3\frac{3}{11} = 15\frac{3}{11}$ km/h and the speed against the wind is $12 - 3\frac{3}{11} = 8\frac{8}{11}$ km/h. The time for the trip against the wind is $\dfrac{8}{8\frac{8}{11}}$, or $\dfrac{11}{12}$ hours. The time for the trip with the wind is $\dfrac{14}{15\frac{3}{11}}$, or $\dfrac{11}{12}$ hours. The times are the same. The values check.

***State*.** The speed of the wind is $3\frac{3}{11}$ mph.

27. *Familiarize*. Because the surface area A, in square inches varies directly as the square of the radius r, in inches, we write $A = kr^2$. We know that $A = 325$ when $r = 5$.

***Translate*.** Find k and the equation of variation.

$$A = kr^2$$
$$325 = k(5)^2$$
$$13 = k$$
$$d = 13r^2 \quad \text{Equation of variation}$$

***Carry out*.** We find the value of d when r is 7.

$$A = 13(7)^2 = 637 \text{ in}^2$$

***Check*.** Reexamine the calculations.

***State*.** The area would be 637 in^2 when the radius is 7 in.

29.
$$\frac{6}{x-15} - \frac{6}{x} = \frac{90}{x^2 - 15x}$$
$$\frac{6}{x-15} - \frac{6}{x} = \frac{90}{x(x-15)}$$

To ensure that none of the denominators is 0, we note at the outset that $x \neq 0$ and $x \neq 15$. Then we multiply both sides by the LCD, $x(x-15)$.

$$x(x-15)\left(\frac{6}{x-15} - \frac{6}{x}\right) = x(x-15) \cdot \frac{90}{x(x-15)}$$
$$6x - 6(x-15) = 90$$
$$6x - 6x + 90 = 90$$
$$0 = 0$$

This is a true statement. Recall that because of the restrictions above, 0 and 15 cannot be a solution, however all other real numbers *can* be a solution. The solution is $\{x | x \text{ is a real number } and\ x \neq 0\ and\ x \neq 15\}$.

31. The ration of the number of lawns Alex mowed to the number of lawns Ryan mowed is $\dfrac{4}{3}$. Let $x =$ the number of lawns Alex mowed, then $98 - x =$ the number of lawns Ryan mowed. Set the ratios equal to one another ans solve for x.

$$\frac{4}{3} = \frac{x}{98-x}, \quad \text{LCD is } 3(98-x)$$
$$3(98-x) \cdot \frac{4}{3} = 3(98-x) \cdot \frac{x}{98-x}$$
$$4(98-x) = 3x$$
$$392 - 4x = 3x$$
$$392 = 7x$$
$$56 = x$$

Then $98 - x = 98 - 56 = 42$.

$\dfrac{56}{42} = \dfrac{4}{3}$ and $56 + 42 = 98$, so the number checks.

Alex mowed 56 lawns and Ryan mowed 42 lawns.

Chapter 7

Exponents and Radicals

1. two; see page 430 in the text.

3. positive; see Example 4.

5. irrational; see page 431 in the text.

7. nonnegative; see page 434 in the text.

9. The square roots of 64 are 8 and -8, because $8^2 = 64$ and $(-8)^2 = 64$.

11. The square roots of 100 are 10 and -10 because $10^2 = 100$ and $(-10)^2 = 100$.

13. The square roots of 400 are 20 and -20 because $20^2 = 400$ and $(-20)^2 = 400$.

15. The square roots of 625 are 25 and -25 because $25^2 = 625$ and $(-25)^2 = 625$.

17. $\sqrt{49} = 7$ Remember, $\sqrt{}$ indicates the principle square root.

19. $-\sqrt{16} = -4$ Since, $\sqrt{16} = 4$, $-\sqrt{16} = -4$

21. $\sqrt{\dfrac{36}{49}} = \dfrac{6}{7}$

23. $-\sqrt{169} = -13$ Since, $\sqrt{169} = 13$, $-\sqrt{169} = -13$

25. $-\sqrt{\dfrac{16}{81}} = -\dfrac{4}{9}$ Since, $\sqrt{\dfrac{16}{81}} = \dfrac{4}{9}$, $-\sqrt{\dfrac{16}{81}} = -\dfrac{4}{9}$

27. $\sqrt{0.04} = 0.2$

29. $\sqrt{0.0081} = 0.09$

31. $5\sqrt{p^2 + 4}$

    The radicand is the expression written under the radical sign, $p^2 + 4$.

    Since the index is not written, we know it is 2.

33. $x^2 y^3 \sqrt[5]{\dfrac{x}{y+4}}$

    The radicand is the expression written under the radical sign, $\dfrac{x}{y+4}$.

    The index is 5.

35. $f(t) = \sqrt{5t - 10}$

    $f(3) = \sqrt{5(3) - 10} = \sqrt{5}$

    $f(2) = \sqrt{5(2) - 10} = \sqrt{0} = 0$

    $f(1) = \sqrt{5(1) - 10} = \sqrt{-5}$

         Since negative numbers do not have real-number square roots, $f(1)$ does not exist.

    $f(-1) = \sqrt{5(-1) - 10} = \sqrt{-15}$

         Since negative numbers do not have real-number square roots, $f(1)$ does not exist.

37. $t(x) = -\sqrt{2x^2 - 1}$

    $t(5) = -\sqrt{2 \cdot 5^2 - 1} = -\sqrt{49} = -7$

    $t(0) = -\sqrt{2 \cdot 0^2 - 1} = \sqrt{-1}$

         $t(0)$ does not exist

    $t(-1) = -\sqrt{2(-1)^2 - 1} = -\sqrt{1} = -1$

    $t\left(-\dfrac{1}{2}\right) = -\sqrt{2\left(-\dfrac{1}{2}\right)^2 - 1} = -\sqrt{-\dfrac{1}{2}}$ does not exist

39. $f(t) = \sqrt{t^2 + 1}$

    $f(0) = \sqrt{0^2 + 1} = \sqrt{1} = 1$

    $f(-1) = \sqrt{(-1)^2 + 1} = \sqrt{2}$

    $f(-10) = \sqrt{(-10)^2 + 1} = \sqrt{101}$

41. $\sqrt{100x^2} = \sqrt{(10x)^2} = |10x| = 10|x|$

    Since x might be negative, absolute-value notation is necessary.

43. $\sqrt{(8-t)^2} = |8-t|$

    Since $8 - t$ might be negative, absolute-value notation is necessary.

45. $\sqrt{y^2 + 16y + 64} = \sqrt{(y+8)^2} = |y+8|$

    Since $y + 8$ might be negative, absolute-value notation is necessary.

47. $\sqrt{4x^2 + 28x + 49} = \sqrt{(2x+7)^2} = |2x+7|$

Since $2x + 7$ might be negative, absolute-value notation is necessary.

49. $-\sqrt[4]{256} = -4$ Since $4^4 = 256$

51. $\sqrt[3]{-1} = -1$ Since $(-1)^3 = -1$

53. $-\sqrt[5]{-\dfrac{32}{243}} = \dfrac{2}{3}$ Since $\left(-\dfrac{2}{3}\right)^5 = -\dfrac{32}{243}$

55. $\sqrt[6]{x^6} = |x|$

The index is even. Use absolute-value notation since x could have a negative value.

57. $\sqrt[9]{t^9} = t$

The index is odd. Absolute-value signs are not necessary.

59. $\sqrt[4]{(6a)^4} = |6a| = 6|a|$

The index is even. Use absolute-value notation since a could have a negative value.

61. $\sqrt[10]{(-6)^{10}} = |-6| = 6$

63. $\sqrt[414]{(a+b)^{414}} = |a+b|$

The index is even. Use absolute-value notation since $a + b$ could have a negative value.

65. $\sqrt{a^{22}} = |a^{11}|$ Note that $\left(a^{11}\right)^2 = a^{22}$; could have a negative value.

67. $\sqrt{-25}$ is not a real number, so $\sqrt{-25}$ cannot be simplified.

69. $\sqrt{16x^2} = \sqrt{(4x)^2} = 4x$ Assuming x is nonnegative

71. $-\sqrt{(3t)^2} = -3t$ Assuming t is nonnegative

73. $\sqrt{(-5b)^2} = 5b$

75. $\sqrt{a^2 + 2a + 1} = \sqrt{(a+1)^2} = a+1$

77. $\sqrt[3]{27} = 3$ $(3^3 = 27)$

79. $\sqrt[4]{16x^4} = \sqrt[4]{(2x)^4} = 2x$

81. $\sqrt[5]{(x-1)^5} = x-1$

83. $-\sqrt[3]{-125y^3} = -(-5y)$ $\left[(-5y^3) = -125y^3\right]$
       $= 5y$

85. $\sqrt{t^{18}} = \sqrt{(t^9)^2} = t^9$

87. $\sqrt{(x-2)^8} = \sqrt{\left[(x-2)^4\right]^2} = (x-2)^4$

89. $f(x) = \sqrt[3]{x+1}$
        $f(7) = \sqrt[3]{7+1} = \sqrt[3]{8} = 2$
        $f(26) = \sqrt[3]{26+1} = \sqrt[3]{27} = 3$
        $f(-9) = \sqrt[3]{-9+1} = \sqrt[3]{-8} = -2$
        $f(-65) = \sqrt[3]{-65+1} = \sqrt[3]{-64} = -4$

91. $g(t) = \sqrt[4]{t-3}$
        $g(19) = \sqrt[4]{19-3} = \sqrt[4]{16} = 2$
        $g(-13) = \sqrt[4]{-13-3} = \sqrt[4]{-16}$
            $g(-13)$ does not exist
        $g(1) = \sqrt[4]{1-3} = \sqrt[4]{-2}$
            $g(1)$ does not exist
        $g(84) = \sqrt[4]{84-3} = \sqrt[4]{81} = 3$

93. $f(x) = \sqrt{x-6}$

Since the index is even, the radicand, $x - 6$, must be nonnegative. We solve the inequality:
$$x - 6 \geq 0$$
$$x \geq 6$$
Domain of $f = \{x | x \geq 6\}$, or $[6, \infty)$

95. $g(t) = \sqrt[4]{t+8}$

Since the index is even, the radicand, $t + 8$, must be nonnegative. We solve the inequality:
$$t + 8 \geq 0$$
$$t \geq -8$$
Domain of $g = \{t | t \geq -8\}$, or $[-8, \infty)$

97. $g(x) = \sqrt[4]{10-2x}$

Since the index is even, the radicand, $10 - 2x$, must be nonnegative. We solve the inequality:
$$10 - 2x \geq 0$$
$$-2x \geq -10$$
$$x \leq 5$$
Domain of $g = \{x | x \leq 5\}$, or $(-\infty, 5]$

99. $f(t) = \sqrt[5]{2t+7}$

Since the index is odd, the radicand can be any real number.

Domain of $f = \{t | t \text{ is a real number}\}$, or $(-\infty, \infty)$

101. $h(z) = -\sqrt[6]{5z+2}$

Since the index is even, the radicand, $5z + 2$, must be nonnegative. We solve the inequality:
$$5z + 2 \geq 0$$
$$5z \geq -2$$
$$z \geq -\tfrac{2}{5}$$

Domain of $h = \left\{z \big| z \geq -\tfrac{2}{5}\right\}$, or $\left[-\tfrac{2}{5},\ \infty\right)$

103. $f(t) = 7 + \sqrt[8]{t^8}$

Since we can compute $7 + \sqrt[8]{t^8}$ for any real number t, the domain is the set of real numbers, or
$\{t | t \text{ is a real number}\}$ or $(-\infty,\ \infty)$.

105. *Writing Exercise.* Write a minus sign in front of the square root notation. For example, for a number n, we can write the negative square root of n as $-\sqrt{n}$.

107. $\left(a^2 b\right)\left(a^4 b\right) = a^{2+4} b^{1+1} = a^6 b^2$

109. $\left(5x^2 y^{-3}\right)^3 = 5^3 x^{2\cdot 3} y^{-3\cdot 3} = 125 x^6 y^{-9} = \dfrac{125x^6}{y^9}$

111. $\left(\dfrac{10x^{-1} y^5}{5x^2 y^{-1}}\right)^{-1} = \left(2x^{-3} y^6\right)^{-1} = 2^{-1} x^3 y^{-6} = \dfrac{x^3}{2y^6}$

113. *Writing Exercise.* If n is an odd number, $\sqrt[n]{x^3}$ exists for all real values of x, because every real number has a real root when n is odd. If n is an even number, then $\sqrt[n]{x^3}$ exists for all nonnegative values of x. This is true because when x is nonnegative, then x^3 is nonnegative and even roots of nonnegative numbers exist. If x is negative, then x^3 is also negative and negative numbers do not have real nth roots when n is even.

115. $S = 88.63 \sqrt[4]{A}$

Substitute 63,000 for A.
$$S = 88.63 \sqrt[4]{63,000}$$
$$S \approx 1404$$
There are about 1404 species of plants.

117. $f(x) = \sqrt{x+5}$

Since the index is even, the radicand, $x + 5$, must be nonnegative. We solve the inequality:
$$x + 5 \geq 0$$
$$x \geq -5$$
Domain of $f = \{x | x \geq -5\}$, or $[-5,\ \infty)$

Make a table of values, keeping in mind that x must be -5 or greater. Plot these points and draw the graph.

x	$f(x)$
-5	0
-4	1
-1	2
1	2.4
3	2.8
4	3

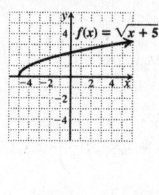

119. $g(x) = \sqrt{x} - 2$

Since the index is even, the radicand, x, must be nonnegative, so we have $x \geq 0$.

Domain of $g = \{x | x \geq 0\}$, or $[0,\ \infty)$

Make a table of values, keeping in mind that x must be 0 or greater. Plot these points and draw the graph.

x	$g(x)$
0	-2
1	-1
4	0
6	0.4
8	0.8

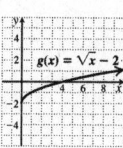

121. $f(x) = \dfrac{\sqrt{x+3}}{\sqrt[4]{2-x}}$

In the numerator we must have $x + 3 \geq 0$, or $x \geq -3$, and in the denominator we must have $2 - x > 0$, or $x < 2$, so Domain of $f = \{x | -3 \leq x < 2\}$, or $[-3,\ 2)$.

123. $F(x) = \dfrac{x}{\sqrt{x^2 - 5x - 6}}$

Since the radical expression in the denominator has an even index, so the radicand, $x^2 - 5x - 6$, must be nonnegative in order for $\sqrt{x^2 - 5x - 6}$ to exist. In addition, the denominator cannot be zero, so the radicand must be positive. We solve the inequality:
$$x^2 - 5x - 6 > 0$$
$$(x+1)(x-6) > 0$$
We have $x < -1$ *and* $x > 6$, so
Domain of $F = \{x | x < -1 \text{ or } x > 6\}$, or $(-\infty, -1) \cup (6,\ \infty)$.

125. *Graphing Calculator Exercise.*

Exercise Set 7.2

1. Choice (g) is correct because $a^{m/n} = \sqrt[n]{a^m}$.

3. $x^{-5/2} = \dfrac{1}{x^{5/2}} = \dfrac{1}{(\sqrt{x})^5}$, so choice (e) is correct.

5. $x^{1/5} \cdot x^{2/5} = x^{1/5+2/5} = x^{3/5}$, so choice (a) is correct.

7. Choice (b) is correct because $\sqrt[n]{a^m}$ and $(\sqrt[n]{a})^m$ are equivalent.

9. $y^{1/3} = \sqrt[3]{y}$

11. $36^{1/2} = \sqrt{36} = 6$

13. $32^{1/5} = \sqrt[5]{32} = 2$

15. $64^{1/2} = \sqrt{64} = 8$

17. $(xyz)^{1/2} = \sqrt{xyz}$

19. $\left(a^2 b^2\right)^{1/5} = \sqrt[5]{a^2 b^2}$

21. $t^{5/6} = \sqrt[6]{t^5}$

23. $16^{3/4} = \sqrt[4]{16^3} = \left(\sqrt[4]{16}\right)^3 = 2^3 = 8$

25. $125^{4/3} = \sqrt[3]{125^4} = \left(\sqrt[3]{125}\right)^4 = 5^4 = 625$

27. $(81x)^{3/4} = \sqrt[4]{(81x)^3} = \sqrt[4]{81^3 x^3}$, or
$$\sqrt[4]{81^3} \cdot \sqrt[4]{x^3} = \left(\sqrt[4]{81}\right)^3 \cdot \left(\sqrt[4]{x^3}\right) = 3^3 \sqrt[4]{x^3} = 27\sqrt[4]{x^3}$$

29. $\left(25x^4\right)^{3/2} = \sqrt{\left(25x^4\right)^3} = \sqrt{25^3 \cdot x^{12}} = \sqrt{25^3} \cdot \sqrt{x^{12}}$
$$= \left(\sqrt{25}\right)^3 x^6 = 5^3 x^6 = 125 x^6$$

31. $\sqrt[3]{18} = 18^{1/3}$

33. $\sqrt{30} = 30^{1/2}$

35. $\sqrt{x^7} = x^{7/2}$

37. $\sqrt[5]{m^2} = m^{2/5}$

39. $\sqrt[4]{pq} = (pq)^{1/4}$

41. $\sqrt[5]{xy^2 z} = \left(xy^2 z\right)^{1/5}$

43. $\left(\sqrt{3mn}\right)^3 = (3mn)^{3/2}$

45. $\left(\sqrt[7]{8x^2 y}\right)^5 = \left(8x^2 y\right)^{5/7}$

47. $\dfrac{2x}{\sqrt[3]{z^2}} = \dfrac{2x}{z^{2/3}}$

49. $a^{-1/4} = \dfrac{1}{a^{1/4}}$

51. $(2rs)^{-3/4} = \dfrac{1}{(2rs)^{3/4}}$

53. $\left(\dfrac{1}{16}\right)^{-3/4} = \left(\dfrac{16}{1}\right)^{3/4} = \left(2^4\right)^{3/4} = 2^{4(3/4)} = 2^3 = 8$

55. $\dfrac{8c}{a^{-3/5}} = 8a^{3/5} c$

57. $2a^{3/4} b^{-1/2} c^{2/3} = 2 \cdot a^{3/4} \cdot \dfrac{1}{b^{1/2}} \cdot c^{2/3} = \dfrac{2a^{3/4} c^{2/3}}{b^{1/2}}$

59. $3^{-5/2} a^3 b^{-7/3} = \dfrac{1}{3^{5/2}} \cdot a^3 \cdot \dfrac{1}{b^{7/3}} = \dfrac{a^3}{3^{5/2} b^{7/3}}$

61. $\left(\dfrac{2ab}{3c}\right)^{-5/6} = \left(\dfrac{3c}{2ab}\right)^{5/6}$ Finding the reciprocal of the base and changing the sign of the exponent

63. $\dfrac{6a}{\sqrt[4]{b}} = \dfrac{6a}{b^{1/4}}$

65. $11^{1/2} \cdot 11^{1/3} = 11^{1/2+1/3} = 11^{3/6+2/6} = 11^{5/6}$

We added exponents after finding a common denominator.

67. $\dfrac{3^{5/8}}{3^{-1/8}} = 3^{5/8-(-1/8)} = 3^{5/8+1/8} = 3^{6/8} = 3^{3/4}$

We subtracted exponents and simplified.

69. $\dfrac{4.3^{-1/5}}{4.3^{-7/10}} = 4.3^{-1/5-(-7/10)} = 4.3^{-1/5+7/10}$
$$= 4.3^{-2/10+7/10} = 4.3^{5/10} = 4.3^{1/2}$$

We subtracted exponents after finding a common denominator. Then we simplified.

71. $\left(10^{3/5}\right)^{2/5} = 10^{3/5 \cdot 2/5} = 10^{6/25}$

We multiplied exponents.

73. $a^{2/3} \cdot a^{5/4} = a^{2/3+5/4} = a^{8/12+15/12} = a^{23/12}$

We added exponents after finding a common denominator.

75. $\left(64^{3/4}\right)^{4/3} = 64^{\frac{3 \cdot 4}{4 \cdot 3}} = 64^1 = 64$

77. $\left(m^{2/3} n^{-1/4}\right)^{1/2} = m^{2/3 \cdot 1/2} n^{-1/4 \cdot 1/2} = m^{1/3} n^{-1/8}$
$$= m^{1/3} \cdot \dfrac{1}{n^{1/8}} = \dfrac{m^{1/3}}{n^{1/8}}$$

79. $\sqrt[9]{x^3} = x^{3/9}$ Converting to exponential notation
$$= x^{1/3} \quad \text{Simplifying the exponent}$$
$$= \sqrt[3]{x} \quad \text{Returning to radical notation}$$

81. $\sqrt[3]{y^{15}} = y^{15/3}$ Converting to exponential notation

$\quad = y^5$ Simplifying

83. $\sqrt[12]{a^6} = a^{6/12}$ Converting to exponential notation

$\quad = a^{1/2}$ Simplifying the exponent

$\quad = \sqrt{a}$ Returning to radical notation

85. $\left(\sqrt[7]{xy}\right)^{14} = (xy)^{14/7}$ Converting to exponential notation

$\quad = (xy)^2$ Simplifying the exponent

$\quad = x^2 y^2$ Using the laws of exponents

87. $\sqrt[4]{(7a)^2} = (7a)^{2/4}$ Converting to exponential notation

$\quad = (7a)^{1/2}$ Simplifying the exponent

$\quad = \sqrt{7a}$ Returning to radical notation

89. $\sqrt[8]{(2x)^6} = (2x)^{6/8}$ Converting to exponential notation

$\quad = (2x)^{3/4}$ Simplifying the exponent

$\quad = \sqrt[4]{(2x)^3}$ Returning to radical notation

$\quad = \sqrt[4]{8x^3}$ Using the laws of exponents

91. $\sqrt{\sqrt[5]{m}} = \sqrt{m^{1/5}}$ Converting to

$\quad = \left(m^{1/5}\right)^{1/2}$ exponential notation

$\quad = m^{1/10}$ Using the laws of exponents

$\quad = \sqrt[10]{m}$ Returning to radical notation

93. $\sqrt[4]{(xy)^{12}} = (xy)^{12/4}$ Converting to exponential notation

$\quad = (xy)^3$ Simplifying the exponent

$\quad = x^3 y^3$ Using the laws of exponents

95. $\left(\sqrt[5]{a^2 b^4}\right)^{15} = \left(a^2 b^4\right)^{15/5}$ Converting to exponential notation

$\quad = \left(a^2 b^4\right)^3$ Simplifying the exponent

$\quad = a^6 b^{12}$ Using the laws of exponents

97. $\sqrt[3]{\sqrt[4]{xy}} = \sqrt[3]{(xy)^{1/4}}$ Converting to

$\quad = \left[(xy)^{1/4}\right]^{1/3}$ exponential notation

$\quad = (xy)^{1/12}$ Using the laws of exponents

$\quad = \sqrt[12]{xy}$ Returning to radical notation

99. *Writing Exercise.* $f(x) = (x+5)^{1/2}(x+7)^{-1/2}$

Consider $(x+5)^{1/2}$. Since the index is $\frac{1}{2}$, $x+5$ must be nonnegative. Then $x+5 \geq 0$, or $x \geq -5$.

Consider $(x+7)^{-1/2}$. Since the index is $-\frac{1}{2}$, $x+7$ must be positive. Then $x+7 > 0$, or $x > -7$.

Then the domain of $f = \{x | x \geq -5 \ and \ x > -7\}$, or $f = \{x | x \geq -5\}$.

101. $(x+5)(x-5) = x^2 - 25$ Difference of squares

103. $4x^2 + 20x + 25 = (2x+5)(2x+5) = (2x+5)^2$

105. $5t^2 - 10t + 5 = 5\left(t^2 - 2t + 1\right) = 5(t-1)(t-1)$

$\quad\quad = 5(t-1)^2$

107. *Writing Exercise.* For any value of x, $x^6 \geq 0$, and $x^2 \geq 0$, so $\sqrt[3]{x^6} = x^2$. For $x \geq 0$, we have $x^6 \geq 0$ and $x^3 \geq 0$, so $\sqrt[2]{x^6} = x^3$; but for $x < 0$, we have $x^6 > 0$ and $x^3 < 0$. Thus $\sqrt[2]{x^6} \neq x^3$ because $\sqrt[2]{x^6}$ must be nonnegative. Then $\sqrt[2]{x^6} = x^3$ only when $x \geq 0$.

109. $\sqrt{x \sqrt[3]{x^2}} = \sqrt{x \cdot x^{2/3}} = \left(x^{5/3}\right)^{1/2} = x^{5/6} = \sqrt[6]{x^5}$

111. $\sqrt[14]{c^2 - 2cd + d^2} = \sqrt[14]{(c-d)^2} = \left[(c-d)^2\right]^{1/14}$

$\quad\quad = (c-d)^{2/14} = (c-d)^{1/7} = \sqrt[7]{c-d}$

113. $2^{7/12} \approx 1.498 \approx 1.5$ so the G that is 7 half steps above middle C has a frequency that is about 1.5 times that of middle C.

115. a) $L = \dfrac{(0.000169)60^{2.27}}{1} \approx 1.8$ m

b) $L = \dfrac{(0.000169)75^{2.27}}{0.9906} \approx 3.1$ m

c) $L = \dfrac{(0.000169)80^{2.27}}{2.4} \approx 1.5$ m

d) $L = \dfrac{(0.000169)100^{2.27}}{1.1} \approx 5.3$ m

117. $T = 0.936 d^{1.97} h^{0.85}$

$\quad = 0.936(3)^{1.97}(80)^{0.85}$

$\quad \approx 338$ cubic feet

119. *Graphing Calculator Exercise*

Exercise Set 7.3

1. True; see page 446 in the text.

3. False; see page 447 in the text.

5. True; see page 447 in the text.

7. $\sqrt{3}\sqrt{10} = \sqrt{3 \cdot 10} = \sqrt{30}$

9. $\sqrt[3]{7} \sqrt[3]{5} = \sqrt[3]{7 \cdot 5} = \sqrt[3]{35}$

11. $\sqrt[4]{6} \sqrt[4]{9} = \sqrt[4]{6 \cdot 9} = \sqrt[4]{54}$

13. $\sqrt{2x}\sqrt{13y} = \sqrt{2x \cdot 13y} = \sqrt{26xy}$

15. $\sqrt[5]{8y^3}\sqrt[5]{10y} = \sqrt[5]{8y^3 \cdot 10y} = \sqrt[5]{80y^4}$

17. $\sqrt{y-b}\sqrt{y+b} = \sqrt{(y-b)(y+b)} = \sqrt{y^2 - b^2}$

19. $\sqrt[3]{0.7y}\sqrt[3]{0.3y} = \sqrt[3]{0.7y \cdot 0.3y} = \sqrt[3]{0.21y^2}$

21. $\sqrt[5]{x-2}\sqrt[5]{(x-2)^2} = \sqrt[5]{(x-2)(x-2)^2} = \sqrt[5]{(x-2)^3}$

23. $\sqrt{\dfrac{2}{t}}\sqrt{\dfrac{3s}{11}} = \sqrt{\dfrac{2}{t}\cdot\dfrac{3s}{11}} = \sqrt{\dfrac{6s}{11t}}$

25. $\sqrt[7]{\dfrac{x-3}{4}}\sqrt[7]{\dfrac{5}{x+2}} = \sqrt[7]{\dfrac{x-3}{4}\cdot\dfrac{5}{x+2}} = \sqrt[7]{\dfrac{5x-15}{4x+8}}$

27. $\sqrt{12}$
$= \sqrt{4\cdot 3}\qquad$ 4 is the largest perfect square factor of 12
$= \sqrt{4}\cdot\sqrt{3}$
$= 2\sqrt{3}$

29. $\sqrt{45}$
$= \sqrt{9\cdot 5}\qquad$ 9 is the largest perfect square factor of 45
$= \sqrt{9}\cdot\sqrt{5}$
$= 3\sqrt{5}$

31. $\sqrt{8x^9}$
$= \sqrt{4x^8 \cdot 2x}\qquad 4x^8$ is a perfect square
$= \sqrt{4x^8}\cdot\sqrt{2x}\qquad$ Factoring into two radicals
$= 2x^4\sqrt{2x}\qquad$ Taking the square root of $4x^8$

33. $\sqrt{120} = \sqrt{4\cdot 30} = \sqrt{4}\cdot\sqrt{30} = 2\sqrt{30}$

35. $\sqrt{36a^4 b}$
$= \sqrt{36a^4 \cdot b}\qquad 36a^4$ is a perfect square
$= \sqrt{36a^4}\cdot\sqrt{b}\qquad$ Factoring into two radicals
$= 6a^2\sqrt{b}\qquad$ Taking the square root of $36a^4$

37. $\sqrt[3]{8x^3 y^2}$
$= \sqrt[3]{8x^3 \cdot y^2}\qquad 8x^3$ is a perfect cube
$= \sqrt[3]{8x^3}\cdot\sqrt[3]{y^2}\qquad$ Factoring into two radicals
$= 2x\sqrt[3]{y^2}\qquad$ Taking the cube root of $8x^3$

39. $\sqrt[3]{-16x^6}$
$= \sqrt[3]{-8x^6 \cdot 2}\qquad -8x^6$ is a perfect cube
$= \sqrt[3]{-8x^6}\cdot\sqrt[3]{2}\qquad$ Factoring into two radicals
$= -2x^2\sqrt[3]{2}\qquad$ Taking the cube root of $-8x^6$

41. $f(x) = \sqrt[3]{40x^6}$
$= \sqrt[3]{8x^6 \cdot 5}$
$= \sqrt[3]{8x^6}\cdot\sqrt[3]{5}$
$= 2x^2\sqrt[3]{5}$

43. $f(x) = \sqrt{49(x-3)^2}\qquad 49(x-3)^2$ is a perfect square.
$= |7(x-3)|,\ \text{or}\ 7|x-3|$

45. $f(x) = \sqrt{5x^2 - 10x + 5}$
$= \sqrt{5(x^2 - 2x + 1)}$
$= \sqrt{5(x-1)^2}$
$= \sqrt{(x-1)^2}\cdot\sqrt{5}$
$= |x-1|\sqrt{5}$

47. $\sqrt{a^{10}b^{11}}$
$= \sqrt{a^{10}\cdot b^{10}\cdot b}\qquad$ Identifying the largest even
$\qquad\qquad\qquad\qquad$ powers of a and b
$= \sqrt{a^{10}}\sqrt{b^{10}}\sqrt{b}\qquad$ Factoring into several radicals
$= a^5 b^5\sqrt{b}$

49. $\sqrt[3]{x^5 y^6 z^{10}}$
$= \sqrt[3]{x^3 \cdot x^2 \cdot y^6 \cdot z^9 \cdot z}\qquad$ Identifying the largest
$\qquad\qquad\qquad\qquad\qquad$ perfect-cube powers of $x,\ y$ and z
$= \sqrt[3]{x^3}\cdot\sqrt[3]{y^6}\cdot\sqrt[3]{z^9}\cdot\sqrt[3]{x^2 z}\quad$ Factoring into several radicals
$= xy^2 z^3\sqrt[3]{x^2 z}$

51. $\sqrt[4]{16x^5 y^{11}} = \sqrt[4]{2^4 \cdot x^4 \cdot x \cdot y^8 \cdot y^3}$
$= \sqrt[4]{2^4}\cdot\sqrt[4]{x^4}\cdot\sqrt[4]{y^8}\cdot\sqrt[4]{xy^3}$
$= 2xy^2\sqrt[4]{xy^3}$

53. $\sqrt[5]{x^{13}y^8 z^{17}} = \sqrt[5]{x^{10}\cdot x^3 \cdot y^5 \cdot y^3 \cdot z^{15}\cdot z^2}$
$= \sqrt[5]{x^{10}}\cdot\sqrt[5]{y^5}\cdot\sqrt[5]{z^{15}}\cdot\sqrt[5]{x^3 y^3 z^2}$
$= x^2 yz^3\sqrt[5]{x^3 y^3 z^2}$

55. $\sqrt[3]{-80a^{14}} = \sqrt[3]{-8\cdot 10\cdot a^{12}\cdot a^2}$
$= \sqrt[3]{-8}\cdot\sqrt[3]{a^{12}}\cdot\sqrt[3]{10a^2}$
$= -2a^4\sqrt[3]{10a^2}$

57. $\sqrt{5}\sqrt{10} = \sqrt{5\cdot 10} = \sqrt{50} = \sqrt{25\cdot 2} = 5\sqrt{2}$

59. $\sqrt{6}\sqrt{33} = \sqrt{6\cdot 33} = \sqrt{198} = \sqrt{9\cdot 22} = 3\sqrt{22}$

61. $\sqrt[3]{9}\sqrt[3]{3} = \sqrt[3]{9\cdot 3} = \sqrt[3]{27} = 3$

63. $\sqrt{24y^5}\sqrt{24y^5} = \sqrt{\left(24y^5\right)^2} = 24y^5$

65. $\sqrt[3]{5a^2}\sqrt[3]{2a} = \sqrt[3]{5a^2 \cdot 2a} = \sqrt[3]{10a^3} = \sqrt[3]{a^3 \cdot 10} = a\sqrt[3]{10}$

67. $\sqrt{2x^5}\sqrt{10x^2} = \sqrt{20x^7} = \sqrt{4x^6 \cdot 5x} = 2x^3\sqrt{5x}$

69. $\sqrt[3]{s^2 t^4}\,\sqrt[3]{s^4 t^6} = \sqrt[3]{s^6 t^{10}} = \sqrt[3]{s^6 t^9 \cdot t} = s^2 t^3 \sqrt[3]{t}$

71. $\sqrt[3]{(x-y)^2}\,\sqrt[3]{(x-y)^{10}} = \sqrt[3]{(x-y)^{12}} = (x-y)^4$

73. $\sqrt[4]{20a^3b^7}\cdot\sqrt[4]{4a^2b^5} = \sqrt[4]{80a^5b^{12}} = \sqrt[4]{16a^4b^{12}\cdot 5a}$
$$= 2ab^3\,\sqrt[4]{5a}$$

75. $\sqrt[5]{x^3(y+z)^6}\,\sqrt[5]{x^3(y+z)^4} = \sqrt[5]{x^6(y+z)^{10}}$
$$= \sqrt[5]{x^5(y+z)^{10}\cdot x} = x(y+z)^2\,\sqrt[5]{x}$$

77. *Writing Exercise.* Suppose that we let $x = 5$. Then we have $\sqrt{x^2 - 16} = \sqrt{5^2 - 16} = \sqrt{25 - 16} = \sqrt{9} = 3$, and $\sqrt{x^2} - \sqrt{16} = \sqrt{5^2} - \sqrt{16} = 5 - 4 = 1$. Thus, $\sqrt{x^2 - 16} \neq \sqrt{x^2} - \sqrt{16}$.

79. $\dfrac{15a^2x}{8b}\cdot\dfrac{24b^2x}{5a} = \dfrac{(5a \cdot 3ax)(8b \cdot 3bx)}{8b \cdot 5a} = 9abx^2$

81. $\dfrac{x-3}{2x-10} - \dfrac{3x-5}{x^2-25} = \dfrac{x-3}{2(x-5)} - \dfrac{3x-5}{(x+5)(x-5)}$
$$= \dfrac{x-3}{2(x-5)}\cdot\dfrac{x+5}{x+5} - \dfrac{3x-5}{(x+5)(x-5)}\cdot\dfrac{2}{2}$$
$$= \dfrac{x^2 + 2x - 15 - (6x - 10)}{2(x+5)(x-5)}$$
$$= \dfrac{x^2 - 4x - 5}{2(x+5)(x-5)}$$
$$= \dfrac{(x-5)(x+1)}{2(x+5)(x-5)}$$
$$= \dfrac{x+1}{2(x+5)}$$

83. $\dfrac{a^{-1}+b^{-1}}{ab} = \dfrac{\frac{1}{a}+\frac{1}{b}}{ab} = \dfrac{\frac{1}{a}+\frac{1}{b}}{ab}\cdot\dfrac{ab}{ab} = \dfrac{b+a}{a^2b^2}$

85. *Writing Exercise.* $\sqrt[n]{ab} = (ab)^{1/n} = a^{1/n}b^{1/n} = \sqrt[n]{a}\,\sqrt[n]{b}$

87. $R(x) = \dfrac{1}{2}\sqrt[4]{\dfrac{x\cdot 3.0\times 10^6}{\pi^2}}$
$$R(5\times 10^4) = \dfrac{1}{2}\sqrt[4]{\dfrac{5\times 10^4 \cdot 3.0\times 10^6}{\pi^2}}$$
$$= \dfrac{1}{2}\sqrt[4]{\dfrac{15\times 10^{10}}{\pi^2}}$$
$$\approx 175.6\text{ mi}$$

89. a) $T_w = 33 - \dfrac{(10.45 + 10\sqrt{8} - 8)(33 - 7)}{22}$
$$\approx -3.3\ ^\circ\text{C}$$

    b) $T_w = 33 - \dfrac{(10.45 + 10\sqrt{12} - 12)(33 - 0)}{22}$
$$\approx -16.6\ ^\circ\text{C}$$

    c) $T_w = 33 - \dfrac{(10.45 + 10\sqrt{14} - 14)(33 - (-5))}{22}$
$$\approx -25.5\ ^\circ\text{C}$$

    d) $T_w = 33 - \dfrac{(10.45 + 10\sqrt{15} - 15)(33 - (-23))}{22}$
$$\approx -54.0\ ^\circ\text{C}$$

91. $\left(\sqrt[3]{25x^4}\right)^4 = \sqrt[3]{(25x^4)^4} = \sqrt[3]{25^4 x^{16}}$
$$= \sqrt[3]{25^3 \cdot 25 \cdot x^{15} \cdot x} = \sqrt[3]{25^3}\,\sqrt[3]{x^{15}}\,\sqrt[3]{25x}$$
$$= 25x^5 \sqrt[3]{25x}$$

93. $\left(\sqrt{a^3b^5}\right)^7 = \sqrt{(a^3b^5)^7} = \sqrt{a^{21}b^{35}}$
$$= \sqrt{a^{20}\cdot a \cdot b^{34}\cdot b} = \sqrt{a^{20}}\,\sqrt{b^{34}}\,\sqrt{ab} = a^{10}b^{17}\sqrt{ab}$$

95.

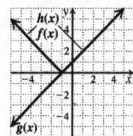

We see that $f(x) = h(x)$ and $f(x) \neq g(x)$.

97. $g(x) = x^2 - 6x + 8$

We must have $x^2 - 6x + 8 \geq 0$, or $(x-2)(x-4) \geq 0$. We graph $y = x^2 - 6x + 8$.

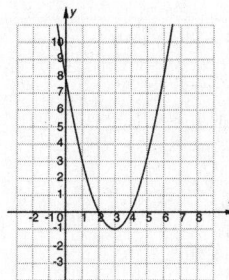

From the graph we see that $y \geq 0$ for $x \leq 2$ or $x \geq 4$, so the domain of g is $\{x \mid x \leq 2 \text{ or } x \geq 4\}$, or $(-\infty,\, 2]\cup[4,\, \infty)$.

99. $\sqrt[5]{4a^{3k+2}}\,\sqrt[5]{8a^{6-k}} = 2a^4$
$$\sqrt[5]{32a^{2k+8}} = 2a^4$$
$$2\sqrt[5]{a^{2k+8}} = 2a^4$$
$$\sqrt[5]{a^{2k+8}} = a^4$$
$$a^{\frac{2k+8}{5}} = a^4$$

Since the base is the same, the exponents must be equal. We have:
$$\dfrac{2k+8}{5} = 4$$
$$2k + 8 = 20$$
$$2k = 12$$
$$k = 6$$

101. *Writing Exercise.* Since $\sqrt{x}$ exists only for $\{x \mid x \geq 0\}$, this is the domain of $y = \sqrt{x} \cdot \sqrt{x}$.

Exercise Set 7.4

1. $\sqrt[4]{\dfrac{16a^6}{a^2}} = \sqrt[4]{16a^4} = 2a$, so choice (g) is correct.

3. $\sqrt[5]{\dfrac{a^6}{b^4}} = \sqrt[5]{\dfrac{a^6}{b^4} \cdot \dfrac{b}{b}} = \sqrt[5]{\dfrac{a^6 b}{b^4 \cdot b}}$, so choice (f) is correct.

5. $\dfrac{\sqrt[5]{a^2}}{\sqrt[5]{b^2}} = \dfrac{\sqrt[5]{a^2}}{\sqrt[5]{b^2}} \cdot \dfrac{\sqrt[5]{b^3}}{\sqrt[5]{b^3}} = \dfrac{\sqrt[5]{a^2 b^3}}{\sqrt[5]{b^5}}$, so choice (h) is correct.

7. $\dfrac{\sqrt[5]{a^2}}{\sqrt[5]{b^3}} = \dfrac{\sqrt[5]{a^2}}{\sqrt[5]{b^3}} \cdot \dfrac{\sqrt[5]{b^2}}{\sqrt[5]{b^2}} = \dfrac{\sqrt[5]{a^2 b^2}}{\sqrt[5]{b^5}}$, so choice (a) is correct.

9. $\sqrt{\dfrac{49}{100}} = \dfrac{\sqrt{49}}{\sqrt{100}} = \dfrac{7}{10}$

11. $\sqrt[3]{\dfrac{125}{8}} = \dfrac{\sqrt[3]{125}}{\sqrt[3]{8}} = \dfrac{5}{2}$

13. $\sqrt{\dfrac{121}{t^2}} = \dfrac{\sqrt{121}}{\sqrt{t^2}} = \dfrac{11}{t}$

15. $\sqrt{\dfrac{36y^3}{x^4}} = \dfrac{\sqrt{36y^3}}{\sqrt{x^4}} = \dfrac{\sqrt{36y^2 \cdot y}}{\sqrt{x^4}} = \dfrac{\sqrt{36y^2}\sqrt{y}}{\sqrt{x^4}} = \dfrac{6y\sqrt{y}}{x^2}$

17. $\sqrt[3]{\dfrac{27a^4}{8b^3}} = \dfrac{\sqrt[3]{27a^4}}{\sqrt[3]{8b^3}} = \dfrac{\sqrt[3]{27a^3 \cdot a}}{\sqrt[3]{8b^3}} = \dfrac{\sqrt[3]{27a^3}\sqrt[3]{a}}{\sqrt[3]{8b^3}} = \dfrac{3a\sqrt[3]{a}}{2b}$

19. $\sqrt[4]{\dfrac{32a^4}{2b^4 c^8}} = \sqrt[4]{\dfrac{16a^4}{b^4 c^8}} = \dfrac{\sqrt[4]{16a^4}}{\sqrt[4]{b^4 c^8}} = \dfrac{2a}{bc^2}$

21. $\sqrt[4]{\dfrac{a^5 b^8}{c^{10}}} = \dfrac{\sqrt[4]{a^5 b^8}}{\sqrt[4]{c^{10}}} = \dfrac{\sqrt[4]{a^4 b^8 \cdot a}}{\sqrt[4]{c^8 \cdot c^2}} = \dfrac{\sqrt[4]{a^4 b^8}\sqrt[4]{a}}{\sqrt[4]{c^8}\sqrt[4]{c^2}} = \dfrac{ab^2 \sqrt[4]{a}}{c^2 \sqrt[4]{c^2}}$,

   or $\dfrac{ab^2}{c^2}\sqrt[4]{\dfrac{a}{c^2}}$

23. $\sqrt[5]{\dfrac{32x^6}{y^{11}}} = \dfrac{\sqrt[5]{32x^6}}{\sqrt[5]{y^{11}}} = \dfrac{\sqrt[5]{32x^5 \cdot x}}{\sqrt[5]{y^{10} \cdot y}}$

   $= \dfrac{\sqrt[5]{32x^5} \cdot \sqrt[5]{x}}{\sqrt[5]{y^{10}}\sqrt[5]{y}} = \dfrac{2x\sqrt[5]{x}}{y^2 \sqrt[5]{y}}$, or $\dfrac{2x}{y^2}\sqrt[5]{\dfrac{x}{y}}$

25. $\sqrt[6]{\dfrac{x^6 y^8}{z^{15}}} = \dfrac{\sqrt[6]{x^6 y^8}}{\sqrt[6]{z^{15}}} = \dfrac{\sqrt[6]{x^6 y^6 \cdot y^2}}{\sqrt[6]{z^{12} \cdot z^3}} = \dfrac{\sqrt[6]{x^6 y^6}\sqrt[6]{y^2}}{\sqrt[6]{z^{12}}\sqrt[6]{z^3}} = \dfrac{xy\sqrt[6]{y^2}}{z^2 \sqrt[6]{z^3}}$,

   or $\dfrac{xy}{z^2}\sqrt[6]{\dfrac{y^2}{z^3}}$

27. $\dfrac{\sqrt{18y}}{\sqrt{2y}} = \sqrt{\dfrac{18y}{2y}} = \sqrt{9} = 3$

29. $\dfrac{\sqrt[3]{26}}{\sqrt[3]{13}} = \sqrt[3]{\dfrac{26}{13}} = \sqrt[3]{2}$

31. $\dfrac{\sqrt{40xy^3}}{\sqrt{8x}} = \sqrt{\dfrac{40xy^3}{8x}} = \sqrt{5y^3} = \sqrt{y^2 \cdot 5y}$

   $= \sqrt{y^2}\sqrt{5y} = y\sqrt{5y}$

33. $\dfrac{\sqrt[3]{96a^4 b^2}}{\sqrt[3]{12a^2 b}} = \sqrt[3]{\dfrac{96a^4 b^2}{12a^2 b}} = \sqrt[3]{8a^2 b} = \sqrt[3]{8}\sqrt[3]{a^2 b} = 2\sqrt[3]{a^2 b}$

35. $\dfrac{\sqrt{100ab}}{5\sqrt{2}} = \dfrac{1}{5}\dfrac{\sqrt{100ab}}{\sqrt{2}} = \dfrac{1}{5}\sqrt{\dfrac{100ab}{2}} = \dfrac{1}{5}\sqrt{50ab}$

   $= \dfrac{1}{5}\sqrt{25 \cdot 2ab} = \dfrac{1}{5} \cdot 5\sqrt{2ab} = \sqrt{2ab}$

37. $\dfrac{\sqrt[4]{48x^9 y^{13}}}{\sqrt[4]{3xy^{-2}}} = \sqrt[4]{\dfrac{48x^9 y^{13}}{3xy^{-2}}} = \sqrt[4]{16x^8 y^{15}} = \sqrt[4]{16x^8 y^{12}}\sqrt[4]{y^3}$

   $= 2x^2 y^3 \sqrt[4]{y^3}$

39. $\dfrac{\sqrt[3]{x^3 - y^3}}{\sqrt[3]{x - y}} = \sqrt[3]{\dfrac{x^3 - y^3}{x - y}} = \sqrt[3]{\dfrac{(x-y)(x^2 + xy + y^2)}{x - y}}$

   $= \sqrt[3]{\dfrac{(x-y)(x^2 + xy + y^2)}{x - y}} = \sqrt[3]{x^2 + xy + y^2}$

41. $\sqrt{\dfrac{2}{5}} = \sqrt{\dfrac{2}{5} \cdot \dfrac{5}{5}} = \sqrt{\dfrac{10}{25}} = \dfrac{\sqrt{10}}{\sqrt{25}} = \dfrac{\sqrt{10}}{5}$

43. $\dfrac{2\sqrt{5}}{7\sqrt{3}} = \dfrac{2\sqrt{5}}{7\sqrt{3}} \cdot \dfrac{\sqrt{3}}{\sqrt{3}} = \dfrac{2\sqrt{15}}{21}$

45. $\sqrt[3]{\dfrac{5}{4}} = \sqrt[3]{\dfrac{5 \cdot 2}{4 \cdot 2}} = \sqrt[3]{\dfrac{10}{8}} = \dfrac{\sqrt[3]{10}}{\sqrt[3]{8}} = \dfrac{\sqrt[3]{10}}{2}$

47. $\dfrac{\sqrt[3]{3a}}{\sqrt[3]{5c}} = \dfrac{\sqrt[3]{3a}}{\sqrt[3]{5c}} \cdot \dfrac{\sqrt[3]{5^2 c^2}}{\sqrt[3]{5^2 c^2}} = \dfrac{\sqrt[3]{75ac^2}}{\sqrt[3]{5^3 c^3}} = \dfrac{\sqrt[3]{75ac^2}}{5c}$

49. $\dfrac{\sqrt[4]{5y^6}}{\sqrt[4]{9x}} = \dfrac{\sqrt[4]{5y^6}}{\sqrt[4]{9x}} \cdot \dfrac{\sqrt[4]{9x^3}}{\sqrt[4]{9x^3}} = \dfrac{\sqrt[4]{5y^6 \cdot 9x^3}}{\sqrt[4]{81x^4}} = \dfrac{\sqrt[4]{45x^3 y^2}}{3x}$

51. $\sqrt[3]{\dfrac{2}{x^2 y}} = \sqrt[3]{\dfrac{2}{x^2 y} \cdot \dfrac{xy^2}{xy^2}} = \sqrt[3]{\dfrac{2xy^2}{x^3 y^3}} = \dfrac{\sqrt[3]{2xy^2}}{\sqrt[3]{x^3 y^3}} = \dfrac{\sqrt[3]{2xy^2}}{xy}$

53. $\sqrt{\dfrac{7a}{18}} = \sqrt{\dfrac{7a}{18} \cdot \dfrac{2}{2}} = \sqrt{\dfrac{14a}{36}} = \dfrac{\sqrt{14a}}{\sqrt{36}} = \dfrac{\sqrt{14a}}{6}$

55. $\sqrt[5]{\dfrac{9}{32x^5 y}} = \sqrt[5]{\dfrac{9}{32x^5 y} \cdot \dfrac{y^4}{y^4}} = \dfrac{\sqrt[5]{9y^4}}{\sqrt[5]{32x^5 y^5}} = \dfrac{\sqrt[5]{9y^4}}{2xy}$

57. $\sqrt{\dfrac{10ab^2}{72a^3 b}} = \sqrt{\dfrac{5b}{36a^2}} = \dfrac{\sqrt{5b}}{6a}$

59. $\sqrt{\dfrac{5}{11}} = \sqrt{\dfrac{5}{11} \cdot \dfrac{5}{5}} = \sqrt{\dfrac{25}{55}} = \dfrac{\sqrt{25}}{\sqrt{55}} = \dfrac{5}{\sqrt{55}}$

61. $\dfrac{2\sqrt{6}}{5\sqrt{7}} = \dfrac{2\sqrt{6}}{5\sqrt{7}} \cdot \dfrac{\sqrt{6}}{\sqrt{6}} = \dfrac{2\sqrt{36}}{5\sqrt{42}} = \dfrac{2 \cdot 6}{5\sqrt{42}} = \dfrac{12}{5\sqrt{42}}$

63. $\dfrac{\sqrt{8}}{2\sqrt{3x}} = \dfrac{\sqrt{8}}{2\sqrt{3x}} \cdot \dfrac{\sqrt{8}}{\sqrt{8}} = \dfrac{\sqrt{64}}{2\sqrt{24x}} = \dfrac{8}{2\sqrt{24x}} = \dfrac{8}{4\sqrt{6x}} = \dfrac{2}{\sqrt{6x}}$

65. $\dfrac{\sqrt[3]{7}}{\sqrt[3]{2}} = \dfrac{\sqrt[3]{7}}{\sqrt[3]{2}} \cdot \dfrac{\sqrt[3]{7^2}}{\sqrt[3]{7^2}} = \dfrac{\sqrt[3]{7^3}}{\sqrt[3]{98}} = \dfrac{7}{\sqrt[3]{98}}$

67. $\sqrt{\dfrac{7x}{3y}} = \sqrt{\dfrac{7x}{3y} \cdot \dfrac{7x}{7x}} = \sqrt{\dfrac{(7x)^2}{21xy}} = \dfrac{7x}{\sqrt{21xy}}$

69. $\sqrt[3]{\dfrac{2a^5}{5b}} = \sqrt[3]{\dfrac{2a^5}{5b} \cdot \dfrac{4a}{4a}} = \sqrt[3]{\dfrac{8a^6}{20ab}} = \dfrac{2a^2}{\sqrt[3]{20ab}}$

71. $\sqrt{\dfrac{x^3y}{2}} = \sqrt{\dfrac{x^3y}{2} \cdot \dfrac{xy}{xy}} = \sqrt{\dfrac{x^4y^2}{2xy}} = \dfrac{\sqrt{x^4y^2}}{\sqrt{2xy}} = \dfrac{x^2y}{\sqrt{2xy}}$

73. *Writing Exercise.* Assuming that a calculator is not available, the calculation is 1.414213562/2 is easier to perform than 1/1.414213562.

75. $3x - 8xy + 2xz = x(3 - 8y + 2z)$

77. $(a + b)(a - b) = a^2 - b^2$ Difference of squares

79. $(8 + 3x)(7 - 4x) = 56 - 32x + 21x - 12x^2$ FOIL
$$= 56 - 11x - 12x^2$$

81. *Writing Exercise.* No; $\dfrac{\sqrt{8}}{\sqrt{2}} = \sqrt{\dfrac{8}{2}} = \sqrt{4} = 2.$

83. a) $T = 2\pi\sqrt{\dfrac{65}{980}} \approx 1.62$ sec

   b) $T = 2\pi\sqrt{\dfrac{98}{980}} \approx 1.99$ sec

   c) $T = 2\pi\sqrt{\dfrac{120}{980}} \approx 2.20$ sec

85. $\dfrac{\left(\sqrt[3]{81mn^2}\right)^2}{\left(\sqrt[3]{mn}\right)^2} = \dfrac{\sqrt[3]{(81mn^2)^2}}{\sqrt[3]{(mn)^2}}$

$\qquad\qquad = \dfrac{\sqrt[3]{6561m^2n^4}}{\sqrt[3]{m^2n^2}}$

$\qquad\qquad = \sqrt[3]{\dfrac{6561m^2n^4}{m^2n^2}}$

$\qquad\qquad = \sqrt[3]{6561n^2}$

$\qquad\qquad = \sqrt[3]{729 \cdot 9n^2}$

$\qquad\qquad = \sqrt[3]{729}\sqrt[3]{9n^2}$

$\qquad\qquad = 9\sqrt[3]{9n^2}$

87. $\sqrt{a^2 - 3} - \dfrac{a^2}{\sqrt{a^2 - 3}} = \sqrt{a^2 - 3} - \dfrac{a^2}{\sqrt{a^2 - 3}} \cdot \dfrac{\sqrt{a^2 - 3}}{\sqrt{a^2 - 3}}$

$\qquad = \sqrt{a^2 - 3} - \dfrac{a^2\sqrt{a^2 - 3}}{a^2 - 3}$

$\qquad = \sqrt{a^2 - 3} \cdot \dfrac{a^2 - 3}{a^2 - 3} - \dfrac{a^2\sqrt{a^2 - 3}}{a^2 - 3}$

$\qquad = \dfrac{a^2\sqrt{a^2 - 3} - 3\sqrt{a^2 - 3} - a^2\sqrt{a^2 - 3}}{a^2 - 3}$

$\qquad = \dfrac{-3\sqrt{a^2 - 3}}{a^2 - 3},$ or $\dfrac{-3}{\sqrt{a^2 - 3}}$

89. Step 1: $\sqrt[n]{x} = x^{1/n},$ by definition;

   Step 2: $\left(\dfrac{x}{y}\right)^n = \dfrac{x^n}{y^n},$ raising a quotient to a power;

   Step 3: $x^{1/n} = \sqrt[n]{x},$ by definition

91. $f(x) = \sqrt{18x^3}, \ g(x) = \sqrt{2x}$

$\left(f / g\right)(x) = \dfrac{f(x)}{g(x)} = \dfrac{\sqrt{18x^3}}{\sqrt{2x}} = \sqrt{\dfrac{18x^3}{2x}} = \sqrt{9x^2} = 3x$

$\sqrt{2x}$ is defined for $2x \geq 0$, or $x \geq 0$. To avoid division by 0, we must exclude 0 from the domain. Thus, the domain of $f / g = \left\{x | x \text{ is a real number and } x > 0\right\},$ or $(0, \infty).$

93. $f(x) = \sqrt{x^2 - 9}, \ g(x) = \sqrt{x - 3}$

$\left(f / g\right)(x) = \dfrac{f(x)}{g(x)} = \dfrac{\sqrt{x^2 - 9}}{\sqrt{x - 3}} = \sqrt{\dfrac{x^2 - 9}{x - 3}}$

$\qquad\qquad = \sqrt{\dfrac{(x + 3)(x - 3)}{x - 3}} = \sqrt{x + 3}$

$\sqrt{x - 3}$ is defined for $x - 3 \geq 0$, or $x \geq 3$. To avoid division by 0, we must exclude 0 from the domain. Thus, the domain of $f / g = \left\{x | x \text{ is a real number and } x > 3\right\},$ or $(3, \infty).$

Exercise Set 7.5

1. To add radical expressions, the <u>indices</u> and the <u>radicands</u> must be the same.

3. To find a product by adding exponents, the <u>bases</u> must be the same.

5. To rationalize the <u>numerator</u> of $\dfrac{\sqrt{c}-\sqrt{a}}{5}$, we multiply

by a form of 1, using the <u>conjugate</u> of $\sqrt{c}-\sqrt{a}$, or

$\sqrt{c}+\sqrt{a}$, to write 1.

7. $4\sqrt{3}+7\sqrt{3}=(4+7)\sqrt{3}=11\sqrt{3}$

9. $7\sqrt[3]{4}-5\sqrt[3]{4}=(7-5)\sqrt[3]{4}=2\sqrt[3]{4}$

11. $\sqrt[3]{y}+9\sqrt[3]{y}=(1+9)\sqrt[3]{y}=10\sqrt[3]{y}$

13. $8\sqrt{2}-\sqrt{2}+5\sqrt{2}=(8-1+5)\sqrt{2}=12\sqrt{2}$

15. $9\sqrt[3]{7}-\sqrt{3}+4\sqrt[3]{7}+2\sqrt{3}$
$=(9+4)\sqrt[3]{7}+(-1+2)\sqrt{3}=13\sqrt[3]{7}+\sqrt{3}$

17. $4\sqrt{27}-3\sqrt{3}$
$=4\sqrt{9\cdot3}-3\sqrt{3}$ Factoring the
$=4\sqrt{9}\cdot\sqrt{3}-3\sqrt{3}$ first radical
$=4\cdot3\sqrt{3}-3\sqrt{3}$ Taking the square root of 9
$=12\sqrt{3}-3\sqrt{3}$
$=9\sqrt{3}$ Combining the radicals

19. $3\sqrt{45}-8\sqrt{20}$
$=3\sqrt{9\cdot5}-8\sqrt{4\cdot5}$ Factoring the
$=3\sqrt{9}\cdot\sqrt{5}-8\sqrt{4}\cdot\sqrt{5}$ radicals
$=3\cdot3\sqrt{5}-8\cdot2\sqrt{5}$ Taking the square roots
$=9\sqrt{5}-16\sqrt{5}$
$=-7\sqrt{5}$ Combining like radicals

21. $3\sqrt[3]{16}+\sqrt[3]{54}=3\sqrt[3]{8\cdot2}+\sqrt[3]{27\cdot2}$
$=3\sqrt[3]{8}\cdot\sqrt[3]{2}+\sqrt[3]{27}\cdot\sqrt[3]{2}=3\cdot2\sqrt[3]{2}+3\sqrt[3]{2}$
$=6\sqrt[3]{2}+3\sqrt[3]{2}=9\sqrt[3]{2}$

23. $\sqrt{a}+3\sqrt{16a^3}=\sqrt{a}+3\sqrt{16a^2\cdot a}=\sqrt{a}+3\sqrt{16a^2}\cdot\sqrt{a}$
$=\sqrt{a}+3\cdot4a\sqrt{a}=\sqrt{a}+12a\sqrt{a}$
$=(1+12a)\sqrt{a}$

25. $\sqrt[3]{6x^4}-\sqrt[3]{48x}=\sqrt[3]{x^3\cdot6x}-\sqrt[3]{8\cdot6x}$
$=\sqrt[3]{x^3}\cdot\sqrt[3]{6x}-\sqrt[3]{8}\cdot\sqrt[3]{6x}=x\sqrt[3]{6x}-2\sqrt[3]{6x}$
$=(x-2)\sqrt[3]{6x}$

27. $\sqrt{4a-4}+\sqrt{a-1}=\sqrt{4(a-1)}+\sqrt{a-1}$
$=\sqrt{4}\sqrt{a-1}+\sqrt{a-1}=2\sqrt{a-1}+\sqrt{a-1}=3\sqrt{a-1}$

29. $\sqrt{x^3-x^2}+\sqrt{9x-9}=\sqrt{x^2(x-1)}+\sqrt{9(x-1)}$
$=\sqrt{x^2}\cdot\sqrt{x-1}+\sqrt{9}\cdot\sqrt{x-1}$
$=x\sqrt{x-1}+3\sqrt{x-1}=(x+3)\sqrt{x-1}$

31. $\sqrt{2}\left(5+\sqrt{2}\right)=\sqrt{2}\cdot5+\sqrt{2}\cdot\sqrt{2}=5\sqrt{2}+2$

33. $3\sqrt{5}\left(\sqrt{6}-\sqrt{7}\right)=3\sqrt{5}\cdot\sqrt{6}-3\sqrt{5}\cdot\sqrt{7}=3\sqrt{30}-3\sqrt{35}$

35. $\sqrt{2}\left(3\sqrt{10}-\sqrt{8}\right)=\sqrt{2}\cdot3\sqrt{10}-\sqrt{2}\cdot\sqrt{8}=3\sqrt{20}-\sqrt{16}$
$=3\sqrt{4\cdot5}-\sqrt{16}=3\cdot2\sqrt{5}-4$
$=6\sqrt{5}-4$

37. $\sqrt[3]{3}\left(\sqrt[3]{9}-4\sqrt[3]{21}\right)=\sqrt[3]{3}\cdot\sqrt[3]{9}-\sqrt[3]{3}\cdot4\sqrt[3]{21}$
$=\sqrt[3]{27}-4\sqrt[3]{63}$
$=3-4\sqrt[3]{63}$

39. $\sqrt[3]{a}\left(\sqrt[3]{a^2}+\sqrt[3]{24a^2}\right)=\sqrt[3]{a}\cdot\sqrt[3]{a^2}+\sqrt[3]{a}\sqrt[3]{24a^2}$
$=\sqrt[3]{a^3}+\sqrt[3]{24a^3}$
$=\sqrt[3]{a^3}+\sqrt[3]{8a^3\cdot3}$
$=a+2a\sqrt[3]{3}$

41. $\left(2+\sqrt{6}\right)\left(5-\sqrt{6}\right)=2\cdot5-2\sqrt{6}+5\sqrt{6}-\sqrt{6}\cdot\sqrt{6}$
$=10+3\sqrt{6}-6=4+3\sqrt{6}$

43. $\left(\sqrt{2}+\sqrt{7}\right)\left(\sqrt{3}-\sqrt{7}\right)=\sqrt{2}\cdot\sqrt{3}-\sqrt{2}\cdot\sqrt{7}+\sqrt{7}\cdot\sqrt{3}-\sqrt{7}\cdot\sqrt{7}$
$=\sqrt{6}-\sqrt{14}+\sqrt{21}-7$

45. $\left(2-\sqrt{3}\right)\left(2+\sqrt{3}\right)=2^2-\left(\sqrt{3}\right)^2=4-3=1$

47. $\left(\sqrt{10}-\sqrt{15}\right)\left(\sqrt{10}+\sqrt{15}\right)=\left(\sqrt{10}\right)^2-\left(\sqrt{15}\right)^2$
$=10-15=-5$

49. $\left(3\sqrt{7}+2\sqrt{5}\right)\left(2\sqrt{7}-4\sqrt{5}\right)$
$=3\sqrt{7}\cdot2\sqrt{7}-3\sqrt{7}\cdot4\sqrt{5}+2\sqrt{5}\cdot2\sqrt{7}-2\sqrt{5}\cdot4\sqrt{5}$
$=6\cdot7-12\sqrt{35}+4\sqrt{35}-8\cdot5=42-8\sqrt{35}-40$
$=2-8\sqrt{35}$

51. $\left(4+\sqrt{7}\right)^2=4^2+2\cdot4\cdot\sqrt{7}+\left(\sqrt{7}\right)^2=16+8\sqrt{7}+7$
$=23+8\sqrt{7}$

53. $\left(\sqrt{3}-\sqrt{2}\right)^2=\left(\sqrt{3}\right)^2-2\cdot\sqrt{3}\cdot\sqrt{2}+\left(\sqrt{2}\right)^2$
$=3-2\sqrt{6}+2=5-2\sqrt{6}$

55. $\left(\sqrt{2t}+\sqrt{5}\right)^2=\left(\sqrt{2t}\right)^2+2\cdot\sqrt{2t}\cdot\sqrt{5}+\left(\sqrt{5}\right)^2$
$=2t+2\sqrt{10t}+5$

57. $\left(3-\sqrt{x+5}\right)^2=3^2-2\cdot3\cdot\sqrt{x+5}+\left(\sqrt{x+5}\right)^2$
$=9-6\sqrt{x+5}+x+5$
$=14-6\sqrt{x+5}+x$

59. $\left(2\sqrt[4]{7}-\sqrt[4]{6}\right)\left(3\sqrt[4]{9}+2\sqrt[4]{5}\right)$
$=2\sqrt[4]{7}\cdot3\sqrt[4]{9}+2\sqrt[4]{7}\cdot2\sqrt[4]{5}-\sqrt[4]{6}\cdot3\sqrt[4]{9}-\sqrt[4]{6}\cdot2\sqrt[4]{5}$
$=6\sqrt[4]{63}+4\sqrt[4]{35}-3\sqrt[4]{54}-2\sqrt[4]{30}$

61. $\dfrac{6}{3-\sqrt{2}}=\dfrac{6}{3-\sqrt{2}}\cdot\dfrac{3+\sqrt{2}}{3+\sqrt{2}}=\dfrac{6\left(3+\sqrt{2}\right)}{\left(3-\sqrt{2}\right)\left(3+\sqrt{2}\right)}$
$=\dfrac{18+6\sqrt{2}}{3^2-\left(\sqrt{2}\right)^2}=\dfrac{18+6\sqrt{2}}{9-2}=\dfrac{18+6\sqrt{2}}{7}$

63. $\dfrac{2+\sqrt{5}}{6+\sqrt{3}} = \dfrac{2+\sqrt{5}}{6+\sqrt{3}} \cdot \dfrac{6-\sqrt{3}}{6-\sqrt{3}}$

$= \dfrac{(2+\sqrt{5})(6-\sqrt{3})}{(6+\sqrt{3})(6-\sqrt{3})} = \dfrac{12-2\sqrt{3}+6\sqrt{5}-\sqrt{15}}{36-3}$

$= \dfrac{12-2\sqrt{3}+6\sqrt{5}-\sqrt{15}}{33}$

65. $\dfrac{\sqrt{a}}{\sqrt{a}+\sqrt{b}} = \dfrac{\sqrt{a}}{\sqrt{a}+\sqrt{b}} \cdot \dfrac{\sqrt{a}-\sqrt{b}}{\sqrt{a}-\sqrt{b}}$

$= \dfrac{\sqrt{a}(\sqrt{a}-\sqrt{b})}{(\sqrt{a}+\sqrt{b})(\sqrt{a}-\sqrt{b})} = \dfrac{a-\sqrt{ab}}{a-b}$

67. $\dfrac{\sqrt{7}-\sqrt{3}}{\sqrt{3}-\sqrt{7}} = \dfrac{-1(\sqrt{3}-\sqrt{7})}{\sqrt{3}-\sqrt{7}} = -1 \cdot \dfrac{\sqrt{3}-\sqrt{7}}{\sqrt{3}-\sqrt{7}} = -1 \cdot 1 = -1$

69. $\dfrac{3\sqrt{2}-\sqrt{7}}{4\sqrt{2}+2\sqrt{5}} = \dfrac{3\sqrt{2}-\sqrt{7}}{4\sqrt{2}+2\sqrt{5}} \cdot \dfrac{4\sqrt{2}-2\sqrt{5}}{4\sqrt{2}-2\sqrt{5}}$

$= \dfrac{(3\sqrt{2}-\sqrt{7})(4\sqrt{2}-2\sqrt{5})}{(4\sqrt{2}+2\sqrt{5})(4\sqrt{2}-2\sqrt{5})}$

$= \dfrac{12 \cdot 2 - 6\sqrt{10} - 4\sqrt{14} + 2\sqrt{35}}{16 \cdot 2 - 4 \cdot 5}$

$= \dfrac{24 - 6\sqrt{10} - 4\sqrt{14} + 2\sqrt{35}}{32-20} = \dfrac{24 - 6\sqrt{10} - 4\sqrt{14} + 2\sqrt{35}}{12}$

$= \dfrac{2(12 - 3\sqrt{10} - 2\sqrt{14} + \sqrt{35})}{2 \cdot 6} = \dfrac{12 - 3\sqrt{10} - 2\sqrt{14} + \sqrt{35}}{6}$

71. $\dfrac{\sqrt{5}+1}{4} = \dfrac{\sqrt{5}+1}{4} \cdot \dfrac{\sqrt{5}-1}{\sqrt{5}-1} = \dfrac{(\sqrt{5}+1)(\sqrt{5}-1)}{4(\sqrt{5}-1)}$

$= \dfrac{(\sqrt{5})^2 - 1}{4(\sqrt{5}-1)} = \dfrac{5-1}{4(\sqrt{5}-1)} = \dfrac{4}{4(\sqrt{5}-1)}$

$= \dfrac{1}{\sqrt{5}-1}$

73. $\dfrac{\sqrt{6}-2}{\sqrt{3}+7} = \dfrac{\sqrt{6}-2}{\sqrt{3}+7} \cdot \dfrac{\sqrt{6}+2}{\sqrt{6}+2} = \dfrac{(\sqrt{6}-2)(\sqrt{6}+2)}{(\sqrt{3}+7)(\sqrt{6}+2)}$

$= \dfrac{6-4}{\sqrt{18}+2\sqrt{3}+7\sqrt{6}+14} = \dfrac{2}{3\sqrt{2}+2\sqrt{3}+7\sqrt{6}+14}$

75. $\dfrac{\sqrt{x}-\sqrt{y}}{\sqrt{x}+\sqrt{y}} = \dfrac{\sqrt{x}-\sqrt{y}}{\sqrt{x}+\sqrt{y}} \cdot \dfrac{\sqrt{x}+\sqrt{y}}{\sqrt{x}+\sqrt{y}}$

$= \dfrac{(\sqrt{x}-\sqrt{y})(\sqrt{x}+\sqrt{y})}{(\sqrt{x}+\sqrt{y})(\sqrt{x}+\sqrt{y})} = \dfrac{x-y}{x+2\sqrt{xy}+y}$

77. $\dfrac{\sqrt{a+h}-\sqrt{a}}{h} = \dfrac{\sqrt{a+h}-\sqrt{a}}{h} \cdot \dfrac{\sqrt{a+h}+\sqrt{a}}{\sqrt{a+h}+\sqrt{a}}$

$= \dfrac{(\sqrt{a+h}-\sqrt{a})(\sqrt{a+h}+\sqrt{a})}{h(\sqrt{a+h}+\sqrt{a})} = \dfrac{a+h-a}{h(\sqrt{a+h}+\sqrt{a})}$

$= \dfrac{h}{h(\sqrt{a+h}+\sqrt{a})} = \dfrac{1}{\sqrt{a+h}+\sqrt{a}}$

79. $\sqrt[3]{a}\,\sqrt[6]{a}$

$= a^{1/3} \cdot a^{1/6}$ Converting to exponential notation

$= a^{1/2}$ Adding exponents

$= \sqrt{a}$ Returning to radical notation

81. $\sqrt{b^3}\,\sqrt[5]{b^4}$

$= b^{3/2} \cdot b^{4/5}$ Converting to exponential notation

$= b^{23/10}$ Adding exponents

$= b^{2+3/10}$ Writing $\frac{23}{10}$ as a mixed number

$= b^2 b^{3/10}$ Factoring

$= b^2 \sqrt[10]{b^3}$ Returning to radical notation

83. $\sqrt{xy^3}\,\sqrt[3]{x^2 y} = (xy^3)^{1/2}(x^2 y)^{1/3}$

$= (xy^3)^{3/6}(x^2 y)^{2/6}$

$= \left[(xy^3)^3 (x^2 y)^2\right]^{1/6}$

$= \sqrt[6]{x^3 y^9 \cdot x^4 y^2}$

$= \sqrt[6]{x^7 y^{11}}$

$= \sqrt[6]{x^6 y^6 \cdot xy^5}$

$= xy\sqrt[6]{xy^5}$

85. $\sqrt[4]{9ab^3}\,\sqrt{3a^4 b} = (9ab^3)^{1/4}(3a^4 b)^{1/2}$

$= (9ab^3)^{1/4}(3a^4 b)^{2/4}$

$= \left[(9ab^3)(3a^4 b)^2\right]^{1/4}$

$= \sqrt[4]{9ab^3 \cdot 9a^8 b^2}$

$= \sqrt[4]{81a^9 b^5}$

$= \sqrt[4]{81a^8 b^4 \cdot ab}$

$= 3a^2 b\sqrt[4]{ab}$

87. $\sqrt{a^4 b^3 c^4}\,\sqrt[3]{ab^2 c} = (a^4 b^3 c^4)^{1/2}(ab^2 c)^{1/3}$

$= (a^4 b^3 c^4)^{3/6}(ab^2 c)^{2/6}$

$= \left[(a^4 b^3 c^4)^3 (ab^2 c)^2\right]^{1/6}$

$= \sqrt[6]{a^{12} b^9 c^{12} \cdot a^2 b^4 c^2}$

$= \sqrt[6]{a^{14} b^{13} c^{14}}$

$= \sqrt[6]{a^{12} b^{12} c^{12} \cdot a^2 bc^2}$

$= a^2 b^2 c^2 \sqrt[6]{a^2 bc^2}$

89. $\dfrac{\sqrt[3]{a^2}}{\sqrt[4]{a}}$

$= \dfrac{a^{2/3}}{a^{1/4}}$ Converting to exponential notation

$= a^{2/3-1/4}$ Subtracting exponents

$= a^{5/12}$ Converting back

$= \sqrt[12]{a^5}$ to radical notation

91. $\dfrac{\sqrt[4]{x^2 y^3}}{\sqrt[3]{xy}}$

$= \dfrac{\left(x^2 y^3\right)^{1/4}}{(xy)^{1/3}}$ Converting to exponential notation

$= \dfrac{x^{2/4} y^{3/4}}{x^{1/3} y^{1/3}}$ Using the power and product rules

$= x^{2/4-1/3} y^{3/4-1/3}$ Subtracting exponents

$= x^{2/12} y^{5/12}$

$= \left(x^2 y^5\right)^{1/12}$ Converting back to radical notation

$= \sqrt[12]{x^2 y^5}$

93. $\dfrac{\sqrt{ab^3}}{\sqrt[5]{a^2 b^3}}$

$= \dfrac{\left(ab^3\right)^{1/2}}{\left(a^2 b^3\right)^{1/5}}$ Converting to exponential notation

$= \dfrac{a^{1/2} b^{3/2}}{a^{2/5} b^{3/5}}$

$= a^{1/10} b^{9/10}$ Subtracting exponents

$= \left(ab^9\right)^{1/10}$ Converting back to radical notation

$= \sqrt[10]{ab^9}$

95. $\dfrac{\sqrt{(7-y)^3}}{\sqrt[3]{(7-y)^2}}$

$= \dfrac{(7-y)^{3/2}}{(7-y)^{2/3}}$ Converting to exponential notation

$= (7-y)^{3/2-2/3}$ Subtracting exponents

$= (7-y)^{5/6}$

$= \sqrt[6]{(7-y)^5}$ Returning to radical notation

97. $\dfrac{\sqrt[4]{(5+3x)^3}}{\sqrt[3]{(5+3x)^2}}$

$= \dfrac{(5+3x)^{3/4}}{(5+3x)^{2/3}}$ Converting to exponential notation

$= (5+3x)^{3/4-2/3}$ Subtracting exponents

$= (5+3x)^{1/12}$ Converting back to radical notation

$= \sqrt[12]{5+3x}$

99. $\sqrt[3]{x^2 y}\left(\sqrt{xy} - \sqrt[5]{xy^3}\right)$

$= \left(x^2 y\right)^{1/3}\left[(xy)^{1/2} - \left(xy^3\right)^{1/5}\right]$

$= x^{2/3} y^{1/3}\left(x^{1/2} y^{1/2} - x^{1/5} y^{3/5}\right)$

$= x^{2/3} y^{1/3} x^{1/2} y^{1/2} - x^{2/3} y^{1/3} x^{1/5} y^{3/5}$

$= x^{2/3+1/2} y^{1/3+1/2} - x^{2/3+1/5} y^{1/3+3/5}$

$= x^{7/6} y^{5/6} - x^{13/15} y^{14/15}$ Writing as a mixed numeral

$= x \cdot x^{1/6} y^{5/6} - x^{13/15} y^{14/15}$

$= x\left(xy^5\right)^{1/6} - \left(x^{13} y^{14}\right)^{1/15}$

$= x\sqrt[6]{xy^5} - \sqrt[15]{x^{13} y^{14}}$

101. $\left(m + \sqrt[3]{n^2}\right)\left(2m + \sqrt[4]{n}\right)$

$= \left(m + n^{2/3}\right)\left(2m + n^{1/4}\right)$ Converting to exponential notation

$= 2m^2 + mn^{1/4} + 2mn^{2/3} + n^{2/3} n^{1/4}$ Using FOIL

$= 2m^2 + mn^{1/4} + 2mn^{2/3} + n^{2/3+1/4}$ Adding exponents

$= 2m^2 + mn^{1/4} + 2mn^{2/3} + n^{11/12}$

$= 2m^2 + m\sqrt[4]{n} + 2m\sqrt[3]{n^2} + \sqrt[12]{n^{11}}$ Converting back to radical notation

103. $f(x) = \sqrt[4]{x}, \;\; g(x) = 2\sqrt{x} - \sqrt[3]{x^2}$

$(f \cdot g)(x) = \sqrt[4]{x}\left(2\sqrt{x} - \sqrt[3]{x^2}\right)$

$= x^{1/4} \cdot 2x^{1/2} - x^{1/4} \cdot x^{2/3}$

$= 2x^{1/4+1/2} - x^{1/4+2/3}$

$= 2x^{1/4+2/4} - x^{3/12+8/12}$

$= 2x^{3/4} - x^{11/12}$

$= 2\sqrt[4]{x^3} - \sqrt[12]{x^{11}}$

105. $f(x) = x + \sqrt{7}, \;\; g(x) = x - \sqrt{7}$

$(f \cdot g)(x) = (x + \sqrt{7})(x - \sqrt{7})$

$= x^2 - \left(\sqrt{7}\right)^2$

$= x^2 - 7$

107. $f(x) = x^2$

$f(3 - \sqrt{2}) = (3 - \sqrt{2})^2 = 3^2 - 2 \cdot 3 \cdot \sqrt{2} + \left(\sqrt{2}\right)^2$

$= 9 - 6\sqrt{2} + 2 = 11 - 6\sqrt{2}$

109. $f(x) = x^2$

$f(\sqrt{6} + \sqrt{21}) = (\sqrt{6} + \sqrt{21})^2$

$= \left(\sqrt{6}\right)^2 + 2 \cdot \sqrt{6} \cdot \sqrt{21} + \left(\sqrt{21}\right)^2$

$= 6 + 2\sqrt{126} + 21 = 27 + 2\sqrt{9 \cdot 14}$

$= 27 + 6\sqrt{14}$

111. *Writing Exercise.* The distributive law is used to combine radical expressions with the same indices and radicands just as it is used to combine monomials with the same variables and exponents.

113. $3x - 1 = 125$

$3x = 126$

$x = 42$

The solution is 42.

115. $x^2 + 2x + 1 = 22 - 2x$

$x^2 + 4x - 21 = 0$

$(x + 7)(x - 3) = 0$

$x + 7 = 0 \quad or \quad x - 3 = 0$

$x = -7 \quad or \qquad x = 3$

The solutions are –7 and 3.

117.
$$\frac{1}{x} + \frac{1}{2} = \frac{1}{6}$$
$$\frac{1}{x} \cdot 6x + \frac{1}{2} \cdot 6x = \frac{1}{6} \cdot 6x$$
$$6 + 3x = x$$
$$2x = -6$$
$$x = -3$$
The solution is –3.

119. *Writing Exercise.* It appears that Ramon multiplied the exponents rather than adding them.

121. $f(x) = \sqrt{x^3 - x^2} + \sqrt{9x^3 - 9x^2} - \sqrt{4x^3 - 4x^2}$
$$= \sqrt{x^2(x-1)} + \sqrt{9x^2(x-1)} - \sqrt{4x^2(x-1)}$$
$$= x\sqrt{x-1} + 3x\sqrt{x-1} - 2x\sqrt{x-1}$$
$$= 2x\sqrt{x-1}$$

123. $f(x) = \sqrt[4]{x^5 - x^4} + 3\sqrt[4]{x^9 - x^8}$
$$= \sqrt[4]{x^4(x-1)} + 3\sqrt[4]{x^8(x-1)}$$
$$= \sqrt[4]{x^4} \cdot \sqrt[4]{x-1} + 3\sqrt[4]{x^8}\sqrt[4]{x-1}$$
$$= x\sqrt[4]{x-1} + 3x^2\sqrt[4]{x-1}$$
$$= (x + 3x^2)\sqrt[4]{x-1}$$

125. $7x\sqrt{(x+y)^3} - 5xy\sqrt{x+y} - 2y\sqrt{(x+y)^3}$
$$= 7x\sqrt{(x+y)^2(x+y)} - 5xy\sqrt{x+y} - 2y\sqrt{(x+y)^2(x+y)}$$
$$= 7x(x+y)\sqrt{x+y} - 5xy\sqrt{x+y} - 2y(x+y)\sqrt{x+y}$$
$$= [7x(x+y) - 5xy - 2y(x+y)]\sqrt{x+y}$$
$$= (7x^2 + 7xy - 5xy - 2xy - 2y^2)\sqrt{x+y}$$
$$= (7x^2 - 2y^2)\sqrt{x+y}$$

127. $\sqrt{8x(y+z)^5}\sqrt[3]{4x^2(y+z)^2}$
$$= \left[8x(y+z)^5\right]^{1/2}\left[4x^2(y+z)^2\right]^{1/3}$$
$$= \left[8x(y+z)^5\right]^{3/6}\left[4x^2(y+z)^2\right]^{2/6}$$
$$= \left\{\left[2^3 x(y+z)^5\right]^3\left[2^2 x^2(y+z)^2\right]^2\right\}^{1/6}$$
$$= \sqrt[6]{2^9 x^3(y+z)^{15} \cdot 2^4 x^4(y+z)^4}$$
$$= \sqrt[6]{2^{13} x^7(y+z)^{19}}$$
$$= \sqrt[6]{2^{12} x^6(y+z)^{18} \cdot 2x(y+z)}$$
$$= 2^2 x(y+z)^3\sqrt[6]{2x(y+z)}, \text{ or}$$
$$4x(y+z)^3\sqrt[6]{2x(y+z)}$$

129. $\dfrac{\dfrac{1}{\sqrt{w}} - \sqrt{w}}{\dfrac{\sqrt{w}+1}{\sqrt{w}}} = \dfrac{\dfrac{1}{\sqrt{w}} - \sqrt{w}}{\dfrac{\sqrt{w}+1}{\sqrt{w}}} \cdot \dfrac{\sqrt{w}}{\sqrt{w}} = \dfrac{1-w}{\sqrt{w}+1}$
$$= \frac{1-w}{\sqrt{w}+1} \cdot \frac{\sqrt{w}-1}{\sqrt{w}-1} = \frac{\sqrt{w}-1-w\sqrt{w}+w}{w-1}$$
$$= \frac{(w-1) - \sqrt{w}(w-1)}{w-1} = \frac{(w-1)(1-\sqrt{w})}{w-1}$$
$$= 1 - \sqrt{w}$$

131. $x - 5 = \left(\sqrt{x}\right)^2 - \left(\sqrt{5}\right)^2 = \left(\sqrt{x} + \sqrt{5}\right)\left(\sqrt{x} - \sqrt{5}\right)$

133. $x - a = \left(\sqrt{x}\right)^2 - \left(\sqrt{a}\right)^2 = \left(\sqrt{x} + \sqrt{a}\right)\left(\sqrt{x} - \sqrt{a}\right)$

135. $\left(\sqrt{x+2} - \sqrt{x-2}\right)^2 = x + 2 - 2\sqrt{(x+2)(x-2)} + x - 2$
$$= x + 2 - 2\sqrt{x^2 - 4} + x - 2$$
$$= 2x - 2\sqrt{x^2 - 4}$$

Connecting the Concepts

1. $\sqrt{(t+5)^2} = t + 5$

3. $\sqrt{6x}\sqrt{15x} = \sqrt{6x \cdot 15x} = \sqrt{90x^2} = \sqrt{9x^2 \cdot 10} = 3x\sqrt{10}$

5. $\sqrt{15t} + 4\sqrt{15t} = (1+4)\sqrt{15t} = 5\sqrt{15t}$

7. $\sqrt{6}\left(\sqrt{10} - \sqrt{33}\right) = \sqrt{6}\sqrt{10} - \sqrt{6}\sqrt{33} = \sqrt{6 \cdot 10} - \sqrt{6 \cdot 33}$
$$= \sqrt{60} - \sqrt{198} = \sqrt{4 \cdot 15} - \sqrt{9 \cdot 22}$$
$$= 2\sqrt{15} - 3\sqrt{22}$$

9. $\dfrac{\sqrt{t}}{\sqrt[8]{t^3}} = \dfrac{t^{1/2}}{t^{3/8}} = t^{1/2 - 3/8} = t^{1/8} = \sqrt[8]{t}$

11. $2\sqrt{3} - 5\sqrt{12} = 2\sqrt{3} - 5\sqrt{4 \cdot 3} = 2\sqrt{3} - 5 \cdot 2\sqrt{3}$
$$= 2\sqrt{3} - 10\sqrt{3} = -8\sqrt{3}$$

13. $\left(\sqrt{15} + \sqrt{10}\right)^2 = \left(\sqrt{15}\right)^2 + 2 \cdot \sqrt{10}\sqrt{15} + \left(\sqrt{10}\right)^2$
$$= 15 + 2\sqrt{25 \cdot 6} + 10 = 25 + 2 \cdot 5\sqrt{6}$$
$$= 25 + 10\sqrt{6}$$

15. $\sqrt{x^3 y}\sqrt[5]{xy^4} = x^{3/2}y^{1/2}x^{1/5}y^{4/5} = x^{3/2 + 1/5}y^{1/2 + 4/5}$
$$= x^{17/10}y^{13/10} = x^{1 + 7/10}y^{1 + 3/10}$$
$$= xy\sqrt[10]{x^7 y^3}$$

17. $\sqrt{\sqrt[5]{x^2}} = \sqrt{x^{2/5}} = \left(x^{2/5}\right)^{1/2} = x^{\frac{2 \cdot 1}{5 \cdot 2}} = x^{1/5} = \sqrt[5]{x}$

19. $\left(\sqrt[4]{a^2 b^3}\right)^2 = \left(a^{2/4}b^{3/4}\right)^2 = a^{\frac{2 \cdot 2}{4}}b^{\frac{3 \cdot 2}{4}}$
$$= ab^{3/2} = ab^1 b^{1/2} = ab\sqrt{b}$$

Exercise Set 7.6

1. False; if $x^2 = 25$, then $x = 5$, or $x = -5$.

3. True by the principle of powers

5. If we add 8 to both sides of $\sqrt{x} - 8 = 7$, we get $\sqrt{x} = 15$, so the statement is true.

7. $\sqrt{5x+1} = 4$

$\left(\sqrt{5x+1}\right)^2 = 4^2$ Principle of powers (squaring)

$5x + 1 = 16$

$5x = 15$

$x = 3$

Check: $\dfrac{\sqrt{5x+1} = 4}{\begin{array}{c|c} \sqrt{5\cdot 3+1} & 4 \\ \sqrt{16} & \end{array}}$

$4 \overset{?}{=} 4$ TRUE

The solution is 3.

9. $\sqrt{3x} + 1 = 5$

$\sqrt{3x} = 4$ Adding to isolate the radical

$\left(\sqrt{3x}\right)^2 = 4^2$ Principle of powers (squaring)

$3x = 16$

$x = \dfrac{16}{3}$

Check: $\dfrac{\sqrt{3x}+1 = 5}{\begin{array}{c|c} \sqrt{3\cdot\frac{16}{3}+1} & 5 \\ \sqrt{16}+1 & \\ 4+1 & \end{array}}$

$5 \overset{?}{=} 5$ TRUE

The solution is $\dfrac{16}{3}$.

11. $\sqrt{y+5} - 4 = 1$

$\sqrt{y+5} = 5$ Adding to isolate the radical

$\left(\sqrt{y+5}\right)^2 = 5^2$ Principle of powers (squaring)

$y + 5 = 25$

$y = 20$

Check: $\dfrac{\sqrt{y+5}-4 = 1}{\begin{array}{c|c} \sqrt{20+5}-4 & 1 \\ \sqrt{25}-4 & \\ 5-4 & \end{array}}$

$1 \overset{?}{=} 1$ TRUE

The solution is 20.

13. $\sqrt{8-x} + 7 = 10$

$\sqrt{8-x} = 3$ Adding to isolate the radical

$\left(\sqrt{8-x}\right)^2 = 3^2$ Principle of powers (squaring)

$8 - x = 9$

$x = -1$

Check: $\dfrac{\sqrt{8-x}+7 = 10}{\begin{array}{c|c} \sqrt{8-(-1)}+7 & 10 \\ \sqrt{9}+7 & \\ 3+7 & \end{array}}$

$10 \overset{?}{=} 10$ TRUE

The solution is -1.

15. $\sqrt[3]{y+3} = 2$

$\left(\sqrt[3]{y+3}\right)^3 = 2^3$ Principle of powers (cubing)

$y + 3 = 8$

$y = 5$

Check: $\dfrac{\sqrt[3]{y+3} = 2}{\begin{array}{c|c} \sqrt[3]{5+3} & 2 \\ \sqrt[3]{8} & \end{array}}$

$2 \overset{?}{=} 2$ TRUE

The solution is 5.

17. $\sqrt[4]{t-10} = 3$

$\left(\sqrt[4]{t-10}\right)^4 = 3^4$

$t - 10 = 81$

$t = 91$

Check: $\dfrac{\sqrt[4]{t-10} = 3}{\begin{array}{c|c} \sqrt[4]{91-10} & 3 \\ \sqrt[4]{81} & \end{array}}$

$3 \overset{?}{=} 3$ TRUE

The solution is 91.

19. $6\sqrt{x} = x$

$\left(6\sqrt{x}\right)^2 = x^2$

$36x = x^2$

$0 = x^2 - 36x$

$0 = x(x - 36)$

$x = 0 \ \ or \ \ x = 36$

Check:

For $x = 0$: $\dfrac{6\sqrt{x} = x}{\begin{array}{c|c} 6\sqrt{0} & 0 \end{array}}$

$0 \overset{?}{=} 0$ TRUE

For $x = 36$: $\dfrac{6\sqrt{x} = x}{\begin{array}{c|c} 6\sqrt{36} & 36 \\ 6\cdot 6 & \end{array}}$

$36 \overset{?}{=} 36$ TRUE

The solutions are 0 and 36.

21. $2y^{1/2} - 13 = 7$

$2\sqrt{y} - 13 = 7$

$2\sqrt{y} = 20$

$\sqrt{y} = 10$

$\left(\sqrt{y}\right)^2 = 10^2$

$y = 100$

Check: $\dfrac{2y^{1/2} - 13 = 7}{\begin{array}{c|c} 2\cdot 100^{1/2} - 13 & 7 \\ 2\cdot 10 - 13 & \\ 20 - 13 & \end{array}}$

$7 \overset{?}{=} 7$ TRUE

The solution is 100.

23. $\sqrt[3]{x} = -5$

$\left(\sqrt[3]{x}\right)^3 = (-5)^3$

$x = -125$

Check: $\dfrac{\sqrt[3]{x} = -5}{\begin{array}{c|c}\sqrt[3]{-125} & -5 \\ \sqrt[3]{(-5)^3} & \end{array}}$

$-5 \overset{?}{=} -5$ TRUE

The solution is -125.

25. $z^{1/4} + 8 = 10$

$z^{1/4} = 2$

$\left(z^{1/4}\right)^4 = 2^4$

$z = 16$

Check: $\dfrac{z^{1/4} + 8 = 10}{\begin{array}{c|c}16^{1/4} + 8 & 10 \\ 2 + 8 & \end{array}}$

$10 \overset{?}{=} 10$ TRUE

The solution is 16.

27. $\sqrt{n} = -2$

This equation has no solution, since the principal square root is never negative.

29. $\sqrt[4]{3x+1} - 4 = -1$

$\sqrt[4]{3x+1} = 3$

$\left(\sqrt[4]{3x+1}\right)^4 = 3^4$

$3x + 1 = 81$

$3x = 80$

$x = \dfrac{80}{3}$

Check: $\dfrac{\sqrt[4]{3x+1} - 4 = -1}{\begin{array}{c|c}\sqrt[4]{3 \cdot \frac{80}{3} + 1} - 4 & -1 \\ \sqrt[4]{81} - 4 & \\ 3 - 4 & \end{array}}$

$-1 \overset{?}{=} -1$ TRUE

The solution is $\dfrac{80}{3}$.

31. $(21x + 55)^{1/3} = 10$

$\left[(21x + 55)^{1/3}\right]^3 = 10^3$

$21x + 55 = 1000$

$21x = 945$

$x = 45$

Check: $\dfrac{(21x + 55)^{1/3} = 10}{\begin{array}{c|c}(21 \cdot 45 + 55)^{1/3} & 10 \\ (945 + 55)^{1/3} & \end{array}}$

$10 \overset{?}{=} 10$ TRUE

The solution is 45.

33. $\sqrt[3]{3y+6} + 7 = 8$

$\sqrt[3]{3y+6} = 1$

$\left(\sqrt[3]{3y+6}\right)^3 = 1^3$

$3y + 6 = 1$

$3y = -5$

$y = -\dfrac{5}{3}$

Check: $\dfrac{\sqrt[3]{3y+6} + 7 = 8}{\begin{array}{c|c}\sqrt[3]{3\left(-\frac{5}{3}\right)+6} + 7 & 8 \\ \sqrt[3]{1} + 7 & \\ 1 + 7 & \end{array}}$

$8 \overset{?}{=} 8$ TRUE

The solution is $-\dfrac{5}{3}$.

35. $\sqrt{3t+4} = \sqrt{4t+3}$

$\left(\sqrt{3t+4}\right)^2 = \left(\sqrt{4t+3}\right)^2$

$3t + 4 = 4t + 3$

$4 = t + 3$

$1 = t$

Check:

$\dfrac{\sqrt{3t+4} = \sqrt{4t+3}}{\begin{array}{c|c}\sqrt{3 \cdot 1 + 4} & \sqrt{4 \cdot 1 + 3}\end{array}}$

$\sqrt{7} \overset{?}{=} \sqrt{7}$ TRUE

The solution is 1.

37. $3(4-t)^{1/4} = 6^{1/4}$

$\left[3(4-t)^{1/4}\right]^4 = \left(6^{1/4}\right)^4$

$81(4-t) = 6$

$324 - 81t = 6$

$-81t = -318$

$t = \dfrac{106}{27}$

The number $\dfrac{106}{27}$ checks and is the solution.

39. $3 + \sqrt{5-x} = x$

$\sqrt{5-x} = x - 3$

$\left(\sqrt{5-x}\right)^2 = (x-3)^2$

$5 - x = x^2 - 6x + 9$

$0 = x^2 - 5x + 4$

$0 = (x-1)(x-4)$

$x - 1 = 0 \quad \text{or} \quad x - 4 = 0$

$x = 1 \quad \text{or} \quad x = 4$

Check:

For 1: $\dfrac{3 + \sqrt{5-x} = x}{\begin{array}{c|c}3 + \sqrt{5-1} & 1 \\ 3 + \sqrt{4} & \\ 3 + 2 & \end{array}}$

$5 \overset{?}{=} 1$ FALSE

For 4: $3+\sqrt{5-x}=x$

$$
\begin{array}{c|c}
3+\sqrt{5-4} & 4 \\
3+\sqrt{1} & \\
3+1 & \\
\end{array}
$$

$$4 \overset{?}{=} 4 \quad \text{TRUE}$$

Since 4 checks but 1 does not, the solution is 4.

41.

$$\sqrt{4x-3}=2+\sqrt{2x-5} \quad \text{One radical is already isolated.}$$

$$\left(\sqrt{4x-3}\right)^2=\left(2+\sqrt{2x-5}\right)^2 \quad \text{Squaring both sides}$$

$$4x-3=4+4\sqrt{2x-5}+2x-5$$

$$2x-2=4\sqrt{2x-5}$$

$$x-1=2\sqrt{2x-5}$$

$$x^2-2x+1=8x-20$$

$$x^2-10x+21=0$$

$$(x-7)(x-3)=0$$

$$x-7=0 \quad \text{or} \quad x-3=0$$

$$x=7 \quad \text{or} \quad x=3$$

Both numbers check. The solutions are 7 and 3.

43. $\sqrt{20-x}+8=\sqrt{9-x}+11$

$$\sqrt{20-x}=\sqrt{9-x}+3 \quad \text{Isolating one radical}$$

$$\left(\sqrt{20-x}\right)^2=\left(\sqrt{9-x}+3\right)^2 \quad \text{Squaring both sides}$$

$$20-x=9-x+6\sqrt{9-x}+9$$

$$2=6\sqrt{9-x} \quad \text{Isolating the remaining radical}$$

$$1=3\sqrt{9-x} \quad \text{Multiplying by}\,\frac{1}{2}$$

$$1^2=\left(3\sqrt{9-x}\right)^2 \quad \text{Squaring both sides}$$

$$1=9(9-x)$$

$$1=81-9x$$

$$-80=-9x$$

$$\frac{80}{9}=x$$

The number $\frac{80}{9}$ checks and is the solution.

45. $\sqrt{x+2}+\sqrt{3x+4}=2$

$$\sqrt{x+2}=2-\sqrt{3x+4} \quad \text{Isolating one radical}$$

$$\left(\sqrt{x+2}\right)^2=\left(2-\sqrt{3x+4}\right)^2$$

$$x+2=4-4\sqrt{3x+4}+3x+4$$

$$-2x-6=-4\sqrt{3x+4} \quad \text{Isolating the remaining radical}$$

$$x+3=2\sqrt{3x+4} \quad \text{Multiplying by}-\frac{1}{2}$$

$$(x+3)^2=\left(2\sqrt{3x+4}\right)^2$$

$$x^2+6x+9=4(3x+4)$$

$$x^2+6x+9=12x+16$$

$$x^2-6x-7=0$$

$$(x-7)(x+1)=0$$

$$x-7=0 \quad \text{or} \quad x+1=0$$

$$x=7 \quad \text{or} \quad x=-1$$

Check:

For 7:

$$
\begin{array}{c|c}
\sqrt{x+2}+\sqrt{3x+4}=2 & \\
\hline
\sqrt{7+2}+\sqrt{3\cdot7+4} & 2 \\
\sqrt{9}+\sqrt{25} & \\
\end{array}
$$

$$8 \overset{?}{=} 2 \quad \text{FALSE}$$

For -1:

$$
\begin{array}{c|c}
\sqrt{x+2}+\sqrt{3x+4}=2 & \\
\hline
\sqrt{-1+2}+\sqrt{3\cdot(-1)+4} & \\
\sqrt{1}+\sqrt{1} & \\
\end{array}
$$

$$2 \overset{?}{=} 2 \quad \text{TRUE}$$

Since -1 checks but 7 does not, the solution is -1.

47. We must have $f(x)=1$, or $\sqrt{x}+\sqrt{x-9}=1$.

$$\sqrt{x}+\sqrt{x-9}=1$$

$$\sqrt{x-9}=1-\sqrt{x} \quad \text{Isolating one radical term}$$

$$\left(\sqrt{x-9}\right)^2=\left(1-\sqrt{x}\right)^2$$

$$x-9=1-2\sqrt{x}+x$$

$$-10=-2\sqrt{x} \quad \text{Isolating the remaining radical term}$$

$$5=\sqrt{x}$$

$$25=x$$

This value does not check. There is no solution, so there is no value of x for which $f(x)=1$.

49. $\sqrt{t-2}-\sqrt{4t+1}=-3$

$$\sqrt{t-2}=\sqrt{4t+1}-3$$
$$\left(\sqrt{t-2}\right)^2=\left(\sqrt{4t+1}-3\right)^2$$
$$t-2=4t+1-6\sqrt{4t+1}+9$$
$$-3t-12=-6\sqrt{4t+1}$$
$$t+4=2\sqrt{4t+1}$$
$$(t+4)^2=\left(2\sqrt{4t+1}\right)^2$$
$$t^2+8t+16=4(4t+1)$$
$$t^2+8t+16=16t+4$$
$$t^2-8t+12=0$$
$$(t-2)(t-6)=0$$
$$t-2=0 \quad or \quad t-6=0$$
$$t=2 \quad or \quad \quad t=6$$

Both numbers check, so we have $f(t)=-3$ when $t=2$ and when $t=6$.

51. We must have $\sqrt{2x-3}=\sqrt{x+7}-2$.

$$\sqrt{2x-3}=\sqrt{x+7}-2$$
$$\left(\sqrt{2x-3}\right)^2=\left(\sqrt{x+7}-2\right)^2$$
$$2x-3=x+7-4\sqrt{x+7}+4$$
$$x-14=-4\sqrt{x+7}$$
$$(x-14)^2=\left(-4\sqrt{x+7}\right)^2$$
$$x^2-28x+196=16(x+7)$$
$$x^2-28x+196=16x+112$$
$$x^2-44x+84=0$$
$$(x-2)(x-42)=0$$
$$x=2 \quad or \quad x=42$$

Since 2 checks but 42 does not, we have $f(x)=g(x)$ when $x=2$.

53. We must have $4-\sqrt{t-3}=(t+5)^{1/2}$.

$$4-\sqrt{t-3}=(t+5)^{1/2}$$
$$\left(4-\sqrt{t-3}\right)^2=\left[(t+5)^{1/2}\right]^2$$
$$16-8\sqrt{t-3}+t-3=t+5$$
$$-8\sqrt{t-3}=-8$$
$$\sqrt{t-3}=1$$
$$\left(\sqrt{t-3}\right)^2=1^2$$
$$t-3=1$$
$$t=4$$

The number 4 checks, so we have $f(t)=g(t)$ when $t=4$.

55. *Writing Exercise.* For an even power n, $a^n=(-a)^n$, but $a\neq-a$ (for $a\neq0$). Thus, we must check the possible solutions found when using the principle of powers.

57. *Familiarize.* Let $w=$ the width of the rectangle, in feet. Then $13w+5$ is the length. Recall the formula for the perimeter of a rectangle with width w and length l is $P=2w+2l$.

Translate. Substitute in the formula.

$$430=2w+2(13w+5)$$

Carry out. We solve the equation.

$$430=2w+2(13w+5)$$
$$430=2w+26w+10$$
$$430=28w+10$$
$$420=28w$$
$$15=w$$

Check. If the width is 15 ft, then the length is $13\cdot15+5$, or 200 ft, and the perimeter is $2\cdot15+2\cdot200$, or 430 ft. The answer checks.

State. The width is 15 ft and the length is 200 ft.

59. *Familiarize.* Let w represent the width of the photo, in inches. Then $w+4$ represents the length. Recall the formula for the area of a rectangle with width w and length l is $A=\text{width}\times\text{length}$.

Translate. The area is 140 in^2.

$$w(w+4)=140$$

Carry out. We solve the equation.

$$w(w+4)=140$$
$$w^2+4w=140$$
$$w^2+4w-140=0$$
$$(w-10)(w+14)=0$$
$$w-10=0 \quad or \quad w+14=0$$
$$w=10 \quad or \quad \quad w=-14$$

Check. We check only 10, since width cannot be negative. If the width is 10 in., the length is $10+4$, or 14 in., and the area is $10\cdot14$, or 140 in^2. We have a solution.

State. The width is 10 in. and the length is 14 in.

61. *Familiarize.* Let x represent the first integer, $x+2$ the second, and $x+4$ the third.

Translate. We use the Pythagorean theorem.

$$x^2+(x+2)^2=(x+4)^2$$

Carry out. We solve the equation.

$$x^2+(x+2)^2=(x+4)^2$$
$$x^2+x^2+4x+4=x^2+8x+16$$
$$2x^2+4x+4=x^2+8x+16$$
$$x^2-4x-12=0$$
$$(x-6)(x+2)=0$$
$$x-6=0 \quad or \quad x+2=0$$
$$x=6 \quad or \quad \quad x=-2$$

Check. We check only 6, since the side cannot be negative. The integers are 6, 8, and 10. The square of 6, or 36 plus the square of 8, or 64, is the square of 10, or 100.

State. The sides are 6, 8, and 10.

63. *Writing Exercise.* Answers may vary. One procedure is to set a radical expression with an even index equal to a negative number This is equivalent to writing an equation similar to those in Exercises 27 and 28.

65. Substitute 100 for $v(p)$ and solve for p.

$$v(p) = 12.1\sqrt{p}$$
$$100 = 12.1\sqrt{p}$$
$$8.2645 \approx \sqrt{p}$$
$$(8.2645)^2 \approx \left(\sqrt{p}\right)^2$$
$$68.3013 \approx p$$

The nozzle pressure is about 68 psi.

67. Let f be the frequency of the string and t be the tension of the string. Substitute 260 for f, 28 for t, and solve for k, the constant of variation.

$$f = k\sqrt{t}$$
$$260 = k\sqrt{28}$$
$$k = \frac{260}{\sqrt{28}} \approx 49.135$$

Then substitute 32 for t and solve for f.

$$f = 49.135\sqrt{t}$$
$$f = 49.135\sqrt{32} \approx 277.952$$

The frequency is about 278 Hz.

69. Substitute 1880 for $S(t)$ and solve for t.

$$1880 = 1087.7\sqrt{\frac{9t + 2617}{2457}}$$
$$1.7284 \approx \sqrt{\frac{9t + 2617}{2457}} \quad \text{Dividing by 1087.7}$$
$$(1.7284)^2 \approx \left(\sqrt{\frac{9t + 2617}{2457}}\right)^2$$
$$2.9874 \approx \frac{9t + 2617}{2457}$$
$$7340.0418 \approx 9t + 2617$$
$$4723.0418 \approx 9t$$
$$524.7824 \approx t$$

The temperature is about 524.8°C.

71.
$$S = 1087.7\sqrt{\frac{9t + 2617}{2457}}$$
$$\frac{S}{1087.7} = \sqrt{\frac{9t + 2617}{2457}}$$
$$\left(\frac{S}{1087.7}\right)^2 = \left(\sqrt{\frac{9t + 2617}{2457}}\right)^2$$
$$\frac{S^2}{1087.7^2} = \frac{9t + 2617}{2457}$$
$$\frac{2457S^2}{1087.7^2} = 9t + 2617$$
$$\frac{2457S^2}{1087.7^2} - 2617 = 9t$$
$$\frac{1}{9}\left(\frac{2457S^2}{1087.7^2} - 2617\right) = t$$

73. $d(n) = 0.75\sqrt{2.8n}$

Substitute 84 for $d(n)$ and solve for n.

$$84 = 0.75\sqrt{2.8n}$$
$$112 = \sqrt{2.8n}$$
$$(112)^2 = \left(\sqrt{2.8n}\right)^2$$
$$12{,}544 = 2.8n$$
$$4480 = n$$

About 4480 rpm will produce peak performance.

75.
$$v = \sqrt{2gr}\sqrt{\frac{h}{r + h}}$$
$$v^2 = 2gr \cdot \frac{h}{r + h} \quad \text{Squaring both sides}$$
$$v^2(r + h) = 2grh \quad \text{Multiplying by } r + h$$
$$v^2 r + v^2 h = 2grh$$
$$v^2 h = 2grh - v^2 r$$
$$v^2 h = r\left(2gh - v^2\right)$$
$$\frac{v^2 h}{2gh - v^2} = r$$

77.
$$\frac{x + \sqrt{x + 1}}{x - \sqrt{x + 1}} = \frac{5}{11}$$
$$11\left(x + \sqrt{x + 1}\right) = 5\left(x - \sqrt{x + 1}\right)$$
$$11x + 11\sqrt{x + 1} = 5x - 5\sqrt{x + 1}$$
$$16\sqrt{x + 1} = -6x$$
$$8\sqrt{x + 1} = -3x$$
$$\left(8\sqrt{x + 1}\right)^2 = (-3x)^2$$
$$64(x + 1) = 9x^2$$
$$64x + 64 = 9x^2$$
$$0 = 9x^2 - 64x - 64$$
$$0 = (9x + 8)(x - 8)$$

$$9x + 8 = 0 \quad \text{or} \quad x - 8 = 0$$
$$9x = -8 \quad \text{or} \quad x = 8$$
$$x = -\tfrac{8}{9} \quad \text{or} \quad x = 8$$

Since $-\tfrac{8}{9}$ checks but 8 does not, the solution is $-\tfrac{8}{9}$.

79.
$$\left(z^2+17\right)^{3/4}=27$$
$$\left[\left(z^2+17\right)^{3/4}\right]^{4/3}=\left(3^3\right)^{4/3}$$
$$z^2+17=3^4$$
$$z^2+17=81$$
$$z^2-64=0$$
$$(z+8)(z-8)=0$$
$$z=-8 \quad\text{or}\quad z=8$$

Both -8 and 8 check. They are the solutions.

81.
$$\sqrt{8-b}=b\sqrt{8-b}$$
$$\left(\sqrt{8-b}\right)^2=\left(b\sqrt{8-b}\right)^2$$
$$(8-b)=b^2(8-b)$$
$$0=b^2(8-b)-(8-b)$$
$$0=(8-b)(b^2-1)$$
$$0=(8-b)(b+1)(b-1)$$
$$8-b=0 \quad\text{or}\quad b+1=0 \quad\text{or}\quad b-1=0$$
$$8=b \quad\text{or}\quad b=-1 \quad\text{or}\quad b=1$$

Since the numbers 8 and 1 check but -1 does not, 8 and 1 are the solutions.

83. We find the values of x for which $g(x)=0$.
$$6x^{1/2}+6x^{-1/2}-37=0$$
$$6\sqrt{x}+\frac{6}{\sqrt{x}}=37$$
$$\left(6\sqrt{x}+\frac{6}{\sqrt{x}}\right)^2=37^2$$
$$36x+72+\frac{36}{x}=1369$$
$$36x^2+72x+36=1369x \quad \text{Multiplying by } x$$
$$36x^2-1297x+36=0$$
$$(36x-1)(x-36)=0$$
$$36x-1=0 \quad\text{or}\quad x-36=0$$
$$36x=1 \quad\text{or}\quad x=36$$
$$x=\tfrac{1}{36} \quad\text{or}\quad x=36$$

Both numbers check. The x-intercepts are $\left(\frac{1}{36},\,0\right)$ and $(36,\,0)$.

85. *Graphing Calculator Exercise*

87. *Graphing Calculator Exercise*

Exercise Set 7.7

1. The correct choice is (d) Right; see page 473 in the text.

3. The correct choice is (e) Square roots; see page 473 in the text.

5. The correct choice is (f) 30°-60°-90°; see page 476 in the text.

7. $a=5$, $b=3$

Find c.
$$c^2=a^2+b^2 \quad \text{Pythagorean theorem}$$
$$c^2=5^2+3^2 \quad \text{Substituting}$$
$$c^2=25+9$$
$$c^2=34$$
$$c=\sqrt{34} \quad \text{Exact answer}$$
$$c\approx5.831 \quad \text{Approximation}$$

9. $a=9$, $b=9$

Observe that the legs have the same length, so this is an isosceles right triangle. Then we know that the length of the hypotenuse is the length of a leg times $\sqrt{2}$, or $9\sqrt{2}$, or approximately 12.728.

11. $b=15$, $c=17$

Find a.
$$a^2+b^2=c^2 \quad \text{Pythagorean theorem}$$
$$a^2+15^2=17^2 \quad \text{Substituting}$$
$$a^2+225=289$$
$$a^2=64$$
$$a=8$$

13.
$$a^2+b^2=c^2 \quad \text{Pythagorean theorem}$$
$$\left(4\sqrt{3}\right)^2+b^2=8^2$$
$$16\cdot3+b^2=64$$
$$48+b^2=64$$
$$b^2=16$$
$$b=4$$

The other leg is 4 m long.

15.
$$a^2+b^2=c^2 \quad \text{Pythagorean theorem}$$
$$1^2+b^2=\left(\sqrt{20}\right)^2 \quad \text{Substituting}$$
$$1+b^2=20$$
$$b^2=19$$
$$b=\sqrt{19}$$
$$b\approx4.359$$

The length of the other leg is $\sqrt{19}$ in., or about 4.359 in.

17. Observe that the length of the hypotenuse, $\sqrt{2}$, is $\sqrt{2}$ times the length of the given leg, 1 m. Thus, we have an isosceles right triangle and the length of the other leg is also 1 m.

19. From the drawing in the text we see that we have a right triangle with legs of 150 ft and 200 ft. Let $d=$ the length of the diagonal, in feet. We use the Pythagorean theorem to find d.

$$150^2 + 200^2 = d^2$$
$$22{,}500 + 40{,}000 = d^2$$
$$62{,}500 = d^2$$
$$250 = d$$

Clare travels 250 ft across the parking lot.

21. We make a drawing and let $d=$ the distance from home plate to second base.

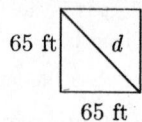

65 ft

d

65 ft

Note that we have an isosceles right triangle. Then the length of the hypotenuse is the length of a leg times $\sqrt{2}$, or $65\sqrt{2}$ ft. This is about 91.924 ft.

(We could also have used the Pythagorean theorem, solving $65^2 + 65^2 = d^2$.)

23. We make a drawing similar to the one in the text.

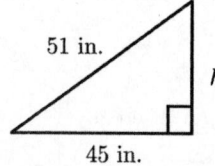

51 in.

h

45 in.

We use the Pythagorean theorem to find h.

$$45^2 + h^2 = 51^2$$
$$2051 + h^2 = 2601$$
$$h^2 = 576$$
$$h = 24$$

The height of the screen is 24 in.

25. First we will find the diagonal distance, d, in feet, across the room. We make a drawing.

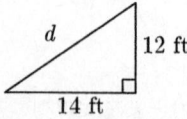

d

12 ft

14 ft

Now we use the Pythagorean theorem.

$$12^2 + 14^2 = d^2$$
$$144 + 196 = d^2$$
$$340 = d^2$$
$$\sqrt{340} = d$$
$$18.439 \approx d$$

Recall that 4 ft of slack is required on each end. Thus, $\sqrt{340} + 2 \cdot 4$, or $(\sqrt{340} + 8)$ ft, of wire should be purchased. This is about 26.439 ft.

27. The diagonal is the hypotenuse of a right triangle with legs of 70 paces and 40 paces. First we use the Pythagorean theorem to find the length d of the diagonal, in paces.

$$70^2 + 40^2 = d^2$$
$$4900 + 1600 = d^2$$
$$6500 = d^2$$
$$\sqrt{6500} = d$$
$$80.623 \approx d$$

If Marissa walks along two sides of the quad she takes $70 + 40$, or 110 paces. Then by using the diagonal she saves $(110 - \sqrt{6500})$ paces. This is approximately $110 - 80.623$, or 29.377 paces.

29. Since one acute angle is $45°$, this is an isosceles right triangle with one leg $= 5$. Then the other leg $= 5$ also. And the hypotenuse is the length of the a leg times $\sqrt{2}$, or $5\sqrt{2}$..

Exact answer: Leg $= 5$, hypotenuse $= 5\sqrt{2}$

Approximation: hypotenuse ≈ 7.071

31. This is a 30-60-90 right triangle with hypotenuse 14. We find the legs:

$2a = 14$, so $a = 7$ and $a\sqrt{3} = 7\sqrt{3}$

Exact answer: shorter leg $= 7$; longer leg $= 7\sqrt{3}$

Approximation: longer leg ≈ 12.124

33. This is a 30-60-90 right triangle with one leg $= 15$. We substitute to find the length of the other leg, a, and the hypotenuse, c.

$$b = a\sqrt{3}$$
$$15 = a\sqrt{3}$$
$$\frac{15}{\sqrt{3}} = a$$
$$\frac{15\sqrt{3}}{3} = a \quad \text{Rationalizing the denominator}$$
$$5\sqrt{3} = a \quad \text{Simplifying}$$
$$c = 2a$$
$$c = 2 \cdot 5\sqrt{3}$$
$$c = 10\sqrt{3}$$

Exact answer: $a = 5\sqrt{3}, c = 10\sqrt{3}$

Approximations: $a \approx 8.660, c \approx 17.321$

35. This is an isosceles right triangle with hypotenuse 13. The two legs have the same length, a.

$$a\sqrt{2} = 13$$
$$a = \frac{13}{\sqrt{2}} = \frac{13\sqrt{2}}{2}$$

Exact answer: $\dfrac{13\sqrt{2}}{2}$

Approximation: 9.192

37. This is a 30-60-90 triangle with the shorter leg $= 14$. We find the longer leg and the hypotenuse.

$a\sqrt{3} = 14\sqrt{3}$, and $2a = 2 \cdot 14 = 28$.

Exact answer: longer leg $= 14\sqrt{3}$, hypotenuse $= 28$

Approximation: longer leg ≈ 24.249

39. h is the longer leg of a 30-60-90 right triangle with shorter leg $= 5$. Then $h = 5\sqrt{3} \approx 8.660$.

41. We make a drawing.

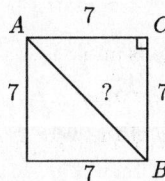

Triangle ABC is an isosceles right triangle with legs of length 7. Then the hypotenuse $= 7\sqrt{2} \approx 9.899$.

43. We make a drawing.

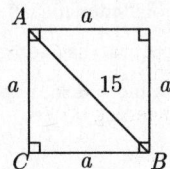

Triangle ABC is an isosceles right triangle with hypotenuse $= 15$. Then $a = \dfrac{15\sqrt{2}}{2} \approx 10.607$.

45. We will express all distances in feet. Recall that $1\,\text{mi} = 5280\,\text{ft}$.

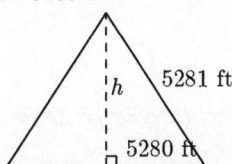

We use the Pythagorean theorem to find h.

$$h^2 + (5280)^2 = (5281)^2$$
$$h^2 + 27,878,400 = 27,888,961$$
$$h^2 = 10,561$$
$$h = \sqrt{10,561}$$
$$h \approx 102.767$$

The height of the bulge is $\sqrt{10,561}$ ft, or about 102.767 ft.

47. We make a drawing.

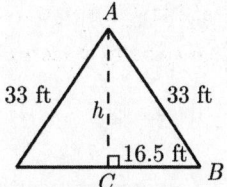

The base of the lodge is an equilateral triangle, so all the angles are 60°. The altitude bisects one angle and one side. Then the triangle ABC is a 30°-60°-90° right triangle with the shorter leg of length $\frac{33}{2}$, or 16.5 ft, and hypotenuse of length 33. Then the height is the length of the shorter leg times $\sqrt{3}$.

Exact answer: $h = \dfrac{33\sqrt{3}}{2}$ ft

Approximation: $h \approx 28.579$ ft

If the height of triangle ABC is $\frac{33\sqrt{3}}{2}$ and the base is 33 ft, the area is $\frac{1}{2} \cdot 33 \cdot \frac{33\sqrt{3}}{2} = \frac{1089}{4}\sqrt{3}$ ft^2, or about 471.551 ft^2.

49. We make a drawing.

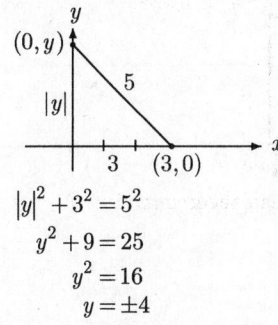

$$|y|^2 + 3^2 = 5^2$$
$$y^2 + 9 = 25$$
$$y^2 = 16$$
$$y = \pm 4$$

The points are $(0, -4)$ and $(0, 4)$.

51. Using the distance formula $d = \sqrt{(x_2 - x_1)^2 + (y_2 - y_1)^2}$ for the points $(4, 5)$ and $(7, 1)$,

$$d = \sqrt{(7-4)^2 + (1-5)^2}$$
$$= \sqrt{3^2 + (-4)^2} = \sqrt{9+16} = \sqrt{25}$$
$$= 5$$

53. Using the distance formula $d = \sqrt{(x_2 - x_1)^2 + (y_2 - y_1)^2}$ for the points $(1, -2)$ and $(0, -5)$,

$$d = \sqrt{(1-0)^2 + (-2-(-5))^2}$$
$$= \sqrt{1^2 + 3^2} = \sqrt{1+9} = \sqrt{10}$$
$$\approx 3.162$$

55. Using the distance formula $d = \sqrt{(x_2 - x_1)^2 + (y_2 - y_1)^2}$

for the points $(6, -6)$ and $(-4, 4)$,

$$d = \sqrt{(6-(-4))^2 + (-6-4)^2}$$
$$= \sqrt{10^2 + (-10)^2} = \sqrt{100 + 100} = \sqrt{200}$$
$$\approx 14.142$$

57. Using the distance formula $d = \sqrt{(x_2 - x_1)^2 + (y_2 - y_1)^2}$

for the points $(-9.2, -3.4)$ and $(8.6, -3.4)$,

$$d = \sqrt{(-9.8 - 8.6)^2 + (-3.4 - (-3.4))^2}$$
$$= \sqrt{(-17.8)^2 + 0^2} = \sqrt{316.84}$$
$$\approx 17.8$$

59. Using the distance formula $d = \sqrt{(x_2 - x_1)^2 + (y_2 - y_1)^2}$

for the points $\left(\frac{5}{6}, -\frac{1}{6}\right)$ and $\left(\frac{1}{2}, \frac{1}{3}\right)$,

$$d = \sqrt{\left(\frac{5}{6} - \frac{1}{2}\right)^2 + \left(-\frac{1}{6} - \frac{1}{3}\right)^2}$$
$$= \sqrt{\left(\frac{2}{6}\right)^2 + \left(-\frac{3}{6}\right)^2} = \sqrt{\frac{4}{36} + \frac{9}{36}} = \sqrt{\frac{13}{36}} = \frac{\sqrt{13}}{6}$$
$$\approx 0.601$$

61. Using the distance formula $d = \sqrt{(x_2 - x_1)^2 + (y_2 - y_1)^2}$

for the points $(0, 0)$ and $\left(-\sqrt{6}, \sqrt{6}\right)$,

$$d = \sqrt{\left(0 - (-\sqrt{6})\right)^2 + (0 - \sqrt{6})^2}$$
$$= \sqrt{(\sqrt{6})^2 + (-\sqrt{6})^2} = \sqrt{6 + 6} = \sqrt{12}$$
$$\approx 3.464$$

63. Using the distance formula $d = \sqrt{(x_2 - x_1)^2 + (y_2 - y_1)^2}$

for the points $(-2, -40)$ and $(-1, -30)$,

$$d = \sqrt{(-2 - (-1))^2 + (-40 - (-30))^2}$$
$$= \sqrt{(-1)^2 + (-10)^2} = \sqrt{1 + 100} = \sqrt{101}$$
$$\approx 10.050$$

65. Using the midpoint formula $\left(\frac{x_1 + x_2}{2}, \frac{y_1 + y_2}{2}\right)$ for the

points $(-2, 5)$ and $(8, 3)$,

$\left(\frac{-2 + 8}{2}, \frac{5 + 3}{2}\right)$, or $\left(\frac{6}{2}, \frac{8}{2}\right)$, or $(3, 4)$

67. Using the midpoint formula $\left(\frac{x_1 + x_2}{2}, \frac{y_1 + y_2}{2}\right)$ for the

points $(2, -1)$ and $(5, 8)$,

$\left(\frac{2 + 5}{2}, \frac{-1 + 8}{2}\right)$, or $\left(\frac{7}{2}, \frac{7}{2}\right)$

69. Using the midpoint formula $\left(\frac{x_1 + x_2}{2}, \frac{y_1 + y_2}{2}\right)$ for the

points $(-8, -5)$ and $(6, -1)$,

$\left(\frac{-8 + 6}{2}, \frac{-5 + (-1)}{2}\right)$, or $\left(-\frac{2}{2}, \frac{-6}{2}\right)$, or $(-1, -3)$

71. Using the midpoint formula $\left(\frac{x_1 + x_2}{2}, \frac{y_1 + y_2}{2}\right)$ for the

points $(-3.4, 8.1)$ and $(4.8, -8.1)$,

$\left(\frac{-3.4 + 4.8}{2}, \frac{8.1 + (-8.1)}{2}\right)$, or $\left(\frac{1.4}{2}, \frac{0}{2}\right)$, or $(0.7, 0)$

73. Using the midpoint formula $\left(\frac{x_1 + x_2}{2}, \frac{y_1 + y_2}{2}\right)$ for the

points $\left(\frac{1}{6}, -\frac{3}{4}\right)$ and $\left(-\frac{1}{3}, \frac{5}{6}\right)$,

$\left(\frac{\frac{1}{6} + \left(-\frac{1}{3}\right)}{2}, \frac{-\frac{3}{4} + \frac{5}{6}}{2}\right)$, or $\left(\frac{-\frac{1}{6}}{2}, \frac{\frac{1}{12}}{2}\right)$, or $\left(-\frac{1}{12}, \frac{1}{24}\right)$

75. Using the midpoint formula $\left(\frac{x_1 + x_2}{2}, \frac{y_1 + y_2}{2}\right)$ for the

points $\left(\sqrt{2}, -1\right)$ and $\left(\sqrt{3}, 4\right)$,

$\left(\frac{\sqrt{2} + \sqrt{3}}{2}, \frac{-1 + 4}{2}\right)$, or $\left(\frac{\sqrt{2} + \sqrt{3}}{2}, \frac{3}{2}\right)$

77. *Writing Exercise.* If a right triangle has consecutive numbers for the lengths of its sides, then a, $a+1$, and $a+2$ represent the lengths of the sides and these numbers must satisfy the equation $a^2 + (a+1)^2 = (a+2)^2$. The solutions of this equation are -1 and 3. Since the length of a side cannot be negative, the only possible sides measure 3, 4, and 5. Thus, the only right triangle that has consecutive numbers for the lengths of its sides has sides measuring 3, 4, and 5.

79. $y = 2x - 3$

Slope is 2; y-intercept is $(0, -3)$.

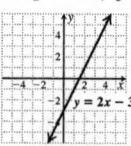

81. $8x - 4y = 8$

To find the y-intercept, let $x = 0$ and solve for y.

$$8 \cdot 0 - 4y = 8$$
$$y = -2$$

The y-intercept is $(0, -2)$.

To find the x-intercept, let $y = 0$ and solve for x.

$$8x - 4 \cdot 0 = 8$$
$$x = 1$$

The x-intercept is $(1, 0)$.

Plot these points and draw the line. A third point could be used as a check.

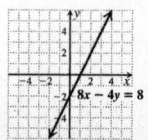

83. $x \geq 1$

Graph the line $x = 1$. Draw the line solid since the inequality symbol is $\geq$. Test the point $(0, 0)$ to determine if it is a solution.

$$x \geq 1$$
$$\overline{0 \;\; 1}$$
$$\overset{?}{0 \geq 1} \quad \text{FALSE}$$

Since $0 \geq 1$ is false, we shade the half plane that does not contains $(0, 0)$ and obtain the graph.

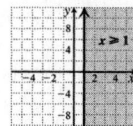

85. *Writing Exercise.* If $P_1(x_1,\ y_1)$, $P_2(x_2,\ y_2)$, and $P_3(x_3,\ y_3)$ are vertices of a right triangle, the distance formula could be used to find the lengths of the legs and the length of the hypotenuse. Then the right triangle can be verified by substituting the lengths for a, b, and c, in the Pythagorean theorem.

87. The length of a side of the hexagon is 72/6, or 12 cm. Then the shaded region is a triangle with base 12 cm. To find the height of the triangle, note that it is the longer leg of a 30°-60°-90° right triangle. Thus its length is the length of the length of the shorter leg times $\sqrt{3}$. The length of the shorter leg is half the length of the base, $\frac{1}{2} \cdot 12$ cm, or 6 cm, so the length of the longer leg is $6\sqrt{3}$ cm. Now we find the area of the triangle.

$$A = \frac{1}{2}bh$$
$$= \frac{1}{2}(12 \text{ cm})\left(6\sqrt{3} \text{ cm}\right)$$
$$= 36\sqrt{3} \text{ cm}^2$$
$$\approx 62.354 \text{ cm}^2$$

89. We make a drawing.

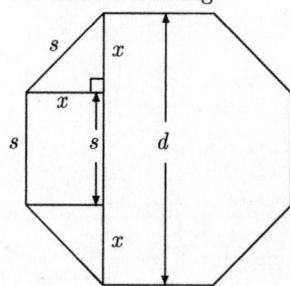

$d = s + 2x$

Use the Pythagoran theorem to find x.

$$x^2 + x^2 = s^2$$
$$2x^2 = s^2$$
$$x^2 = \frac{s^2}{2}$$
$$x = \frac{s}{\sqrt{2}} = \frac{s}{\sqrt{2}} \cdot \frac{\sqrt{2}}{2} = \frac{s\sqrt{2}}{2}$$

Then $d = s + 2x = s + 2\left(\frac{s\sqrt{2}}{2}\right) = s + s\sqrt{2}$.

91. We make a drawing.

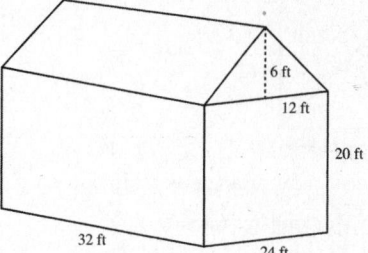

The area to be painted consists of two 20 ft by 24 ft rectangles, two 20 ft by 32 ft rectangles, and two triangles with height 6 ft and base 24 ft. The area of the two 20 ft by 24 ft rectangles is $2 \cdot 20 \text{ ft} \cdot 24 \text{ ft} = 960 \text{ ft}^2$. The area of the two 20 ft by 32 ft rectangles is $2 \cdot 20 \text{ ft} \cdot 32 \text{ ft} = 1280 \text{ ft}^2$. The area of the two triangles is $2 \cdot \frac{1}{2} \cdot 24 \text{ ft} \cdot 6 \text{ ft} = 144 \text{ ft}^2$. Thus, the total area to be painted is $960 \text{ ft}^2 + 1280 \text{ ft}^2 + 144 \text{ ft}^2 = 2384 \text{ ft}^2$.

One gallon of paint covers a minimum of 450 ft^2, so we divide to determine how many gallons of paint are required: $\frac{2384}{450} \approx 5.3$. Thus, 5 gallons of paint should be bought to paint the house. This answer assumes that the total area of the doors and windows is 134 ft^2 or more. ($5 \cdot 450 = 2250$ and $2384 = 2250 + 134$)

93. First we find the radius of a circle with an area of 6160 ft^2. This is the length of the hose.

$$A = \pi r^2$$
$$6160 = \pi r^2$$
$$\frac{6160}{\pi} = r^2$$
$$\sqrt{\frac{6160}{\pi}} = r$$
$$44.28 \approx r$$

Now we make a drawing of the room.

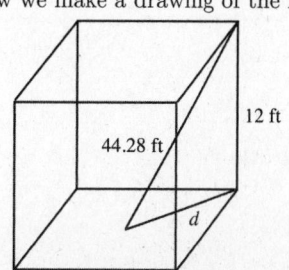

We use the Pythagorean theorem to find d.

$$d^2 + 12^2 = 44.28^2$$
$$d^2 + 144 = 1960.7184$$
$$d^2 = 1816.7184$$
$$d \approx 42.623$$

Now we make a drawing of the floor of the room.

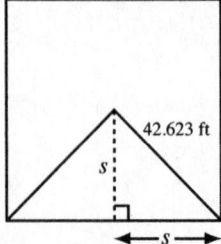

We have an isosceles right triangle with hypotenuse 42.623 ft. We find the length of a side s.

$$s\sqrt{2} = 42.623$$
$$s = \frac{42.623}{\sqrt{2}} \approx 30.14 \text{ ft}$$

Then the length of a side of the room is

$2s = 2(30.14 \text{ ft}) = 60.28$ ft; so the dimensions of the largest square room that meets the given conditions is 60.28 ft by 60.28 ft.

95. We make a drawing.

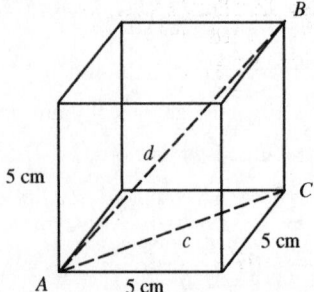

First find the length of a diagonal of the base of the cube. It is the hypotenuse of an isosceles right triangle with legs 5 cm. Then $c = 5\sqrt{2}$ cm.

Triangle ABC is a right triangle with legs of $5\sqrt{2}$ cm and 5 cm and hypotenuse d. Use the Pythagorean theorem to find d, the length of the diagonal that connects two opposite corners of the cube.

$$d^2 = (5\sqrt{2})^2 + 5^2$$
$$d^2 = 25 \cdot 2 + 25$$
$$d^2 = 50 + 25$$
$$d^2 = 75$$
$$d = \sqrt{75}$$

Exact answer: $d = \sqrt{75}$ cm

Exercise Set 7.8

1. False; see page 484 in the text.

3. True; see page 485 in the text.

5. True; see page 486 in the text.

7. False; see Exercises 59-64.

9. $\sqrt{-100} = \sqrt{-1 \cdot 100} = \sqrt{-1} \cdot \sqrt{100} = i \cdot 10 = 10i$

11. $\sqrt{-5} = \sqrt{-1 \cdot 5} = \sqrt{-1} \cdot \sqrt{5} = i \cdot \sqrt{5}$, or $\sqrt{5}\,i$

13. $\sqrt{-8} = \sqrt{-1} \cdot \sqrt{4} \cdot \sqrt{2} = i \cdot 2 \cdot \sqrt{2} = 2i\sqrt{2}$, or $2\sqrt{2}\,i$

15. $-\sqrt{-11} = -\sqrt{-1} \cdot \sqrt{11} = -i \cdot \sqrt{11} = -i\sqrt{11}$, or $-\sqrt{11}\,i$

17. $-\sqrt{-49} = -\sqrt{-1 \cdot 49} = -\sqrt{-1} \cdot \sqrt{49} = -i \cdot 7 = -7i$

19. $-\sqrt{-300} = -\sqrt{-1} \cdot \sqrt{100} \cdot \sqrt{3} = -i \cdot 10 \cdot \sqrt{3} = -10i\sqrt{3}$, or $-10\sqrt{3}i$

21. $6 - \sqrt{-84} = 6 - \sqrt{-1 \cdot 4 \cdot 21} = 6 - i \cdot 2\sqrt{21} = 6 - 2i\sqrt{21}$, or $6 - 2\sqrt{21}i$

23. $-\sqrt{-76} + \sqrt{-125} = -\sqrt{-1 \cdot 4 \cdot 19} + \sqrt{-1 \cdot 25 \cdot 5}$
$= -i \cdot 2\sqrt{19} + i \cdot 5\sqrt{5} = -2i\sqrt{19} + 5i\sqrt{5} = (-2\sqrt{19} + 5\sqrt{5})i$

25. $\sqrt{-18} - \sqrt{-64} = \sqrt{-1 \cdot 9 \cdot 2} - \sqrt{-1 \cdot 64}$
$= i \cdot 3 \cdot \sqrt{2} - i \cdot 8 = 3i\sqrt{2} - 8i$, or $(3\sqrt{2} - 8)i$

27. $(3 + 4i) + (2 - 7i)$
$= (3 + 2) + (4 - 7)i$ Combining the real and
                     the imaginary parts
$= 5 - 3i$

29. $(9 + 5i) - (2 + 3i) = (9 - 2) + (5 - 3)i$
$= 7 + 2i$

31. $(7 - 4i) - (5 - 3i) = (7 - 5) + [-4 - (-3)]i = 2 - i$

33. $(-5 - i) - (7 + 4i) = (-5 - 7) + (-1 - 4)i = -12 - 5i$

35. $5i \cdot 8i = 40 \cdot i^2 = 40(-1) = -40$

37. $(-4i)(-6i) = 24 \cdot i^2 = 24(-1) = -24$

39. $\sqrt{-36}\sqrt{-9} = \sqrt{-1} \cdot \sqrt{36} \cdot \sqrt{-1} \cdot \sqrt{9} = i \cdot 6 \cdot i \cdot 3$
$= i^2 \cdot 18 = -1 \cdot 18 = -18$

41. $\sqrt{-3}\sqrt{-10} = \sqrt{-1} \cdot \sqrt{3} \cdot \sqrt{-1} \cdot \sqrt{10}$
$= i \cdot \sqrt{3} \cdot i \cdot \sqrt{10}$
$= \sqrt{30} \cdot i^2 = \sqrt{30} \cdot (-1)$
$= -\sqrt{30}$

43. $\sqrt{-6}\sqrt{-21} = \sqrt{-1}\cdot\sqrt{6}\cdot\sqrt{-1}\cdot\sqrt{21}$
$\qquad = i\cdot\sqrt{6}\cdot i\cdot\sqrt{21} = i^2\sqrt{126} = -1\cdot\sqrt{9\cdot14}$
$\qquad = -3\sqrt{14}$

45. $5i(2+6i) = 5i\cdot2 + 5i\cdot6i = 10i + 30i^2$
$\qquad = 10i - 30 = -30 + 10i$

47. $-7i(3+4i) = -7i\cdot3 - 7i\cdot4i$
$\qquad = -21i - 28i^2$
$\qquad = -21i + 28 = 28 - 21i$

49. $(1+i)(3+2i) = 3 + 2i + 3i + 2i^2$
$\qquad = 3 + 2i + 3i - 2 = 1 + 5i$

51. $(6-5i)(3+4i) = 18 + 24i - 15i - 20i^2$
$\qquad = 18 + 24i - 15i + 20 = 38 + 9i$

53. $(7-2i)(2-6i) = 14 - 42i - 4i + 12i^2$
$\qquad = 14 - 42i - 4i - 12 = 2 - 46i$

55. $(3+8i)(3-8i) = 3^2 - (8i)^2$ Difference of squares
$\qquad = 9 - 64i^2 = 9 - 64(-1)$
$\qquad = 9 + 64 = 73$

57. $(-7+i)(-7-i) = (-7)^2 - i^2$ Difference of squares
$\qquad = 49 - i^2 = 49 - (-1)$
$\qquad = 49 + 1 = 50$

59. $(4-2i)^2 = 4^2 - 2\cdot4\cdot2i + (2i)^2 = 16 - 16i + 4i^2$
$\qquad = 16 - 16i - 4 = 12 - 16i$

61. $(2+3i)^2 = 2^2 + 2\cdot2\cdot3i + (3i)^2 = 4 + 12i + 9i^2$
$\qquad = 4 + 12i - 9 = -5 + 12i$

63. $(-2+3i)^2 = (-2)^2 + 2(-2)(3i) + (3i)^2$
$\qquad = 4 - 12i + 9i^2 = 4 - 12i - 9 = -5 - 12i$

65. $\dfrac{10}{3+i} = \dfrac{10}{3+i}\cdot\dfrac{3-i}{3-i}$ Multiplying by 1, using the conjugate
$\qquad = \dfrac{30-10i}{9-i^2}$
$\qquad = \dfrac{30-10i}{9-(-1)}$
$\qquad = \dfrac{30-10i}{10}$
$\qquad = 3 - i$

67. $\dfrac{2}{3-2i} = \dfrac{2}{3-2i}\cdot\dfrac{3+2i}{3+2i}$ Multiplying by 1, using the conjugate
$\qquad = \dfrac{6+4i}{9-4i^2}$
$\qquad = \dfrac{6+4i}{9-4(-1)}$
$\qquad = \dfrac{6+4i}{13}$
$\qquad = \dfrac{6}{13} + \dfrac{4}{13}i$

69. $\dfrac{2i}{5+3i} = \dfrac{2i}{5+3i}\cdot\dfrac{5-3i}{5-3i} = \dfrac{10i-6i^2}{25-9i^2} = \dfrac{10i+6}{25+9}$
$\qquad = \dfrac{10i+6}{34} = \dfrac{6}{34} + \dfrac{10}{34}i = \dfrac{3}{17} + \dfrac{5}{17}i$

71. $\dfrac{5}{6i} = \dfrac{5}{6i}\cdot\dfrac{i}{i} = \dfrac{5i}{6i^2} = \dfrac{5i}{-6} = -\dfrac{5}{6}i$

73. $\dfrac{5-3i}{4i} = \dfrac{5-3i}{4i}\cdot\dfrac{i}{i} = \dfrac{5i-3i^2}{4i^2} = \dfrac{5i+3}{-4}$
$\qquad = -\dfrac{3}{4} - \dfrac{5}{4}i$

75. $\dfrac{7i+14}{7i} = \dfrac{7i}{7i} + \dfrac{14}{7i} = 1 + \dfrac{2}{i} = 1 + \dfrac{2}{i}\cdot\dfrac{i}{i}$
$\qquad = 1 + \dfrac{2i}{i^2} = 1 + \dfrac{2i}{-1} = 1 - 2i$

77. $\dfrac{4+5i}{3-7i} = \dfrac{4+5i}{3-7i}\cdot\dfrac{3+7i}{3+7i} = \dfrac{12+28i+15i+35i^2}{9-49i^2}$
$\qquad = \dfrac{12+28i+15i-35}{9+49} = \dfrac{-23+43i}{58}$
$\qquad = -\dfrac{23}{58} + \dfrac{43}{58}i$

79. $\dfrac{2+3i}{2+5i} = \dfrac{2+3i}{2+5i}\cdot\dfrac{2-5i}{2-5i} = \dfrac{4-10i+6i-15i^2}{4-25i^2}$
$\qquad = \dfrac{4-10i+6i+15}{4+25} = \dfrac{19-4i}{29}$
$\qquad = \dfrac{19}{29} - \dfrac{4}{29}i$

81. $\dfrac{3-2i}{4+3i} = \dfrac{3-2i}{4+3i}\cdot\dfrac{4-3i}{4-3i} = \dfrac{12-9i-8i+6i^2}{16-9i^2}$
$\qquad = \dfrac{12-9i-8i-6}{16+9} = \dfrac{6-17i}{25}$
$\qquad = \dfrac{6}{25} - \dfrac{17}{25}i$

83. $i^{32} = (i^2)^{16} = (-1)^{16} = 1$

85. $i^{15} = i^{14}\cdot i = (i^2)^7\cdot i = (-1)^7\cdot i = -i$

87. $i^{42} = (i^2)^{21} = (-1)^{21} = -1$

89. $i^9 = (i^2)^4\cdot i = (-1)^4\cdot i = 1\cdot i = i$

91. $(-i)^6 = (-1\cdot i)^6 = (-1)^6\cdot i^6 = 1\cdot i^6 = (i^2)^3 = (-1)^3 = -1$

93. $(5i)^3 = 5^3\cdot i^3 = 125\cdot i^2\cdot i = 125(-1)(i) = -125i$

95. $i^2 + i^4 = -1 + (i^2)^2 = -1 + (-1)^2 = -1 + 1 = 0$

97. *Writing Exercise.* No; the product of two pure imaginary numbers is a real number.

99. $x^2 - x - 6 = 0$
$(x + 2)(x - 3) = 0$
$x + 2 = 0 \quad or \quad x - 3 = 0$
$x = -2 \quad or \quad x = 3$

The solutions are –2 and 3.

101. $t^2 = 100$
$t^2 - 100 = 0$
$(t + 10)(t - 10) = 0$
$t + 10 = 0 \quad or \quad t - 10 = 0$
$t = -10 \quad or \quad t = 10$

The solutions are –10 and 10.

103. $15x^2 = 14x + 8$
$15x^2 - 14x - 8 = 0$
$(5x + 2)(3x - 4) = 0$
$5x + 2 = 0 \quad or \quad 3x - 4 = 0$
$x = -\frac{2}{5} \quad or \quad x = \frac{4}{3}$

The solutions are $-\frac{2}{5}$ and $\frac{4}{3}$.

105. *Writing Exercise.* Yes; every real number a is a complex number $a + bi$ with $b = 0$.

107.

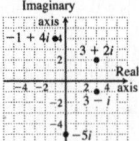

109. $|3 + 4i| = \sqrt{3^2 + 4^2} = \sqrt{9 + 16} = \sqrt{25} = 5$

111. $|-1 + i| = \sqrt{(-1)^2 + 1^2} = \sqrt{1 + 1} = \sqrt{2}$

113. $g(3i) = \dfrac{(3i)^4 - (3i)^2}{3i - 1} = \dfrac{81i^4 - 9i^2}{-1 + 3i} = \dfrac{81 + 9}{-1 + 3i}$
$= \dfrac{90}{-1 + 3i} = \dfrac{90}{-1 + 3i} \cdot \dfrac{-1 - 3i}{-1 - 3i} = \dfrac{90(-1 - 3i)}{1 - 9i^2}$
$= \dfrac{90(-1 - 3i)}{1 + 9} = \dfrac{90(-1 - 3i)}{10} = \dfrac{9 \cdot 10(-1 - 3i)}{10}$
$= 9(-1 - 3i) = -9 - 27i$

115. First we simplify $g(z)$.

$g(z) = \dfrac{z^4 - z^2}{z - 1} = \dfrac{z^2(z^2 - 1)}{z - 1} = \dfrac{z^2(z + 1)(z - 1)}{z - 1}$
$= \dfrac{z^2(z + 1)(z - 1)}{z - 1} = z^2(z + 1)$

Now we substitute.

$g(5i - 1) = (5i - 1)^2(5i - 1 + 1)$
$= (25i^2 - 10i + 1)(5i)$
$= (-25 - 10i + 1)(5i) = (-24 - 10i)(5i)$
$= -120i - 50i^2 = 50 - 120i$

117. $\dfrac{1}{w - w^2} = \dfrac{1}{\frac{1-i}{10} - \left(\frac{1-i}{10}\right)^2}$
$= \dfrac{1}{\frac{1-i}{10} - \frac{1 - 2i + i^2}{100}} = \dfrac{1}{\frac{10 - 10i - (1 - 2i - 1)}{100}}$
$= \dfrac{1}{\frac{10 - 8i}{100}} = \dfrac{100}{10 - 8i} = \dfrac{50}{5 - 4i}$
$= \dfrac{50}{5 - 4i} \cdot \dfrac{5 + 4i}{5 + 4i} = \dfrac{250 + 200i}{25 + 16}$
$= \dfrac{250 + 200i}{41} = \dfrac{250}{41} + \dfrac{200}{41}i$

119. $(1 - i)^3(1 + i)^3$
$= (1 - i)(1 + i) \cdot (1 - i)(1 + i) \cdot (1 - i)(1 + i)$
$= (1 - i^2)(1 - i^2)(1 - i^2) = (1 + 1)(1 + 1)(1 + 1)$
$= 2 \cdot 2 \cdot 2 = 8$

121. $\dfrac{6}{1 + \frac{3}{i}} = \dfrac{6}{\frac{i + 3}{i}} = \dfrac{6i}{i + 3} = \dfrac{6i}{i + 3} \cdot \dfrac{-i + 3}{-i + 3}$
$= \dfrac{-6i^2 + 18i}{-i^2 + 9} = \dfrac{6 + 18i}{10} = \dfrac{6}{10} + \dfrac{18}{10}i$
$= \dfrac{3}{5} + \dfrac{9}{5}i$

123. $\dfrac{i - i^{38}}{1 + i} = \dfrac{i - (i^2)^{19}}{1 + i} = \dfrac{i - (-1)^{19}}{1 + i} = \dfrac{i - (-1)}{1 + i} = \dfrac{i + 1}{1 + i} = 1$

Chapter 7 Review

1. True, see page 446 of the text.

3. False, see page 432 of the text.

5. True, see page 440 of the text.

7. True, see page 466 of the text.

9. $\sqrt{\dfrac{100}{121}} = \dfrac{\sqrt{100}}{\sqrt{121}} = \dfrac{10}{11}$

11. $f(x) = \sqrt{x + 10}$
$f(15) = \sqrt{15 + 10} = \sqrt{25} = 5$

13. $\sqrt{64t^2} = \sqrt{(8t)^2} = |8t| = 8|t|$

15. $\sqrt{4x^2 + 4x + 1} = \sqrt{(2x + 1)^2} = |2x + 1|$

17. $\left(\sqrt[3]{5ab}\right)^4 = (5ab)^{4/3}$

19. $\sqrt{x^6 y^{10}} = \left(x^6 y^{10}\right)^{1/2} = x^{6/2} y^{10/2} = x^3 y^5$

21. $\left(x^{-2/3}\right)^{3/5} = x^{-\frac{2}{3} \cdot \frac{3}{5}} = x^{-2/5} = \dfrac{1}{x^{2/5}}$

23. $f(x) = \sqrt{25}\sqrt{(x-6)^2} = 5|x-6|$

25. $\sqrt{250x^3y^2} = \sqrt{25x^2y^2 \cdot 10x} = 5xy\sqrt{10x}$

27. $\sqrt[3]{3x^4b}\sqrt[3]{9xb^2} = \sqrt[3]{3x^4b \cdot 9xb^2} = \sqrt[3]{27x^3b^3 \cdot x^2} = 3xb\sqrt[3]{x^2}$

29. $\sqrt[3]{\dfrac{-27y^{12}}{64}} = \dfrac{\sqrt[3]{(-3y^4)^3}}{\sqrt[3]{4^3}} = \dfrac{-3y^4}{4}$

31. $\dfrac{\sqrt{75x}}{2\sqrt{3}} = \dfrac{1}{2}\sqrt{\dfrac{75x}{3}} = \dfrac{1}{2}\sqrt{25x} = \dfrac{5\sqrt{x}}{2}$

33. $5\sqrt[3]{4y} + 2\sqrt[3]{4y} = (5+2)\sqrt[3]{4y} = 7\sqrt[3]{4y}$

35. $\sqrt[3]{8x^4} + \sqrt[3]{xy^6} = \sqrt[3]{8x^3 \cdot x} + \sqrt[3]{y^6 \cdot x} = 2x\sqrt[3]{x} + y^2\sqrt[3]{x}$
$$= (2x + y^2)\sqrt[3]{x}$$

37. $(3+\sqrt{10})(3-\sqrt{10}) = 3^2 - (\sqrt{10})^2 = 9 - 10 = -1$

39. $\sqrt[4]{x}\sqrt{x} = x^{1/4} \cdot x^{1/2} = x^{1/4+2/4} = x^{3/4} = \sqrt[4]{x^3}$

41. $f(2-\sqrt{a}) = (2-\sqrt{a})^2 = 2^2 - 2\cdot 2\sqrt{a} + (\sqrt{a})^2$
$$= 4 - 4\sqrt{a} + a$$

43. $\dfrac{4\sqrt{5}}{\sqrt{2}+\sqrt{3}} = \dfrac{4\sqrt{5}}{\sqrt{2}+\sqrt{3}} \cdot \dfrac{\sqrt{5}}{\sqrt{5}} = \dfrac{4\sqrt{25}}{\sqrt{10}+\sqrt{15}} = \dfrac{20}{\sqrt{10}+\sqrt{15}}$

45. $(x+1)^{1/3} = -5$
$$\sqrt[3]{x+1} = -5$$
$$(\sqrt[3]{x+1})^3 = (-5)^3$$
$$x+1 = -125$$
$$x = -126$$

Check: $\dfrac{(x+1)^{1/3} = -5}{(-126+1)^{1/3}\ \Big|\ -5}$
$$(-125)\ \Big|$$
$$-5 \overset{?}{=} -5 \quad \text{TRUE}$$
The solution is –126.

47. $f(a) = \sqrt{a+2} + a = 4$
$$\sqrt{a+2} + a = 4$$
$$\sqrt{a+2} = 4 - a$$
$$(\sqrt{a+2})^2 = (4-a)^2$$
$$a+2 = 16 - 8a + a^2$$
$$0 = a^2 - 9a + 14$$
$$0 = (a-2)(a-7)$$
$$a = 2 \ \text{ or } \ a = 7$$

Check:

For $a = 2$: $\dfrac{\sqrt{a+2}+a = 4}{\sqrt{2+2}+2\ \Big|\ 4}$
$$2+2\ \Big|$$
$$4 \overset{?}{=} 4 \quad \text{TRUE}$$

For $a = 7$: $\dfrac{\sqrt{a+2}+a = 4}{\sqrt{7+2}+7\ \Big|\ 4}$
$$3+7\ \Big|$$
$$10 \overset{?}{=} 4 \quad \text{FALSE}$$
The solution is 2.

49. Let b represent the base. We use the Pythagorean theorem to find b.
$$b^2 + 2^2 = 6^2$$
$$b^2 + 4 = 36$$
$$b^2 = 32$$
$$b = \sqrt{32} \approx 5.657$$
The base is $\sqrt{32}$ ft or about 5.657 ft.

51. Using the distance formula $d = \sqrt{(x_2-x_1)^2 + (y_2-y_1)^2}$ for the points $(-6, 4)$ and $(-1, 5)$,
$$d = \sqrt{(-1-(-6))^2 + (5-4)^2}$$
$$= \sqrt{5^2 + 1^2} = \sqrt{26+1} = \sqrt{26}$$
$$\approx 5.099$$

53. $\sqrt{-45} = \sqrt{-1}\cdot\sqrt{9}\cdot\sqrt{5} = 3i\sqrt{5}$ or $3\sqrt{5}\,i$

55. $(9-7i) - (3-8i) = (9-3) + (-7+8)i = 6 + i$

57. $i^{34} = (i^2)^{17} = (-1)^{17} = -1$

59. $\dfrac{7-2i}{3+4i} = \dfrac{7-2i}{3+4i} \cdot \dfrac{3-4i}{3-4i} = \dfrac{21 - 28i - 6i + 8i^2}{9 - 16i^2}$
$$= \dfrac{21 - 34i - 8}{9 + 16} = \dfrac{13 - 34i}{25} = \dfrac{13}{25} - \dfrac{34}{25}i$$

61. *Writing Exercise.* An absolute-value sign must be used to simplify $\sqrt[n]{x^n}$ when n is even, since x may be negative. If x is negative while n is even, the radical expression cannot be simplified to x, since $\sqrt[n]{x^n}$ represents the principal, or nonnegative, root. When n is odd, there is only one root, and it will be positive or negative depending on the sign of x. Thus, there is no absolute-value sign when n is odd.

63. $\sqrt{11x + \sqrt{6+x}} = 6$
$$\left(\sqrt{11x + \sqrt{6+x}}\right)^2 = (6)^2$$
$$11x + \sqrt{6+x} = 36$$
$$\sqrt{6+x} = 36 - 11x$$
$$(\sqrt{6+x})^2 = (36 - 11x)^2$$
$$6 + x = 1296 - 792x + 121x^2$$
$$0 = 121x^2 - 793x + 1290$$
$$0 = (x-3)(121x - 430)$$
$$x = 3 \ \text{ or } \ x = \tfrac{430}{121}$$

Check:

For $x = 3$: $\sqrt{11x + \sqrt{6+x}} = 6$

$$\frac{\sqrt{11 \cdot 3 + \sqrt{6+3}}}{\sqrt{33+3}} \bigg| 6$$

$$6 \overset{?}{=} 6 \quad \text{TRUE}$$

For $x = \frac{430}{121}$: $\sqrt{11x + \sqrt{6+x}} = 6$

$$\frac{\sqrt{11 \cdot \frac{430}{121} + \sqrt{6 + \frac{430}{121}}}}{\sqrt{\frac{430}{11} + \frac{34}{11}}} \bigg| 6$$

$$\sqrt{\frac{464}{11}} \overset{?}{=} 6 \quad \text{FALSE}$$

The solution is 3.

65. The isosceles right triangle has hypotenuse 6. Let x represent the leg of the triangle. Using the Pythagorean theorem,

$$x^2 + x^2 = 6^2$$
$$2x^2 = 36$$
$$x^2 = 18$$
$$x = \sqrt{18} \text{ ft}$$

The area of the isosceles right triangle is

$$A = \frac{1}{2}x^2 = \frac{1}{2}\left(\sqrt{18}\right)^2 = 9 \text{ ft}^2$$

Then in the 30°-60°-90° triangle, a is the shorter leg and we have

$$6 = 2a$$
$$3 = a$$
$$b = a\sqrt{3}$$
$$b = 3\sqrt{3}$$

The area of the 30-60-90 triangle is

$$A = \frac{1}{2}(3\sqrt{3})(3) = \frac{9\sqrt{3}}{2} \text{ ft}^2 \approx 7.794 \text{ ft}^2$$

The area of the isosceles right triangle is larger by about 1.206 ft^2.

Chapter 7 Test

1. $\sqrt{50} = \sqrt{25 \cdot 2} = \sqrt{25} \cdot \sqrt{2} = 5\sqrt{2}$

3. $\sqrt{81a^2} = \sqrt{(9a)^2} = |9a| = 9|a|$

5. $\sqrt{7xy} = (7xy)^{1/2}$

7. $f(x) = \sqrt{2x - 10}$

Since the index is even, the radicand, $2x - 10$, must be non-negative. We solve the inequality:

$$2x - 10 \geq 0$$
$$x \geq 5$$

Domain of $f = \{x | x \geq 5\}$, or $[5, \infty)$

9. $\sqrt[5]{32x^{16}y^{10}} = \sqrt[5]{2^5 x^{15} y^{10} \cdot x} = \sqrt[5]{2^5 x^{15} y^{10}} \cdot \sqrt[5]{x} = 2x^3 y^2 \sqrt[5]{x}$

11. $\sqrt{\frac{100a^4}{9b^6}} = \frac{\sqrt{100a^4}}{\sqrt{9b^6}} = \frac{10a^2}{3b^3}$

13. $\sqrt[4]{x^3}\sqrt{x} = x^{3/4}x^{1/2} = x^{3/4 + 1/2}$
$$= x^{5/4} = x^{1+1/4} = x\sqrt[4]{x}$$

15. $8\sqrt{2} - 2\sqrt{2} = (8-2)\sqrt{2} = 6\sqrt{2}$

17. $(7 + \sqrt{x})(2 - 3\sqrt{x}) = 14 - 7 \cdot 3\sqrt{x} + 2\sqrt{x} - \sqrt{x} \cdot 3\sqrt{x}$
$$= 14 - 21\sqrt{x} + 2\sqrt{x} - 3(\sqrt{x})^2$$
$$= 14 - 19\sqrt{x} - 3x$$

19. $6 = \sqrt{x-3} + 5$
$$1 = \sqrt{x-3}$$
$$1^2 = \left(\sqrt{x-3}\right)^2$$
$$1 = x - 3$$
$$4 = x$$

Check: $6 = \sqrt{x-3} + 5$

$$\frac{6}{6} \bigg| \frac{\sqrt{4-3} + 5}{\sqrt{1} + 5}$$

$$6 \overset{?}{=} 6 \quad \text{TRUE}$$

The solution is 4.

21. $$\sqrt{2x} = \sqrt{x+1} + 1$$
$$\sqrt{2x} - 1 = \sqrt{x+1}$$
$$\left(\sqrt{2x} - 1\right)^2 = \left(\sqrt{x+1}\right)^2$$
$$2x - 2\sqrt{2x} + 1 = x + 1$$
$$-2\sqrt{2x} = -x$$
$$\left(-2\sqrt{2x}\right)^2 = (-x)^2$$
$$4(2x) = x^2$$
$$0 = x^2 - 8x$$
$$0 = x(x - 8)$$

$$x = 0 \quad \text{or} \quad x = 8$$

Check:

For $x = 0$: $\sqrt{2x} = \sqrt{x+1} + 1$

$$\frac{\sqrt{2 \cdot 0}}{} \bigg| \frac{\sqrt{0+1} + 1}{1 + 1}$$

$$0 \overset{?}{=} 2 \quad \text{FALSE}$$

For $x = 8$: $\sqrt{2x} = \sqrt{x+1} + 1$

$$\frac{\sqrt{2 \cdot 8}}{\sqrt{16}} \bigg| \frac{\sqrt{8+1} + 1}{3 + 1}$$

$$4 \overset{?}{=} 4 \quad \text{TRUE}$$

The solution is 8.

23. This is a 30°-60°-90° right triangle with hypotenuse 10.

Let $a =$ the shorter leg and $b =$ the longer leg.

$2a = 10$, so $a = 5$ cm, and

$b = a\sqrt{3} = 5\sqrt{3} \approx 8.660$ cm .

25. Using the midpoint formula $\left(\dfrac{x_1 + x_2}{2}, \ \dfrac{y_1 + y_2}{2}\right)$ for the

points $(2, -5)$ and $(1, -7)$,

$\left(\dfrac{2+1}{2}, \ \dfrac{-5+(-7)}{2}\right)$, or $\left(\dfrac{3}{2}, \ \dfrac{-12}{2}\right)$, or $\left(\dfrac{3}{2}, -6\right)$

27. $(9 + 8i) - (-3 + 6i) = (9 + 3) + (8 - 6)i = 12 + 2i$

29. $\dfrac{-2+i}{3-5i} = \dfrac{-2+i}{3-5i} \cdot \dfrac{3+5i}{3+5i} = \dfrac{-6 - 10i + 3i + 5i^2}{9 - 25i^2}$

$\qquad = \dfrac{-6 - 7i - 5}{9 + 25} = \dfrac{-11 - 7i}{34}$

$\qquad = -\dfrac{11}{34} - \dfrac{7}{34}i$

31.

$$\sqrt{2x - 2} + \sqrt{7x + 4} = \sqrt{13x + 10}$$

$$\left(\sqrt{2x - 2} + \sqrt{7x + 4}\right)^2 = \left(\sqrt{13x + 10}\right)^2$$

$$2x - 2 + 2\sqrt{2x - 2}\sqrt{7x + 4} + 7x + 4 = 13x + 10$$

$$2\sqrt{2x - 2}\sqrt{7x + 4} = 4x + 8$$

$$\left(\sqrt{2x - 2}\sqrt{7x + 4}\right)^2 = (2x + 4)^2$$

$$(2x - 2)(7x + 4) = 4x^2 + 16x + 16$$

$$14x^2 - 6x - 8 = 4x^2 + 16x + 16$$

$$10x^2 - 22x - 24 = 0$$

$$2\left(5x^2 - 11x - 12\right) = 0$$

$$2(5x + 4)(x - 3) = 0$$

$$x = -\dfrac{4}{5} \quad or \quad x = 3$$

Check:

For $x = -\dfrac{4}{5}$:

$$\dfrac{\sqrt{2x - 2} + \sqrt{7x + 4} = \sqrt{13x + 10}}{\sqrt{2 \cdot \left(-\frac{4}{5}\right) + 2} + \sqrt{7\left(-\frac{4}{5}\right) + 4} \ \bigg| \ \sqrt{13\left(-\frac{4}{5}\right) + 10}}$$

$$\sqrt{\tfrac{2}{5}} + \sqrt{-\tfrac{8}{5}} \ \bigg| \ \sqrt{-\tfrac{2}{5}}$$

Since the values in the check are not real, $-\dfrac{4}{5}$ is not a solution.

For $x = 3$:

$$\dfrac{\sqrt{2x - 2} + \sqrt{7x + 4} = \sqrt{13x + 10}}{\sqrt{2 \cdot 3 - 2} + \sqrt{7 \cdot 3 + 4} \ \bigg| \ \sqrt{13 \cdot 3 + 10}}$$

$$\sqrt{4} + \sqrt{25} \ \bigg| \ \sqrt{49}$$

$$2 + 5 \ \bigg|$$

$$7 \overset{?}{=} 7 \quad \text{TRUE}$$

The solution is 3.

33. Substitute 180 for $D(h)$, and solve for h.

$$D(h) = 1.2\sqrt{h}$$

$$180 = 1.2\sqrt{h}$$

$$150 = \sqrt{h}$$

$$(150)^2 = \left(\sqrt{h}\right)^2$$

$$22{,}500 = h$$

The pilot must be above 22,500 ft.

Chapter 8

Quadratic Functions and Equations

Exercise Set 8.1

1. $\sqrt{k}; \ -\sqrt{k}$

3. $t+3; \ t+3$

5. $25; \ 5$

7. $x^2 = 100$

$x = 10 \quad or \quad x = -10$ Using the principle of square roots

The solutions are -10 and 10, or ± 10.

9. $p^2 - 50 = 0$

$p^2 = 50$ Isolating p^2

$p = \sqrt{50} \quad or \quad p = -\sqrt{50}$ Principle of square roots

$p = 5\sqrt{2} \quad or \quad p = -5\sqrt{2}$

The solutions are $5\sqrt{2}$ and $-5\sqrt{2}$ or $\pm 5\sqrt{2}$.

11. $5y^2 = 30$

$y^2 = 6$ Isolating y^2

$y = \sqrt{6} \quad or \quad y = -\sqrt{6}$ Principle of square roots

The solutions are $\sqrt{6}$ and $-\sqrt{6}$ or $\pm\sqrt{6}$.

13. $9x^2 - 49 = 0$

$x^2 = \dfrac{49}{9}$ Isolating x^2

$x = \sqrt{\dfrac{49}{9}} \quad or \quad x = -\sqrt{\dfrac{49}{9}}$ Principle of square roots

$x = \dfrac{7}{3} \quad or \quad x = -\dfrac{7}{3}$

The solutions are $\dfrac{7}{3}$ and $-\dfrac{7}{3}$ or $\pm\dfrac{7}{3}$.

15. $6t^2 - 5 = 0$

$t^2 = \dfrac{5}{6}$

$t = \sqrt{\dfrac{5}{6}} \quad or \quad t = -\sqrt{\dfrac{5}{6}}$ Principle of square roots

$t = \sqrt{\dfrac{5}{6}\cdot\dfrac{6}{6}} \quad or \quad t = -\sqrt{\dfrac{5}{6}\cdot\dfrac{6}{6}}$ Rationalizing denominators

$t = \dfrac{\sqrt{30}}{6} \quad or \quad t = -\dfrac{\sqrt{30}}{6}$

The solutions are $\sqrt{\dfrac{5}{6}}$ and $-\sqrt{\dfrac{5}{6}}$. This can also be

written as $\pm\sqrt{\dfrac{5}{6}}$ or, if we rationalize the denominator,

$\pm\dfrac{\sqrt{30}}{6}$.

17. $a^2 + 1 = 0$

$a^2 = -1$

$a = \sqrt{-1} \quad or \quad a = -\sqrt{-1}$

$a = i \quad\quad or \quad a = -i$

The solutions are i and $-i$ or $\pm i$.

19. $4d^2 + 81 = 0$

$d^2 = -\dfrac{81}{4}$

$d = \sqrt{-\dfrac{81}{4}} \quad or \quad d = -\sqrt{-\dfrac{81}{4}}$

$d = \dfrac{9}{2}i \quad\quad or \quad d = -\dfrac{9}{2}i$

The solutions are $\dfrac{9}{2}i$ and $-\dfrac{9}{2}i$ or $\pm\dfrac{9}{2}i$.

21. $(x-3)^2 = 16$

$x - 3 = \sqrt{16} \quad or \quad x - 3 = -\sqrt{16}$

$x - 3 = 4 \quad\quad or \quad x - 3 = -4$

$x = 7 \quad\quad\quad or \quad\quad x = -1$

The solutions are -1 and 7.

23. $(t+5)^2 = 12$

$t + 5 = \sqrt{12} \quad\quad or \quad t + 5 = -\sqrt{12}$

$t + 5 = 2\sqrt{3} \quad\quad or \quad t + 5 = -2\sqrt{3}$

$t = -5 + 2\sqrt{3} \quad or \quad t = -5 - 2\sqrt{3}$

The solutions are $-5 + 2\sqrt{3}$ and $-5 - 2\sqrt{3}$, or $-5 \pm 2\sqrt{3}$.

25. $(x+1)^2 = -9$

$x + 1 = \sqrt{-9} \quad\quad or \quad x + 1 = -\sqrt{-9}$

$x + 1 = 3i \quad\quad\quad or \quad x + 1 = -3i$

$x = -1 + 3i \quad or \quad\quad x = -1 - 3i$

The solutions are $-1 + 3i$ and $-1 - 3i$, or $-1 \pm 3i$.

27. $\left(y + \dfrac{3}{4}\right)^2 = \dfrac{17}{16}$

$y + \dfrac{3}{4} = \pm\dfrac{\sqrt{17}}{4}$

$y = -\dfrac{3}{4} \pm \dfrac{\sqrt{17}}{4}, \ \text{or} \ \dfrac{-3 \pm \sqrt{17}}{4}$

The solutions are $-\dfrac{3}{4} \pm \dfrac{\sqrt{17}}{4}$, or $\dfrac{-3 \pm \sqrt{17}}{4}$.

29. $x^2 - 10x + 25 = 64$

$$(x-5)^2 = 64$$
$$x - 5 = \pm 8$$
$$x = 5 \pm 8$$
$$x = 13 \quad or \quad x = -3$$

The solutions are 13 and –3.

31. $f(x) = x^2$

$$19 = x^2 \qquad \text{Substituting}$$
$$\sqrt{19} = x \quad or \quad -\sqrt{19} = x$$

The solutions are $\sqrt{19}$ and $-\sqrt{19}$ or $\pm\sqrt{19}$.

33. $\qquad f(x) = 16$

$$(x-5)^2 = 16 \qquad \text{Substituting}$$
$$x - 5 = 4 \quad or \quad x - 5 = -4$$
$$x = 9 \quad or \qquad x = 1$$

The solutions are 9 and 1.

35. $\qquad F(t) = 13$

$$(t+4)^2 = 13 \qquad \text{Substituting}$$
$$t + 4 = \sqrt{13} \qquad or \quad t + 4 = -\sqrt{13}$$
$$t = -4 + \sqrt{13} \quad or \qquad t = -4 - \sqrt{13}$$

The solutions are $-4 + \sqrt{13}$ and $-4 - \sqrt{13}$, or $-4 \pm \sqrt{13}$.

37. $g(x) = x^2 + 14x + 49$

Observe first that $g(0) = 49$. Also observe that when $x = -14$, then

$$x^2 + 14x = (-14)^2 - (14)(14) = (14)^2 - (14)^2 = 0, \text{ so}$$

$g(-14) = 49$ as well. Thus, we have $x = 0$ or $x = 14$.

We can also do this problem as follows.

$$g(x) = 49$$
$$x^2 + 14x + 49 = 49 \qquad \text{Substituting}$$
$$(x+7)^2 = 49$$
$$x + 7 = 7 \quad or \quad x + 7 = -7$$
$$x = 0 \quad or \qquad x = -14$$

The solutions are 0 and -14.

39. $x^2 + 16x$

We take half the coefficient of x and square it: Half of 16 is 8, and $8^2 = 64$. We add 64.

$$x^2 + 16x + 64 = (x+8)^2$$

41. $t^2 - 10t$

We take half the coefficient of t and square it:

Half of –10 is –5, and $(-5)^2 = 25$. We add 25.

$$t^2 - 10t + 25 = (t-5)^2$$

43. $t^2 - 2t$

We take half the coefficient of t and square it:

$\frac{1}{2}(-2) = -1$, and $(-1)^2 = 1$. We add 1.

$$t^2 - 2t + 1 = (t-1)^2$$

45. $x^2 + 3x$

We take half the coefficient of t and square it:

$\frac{1}{2}(3) = \frac{3}{2}$, and $\left(\frac{3}{2}\right)^2 = \frac{9}{4}$. We add $\frac{9}{4}$.

$$x^2 + 3x + \frac{9}{4} = \left(x + \frac{3}{2}\right)^2$$

47. $x^2 + \frac{2}{5}x$

$\frac{1}{2} \cdot \frac{2}{5} = \frac{1}{5}$, and $\left(\frac{1}{5}\right)^2 = \frac{1}{25}$. We add $\frac{1}{25}$.

$$x^2 + \frac{2}{5}x + \frac{1}{25} = \left(x + \frac{1}{5}\right)^2$$

49. $t^2 - \frac{5}{6}t$

$\frac{1}{2}\left(-\frac{5}{6}\right) = -\frac{5}{12}$, and $\left(-\frac{5}{12}\right)^2 = \frac{25}{144}$. We add $\frac{25}{144}$.

$$t^2 - \frac{5}{6}t + \frac{25}{144} = \left(t - \frac{5}{12}\right)^2$$

51. $\qquad x^2 + 6x = 7$

$$x^2 + 6x + 9 = 7 + 9 \qquad \text{Adding 9 to both sides to complete the square}$$
$$(x+3)^2 = 16 \qquad \text{Factoring}$$
$$x + 3 = \pm 4 \qquad \text{Principle of square roots}$$
$$x = -3 \pm 4$$
$$x = -3 + 4 \quad or \quad x = -3 - 4$$
$$x = 1 \qquad or \quad x = -7$$

The solutions are 1 and –7.

53. $\qquad t^2 - 10t = -23$

$$t^2 - 10t + 25 = -23 + 25 \qquad \text{Adding 25 to both sides to complete the square}$$
$$(t-5)^2 = 2 \qquad \text{Factoring}$$
$$t - 5 = \pm\sqrt{2} \qquad \text{Principle of square roots}$$
$$t = 5 \pm \sqrt{2}$$

The solutions are $5 \pm \sqrt{2}$.

55. $x^2 + 12x + 32 = 0$

$$x^2 + 12x = -32$$
$$x^2 + 12x + 36 = -32 + 36$$
$$(x+6)^2 = 4$$
$$x + 6 = \pm 2$$
$$x = -6 \pm 2$$
$$x = -6 + 2 \quad or \quad x = -6 - 2$$
$$x = -4 \qquad or \quad x = -8$$

The solutions are –8 and –4.

57. $t^2 + 8t - 3 = 0$

$t^2 + 8t = 3$

$t^2 + 8t + 16 = 3 + 16$

$(t + 4)^2 = 19$

$t + 4 = \pm\sqrt{19}$

$t = -4 \pm \sqrt{19}$

The solutions are $-4 \pm \sqrt{19}$.

59. The value of $f(x)$ must be 0 at any x-intercepts.

$f(x) = 0$

$x^2 + 6x + 7 = 0$

$x^2 + 6x = -7$

$x^2 + 6x + 9 = -7 + 9$

$(x + 3)^2 = 2$

$x + 3 = \pm\sqrt{2}$

$x = -3 \pm \sqrt{2}$

The x-intercepts are $(-3 - \sqrt{2},\ 0)$ and $(-3 + \sqrt{2},\ 0)$.

61. The value of $g(x)$ must be 0 at any x-intercepts.

$g(x) = 0$

$x^2 + 9x - 25 = 0$

$x^2 + 9x = 25$

$x^2 + 9x + \dfrac{81}{4} = 25 + \dfrac{81}{4}$

$\left(x + \dfrac{9}{2}\right)^2 = \dfrac{181}{4}$

$x + \dfrac{9}{2} = \pm\dfrac{\sqrt{181}}{2}$

$x = -\dfrac{9}{2} \pm \dfrac{\sqrt{181}}{2}$

The x-intercepts are $\left(-\dfrac{9}{2} - \dfrac{\sqrt{181}}{2},\ 0\right)$ and

$\left(-\dfrac{9}{2} + \dfrac{\sqrt{181}}{2},\ 0\right)$.

63. The value of $f(x)$ must be 0 at any x-intercepts.

$f(x) = 0$

$x^2 - 10x - 22 = 0$

$x^2 - 10x = 22$

$x^2 - 10x + 25 = 22 + 25$

$(x - 5)^2 = 47$

$x - 5 = \pm\sqrt{47}$

$x = 5 \pm \sqrt{47}$

The x-intercepts are $(5 - \sqrt{47},\ 0)$ and $(5 + \sqrt{47},\ 0)$.

65. $9x^2 + 18x = -8$

$x^2 + 2x = -\dfrac{8}{9}$ Dividing both sides by 9

$x^2 + 2x + 1 = -\dfrac{8}{9} + 1$

$(x + 1)^2 = \dfrac{1}{9}$

$x + 1 = \pm\dfrac{1}{3}$

$x = -1 \pm \dfrac{1}{3}$

$x = -1 - \dfrac{1}{3}$ or $x = -1 + \dfrac{1}{3}$

$x = -\dfrac{4}{3}$ or $x = -\dfrac{2}{3}$

The solutions are $-\dfrac{4}{3}$ and $-\dfrac{2}{3}$.

67. $3x^2 - 5x - 2 = 0$

$3x^2 - 5x = 2$

$x^2 - \dfrac{5}{3}x = \dfrac{2}{3}$ Dividing both sides by 3

$x^2 - \dfrac{5}{3}x + \dfrac{25}{36} = \dfrac{2}{3} + \dfrac{25}{36}$

$\left(x - \dfrac{5}{6}\right)^2 = \dfrac{49}{36}$

$x - \dfrac{5}{6} = \pm\dfrac{7}{6}$

$x = \dfrac{5}{6} \pm \dfrac{7}{6}$

$x = \dfrac{5}{6} - \dfrac{7}{6}$ or $x = \dfrac{5}{6} + \dfrac{7}{6}$

$x = -\dfrac{1}{3}$ or $x = 2$

The solutions are $-\dfrac{1}{3}$ and 2.

69. $5x^2 + 4x - 3 = 0$

$5x^2 + 4x = 3$

$x^2 + \dfrac{4}{5}x = \dfrac{3}{5}$ Dividing both sides by 5

$x^2 + \dfrac{4}{5}x + \dfrac{4}{25} = \dfrac{3}{5} + \dfrac{4}{25}$

$\left(x + \dfrac{2}{5}\right)^2 = \dfrac{19}{25}$

$x + \dfrac{2}{5} = \pm\dfrac{\sqrt{19}}{5}$

$x = -\dfrac{2}{5} \pm \dfrac{\sqrt{19}}{5}$, or $\dfrac{-2 \pm \sqrt{19}}{5}$

The solutions are $-\dfrac{2}{5} \pm \dfrac{\sqrt{19}}{5}$, or $\dfrac{-2 \pm \sqrt{19}}{5}$.

71. The value of $f(x)$ must be 0 at any x-intercepts.

$$f(x) = 0$$
$$4x^2 + 2x - 3 = 0$$
$$4x^2 + 2x = 3$$
$$x^2 + \frac{1}{2}x = \frac{3}{4} \qquad \text{Dividing both sides by 4}$$
$$x^2 + \frac{1}{2}x + \frac{1}{16} = \frac{3}{4} + \frac{1}{16}$$
$$\left(x + \frac{1}{4}\right)^2 = \frac{13}{16}$$
$$x + \frac{1}{4} = \pm\frac{\sqrt{13}}{4}$$
$$x = -\frac{1}{4} \pm \frac{\sqrt{13}}{4}, \text{ or } \frac{-1 \pm \sqrt{13}}{4}$$

The x-intercepts are $\left(-\frac{1}{4} - \frac{\sqrt{13}}{4}, 0\right)$ and $\left(-\frac{1}{4} + \frac{\sqrt{13}}{4}, 0\right)$, or $\left(\frac{-1-\sqrt{13}}{4}, 0\right)$ and $\left(\frac{-1+\sqrt{13}}{4}, 0\right)$.

73. The value of $g(x)$ must be 0 at any x-intercepts.

$$g(x) = 0$$
$$2x^2 - 3x - 1 = 0$$
$$2x^2 - 3x = 1$$
$$x^2 - \frac{3}{2}x = \frac{1}{2} \qquad \text{Dividing both sides by 2}$$
$$x^2 - \frac{3}{2}x + \frac{9}{16} = \frac{1}{2} + \frac{9}{16}$$
$$\left(x - \frac{3}{4}\right)^2 = \frac{17}{16}$$
$$x - \frac{3}{4} = \pm\frac{\sqrt{17}}{4}$$
$$x = \frac{3}{4} \pm \frac{\sqrt{17}}{4}, \text{ or } \frac{3 \pm \sqrt{17}}{4}$$

The x-intercepts are $\left(\frac{3}{4} - \frac{\sqrt{17}}{4}, 0\right)$ and $\left(\frac{3}{4} + \frac{\sqrt{17}}{4}, 0\right)$, or $\left(\frac{3-\sqrt{17}}{4}, 0\right)$ and $\left(\frac{3+\sqrt{17}}{4}, 0\right)$.

75. *Familiarize.* We are already familiar with the compound-interest formula.

Translate. We substitute into the formula.

$$A = P(1+r)^t$$
$$2420 = 2000(1+r)^2$$

Carry out. We solve for r.

$$2420 = 2000(1+r)^2$$
$$\frac{2420}{2000} = (1+r)^2$$
$$\frac{121}{100} = (1+r)^2$$
$$\pm\sqrt{\frac{121}{100}} = 1 + r$$
$$\pm\frac{11}{10} = 1 + r$$
$$-\frac{10}{10} \pm \frac{11}{10} = r$$
$$\frac{1}{10} = r \text{ or } -\frac{21}{10} = r$$

Check. Since the interest rate cannot be negative, we need only check $\frac{1}{10}$, or 10%. If $2000 were invested at 10% interest, compounded annually, then in 2 years it would grow to $2000(1.1)^2$, or $2420. The number 10% checks.

State. The interest rate is 10%.

77. *Familiarize.* We are already familiar with the compound-interest formula.

Translate. We substitute into the formula.

$$A = P(1+r)^t$$
$$6760 = 6250(1+r)^2$$

Carry out. We solve for r.

$$\frac{6760}{6250} = (1+r)^2$$
$$\frac{676}{625} = (1+r)^2$$
$$\pm\frac{26}{25} = 1 + r$$
$$-\frac{25}{25} \pm \frac{26}{25} = r$$
$$\frac{1}{25} = r \text{ or } -\frac{51}{25} = r$$

Check. Since the interest rate cannot be negative, we need only check $\frac{1}{25}$, or 4%. If $6250 were invested at 4% interest, compounded annually, then in 2 years it would grow to $6250(1.04)^2$, or $6760. The number 4% checks.

State. The interest rate is 4%.

79. *Familiarize.* We will use the formula $s = 16t^2$.

Translate. We substitute into the formula.

$$s = 16t^2$$
$$290 = 16t^2$$

Carry out. We solve for t.

$$290 = 16t^2$$
$$\frac{290}{16} = t^2$$
$$\sqrt{\frac{290}{16}} = t \qquad \text{Principle of square roots;}$$
$$4.3 \approx t \qquad \text{rejecting the negative square root}$$

Check. Since $16(4.3)^2 = 295.84 \approx 290$, our answer checks.

State. It would take an object about 4.3 sec to fall freely from the bridge.

81. *Familiarize.* We will use the formula $s = 16t^2$.

Translate. We substitute into the formula.

$$s = 16t^2$$
$$2063 = 16t^2$$

Carry out. We solve for t.

$$2063 = 16t^2$$
$$\frac{2063}{16} = t^2$$
$$\sqrt{\frac{2063}{16}} = t \qquad \text{Principle of square roots;}$$
$$\qquad\qquad\qquad \text{rejecting the negative square root}$$
$$11.4 \approx t$$

Check. Since $16(11.4)^2 = 2079.63 \approx 2063$, our answer checks.

State. It would take an object about 11.4 sec to fall freely from the top.

83. *Writing Exercise.*

1) If the quadratic equation is of the type $x^2 = k$, use the principle of square roots.

2) If the quadratic equation is of the type $ax^2 + bx + c = 0$, $b \neq 0$, use the principle of zero products, if possible.

3) If the quadratic equation is of the type $ax^2 + bx + c = 0$, $b \neq 0$, and factoring is difficult or impossible, solve by completing the square.

85. $b^2 - 4ac = 2^2 - 4 \cdot 3 \cdot (-5)$
$$= 4 + 60$$
$$= 64$$

87. $\sqrt{200} = \sqrt{100} \cdot \sqrt{2} = 10\sqrt{2}$

89. $\sqrt{-4} = \sqrt{4} \cdot \sqrt{-1} = 2i$

91. $\sqrt{-8} = \sqrt{4} \cdot \sqrt{-1} \cdot \sqrt{2} = 2i\sqrt{2}$, or $2\sqrt{2}i$

93. *Writing Exercise.* It would be better to receive 3% interest every 6 months, because the interest compounds faster in this situation.

95. In order for $x^2 + bx + 81$ to be a square, the following must be true:

$$\left(\frac{b}{2}\right)^2 = 81$$
$$\frac{b^2}{4} = 81$$
$$b^2 = 324$$
$$b = 18 \quad \text{or} \quad b = -18$$

97. We see that x is a factor of each term, so x is also a factor of $f(x)$. We have $f(x) = x(2x^4 - 9x^3 - 66x^2 + 45x + 280)$. Since $x^2 - 5$ is a factor of $f(x)$ it is also a factor of $2x^4 - 9x^3 - 66x^2 + 45x + 280$. We divide to find another factor.

$$
\begin{array}{r}
2x^2 \quad -9x \quad -56 \\
x^2 - 5 \overline{)2x^4 - 9x^3 - 66x^2 + 45x + 280} \\
\underline{2x^4 \qquad\quad -10x^2} \\
-9x^3 - 56x^2 + 45x \\
\underline{-9x^3 \qquad\quad +45x} \\
-56x^2 \qquad\quad + 280 \\
\underline{-56x^2 \qquad\quad + 280} \\
0
\end{array}
$$

Then we have $f(x) = x(x^2 - 5)(2x^2 - 9x - 56)$, or $f(x) = x(x^2 - 5)(2x + 7)(x - 8)$. Now we find the values of a for which $f(a) = 0$.

$$f(a) = 0$$
$$a(a^2 - 5)(2a + 7)(a - 8) = 0$$

$a = 0$ or $a^2 - 5 = 0$ or $2a + 7 = 0$ or $a - 8 = 0$
$a = 0$ or $\quad a^2 = 5$ or $\quad 2a = -7$ or $\quad a = 8$
$a = 0$ or $\quad a = \pm\sqrt{5}$ or $\quad a = -\frac{7}{2}$ or $\quad a = 8$

The solutions are 0, $\sqrt{5}$, $-\sqrt{5}$, $-\frac{7}{2}$, and 8.

99. ***Familiarize***. It is helpful to list information in a chart and make a drawing. Let r represent the speed of the fishing boat. Then $r - 7$ represents the speed of the barge.

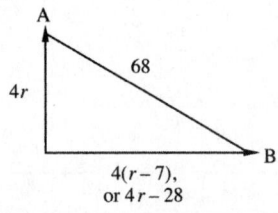

$$4(r-7),$$
$$\text{or } 4r - 28$$

Boat	r	t	d
Fishing	r	4	$4r$
Barge	$r - 7$	4	$4(r-7)$

Translate. We use the Pythagorean equation:

$$a^2 + b^2 = c^2$$
$$(4r - 28)^2 + (4r)^2 = 68^2$$

Carry out.

$$(4r - 28)^2 + (4r)^2 = 68^2$$
$$16r^2 - 224r + 784 + 16r^2 = 4624$$
$$32r^2 - 224r - 3840 = 0$$
$$r^2 - 7r - 120 = 0$$
$$(r + 8)(r - 15) = 0$$
$$r + 8 = 0 \quad \text{or} \quad r - 15 = 0$$
$$r = -8 \quad \text{or} \quad\quad r = 15$$

Check. We check only 15 since the speeds of the boats cannot be negative. If the speed of the fishing boat is 15 km/h, then the speed of the barge is $15 - 7$, or 8 km/h, and the distances they travel are $4 \cdot 15$ (or 60) and $4 \cdot 8$ (or 32).

$60^2 + 32^2 = 3600 + 1024 = 4624 = 68^2$ The values check.

State. The speed of the fishing boat is 15 km/h, and the speed of the barge is 8 km/h.

101. *Graphing Calculator Exercise*

103. *Writing Exercise.* From a reading of the problem we know that we are interested only in positive values of r and it is safe to assume $r \leq 1$. We also know that we want to find the value of r for which $4410 = 4000(1+r)^2$, so the window must include the y-value 4410. A suitable viewing window might be $[0, 1, 4000, 4500]$, Xscl = 0.1, Yscl = 100.

Exercise Set 8.2

1. True; see page 512 in the text.

3. False; see Example 3.

5. False; the quadratic formula yields at most two solutions.

7.
$$2x^2 + 3x - 5 = 0$$
$$(2x+5)(x-1) = 0 \qquad \text{Factoring}$$
$$2x+5 = 0 \quad or \quad x-1 = 0$$
$$x = -\frac{5}{2} \quad or \quad x = 1$$

The solutions are $-\frac{5}{2}$ and 1.

9.
$$u^2 + 2u - 4 = 0$$
$$u^2 + 2u = 4$$
$$u^2 + 2u + 1 = 4 + 1 \qquad \text{Completing the square}$$
$$(u+1)^2 = 5$$
$$u+1 = \pm\sqrt{5} \qquad \text{Principle of square roots}$$
$$u = -1 \pm \sqrt{5}$$

The solutions are $-1+\sqrt{5}$ and $-1-\sqrt{5}$.

11.
$$t^2 + 3 = 6t$$
$$t^2 - 6t = -3$$
$$t^2 - 6t + 9 = -3 + 9$$
$$(t-3)^2 = 6$$
$$t - 3 = \pm\sqrt{6}$$
$$t = 3 \pm \sqrt{6}$$

The solutions are $3+\sqrt{6}$ and $3-\sqrt{6}$.

13.
$$x^2 = 3x + 5$$
$$x^2 - 3x - 5 = 0$$
$$a = 1, \ b = -3, \ c = -5$$
$$x = \frac{-b \pm \sqrt{b^2 - 4ac}}{2a}$$
$$x = \frac{-(-3) \pm \sqrt{(-3)^2 - 4 \cdot 1 \cdot (-5)}}{2 \cdot 1} = \frac{3 \pm \sqrt{9+20}}{2}$$
$$x = \frac{3 \pm \sqrt{29}}{2} = \frac{3}{2} \pm \frac{\sqrt{29}}{2}$$

The solutions are $\frac{3}{2} + \frac{\sqrt{29}}{2}$ and $\frac{3}{2} - \frac{\sqrt{29}}{2}$.

15.
$$3t(t+2) = 1$$
$$3t^2 + 6t = 1$$
$$3t^2 + 6t - 1 = 0$$
$$a = 3, \ b = 6, \ c = -1$$
$$t = \frac{-b \pm \sqrt{b^2 - 4ac}}{2a}$$
$$t = \frac{-6 \pm \sqrt{6^2 - 4 \cdot 3 \cdot (-1)}}{2 \cdot 3} = \frac{-6 \pm \sqrt{36+12}}{6}$$
$$t = \frac{-6 \pm \sqrt{48}}{6} = \frac{-6 \pm 4\sqrt{3}}{6}$$
$$t = -\frac{6}{6} \pm \frac{4\sqrt{3}}{6} = -1 \pm \frac{2\sqrt{3}}{3}$$

The solutions are $-1 + \frac{2\sqrt{3}}{3}$ and $-1 - \frac{2\sqrt{3}}{3}$.

17.
$$\frac{1}{x^2} - 3 = \frac{8}{x}, \qquad \text{LCD is } x^2$$
$$x^2\left(\frac{1}{x^2} - 3\right) = x^2 \cdot \frac{8}{x}$$
$$x^2 \cdot \frac{1}{x^2} - x^2 \cdot 3 = 8x$$
$$1 - 3x^2 = 8x$$
$$0 = 3x^2 + 8x - 1$$
$$a = 3, \ b = 8, \ c = -1$$
$$x = \frac{-8 \pm \sqrt{8^2 - 4 \cdot 3 \cdot (-1)}}{2 \cdot 3} = \frac{-8 \pm \sqrt{64+12}}{6}$$
$$x = \frac{-8 \pm \sqrt{76}}{6} = \frac{-8 \pm \sqrt{4 \cdot 19}}{6} = \frac{-8 \pm 2\sqrt{19}}{6}$$
$$x = \frac{-4 \pm \sqrt{19}}{3} = -\frac{4}{3} \pm \frac{\sqrt{19}}{3}$$

The solutions are $-\frac{4}{3} - \frac{\sqrt{19}}{3}$ and $-\frac{4}{3} + \frac{\sqrt{19}}{3}$.

19. $t^2 + 10 = 6t$

$t^2 - 6t + 10 = 0$

$a = 1, \ b = -6, \ c = 10$

$t = \dfrac{-b \pm \sqrt{b^2 - 4ac}}{2a}$

$t = \dfrac{-(-6) \pm \sqrt{(-6)^2 - 4 \cdot 1 \cdot 10}}{2 \cdot 1} = \dfrac{6 \pm \sqrt{36 - 40}}{6}$

$t = \dfrac{6 \pm \sqrt{-4}}{2} = \dfrac{6 \pm 2i}{2}$

$t = \dfrac{6}{2} \pm \dfrac{2i}{2} = 3 \pm i$

The solutions are $3 + i$ and $3 - i$.

21. $p^2 - p + 1 = 0$

$a = 1, \ b = -1, \ c = 1$

$p = \dfrac{-b \pm \sqrt{b^2 - 4ac}}{2a}$

$p = \dfrac{-(-1) \pm \sqrt{(-1)^2 - 4 \cdot 1 \cdot 1}}{2 \cdot 1} = \dfrac{1 \pm \sqrt{1 - 4}}{2}$

$p = \dfrac{1 \pm \sqrt{-3}}{2} = \dfrac{1}{2} \pm \dfrac{\sqrt{3}}{2}i$

The solutions are $\dfrac{1}{2} + \dfrac{\sqrt{3}}{2}i$ and $\dfrac{1}{2} - \dfrac{\sqrt{3}}{2}i$.

23. $x^2 + 4x + 6 = 0$

$x = \dfrac{-b \pm \sqrt{b^2 - 4ac}}{2a}$

$x = \dfrac{-4 \pm \sqrt{4^2 - 4 \cdot 1 \cdot 6}}{2 \cdot 1} = \dfrac{-4 \pm \sqrt{16 - 24}}{2}$

$x = \dfrac{-4 \pm \sqrt{-8}}{2} = -\dfrac{4}{2} \pm \dfrac{2\sqrt{2}}{2}i$

$x = -2 \pm \sqrt{2}i$

The solutions are $-2 + \sqrt{2}i$ and $-2 - \sqrt{2}i$.

25. $12t^2 + 17t = 40$

$12t^2 + 17t - 40 = 0$

$(3t + 8)(4t - 5) = 0$

$3t + 8 = 0 \quad or \quad 4t - 5 = 0$

$t = -\dfrac{8}{3} \quad or \quad t = \dfrac{5}{4}$

The solutions are $-\dfrac{8}{3}$ and $\dfrac{5}{4}$.

27. $25x^2 - 20x + 4 = 0$

$(5x - 2)(5x - 2) = 0$

$5x - 2 = 0 \quad or \quad 5x - 2 = 0$

$5x = 2 \quad or \quad 5x = 2$

$x = \dfrac{2}{5} \quad or \quad x = \dfrac{2}{5}$

The solution is $\dfrac{2}{5}$.

29. $7x(x + 2) + 5 = 3x(x + 1)$

$7x^2 + 14x + 5 = 3x^2 + 3x$

$4x^2 + 11x + 5 = 0$

$a = 4, \ b = 11, \ c = 5$

$x = \dfrac{-11 \pm \sqrt{11^2 - 4 \cdot 4 \cdot 5}}{2 \cdot 4} = \dfrac{-11 \pm \sqrt{121 - 80}}{8}$

$x = \dfrac{-11 \pm \sqrt{41}}{8} = -\dfrac{11}{8} \pm \dfrac{\sqrt{41}}{8}$

The solutions are $-\dfrac{11}{8} - \dfrac{\sqrt{41}}{8}$ and $-\dfrac{11}{8} + \dfrac{\sqrt{41}}{8}$.

31. $14(x - 4) - (x + 2) = (x + 2)(x - 4)$

$14x - 56 - x - 2 = x^2 - 2x - 8 \quad$ Removing parentheses

$13x - 58 = x^2 - 2x - 8$

$0 = x^2 - 15x + 50$

$0 = (x - 10)(x - 5)$

$x - 10 = 0 \quad or \quad x - 5 = 0$

$x = 10 \quad or \quad x = 5$

The solutions are 10 and 5.

33. $51p = 2p^2 + 72$

$0 = 2p^2 - 51p + 72$

$0 = (2p - 3)(p - 24)$

$2p - 3 = 0 \quad or \quad p - 24 = 0$

$p = \dfrac{3}{2} \quad or \quad p = 24$

The solutions are $\dfrac{3}{2}$ and 24.

35. $x(x - 3) = x - 9$

$x^2 - 3x = x - 9 \quad$ Removing parentheses

$x^2 - 4x = -9$

$x^2 - 4x + 4 = -9 + 4 \quad$ Completing the square

$(x - 2)^2 = -5$

$x - 2 = \pm\sqrt{-5}$

$x = 2 \pm \sqrt{5}i$

The solutions are $2 + \sqrt{5}i$ and $2 - \sqrt{5}i$.

37. $x^3 - 8 = 0$

$x^3 - 2^3 = 0$

$(x - 2)(x^2 + 2x + 4) = 0$

$x - 2 = 0 \quad or \quad x^2 + 2x + 4 = 0$

$x = 2 \quad or \quad x = \dfrac{-2 \pm \sqrt{2^2 - 4 \cdot 1 \cdot 4}}{2 \cdot 1}$

$x = 2 \quad or \quad x = \dfrac{-2 \pm \sqrt{-12}}{2} = \dfrac{-2 \pm 2i\sqrt{3}}{2}$

$x = 2 \quad or \quad x = -\dfrac{2}{2} \pm \dfrac{2\sqrt{3}}{2}i$

$x = 2 \quad or \quad x = -1 \pm \sqrt{3}i$

The solutions are 2, $-1 + \sqrt{3}i$, and $-1 - \sqrt{3}i$.

39.
$$f(x) = 0$$
$$6x^2 - 7x - 20 = 0$$
$$(3x + 4)(2x - 5) = 0$$
$$3x + 4 = 0 \quad or \quad 2x - 5 = 0$$
$$x = -\frac{4}{3} \quad or \quad x = \frac{5}{2}$$
$$f(x) = 0 \text{ for } x = -\frac{4}{3} \text{ and } x = \frac{5}{2}.$$

41.
$$f(x) = 1 \qquad \text{Substituting}$$
$$\frac{7}{x} + \frac{7}{x+4} = 1$$
$$x(x+4)\left(\frac{7}{x} + \frac{7}{x+4}\right) = x(x+4) \cdot 1$$
$$\qquad \text{Multiplying by the LCD}$$
$$7(x+4) + 7x = x^2 + 4x$$
$$7x + 28 + 7x = x^2 + 4x$$
$$14x + 28 = x^2 + 4x$$
$$0 = x^2 - 10x - 28$$
$$a = 1, \ b = -10, \ c = -28$$
$$x = \frac{-(-10) \pm \sqrt{(-10)^2 - 4 \cdot 1 \cdot (-28)}}{2 \cdot 1}$$
$$x = \frac{10 \pm \sqrt{100 + 112}}{2} = \frac{10 \pm \sqrt{212}}{2}$$
$$x = \frac{10 \pm \sqrt{4 \cdot 53}}{2} = \frac{10 \pm 2\sqrt{53}}{2}$$
$$x = 5 \pm \sqrt{53}$$
$$f(x) = 1 \text{ for } x = 5 + \sqrt{53} \text{ and } x = 5 - \sqrt{53}.$$

43.
$$F(x) = G(x)$$
$$\frac{3 - x}{4} = \frac{1}{4x}$$
$$4x \cdot \frac{3 - x}{4} = 4x \cdot \frac{1}{4x}$$
$$3x - x^2 = 1$$
$$0 = x^2 - 3x + 1$$
$$x = \frac{-(-3) \pm \sqrt{(-3)^2 - 4 \cdot 1 \cdot 1}}{2 \cdot 1} = \frac{3 \pm \sqrt{5}}{2}$$
$$x = \frac{3}{2} \pm \frac{\sqrt{5}}{2}$$

45. $x^2 + 4x - 7 = 0$
$$a = 1, \ b = 4, \ c = -7$$
$$x = \frac{-4 \pm \sqrt{4^2 - 4 \cdot 1 \cdot (-7)}}{2 \cdot 1} = \frac{-4 \pm \sqrt{16 + 28}}{2}$$
$$x = \frac{-4 \pm \sqrt{44}}{2}$$

Using a calculator we find that $\frac{-4 + \sqrt{44}}{2} \approx 1.317$ and

$\frac{-4 - \sqrt{44}}{2} \approx -5.317$.

The solutions are approximately 1.317 and –5.317.

47. $x^2 - 6x + 4 = 0$
$$a = 1, \ b = -6, \ c = 4$$

$$x = \frac{-(-6) \pm \sqrt{(-6)^2 - 4 \cdot 1 \cdot 4}}{2 \cdot 1} = \frac{6 \pm \sqrt{36 - 16}}{2}$$
$$x = \frac{6 \pm \sqrt{20}}{2}$$

Using a calculator we find that $\frac{6 + \sqrt{20}}{2} \approx 5.236$ and

$\frac{6 - \sqrt{20}}{2} \approx 0.764$.

The solutions are approximately 5.236 and 0.764.

49. $2x^2 - 3x - 7 = 0$
$$a = 2, \ b = -3, \ c = -7$$
$$x = \frac{-(-3) \pm \sqrt{(-3)^2 - 4 \cdot 2 \cdot (-7)}}{2 \cdot 2}$$
$$x = \frac{3 \pm \sqrt{9 + 56}}{4} = \frac{3 \pm \sqrt{65}}{4}$$

Using a calculator we find that $\frac{3 + \sqrt{65}}{4} \approx 2.766$ and

$\frac{3 - \sqrt{65}}{4} \approx -1.266$.

The solutions are approximately 2.766 and –1.266.

51. *Writing Exercise.* No; the quadratic formula is derived by solving the equation $ax^2 + bx + c = 0$ by completing the square, so, any equation that can be solved using the quadratic formula can also be solved by completing the square.

53. $(x - 2i)(x + 2i) = x^2 - (2i)^2 = x^2 - 4i^2$
$$= x^2 + 4$$

55. $\left[x - (2 - \sqrt{7})\right]\left[x - (2 + \sqrt{7})\right]$
$$= \left[(x - 2) + \sqrt{7}\right]\left[(x - 2) - \sqrt{7}\right] \qquad \text{Regrouping}$$
$$= (x - 2)^2 - \left(\sqrt{7}\right)^2 \qquad \text{Difference of squares}$$
$$= x^2 - 4x + 4 - 7$$
$$= x^2 - 4x - 3$$

57. $\dfrac{-6 \pm \sqrt{(-4)^2 - 4(2)(2)}}{2(2)} = \dfrac{-6 \pm \sqrt{16 - 16}}{4} = \dfrac{-6 \pm 0}{4} = -\dfrac{3}{2}$

59. *Writing Exercise.* A quadratic polynomial in the form $ax^2 + bx + c$ can be written in factored form
$$\left(x - \frac{-b + \sqrt{b^2 - 4ac}}{2a}\right)\left(x + \frac{-b - \sqrt{b^2 - 4ac}}{2a}\right).$$

61. $f(x) = \dfrac{x^2}{x - 2} + 1$

To find the x-coordinates of the x-intercepts of the graph of f, we solve $f(x) = 0$.
$$\frac{x^2}{x - 2} + 1 = 0$$
$$x^2 + x - 2 = 0 \quad \text{Multiplying by } x - 2$$
$$(x + 2)(x - 1) = 0$$

$x = -2$ *or* $x = 1$

The x-intercepts are $(-2, 0)$ and $(1, 0)$.

63.
$$f(x) = g(x)$$
$$\frac{x^2}{x-2} + 1 = \frac{4x-2}{x-2} + \frac{x+4}{2} \quad \text{Substituting}$$
$$2(x-2)\left(\frac{x^2}{x-2} + 1\right) = 2(x-2)\left(\frac{4x-2}{x-2} + \frac{x+4}{2}\right)$$
$$\text{Multiplying by the LCD}$$
$$2x^2 + 2(x-2) = 2(4x-2) + (x-2)(x+4)$$
$$2x^2 + 2x - 4 = 8x - 4 + x^2 + 2x - 8$$
$$2x^2 + 2x - 4 = x^2 + 10x - 12$$
$$x^2 - 8x + 8 = 0$$
$$a = 1, \ b = -8, \ c = 8$$
$$x = \frac{-(-8) \pm \sqrt{(-8)^2 - 4 \cdot 1 \cdot 8}}{2 \cdot 1} = \frac{8 \pm \sqrt{64 - 32}}{2}$$
$$x = \frac{8 \pm \sqrt{32}}{2} = \frac{8 \pm \sqrt{16 \cdot 2}}{2} = \frac{8 \pm 4\sqrt{2}}{2}$$
$$x = \frac{8}{2} \pm \frac{4\sqrt{2}}{2} = 4 \pm 2\sqrt{2}$$

The solutions are $4 + 2\sqrt{2}$ and $4 - 2\sqrt{2}$.

65. $z^2 + 0.84z - 0.4 = 0$

$a = 1, \ b = 0.84, \ c = -0.4$
$$z = \frac{-0.84 \pm \sqrt{(0.84)^2 - 4 \cdot 1 \cdot (-0.4)}}{2 \cdot 1}$$
$$z = \frac{-0.84 \pm \sqrt{2.3056}}{2}$$
$$z = \frac{-0.84 + \sqrt{2.3056}}{2} \approx 0.339$$
$$z = \frac{-0.84 - \sqrt{2.3056}}{2} \approx -1.179$$

The solutions are approximately 0.339 and -1.179.

67. $\sqrt{2}x^2 + 5x + \sqrt{2} = 0$
$$x = \frac{-5 \pm \sqrt{5^2 - 4 \cdot \sqrt{2} \cdot \sqrt{2}}}{2\sqrt{2}} = \frac{-5 \pm \sqrt{17}}{2\sqrt{2}}, \text{ or}$$
$$x = \frac{-5 \pm \sqrt{17}}{2\sqrt{2}} \cdot \frac{\sqrt{2}}{\sqrt{2}} = \frac{-5\sqrt{2} \pm \sqrt{34}}{4}$$

The solutions are $\dfrac{-5\sqrt{2} \pm \sqrt{34}}{4}$.

69.
$$kx^2 + 3x - k = 0$$
$$k(-2)^2 + 3(-2) - k = 0 \quad \text{Substituting } -2 \text{ for } x$$
$$4k - 6 - k = 0$$
$$3k = 6$$
$$k = 2$$
$$2x^2 + 3x - 2 = 0 \quad \text{Substituting 2 for } k$$
$$(2x - 1)(x + 2) = 0$$
$$2x - 1 = 0 \quad \text{or} \quad x + 2 = 0$$
$$x = \frac{1}{2} \quad \text{or} \quad x = -2$$

The other solution is $\dfrac{1}{2}$.

71. *Graphing Calculator Exercise*

Exercise Set 8.3

1. Since the discriminant is 9, a perfect square, choice (b) two different rational solutions, is correct.

3. Since the discriminant is −1, a negative number, choice (d) two different imaginary-number solutions, is correct.

5. Since the discriminant is 8, a positive number that is not a perfect square, choice (c) two different irrational solutions, is correct.

7. $x^2 - 7x + 5 = 0$

$a = 1, \ b = -7, \ c = 5$

We substitute and compute the discriminant.
$$b^2 - 4ac = (-7)^2 - 4 \cdot 1 \cdot 5$$
$$= 49 - 20$$
$$= 29$$

Since the discriminant is a positive number that is not a perfect square, there are two irrational solutions.

9. $x^2 + 11 = 0$

$a = 1, \ b = 0, \ c = 11$

We substitute and compute the discriminant.
$$b^2 - 4ac = 0^2 - 4 \cdot 1 \cdot 11 = -44$$

Since the discriminant is negative, there are two imaginary-number solutions.

11. $x^2 - 11 = 0$

$a = 1, \ b = 0, \ c = -11$

We substitute and compute the discriminant.
$$b^2 - 4ac = 0^2 - 4 \cdot 1 \cdot (-11) = 44$$

Since the discriminant is a positive number that is not a perfect square, there are two irrational solutions.

13. $4x^2 + 8x - 5 = 0$

$a = 4, \ b = 8, \ c = -5$

We substitute and compute the discriminant.
$$b^2 - 4ac = 8^2 - 4 \cdot 4 \cdot (-5)$$
$$= 64 + 80$$
$$= 144$$

Since the discriminant is a positive number and a perfect square, there are two rational solutions.

15. $x^2 + 4x + 6 = 0$

$a = 1, \ b = 4, \ c = 6$

We substitute and compute the discriminant.

$$b^2 - 4ac = 4^2 - 4 \cdot 1 \cdot 6$$
$$= 16 - 24$$
$$= -8$$

Since the discriminant is negative, there are two imaginary-number solutions.

17. $9t^2 - 48t + 64 = 0$

$a = 9, \ b = -48, \ c = 64$

We substitute and compute the discriminant.

$$b^2 - 4ac = (-48)^2 - 4 \cdot 9 \cdot 64$$
$$= 2304 - 2304$$
$$= 0$$

Since the discriminant is 0, there is just one solution and it is a rational number.

19. $9t^2 + 3t = 0$

Observe that we can factor $9t^2 + 3t$. This tells us that there are two rational solutions. We could also do this problem as follows.

$$b^2 - 4ac = 3^2 - 4 \cdot 9 \cdot 0 = 9$$

Since the discriminant is a positive number and a perfect square, there are two rational solutions.

21. $x^2 + 4x = 8$

$x^2 + 4x - 8 = 0 \quad$ Standard form

$a = 1, \ b = 4, \ c = -8$

We substitute and compute the discriminant.

$$b^2 - 4ac = 4^2 - 4 \cdot 1 \cdot (-8)$$
$$= 16 + 32 = 48$$

Since the discriminant is a positive number that is not a perfect square, there are two irrational solutions.

23. $2a^2 - 3a = -5$

$2a^2 - 3a + 5 = 0 \quad$ Standard form

$a = 2, \ b = -3, \ c = 5$

We substitute and compute the discriminant.

$$b^2 - 4ac = (-3)^2 - 4 \cdot 2 \cdot 5$$
$$= 9 - 40$$
$$= -31$$

Since the discriminant is negative, there are two imaginary-number solutions.

25. $7x^2 = 19x$

$7x^2 - 19x = 0 \quad$ Standard form

$a = 7, \ b = -19, \ c = 0$

We substitute and compute the discriminant.

$$b^2 - 4ac = (-19)^2 - 4 \cdot 7 \cdot 0 = 361$$

Since the discriminant is a positive number and a perfect square, there are two different rational solutions.

27. $y^2 + \dfrac{9}{4} = 4y$

$y^2 - 4y + \dfrac{9}{4} = 0 \quad$ Standard form

$a = 1, \ b = -4, \ c = \dfrac{9}{4}$

We substitute and compute the discriminant.

$$b^2 - 4ac = (-4)^2 - 4 \cdot 1 \cdot \dfrac{9}{4}$$
$$= 16 - 9$$
$$= 7$$

The discriminant is a positive number that is not a perfect square. There are two irrational solutions.

29. The solutions are -5 and 4.

$$x = -5 \quad or \quad x = 4$$
$$x + 5 = 0 \quad or \quad x - 4 = 0$$
$$(x + 5)(x - 4) = 0 \quad \text{Principle of zero products}$$
$$x^2 + x - 20 = 0 \quad \text{FOIL}$$

31. The only solution is 3. It must be a repeated solution.

$$x = 3 \quad or \quad x = 3$$
$$x - 3 = 0 \quad or \quad x - 3 = 0$$
$$(x - 3)(x - 3) = 0 \quad \text{Principle of zero products}$$
$$x^2 - 6x + 9 = 0 \quad \text{FOIL}$$

33. The solutions are -1 and -3.

$$x = -1 \quad or \quad x = -3$$
$$x + 1 = 0 \quad or \quad x + 3 = 0$$
$$(x + 1)(x + 3) = 0$$
$$x^2 + 4x + 3 = 0$$

35. The solutions are 5 and $\dfrac{3}{4}$.

$$x = 5 \quad or \quad x = \dfrac{3}{4}$$
$$x - 5 = 0 \quad or \quad x - \dfrac{3}{4} = 0$$
$$(x - 5)\left(x - \dfrac{3}{4}\right) = 0$$
$$x^2 - \dfrac{3}{4}x - 5x + \dfrac{15}{4} = 0$$
$$x^2 - \dfrac{23}{4}x + \dfrac{15}{4} = 0$$
$$4x^2 - 23x + 15 = 0 \quad \text{Multiplying by 4}$$

37. The solutions are $-\dfrac{1}{4}$ and $-\dfrac{1}{2}$.

$$x = -\dfrac{1}{4} \quad or \quad x = -\dfrac{1}{2}$$
$$x + \dfrac{1}{4} = 0 \quad or \quad x + \dfrac{1}{2} = 0$$
$$\left(x + \dfrac{1}{4}\right)\left(x + \dfrac{1}{2}\right) = 0$$
$$x^2 + \dfrac{1}{2}x + \dfrac{1}{4}x + \dfrac{1}{8} = 0$$
$$x^2 + \dfrac{3}{4}x + \dfrac{1}{8} = 0$$
$$8x^2 + 6x + 1 = 0 \quad \text{Multiplying by 8}$$

39. The solutions are 2.4 and –0.4.

$$x = 2.4 \quad or \quad x = -0.4$$
$$x - 2.4 = 0 \quad or \quad x + 0.4 = 0$$
$$(x - 2.4)(x + 0.4) = 0$$
$$x^2 + 0.4x - 2.4x - 0.96 = 0$$
$$x^2 - 2x - 0.96 = 0$$

41. The solutions are $-\sqrt{3}$ and $\sqrt{3}$.

$$x = -\sqrt{3} \quad or \quad x = \sqrt{3}$$
$$x + \sqrt{3} = 0 \quad or \quad x - \sqrt{3} = 0$$
$$(x + \sqrt{3})(x - \sqrt{3}) = 0$$
$$x^2 - 3 = 0$$

43. The solutions are $2\sqrt{5}$ and $-2\sqrt{5}$.

$$x = 2\sqrt{5} \quad or \quad x = -2\sqrt{5}$$
$$x - 2\sqrt{5} = 0 \quad or \quad x + 2\sqrt{5} = 0$$
$$(x - 2\sqrt{5})(x + 2\sqrt{5}) = 0$$
$$x^2 - (2\sqrt{5})^2 = 0$$
$$x^2 - 4 \cdot 5 = 0$$
$$x^2 - 20 = 0$$

45. The solutions are $4i$ and $-4i$.

$$x = 4i \quad or \quad x = -4i$$
$$x - 4i = 0 \quad or \quad x + 4i = 0$$
$$(x - 4i)(x + 4i) = 0$$
$$x^2 - (4i)^2 = 0$$
$$x^2 + 16 = 0$$

47. The solutions are $2 - 7i$ and $2 + 7i$.

$$x = 2 - 7i \quad or \quad x = 2 + 7i$$
$$x - 2 + 7i = 0 \quad or \quad x - 2 - 7i = 0$$
$$(x - 2) + 7i = 0 \quad or \quad (x - 2) - 7i = 0$$
$$[(x - 2) + 7i][(x - 2) - 7i] = 0$$
$$(x - 2)^2 - (7i)^2 = 0$$
$$x^2 - 4x + 4 - 49i^2 = 0$$
$$x^2 - 4x + 4 + 49 = 0$$
$$x^2 - 4x + 53 = 0$$

49. The solutions are $3 - \sqrt{14}$ and $3 + \sqrt{14}$.

$$x = 3 - \sqrt{14} \quad or \quad x = 3 + \sqrt{14}$$
$$x - 3 + \sqrt{14} = 0 \quad or \quad x - 3 - \sqrt{14} = 0$$
$$(x - 3) + \sqrt{14} = 0 \quad or \quad (x - 3) - \sqrt{14} = 0$$
$$[(x - 3) + \sqrt{14}][(x - 3) - \sqrt{14}] = 0$$
$$(x - 3)^2 - (\sqrt{14})^2 = 0$$
$$x^2 - 6x + 9 - 14 = 0$$
$$x^2 - 6x - 5 = 0$$

51. The solutions are $1 - \dfrac{\sqrt{21}}{3}$ and $1 + \dfrac{\sqrt{21}}{3}$.

$$x = 1 - \frac{\sqrt{21}}{3} \quad or \quad x = 1 + \frac{\sqrt{21}}{3}$$
$$x - 1 + \frac{\sqrt{21}}{3} = 0 \quad or \quad x - 1 - \frac{\sqrt{21}}{3} = 0$$
$$(x - 1) + \frac{\sqrt{21}}{3} = 0 \quad or \quad (x - 1) - \frac{\sqrt{21}}{3} = 0$$
$$\left[(x-1) + \frac{\sqrt{21}}{3}\right]\left[(x-1) - \frac{\sqrt{21}}{3}\right] = 0$$
$$(x - 1)^2 - \left(\frac{\sqrt{21}}{3}\right)^2 = 0$$
$$x^2 - 2x + 1 - \frac{21}{9} = 0$$
$$x^2 - 2x + 1 - \frac{7}{3} = 0$$
$$x^2 - 2x - \frac{4}{3} = 0$$
$$3x^2 - 6x - 4 = 0 \quad \text{Multiplying by 3}$$

53. The solutions are –2, 1, and 5.

$$x = -2 \quad or \quad x = 1 \quad or \quad x = 5$$
$$x + 2 = 0 \quad or \quad x - 1 = 0 \quad or \quad x - 5 = 0$$
$$(x + 2)(x - 1)(x - 5) = 0$$
$$(x^2 + x - 2)(x - 5) = 0$$
$$x^3 + x^2 - 2x - 5x^2 - 5x + 10 = 0$$
$$x^3 - 4x^2 - 7x + 10 = 0$$

55. The solutions are –1, 0, and 3.

$$x = -1 \quad or \quad x = 0 \quad or \quad x = 3$$
$$x + 1 = 0 \quad or \quad x = 0 \quad or \quad x - 3 = 0$$
$$(x + 1)(x)(x - 3) = 0$$
$$(x^2 + x)(x - 3) = 0$$
$$x^3 - 3x^2 + x^2 - 3x = 0$$
$$x^3 - 2x^2 - 3x = 0$$

57. *Writing Exercise.* When the discriminant, $b^2 - 4ac$ is 0, then the quadratic formula can be simplified to

$$x = \frac{-b \pm \sqrt{b^2 - 4ac}}{2a} = \frac{-b \pm \sqrt{0}}{2a} = -\frac{b}{2a}.$$

There is only one solution, $-\dfrac{b}{2a}$.

59.
$$\frac{c}{d} = c + d$$
$$d \cdot \frac{c}{d} = d(c + d)$$
$$c = cd + d^2$$
$$c - cd = d^2$$
$$c(1 - d) = d^2$$
$$c = \frac{d^2}{1 - d}$$

61.
$$x = \frac{3}{1-y}$$
$$(1-y)x = (1-y)\frac{3}{1-y}$$
$$x - xy = 3$$
$$x - 3 = xy$$
$$\frac{x-3}{x} = y \quad \text{or} \quad y = 1 - \frac{3}{x}$$

63. *Familiarize*. Let $r =$ Jamal's speed in mph. Then $r - 1.5 =$ Kade's speed in mph. We organize the information in a table. The time is the same for both so we use t for each time.

	Distance	Speed	Time
Jamal	7	r	t
Kade	4	$r-1.5$	t

Translate. Using the formula Time = Distance/Rate in each row of the table and the fact that the times are the same, we can write an equation.
$$\frac{7}{r} = \frac{4}{r-1.5}$$

Carry out. We solve the equation.
$$\frac{7}{r} = \frac{4}{r-1.5}, \quad \text{LCD is } r(r-1.5)$$
$$r(r-1.5)\cdot\frac{7}{r} = r(r-1.5)\cdot\frac{4}{r-1.5}$$
$$7(r-1.5) = 4r$$
$$7r - 10.5 = 4r$$
$$-10.5 = -3r$$
$$3.5 = r$$

Check. If the Jamal's speed is 3.5 mph, then Kade's speed is $3.5 - 1.5$, or 2 mph. Traveling 7 mi at 3.5 mph takes $\frac{7}{3.5} = 2\,\text{hr}$. Traveling 4 mi at 2 mph takes $\frac{4}{2} = 2\,\text{hr}$. Since the times are the same, the answer checks.

State. The Jamal's speed is 3.5 mph and Kade's speed is 2 mph.

65. *Writing Exercise*. Consider a quadratic equation in standard form, $ax^2 + bx + c = 0$. The solutions are
$$\frac{-b \pm \sqrt{b^2 - 4ac}}{2a}.$$
The product of the solutions is
$$\left(\frac{-b + \sqrt{b^2-4ac}}{2a}\right)\left(\frac{-b - \sqrt{b^2-4ac}}{2a}\right)$$
$$= \frac{(-b)^2 - \left(\sqrt{b^2-4ac}\right)^2}{(2a)^2} = \frac{b^2 - (b^2 - 4ac)}{4a^2} = \frac{4ac}{4a^2} = \frac{c}{a}$$
For integers c and a, this is a real number.

67. The graph includes the points $(-3, 0)$, $(0, -3)$, and $(1, 0)$. Substituting in $y = ax^2 + bx + c$, we have three equations.

$$0 = 9a - 3b + c,$$
$$-3 = \qquad\qquad c,$$
$$0 = \quad a + \; b + c$$

The solution of this system of equations is $a = 1$, $b = 2$, $c = -3$.

69. a) $kx^2 - 2x + k = 0$; one solution is -3

We first find k by substituting -3 for x.
$$k(-3)^2 - 2(-3) + k = 0$$
$$9k + 6 + k = 0$$
$$10k = -6$$
$$k = -\frac{6}{10}$$
$$k = -\frac{3}{5}$$

b) Now substitute $-\frac{3}{5}$ for k in the original equation.
$$-\frac{3}{5}x^2 - 2x + \left(-\frac{3}{5}\right) = 0$$
$$3x^2 + 10x + 3 = 0 \quad \text{Multiplying by } -5$$
$$(3x+1)(x+3) = 0$$
$$x = -\frac{1}{3} \quad \text{or} \quad x = -3$$

The other solution is $-\frac{1}{3}$.

71. a) $x^2 - (6+3i)x + k = 0$; one solution is 3.

We first find k by substituting 3 for x.
$$3^2 - (6+3i)3 + k = 0$$
$$9 - 18 - 9i + k = 0$$
$$-9 - 9i + k = 0$$
$$k = 9 + 9i$$

b) Now we substitute $9 + 9i$ for k in the original equation.
$$x^2 - (6+3i)x + (9+9i) = 0$$
$$x^2 - (6+3i)x + 3(3+3i) = 0$$
$$[x - (3+3i)][x - 3] = 0$$

The other solution is $3 + 3i$.

73. The solutions of $ax^2 + bx + c = 0$ are $x = \frac{-b \pm \sqrt{b^2 - 4ac}}{2a}$.

When there is just one solution, $b^2 - 4ac = 0$, so
$$x = \frac{-b \pm 0}{2a} = -\frac{b}{2a}.$$

75. We substitute $(-3, 0)$, $\left(\frac{1}{2}, 0\right)$, and $(0, -12)$ in $f(x) = ax^2 + bx + c$ and get three equations.
$$0 = 9a - 3b + c,$$
$$0 = \frac{1}{4}a + \frac{1}{2}b + c,$$
$$-12 = c$$

The solution of this system of equations is $a = 8$, $b = 20$, $c = -12$.

77. If $-\sqrt{2}$ is one solution then $\sqrt{2}$ is another solution. Then

$$x = -\sqrt{2} \quad or \quad x = \sqrt{2}$$
$$x = \pm\sqrt{2}$$
$$x^2 = 2 \quad \text{Principle of square roots}$$
$$x^2 - 2 = 0$$

79. If $1 - \sqrt{5}$ and $3 + 2i$ are two solutions, then $1 + \sqrt{5}$ and $3 - 2i$ are also solutions. The equation of lowest degree that has these solutions is found as follows.

$$[x - (1 - \sqrt{5})][x - (1 + \sqrt{5})][x - (3 + 2i)][x - (3 - 2i)] = 0$$
$$(x^2 - 2x - 4)(x^2 - 6x + 13) = 0$$
$$x^4 - 8x^3 + 21x^2 - 2x - 52 = 0$$

81. *Writing Exercise.* If the discriminant indicates that there are two imaginary-number solutions, or only one rational solution, Keisha would know that her graph is not correct.

Exercise Set 8.4

1. *Familiarize.* We first make a drawing, labeling it with the known and unknown information. We can also organize the information in a table. We let r represent the speed and t the time for the first part of the trip.

r mph	t hr		$r - 10$ mph	$4 - t$ hr
120 mi			100 mi	

Trip	Distance	Speed	Time
1st part	120	r	t
2nd part	100	$r - 10$	$4 - t$

Translate. Using $r = \frac{d}{t}$, we get two equations from the table, $r = \frac{120}{t}$ and $r - 10 = \frac{100}{4 - t}$.

Carry out. We substitute $\frac{120}{t}$ for r in the second equation and solve for t.

$$\frac{120}{t} - 10 = \frac{100}{4 - t}, \quad \text{LCD is } t(4 - t)$$
$$t(4 - t)\left(\frac{120}{t} - 10\right) = t(4 - t) \cdot \frac{100}{4 - t}$$
$$120(4 - t) - 10t(4 - t) = 100t$$
$$480 - 120t - 40t + 10t^2 = 100t$$
$$10t^2 - 260t + 480 = 0 \quad \text{Standard form}$$
$$t^2 - 26t + 48 = 0 \quad \text{Multiplying by } \frac{1}{10}$$
$$(t - 2)(t - 24) = 0$$

$t = 2 \quad or \quad t = 24$

Check. Since the time cannot be negative (If $t = 24$, $4 - t = -20$.), we check only 2 hr. If $t = 2$, then

$4 - t = 2$. The speed of the first part is $\frac{120}{2}$, or 60 mph. The speed of the second part is $\frac{100}{2}$, or 50 mph. The speed of the second part is 10 mph slower than the first part. The value checks.

State. The speed of the first part was 60 mph, and the speed of the second part was 50 mph.

3. *Familiarize.* We first make a drawing. We also organize the information in a table. We let r = the speed and t = the time of the slower trip.

200 mi	r mph	t hr
200 mi	$r + 10$ mph	$t - 1$ hr

Trip	Distance	Speed	Time
Slower	200	r	t
Faster	200	$r + 10$	$t - 1$

Translate. Using $t = d/r$, we get two equations from the table:

$$t = \frac{200}{r} \quad \text{and} \quad t - 1 = \frac{200}{r + 10}$$

Carry out. We substitute $\frac{200}{r}$ for t in the second equation and solve for r.

$$\frac{200}{r} - 1 = \frac{200}{r + 10}, \qquad \text{LCD is } r(r + 10)$$
$$r(r + 10)\left(\frac{200}{r} - 1\right) = r(r + 10) \cdot \frac{200}{r + 10}$$
$$200(r + 10) - r(r + 10) = 200r$$
$$200r + 2000 - r^2 - 10r = 200r$$
$$0 = r^2 + 10r - 2000$$
$$0 = (r + 50)(r - 40)$$

$r = -50 \quad or \quad r = 40$

Check. Since negative speed has no meaning in this problem, we check only 40. If $r = 40$, then the time for the slower trip is $\frac{200}{40}$, or 5 hours. If $r = 40$, then $r + 10 = 50$ and the time for the faster trip is $\frac{200}{50}$, or 4 hours. This is 1 hour less time than the slower trip took, so we have an answer to the problem.

State. The speed is 40 mph.

5. *Familiarize.* We make a drawing and then organize the information in a table. We let r = the speed and t = the time of the Cessna.

600 mi	r mph	t hr
1000 mi	$r + 50$ mph	$t + 1$ hr

Plane	Distance	Speed	Time
Cessna	600	r	t
Beechcraft	1000	$r + 50$	$t + 1$

Translate. Using $t = d/r$, we get two equations from the table:

$$t = \frac{600}{r} \text{ and } t + 1 = \frac{1000}{r + 50}$$

Carry out. We substitute $\frac{600}{r}$ for t in the second equation and solve for r.

$$\frac{600}{r} + 1 = \frac{1000}{r + 50}, \qquad \text{LCD is } r(r + 50)$$

$$r(r + 50)\left(\frac{600}{r} + 1\right) = r(r + 50) \cdot \frac{1000}{r + 50}$$

$$600(r + 50) + r(r + 50) = 1000r$$

$$600r + 30{,}000 + r^2 + 50r = 1000r$$

$$r^2 - 350r + 30{,}000 = 0$$

$$(r - 150)(r - 200) = 0$$

$$r = 150 \quad or \quad r = 200$$

Check. If $r = 150$, then the Cessna's time is $\frac{600}{150}$, or 4 hr and the Beechcraft's time is $\frac{1000}{150 + 50}$, or $\frac{1000}{200}$, or 5 hr. If $r = 200$, then the Cessna's time is $\frac{600}{200}$, or 3 hr and the Beechcraft's time is $\frac{1000}{200 + 50}$, or $\frac{1000}{250}$, or 4 hr. Since the Beechcraft's time is 1 hr longer in each case, both values check. There are two solutions.

State. The speed of the Cessna is 150 mph and the speed of the Beechcraft is 200 mph; or the speed of the Cessna is 200 mph and the speed of the Beechcraft is 250 mph.

7. Familiarize. We make a drawing and then organize the information in a table. We let r represent the speed and t the time of the trip to Hillsboro.

			Hillsboro
36 mi	r mph	t hr	→
36 mi	$r - 3$ mph	$7 - t$ hr	

Trip	Distance	Speed	Time
To Hillsboro	36	r	t
Return	36	$r - 3$	$7 - t$

Translate. Using $t = \frac{d}{r}$, we get two equations from the table,

$$t = \frac{36}{r} \text{ and } 7 - t = \frac{36}{r - 3}.$$

Carry out. We substitute $\frac{36}{r}$ for t in the second equation and solve for r.

$$7 - \frac{36}{r} = \frac{36}{r - 3}, \qquad \text{LCD is } r(r - 3)$$

$$r(r - 3)\left(7 - \frac{36}{r}\right) = r(r - 3) \cdot \frac{36}{r - 3}$$

$$7r(r - 3) - 36(r - 3) = 36r$$

$$7r^2 - 21r - 36r + 108 = 36r$$

$$7r^2 - 93r + 108 = 0$$

$$(7r - 9)(r - 12) = 0$$

$$r = \frac{9}{7} \quad or \quad r = 12$$

Check. Since negative speed has no meaning in this problem (If $r = \frac{9}{7}$, then $r - 3 = -\frac{12}{7}$.), we check only 12 mph. If $r = 12$, then the time of the trip to Hillsboro is $\frac{36}{12}$, or 3 hr. The speed of the return trip is $12 - 3$, or 9 mph, and the time is $\frac{36}{9}$, or 4 hr. The total time for the round trip is $3 \text{ hr} + 4 \text{ hr}$, or 7 hr. The value checks.

State. Naoki's speed on the trip to Hillsboro was 12 mph and it was 9 mph on the return trip.

9. Familiarize. We make a drawing and organize the information in a table. Let r represent the speed of the boat in still water, and let t represent the time of the trip upriver.

60 mi	$r - 3$ mph	t hr	Upriver
Downriver	60 mi	$r + 3$ mph	$9 - t$ hr

Trip	Distance	Speed	Time
Upriver	60	$r - 3$	t
Downriver	60	$r + 3$	$9 - t$

Translate. Using $t = \frac{d}{r}$, we get two equations from the table,

$$t = \frac{60}{r - 3} \text{ and } 9 - t = \frac{60}{r + 3}.$$

Carry out. We substitute $\frac{60}{r - 3}$ for t in the second equation and solve for r.

$$9 - \frac{60}{r - 3} = \frac{60}{r + 3}$$

$$(r - 3)(r + 3)\left(9 - \frac{60}{r - 3}\right) = (r - 3)(r + 3) \cdot \frac{60}{r + 3}$$

$$9(r - 3)(r + 3) - 60(r + 3) = 60(r - 3)$$

$$9r^2 - 81 - 60r - 180 = 60r - 180$$

$$9r^2 - 120r - 81 = 0$$

$$3r^2 - 40r - 27 = 0 \quad \text{Dividing by 3}$$

We use the quadratic formula.

$$r = \frac{-(-40) \pm \sqrt{(-40)^2 - 4 \cdot 3 \cdot (-27)}}{2 \cdot 3}$$

$$r = \frac{40 \pm \sqrt{1924}}{6}$$

$$r \approx 14 \quad or \quad r \approx -0.6$$

Check. Since negative speed has no meaning in this problem, we check only 14 mph. If $r \approx 14$, then the speed upriver is about $14 - 3$, or 11 mph, and the time is about $\frac{60}{11}$, or 5.5 hr. The speed downriver is about $14 + 3$, or 17 mph, and the time is about $\frac{60}{17}$, or 3.5 hr. The total time of the round trip is $5.5 + 3.5$, or 9 hr. The value checks.

State. The speed of the boat in still water is about 14 mph.

11. *Familiarize.* Let x represent the time it take the spring to fill the pool. Then $x - 8$ represents the time it takes the well to fill the pool. It takes them 3 hr to fill the pool working together, so they can fill $\frac{1}{3}$ of the pool in 1 hr.

The spring will fill $\frac{1}{x}$ of the pool in 1 hr, and the well will fill $\frac{1}{x-8}$ of the pool in 1 hr.

Translate. We have an equation.
$$\frac{1}{x} + \frac{1}{x-8} = \frac{1}{3}$$

Carry out. We solve the equation.
We multiply by the LCD, $3x(x-8)$.
$$3x(x-8)\left(\frac{1}{x} + \frac{1}{x-8}\right) = 3x(x-8) \cdot \frac{1}{3}$$
$$3(x-8) + 3x = x(x-8)$$
$$3x - 24 + 3x = x^2 - 8x$$
$$0 = x^2 - 14x + 24$$
$$0 = (x-2)(x-12)$$

Check. Since negative time has no meaning in this problem, 2 is not a solution $(2 - 8 = -6)$. We check only 12 hr. This is the time it would take the spring working alone. Then the well would take $12 - 8$, or 4 hr working alone. The well would fill $3\left(\frac{1}{4}\right)$, or $\frac{3}{4}$ of the pool in 3 hr, and the spring would fill $3\left(\frac{1}{12}\right)$, or $\frac{1}{4}$ of the pool in 3 hr. Thus, in 3 hr they would fill $\frac{3}{4} + \frac{1}{4}$ of the pool. This is all of it, so the numbers check.

State. It takes the spring, working alone, 12 hr to fill the pool.

13. We make a drawing and then organize the information in a table. We let r represent Kofi's speed in still water. Then $r - 2$ is the speed upstream and $r + 2$ is the speed downstream. Using $t = \frac{d}{r}$, we let $\frac{1}{r-2}$ represent the time upstream and $\frac{1}{r+2}$ represent the time downstream.

$$\overset{\text{1 mi}}{\underset{}{\xrightarrow{\hspace{3cm}}}} \quad r - 2 \text{ mph}$$
Upstream

Downstream $\xleftarrow{\hspace{2cm}}$ 1 mi $\quad r + 2$ mph

Trip	Distance	Speed	Time
Upstream	1	$r-2$	$\frac{1}{r-2}$
Downstream	1	$r+2$	$\frac{1}{r+2}$

Translate. The time for the round trip is 1 hour. We now have an equation.
$$\frac{1}{r-2} + \frac{1}{r+2} = 1$$

Carry out. We solve the equation. We multiply by the LCD, $(r-2)(r+2)$.
$$(r-2)(r+2)\left(\frac{1}{r-2} + \frac{1}{r+2}\right) = (r-2)(r+2) \cdot 1$$
$$(r+2) + (r-2) = (r-2)(r+2)$$
$$2r = r^2 - 4$$
$$0 = r^2 - 2r - 4$$
$$a = 1, \ b = -2, \ c = -4$$
$$r = \frac{-(-2) \pm \sqrt{(-2)^2 - 4 \cdot 1 \cdot (-4)}}{2 \cdot 1}$$
$$r = \frac{2 \pm \sqrt{4 + 16}}{2} = \frac{2 \pm \sqrt{20}}{2}$$
$$r = \frac{2 \pm 2\sqrt{5}}{2} = 1 \pm \sqrt{5}$$
$$1 + \sqrt{5} \approx 1 + 2.236 \approx 3.24$$
$$1 - \sqrt{5} \approx 1 - 2.236 \approx -1.24$$

Check. Since negative speed has no meaning in this problem, we check only 3.24 mph. If $r \approx 3.24$, then $r - 2 \approx 1.24$ and $r + 2 \approx 5.24$. The time it takes to travel upstream is approximately $\frac{1}{1.24}$, or 0.806 hr, and the time it takes to travel downstream is approximately $\frac{1}{5.24}$, or 0.191 hr. The total time is 0.997 which is approximately 1 hour. The value checks.

State. Kofi's speed in still water is approximately 3.24 mph.

15.
$$A = 4\pi r^2$$
$$\frac{A}{4\pi} = r^2 \qquad \text{Dividing by } 4\pi$$
$$\frac{1}{2}\sqrt{\frac{A}{\pi}} = r \qquad \text{Taking the positive square root}$$

17. $A = 2\pi r^2 + 2\pi rh$

$0 = 2\pi r^2 + 2\pi rh - A$ Standard form

$a = 2\pi, \ b = 2\pi h, \ c = -A$

$r = \dfrac{-2\pi h \pm \sqrt{(2\pi h)^2 - 4 \cdot 2\pi \cdot (-A)}}{2 \cdot 2\pi}$ Using the quadratic formula

$r = \dfrac{-2\pi h \pm \sqrt{4\pi^2 h^2 + 8\pi A}}{4\pi}$

$r = \dfrac{-2\pi h \pm 2\sqrt{\pi^2 h^2 + 2\pi A}}{4\pi}$

$r = \dfrac{-\pi h \pm \sqrt{\pi^2 h^2 + 2\pi A}}{2\pi}$

Since taking the negative square root would result in a negative answer, we take the positive one.

$r = \dfrac{-\pi h + \sqrt{\pi^2 h^2 + 2\pi A}}{2\pi}$

19. $F = \dfrac{Gm_1 m_2}{r^2}$

$Fr^2 = Gm_1 m_2$

$r^2 = \dfrac{Gm_1 m_2}{F}$

$r = \sqrt{\dfrac{Gm_1 m_2}{F}}$

21. $c = \sqrt{gH}$

$c^2 = gH$ Squaring

$\dfrac{c^2}{g} = H$

23. $a^2 + b^2 = c^2$

$b^2 = c^2 - a^2$

$b = \sqrt{c^2 - a^2}$

25. $s = v_0 t + \dfrac{gt^2}{2}$

$0 = \dfrac{gt^2}{2} + v_0 t - s$ Standard form

$a = \dfrac{g}{2}, \ b = v_0, \ c = -s$

$t = \dfrac{-v_0 \pm \sqrt{v_0^2 - 4\left(\dfrac{g}{2}\right)(-s)}}{2\left(\dfrac{g}{2}\right)}$

$t = \dfrac{-v_0 \pm \sqrt{v_0^2 + 2gs}}{g}$

Since taking the negative square root would result in a negative answer, we take the positive one.

$t = \dfrac{-v_0 + \sqrt{v_0^2 + 2gs}}{g}$

27. $N = \dfrac{1}{2}\left(n^2 - n\right)$

$N = \dfrac{1}{2}n^2 - \dfrac{1}{2}n$

$0 = \dfrac{1}{2}n^2 - \dfrac{1}{2}n - N$

$a = \dfrac{1}{2}, \ b = -\dfrac{1}{2}, \ c = -N$

$n = \dfrac{-\left(-\dfrac{1}{2}\right) \pm \sqrt{\left(-\dfrac{1}{2}\right)^2 - 4 \cdot \dfrac{1}{2} \cdot (-N)}}{2\left(\dfrac{1}{2}\right)}$

$n = \dfrac{1}{2} \pm \sqrt{\dfrac{1}{4} + 2N}$

$n = \dfrac{1}{2} \pm \sqrt{\dfrac{1 + 8N}{4}}$

$n = \dfrac{1}{2} \pm \dfrac{1}{2}\sqrt{1 + 8N}$

Since taking the negative square root would result in a negative answer, we take the positive one.

$n = \dfrac{1}{2} + \dfrac{1}{2}\sqrt{1 + 8N}$, or $\dfrac{1 + \sqrt{1 + 8N}}{2}$

29. $T = 2\pi\sqrt{\dfrac{l}{g}}$

$\dfrac{T}{2\pi} = \sqrt{\dfrac{l}{g}}$ Multiplying by $\dfrac{1}{2\pi}$

$\dfrac{T^2}{4\pi^2} = \dfrac{l}{g}$ Squaring

$gT^2 = 4\pi^2 l$ Multiplying by $4\pi^2 g$

$g = \dfrac{4\pi^2 l}{T^2}$ Multiplying by $\dfrac{1}{T^2}$

31. $at^2 + bt + c = 0$

The quadratic formula gives the result.

$t = \dfrac{-b \pm \sqrt{b^2 - 4ac}}{2a}$

33. a)   ***Familiarize and Translate***. From Example 4, we know

$t = \dfrac{-v_0 + \sqrt{v_0^2 + 19.6s}}{9.8}$.

Carry out. Substituting 500 for s and 0 for v_0, we have

$t = \dfrac{0 + \sqrt{0^2 + 19.6(500)}}{9.8}$

$t \approx 10.1$

Check. Substitute 10.1 for t and 0 for v_0 in the original formula. (See Example 4.)

$s = 4.9t^2 + v_0 t = 4.9(10.1)^2 + 0 \cdot (10.1)^2 \approx 500$

The answer checks.

State. It takes the bolt about 10.1 sec to reach the ground.

b) *Familiarize and Translate*. From Example 4, we know

$$t = \frac{-v_0 + \sqrt{v_0^2 + 19.6s}}{9.8}$$

Carry out. Substitute 500 for s and 30 for v_0.

$$t = \frac{-30 + \sqrt{30^2 + 19.6(500)}}{9.8}$$
$$t \approx 7.49$$

Check. Substitute 30 for v_0 and 7.49 for t in the original formula. (See Example 4.)

$$s = 4.9t^2 + v_0 t = 4.9(7.49)^2 + (30)(7.49) \approx 500$$

The answer checks.

State. It takes the ball about 7.49 sec to reach the ground.

c) *Familiarize and Translate*. We will use the formula in Example 4, $s = 4.9t^2 + v_0 t$.

Carry out. Substitute 5 for t and 30 for v_0.

$$s = 4.9(5)^2 + 30(5) = 272.5$$

Check. We can substitute 30 for v_0 and 272.5 for s in the form of the formula we used in part (b).

$$t = \frac{-v_0 + \sqrt{v_0^2 + 19.6s}}{9.8}$$
$$= \frac{-30 + \sqrt{(30)^2 + 19.6(272.5)}}{9.8} = 5$$

The answer checks.

State. The object will fall 272.5 m.

35. *Familiarize*. We will use the formula $4.9t^2 = s$.

Translate. Substitute 40 for s.

$$4.9t^2 = 40$$

Carry out. We solve the equation.

$$4.9t^2 = 40$$
$$t^2 = \frac{40}{4.9}$$
$$t = \sqrt{\frac{40}{4.9}}$$
$$t \approx 2.9$$

Check. Substitute 2.9 for t in the formula.

$$s = 4.9(2.9)^2 = 41.209 \approx 40$$

The answer checks.

State. Chad will fall for about 2.9 sec before the cord begins to stretch.

37. *Familiarize*. We will use the formula $V = 48T^2$.

Translate. Substitute 38 for V.

$$38 = 48T^2$$

Carry out. We solve the equation.

$$38 = 48T^2$$
$$\frac{38}{48} = T^2$$
$$T = \sqrt{\frac{38}{48}}$$
$$T \approx 0.890$$

Check. Substitute 0.890 for T in the formula.

$$V = 48(0.890)^2 \approx 38$$

The answer checks.

State. Dwight Howard's hang time is about 0.890 sec.

39. *Familiarize and Translate*. We will use the formula in Example 4, $s = 4.9t^2 + v_0 t$.

Carry out. Solve the formula for v_0.

$$s - 4.9t^2 = v_0 t$$
$$\frac{s - 4.9t^2}{t} = v_0$$

Now substitute 51.6 for s and 3 for t.

$$\frac{51.6 - 4.9(3)^2}{3} = v_0$$
$$2.5 = v_0$$

Check. Substitute 3 for t and 2.5 for v_0 in the original formula.

$$s = 4.9(3)^2 + 2.5(3) = 51.6$$

The solution checks.

State. The initial velocity is 2.5 m/sec.

41. *Familiarize and Translate*. From Exercise 32 we know that $r = -1 + \dfrac{-P_2 + \sqrt{P_2^2 + 4AP_1}}{2P_1}$

where A is the total amount in the account after two years, P_1 is the amount of the original deposit, P_2 is deposited at the beginning of the second year, and r is the annual interest rate.

Carry out. Substitute 3200 for P_1, 1800 for P_2, and 5375.48 for A.

$$r = -1 + \frac{-1800 + \sqrt{(1800)^2 + 4(5375.48)(3200)}}{2(3200)}$$

Using a calculator we have $r = 0.045$.

Check. Substitute in the original formula in Exercise 32.

$$A = P_1(1+r)^2 + P_2(1+r)$$
$$A = 3200(1.045)^2 + 1800(1.045) = 5375.48$$

The solution checks.

State. The annual interest rate is 0.045 or 4.5%.

43. *Writing Exercise*. Let the length of Rafe's cord be s, in meters. Then the length of Marti's cord is $2s$. Then, using

the formula in Example 4, we find that the time t_1 that Rafe will fall before his cord begins to stretch is given by

$$t_1 = \frac{0 \pm \sqrt{0^2 + 19.6s}}{9.8} \approx 0.5\sqrt{s} \text{ sec. Similarly, the time } t_2$$

that Marti will fall before her cord begins to stretch is

given by $t_2 = \dfrac{0 \pm \sqrt{0^2 + 19.6(2s)}}{9.8} \approx 0.6\sqrt{s}$ sec. Since

$t_2 \ne 2t_1$, Marti's fall will not take twice as long as Rafe's.

45. $\left(m^{-1}\right)^2 = m^{-1 \cdot 2} = m^{-2}$ or $\dfrac{1}{m^2}$

47. $\left(y^{1/6}\right)^2 = y^{\frac{1}{6} \cdot 2} = y^{2/6} = y^{1/3}$

49. $t^{-1} = \dfrac{1}{2}$

$\dfrac{1}{t} = \dfrac{1}{2}$

$2t \cdot \dfrac{1}{t} = 2t \cdot \dfrac{1}{2}$

$2 = t$

51. *Writing Exercise.* Answers may vary.

An express train travels 30 mph faster than a local train. The express train travels 450 mi in 2 hr less time than it takes the local train to travel 420 mi. Find the speed of the express train.

53. $A = 6.5 - \dfrac{20.4t}{t^2 + 36}$

$(t^2 + 36)A = (t^2 + 36)\left(6.5 - \dfrac{20.4t}{t^2 + 36}\right)$

$At^2 + 36A = (t^2 + 36)(6.5) - (t^2 + 36)\left(\dfrac{20.4t}{t^2 + 36}\right)$

$At^2 + 36A = 6.5t^2 + 234 - 20.4t$

$At^2 - 6.5t^2 + 20.4t + 36A - 234 = 0$

$(A - 6.5)t^2 + 20.4t + (36A - 234) = 0$

$a = A - 6.5,\ b = 20.4,\ c = 36A - 234$

$t = \dfrac{-20.4 + \sqrt{(20.4)^2 - 4(A - 6.5)(36A - 234)}}{2(A - 6.5)}$

$t = \dfrac{-20.4 + \sqrt{416.16 - 144A^2 + 1872A - 6084}}{2(A - 6.5)}$

$t = \dfrac{-20.4 + \sqrt{-144A^2 + 1872A - 5667.84}}{2(A - 6.5)}$

$t = \dfrac{-20.4 + \sqrt{144(-A^2 + 13A - 39.36)}}{2(A - 6.5)}$

$t = \dfrac{-20.4 + 12\sqrt{-A^2 + 13A - 39.36}}{2(A - 6.5)}$

$t = \dfrac{2(-10.2 + 6\sqrt{-A^2 + 13A - 39.36})}{2(A - 6.5)}$

$t = \dfrac{-10.2 + 6\sqrt{-A^2 + 13A - 39.36}}{A - 6.5}$

55. *Familiarize.* Let $a =$ the number. Then $a - 1$ is 1 less than a and the reciprocal of that number is $\dfrac{1}{a-1}$. Also, 1 more than the number is $a + 1$.

Translate.

The reciprocal of 1 less than a number	is	1 more than the number.
$\downarrow$	$\downarrow$	$\downarrow$
$\dfrac{1}{(a-1)}$	$=$	$a+1$

Carry out. We solve the equation.

$\dfrac{1}{a-1} = a + 1,$ LCD is $a - 1$

$(a-1) \cdot \dfrac{1}{a-1} = (a-1)(a+1)$

$1 = a^2 - 1$

$2 = a^2$

$\pm\sqrt{2} = a$

Check. $\dfrac{1}{\sqrt{2}-1} \approx 2.4142 \approx \sqrt{2} + 1$ and

$\dfrac{1}{-\sqrt{2}-1} \approx -0.4142 \approx -\sqrt{2} + 1$. The answers check.

State. The numbers are $\sqrt{2}$ and $-\sqrt{2}$, or $\pm\sqrt{2}$.

57. $\dfrac{w}{l} = \dfrac{l}{w+l}$

$l(w+l) \cdot \dfrac{w}{l} = l(w+l) \cdot \dfrac{l}{w+l}$

$w(w+l) = l^2$

$w^2 + lw = l^2$

$0 = l^2 - lw - w^2$

Use the quadratic formula with $a = 1$, $b = -w$, and $c = -w^2$.

$l = \dfrac{-(-w) \pm \sqrt{(-w)^2 - 4 \cdot 1 \cdot (-w^2)}}{2 \cdot 1}$

$l = \dfrac{w \pm \sqrt{w^2 + 4w^2}}{2} = \dfrac{w \pm \sqrt{5w^2}}{2}$

$l = \dfrac{w \pm w\sqrt{5}}{2}$

Since $\dfrac{w - w\sqrt{5}}{2}$ is negative we use the positive square

root: $l = \dfrac{w + w\sqrt{5}}{2}$

59. $mn^4 - r^2pm^3 - r^2n^2 + p = 0$

Let $u = n^2$. Substitute and rearrange.

$mu^2 - r^2u - r^2pm^3 + p = 0$

$a = m,\ b = -r^2,\ c = -r^2pm^3 + p$

$$u = \frac{-(-r^2) \pm \sqrt{(-r^2)^2 - 4 \cdot m(-r^2 pm^3 + p)}}{2 \cdot m}$$

$$u = \frac{r^2 \pm \sqrt{r^4 + 4m^4 r^2 p - 4mp}}{2m}$$

$$n^2 = \frac{r^2 \pm \sqrt{r^4 + 4m^4 r^2 p - 4mp}}{2m}$$

$$n = \pm\sqrt{\frac{r^2 \pm \sqrt{r^4 + 4m^4 r^2 p - 4mp}}{2m}}$$

61. Let s represent a length of a side of the cube, let S represent the surface area of the cube, and let A represent the surface area of the sphere. Then the diameter of the sphere is s, so the radius r is $s/2$. From Exercise 15, we know, $A = 4\pi r^2$, so when $r = s/2$ we have

$$A = 4\pi\left(\frac{s}{2}\right)^2 = 4\pi \cdot \frac{s^2}{4} = \pi s^2 .$$

From the formula for the surface area of a cube (See Exercise 16.) we know that $S = 6s^2$, so $\frac{S}{6} = s^2$ and then

$$A = \pi \cdot \frac{S}{6}, \text{ or } A(S) = \frac{\pi S}{6} .$$

Exercise Set 8.5

1. $x^6 = (x^3)^2$, so (f) is an appropriate choice.

3. $x^8 = (x^4)^2$, so (h) is an appropriate choice.

5. $x^{4/3} = (x^{2/3})^2$, so (g) is an appropriate choice.

7. $x^{-4/3} = (x^{-2/3})^2$, so (e) is an appropriate choice.

9. Since $\left(\sqrt{p}\right)^2 = p$, use $u = \sqrt{p}$.

11. Since $\left(x^2 + 3\right)^2 = \left(x^2 + 3\right)^2$, use $u = x^2 + 3$.

13. Since $\left[(1+t)^2\right]^2 = (1+t)^2$, use $u = (1+t)^2$.

15. $x^4 - 13x^2 + 36 = 0$

Let $u = x^2$ and $u^2 = x^4$.

$u^2 - 13u + 36 = 0$ Substituting u for x^2
$(u-4)(u-9) = 0$
$u - 4 = 0$ or $u - 9 = 0$
$\quad u = 4$ or $\quad u = 9$

Now replace u with x^2 and solve these equations.

$\quad x^2 = 4$ or $x^2 = 9$
$\quad x = \pm 2$ or $x = \pm 3$

The numbers 2, –2, 3, and –3 check. They are the solutions.

17. $t^4 - 7t^2 + 12 = 0$

Let $u = t^2$ and $u^2 = t^4$.

$u^2 - 7u + 12 = 0$ Substituting u for t^2
$(u-3)(u-4) = 0$
$u - 3 = 0$ or $u - 4 = 0$
$\quad u = 3$ or $\quad u = 4$

Now replace u with t^2 and solve these equations.

$\quad t^2 = 3$ or $t^2 = 4$
$\quad t = \pm\sqrt{3}$ or $t = \pm 2$

The numbers $\sqrt{3}, -\sqrt{3}$, 2, and –2 check. They are the solutions.

19. $4x^4 - 9x^2 + 5 = 0$

Let $u = x^2$ and $u^2 = x^4$.

$\quad 4u^2 - 9u + 5 = 0$ Substituting u for x^2
$(4u-5)(u-1) = 0$
$4u - 5 = 0$ or $u - 1 = 0$
$\quad u = \frac{5}{4}$ or $\quad u = 1$

Now replace u with x^2 and solve these equations.

$\quad x^2 = \frac{5}{4}$ or $x^2 = 1$
$\quad x = \pm\sqrt{\frac{5}{4}} = \pm\frac{\sqrt{5}}{2}$ or $x = \pm 1$

The numbers $\frac{\sqrt{5}}{2}, -\frac{\sqrt{5}}{2}$, 1, and –1 check. They are the solutions.

21. $w + 4\sqrt{w} - 12 = 0$

Let $u = \sqrt{w}$ and $u^2 = \left(\sqrt{w}\right)^2 = w$.

$\quad u^2 + 4u - 12 = 0$
$(u+6)(u-2) = 0$
$u + 6 = 0$ or $u - 2 = 0$
$\quad u = -6$ or $\quad u = 2$

Now replace u with $\sqrt{w}$ and solve these equations.

$\quad \sqrt{w} = -6$ or $\sqrt{w} = 2$
$\quad\quad\quad\quad$ or $w = 4$

Since the principal square root cannot be negative, only 4 checks as a solution.

23. $(x^2 - 7)^2 - 3(x^2 - 7) + 2 = 0$

Let $u = x^2 - 7$ and $u^2 = (x^2 - 7)^2$.

$\quad u^2 - 3u + 2 = 0$ Substituting
$(u-1)(u-2) = 0$
$\quad u = 1$ or $\quad u = 2$
$x^2 - 7 = 1$ or $x^2 - 7 = 2$ Replacing u with $x^2 - 7$
$\quad x^2 = 8$ or $\quad x^2 = 9$
$x = \pm\sqrt{8} = \pm 2\sqrt{2}$ or $x = \pm 3$

The numbers $2\sqrt{2}$, $-2\sqrt{3}$, 3, and –3 check. They are the solutions.

25. $r - 2\sqrt{r} - 6 = 0$

Let $u = \sqrt{r}$ and $u^2 = r$.

$u^2 - 2u - 6 = 0$

$$u = \frac{-(-2) \pm \sqrt{(-2)^2 - 4 \cdot 1 \cdot (-6)}}{2 \cdot 1}$$

$$u = \frac{2 \pm \sqrt{28}}{2} = \frac{2 + 2\sqrt{7}}{2}$$

$$u = 1 \pm \sqrt{7}$$

Replace u with $\sqrt{r}$ and solve these equations:

$$\sqrt{r} = 1 + \sqrt{7} \quad or \quad \sqrt{r} = 1 - \sqrt{7}$$
$$\left(\sqrt{r}\right)^2 = \left(1 + \sqrt{7}\right)^2$$
$$r = 1 + 2\sqrt{7} + 7$$
$$r = 8 + 2\sqrt{7}$$

The number $1 - \sqrt{7}$ is not a solution since it is negative. The number $8 + 2\sqrt{7}$ checks. It is the solution.

27. $(1 + \sqrt{x})^2 + 5(1 + \sqrt{x}) + 6 = 0$

Let $u = 1 + \sqrt{x}$ and $u^2 = (1 + \sqrt{x})^2$.

$u^2 + 5u + 6 = 0$ Substituting
$(u + 3)(u + 2) = 0$

$$u = -3 \quad or \quad u = -2$$
$$1 + \sqrt{x} = -3 \quad or \quad 1 + \sqrt{x} = -2 \quad \text{Replacing } u \text{ with } 1 + \sqrt{x}$$
$$\sqrt{x} = -4 \quad or \quad \sqrt{x} = -3$$

Since the principal square root cannot be negative, this equation has no solution.

29. $x^{-2} - x^{-1} - 6 = 0$

Let $u = x^{-1}$ and $u^2 = x^{-2}$.

$u^2 - u - 6 = 0$ Substituting
$(u - 3)(u + 2) = 0$

$u = 3 \ or \ u = -2$

Now we replace u with x^{-1} and solve these equations:

$$x^{-1} = 3 \quad or \quad x^{-1} = -2$$
$$\frac{1}{x} = 3 \quad or \quad \frac{1}{x} = -2$$
$$\frac{1}{3} = x \quad or \quad -\frac{1}{2} = x$$

Both $\frac{1}{3}$ and $-\frac{1}{2}$ check. They are the solutions.

31. $4t^{-2} - 3t^{-1} - 1 = 0$

Let $u = t^{-1}$ and $u^2 = t^{-2}$.

$4u^2 - 3u - 1 = 0$ Substituting
$(4u + 1)(u - 1) = 0$

$$4u + 1 = 0 \quad or \quad u - 1 = 0$$
$$u = -\frac{1}{4} \quad or \quad u = 1$$

Now replace u with t^{-1} and solve these equations.

$$t^{-1} = -\frac{1}{4} \quad or \quad t^{-1} = 1$$
$$\frac{1}{t} = -\frac{1}{4} \quad or \quad \frac{1}{t} = 1$$
$$-4 = t \quad or \quad 1 = t$$

Both -4 and 1 check. They are the solutions.

33. $t^{2/3} + t^{1/3} - 6 = 0$

Let $u = t^{1/3}$ and $u^2 = t^{2/3}$.

$u^2 + u - 6 = 0$ Substituting
$(u + 3)(u - 2) = 0$

$u = -3 \ or \ u = 2$

Now we replace u with $t^{1/3}$ and solve these equations:

$$t^{1/3} = -3 \quad or \quad t^{1/3} = 2$$
$$t = (-3)^3 \quad or \quad t = 2^3 \quad \text{Raising to the third power}$$
$$t = -27 \quad or \quad t = 8$$

Both -27 and 8 check. They are the solutions.

35. $y^{1/3} - y^{1/6} - 6 = 0$

Let $u = y^{1/6}$ and $u^2 = y^{2/3}$.

$u^2 - u - 6 = 0$ Substituting
$(u - 3)(u + 2) = 0$

$u = 3 \ or \ u = -2$

Now we replace u with $y^{1/6}$ and solve these equations:

$$y^{1/6} = 3 \quad or \quad y^{1/6} = -2$$
$$\sqrt[6]{y} = 3 \quad or \quad \sqrt[6]{y} = -2$$
$$y = 3^6$$
$$y = 729$$

The equation $\sqrt[6]{y} = -2$ has no solution since principal sixth roots are never negative. The number 729 checks and is the solution.

37. $t^{1/3} + 2t^{1/6} = 3$
$t^{1/3} + 2t^{1/6} - 3 = 0$

Let $u = t^{1/6}$ and $u^2 = t^{2/6} = t^{1/3}$.

$u^2 + 2u - 3 = 0$ Substituting
$(u + 3)(u - 1) = 0$

$$u = -3 \quad or \quad u = 1$$
$$t^{1/6} = -3 \quad or \quad t^{1/6} = 1 \quad \text{Substituting } t^{1/6} \text{ for } u$$
$$t = 1$$

Since principal sixth roots are never negative, the equation $t^{1/6} = -3$ has no solution.

The number 1 checks and is the solution.

39. $(3 - \sqrt{x})^2 - 10(3 - \sqrt{x}) + 23 = 0$

Let $u = 3 - \sqrt{x}$ and $u^2 = (3 - \sqrt{x})^2$.

$u^2 - 10u + 23 = 0$ Substituting

$u = \dfrac{-(-10) \pm \sqrt{(-10)^2 - 4 \cdot 1 \cdot 23}}{2 \cdot 1}$

$u = \dfrac{10 \pm \sqrt{8}}{2} = \dfrac{2 \cdot 5 \pm 2\sqrt{2}}{2}$

$u = 5 \pm \sqrt{2}$

$u = 5 + \sqrt{2}$ or $u = 5 - \sqrt{2}$

Now we replace u with $3 - \sqrt{x}$ and solve these equations:

$\begin{array}{llll} 3 - \sqrt{x} = 5 + \sqrt{2} & or & 3 - \sqrt{x} = 5 - \sqrt{2} \\ -\sqrt{x} = 2 + \sqrt{2} & or & -\sqrt{x} = 2 - \sqrt{2} \\ \sqrt{x} = -2 - \sqrt{2} & or & \sqrt{x} = -2 + \sqrt{2} \end{array}$

Since both $-2 - \sqrt{2}$ and $-2 + \sqrt{2}$ are negative and principal square roots are never negative, the equation has no solution.

41. $16\left(\dfrac{x-1}{x-8}\right)^2 + 8\left(\dfrac{x-1}{x-8}\right) + 1 = 0$

Let $u = \dfrac{x-1}{x-8}$ and $u^2 = \left(\dfrac{x-1}{x-8}\right)^2$.

$16u^2 + 8u + 1 = 0$ Substituting
$(4u+1)(4u+1) = 0$

$u = -\dfrac{1}{4}$

Now we replace u with $\dfrac{x-1}{x-8}$ and solve this equation:

$\dfrac{x-1}{x-8} = -\dfrac{1}{4}$

$4x - 4 = -x + 8$ Multiplying by $4(x-8)$

$5x = 12$

$x = \dfrac{12}{5}$

The number $\dfrac{12}{5}$ checks and is the solution.

43. $x^4 + 5x^2 - 36 = 0$

Let $u = x^2$ and $u^2 = x^4$.

$u^2 + 5u - 36 = 0$
$(u-4)(u+9) = 0$

$\begin{array}{llll} u - 4 = 0 & or & u + 9 = 0 \\ u = 4 & or & u = -9 \end{array}$

Now replace u with x^2 and solve these equations.

$\begin{array}{llll} x^2 = 4 & or & x^2 = -9 \\ x = \pm 2 & or & x = \pm\sqrt{-9} = \pm 3i \end{array}$

The numbers 2, -2, $3i$, and $-3i$ check. They are the solutions.

45. $\left(n^2 + 6\right)^2 - 7\left(n^2 + 6\right) + 10 = 0$

Let $u = n^2 + 6$ and $u^2 = \left(n^2 + 6\right)^2$.

$u^2 - 7u + 10 = 0$
$(u-2)(u-5) = 0$

$\begin{array}{llll} u - 2 = 0 & or & u - 5 = 0 \\ u = 2 & or & u = 5 \end{array}$

Now replace u with $n^2 + 6$ and solve these equations.

$\begin{array}{llll} n^2 + 6 = 2 & or & n^2 + 6 = 5 \\ n^2 = -4 & or & n^2 = -1 \\ n = \pm\sqrt{-4} = \pm 2i & or & n = \pm\sqrt{-1} = \pm i \end{array}$

$2i$, $-2i$, i, and $-i$ check. They are the solutions.

47. The x-intercepts occur where $f(x) = 0$. Thus, we must have $5x + 13\sqrt{x} - 6 = 0$.

Let $u = \sqrt{x}$ and $u^2 = x$.

$5u^2 + 13u - 6 = 0$ Substituting
$(5u - 2)(u + 3) = 0$

$u = \dfrac{2}{5}$ or $u = -3$

Now replace u with $\sqrt{x}$ and solve these equations:

$\sqrt{x} = \dfrac{2}{5}$ or $\sqrt{x} = -3$ has no solution

$x = \dfrac{4}{25}$

The number $\dfrac{4}{25}$ checks. Thus, the x-intercept is $\left(\dfrac{4}{25}, 0\right)$.

49. The x-intercepts occur where $f(x) = 0$. Thus, we must have $(x^2 - 3x)^2 - 10(x^2 - 3x) + 24 = 0$.

Let $u = x^2 - 3x$ and $u^2 = (x^2 - 3x)^2$.

$u^2 - 10u + 24 = 0$ Substituting
$(u - 6)(u - 4) = 0$

$u = 6$ or $u = 4$

Now replace u with $x^2 - 3x$ and solve these equations:

$\begin{array}{llll} x^2 - 3x = 6 & or & x^2 - 3x = 4 \\ x^2 - 3x - 6 = 0 & or & x^2 - 3x - 4 = 0 \end{array}$

$x = \dfrac{-(-3) \pm \sqrt{(-3)^2 - 4(1)(-6)}}{2 \cdot 1}$ or $(x-4)(x+1) = 0$

$x = \dfrac{3}{2} \pm \dfrac{\sqrt{33}}{2}$ $\qquad or \quad x = 4 \ or \ x = -1$

All four numbers check. Thus, the x-intercepts are $\left(\dfrac{3}{2} + \dfrac{\sqrt{33}}{2}, 0\right)$, $\left(\dfrac{3}{2} - \dfrac{\sqrt{33}}{2}, 0\right)$, $(4, 0)$, and $(-1, 0)$.

51. The x-intercepts occur where $f(x) = 0$. Thus, we must have $x^{2/5} + x^{1/5} - 6 = 0$.

Let $u = x^{1/5}$ and $u^2 = x^{2/5}$.

$u^2 + u - 6 = 0$ Substituting
$(u + 3)(u - 2) = 0$

$\begin{array}{llll} u = -3 & or & u = 2 \\ x^{1/5} = -3 & or & x^{1/5} = 2 & \text{Replacing } u \text{ with } x^{1/5} \\ x = -243 & & x = 32 & \text{Raising to the fifth power} \end{array}$

Both -243 and 32 check. Thus, the x-intercepts are $(-243, 0)$ and $(32, 0)$.

53. $f(x) = \left(\dfrac{x^2+2}{x}\right)^4 + 7\left(\dfrac{x^2+2}{x}\right)^2 + 5$

Observe that, for all real numbers x, each term is positive. Thus, there are no real-number values of x for which $f(x) = 0$ and hence no x-intercepts.

55. *Writing Exercise.* They can both be correct. Jose's substitution reduces the original equation to the quadratic equation $u^2 - 2u + 1 = 0$ while Robin's produces the equation $25u^2 - 10u + 1 = 0$. Both equations can be solved and they yield the same result.

57. Graph $f(x) = x$.

We find some ordered pairs, plot points, and draw the graph.

x	y
-2	-2
-1	-1
0	0
1	1
2	2

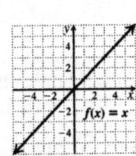

59. Graph $h(x) = x - 2$.

We find some ordered pairs, plot points, and draw the graph.

x	y
-1	-3
0	-2
1	-1
2	0
3	1

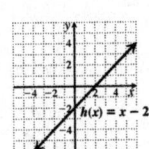

61. Graph $g(x) = x^2 + 2$.

We find some ordered pairs, plot points, and draw the graph.

x	y
-2	6
-1	3
0	2
1	3
2	6

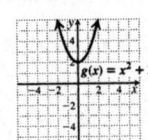

63. *Writing Exercise.* Substitute u for x^2 and solve $au^2 + bu + c = 0$. Then replace u with x^2 and solve for x.

65. $3x^4 + 5x^2 - 1 = 0$

Let $u = x^2$ and $u^2 = x^4$.

$3u^2 + 5u - 1 = 0$ Substituting

$u = \dfrac{-5 \pm \sqrt{5^2 - 4 \cdot 3 \cdot (-1)}}{2 \cdot 3}$

$u = \dfrac{-5 \pm \sqrt{37}}{6}$

$x^2 = \dfrac{-5 \pm \sqrt{37}}{6}$ Replacing u with x^2

$x = \pm\sqrt{\dfrac{-5 \pm \sqrt{37}}{6}}$

All four numbers check and are the solutions.

67. $(x^2 - 5x - 1)^2 - 18(x^2 - 5x - 1) + 65 = 0$

Let $u = x^2 - 5x - 1$ and $u^2 = (x^2 - 5x - 1)^2$.

$u^2 - 18u + 65 = 0$ Substituting
$(u - 5)(u - 13) = 0$

$\qquad u = 5 \quad or \qquad\qquad u = 13$
$x^2 - 5x - 1 = 5 \quad or \quad x^2 - 5x - 1 = 13$
$\qquad\qquad$ Replacing u with $x^2 - 4x - 2$
$x^2 - 5x - 6 = 0 \quad or \quad x^2 - 5x - 14 = 0$
$(x - 6)(x + 1) = 0 \quad or \quad (x - 7)(x + 2) = 0$

$x = 6 \quad or \quad x = -1 \quad or \quad x = 7 \quad or \quad x = -2$

The numbers 6, –1, 7 and –2 check and are the solutions.

69. $\dfrac{x}{x-1} - 6\sqrt{\dfrac{x}{x-1}} - 40 = 0$

Let $u = \sqrt{\dfrac{x}{x-1}}$ and $u^2 = \dfrac{x}{x-1}$.

$u^2 - 6u - 40 = 0$ Substituting
$(u - 10)(u + 4) = 0$

$\qquad u = 10 \quad or \qquad u = -4$
$\sqrt{\dfrac{x}{x-1}} = 10 \quad or \quad \sqrt{\dfrac{x}{x-1}} = -4$ has no solution

$\dfrac{x}{x-1} = 100$

$x = 100x - 100$ Multiplying by $(x-1)$
$100 = 99x$
$\dfrac{100}{99} = x$

The number $\dfrac{100}{99}$ checks. It is the solution.

71. $a^5(a^2 - 25) + 13a^3(25 - a^2) + 36a(a^2 - 25) = 0$
$a^5(a^2 - 25) - 13a^3(a^2 - 25) + 36a(a^2 - 25) = 0$
$a(a^2 - 25)(a^4 - 13a^2 + 36) = 0$
$a(a^2 - 25)(a^2 - 4)(a^2 - 9) = 0$

$a = 0 \quad or \quad a^2 - 25 = 0 \quad or \quad a^2 - 4 = 0 \quad or \quad a^2 - 9 = 0$
$a = 0 \quad or \qquad a^2 = 25 \quad or \qquad a^2 = 4 \quad or \qquad a^2 = 9$
$a = 0 \quad or \qquad a = \pm 5 \quad or \qquad a = \pm 2 \quad or \qquad a = \pm 3$

All seven numbers check. The solutions are 0, 5, –5, 2, –2, 3, and –3.

73. $x^6 - 28x^3 + 27 = 0$

Let $u = x^3$.

$u^2 - 28u + 27 = 0$
$(u - 27)(u - 1) = 0$
$u = 27 \quad or \quad u = 1$
$x^3 = 27 \quad or \quad x^3 = 1$
$x^3 - 27 = 0 \quad or \quad x^3 - 1 = 0$

First we solve $x^3 - 27 = 0$.

$x^3 - 27 = 0$
$(x - 3)(x^2 + 3x + 9) = 0$
$x - 3 = 0 \quad or \quad x^2 + 3x + 9 = 0$

$x = 3 \quad or \quad x = \dfrac{-3 \pm \sqrt{3^2 - 4 \cdot 1 \cdot 9}}{2 \cdot 1}$

$x = 3 \quad or \quad x = \dfrac{-3 \pm \sqrt{-27}}{2}$

$x = 3 \quad or \quad x = -\dfrac{3}{2} \pm \dfrac{3\sqrt{3}}{2}i$

Next we solve $x^3 - 1 = 0$.

$x^3 - 1 = 0$
$(x - 1)(x^2 + x + 1) = 0$
$x - 1 = 0 \quad or \quad x^2 + x + 1 = 0$

$x = 1 \quad or \quad x = \dfrac{-1 \pm \sqrt{1^2 - 4 \cdot 1 \cdot 1}}{2 \cdot 1}$

$x = 1 \quad or \quad x = \dfrac{-1 \pm \sqrt{-3}}{2}$

$x = 1 \quad or \quad x = -\dfrac{1}{2} \pm \dfrac{\sqrt{3}}{2}i$

All six numbers check.

75. *Graphing Calculator Exercise*

77. *Writing Exercise.* Salam can find only two real-number solutions using a graphing calculator. There are also two imaginary-number solutions which must be found algebraically.

Connecting the Concepts

1. $x^2 - 3x - 10 = 0$
$(x + 2)(x - 5) = 0$
$x + 2 = 0 \quad or \quad x - 5 = 0$
$x = -2 \quad or \quad x = 5$

3. $x^2 + 6x = 10$
$x^2 + 6x + 9 = 10 + 9 \quad$ Completing the square
$(x + 3)^2 = 19$
$x + 3 = \pm\sqrt{19}$
$x = -3 \pm \sqrt{19}$

5. $(x + 1)^2 = 2$
$x + 1 = \pm\sqrt{2}$
$x = -1 \pm \sqrt{2}$

7. $x^2 - x - 1 = 0$

$a = 1, \, b = -1, \, c = -1$

$x = \dfrac{-(-1) \pm \sqrt{(-1)^2 - 4 \cdot 1 \cdot (-1)}}{2 \cdot 1} = \dfrac{1 \pm \sqrt{1 + 4}}{2}$

$= \dfrac{1 \pm \sqrt{5}}{2} = \dfrac{1}{2} \pm \dfrac{\sqrt{5}}{2}$

9. $4t^2 = 11$
$t^2 = \dfrac{11}{4}$
$t = \pm\sqrt{\dfrac{11}{4}} = \pm\dfrac{\sqrt{11}}{2}$

11. $c^2 + c + 1 = 0$

$a = 1, \, b = 1, \, c = 1$

$c = \dfrac{-1 \pm \sqrt{1^2 - 4 \cdot 1 \cdot 1}}{2 \cdot 1} = \dfrac{-1 \pm \sqrt{1 - 4}}{2}$

$= \dfrac{-1 \pm \sqrt{-3}}{2} = -\dfrac{1}{2} \pm \dfrac{\sqrt{3}}{2}i$

13. $6y^2 - 7y - 10 = 0$
$(6y + 5)(y - 2) = 0$
$6y + 5 = 0 \quad or \quad y - 2 = 0$
$y = -\dfrac{5}{6} \quad or \quad y = 2$

15. $x^4 - 10x^2 + 9 = 0$

Let $u = x^2$ and $u^2 = x^4$.

$u^2 - 10u + 9 = 0$
$(u - 1)(u - 9) = 0$
$u - 1 = 0 \quad or \quad u - 9 = 0$
$u = 1 \quad or \quad u = 9$

Replace u with x^2.

$x^2 = 1 \quad or \quad x^2 = 9$
$x = \pm 1 \quad or \quad x = \pm 3$

17. $t(t - 3) = 2t(t + 1)$
$t^2 - 3t = 2t^2 + 2t$
$0 = t^2 + 5t$
$0 = t(t + 5)$
$t = 0 \quad or \quad t + 5 = 0$
$t = 0 \quad or \quad t = -5$

19. $(m^2 + 3)^2 - 4(m^2 + 3) - 5 = 0$

Let $u = m^2 + 3$ and $u^2 = (m^2 + 3)^2$.

$u^2 - 4u - 5 = 0$
$(u - 5)(u + 1) = 0$
$u - 5 = 0 \quad or \quad u + 1 = 0$
$u = 5 \quad or \quad u = -1$

Replace u with $m^2 + 3$.

$$m^2 + 3 = 5 \quad or \quad m^2 + 3 = -1$$
$$m^2 = 2 \quad or \quad m^2 = -4$$
$$m = \pm\sqrt{2} \quad or \quad m = \pm\sqrt{-4} = \pm 2i$$

Exercise Set 8.6

1. The graph of $f(x) = 2(x-1)^2 + 3$ has vertex $(1, 3)$ and opens up. Choice (h) is correct.

3. The graph of $f(x) = 2(x+1)^2 + 3$ has vertex $(-1, 3)$ and opens up. Choice (f) is correct.

5. The graph of $f(x) = -2(x+1)^2 + 3$ has vertex $(-1, 3)$ and opens down. Choice (b) is correct.

7. The graph of $f(x) = 2(x+1)^2 - 3$ has vertex $(-1, -3)$ and opens up. Choice (e) is correct.

9. $f(x) = x^2$

See Example 1 in the text.

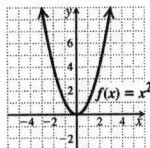

11. $f(x) = -2x^2$

We choose some numbers for x and compute $f(x)$ for each one. Then we plot the ordered pairs $(x, f(x))$ and connect them with a smooth curve.

x	$f(x) = -2x^2$
0	0
1	-2
2	-8
-1	-2
-2	-8

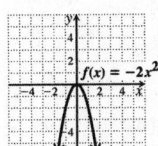

13. $g(x) = \frac{1}{3}x^2$

x	$g(x) = \frac{1}{3}x^2$
0	0
1	$\frac{1}{3}$
2	$\frac{4}{3}$
3	3
-1	$\frac{1}{3}$
-2	$\frac{4}{3}$
-3	3

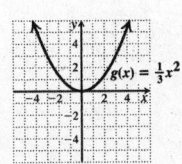

15. $h(x) = -\frac{1}{3}x^2$

Observe that the graph of $h(x) = -\frac{1}{3}x^2$ is the reflection of the graph of $g(x) = \frac{1}{3}x^2$ across the x-axis. We graphed $g(x)$ in Exercise 13, so we can use it to graph $h(x)$. If we did not make this observation we could find some ordered pairs, plot points, and connect them with a smooth curve.

x	$h(x) = -\frac{1}{3}x^2$
0	0
1	$-\frac{1}{3}$
2	$-\frac{4}{3}$
3	-3
-1	$-\frac{1}{3}$
-2	$-\frac{4}{3}$
-3	-3

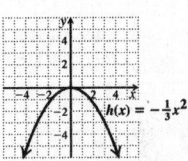

17. $f(x) = \frac{5}{2}x^2$

x	$f(x) = \frac{5}{2}x^2$
0	0
1	$\frac{5}{2}$
2	10
-1	$\frac{5}{2}$
-2	10

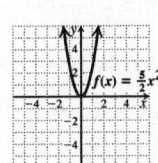

19. $g(x) = (x+1)^2 = [x - (-1)]^2$

We know that the graph of $g(x) = (x+1)^2$ looks like the graph of $f(x) = x^2$ (see Exercise 9) but moved to the left 1 unit.

Vertex: $(-1, 0)$, axis of symmetry: $x = -1$

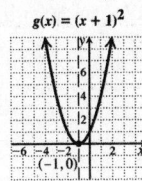

21. $f(x) = (x-2)^2$

The graph of $f(x) = (x-2)^2$ looks like the graph of

$f(x) = x^2$ (see Exercise 9) but moved to the right 2 units.

Vertex: $(2, 0)$, axis of symmetry: $x = 2$

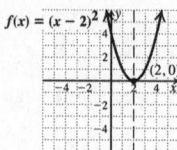

23. $g(x) = -(x+1)^2$

The graph of $g(x) = -(x+1)^2$ looks like the graph of

$f(x) = x^2$ (see Exercise 9) but moved to the left 1 unit. It

will also open downward because of the negative

coefficient, -1.

Vertex: $(-1, 0)$, axis of symmetry: $x = -1$

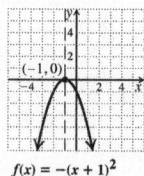

25. $f(x) = -(x-2)^2$

The graph of $f(x) = -(x-2)^2$ looks like the graph of

$f(x) = x^2$ (see Exercise 9) but moved to the right 2 units.

It will also open downward because of the negative

coefficient, -1.

Vertex: $(2, 0)$, axis of symmetry: $x = 2$

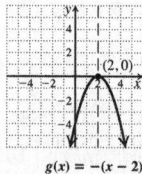

27. $f(x) = 2(x+1)^2$

The graph of $f(x) = 2(x+1)^2$ looks like the graph of

$h(x) = 2x^2$ (see graph following Example 1) but moved to

the left 1 unit.

Vertex: $(-1, 0)$, axis of symmetry: $x = -1$

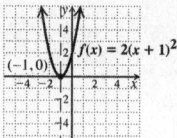

29. $g(x) = 3(x-4)^2$

The graph of $g(x) = 3(x-4)^2$ looks like the graph of

$g(x) = 3x^2$ but moved to the right 4 units.

Vertex: $(4, 0)$, axis of symmetry: $x = 4$

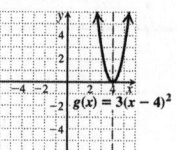

31. $h(x) = -\frac{1}{2}(x-4)^2$

The graph of $h(x) = -\frac{1}{2}(x-4)^2$ looks like the graph of

$g(x) = \frac{1}{2}x^2$ (see graph following Example 1) but moved to

the right 4 units. It will also open downward because of

the negative coefficient, $-\frac{1}{2}$.

Vertex: $(4, 0)$, axis of symmetry: $x = 4$

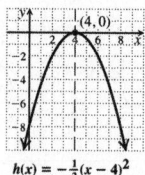

33. $f(x) = \frac{1}{2}(x-1)^2$

The graph of $f(x) = \frac{1}{2}(x-1)^2$ looks like the graph of

$g(x) = \frac{1}{2}x^2$ (see graph following Example 1) but moved

to the right 1 unit.

Vertex: $(1, 0)$, axis of symmetry: $x = 1$

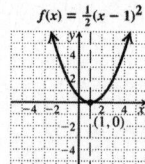

35. $f(x) = -2(x+5)^2 = -2[x-(-5)]^2$

The graph of $f(x) = -2(x+5)^2$ looks like the graph of

$h(x) = 2x^2$ (see the graph following Example 1) but

moved to the left 5 units. It will also open downward

because of the negative coefficient, -2.

Vertex: $(-5, 0)$, axis of symmetry: $x = -5$

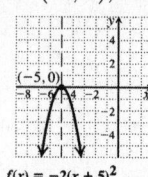

37. $h(x) = -3\left(x - \frac{1}{2}\right)^2$

The graph of $h(x) = -3\left(x - \frac{1}{2}\right)^2$ looks like the graph of

$f(x) = -3x^2$ (see Exercise 12) but moved to the right $\frac{1}{2}$

unit.

Vertex: $\left(\frac{1}{2}, 0\right)$, axis of symmetry: $x = \frac{1}{2}$

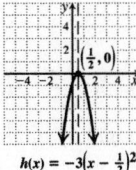

39. $f(x) = (x - 5)^2 + 2$

We know that the graph looks like the graph of $f(x) = x^2$ (see Example 1) but moved to the right 5 units and up 2 units. The vertex is (5, 2), and the axis of symmetry is $x = 5$. Since the coefficient of $(x - 5)^2$ is positive $(1 > 0)$, there is a minimum function value, 2.

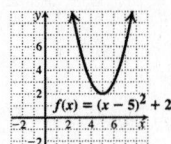

41. $f(x) = (x + 1)^2 - 3$

We know that the graph looks like the graph of $f(x) = x^2$ (see Example 1) but moved to the left 1 unit and down 3 units. The vertex is (−1, −3), and the axis of symmetry is $x = -1$. Since the coefficient of $(x + 1)^2$ is positive $(1 > 0)$, there is a minimum function value, −3.

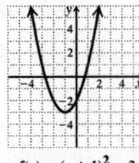

43. $g(x) = \frac{1}{2}(x + 4)^2 + 1$

We know that the graph looks like the graph of

$f(x) = \frac{1}{2}x^2$ (see graph following Example 1) but moved

to the left 4 units and up 1 unit. The vertex is (−4, 1), and the axis of symmetry is $x = -4$, and the minimum

function value is 1.

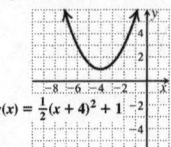

45. $h(x) = -2(x - 1)^2 - 3$

We know that the graph looks like the graph of

$h(x) = 2x^2$ (see graph following Example 1) but moved to the right 1 unit and down 3 units and turned upside down. The vertex is (1, −3), and the axis of symmetry is $x = 1$. The maximum function value is −3.

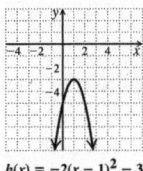

47. $f(x) = 2(x + 3)^2 + 1$

We know that the graph looks like the graph of

$f(x) = 2x^2$ (see graph following Example 1) but moved to the left 3 units and up 1 unit. The vertex is (−3, 1), and the axis of symmetry is $x = -3$. The minimum function value is 1.

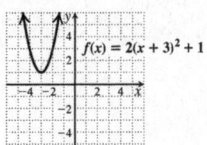

49. $g(x) = -\frac{3}{2}(x - 2)^2 + 4$

We know that the graph looks like the graph of

$f(x) = \frac{3}{2}x^2$ (see Exercise 18) but moved to the right 2

units and up 4 units and turned upside down. The vertex is (2, 4), and the axis of symmetry is $x = 2$, and the maximum function value is 4.

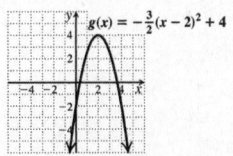

51. $f(x) = 5(x - 3)^2 + 9$

The function is of the form $f(x) = a(x - h)^2 + k$ with $a = 5$, $h = 3$, and $k = 9$. The vertex is (h, k), or (3, 9). The axis of symmetry is $x = h$, or $x = 3$. Since $a > 0$, then k, or 9, is the minimum function value.

53. $f(x) = -\frac{3}{7}(x+8)^2 + 2$

The function is of the form $f(x) = a(x-h)^2 + k$ with $a = -\frac{3}{7}$, $h = -8$, and $k = 2$. The vertex is (h, k), or $(-8, 2)$. The axis of symmetry is $x = h$, or $x = -8$. Since $a < 0$, then k, or 2, is the maximum function value.

55. $f(x) = \left(x - \frac{7}{2}\right)^2 - \frac{29}{4}$

The function is of the form $f(x) = a(x-h)^2 + k$ with $a = 1$, $h = \frac{7}{2}$, and $k = -\frac{29}{4}$. The vertex is (h, k), or $\left(\frac{7}{2}, -\frac{29}{4}\right)$. The axis of symmetry is $x = h$, or $x = \frac{7}{2}$. Since $a > 0$, then k, or $-\frac{29}{4}$, is the minimum function value.

57. $f(x) = -\sqrt{2}(x + 2.25)^2 - \pi$

The function is of the form $f(x) = a(x-h)^2 + k$ with $a = -\sqrt{2}$, $h = -2.25$, and $k = -\pi$. The vertex is (h, k), or $(-2.25, -\pi)$. The axis of symmetry is $x = h$, or $x = -2.25$. Since $a < 0$, then k, or $-\pi$, is the maximum function value.

59. *Writing Exercise.* For any input, the output of $y = x^2 - 4$ is 4 less than (or 4 units down from) the output of $y = x^2$.

61. $8x - 6y = 24$

Find the x-intercept.
$$8x - 6 \cdot 0 = 24$$
$$8x = 24$$
$$x = 3$$
The x-intercept is $(3, 0)$.

Find the y-intercept.
$$8 \cdot 0 - 6y = 24$$
$$-6y = 24$$
$$y = -4$$
The y-intercept is $(0, -4)$.

63. $f(x) = x^2 + 8x + 15$

Find the x-intercepts.
$$x^2 + 8x + 15 = 0$$
$$(x+5)(x+3) = 0$$
$$x + 5 = 0 \quad or \quad x + 3 = 0$$
$$x = -5 \quad or \quad x = -3$$
The x-intercepts are $(-5, 0)$ and $(-3, 0)$.

65. $x^2 - 14x$

We take half the coefficient of x and square it.
$$\frac{1}{2}(-14) = -7, \quad (-7)^2 = 49$$
Then we have $x^2 - 14x + 49 = (x-7)^2$.

67. *Writing Exercise.* She uses symmetry to find the mirror images of the two ordered pairs she calculates after plotting the vertex.

69. The equation will be of the form $f(x) = \frac{3}{5}(x-h)^2 + k$ with $h = 1$ and $k = 3$:
$$f(x) = \frac{3}{5}(x-1)^2 + 3$$

71. The equation will be of the form $f(x) = \frac{3}{5}(x-h)^2 + k$ with $h = 4$ and $k = -7$:
$$f(x) = \frac{3}{5}(x-4)^2 - 7$$

73. The equation will be of the form $f(x) = \frac{3}{5}(x-h)^2 + k$ with $h = -2$ and $k = -5$:
$$f(x) = \frac{3}{5}[x - (-2)]^2 + (-5), or$$
$$f(x) = \frac{3}{5}(x+2)^2 - 5$$

75. Since there is a minimum at $(2, 0)$, the parabola will have the same shape as $f(x) = 2x^2$. It will be of the form $f(x) = 2(x-h)^2 + k$ with $h = 2$ and $k = 0$:
$$f(x) = 2(x-2)^2$$

77. Since there is a maximum at $(0, -5)$, the parabola will have the same shape as $g(x) = -2x^2$. It will be of the form $g(x) = -2(x-h)^2 + k$ with $h = 0$ and $k = -5$:
$$g(x) = -2(x-0)^2 - 5, or \quad g(x) = -2x^2 - 5.$$

79. If h is increased, the graph will move to the right.

81. If a is replaced with $-a$, the graph will be reflected across the x-axis.

83. The maximum value of $g(x)$ is 1 and occurs at the point $(5, 1)$, so for $F(x)$ we have $h = 5$ and $k = 1$. $F(x)$ has the same shape as $f(x)$ and has a minimum, so $a = 3$. Thus, $F(x) = 3(x-5)^2 + 1$.

85. The graph of $y = f(x-1)$ looks like the graph of $y = f(x)$ moved 1 unit to the right.

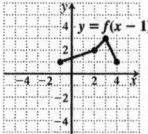

87. The graph of $y = f(x) + 2$ looks like the graph of $y = f(x)$ moved up 2 units.

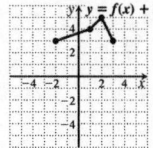

89. The graph of $y = f(x+3) - 2$ looks like the graph of $y = f(x)$ moved 3 units to the left and also moved down 2 units.

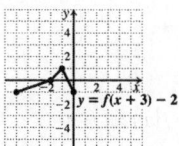

91. *Graphing Calculator Exercise*

93. *Writing Exercise.* The coefficient of x^2 is negative, so the parabola should open down.

Exercise Set 8.7

1. True; since $a = 3 > 0$, the graph opens upward.

3. True

5. False; the axis of symmetry is $x = \dfrac{3}{2}$.

7. False; the y-intercept is $(0, 7)$.

9. $\dfrac{1}{2} \cdot (-8) = -4; \quad (-4)^2 = 16$

$\begin{aligned} f(x) &= x^2 - 8x + 2 \\ &= \left(x^2 - 8x + 16\right) - 16 + 2 \\ &= (x-4)^2 + (-14) \end{aligned}$

11. $\dfrac{1}{2} \cdot 3 = \dfrac{3}{2}; \quad \left(\dfrac{3}{2}\right)^2 = \dfrac{9}{4}$

$\begin{aligned} f(x) &= x^2 + 3x - 5 \\ &= \left(x^2 + 3x + \dfrac{9}{4}\right) - \dfrac{9}{4} - 5 \\ &= \left[x - \left(-\dfrac{3}{2}\right)\right]^2 + \left(-\dfrac{29}{4}\right) \end{aligned}$

13. $\dfrac{1}{2} \cdot 2 = 1; \quad 1^2 = 1$

$\begin{aligned} f(x) &= 3x^2 + 6x - 2 \\ &= 3\left(x^2 + 2x\right) - 2 \\ &= 3\left(x^2 + 2x + 1\right) + 3(-1) - 2 \\ &= 3[x - (-1)]^2 + (-5) \end{aligned}$

15. $\dfrac{1}{2} \cdot 4 = 2; \quad 2^2 = 4$

$\begin{aligned} f(x) &= -x^2 - 4x - 7 \\ &= -1\left(x^2 + 4x\right) - 7 \\ &= -\left(x^2 + 4x + 4\right) + -1(-4) - 7 \\ &= -[x - (-2)]^2 + (-3) \end{aligned}$

17. $\dfrac{1}{2} \cdot \left(-\dfrac{5}{2}\right) = -\dfrac{5}{4}; \quad \left(-\dfrac{5}{4}\right)^2 = \dfrac{25}{16}$

$\begin{aligned} f(x) &= 2x^2 - 5x + 10 \\ &= 2\left(x^2 - \dfrac{5}{2}x\right) + 10 \\ &= 2\left(x^2 - \dfrac{5}{2}x + \dfrac{25}{16}\right) + 2\left(-\dfrac{25}{16}\right) + 10 \\ &= 2\left(x - \dfrac{5}{4}\right)^2 + \dfrac{55}{8} \end{aligned}$

19. a) $\begin{aligned} f(x) &= x^2 + 4x + 5 \\ &= (x^2 + 4x + 4 - 4) + 5 \quad \text{Adding } 4 - 4 \\ &= (x^2 + 4x + 4) - 4 + 5 \quad \text{Regrouping} \\ &= (x+2)^2 + 1 \end{aligned}$

The vertex is $(-2, 1)$, the axis of symmetry is $x = -2$, and the graph opens upward since the coefficient 1 is positive. We plot a few points as a check and draw the curve.

b)

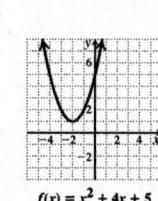

$f(x) = x^2 + 4x + 5$

21. a) $\begin{aligned} f(x) &= x^2 + 8x + 20 \\ &= (x^2 + 8x + 16 - 16) + 20 \quad \text{Adding } 16 - 16 \\ &= (x^2 + 8x + 16) - 16 + 20 \quad \text{Regrouping} \\ &= (x+4)^2 + 4 \end{aligned}$

The vertex is $(-4, 4)$, the axis of symmetry is

$x = -4$, and the graph opens upward since the coefficient 1 is positive.

b)

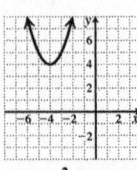

$f(x) = x^2 + 8x + 20$

23. a) $h(x) = 2x^2 - 16x + 25$

$= 2(x^2 - 8x) + 25$ Factoring 2 from the first two terms

$= 2(x^2 - 8x + 16 - 16) + 25$ Adding $16 - 16$ inside the parentheses

$= 2(x^2 - 8x + 16) + 2(-16) + 25$ Distributing to obtain a trinomial square

$= 2(x - 4)^2 - 7$

The vertex is $(4, -7)$, the axis of symmetry is $x = 4$, and the graph opens upward since the coefficient 2 is positive.

b)

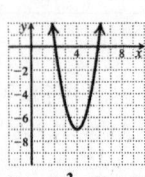

$h(x) = 2x^2 - 16x + 25$

25. a) $f(x) = -x^2 + 2x + 5$

$= -(x^2 - 2x) + 5$ Factoring -1 from the first two terms

$= -(x^2 - 2x + 1 - 1) + 5$ Adding $1 - 1$ inside the parentheses

$= -(x^2 - 2x + 1) - (-1) + 5$

$= -(x - 1)^2 + 6$

The vertex is $(1, 6)$, the axis of symmetry is $x = 1$, and the graph opens downward since the coefficient -1 is negative.

b)

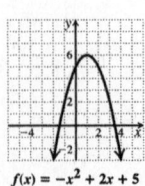

$f(x) = -x^2 + 2x + 5$

27. a) $g(x) = x^2 + 3x - 10$

$= \left(x^2 + 3x + \dfrac{9}{4} - \dfrac{9}{4}\right) - 10$

$= \left(x^2 + 3x + \dfrac{9}{4}\right) - \dfrac{9}{4} - 10$

$= \left(x + \dfrac{3}{2}\right)^2 - \dfrac{49}{4}$

The vertex is $\left(-\dfrac{3}{2}, -\dfrac{49}{4}\right)$, the axis of symmetry is $x = -\dfrac{3}{2}$, and the graph opens upward since the coefficient 1 is positive.

b)

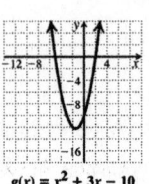

$g(x) = x^2 + 3x - 10$

29. a) $h(x) = x^2 + 7x$

$= \left(x^2 + 7x + \dfrac{49}{4}\right) - \dfrac{49}{4}$

$= \left(x + \dfrac{7}{2}\right)^2 - \dfrac{49}{4}$

The vertex is $\left(-\dfrac{7}{2}, -\dfrac{49}{4}\right)$, the axis of symmetry is $x = -\dfrac{7}{2}$, and the graph opens upward since the coefficient 1 is positive.

b)

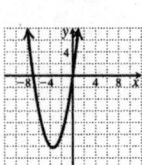

$h(x) = x^2 + 7x$

31. a) $f(x) = -2x^2 - 4x - 6$

$= -2(x^2 + 2x) - 6$ Factoring

$= -2(x^2 + 2x + 1 - 1) - 6$ Adding $1 - 1$ inside the parentheses

$= -2(x^2 + 2x + 1) - 2(-1) - 6$

$= -2(x + 1)^2 - 4$

The vertex is $(-1, -4)$, the axis of symmetry is $x = -1$, and the graph opens downward since the coefficient -2 is negative.

b)

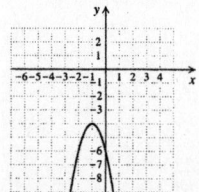

33. a) $g(x) = x^2 - 6x + 13$

$\qquad = (x^2 - 6x + 9 - 9) + 13 \quad$ Adding $9 - 9$

$\qquad = (x^2 - 6x + 9) - 9 + 13 \quad$ Regrouping

$\qquad = (x - 3)^2 + 4$

The vertex is $(3, 4)$, the axis of symmetry is $x = 3$, and the graph opens upward since the coefficient 1 is positive. The minimum is 4.

b)

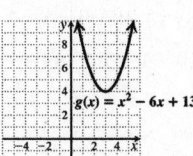

35. a) $g(x) = 2x^2 - 8x + 3$

$\qquad = 2(x^2 - 4x) + 3 \qquad$ Factoring

$\qquad = 2(x^2 - 4x + 4 - 4) + 3 \quad$ Adding $4 - 4$

$\qquad\qquad\qquad\qquad$ inside the parentheses

$\qquad = 2(x^2 - 4x + 4) + 2(-4) + 3$

$\qquad = 2(x - 2)^2 - 5$

The vertex is $(2, -5)$, the axis of symmetry is $x = 2$, and the graph opens upward since the coefficient 2 is positive. The minimum is -5.

b)

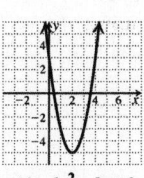

$g(x) = 2x^2 - 8x + 3$

37. a) $f(x) = 3x^2 - 24x + 50$

$\qquad = 3(x^2 - 8x) + 50 \qquad$ Factoring

$\qquad = 3(x^2 - 8x + 16 - 16) + 50 \quad$ Adding $16 - 16$

$\qquad\qquad\qquad\qquad$ inside the parentheses

$\qquad = 3(x^2 - 8x + 16) - 3 \cdot 16 + 50$

$\qquad = 3(x - 4)^2 + 2$

The vertex is $(4, 2)$, the axis of symmetry is $x = 4$, and the graph opens upward since the coefficient 3 is positive. The minimum is 2.

b)

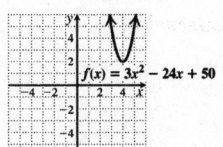

$f(x) = 3x^2 - 24x + 50$

39. a) $f(x) = -3x^2 + 5x - 2$

$\qquad = -3\left(x^2 - \frac{5}{3}x\right) - 2 \qquad$ Factoring

$\qquad = -3\left(x^2 - \frac{5}{3}x + \frac{25}{36} - \frac{25}{36}\right) - 2 \quad$ Adding $\frac{25}{36} - \frac{25}{36}$

$\qquad\qquad\qquad\qquad$ inside the parentheses

$\qquad = -3\left(x^2 - \frac{5}{3}x + \frac{25}{36}\right) - 3\left(-\frac{25}{36}\right) - 2$

$\qquad = -3\left(x - \frac{5}{6}\right)^2 + \frac{1}{12}$

The vertex is $\left(\frac{5}{6}, \frac{1}{12}\right)$, the axis of symmetry is $x = \frac{5}{6}$, and the graph opens downward since the coefficient -3 is negative. The maximum is $\frac{1}{12}$.

b)

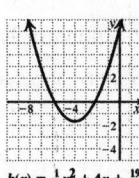

$f(x) = -3x^2 + 5x - 2$

41. a) $h(x) = \frac{1}{2}x^2 + 4x + \frac{19}{3}$

$\qquad = \frac{1}{2}(x^2 + 8x) + \frac{19}{3} \qquad$ Factoring

$\qquad = \frac{1}{2}(x^2 + 8x + 16 - 16) + \frac{19}{3} \quad$ Adding $16 - 16$

$\qquad\qquad\qquad\qquad$ inside parentheses

$\qquad = \frac{1}{2}(x^2 + 8x + 16) + \frac{1}{2}(-16) + \frac{19}{3}$

$\qquad = \frac{1}{2}(x + 4)^2 - \frac{5}{3}$

The vertex is $\left(-4, -\frac{5}{3}\right)$, the axis of symmetry is $x = -4$, and the graph opens upward since the coefficient $\frac{1}{2}$ is positive. The minimum is $-\frac{5}{3}$.

b)

$h(x) = \frac{1}{2}x^2 + 4x + \frac{19}{3}$

43. $f(x) = x^2 - 6x + 3$

To find the x-intercepts, solve the equation $0 = x^2 - 6x + 3$. Use the quadratic formula.

$$x = \frac{-(-6) \pm \sqrt{(-6)^2 - 4 \cdot 1 \cdot 3}}{2 \cdot 1}$$

$$x = \frac{6 \pm \sqrt{24}}{2} = \frac{6 \pm 2\sqrt{6}}{2} = 3 \pm \sqrt{6}$$

The x-intercepts are $(3 - \sqrt{6}, 0)$ and $(3 + \sqrt{6}, 0)$.

The y-intercept is $(0, f(0))$, or $(0, 3)$.

45. $g(x) = -x^2 + 2x + 3$

To find the x-intercepts, solve the equation

$0 = -x^2 + 2x + 3$. We factor.

$$0 = -x^2 + 2x + 3$$
$$0 = x^2 - 2x - 3 \qquad \text{Multiplying by } -1$$
$$0 = (x-3)(x+1)$$
$$x = 3 \text{ or } x = -1$$

The x-intercepts are $(-1, 0)$ and $(3, 0)$.

The y-intercept is $(0, g(0))$, or $(0, 3)$.

47. $f(x) = x^2 - 9x$

To find the x-intercepts, solve the equation $0 = x^2 - 9x$.

We factor.

$$0 = x^2 - 9x$$
$$0 = x(x-9)$$
$$x = 0 \text{ or } x = 9$$

The x-intercepts are $(0, 0)$ and $(9, 0)$.

Since $(0, 0)$ is an x-intercept, we observe that $(0, 0)$ is also the y-intercept.

49. $h(x) = -x^2 + 4x - 4$

To find the x-intercepts, solve the equation

$0 = -x^2 + 4x - 4$. We factor.

$$0 = -x^2 + 4x - 4$$
$$0 = x^2 - 4x + 4 \qquad \text{Multiplying by } -1$$
$$0 = (x-2)(x-2)$$
$$x = 2 \text{ or } x = 2$$

The x-intercept is $(2, 0)$.

The y-intercept is $(0, h(0))$, or $(0, -4)$.

51. $g(x) = x^2 + x - 5$

To find the x-intercepts, solve the equation

$0 = x^2 + x - 5$. Use the quadratic formula.

$$x = \frac{-1 \pm \sqrt{1^2 - 4 \cdot 1 \cdot (-5)}}{2 \cdot 1}$$
$$x = \frac{-1 \pm \sqrt{21}}{2} = -\frac{1}{2} \pm \frac{\sqrt{21}}{2}$$

The x-intercepts are $\left(-\frac{1}{2} - \frac{\sqrt{21}}{2}, 0\right)$ and $\left(-\frac{1}{2} + \frac{\sqrt{21}}{2}, 0\right)$.

The y-intercept is $(0, g(0))$, or $(0, -5)$.

53. $f(x) = 2x^2 - 4x + 6$

To find the x-intercepts, solve the equation

$0 = 2x^2 - 4x + 6$. We use the quadratic formula.

$$x = \frac{-(-4) \pm \sqrt{(-4)^2 - 4 \cdot 2 \cdot 6}}{2 \cdot 2}$$
$$x = \frac{4 \pm \sqrt{-32}}{4} = \frac{4 \pm 4i\sqrt{2}}{2} = 2 \pm 2i\sqrt{2}$$

There are no real-number solutions, so there is no

x-intercept.

The y-intercept is $(0, f(0))$, or $(0, 6)$.

55. *Writing Exercise.* If the quadratic function opens downward and has no x-intercepts it must lie either in quadrant III or IV.

57. $x + y + z = 3,$ (1)
          $x - y + z = 1,$ (2)
         $-x - y + z = -1$ (3)

We eliminate y from two different pairs of equations.

$$\begin{array}{ll} x + y + z = 3 & (1) \\ \underline{x - y + z = 1} & (2) \\ 2x \quad + 2z = 4 & \\ x + z = 2 & (4) \end{array}$$

$$\begin{array}{ll} x + y + z = 3 & (1) \\ \underline{-x - y + z = -1} & (3) \\ 2z = 2 & \\ z = 1 & \end{array}$$

We eliminate not only y, but also x and found $z = 1$.

Substitute 1 for z in Equation (4) to find x.

$$\begin{array}{l} x + 1 = 2 \quad \text{Substituting 1 for } z \text{ in (4)} \\ x = 1 \end{array}$$

Substitute in one of the original equations to find y.

$$\begin{array}{l} 1 + y + 1 = 3 \quad \text{Substituting 1 for } z \text{ and} \\ \qquad\qquad\qquad 1 \text{ for } x \text{ in (1)} \\ y = 1 \end{array}$$

The solution is $(1, 1, 1)$.

59. $z = 8,$ (1)
         $x + y + z = 23,$ (2)
        $2x + y - z = 17$ (3)

We eliminate y from equations (2) and (3).

$$\begin{array}{ll} x + y + z = 23 & (2) \\ \underline{-2x - y + z = -17} & \text{Multiply (3) by } -1 \\ -x \quad + 2z = 6 & (4) \end{array}$$

Substitute 8 for z in Equation (4) to find x.

$$\begin{array}{l} -x + 2(8) = 6 \qquad \text{Substituting 8 for } z \text{ in (4)} \\ -x = -10 \\ x = 10 \end{array}$$

Substitute in one of the original equations to find y.

$$\begin{array}{l} 10 + y + 8 = 23 \quad \text{Substituting 8 for } z \text{ and} \\ \qquad\qquad\qquad\quad 10 \text{ for } x \text{ in (2)} \\ y = 5 \end{array}$$

The solution is $(10, 5, 8)$.

61. $1.5 = c,$ (1)
        $52.5 = 25a + 5b + c,$ (2)
         $7.5 = 4a + 2b + c$ (3)

We eliminate b from equations (2) and (3).

$$\begin{array}{ll} 50a + 10b + 2c = 105 & \text{Multiply (2) by 2} \\ \underline{-20a - 10b - 5c = -37.5} & \text{Multiply (3) } -5 \\ 30a \quad - 3c = 67.5 & (4) \end{array}$$

Substitute 1.5 for c in Equation (4) to find a.

$30a - 3(1.5) = 67.5$ Substituting 1.5 for c in (4)
$30a = 72$
$a = 2.4$

Substitute in one of the original equations to find b.

$7.5 = 4(2.4) + 2b + 1.5$ Substituting 2.4 for a and
1.5 for c in (3)

$-1.8 = b$

The solution is $(2.4, -1.8, 1.5)$.

63. *Writing Exercise.* No; the graphs could open in different directions and have different vertices. Consider the graphs of $f(x) = x^2 - 4$ and $g(x) = -x^2 + 4$, for example. Both have x-intercepts $(-2, 0)$ and $(2, 0)$, but the vertex of $f(x) = (0, -4)$ while the vertex of $g(x)$ is $(0, 4)$.

65. a) $f(x) = 2.31x^2 - 3.135x - 5.89$
$= 2.31(x^2 - 1.357142857x) - 5.89$
$= 2.31(x^2 - 1.357142857x$
$\quad + 0.460459183 - 0.460459183) - 5.89$
$= 2.31(x^2 - 1.357142857x + 0.460459183)$
$\quad + 2.31(-0.460459183) - 5.89$
$= 2.31(x - 0.678571428)^2 - 6.953660714$

Since the coefficient 2.31 is positive, the function has a minimum value. It is -6.953660714.

b) To find the x-intercepts, solve
$0 = 2.31x^2 - 3.135x - 5.89$.

$x = \dfrac{-(-3.135) \pm \sqrt{(-3.135)^2 - 4(2.31)(-5.89)}}{2(2.31)}$

$x \approx \dfrac{3.135 \pm 8.015723611}{4.62}$

$x \approx -1.056433682 \quad or \quad x \approx 2.413576539$

The x-intercepts are $(-1.056433682, 0)$ and $(2.413576539, 0)$.

The y-intercept is $(0, f(0))$, or $(0, -5.89)$.

67. $f(x) = x^2 - x - 6$

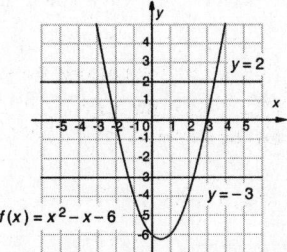

a) The solutions of $x^2 - x - 6 = 2$ are the first coordinates of the points of intersection of the graphs of $f(x) = x^2 - x - 6$ and $y = 2$. From the graph we see that the solutions are approximately -2.4 and 3.4.

b) The solutions of $x^2 - x - 6 = -3$ are the first

coordinates of the points of intersection of the graphs of $f(x) = x^2 - x - 6$ and $y = -3$. From the graph we see that the solutions are approximately -1.3 and 2.3.

69. $f(x) = mx^2 - nx + p$
$= m\left(x^2 - \dfrac{n}{m}x\right) + p$
$= m\left(x^2 - \dfrac{n}{m}x + \dfrac{n^2}{4m^2} - \dfrac{n^2}{4m^2}\right) + p$
$= m\left(x - \dfrac{n}{2m}\right)^2 - \dfrac{n^2}{4m} + p$
$= m\left(x - \dfrac{n}{2m}\right)^2 + \dfrac{-n^2 + 4mp}{4m}$, or
$m\left(x - \dfrac{n}{2m}\right)^2 + \dfrac{4mp - n^2}{4m}$

71. The horizontal distance from $(-1, 0)$ to $(3, -5)$ is $|3 - (-1)|$, or 4, so by symmetry the other x-intercept is $(3 + 4, 0)$, or $(7, 0)$. Substituting the three ordered pairs $(-1, 0)$, $(3, -5)$, and $(7, 0)$ in the equation $f(x) = ax^2 + bx + c$ yields a system of equations:
$0 = a - b + c,$
$-5 = 9a + 3b + c,$
$0 = 49a + 7b + c$

The solution of this system of equations is $\left(\dfrac{5}{16}, -\dfrac{15}{8}, -\dfrac{35}{16}\right)$, so $f(x) = \dfrac{5}{16}x^2 - \dfrac{15}{8}x - \dfrac{35}{16}$.

If we complete the square we find that this function can also be expressed as $f(x) = \dfrac{5}{16}(x - 3)^2 - 5$.

73. $f(x) = |x^2 - 1|$

We plot some points and draw the curve. Note that it will lie entirely on or above the x-axis since absolute value is never negative.

x	$f(x)$
-3	8
-2	3
-1	0
0	1
1	0
2	3
3	8

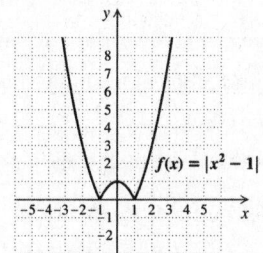

75. $f(x) = \left|2(x-3)^2 - 5\right|$

We plot some points and draw the curve. Note that it will lie entirely on or above the $x-$axis since absolute value is never negative.

x	$f(x)$
-1	27
0	13
1	3
2	3
3	5
4	3
5	3
6	13

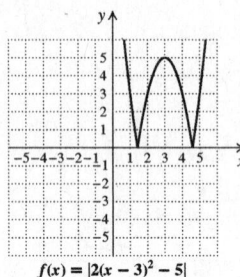

$f(x) = |2(x-3)^2 - 5|$

Section 8.8

1. e

3. c

5. d

7. *Familiarize and Translate.* We are given the formula $p(x) = -0.2x^2 + 1.3x + 6.2$.

Carry out. To find the value of x for which $p(x)$ is a maximum, we first find $-\dfrac{b}{2a}$:

$$-\frac{b}{2a} = -\frac{1.3}{2(-0.2)} = 3.25, \text{ or } 3\frac{1}{4}$$

Now we find the maximum value of the function $p(3.25)$:

$$p(3.25) = -0.2(3.25)^2 + 1.3(3.25) + 6.2 = 8.3125$$

The minimum function value of about 8.3 occurs when $x = 3.25$.

Check. We can go over the calculations again. We could also solve the problem again by completing the square. The answer checks.

State. A calf's daily milk consumption is greatest at 3.25 weeks at about 8.3 lb of milk per day.

9. *Familiarize and Translate.* We want to find the value of x for which $C(x) = 0.1x^2 - 0.7x + 2.425$ is a minimum.

Carry out. We complete the square.

$$C(x) = 0.1(x^2 - 7x + 12.25) + 2.425 - 1.225$$
$$C(x) = 0.1(x - 3.5)^2 + 1.2$$

The minimum function value of 1.2 occurs when $x = 3.5$.

Check. Check a function value for x less than 3.5 and for x greater than 3.5.

$$C(3) = 0.1(3)^2 - 0.7(3) + 2.425 = 1.225$$
$$C(4) = 0.1(4)^2 - 0.7(4) + 2.425 = 1.225$$

Since 1.2 is less than these numbers, it looks as though we have a minimum.

State. The minimum average cost is $1.2 hundred, or $120. To achieve the minimum cost, 3.5 hundred, or 350 dulcimers should be built.

11. *Familiarize.* We make a drawing and label it.

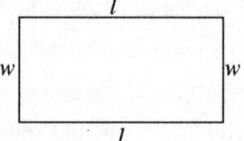

Perimeter: $2l + 2w = 720$ ft

Area: $A = l \cdot w$

Translate. We have a system of equations.

$$2l + 2w = 720,$$
$$A = lw$$

Carry out. Solving the first equation for l, we get $l = 360 - w$. Substituting for l in the second equation we get a quadratic function A:

$$A = (360 - w)w$$
$$A = -w^2 + 360w$$

Completing the square, we get

$$A = -(w - 180)^2 + 32{,}400$$

The maximum function value is 32,400. It occurs when w is 180. When $w = 180$, $l = 360 - 180$, or 180.

Check. We check a function value for w less than 180 and for w greater than 180.

$$A(179) = -179^2 + 360 \cdot 179 = 32{,}399$$
$$A(181) = -181^2 + 360 \cdot 181 = 32{,}399$$

Since 32,400 is greater than these numbers, it looks as though we have a maximum.

State. The maximum area occurs when the dimensions are 180 ft by 180 ft.

13. *Familiarize.* We make a drawing and label it.

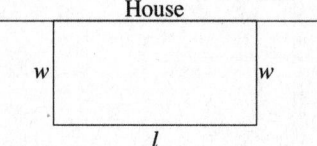

Translate. We have two equations.

$$l + 2w = 60,$$
$$A = lw$$

Carry out. Solve the first equation for l.

$$l = 60 - 2w$$

Substitute for l in the second equation.

$$A = (60 - 2w)w$$
$$A = -2w^2 + 60w$$

Completing the square, we get

$$A = -2(w - 15)^2 + 450 \, .$$

The maximum function value of 450 occurs when $w = 15$.

When $w = 15$, $l = 60 - 2 \cdot 15 = 30$.

Check. Check a function value for w less than 15 and for w greater than 15.

$$A(14) = -2 \cdot 14^2 + 60 \cdot 14 = 448$$
$$A(16) = -2 \cdot 16^2 + 60 \cdot 16 = 448$$

Since 450 is greater than these numbers, it looks as though we have a maximum.

State. The maximum area of 450 ft^2 will occur when the dimensions are 15 ft by 30 ft.

15. Familiarize. Let x represent the height of the file and y represent the width. We make a drawing.

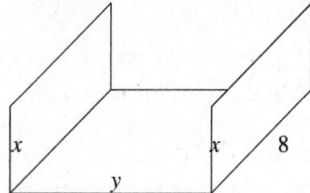

Translate. We have two equations.

$$2x + y = 14$$
$$V = 8xy$$

Carry out. Solve the first equation for y.

$$y = 14 - 2x$$

Substitute for y in the second equation.

$$V = 8x(14 - 2x)$$
$$V = -16x^2 + 112x$$

Completing the square, we get

$$V = -16\left(x - \frac{7}{2}\right)^2 + 196 \, .$$

The maximum function value of 196 occurs when $x = \frac{7}{2}$.

When $x = \frac{7}{2}$, $y = 14 - 2 \cdot \frac{7}{2} = 7$.

Check. Check a function value for x less than $\frac{7}{2}$ and for x greater than $\frac{7}{2}$.

$$V(3) = -16 \cdot 3^2 + 112 \cdot 3 = 192$$
$$V(4) = -16 \cdot 4^2 + 112 \cdot 4 = 192$$

Since 196 is greater than these numbers, it looks as though we have a maximum.

State. The file should be $\frac{7}{2}$ in., or 3.5 in. tall.

17. Familiarize. We let x and y represent the numbers, and we let P represent their product.

Translate. We have two equations.

$$x + y = 18,$$
$$P = xy$$

Carry out. Solving the first equation for y, we get $y = 18 - x$. Substituting for y in the second equation we get a quadratic function P:

$$P = x(18 - x)$$
$$P = -x^2 + 18x$$

Completing the square, we get

$$P = -(x - 9)^2 + 81 \, .$$

The maximum function value is 81. It occurs when $x = 9$.

When $x = 9$, $y = 18 - 9$, or 9.

Check. We can check a function value for x less than 9 and for x greater than 9.

$$P(10) = -10^2 + 18 \cdot 10 = 80$$
$$P(8) = -8^2 + 18 \cdot 8 = 80$$

Since 81 is greater than these numbers, it looks as though we have a maximum.

State. The maximum product of 81 occurs for the numbers 9 and 9.

19. Familiarize. We let x and y represent the two numbers, and we let P represent their product.

Translate. We have two equations.

$$x - y = 8,$$
$$P = xy$$

Carry out. Solve the first equation for x.

$$x = 8 + y$$

Substitute for x in the second equation.

$$P = (8 + y)y$$
$$P = y^2 + 8y$$

Completing the square, we get

$$P = (y + 4)^2 - 16 \, .$$

The minimum function value is –16. It occurs when $y = -4$. When $y = -4$, $x = 8 + (-4)$, or 4.

Check. Check a function value for y less than –4 and for y greater than –4.

$$P(-5) = (-5)^2 + 8(-5) = -15$$
$$P(-3) = (-3)^2 + 8(-3) = -15$$

Since –16 is less than these numbers, it looks as though we have a minimum.

State. The minimum product of –16 occurs for the numbers 4 and –4.

21. From the results of Exercises 17 and 18, we might observe that the numbers are –5 and –5 and that the maximum product is 25. We could also solve this problem as follows.

Familiarize. We let x and y represent the two numbers, and we let P represent their product.

Translate. We have two equations.
$$x + y = -10,$$
$$P = xy$$

Carry out. Solve the first equation for y.
$$y = -10 - x$$

Substitute for y in the second equation.
$$P = x(-10 - x)$$
$$P = -x^2 - 10x$$

Completing the square, we get
$$P = -(x + 5)^2 + 25$$

The maximum function value is 25. It occurs when $x = -5$. When $x = -5$, $y = -10 - (-5)$, or -5.

Check. Check a function value for x less than -5 and for x greater than -5.
$$P(-6) = -(-6)^2 - 10(-6) = 24$$
$$P(-4) = -(-4)^2 - 10(-4) = 24$$

Since 25 is greater than these numbers, it looks as though we have a maximum.

State. The maximum product of 25 occurs for the numbers –5 and –5.

23. The data points rise and then fall. The graph appears to represent a quadratic function that opens downward. Thus a quadratic function $f(x) = ax^2 + bx + c$, $a < 0$, might be used to model the data.

25. The data points rise. The graph does not appear to represent a quadratic function in which the data points would rise and then fall or vice versa. Thus a linear function $f(x) = mx + b$ might be used to model the data.

27. The data points do not represent a linear or quadratic pattern. Thus, it does not appear that the data can be modeled with either a quadratic or a linear function.

29. The data points fall and then rise. The graph appears to represent a quadratic function that opens upward. Thus a quadratic function $f(x) = ax^2 + bx + c$, $a > 0$, might be used to model the data.

31. The data points appear to represent the right half of a quadratic function that opens upward. Thus a quadratic function $f(x) = ax^2 + bx + c$, $a > 0$, might be used to model the data.

33. The data points fall. The graph does not appear to represent a quadratic function in which the data points would rise and then fall or vice versa. Thus a linear function $f(x) = mx + b$ might be used to model the data.

35. We look for a function of the form $f(x) = ax^2 + bx + c$. Substituting the data points, we get
$$4 = a(1)^2 + b(1) + c,$$
$$-2 = a(-1)^2 + b(-1) + c,$$
$$13 = a(2)^2 + b(2) + c,$$
or
$$4 = a + b + c,$$
$$-2 = a - b + c,$$
$$13 = 4a + 2b + c.$$
Solving this system, we get
$$a = 2, \ b = 3, \text{ and } c = -1.$$
Therefore the function we are looking for is
$$f(x) = 2x^2 + 3x - 1.$$

37. We look for a function of the form $f(x) = ax^2 + bx + c$. Substituting the data points, we get
$$0 = a(2)^2 + b(2) + c,$$
$$3 = a(4)^2 + b(4) + c,$$
$$-5 = a(12)^2 + b(12) + c,$$
or
$$0 = 4a + 2b + c,$$
$$3 = 16a + 4b + c,$$
$$-5 = 144a + 12b + c.$$
Solving this system, we get
$$a = -\frac{1}{4}, \ b = 3, \ c = -5.$$
Therefore the function we are looking for is
$$f(x) = -\frac{1}{4}x^2 + 3x - 5.$$

39. a) *Familiarize.* We look for a function of the form $A(s) = as^2 + bs + c$, where $A(s)$ represents the number of nighttime accidents (for every 200 million km) and s represents the travel speed (in km/h).

Translate. We substitute the given values of s and $A(s)$.

$$400 = a(60)^2 + b(60) + c,$$
$$250 = a(80)^2 + b(80) + c,$$
$$250 = a(100)^2 + b(100) + c,$$

or

$$400 = 3600a + 60b + c,$$
$$250 = 6400a + 80b + c,$$
$$250 = 10{,}000a + 100b + c.$$

Carry out. Solving the system of equations, we get

$$a = \frac{3}{16}, \quad b = -\frac{135}{4}, \quad c = 1750.$$

Check. Recheck the calculations.

State. The function

$$A(s) = \frac{3}{16}s^2 - \frac{135}{4}s + 1750 \text{ fits the data.}$$

b) Find $A(50)$.

$$A(50) = \frac{3}{16}(50)^2 - \frac{135}{4}(50) + 1750 = 531.25$$

About 531 accidents occur at 50 km/h.

41. **Familiarize**. Think of a coordinate system placed on the drawing in the text with the origin at the point where the arrow is released. Then three points on the arrow's parabolic path are $(0, 0)$, $(63, 27)$, and $(126, 0)$. We look for a function of the form $h(d) = ad^2 + bd + c$, where $h(d)$ represents the arrow's height and d represents the distance the arrow has traveled horizontally.

Translate. We substitute the values given above for d and $h(d)$.

$$0 = a \cdot 0^2 + b \cdot 0 + c,$$
$$27 = a \cdot 63^2 + b \cdot 63 + c,$$
$$0 = a \cdot 126^2 + b \cdot 126 + c$$

or

$$0 = c,$$
$$27 = 3969a + 63b + c,$$
$$0 = 15{,}876a + 126b + c$$

Carry out. Solving the system of equations, we get $a \approx -0.0068$, $b \approx 0.8571$, and $c = 0$.

Check. Recheck the calculations.

State. The function $h(d) = -0.0068d^2 + 0.8571d$ expresses the arrow's height as a function of the distance it has traveled horizontally.

43. *Writing Exercise*. The graph of a nonlinear function could extend without bound in both the positive and negative directions and thus have neither a minimum nor a maximum value.

45. $2x - 3 > 5$
$$2x > 8$$
$$x > 4$$

The solution set is $\{x | x > 4\}$, or $(4, \infty)$.

47. $|9 - x| \geq 2$

$9 - x \leq -2$	*or*	$2 \leq 9 - x$
$-x \leq -11$	*or*	$-7 \leq -x$
$x \geq 11$	*or*	$7 \geq x$

The solution set is $\{x | x \leq 7 \ or \ x \geq 11\}$, or $(-\infty, 7] \cup [11, \infty)$.

49. $f(x) = \dfrac{x-3}{x+4} - 5$

$$= \frac{x-3}{x+4} - 5$$
$$\text{Note that } x \neq -4.$$
$$= \frac{x-3}{x+4} - 5 \cdot \frac{x+4}{x+4}$$
$$= \frac{x-3-5(x+4)}{x+4}$$
$$= \frac{x-3-5x-20}{x+4}$$
$$= \frac{-4x-23}{x+4}, \quad x \neq -4$$

51. Note the restriction that $x \neq -4$.

$$\frac{x-3}{x+4} = 5$$
$$(x+4) \cdot \frac{x-3}{x+4} = (x+4) \cdot 5$$
$$x - 3 = 5x + 20$$
$$-23 = 4x$$
$$-\frac{23}{4} = x$$

The solution is $-\dfrac{23}{4}$.

53. Note the restriction that $x \neq -7, 3$.

$$\frac{x}{(x-3)(x+7)} = 0$$
$$(x-3)(x+7) \cdot \frac{x}{(x-3)(x+7)} = (x-3)(x+7) \cdot 0$$
$$x = 0$$

The solution is 0.

55. *Writing Exercise*. The graph for the other pitchers appears to be quadratic. It starts at 3.5 at age 20, seems to have a minimum of 3.3 around the age of 30, and a maximum of 4.5 at age 46. For Clemens, the graph appears to be more linear, having a minimum of 3.1 at 23 and continuing to a maximum of 3.3 at age 46. Clemens has no fall and rise with age, just a small rise.

57. *Familiarize.* Position the bridge on a coordinate system as shown with the vertex of the parabola at $(0, 30)$.

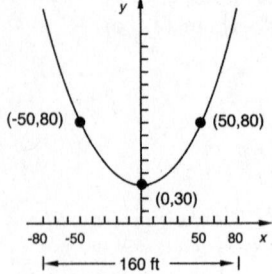

We find a function of the form $y = ax^2 + bx + c$ which represents the parabola containing the points $(0, 30)$, $(-50, 80)$, and $(50, 80)$.

Translate. Substitute for x and y.

$$30 = a \cdot 0^2 + b \cdot 0 + c,$$
$$80 = a(-50)^2 + b(-50) + c,$$
$$80 = a(50)^2 + b(50) + c,$$

or

$$30 = c,$$
$$80 = 2500a - 50b + c,$$
$$80 = 2500a + 50b + c.$$

Carry out. Solving the system of equations, we get

$$a = 0.02, \ b = 0, \ c = 30.$$

The function $y = 0.02x^2 + 30$ represents the parabola. Because the cable supports are 160 ft apart, the tallest supports are positioned 160/2, or 80 ft, to the left and right of the midpoint. This means that the longest vertical cables occur at $x = -80$ and $x = 80$. For $x = \pm 80$,

$$y = 0.02(\pm 80)^2 + 30$$
$$= 128 + 30$$
$$= 158 \text{ ft}$$

Check. We go over the calculations.

State. The longest vertical cables are 158 ft long.

59. *Familiarize.* Let x represent the number of 25 increases in the admission price. Then $10 + 0.25x$ represents the admission price, and $80 - x$ represents the corresponding average attendance. Let R represent the total revenue.

Translate. Since the total revenue is the product of the cover charge and the number attending a show, we have the following function for the amount of money the owner makes.

$$R(x) = (10 + 0.25x)(80 - x), \text{ or}$$
$$R(x) = -0.25x^2 + 10x + 800$$

Carry out. Completing the square, we get

$$R(x) = -0.25(x - 20)^2 + 900$$

The maximum function value of 900 occurs when $x = 20$. The owner should charge $\$10 + \$0.25(20)$, or $\$15$.

Check. We check a function value for x less than 20 and for x greater than 20.

$$R(19) = -0.25(19)^2 + 10 \cdot 19 + 800 = 899.75$$
$$R(21) = -0.25(21)^2 + 10 \cdot 21 + 800 = 899.75$$

Since 900 is greater than these numbers, it looks as though we have a maximum.

State. The owner should charge $\$15$.

61. *Familiarize.* We add labels to the drawing in the text.

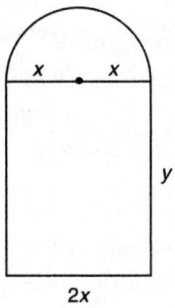

The perimeter of the semicircular portion of the window is $\frac{1}{2} \cdot 2\pi x$, or πx. The perimeter of the rectangular portion is $y + 2x + y$, or $2x + 2y$. The area of the semicircular portion of the window is $\frac{1}{2} \cdot \pi x^2$, or $\frac{\pi}{2}x^2$. The area of the rectangular portion is $2xy$.

Translate. We have two equations, one giving the perimeter of the window and the other giving the area.

$$\pi x + 2x + 2y = 24,$$
$$A = \frac{\pi}{2}x^2 + 2xy$$

Carry out. Solve the first equation for y.

$$\pi x + 2x + 2y = 24$$
$$2y = 24 - \pi x - 2x$$
$$y = 12 - \frac{\pi x}{2} - x$$

Substitute for y in the second equation.

$$A = \frac{\pi}{2}x^2 + 2x\left(12 - \frac{\pi x}{2} - x\right)$$
$$A = \frac{\pi}{2}x^2 + 24x - \pi x^2 - 2x^2$$
$$A = -2x^2 - \frac{\pi}{2}x^2 + 24x$$
$$A = -\left(2x + \frac{\pi}{2}\right)x^2 + 24x$$

Completing the square, we get

$$A = -\left(2 + \frac{\pi}{2}\right)\left(x^2 + \frac{24}{-\left(2 + \frac{\pi}{2}\right)}x\right)$$
$$A = -\left(2 + \frac{\pi}{2}\right)\left(x^2 - \frac{48}{4 + \pi}x\right)$$
$$A = -\left(2 + \frac{\pi}{2}\right)\left(x - \frac{24}{4 + \pi}\right)^2 + \left(\frac{24}{4 + \pi}\right)^2$$

The maximum function value occurs when

$x = \dfrac{24}{4+\pi}$. When $x = \dfrac{24}{4+\pi}$,

$y = 12 - \dfrac{\pi}{2}\left(\dfrac{24}{4+\pi}\right) - \dfrac{24}{4+\pi}$

$= \dfrac{48 + 12\pi}{4+\pi} - \dfrac{12\pi}{4+\pi} - \dfrac{24}{4+\pi} = \dfrac{24}{4+\pi}$

Check. Recheck the calculations.

State. The radius of the circular portion of the window and the height of the rectangular portion should each be $\dfrac{24}{4+\pi}$ ft.

63. a) Enter the data and use the quadratic regression operation on a graphing calculator. We get
$h(x) = 11{,}090.60714x^2 - 29{,}069.62143x + 39{,}983.8$,
where x is the number of years after 2000.

b) In 2010, $x = 2010 - 2000 = 10$.
$h(10) \approx 858{,}348$ vehicles

Section 8.9

1. The solutions of $(x-3)(x+2) = 0$ are 3 and –2 and for a test value in $[-2, 3]$, say 0, $(x-3)(x+2)$ is negative so the statement is true. (Note that the endpoints must be included in the solution set because the inequality symbol is $\leq$.)

3. The solutions of $(x-1)(x-6) = 0$ are 1 and 6. For a value of x less than 1, say 0, $(x-1)(x-6)$ is positive; for a value of x greater than 6, say 7, $(x-1)(x-6)$ is also positive. Thus, the statement is true. (Note that the endpoints of the intervals are not included because the inequality symbol is $>$.)

5. Since $x + 2 = 0$ when $x = -2$ and $x - 3 = 0$ when $x = 3$, the statement is false.

7. $p(x) \leq 0$ when $-4 \leq x \leq \dfrac{3}{2}$,

$\left[-4, \dfrac{3}{2}\right]$ or $\left\{x \middle| -4 \leq x \leq \dfrac{3}{2}\right\}$

9. $x^4 + 12x > 3x^2 + 4x^2$ is equivalent to

$x^4 - 3x^2 - 4x^2 + 12x > 0$, which is the graph in the text.

$p(x) > 0$ when $(-\infty, -2) \cup (0,\ 2) \cup (3,\ \infty)$ or

$\{x | x < -2 \ or \ 0 < x < 2 \ or \ x > 3\}$

11. $\dfrac{x-1}{x+2} < 3$ is equivalent to finding the values of x for

which the graph $r(x)$ is less than 3, or below $g(x)$.

$\left(-\infty, -\dfrac{7}{2}\right) \cup (-2,\ \infty)$ or $\left\{x \middle| x < -\dfrac{7}{2} \ or \ x > -2\right\}$

13. $(x-6)(x-5) < 0$

The solutions of $(x-6)(x-5) = 0$ are 5 and 6. They are not solutions of the inequality, but they divide the real number line in a natural way. The product $(x-6)(x-5)$ is positive or negative, for values other than 5 and 6., depending on the signs of the factors $x-6$ and $x-5$.

$x - 6 > 0$ when $x > 6$ and $x - 6 < 0$ when $x < 6$.

$x - 5 > 0$ when $x > 5$ and $x - 5 < 0$ when $x < 5$

We make a diagram.

Sign of $x-6$	$-$	$-$	$+$
Sign of $x-5$	$-$	$+$	$+$
Sign of product	$+$	$-$	$+$

$$\xleftarrow{\hspace{3cm}}\underset{5}{|}\hspace{1cm}\underset{6}{|}\xrightarrow{\hspace{3cm}}$$

For the product $(x-6)(x-5)$ to be negative, one factor must be positive and the other negative. We see from the diagram that numbers satisfying $5 < x < 6$ are solutions. The solution set of the inequality is $(5, 6)$ or $\{x | 5 < x < 6\}$.

15. $(x+7)(x-2) \geq 0$

The solutions of $(x+7)(x-2) = 0$ are –7 and 2. They divide the number line into three intervals as shown:

$$\overset{A}{\underset{-7}{\rule{0pt}{0pt}}}\quad\overset{B}{\underset{2}{\rule{0pt}{0pt}}}\quad\overset{C}{\rule{0pt}{0pt}}$$

We try test numbers in each interval.

A: Test –8, $f(-8) = (-8+7)(-8-2) = 10$

B: Test 0, $f(0) = (0+7)(0-2) = -14$

C: Test 3, $f(3) = (3+7)(3-2) = 10$

Since $f(-8)$ and $f(3)$ are positive, the function value will be positive for all numbers in the intervals containing –8 and 3. The inequality symbol is $\leq$, so we need to include the endpoints. The solution set is

$(-\infty, -7] \cup [2,\ \infty)$, or $\{x | x \leq -7 \ or \ x \geq 2\}$.

17. $x^2 - x - 2 > 0$
$(x+1)(x-2) > 0$ Factoring

The solutions of $(x+1)(x-2) = 0$ are –1 and 2. They divide the number line into three intervals as shown:

$$\overset{A}{\underset{-1}{\rule{0pt}{0pt}}}\quad\overset{B}{\underset{2}{\rule{0pt}{0pt}}}\quad\overset{C}{\rule{0pt}{0pt}}$$

We try test numbers in each interval.

A: Test -2, $f(-2)=(-2+1)(-2-2)=4$

B: Test 0, $f(0)=(0+1)(0-2)=-2$

C: Test 3, $f(3)=(3+1)(3-2)=4$

Since $f(-2)$ and $f(3)$ are positive, the function value will be positive for all numbers in the intervals containing -2 and 3. The solution set is $(-\infty,-1)\cup(2,\infty)$, or $\{x\,|\,x<-1\ or\ x>2\}$.

19. $x^2+4x+4<0$

$(x+2)^2<0$

Observe that $(x+2)^2\geq0$ for all values of x. Thus, the solution set is $\varnothing$.

21.
$$x^2-4x\leq3$$
$$x^2-4x+4\leq3+4$$
$$(x-2)^2\leq7$$
$$x-2\leq\pm\sqrt{7}$$
$$x\leq2\pm\sqrt{7}$$

The solutions of $x^2-4x-3\leq0$ are $2\pm\sqrt{7}$. They divide the number line into three intervals as shown:

```
        A       B       C
    ──────┬─────┬─────┬──────
        2−√7    2+√7
```

We try test numbers in each interval.

A: Test -1, $f(-1)=(-1)^2-4(-1)-3=2$

B: Test 0, $f(0)=0^2-4(0)-3=-3$

C: Test 5, $f(5)=5^2-4(5)-3=2$

Since $f(0)$ is negative, the function value will be negative for all numbers in the interval containing 0. The solution set is $\left[2-\sqrt{7},\ 2+\sqrt{7}\right]$, or $\{x\,|\,2-\sqrt{7}\leq x\leq2+\sqrt{7}\}$.

23. $3x(x+2)(x-2)<0$

The solutions of $3x(x+2)(x-2)=0$ are 0, -2, and 2. They divide the real-number line into four intervals as shown:

```
      A     B     C     D
    ──┬─────┬─────┬─────┬──
     −2     0     2
```

We try test numbers in each interval.

A: Test -3, $f(-3)=3(-3)(-3+2)(-3-2)=-45$

B: Test -1, $f(-1)=3(-1)(-1+2)(-1-2)=9$

C: Test 1, $f(1)=3(1)(1+2)(1-2)=-9$

D: Test 3, $f(3)=3(3)(3+2)(3-2)=45$

Since $f(-3)$ and $f(1)$ are negative, the function value will be negative for all numbers in the intervals containing -3 and 1. The solution set is $(-\infty,-2)\cup(0,\ 2)$, or $\{x\,|\,x<-2\ or\ 0<x<2\}$.

25. $(x-1)(x+2)(x-4)\geq0$

The solutions of $(x-1)(x+2)(x-4)=0$ are 1, -2, and 4. They divide the real-number line in a natural way. The product $(x-1)(x+2)(x-4)$ is positive or negative depending on the signs of $x-1$, $x+2$, and $x-4$.

Sign of $x-1$	$-$	$-$	$+$	$+$
Sign of $x+2$	$-$	$+$	$+$	$+$
Sign of $x-4$	$-$	$-$	$-$	$+$
Sign of product	$-$	$+$	$-$	$+$

```
    ──────┬─────┬─────┬──────
         −2     1     4
```

A product of three numbers is positive when all three factors are positive or when two are negative and one is positive. Since the $\geq$ symbol allows for equality, the endpoints -2, 1, and 4 are solutions. From the chart we see that the solution set is $[-2,\ 1]\cup[4,\ \infty)$, or $\{x\,|-2\leq x\leq1\ or\ x\geq4\}$.

27.
$$f(x)\geq3$$
$$7-x^2\geq3$$
$$-x^2+4\geq0$$
$$x^2-4\leq0$$
$$(x-2)(x+2)\leq0$$

The solutions of $(x-2)(x+2)=0$ are 2 and -2. They divide the real-number line as shown below.

Sign of $x-2$	$-$	$-$	$+$
Sign of $x+2$	$-$	$+$	$+$
Sign of product	$+$	$-$	$+$

```
    ──────┬───────┬──────
         −2       2
```

Because the inequality symbol is $\leq$, we must include the endpoints in the solution set. From the chart, we see that the solution set is $[-2,\ 2]$, or $\{x\,|-2\leq x\leq2\}$.

29.
$$g(x)>0$$
$$(x-2)(x-3)(x+1)>0$$

The solutions of $(x-2)(x-3)(x+1)=0$ are 2, 3, and -1. They divide the real-number line into four intervals as shown below.

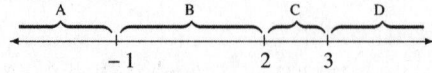

```
      A       B     C   D
    ──┬───────┬─────┬───┬──
     −1       2     3
```

We try test numbers in each interval.

A: Test -2, $f(-2)=(-2-2)(-2-3)(-2+1)=-20$

B: Test 0, $f(0)=(0-2)(0-3)(0+1)=6$

C: Test $\frac{5}{2}$, $f\left(\frac{5}{2}\right)=\left(\frac{5}{2}-2\right)\left(\frac{5}{2}-3\right)\left(\frac{5}{2}+1\right)=-\frac{7}{8}$

D: Test 4, $f(4)=(4-2)(4-3)(4+1)=10$

The function value will be positive for all numbers in intervals B and D. The solution set is $(-1,\ 2)\cup(3,\ \infty)$, or $\{x\,|-1<x<2\ or\ x>3\}$.

31.
$$F(x) \leq 0$$
$$x^3 - 7x^2 + 10x \leq 0$$
$$x(x^2 - 7x + 10) \leq 0$$
$$x(x-2)(x-5) \leq 0$$

The solutions of $x(x-2)(x-5) = 0$ are 0, 2, and 5. They divide the real-number line as shown below.

Sign of x	$-$	$\|$	$+$	$\|$	$+$	$\|$	$+$
Sign of $x - 2$	$-$	$\|$	$-$	$\|$	$+$	$\|$	$+$
Sign of $x - 5$	$-$	$\|$	$-$	$\|$	$-$	$\|$	$+$
Sign of product	$-$	$\|$	$+$	$\|$	$-$	$\|$	$+$

Because the inequality symbol is $\leq$ we must include the endpoints in the solution set. From the chart we see that the solution set is $(-\infty, 0] \cup [2, 5]$ or $\{x \mid x \leq 0 \ or \ 2 \leq x \leq 5\}$.

33. $\dfrac{1}{x-5} < 0$

We write the related equation by changing the $<$ symbol to $=$:.

$$\frac{1}{x-5} = 0$$

We solve the related equation.

$$(x-5) \cdot \frac{1}{x-5} = (x-5) \cdot 0$$
$$1 = 0$$

The related equation has no solution.

Next we find the values that make the denominator 0 by setting the denominator equation to 0 and solving:

$$x - 5 = 0$$
$$x = 5$$

We use 5 to divide the number line into two intervals as shown:

A: Test 0, $\dfrac{1}{0-5} = \dfrac{1}{-5} = -\dfrac{1}{5} < 0$

The number 0 is a solution of the inequality, so the interval A is part of the solution set.

B: Test 6, $\dfrac{1}{6-5} = 1 \not< 0$

The number 6 is not a solution of the inequality, so the interval B is part of the solution set.

The solution set is $(-\infty, 5)$, or $\{x \mid x < 5\}$.

35. $\dfrac{x+1}{x-3} \geq 0$

Solve the related equation.

$$\frac{x+1}{x-3} = 0$$
$$x + 1 = 0$$
$$x = -1$$

Find the values that make the denominator 0.

$$x - 3 = 0$$
$$x = 3$$

Use the numbers -1 and 3 to divide the number line into intervals as shown:

Try test numbers in each interval.

A: Test -2, $\dfrac{-2+1}{-2-3} = \dfrac{-1}{-5} = \dfrac{1}{5} > 0$

The number -2 is a solution of the inequality, so the interval A is part of the solution set.

B: Test 0, $\dfrac{0+1}{0-3} = \dfrac{1}{-3} = -\dfrac{1}{3} \not> 0$

The number 0 is not a solution of the inequality, so the interval B is not part of the solution set.

C: Test 4, $\dfrac{4+1}{4-3} = \dfrac{5}{1} = 5 > 0$

The number 4 is a solution of the inequality, so the interval C is part of the solution set.

The solution set includes intervals A and C. The number -1 is also included since the inequality symbol is $\geq$ and -1 is the solution of the related equation. The number 3 is not included since $\dfrac{x+1}{x-3}$ is undefined for $x = 3$. The solution set is $(-\infty, -1] \cup (3, \infty)$, or $\{x \mid x \leq -1 \ or \ x > 3\}$.

37. $\dfrac{x+1}{x+6} \geq 1$

Solve the related equation.

$$\frac{x+1}{x+6} = 1$$
$$x + 1 = x + 6$$
$$1 = 6$$

The related equation has no solution.

Find the values that make the denominator 0.

$$x + 6 = 0$$
$$x = -6$$

Use the number -6 to divide the number line into two intervals.

Try test numbers in each interval.

A: Test -7, $\dfrac{-7+1}{-7+6} = \dfrac{-6}{-1} = 6 > 1$.

The number -7 is a solution of the inequality, so the interval A is part of the solution set.

B: Test 0, $\dfrac{0+1}{0+6} = \dfrac{1}{6} \not> 1$

The number 0 is not a solution of the inequality, so the interval B is not part of the solution set. The number –6 is not included in the solution set since $\frac{x+1}{x+6}$ is undefined for $x=-6$. The solution set is $(-\infty,-6)$, or $\{x\,|\,x<-6\}$.

39. $\frac{(x-2)(x+1)}{x-5}\leq 0$

Solve the related equation.

$$\frac{(x-2)(x+1)}{x-5}=0$$
$$(x-2)(x+1)=0$$
$$x=2 \ or \ x=-1$$

Find the values that make the denominator 0.

$$x-5=0$$
$$x=5$$

Use the numbers 2, –1, and 5 to divide the number line into intervals as shown:

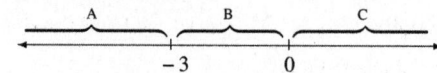

Try test numbers in each interval.

A: Test –2, $\frac{(-2-2)(-2+1)}{-2-5}=\frac{-4(-1)}{-7}=-\frac{4}{7}\leq 0$

Interval A is part of the solution set.

B: Test 0, $\frac{(0-2)(0+1)}{0-5}=\frac{-2\cdot 1}{-5}=\frac{2}{5}\nleq 0$

Interval B is not part of the solution set.

C: Test 3, $\frac{(3-2)(3+1)}{3-5}=\frac{1\cdot 4}{-2}=-2\leq 0$

Interval C is part of the solution set.

D: Test 6, $\frac{(6-2)(6+1)}{6-5}=\frac{4\cdot 7}{1}=28\nleq 0$

Interval D is not part of the solution set.

The solution set includes intervals A and C. The numbers –1 and 2 are also included since the inequality symbol is $\leq$ and –1 and 2 are the solutions of the related equation. The number 5 is not included since $\frac{(x-2)(x+1)}{x-5}$ is undefined for $x=5$. The solution set is $(-\infty,-1]\cup[2, 5)$, or $\{x\,|\,x\leq -1 \ or \ 2\leq x<5\}$.

41. $\frac{x}{x+3}\geq 0$

Solve the related equation.

$$\frac{x}{x+3}=0$$
$$x=0$$

Find the values that make the denominator 0.

$$x+3=0$$
$$x=-3$$

Use the numbers 0 and –3 to divide the number line into

intervals as shown.

Try test numbers in each interval.

A: Test –4, $\frac{-4}{-4+3}=\frac{-4}{-1}=4\geq 0$

Interval A is part of the solution set.

B: Test –1, $\frac{-1}{-1+3}=\frac{-1}{2}=-\frac{1}{2}\ngeq 0$

Interval B is not part of the solution set.

C: Test 1, $\frac{1}{1+3}=\frac{1}{4}\geq 0$

The interval C is part of the solution set.

The solution set includes intervals A and C. The number 0 is also included since the inequality symbol is $\geq$ and 0 is the solution of the related equation. The number –3 is not included since $\frac{x}{x+3}$ is undefined for $x=-3$. The solution set is $(-\infty,-3)\cup[0, \infty)$, or $\{x\,|\,x<-3 \ or \ x\geq 0\}$.

43. $\frac{x-5}{x}<1$

Solve the related equation.

$$\frac{x-5}{x}=1$$
$$x-5=x$$
$$-5=0$$

The related equation has no solution.

Find the values that make the denominator 0.

$$x=0$$

Use the number 0 to divide the number line into two intervals as shown.

Try test numbers in each interval.

A: Test –1, $\frac{-1-5}{-1}=\frac{-6}{-1}=6\nless 1$

Interval A is not part of the solution set.

B: Test 1, $\frac{1-5}{1}=\frac{-4}{1}=-4<1$

Interval B is part of the solution set.

The solution set is $(0, \infty)$ or $\{x\,|\,x>0\}$.

45. $\frac{x-1}{(x-3)(x+4)}\leq 0$

Solve the related equation.

$$\frac{x-1}{(x-3)(x+4)}=0$$
$$x-1=0$$
$$x=1$$

Find the values that make the denominator 0.

$$(x-3)(x+4)=0$$

$x = 3$ or $x = -4$

Use the numbers 1, 3, and –4 to divide the number line into intervals as shown:

Try test numbers in each interval.

A: Test –5, $\dfrac{-5-1}{(-5-3)(-5+4)} = \dfrac{-6}{-8(-1)} = -\dfrac{3}{4} < 0$

Interval A is part of the solution set.

B: Test 0, $\dfrac{0-1}{(0-3)(0+4)} = \dfrac{-1}{-3 \cdot 4} = \dfrac{1}{12} \not< 0$

Interval B is not part of the solution set.

C: Test 2, $\dfrac{2-1}{(2-3)(2+4)} = \dfrac{1}{-1 \cdot 6} = -\dfrac{1}{6} < 0$

Interval C is part of the solution set.

D: Test 4, $\dfrac{4-1}{(4-3)(4+4)} = \dfrac{3}{1 \cdot 8} = \dfrac{3}{8} \not< 0$

Interval D is not part of the solution set.

The solution set includes intervals A and C. The number 1 is also included since the inequality symbol is $\le$ and 1 is the solution of the related equation. The numbers –4 and 3 are not included since $\dfrac{x-1}{(x-3)(x+4)}$ is undefined for $x = -4$ and for $x = 3$.

The solution set is $(-\infty, -4) \cup [1, 3)$, or $\{x \mid x < -4 \text{ or } 1 \le x < 3\}$.

47. $f(x) \ge 0$

$\dfrac{5-2x}{4x+3} \ge 0$

Solve the related equation.

$\dfrac{5-2x}{4x+3} = 0$

$5 - 2x = 0$

$5 = 2x$

$\dfrac{5}{2} = x$

Find the values that make the denominator 0.

$4x + 3 = 0$

$4x = -3$

$x = -\dfrac{3}{4}$

Use the numbers $\dfrac{5}{2}$ and $-\dfrac{3}{4}$ to divide the number line as shown:

Try test numbers in each interval.

A: Test –1, $\dfrac{5-2(-1)}{4(-1)+3} = -7 \not> 0$

Interval A is not part of the solution set.

B: Test 0, $\dfrac{5-2 \cdot 0}{4 \cdot 0 + 3} = \dfrac{5}{3} > 0$

Interval B is part of the solution set.

C: Test 3, $\dfrac{5-2 \cdot 3}{4 \cdot 3 + 3} = -\dfrac{1}{15} \not> 0$

Interval C is not part of the solution set.

The solution set includes interval B. The number $\dfrac{5}{2}$ is also included since the inequality symbol is $\ge$ and $\dfrac{5}{2}$ is the solution of the related equation. The number $-\dfrac{3}{4}$ is not included since $\dfrac{5-2x}{4x+3}$ is undefined for $x = -\dfrac{3}{4}$. The solution set is $\left(-\dfrac{3}{4}, \dfrac{5}{2}\right]$, or $\left\{x \mid -\dfrac{3}{4} < x \le \dfrac{5}{2}\right\}$.

49. $G(x) \le 1$

$\dfrac{1}{x-2} \le 1$

Solve the related equation.

$\dfrac{1}{x-2} = 1$

$1 = x - 2$

$3 = x$

Find the values of x that make the denominator 0.

$x - 2 = 0$

$x = 2$

Use the numbers 2 and 3 to divide the number line as shown.

Try a test number in each interval.

A: Test 0, $\dfrac{1}{0-2} = -\dfrac{1}{2} \le 1$

Interval A is part of the solution set.

B: Test $\dfrac{5}{2}$, $\dfrac{1}{\frac{5}{2}-2} = \dfrac{1}{\frac{1}{2}} = 2 \not\le 1$

Interval B is not part of the solution set.

C: Test 4, $\dfrac{1}{4-2} = \dfrac{1}{2} \le 1$

Interval C is part of the solution set.

The solution set includes intervals A and B. The number 3 is also included since the inequality symbol is $\le$ and 3 is the solution of the related equation. The number 2 is not included since $\dfrac{1}{x-2}$ is undefined for $x = 2$. The solution set is $(-\infty, 2) \cup [3, \infty)$, or $\{x \mid x < 2 \text{ or } x \ge 3\}$.

51. *Writing Exercise.* Consider the quadratic portion of the inequality, $ax^2 + bx + c$. The graph of $f(x) = ax^2 + bx + c$ is a parabola and we can solve the inequality as in Example 1.

53. Graph $f(x) = x^3 - 2$.

x	y
-2	-10
-1	-3
0	-2
1	-1
2	6

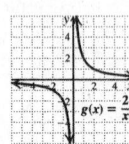

54. Graph $g(x) = \dfrac{2}{x}$.   Note $x \neq 0$.

x	y
-2	-1
-1	-2
0	undefined
1	2
2	1

55. $f(x) = x + 7$
$$f\left(\frac{1}{a^2}\right) = \frac{1}{a^2} + 7$$

57. $g(x) = x^2 + 2$
$$g(2a+5) = (2a+5)^2 + 2$$
$$= 4a^2 + 20a + 25 + 2$$
$$= 4a^2 + 20a + 27$$

59. *Writing Exercise.* If any solutions from step (1) are also replacements for which the rational expression is undefined, they must be excluded from the solution set.

61. $x^2 + 2x < 5$
$$x^2 + 2x - 5 < 0$$
Using the quadratic formula, we find that the solutions of the related equation are $x = -1 \pm \sqrt{6}$. These numbers divide the real-number line into three intervals as shown:

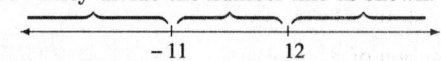

We try test numbers in each interval.

A: Test -4, $f(-4) = (-4)^2 + 2(-4) - 5 = 3$

B: Test 0, $f(0) = 0^2 + 2 \cdot 0 - 5 = -5$

C: Test 2, $f(2) = 2^2 + 2 \cdot 2 - 5 = 3$

The function value will be negative for all numbers in interval B. The solution set is $\left(-1-\sqrt{6}, -1+\sqrt{6}\right)$, or $\left\{x \mid -1-\sqrt{6} < x < -1+\sqrt{6}\right\}$.

63. $x^4 + 3x^2 \leq 0$
$$x^2(x^2 + 3) \leq 0$$
$x^2 = 0$ for $x = 0$, $x^2 > 0$ for $x \neq 0$, $x^2 + 3 > 0$ for all x
The solution set is $\{0\}$.

65. a) $-3x^2 + 630x - 6000 > 0$
$$x^2 - 210x + 2000 < 0 \quad \text{Multiplying by } -\frac{1}{3}$$
$$(x - 200)(x - 10) < 0$$
The solutions of $f(x) = (x - 200)(x - 10) = 0$ are 200 and 10. They divide the number line as shown:

A: Test 0, $f(0) = 0^2 - 210 \cdot 0 + 2000 = 2000$

B: Test 20, $f(20) = 20^2 - 210 \cdot 20 + 2000 = -1800$

C: Test 300, $f(300) = 300^2 - 210 \cdot 300 + 2000 = 29{,}000$

The company makes a profit for values of x such that $10 < x < 200$, or for values of x in the interval $(10, \ 200)$.

b)   See part (a). Keep in mind that x must be nonnegative since negative numbers have no meaning in this application.

The company loses money for values of x such that $0 \leq x < 10$ or $x > 200$, or for values of x in the interval $[0, \ 10) \cup (200, \ \infty)$.

67. We find values of n such that $N \geq 66$ *and* $N \leq 300$.

For $N \geq 66$:
$$\frac{n(n-1)}{2} \geq 66$$
$$n(n-1) \geq 132$$
$$n^2 - n - 132 \geq 0$$
$$(n - 12)(n + 11) \geq 0$$

The solutions of $f(n) = (n-12)(n+11) = 0$ are 12 and -11. They divide the number line as shown:

However, only positive values of n have meaning in this exercise so we need only consider the intervals shown below:

A: Test 1, $f(1) = 1^2 - 1 - 132 = -132$

B: Test 20, $f(20) = 20^2 - 20 - 132 = 248$

Thus, $N \geq 66$ for $\{n \mid n \geq 12\}$.

For $N \leq 300$:
$$\frac{n(n-1)}{2} \leq 300$$
$$n(n-1) \leq 600$$
$$n^2 - n - 600 \leq 0$$
$$(n - 25)(n + 24) \leq 0$$

The solutions of $f(n) = (n-25)(n+24) = 0$ are 25 and

-24. They divide the number line as shown:

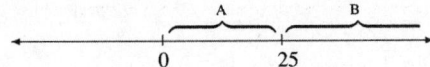

However, only positive values of n have meaning in this exercise so we need only consider the intervals shown below:

A: Test 1, $f(1)=1^2-1-600=-600$

B: Test 30, $f(30)=30^2-30-600=270$

Thus, $N\le 300$ (and $n>0$) for $\{n\mid 0<n\le 25\}$.

Then $66\le N\le 300$ for

$\{n\mid n \text{ is an integer } and \ 12\le n\le 25\}$.

69. From the graph we determine the following:

The solutions of $f(x)=0$ are -2, 1, and 3.

The solution of $f(x)<0$ is $(-\infty,-2)\cup(1,\ 3)$, or

$\{x\mid x<-2 \ or \ 1<x<3\}$.

The solution of $f(x)>0$ is $(-2,\ 1)\cup(3,\ \infty)$, or

$\{x\mid -2<x<1 \ or \ x>3\}$.

71. From the graph we determine the following:

$f(x)$ has no zeros.

The solutions of $f(x)<0$ are $(-\infty,\ 0)$, or $\{x\mid x<0\}$;

The solutions of $f(x)>0$ are $(0,\ \infty)$, or $\{x\mid x>0\}$.

73. From the graph we determine the following:

The solutions of $f(x)=0$ are -1 and 0.

The solution of $f(x)<0$ is $(-\infty,-3)\cup(-1,\ 0)$, or

$\{x\mid x<-3 \ or \ -1<x<0\}$.

The solution of $f(x)>0$ is $(-3,\ -1)\cup(0,\ 2)\cup(2,\ \infty)$, or

$\{x\mid -3<x<-1 \ or \ 0<x<2 \ or \ x>2\}$.

75. For $f(x)=\sqrt{x^2-4x-45}$, we find the domain:

$$x^2-4x-45\ge 0$$
$$(x+5)(x-9)\ge 0$$

The quadratic is nonnegative when $(-\infty,-5]\cup[9,\ \infty)$, or

$\{x\mid x\le -5 \ or \ x\ge 9\}$.

77. For $f(x)=\sqrt{x^2+8x}$, we find the domain:

$$x^2+8x\ge 0$$
$$x(x+8)\ge 0$$

The quadratic is nonnegative when $(-\infty,-8]\cup[0,\ \infty)$, or

$\{x\mid x\le -8 \ or \ x\ge 0\}$.

79. *Writing Exercise.* Answers may vary.

For $a < b$, write the inequality $(x-a)(x-b)\ge 0$, or

$x^2-(a+b)x+ab\ge 0$.

Chapter 8 Review

1. False; see page 518 in the text.

3. True

5. False; the vertex is $(-3,\ -4)$.

7. True; since the coefficient of x^2 is -2, the graph opens down and therefore has no minimum.

9. False; see page 518 in the text.

11. $9x^2-2=0$
$$9x^2=2$$
$$x^2=\frac{2}{9}$$
$$x=\pm\sqrt{\frac{2}{9}}=\pm\frac{\sqrt{2}}{3}$$

The solutions are $-\dfrac{\sqrt{2}}{3}$ and $\dfrac{\sqrt{2}}{3}$.

13. $x^2-12x+36=9$
$$(x-6)^2=9$$
$$x-6=\pm 3$$
$$x=6\pm 3$$

The solutions are 3 and 9.

15. $x(3x+4)=4x(x-1)+15$
$$3x^2+4x=4x^2-4x+15$$
$$0=x^2-8x+15$$
$$0=(x-3)(x-5)$$
$$x-3=0 \quad or \quad x-5=0$$
$$x=3 \quad or \qquad x=5$$

The solutions are 3 and 5.

17. $x^2-5x-2=0$
$$a=1,\ b=-5,\ c=-2$$
$$x=\frac{-(-5)\pm\sqrt{(-5)^2-4\cdot 1\cdot(-2)}}{2\cdot 1}=\frac{5\pm\sqrt{25+8}}{2}$$
$$x=\frac{5\pm\sqrt{33}}{2}$$
$$x\approx -0.372,\ 5.372$$

19. $\dfrac{1}{2}\cdot(-18)=-9;\ (-9)^2=81$
$$x^2-18x+81=(x-9)^2$$

21. $x^2 - 6x + 1 = 0$

$x^2 - 6x = -1$

$x^2 - 6x + 9 = 9 - 1$

$(x - 3)^2 = 8$

$x - 3 = \pm\sqrt{8}$

$x = 3 \pm \sqrt{8}$

$x = 3 \pm 2\sqrt{2}$

23. $s = 16t^2$

$1018 = 16t^2$

$\dfrac{509}{8} = t^2$

$\sqrt{\dfrac{509}{8}} = t$ Principle of square roots; rejecting the negative square root.

$8.0 \approx t$

It will take an object about 8.0 sec to fall.

25. $x^2 + 2x + 5 = 0$

$b^2 - 4ac = 2^2 - 4 \cdot 1 \cdot 5 = -16$

There are two imaginary numbers.

27. The only solution is –5. It must be a repeated solution.

$x = -5 \quad or \quad x = -5$

$x + 5 = 0 \quad or \quad x + 5 = 0$

$(x + 5)(x + 5) = 0$

$x^2 + 10x + 25 = 0$

29. *Familiarize.* Let x represent the time it takes Cheri to reply. Then $x + 6$ represents the time it takes Dani to reply. It takes them 4 hr to reply working together, so they can reply to $\dfrac{1}{4}$ of the emails in 1 hr. Cheri will reply to $\dfrac{1}{x}$ of the emails in 1 hr, and Dani will reply to $\dfrac{1}{x+6}$ of the emails in 1 hr.

Translate. We have an equation.

$$\dfrac{1}{x} + \dfrac{1}{x+6} = \dfrac{1}{4}$$

Carry out. We solve the equation.

We multiply by the LCD, $4x(x+6)$.

$$4x(x+6)\left(\dfrac{1}{x} + \dfrac{1}{x+6}\right) = 4x(x+6) \cdot \dfrac{1}{4}$$

$$4(x+6) + 4x = x(x+6)$$

$$4x + 24 + 4x = x^2 + 6x$$

$$0 = x^2 - 2x - 24$$

$$0 = (x - 6)(x + 4)$$

Check. Since negative time has no meaning in this problem, –4 is not a solution. We check only 6 hr. This is the time it would take Cheri working alone. Then Dani would take 6 + 6, or 12 hr working alone. Cheri would reply to $4\left(\dfrac{1}{6}\right)$, or $\dfrac{2}{3}$ of the emails in 4 hr, and Dani

would reply to $4\left(\dfrac{1}{12}\right)$, or $\dfrac{1}{3}$ of the emails in 4 hr. Thus, in 4 hr they would reply to $\dfrac{2}{3} + \dfrac{1}{3}$ of the emails. This is all of it, so the numbers check.

State. It Cheri, working alone, 6 hr to reply to the emails.

31. $15x^{-2} - 2x^{-1} - 1 = 0$

Let $u = x^{-1}$ and $u^2 = x^{-2}$.

$15u^2 - 2u - 1 = 0$

$(5u + 1)(3u - 1) = 0$

$5u + 1 = 0 \quad or \quad 3u - 1 = 0$

$u = -\dfrac{1}{5} \quad or \quad u = \dfrac{1}{3}$

Replace u with x^{-1}.

$x^{-1} = -\dfrac{1}{5} \quad or \quad x^{-1} = \dfrac{1}{3}$

$x = -5 \quad or \quad x = 3$

The numbers –5 and 3 check. They are the solutions.

33. $f(x) = -3(x + 2)^2 + 4$

We know that the graph looks like the graph of $h(x) = 3x^2$ but moved to the left 2 units and up 4 units and turned upside down. The vertex is (–2, 4), and the axis of symmetry is $x = -2$. The maximum function value is 4.

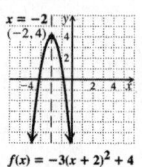

$f(x) = -3(x + 2)^2 + 4$
Maximum: 4

35. $f(x) = x^2 - 9x + 14$

To find the x-intercepts, solve the equation $0 = x^2 - 9x + 14$. We factor.

$0 = x^2 - 9x + 14$

$0 = (x - 2)(x - 7)$

$x = 2 \ or \ x = 7$

The x-intercepts are (2, 0) and (7, 0).

The y-intercept is $(0, f(0))$, or (0, 14).

37. $2A + T = 3T^2$

$3T^2 - T - 2A = 0$

$a = 3, \ b = -1, \ c = -2A$

$T = \dfrac{-(-1) \pm \sqrt{(-1)^2 - 4 \cdot 3 \cdot (-2A)}}{2 \cdot 3}$

$T = \dfrac{1 \pm \sqrt{1 + 24A}}{6}$

39. The data points fall. The graph does not appear to represent a quadratic function in which the data points would rise and then fall or vice versa. Thus a linear function $f(x) = mx + b$ might be used to model the data.

41. a) *Familiarize*. We look for a function of the form

$f(x) = ax^2 + bx + c$, where $f(x)$ represents the percent increase in premiums and x represents the years after 2000.

Translate. We substitute the given values of x and $f(x)$.

$$8 = a(0)^2 + b(0) + c,$$
$$11 = a(2)^2 + b(2) + c,$$
$$8 = a(6)^2 + b(6) + c,$$

or

$$8 = 0a + 0b + c,$$
$$11 = 4a + 2b + c,$$
$$8 = 36a + 6b + c.$$

Carry out. Solving the system of equations, we get

$a = -\dfrac{3}{8}$, $b = \dfrac{9}{4}$, $c = 8$.

Check. Recheck the calculations.

State. The function

$f(x) = -\dfrac{3}{8}x^2 + \dfrac{9}{4}x + 8$ fits the data.

b) Find $f(5)$.

$f(5) = -\dfrac{3}{8}(5)^2 + \dfrac{9}{4}(5) + 8 = 9.875 \approx 10$

About 10% increase for premiums in 2005 is estimated.

43. $\dfrac{x-5}{x+3} \le 0$

Solve the related equation.

$$\dfrac{x-5}{x+3} = 0$$
$$x - 5 = 0$$
$$x = 5$$

Find the values that make the denominator 0.

$$x + 3 = 0$$
$$x = -3$$

Use the numbers 5 and −3 to divide the number line into intervals as shown:

Try test numbers in each interval.

A: Test −4, $\dfrac{-4-5}{-5+3} = \dfrac{-9}{-2} = \dfrac{9}{2} \not\le 0$

The number −4 is a not solution of the inequality, so the interval A is not part of the solution set.

B: Test 0, $\dfrac{0-5}{0+3} = \dfrac{-5}{3} = -\dfrac{5}{3} < 0$

The number 0 is a solution of the inequality, so the interval B is part of the solution set.

C: Test 6, $\dfrac{6-5}{6+3} = \dfrac{1}{9} \not< 0$

The number 6 is a not solution of the inequality, so the interval C is not part of the solution set.

The solution set includes interval B. The number 5 is also included since the inequality symbol is $\le$ and 5 is the solution of the related equation. The number −3 is not included since $\dfrac{x-5}{x+3}$ is undefined for $x = -3$. The solution set is $(-3, -5]$, or $\{x \mid -3 < x \le 5\}$.

45. *Writing Exercise*. Yes; if the discriminant is a perfect square, then the solutions are rational numbers, p/q and r/s. (Note that if the discriminant is 0, then $p/q = r/s$.) Then the equation can be written in factored form, $(qx - p)(sx - r) = 0$.

47. *Writing Exercise*. Completing the square was used to solve quadratic equations and to graph quadratic functions by rewriting the function in the form

$f(x) = a(x - h)^2 + k$.

49. From Section 8.4, we know the sum of the solutions of $ax^2 + bx + c = 0$ is $-\dfrac{b}{a}$, and the product is $\dfrac{c}{a}$.

$$3x^2 - hx + 4k = 0$$
$$a = 3, \ b = -h, \ c = 4k$$

Substituting

For $-\dfrac{b}{a}$: $\ \dfrac{-h}{3} = 20$

$\dfrac{h}{3} = 20$

$h = 60$

For $\dfrac{c}{a}$: $\ \dfrac{4k}{3} = 80$

$4k = 240$

$k = 60$

Chapter 8 Test

1. $25x^2 - 7 = 0$

$$25x^2 = 7$$

$$x^2 = \frac{7}{25}$$

$$x = \pm\sqrt{\frac{7}{25}} = \pm\frac{\sqrt{7}}{5}$$

The solutions are $-\dfrac{\sqrt{7}}{5}$ and $\dfrac{\sqrt{7}}{5}$.

3. $x^2 + 2x + 3 = 0$

$a = 1,\ b = 2,\ c = 3$

$$x = \frac{-2 \pm \sqrt{2^2 - 4(1)(3)}}{2(1)} = \frac{-2 \pm \sqrt{4 - 12}}{2}$$

$$= \frac{-2 \pm \sqrt{-8}}{2} = \frac{-2 \pm 2i\sqrt{2}}{2} = \frac{-2}{2} \pm \frac{2i\sqrt{2}}{2}$$

$$= -1 \pm i\sqrt{2} \quad \text{or} \quad -1 \pm \sqrt{2}i$$

The solutions are $-1 + \sqrt{2}i$ and $-1 - \sqrt{2}i$.

5.
$$x^{-2} - x^{-1} = \frac{3}{4}$$

$$x^{-2} - x^{-1} - \frac{3}{4} = 0$$

$$4x^{-2} - 4x^{-1} - 3 = 0 \quad \text{Clearing fractions}$$

Let $u = x^{-1}$ and $u^2 = x^{-2}$.

$$4u^2 - 4u - 3 = 0$$

$$(2u - 3)(2u + 1) = 0$$

$$2u - 3 = 0 \quad or \quad 2u + 1 = 0$$

$$u = \frac{3}{2} \quad or \qquad u = -\frac{1}{2}$$

Now we replace u with x^{-1} and solve these equations:

$$x^{-1} = \frac{3}{2} \quad or \quad x^{-1} = -\frac{1}{2}$$

$$\frac{1}{x} = \frac{3}{2} \quad or \quad \frac{1}{x} = -\frac{1}{2}$$

$$2 = 3x \quad or \quad 2 = -x$$

$$\frac{2}{3} = x \qquad\qquad -2 = x$$

The solutions are -2 and $\dfrac{2}{3}$.

7. Let $f(x) = 0$ and solve for x.

$$0 = 12x^2 - 19x - 21$$

$$0 = (4x + 3)(3x - 7)$$

$$x = -\frac{3}{4} \quad or \quad x = \frac{7}{3}$$

The solutions are $-\dfrac{3}{4}$ and $\dfrac{7}{3}$.

9. $\dfrac{1}{2} \cdot \dfrac{2}{7} = \dfrac{1}{7};\ \left(\dfrac{1}{7}\right)^2 = \dfrac{1}{49}$

$$x^2 + \frac{2}{7}x + \frac{1}{49} = \left(x + \frac{1}{7}\right)^2$$

11. $x^2 + 2x + 5 = 0$

$b^2 - 4ac = 2^2 - 4(1)(5) = -16$

Two imaginary numbers

13. *Familiarize.* Let r represent the cruiser's speed in still water. Then $r - 4$ is the speed upriver and $r + 4$ is the speed downriver. Using $t = \dfrac{d}{r}$, we let $\dfrac{60}{r - 4}$ represent the time upriver and $\dfrac{60}{r + 4}$ represent the time downriver.

Trip	Distance	Speed	Time
Upriver	60	$r - 4$	$\dfrac{60}{r-4}$
Downriver	60	$r + 4$	$\dfrac{60}{r+4}$

Translate. We have an equation.

$$\frac{60}{r - 4} + \frac{60}{r + 4} = 8$$

Carry out. We solve the equation.

We multiply by the LCD, $(r - 4)(r + 4)$.

$$(r - 4)(r + 4) \cdot \left(\frac{60}{r - 4} + \frac{60}{r + 4}\right) = (r - 4)(r + 4) \cdot 8$$

$$60(r + 4) + 60(r - 4) = 8(r - 4)(r + 4)$$

$$60r + 240 + 60r - 240 = 8r^2 - 128$$

$$0 = 8r^2 - 120r - 128$$

$$0 = 8(r^2 - 15r - 16)$$

$$0 = 8(r - 16)(r + 1)$$

$$r = 16 \quad or \quad r = -1$$

Check. Since negative time has no meaning in this problem, -1 is not a solution. We check only 16 hr. If $r = 16$, then the speed upriver is $16 - 4$, or 12 km/h, and the time is $\dfrac{60}{12}$, or 5 hr. The speed downriver is $16 + 4$, or 20 km/h, and the time is $\dfrac{60}{20}$, or 3 hr. The total time of the round trip is $5 + 3$, or 8 hr. The value checks.

State. The speed of the cruiser in still water is 16 km/h.

15. $f(x) = x^4 - 15x^2 - 16$

To find the x-intercepts, solve the equation

$$0 = x^4 - 15x^2 - 16.$$

Let $u = x^2$ and $u^2 = x^4$.

$$0 = u^2 - 15u - 16$$

$$0 = (u - 16)(u + 1)$$

$$u = 16 \quad or \quad u = -1$$

Replace u with x^2.

$$x^2 = 16 \quad or \quad x^2 = -1 \quad \text{Has no real solutions}$$

$$x = \pm 4$$

The x-intercepts are $(-4, 0)$ and $(4, 0)$.

17. $f(x) = 2x^2 + 4x - 6$

$ = 2(x^2 + 2x) - 6$

$ = 2(x^2 + 2x + 1) - 6 - 2$

$ = 2(x+1)^2 - 8$

We know that the graph looks like the graph of $h(x) = 2x^2$ but moved to the left 1 unit and down 8 units. The vertex is $(-1, -8)$, and the axis of symmetry is $x = -1$.

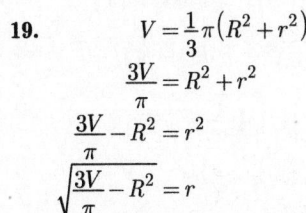

$f(x) = 2x^2 + 4x - 6$

19. $\quad V = \frac{1}{3}\pi(R^2 + r^2)$

$\quad \dfrac{3V}{\pi} = R^2 + r^2$

$\quad \dfrac{3V}{\pi} - R^2 = r^2$

$\quad \sqrt{\dfrac{3V}{\pi} - R^2} = r$

We only consider the positive square root as instructed.

21. $C(x) = 0.2x^2 - 1.3x + 3.4025$

$C(x) = 0.2(x^2 - 6.5x) + 3.4025$

$C(x) = 0.2(x^2 - 6.5x + 10.5625) - 0.2(10.5625) + 3.4025$

$C(x) = 0.2(x - 3.25)^2 + 1.29$

3.25 hundred or 324 cabinets should be built to have a minimum at \$1.29 hundred, or \$129 per cabinet.

23. $\quad x^2 + 5x < 6$

$\quad x^2 + 5x - 6 < 0$

$\quad (x+6)(x-1) < 0$

The solutions of $(x+6)(x-1) = 0$ are –6 and 1. They divide the number line into three intervals as shown:

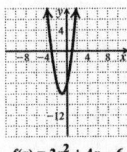

We try test numbers in each interval.

A: Test –7, $f(-7) = (-7+6)(-7-1) = 8$

B: Test 0, $f(0) = (0+6)(0-1) = -6$

C: Test 2, $f(2) = (2+6)(2-1) = 8$

Since $f(0)$ is negative, the function value will be negative for all numbers in the interval containing 0. Because the symbol is $<$, we do not include the endpoints in the solution. The solution set is $(-6, 1)$, or $\{x \mid -6 < x < 1\}$.

25. $\quad kx^2 + 3x - k = 0$

$\quad k(-2)^2 + 3(-2) - k = 0 \quad$ Substitute -2 for x

$\quad 4k - 6 - k = 0$

$\quad 3k = 6$

$\quad k = 2$

Then $\quad 2x^2 + 3x - 2 = 0$

$\quad (x+2)(2x-1) = 0$

The other solution is $\frac{1}{2}$.

27. $x^4 - 4x^2 - 1 = 0$

Let $u = x^2$ and $u^2 = x^4$.

$\quad u^2 - 4u - 1 = 0$

$\quad u^2 - 4u = 1$

$\quad u^2 - 4u + 4 = 1 + 4$

$\quad (u-2)^2 = 5$

$\quad u - 2 = \pm\sqrt{5}$

$\quad u = 2 \pm \sqrt{5}$

Replace u with x^2.

$\quad x^2 = 2 + \sqrt{5} \quad or \quad x^2 = 2 - \sqrt{5}$

$\quad x = \pm\sqrt{2+\sqrt{5}} \quad or \quad x = \pm\sqrt{2-\sqrt{5}}$

Since $2 - \sqrt{5}$ is negative, we can rewrite it as follows:

$$\pm\sqrt{2-\sqrt{5}} = \pm\sqrt{\sqrt{5}-2}\,i$$

The solutions are $\pm\sqrt{\sqrt{5}+2}$ and $\pm\sqrt{\sqrt{5}-2}\,i$.

Chapter 9

Exponential and Logarithmic Functions

1. True; see page 581 in the text.

3. $(g \circ f) = g(f(x)) = x^2 + 3 \neq (x+3)^2$, so the statement is false.

5. False; see page 583 in the text.

7. True; see page 584 in the text.

9. a) $\begin{aligned}(f \circ g)(1) &= f(g(1)) = f(1-3) \\ &= f(-2) = (-2)^2 + 1 \\ &= 4 + 1 = 5\end{aligned}$

b) $\begin{aligned}(g \circ f)(1) &= g(f(1)) = g(1^2 + 1) \\ &= g(2) = 2 - 3 = -1\end{aligned}$

c) $\begin{aligned}(f \circ g) &= f(g(x)) = f(x-3) \\ &= (x-3)^2 + 1 = x^2 - 6x + 9 + 1 \\ &= x^2 - 6x + 10\end{aligned}$

d) $\begin{aligned}(g \circ f)(x) &= g(f(x)) = g(x^2 + 1) \\ &= x^2 + 1 - 3 = x^2 - 2\end{aligned}$

11. a) $\begin{aligned}(f \circ g)(1) &= f(g(1)) = f(2 \cdot 1^2 - 7) \\ &= f(-5) = 5(-5) + 1 = -24\end{aligned}$

b) $\begin{aligned}(g \circ f)(1) &= g(f(1)) = g(5 \cdot 1 + 1) \\ &= g(6) = 2 \cdot 6^2 - 7 = 65\end{aligned}$

c) $\begin{aligned}(f \circ g)(x) &= f(g(x)) = f(2x^2 - 7) \\ &= 5(2x^2 - 7) + 1 = 10x^2 - 34\end{aligned}$

d) $\begin{aligned}(g \circ f)(x) &= g(f(x)) = g(5x + 1) \\ &= 2(5x + 1)^2 - 7 = 2(25x^2 + 10x + 1) - 7 \\ &= 50x^2 + 20x - 5\end{aligned}$

13. a) $\begin{aligned}(f \circ g)(1) &= f(g(1)) = f\left(\frac{1}{1^2}\right) \\ &= f(1) = 1 + 7 = 8\end{aligned}$

b) $(g \circ f)(1) = g(f(1)) = g(1 + 7) = g(8) = \frac{1}{8^2} = \frac{1}{64}$

c) $(f \circ g)(x) = f(g(x)) = f\left(\frac{1}{x^2}\right) = \frac{1}{x^2} + 7$

d) $(g \circ f)(x) = g(f(x)) = g(x + 7) = \frac{1}{(x+7)^2}$

15. a) $\begin{aligned}(f \circ g)(1) &= f(g(1)) = f(1 + 3) \\ &= f(4) = \sqrt{4} = 2\end{aligned}$

b) $\begin{aligned}(g \circ f)(1) &= g(f(1)) = g(\sqrt{1}) \\ &= g(1) = 1 + 3 = 4\end{aligned}$

c) $(f \circ g)(x) = f(g(x)) = f(x + 3) = \sqrt{x + 3}$

d) $(g \circ f)(x) = g(f(x)) = g(\sqrt{x}) = \sqrt{x} + 3$

17. a) $(f \circ g)(1) = f(g(1)) = f\left(\frac{1}{1}\right) = f(1) = \sqrt{4 \cdot 1} = \sqrt{4} = 2$

b) $(g \circ f)(1) = g(f(1)) = g(\sqrt{4 \cdot 1}) = g(\sqrt{4}) = g(2) = \frac{1}{2}$

c) $(f \circ g)(x) = f(g(x)) = f\left(\frac{1}{x}\right) = \sqrt{4 \cdot \frac{1}{x}} = \sqrt{\frac{4}{x}}$

d) $(g \circ f)(x) = g(f(x)) = g(\sqrt{4x}) = \frac{1}{\sqrt{4x}}$

19. a) $\begin{aligned}(f \circ g)(1) &= f(g(1)) = f(\sqrt{1 - 1}) \\ &= f(\sqrt{0}) = f(0) = 0^2 + 4 = 4\end{aligned}$

b) $\begin{aligned}(g \circ f)(1) &= g(f(1)) = g(1^2 + 4) \\ &= g(5) = \sqrt{5 - 1} = \sqrt{4} = 2\end{aligned}$

c) $\begin{aligned}(f \circ g)(x) &= f(g(x)) = f(\sqrt{x - 1}) \\ &= (\sqrt{x - 1})^2 + 4 = x - 1 + 4 = x + 3\end{aligned}$

d) $\begin{aligned}(g \circ f)(x) &= g(f(x)) = g(x^2 + 4) \\ &= \sqrt{x^2 + 4 - 1} = \sqrt{x^2 + 3}\end{aligned}$

21. $h(x) = (3x - 5)^4$

This is $3x - 5$ raise to the fourth power, so the two most obvious functions are $f(x) = x^4$ and $g(x) = 3x - 5$.

23. $h(x) = \sqrt{9x + 1}$

We have $9x + 1$ and take the square root of their expression, so the two most obvious functions are $f(x) = \sqrt{x}$ and $g(x) = 9x + 1$.

25. $h(x) = \frac{6}{5x - 2}$

This is 6 divided by $5x - 2$, so two functions that can be used are $f(x) = \frac{6}{x}$ and $g(x) = 5x - 2$.

27. The graph of $f(x) = -x$ is shown below.

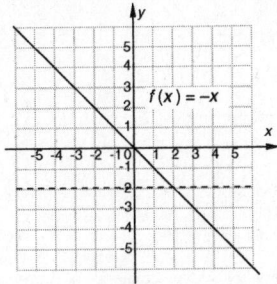

Since there is no horizontal line that crosses the graph more than once, the function is one-to-one.

29. $f(x) = x^2 + 3$

Observe that the graph of this function is a parabola that opens up. Thus, there are many horizontal lines that cross the graph more than once, so the function is not one-to-one. We can also draw the graph as shown below.

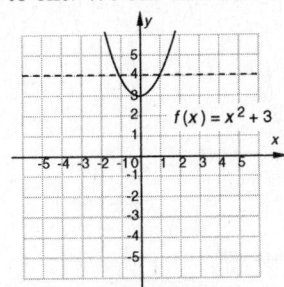

There are many horizontal lines that cross the graph more than once. In particular, the line $y = 4$ crosses the graph more than once. The function is not one-to-one.

31. Since there is no horizontal line that crosses the graph more than once, the function is one-to-one.

33. There are many horizontal lines that cross the graph more than once, the function is not one-to-one.

35. a) The function $f(x) = x + 3$ is a linear function that is not constant, so it passes the horizontal-line test. Thus, f is one-to-one.

     b) Replace $f(x)$ by y: $y = x + 3$
        Interchange x and y: $x = y + 3$
        Solve for y: $x - 3 = y$
        Replace y by $f^{-1}(x)$: $f^{-1}(x) = x - 3$

37. a) The function $f(x) = 2x$ is a linear function that is not constant, so it passes the horizontal-line test. Thus, f is one-to-one.

     b) Replace $f(x)$ by y: $y = 2x$
        Interchange x and y: $x = 2y$
        Solve for y: $\dfrac{x}{2} = y$
        Replace y by $f^{-1}(x)$: $f^{-1}(x) = \dfrac{x}{2}$

39. a) The function $g(x) = 3x - 1$ is a linear function that is not constant, so it passes the horizontal-line test. Thus, g is one-to-one.

     b) Replace $g(x)$ by y: $y = 3x - 1$
        Interchange x and y: $x = 3y - 1$
        Solve for y: $x + 1 = 3y$
                         $\dfrac{x+1}{3} = y$
        Replace y by $g^{-1}(x)$: $f^{-1}(x) = \dfrac{x+1}{3}$

41. a) The function $f(x) = \frac{1}{2}x + 1$ is a linear function that is not constant, so it passes the horizontal-line test. Thus, f is one-to-one.

     b) Replace $f(x)$ by y: $y = \frac{1}{2}x + 1$
        Interchange variables: $x = \frac{1}{2}y + 1$
        Solve for y: $x - 1 = \frac{1}{2}y$
                        $2x - 2 = y$
        Replace y by $f^{-1}(x)$: $f^{-1}(x) = 2x - 2$

43. a) The graph of $g(x) = x^2 + 5$ is shown below. There are many horizontal lines that cross the graph more than once. For example, the line $y = 8$ crosses the graph more than once. The function is not one-to-one.

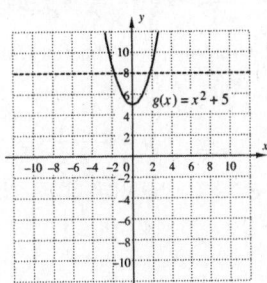

45. a) The function $h(x) = -10 - x$ is a linear function that is not constant, so it passes the horizontal-line test. Thus, h is one-to-one.

     b) Replace $h(x)$ by y: $y = -10 - x$
        Interchange variables: $x = -10 - y$
        Solve for y: $x + 10 = -y$
                       $-x - 10 = y$
        Replace y by $h^{-1}(x)$: $h^{-1}(x) = -x - 10$

47. a) The graph of $f(x)=\frac{1}{x}$ is shown below. It passes the horizontal-line test, so the function is one-to-one.

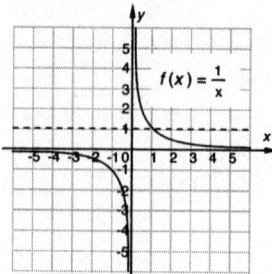

b) Replace $f(x)$ by y:
$$y=\frac{1}{x}$$
Interchange x and y:
$$x=\frac{1}{y}$$
Solve for y:
$$xy=1$$
$$y=\frac{1}{x}$$
Replace y by $f^{-1}(x)$: $\quad f^{-1}(x)=\frac{1}{x}$

49. a) The graph of $g(x)=1$ is shown below. The horizontal line $y=1$ crosses the graph more than once, so the function is not one-to-one.

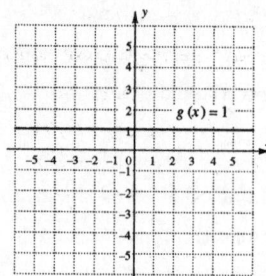

51. a) The function $f(x)=\frac{2x+1}{3}=\frac{2}{3}x+\frac{1}{3}$ is a linear function that is not constant, so it passes the horizontal-line test. Thus, f is one-to-one.

b) Replace $f(x)$ by y:
$$y=\frac{2x+1}{3}$$
Interchange x and y:
$$x=\frac{2y+1}{3}$$
Solve for y:
$$3x=2y+1$$
$$3x-1=2y$$
$$\frac{3x-1}{2}=y$$
Replace y by $f^{-1}(x)$: $\quad f^{-1}(x)=\frac{3x-1}{2}$

53. a) The graph of $f(x)=x^3+5$ is shown below. It passes the horizontal-line test, so the function is one-to-one.

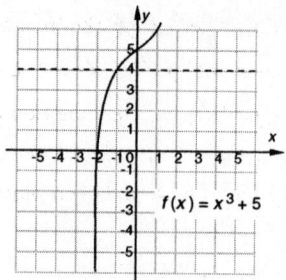

b) Replace $f(x)$ by y: $\quad\quad y=x^3+5$
Interchange x and y: $\quad\quad x=y^3+5$
Solve for y: $\quad\quad x-5=y^3$
$$\sqrt[3]{x-5}=y$$
Replace y by $f^{-1}(x)$: $\quad f^{-1}(x)=\sqrt[3]{x-5}$

55. a) The graph of $g(x)=(x-2)^3$ is shown below. It passes the horizontal-line test, so the function is one-to-one.

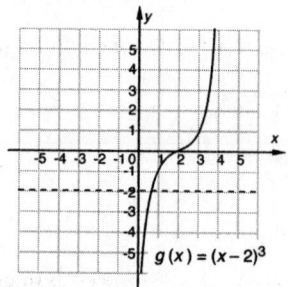

b) Replace $g(x)$ by y: $\quad\quad y=(x-2)^3$
Interchange x and y: $\quad\quad x=(y-2)^3$
Solve for y: $\quad\quad \sqrt[3]{x}=y-2$
$$\sqrt[3]{x}+2=y$$
Replace y by $g^{-1}(x)$: $\quad g^{-1}(x)=\sqrt[3]{x}+2$

57. a) The graph of $f(x)=\sqrt{x}$ is shown below. It passes the horizontal-line test, so the function is one-to-one.

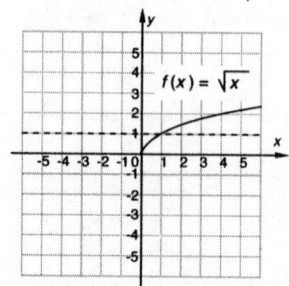

b) Replace $f(x)$ by y: $\quad\quad y=\sqrt{x}$ (Note that $f(x)\geq0$)
Interchange x and y: $\quad\quad x=\sqrt{y}$
Solve for y: $\quad\quad x^2=y$
Replace y by $f^{-1}(x)$: $\quad f^{-1}(x)=x^2; x\geq0$

59. First graph $f(x) = \frac{2}{3}x + 4$. Then graph the inverse function by reflecting the graph of $f(x) = \frac{2}{3}x + 4$ across the line $y = x$. The graph of the inverse function can also be found by first finding a formula for the inverse, substituting to find function values, and then plotting points.

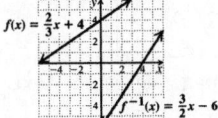

61. Follow the procedure described in Exercise 59 to graph the function and its inverse.

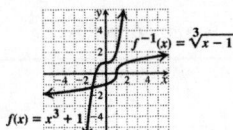

63. Follow the procedure described in Exercise 59 to graph the function and its inverse.

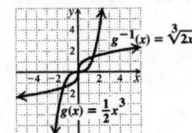

65. Follow the procedure described in Exercise 59 to graph the function and its inverse.

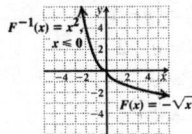

67. Follow the procedure described in Exercise 59 to graph the function and its inverse.

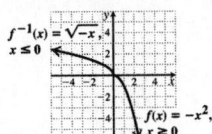

69. We check to see that $\left(f^{-1} \circ f\right)(x) = x$ and $\left(f \circ f^{-1}\right)(x) = x$.

$$(f^{-1} \circ f)(x) = f^{-1}(f(x)) = f^{-1}(\sqrt[3]{x-4})$$
$$= \left(\sqrt[3]{x-4}\right)^3 + 4 = x - 4 + 4 = x$$
$$(f \circ f^{-1})(x) = f(f^{-1}(x)) = f(x^3 + 4)$$
$$= \sqrt[3]{x^3 + 4 - 4} = \sqrt[3]{x^3} = x$$

71. We check to see that $f^{-1} \circ f(x) = x$ and $f \circ f^{-1}(x) = x$.

$$f^{-1} \circ f(x) = f^{-1}(f(x)) = f^{-1}\left(\frac{1-x}{x}\right) = \frac{1}{\frac{1-x}{x}+1}$$
$$= \frac{1}{\frac{1-x}{x}+1} \cdot \frac{x}{x} = \frac{x}{1-x+x} = \frac{x}{1} = x$$

$$f \circ f^{-1}(x) = f(f^{-1}(x)) = f\left(\frac{1}{x+1}\right) = \frac{1 - \frac{1}{x+1}}{\frac{1}{x+1}}$$
$$= \frac{1 - \frac{1}{x+1}}{\frac{1}{x+1}} \cdot \frac{x+1}{x+1} = \frac{x+1-1}{1} = \frac{x}{1} = x$$

73. a) $f(8) = 2(8+12) = 2 \cdot 20 = 40$

Size 40 in Italy corresponds to size 8 in the U.S.

$f(10) = 2(10+12) = 2 \cdot 22 = 44$

Size 44 in Italy corresponds to size 10 in the U.S.

$f(14) = 2(14+12) = 2 \cdot 26 = 52$

Size 52 in Italy corresponds to size 14 in the U.S.

$f(18) = 2(18+12) = 2 \cdot 30 = 60$

Size 60 in Italy corresponds to size 18 in the U.S.

b) The function $f(x) = 2(x+12)$ is a linear function that is not constant, so it passes the horizontal-line test and has an inverse that is a function.

Replace $f(x)$ by y: $\qquad y = 2(x+12)$
Interchange x and y: $\qquad x = 2(y+12)$
Solve for y: $\qquad\qquad x = 2y + 24$
$$x - 24 = 2y$$
$$\frac{x-24}{2} = y$$

Replace y by $f^{-1}(x)$: $\quad f^{-1}(x) = \frac{x-24}{2}$ or $\frac{x}{2} - 12$

c) $f^{-1}(40) = \frac{40-24}{2} = \frac{16}{2} = 8$

Size 8 in the U.S. corresponds to size 40 in Italy.

$f^{-1}(44) = \frac{44-24}{2} = \frac{20}{2} = 10$

Size 10 in the U.S. corresponds to size 44 in Italy.

$f^{-1}(52) = \frac{52-24}{2} = \frac{28}{2} = 14$

Size 14 in the U.S. corresponds to size 52 in Italy.

$f^{-1}(60) = \frac{60-24}{2} = \frac{36}{2} = 18$

Size 18 in the U.S. corresponds to size 60 in Italy.

75. *Writing Exercise.* No; several items can have the same price.

77. $2^{-3} = \frac{1}{2^3} = \frac{1}{8}$

79. $4^{5/2} = \left(\sqrt{4}\right)^5 = 2^5 = 32$

81. Graph $y = x^3$.

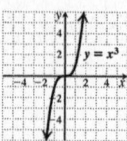

83. *Writing Exercise.* $V^{-1}(t)$ could be used to determine the number of years from 2008 where t is the value of the stamp.

85. Reflect the graph of f across the line $y = x$.

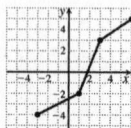

87. From Exercise 73(b), we know that a function that converts dress sizes in Italy to those in the United States is $g(x) = \frac{x-24}{2}$. From Exercise 74, we know that a function that converts dress sizes in the United States to those in France is $f(x) = x + 32$. Then a function that converts dress sizes in Italy to those in France is

$$h(x) = (f \circ g)(x)$$
$$h(x) = f\left(\frac{x-24}{2}\right)$$
$$h(x) = \frac{x-24}{2} + 32$$
$$h(x) = \frac{x}{2} - 12 + 32$$
$$h(x) = \frac{x}{2} + 20.$$

89. *Writing Exercise.* The functions found in Exercises 87 and 88 are inverses. We can show that $(h \circ d)(x) = x$ and $(d \circ h)(x) = x$.

91. Suppose that $h(x) = (f \circ g)(x)$. First note that for $I(x) = x$, $(f \circ I)(x) = f(I(x))$ for any function f.

i) $((g^{-1} \circ f^{-1}) \circ h)(x) = ((g^{-1} \circ f^{-1}) \circ (f \circ g))(x)$
$$= ((g^{-1} \circ (f^{-1} \circ f)) \circ g)(x)$$
$$= ((g^{-1} \circ I) \circ g)(x)$$
$$= (g^{-1} \circ g)(x) = x$$

ii) $(h \circ (g^{-1} \circ f^{-1}))(x) = ((f \circ g) \circ (g^{-1} \circ f^{-1}))(x)$
$$= ((f \circ (g \circ g^{-1})) \circ f^{-1})(x)$$
$$= ((f \circ I) \circ f^{-1})(x)$$
$$= (f \circ f^{-1})(x) = x$$

Therefore, $(g^{-1} \circ f^{-1})(x) = h^{-1}(x)$.

93. $(f \circ g)(x) = x$ and $(g \circ f)(x) = x$, so the functions are inverses.

95. $(f \circ g)(x) \neq x$, so the functions are not inverses. (It is also true that $(g \circ f)(x) \neq x$.)

97. (1) C; (2) A; (3) B; (4) D

99. *Writing Exercise.*

a)   For $x = 2$ and $x = 4$, $y_2(y_1(x)) = x$ and $y_1(y_2(x)) = x$.

b) $Y_1 = 0.5x + 1$; $Y_2 = 2x - 2$

c)   Following the procedure on page 589 in the text, we find that $Y_1^{-1} = 2x - 2$, so the functions are inverses.

Exercise Set 9.2

1. True; see page 593 in the text.

3. True; the graph of $y = f(x - 3)$ is a translation of the graph of $y = f(x)$, 3 units to the right.

5. False; the graph of $y = 3^x$ crosses the y-axis at $(0, 1)$.

7. Graph: $y = f(x) = 3^x$

We compute some function values, thinking of y as $f(x)$, and keep the results in a table.

$$f(0) = 3^0 = 1$$
$$f(1) = 3^1 = 3$$
$$f(2) = 3^2 = 9$$
$$f(-1) = 3^{-1} = \frac{1}{3^1} = \frac{1}{3}$$
$$f(-2) = 3^{-2} = \frac{1}{3^2} = \frac{1}{9}$$

x	y, or $f(x)$
0	1
1	3
2	9
-1	$\frac{1}{3}$
-2	$\frac{1}{9}$

Next we plot these points and connect them with a smooth curve.

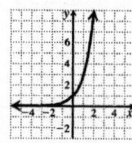

$y = f(x) = 3^x$

9. Graph: $y = 6^x$

We compute some function values, thinking of y as $f(x)$, and keep the results in a table.

$f(0) = 6^0 = 1$

$f(1) = 6^1 = 6$

$f(2) = 6^2 = 36$

$f(-1) = 6^{-1} = \dfrac{1}{6^1} = \dfrac{1}{6}$

$f(-2) = 6^{-2} = \dfrac{1}{6^2} = \dfrac{1}{36}$

x	y, or $f(x)$
0	1
1	6
2	36
-1	$\dfrac{1}{6}$
-2	$\dfrac{1}{36}$

Next we plot these points and connect them with a smooth curve.

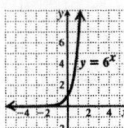

11. Graph: $y = 2^x + 1$

We compute some function values, thinking of y as $f(x)$, and keep the results in a table.

$f(-4) = 2^{-4} + 1 = \dfrac{1}{2^4} + 1 = \dfrac{1}{16} + 1 = 1\dfrac{1}{16}$

$f(-2) = 2^{-2} + 1 = \dfrac{1}{2^2} + 1 = \dfrac{1}{4} + 1 = 1\dfrac{1}{4}$

$f(0) = 2^0 + 1 = 1 + 1 = 2$

$f(1) = 2^1 + 1 = 2 + 1 = 3$

$f(2) = 2^2 + 1 = 4 + 1 = 5$

x	y, or $f(x)$
-4	$1\dfrac{1}{16}$
-2	$1\dfrac{1}{4}$
0	2
1	3
2	5

Next we plot these points and connect them with a smooth curve.

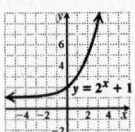

13. Graph: $y = 3^x - 2$

We compute some function values, thinking of y as $f(x)$, and keep the results in a table.

$f(-3) = 3^{-3} - 2 = \dfrac{1}{3^3} - 2 = \dfrac{1}{27} - 2 = -\dfrac{53}{27}$

$f(-1) = 3^{-1} - 2 = \dfrac{1}{3} - 2 = -\dfrac{5}{3}$

$f(0) = 3^0 - 2 = 1 - 2 = -1$

$f(1) = 3^1 - 2 = 3 - 2 = 1$

$f(2) = 3^2 - 2 = 9 - 2 = 7$

x	y, or $f(x)$
-3	$-\dfrac{53}{27}$
-1	$-\dfrac{5}{3}$
0	-1
1	1
2	7

Next we plot these points and connect them with a smooth curve.

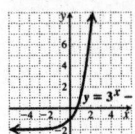

15. Graph: $y = 2^x - 5$

We construct a table of values, thinking of y as $f(x)$. Then we plot the points and connect them with a smooth curve.

$f(0) = 2^0 - 5 = 1 - 5 = -4$

$f(1) = 2^1 - 5 = 2 - 5 = -3$

$f(2) = 2^2 - 5 = 4 - 5 = -1$

$f(3) = 2^3 - 5 = 8 - 5 = 3$

$f(-1) = 2^{-1} - 5 = \dfrac{1}{2} - 5 = -\dfrac{9}{2}$

$f(-2) = 2^{-2} - 5 = \dfrac{1}{4} - 5 = -\dfrac{19}{4}$

$f(-4) = 2^{-4} - 5 = \dfrac{1}{16} - 5 = -\dfrac{79}{16}$

x	y, or $f(x)$
0	-4
1	-3
2	-1
3	3
-1	$-\dfrac{9}{2}$
-2	$-\dfrac{19}{4}$
-4	$-\dfrac{79}{16}$

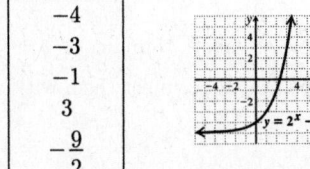

17. Graph: $y = 2^{x-3}$

We construct a table of values, thinking of y as $f(x)$.

Then we plot the points and connect them with a smooth curve.

$$f(0) = 2^{0-3} = 2^{-3} = \frac{1}{8}$$
$$f(-1) = 2^{-1-3} = 2^{-4} = \frac{1}{16}$$
$$f(1) = 2^{1-3} = 2^{-2} = \frac{1}{4}$$
$$f(2) = 2^{2-3} = 2^{-1} = \frac{1}{2}$$
$$f(3) = 2^{3-3} = 2^{0} = 1$$
$$f(4) = 2^{4-3} = 2^{1} = 2$$
$$f(5) = 2^{5-3} = 2^{2} = 4$$

x	y, or $f(x)$
0	$\frac{1}{8}$
−1	$\frac{1}{16}$
1	$\frac{1}{4}$
2	$\frac{1}{2}$
3	1
4	2
5	3

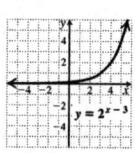

19. Graph: $y = 2^{x+1}$

We construct a table of values, thinking of y as $f(x)$.

Then we plot the points and connect them with a smooth curve.

$$f(-3) = 2^{-3+1} = 2^{-2} = \frac{1}{4}$$
$$f(-1) = 2^{-1+1} = 2^{0} = 1$$
$$f(0) = 2^{0+1} = 2^{1} = 2$$
$$f(1) = 2^{1+1} = 2^{2} = 4$$

x	y, or $f(x)$
−3	$\frac{1}{4}$
−1	1
0	2
1	4

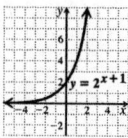

21. Graph: $y = \left(\frac{1}{4}\right)^{x}$

We construct a table of values, thinking of y as $f(x)$.

Then we plot the points and connect them with a smooth curve.

$$f(0) = \left(\frac{1}{4}\right)^{0} = 1$$
$$f(1) = \left(\frac{1}{4}\right)^{1} = \frac{1}{4}$$
$$f(2) = \left(\frac{1}{4}\right)^{2} = \frac{1}{16}$$

$$f(-1) = \left(\frac{1}{4}\right)^{-1} = \frac{1}{\frac{1}{4}} = 4$$
$$f(-2) = \left(\frac{1}{4}\right)^{-2} = \frac{1}{\frac{1}{16}} = 16$$

x	y, or $f(x)$
0	1
1	$\frac{1}{4}$
2	$\frac{1}{16}$
−1	4
−2	16

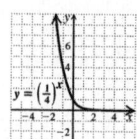

23. Graph: $y = \left(\frac{1}{3}\right)^{x}$

We construct a table of values, thinking of y as $f(x)$.

Then we plot the points and connect them with a smooth curve.

$$f(0) = \left(\frac{1}{3}\right)^{0} = 1$$
$$f(1) = \left(\frac{1}{3}\right)^{1} = \frac{1}{3}$$
$$f(2) = \left(\frac{1}{3}\right)^{2} = \frac{1}{9}$$
$$f(3) = \left(\frac{1}{3}\right)^{3} = \frac{1}{27}$$
$$f(-1) = \left(\frac{1}{3}\right)^{-1} = \frac{1}{\left(\frac{1}{3}\right)^{1}} = \frac{1}{\frac{1}{3}} = 3$$
$$f(-2) = \left(\frac{1}{3}\right)^{-2} = \frac{1}{\left(\frac{1}{3}\right)^{2}} = \frac{1}{\frac{1}{9}} = 9$$
$$f(-3) = \left(\frac{1}{3}\right)^{-3} = \frac{1}{\left(\frac{1}{3}\right)^{3}} = \frac{1}{\frac{1}{27}} = 27$$

x	y, or $f(x)$
0	1
1	$\frac{1}{3}$
2	$\frac{1}{9}$
3	$\frac{1}{27}$
−1	3
−2	9
−3	27

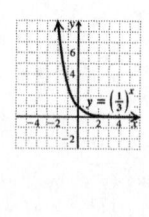

25. Graph: $y = 2^{x+1} - 3$

We construct a table of values, thinking of y as $f(x)$.

Then we plot the points and connect them with a smooth curve.

$f(0) = 2^{0+1} - 3 = 2 - 3 = -1$

$f(1) = 2^{1+1} - 3 = 4 - 3 = 1$

$f(2) = 2^{2+1} - 3 = 8 - 3 = 5$

$f(-1) = 2^{-1+1} - 3 = 1 - 3 = -2$

$f(-2) = 2^{-2+1} - 3 = \frac{1}{2} - 3 = -\frac{5}{2}$

$f(-3) = 2^{-3+1} - 3 = \frac{1}{4} - 3 = -\frac{11}{4}$

x	y, or $f(x)$
0	-1
1	1
2	5
-1	-2
-2	$-\frac{5}{2}$
-3	$-\frac{11}{4}$

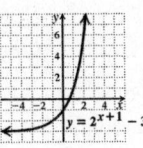

27. Graph: $x = 6^y$

We can find ordered pairs by choosing values for y and then computing values for x.

For $y = 0, x = 6^0 = 1$.

For $y = 1, x = 6^1 = 6$.

For $y = -1, x = 6^{-1} = \frac{1}{6^1} = \frac{1}{6}$.

For $y = -2, x = 6^{-2} = \frac{1}{6^2} = \frac{1}{36}$.

x	y
1	0
6	1
$\frac{1}{6}$	-1
$\frac{1}{36}$	-2

$\uparrow$ $\quad$ $\uparrow$ (1) Choose values for y.

(2) Compute values for x.

We plot the points and connect them with a smooth curve.

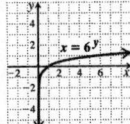

29. Graph: $x = 3^{-y} = \left(\frac{1}{3}\right)^y$

We can find ordered pairs by choosing values for y and then computing values for x. Then we plot these points and connect them with a smooth curve.

For $y = 0, x = \left(\frac{1}{3}\right)^0 = 1$.

For $y = 1, x = \left(\frac{1}{3}\right)^1 = \frac{1}{3}$.

For $y = 2, x = \left(\frac{1}{3}\right)^2 = \frac{1}{9}$.

For $y = -1, x = \left(\frac{1}{3}\right)^{-1} = \frac{1}{\frac{1}{3}} = 3$.

For $y = -2, x = \left(\frac{1}{3}\right)^{-2} = \frac{1}{\frac{1}{9}} = 9$.

x	y
1	0
$\frac{1}{3}$	1
$\frac{1}{9}$	2
3	-1
9	-2

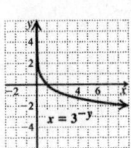

31. Graph: $x = 4^y$

We can find ordered pairs by choosing values for y and then computing values for x. Then we plot these points and connect them with a smooth curve.

For $y = 0, x = 4^0 = 1$.

For $y = 1, x = 4^1 = 4$.

For $y = 2, x = 4^2 = 16$.

For $y = -1, x = 4^{-1} = \frac{1}{4}$.

For $y = -2, x = 4^{-2} = \frac{1}{16}$.

x	y
1	0
4	1
16	2
$\frac{1}{4}$	-1
$\frac{1}{16}$	-2

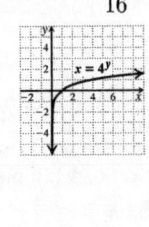

33. Graph: $x = \left(\frac{4}{3}\right)^y$

We can find ordered pairs by choosing values for y and then computing values for x. Then we plot these points and connect them with a smooth curve.

For $y = 0, x = \left(\frac{4}{3}\right)^0 = 1$.

For $y = 1, x = \left(\frac{4}{3}\right)^1 = \frac{4}{3}$.

For $y = 2, x = \left(\frac{4}{3}\right)^2 = \frac{16}{9}$.

For $y = 3, x = \left(\frac{4}{3}\right)^3 = \frac{64}{27}$.

For $y = -1, x = \left(\frac{4}{3}\right)^{-1} = \frac{3}{4}$.

For $y = -2, x = \left(\frac{4}{3}\right)^{-2} = \left(\frac{3}{4}\right)^2 = \frac{9}{16}$.

For $y = -3, x = \left(\frac{4}{3}\right)^{-3} = \left(\frac{3}{4}\right)^3 = \frac{27}{64}$.

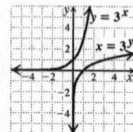

x	y
1	0
$\frac{4}{3}$	1
$\frac{16}{9}$	2
$\frac{64}{27}$	3
$\frac{3}{4}$	-1
$\frac{9}{16}$	-2
$\frac{27}{64}$	-3

35. Graph $y = 3^x$ (see Exercise 8) and $x = 3^y$ (see Exercise 28) using the same set of axes.

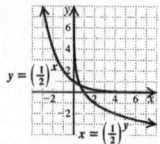

37. Graph $y = \left(\frac{1}{2}\right)^x$ (see Exercise 24) and $x = \left(\frac{1}{2}\right)^y$ (see Exercise 30) using the same set of axes.

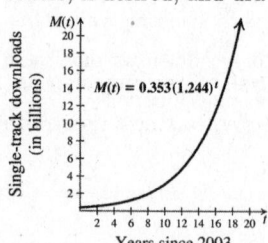

39. a) In 2006, $t = 2006 - 2003 = 3$

$M(3) = 0.353(1.244)^3 \approx 0.680$ billion tracks

In 2008, $t = 2008 - 2003 = 5$

$M(5) = 0.353(1.244)^5 \approx 1.052$ billion tracks

In 2012, $t = 2012 - 2003 = 9$

$M(9) = 0.353(1.244)^9 \approx 2.519$ billion tracks

b) Use the function values computed in part (a) and others, if desired, and draw the graph.

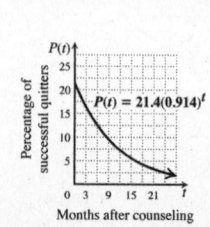

41. a) $P(1) = 21.4(0.914)^1 \approx 19.6\%$

$P(3) = 21.4(0.914)^3 \approx 16.3\%$

1 yr = 12 months; $P(12) = 21.4(0.914)^{12} \approx 7.3\%$

b)

43. a) In 1930, $t = 1930 - 1900 = 30$.

$P(t) = 150(0.960)^t$
$P(30) = 150(0.960)^{30}$
≈ 44.079

In 1930, about 44.079 thousand, or 44,079, humpback whales were alive.

In 1960, $t = 1960 - 1900 = 60$.

$P(t) = 150(0.960)^t$
$P(60) = 150(0.960)^{60}$
≈ 12.953

In 1960, about 12.953 thousand, or 12,953, humpback whales were alive.

b) Plot the points found in part (a), $(30, 44{,}079)$ and $(60, 12{,}953)$ and additional points as needed and graph the function.

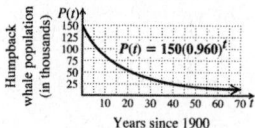

45. a) In 1992, $t = 1992 - 1982 = 10$.

$P(10) = 5.5(1.047)^{10} \approx 8.706$

In 1992, about 8.706 thousand, or 8706, humpback whales were alive.

In 2004, $t = 2004 - 1982 = 22$.

$P(22) = 5.5(1.047)^{22} \approx 15.107$

In 2004, about 15.107 thousand, or 15,107, humpback whales were alive.

b) Use the function values computed in part (a) and others, if desired, and draw the graph.

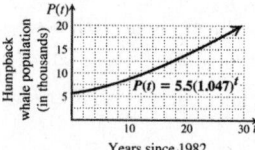

47. a) $A(5) = 10 \cdot 34^5 = 454{,}354{,}240 \text{ cm}^2$

$A(7) = 10 \cdot 34^7 = 525{,}233{,}501{,}400 \text{ cm}^2$

b) Use the function values computed in part (a) and others, if desired, and draw the graph.

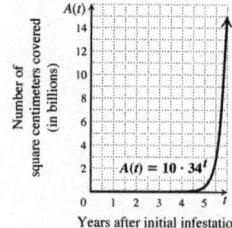

49. *Writing Exercise.* Since $3 < \pi < 4$, $2^3 < 2^\pi < 2^4$; that is, 2^π is greater than 2^3, or 8, but less than 2^4, or 16.

51. $3x^2 - 48 = 3(x^2 - 16) = 3(x+4)(x-4)$

53. $6x^2 + x - 12 = (2x+3)(3x-4)$

55. $t^2 - y^2 + 2y - 1$
$= t^2 - (y^2 - 2y + 1)$
$= t^2 - (y-1)^2$ $\quad$ Difference of squares
$= [t - (y-1)][t + (y-1)]$
$= (t - y + 1)(t + y - 1)$

57. *Writing Exercise.* No; the number computed for 2010 far exceeds the U.S. population, so the number that would be computed for 20 years from now would not be realistic.

59. Since the bases are the same, the one with the larger exponent is the larger number. Thus $\pi^{2.4}$ is larger.

61. Graph: $f(x) = 2.5^x$

Use a calculator with a power key to construct a table of values. (We will round values of $f(x)$ to the nearest hundredth.) Then plot these points and connect them with a smooth curve.

x	y
0	1
1	2.5
2	6.25
3	15.63
-1	0.4
-2	0.16

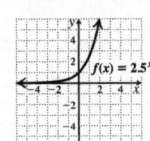

63. Graph: $y = 2^x + 2^{-x}$

Construct a table of values, thinking of y as $f(x)$. Then plot these points and connect them with a curve.

$f(0) = 2^0 + 2^{-0} = 1 + 1 = 2$
$f(1) = 2^1 + 2^{-1} = 2 + \frac{1}{2} = 2\frac{1}{2}$
$f(2) = 2^2 + 2^{-2} = 4 + \frac{1}{4} = 4\frac{1}{4}$
$f(3) = 2^3 + 2^{-3} = 8 + \frac{1}{8} = 8\frac{1}{8}$
$f(-1) = 2^{-1} + 2^{-(-1)} = \frac{1}{2} + 2 = 2\frac{1}{2}$
$f(-2) = 2^{-2} + 2^{-(-2)} = \frac{1}{4} + 4 = 4\frac{1}{4}$
$f(-3) = 2^{-3} + 2^{-(-3)} = \frac{1}{8} + 8 = 8\frac{1}{8}$

x	y, or $f(x)$
0	2
1	$2\frac{1}{2}$
2	$4\frac{1}{4}$
3	$8\frac{1}{8}$
-1	$2\frac{1}{2}$
-2	$4\frac{1}{4}$
-3	$8\frac{1}{8}$

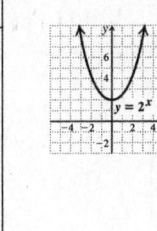

65. Graph: $y = |2^x - 2|$

We construct a table of values, thinking of y as $f(x)$. Then plot these points and connect them with a curve.

$f(0) = |2^0 - 2| = |1 - 2| = |-1| = 1$
$f(1) = |2^1 - 2| = |2 - 2| = |0| = 0$
$f(2) = |2^2 - 2| = |4 - 2| = |2| = 2$
$f(3) = |2^3 - 2| = |8 - 2| = |6| = 6$
$f(-1) = |2^{-1} - 2| = \left|\frac{1}{2} - 2\right| = \left|-\frac{3}{2}\right| = \frac{3}{2}$
$f(-3) = |2^{-3} - 2| = \left|\frac{1}{8} - 2\right| = \left|-\frac{15}{8}\right| = \frac{15}{8}$
$f(-5) = |2^{-5} - 2| = \left|\frac{1}{32} - 2\right| = \left|-\frac{63}{32}\right| = \frac{63}{32}$

x	y, or $f(x)$
0	1
1	0
2	2
3	6
-1	$\frac{3}{2}$
-3	$\frac{15}{8}$
-5	$\frac{63}{32}$

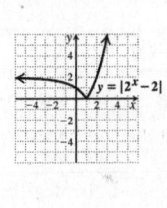

67. Graph: $y = |2^{x^2} - 1|$

We construct a table of values, thinking of y as $f(x)$. Then we plot these points and connect them with a curve.

$f(0) = |2^{0^2} - 1| = |1 - 1| = 0$
$f(1) = |2^{1^2} - 1| = |2 - 1| = 1$
$f(2) = |2^{2^2} - 1| = |16 - 1| = 15$
$f(-1) = |2^{(-1)^2} - 1| = |2 - 1| = 1$
$f(-2) = |2^{(-2)^2} - 1| = |16 - 1| = 15$

x	y, or $f(x)$
0	0
1	1
2	15
-1	1
-2	15

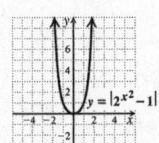

69. $y = 3^{-(x-1)}$ $x = 3^{-(y-1)}$

x	y
0	3
1	1
2	$\frac{1}{3}$
3	$\frac{1}{9}$
-1	9

x	y
3	0
1	1
$\frac{1}{3}$	2
$\frac{1}{9}$	3
9	-1

71. Enter the data points (0, 0.5), (4, 4) and (8, 50) and then use the ExpReg option from the STAT CALC menu of a graphing calculator to find an exponential function that models the data:

$N(t) = 0.464(1.778)^t$, where $N(t)$ is the number of navigational devices in use, in millions, t years after 2000. In 2012, $t = 2012 - 2000 = 12$

$N(12) \approx 463$ million devices.

73. *Writing Exercise.*

a) Section A appears to grow at a Linear rate.
    Section B appears to increase at a Exponential rate.
    Section C appears to have nearly stopped growth so it should be labeled Saturation.

b) Each of the forces affect wave height in different ways. Gravity is a constant force, while wind shows linear growth, and wave height can grow exponentially due to these forces and surface roughness.

75. *Graphing Calculator Exercise*

Exercise Set 9.3

1. $5^2 = 25$, so choice (g) is correct.

3. $5^1 = 5$, so choice (a) is correct.

5. The exponent to which we raise 5 to get 5^x is x, so choice (b) is correct.

7. $5 = 2^x$ is equivalent to $\log_2 5 = x$, so choice (e) is correct.

9. $\log_{10} 1000$ is the exponent to which we raise 10 to get 1000. Since $10^3 = 1000$, $\log_{10} 1000 = 3$.

11. $\log_7 49$ is the exponent to which we raise 7 to get 49. Since $7^2 = 49$, $\log_7 49 = 2$.

13. $\log_3 81$ is the exponent to which we raise 3 to get 81. Since $3^4 = 81$, $\log_3 81 = 4$.

15. $\log_5 \frac{1}{25}$ is the exponent to which we raise 5 to get $\frac{1}{25}$. Since $5^{-2} = \frac{1}{25}$, $\log_5 \frac{1}{25} = -2$.

17. Since $8^{-1} = \frac{1}{8}$, $\log_8 \frac{1}{8} = -1$.

19. Since $5^4 = 625$, $\log_5 625 = 4$.

21. Since $7^1 = 7$, $\log_7 7 = 1$.

23. Since $3^0 = 1$, $\log_3 1 = 0$.

25. $\log_6 6^5$ is the exponent to which we raise 6 to get 6^5. Clearly, this power is 5, so $\log_6 6^5 = 5$.

27. Since $10^{-2} = \frac{1}{100} = 0.01$, $\log_{10} 0.01 = -2$.

29. Since $16^{1/2} = 4$, $\log_{16} 4 = \frac{1}{2}$.

31. Since $9 = 3^2$ and $(3^2)^{3/2} = 3^3 = 27$, $\log_9 27 = \frac{3}{2}$.

33. Since $1000 = 10^3$ and $(10^3)^{2/3} = 10^2 = 100$, $\log_{1000} 100 = \frac{2}{3}$.

35. Since $\log_3 29$ is the power to which we raise 3 to get 29, then 3 raised to this power is 29. That is $3^{\log_3 29} = 29$.

37. Graph: $y = \log_{10} x$

The equation $y = \log_{10} x$ is equivalent to $10^y = x$. We can find ordered pairs by choosing values for y and computing the corresponding x-values.

For $y = 0$, $x = 10^0 = 1$.
For $y = 1$, $x = 10^1 = 10$.
For $y = 2$, $x = 10^2 = 100$.
For $y = -1$, $x = 10^{-1} = \frac{1}{10}$.
For $y = -2$, $x = 10^{-2} = \frac{1}{100}$.

x, or 10^y	y
1	0
10	1
100	2
$\frac{1}{10}$	-1
$\frac{1}{100}$	-2

↑　　↑ (1) Select y.

(2) Compute x.

We plot the set of ordered pairs and connect the points with a smooth curve.

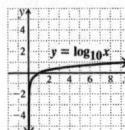

39. Graph: $y = \log_3 x$

The equation $y = \log_3 x$ is equivalent to $3^y = x$. We can find ordered pairs by choosing values for y and computing the corresponding x-values.

For $y = 0$, $x = 3^0 = 1$.
For $y = 1$, $x = 3^1 = 3$.
For $y = 2$, $x = 3^2 = 9$.
For $y = -1$, $x = 3^{-1} = \frac{1}{3}$.
For $y = -2$, $x = 3^{-2} = \frac{1}{9}$.

x, or 3^y	y
1	0
3	1
9	2
$\frac{1}{3}$	-1
$\frac{1}{9}$	-2

We plot the set of ordered pairs and connect the points with a smooth curve.

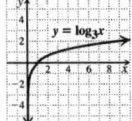

41. Graph: $f(x) = \log_6 x$

Think of $f(x)$ as y. Then $y = \log_6 x$ is equivalent to $6^y = x$. We find ordered pairs by choosing values for y and computing the corresponding x-values. Then we plot the points and connect them with a smooth curve.

For $y = 0$, $x = 6^0 = 1$.
For $y = 1$, $x = 6^1 = 6$.
For $y = 2$, $x = 6^2 = 36$.
For $y = -1$, $x = 6^{-1} = \frac{1}{6}$.
For $y = -2$, $x = 6^{-2} = \frac{1}{36}$.

x, or 6^y	y
1	0
6	1
36	2
$\frac{1}{6}$	-1
$\frac{1}{36}$	-2

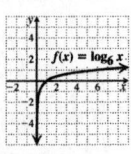

43. Graph: $f(x) = \log_{2.5} x$

Think of $f(x)$ as y. Then $y = \log_{2.5} x$ is equivalent to $2.5^y = x$. We construct a table of values, plot these points and connect them with a smooth curve.

For $y = 0$, $x = 2.5^0 = 1$.
For $y = 1$, $x = 2.5^1 = 2.5$.
For $y = 2$, $x = 2.5^2 = 6.25$.
For $y = 3$, $x = 2.5^3 = 15.625$.
For $y = -1$, $x = 2.5^{-1} = 0.4$.
For $y = -2$, $x = 2.5^{-2} = 0.16$.

x, or 2.5^y	y
1	0
2.5	1
6.25	2
15.625	3
0.4	-1
0.16	-2

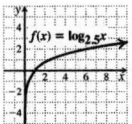

45. Graph $f(x) = 3^x$ (see Exercise Set 9.2, Exercise 7) and $f^{-1}(x) = \log_3 x$ (see Exercise 39 above) on the same set of axes.

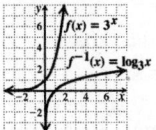

47.

$x = \log_{10} 8 \Rightarrow 10^x = 8$

The base remains the same.

The logarithm is the exponent.

49.

$\log_9 9 = 1 \Rightarrow 9^1 = 9$

The logarithm is the exponent.

The base remains the same.

51. $\log_{10} 0.1 = -1$ is equivalent to $10^{-1} = 0.1$.

53. $\log_{10} 7 = 0.845$ is equivalent to $10^{0.845} = 7$.

55. $\log_c m = 8$ is equivalent to $c^8 = m$.

57. $\log_r C = t$ is equivalent to $r^t = C$.

59. $\log_e 0.25 = -1.3863$ is equivalent to $e^{-1.3863} = 0.25$.

61. $\log_r T = -x$ is equivalent to $r^{-x} = T$.

63.
$$10^2 = 100 \Rightarrow 2 = \log_{10} 100$$
The exponent is the logarithm.
The base remains the same.

65.
$$5^{-3} = \frac{1}{125} \Rightarrow -3 = \log_5 \frac{1}{125}$$
The logarithm is the exponent.
The base remains the same.

67. $16^{1/4} = 2$ is equivalent to $\frac{1}{4} = \log_{16} 2$.

69. $10^{0.4771} = 3$ is equivalent to $0.4771 = \log_{10} 3$.

71. $z^m = 6$ is equivalent to $m = \log_z 6$.

73. $p^t = q$ is equivalent to $t = \log_p q$.

75. $e^3 = 20.0855$ is equivalent to $3 = \log_e 20.0855$.

77. $e^{-4} = 0.0183$ is equivalent to $-4 = \log_e 0.0183$.

79. $\log_6 x = 2$
$$6^2 = x \quad \text{Converting to an exponential equation}$$
$$36 = x \quad \text{Computing } 6^2$$

81. $\log_2 32 = x$
$$2^x = 32 \quad \text{Converting to an exponential equation}$$
$$2^x = 2^5$$
$$x = 5 \quad \text{The exponents must be the same.}$$

83. $\log_x 9 = 1$
$$x^1 = 9 \quad \text{Converting to an exponential equation}$$
$$x = 9 \quad \text{Simplifying } x^1$$

85. $\log_x 7 = 1$
$$x^{1/2} = 7 \quad \text{Converting to an exponential equation}$$
$$x = 49 \quad \text{Simplifying } x^{1/2}$$

87. $\log_3 x = -2$
$$3^{-2} = x \quad \text{Converting to an exponential equation}$$
$$\frac{1}{9} = x \quad \text{Simplifying}$$

89. $\log_{32} x = \frac{2}{5}$
$$32^{2/5} = x \quad \text{Converting to an exponential equation}$$
$$\left(2^5\right)^{2/5} = x$$
$$4 = x$$

91. *Writing Exercise.* By definition, $m = \log_a x$ is equivalent to $a^m = x$. So the number $\log_a x$ is the exponent, m.

93. $\sqrt{18a^3b}\sqrt{50ab^7} = \sqrt{18a^3b \cdot 50ab^7} = \sqrt{900a^4b^8} = 30a^2b^4$

95. $\sqrt{192x} - \sqrt{75x} = 8\sqrt{3x} - 5\sqrt{3x}$
$$= (8-5)\sqrt{3x} = 3\sqrt{3x}$$

97. $\dfrac{\dfrac{3}{x} - \dfrac{2}{xy}}{\dfrac{2}{x^2} + \dfrac{1}{xy}}$

The LCD of all the denominators is x^2y. We multiply numerator and denominator by the LCD.

$$\frac{\dfrac{3}{x} - \dfrac{2}{xy}}{\dfrac{2}{x^2} + \dfrac{1}{xy}} \cdot \frac{x^2y}{x^2y} = \frac{\left(\dfrac{3}{x} - \dfrac{2}{xy}\right)x^2y}{\left(\dfrac{2}{x^2} + \dfrac{1}{xy}\right)x^2y}$$

$$= \frac{\dfrac{3}{x} \cdot x^2y - \dfrac{2}{xy} \cdot x^2y}{\dfrac{2}{x^2} \cdot x^2y + \dfrac{1}{xy} \cdot x^2y}$$

$$= \frac{3xy - 2x}{2y + x}, \text{ or } \frac{x(3y-2)}{2y+x}$$

99. *Writing Exercise.* The graph of a logarithmic function $f(x) = \log_a x$ increases slowly as x increases. Thus, although the manufacturer would be pleased that sales were growing, he or she would probably prefer that they were growing more rapidly.

101. Graph: $y = \left(\frac{3}{2}\right)^x$ Graph: $y = \log_{3/2} x$, or $x = \left(\frac{3}{2}\right)^y$

x	y, or $\left(\dfrac{3}{2}\right)^x$	x, or $\left(\dfrac{3}{2}\right)^y$	y
0	1	1	0
1	$\dfrac{3}{2}$	$\dfrac{3}{2}$	1
2	$\dfrac{9}{4}$	$\dfrac{9}{4}$	2
3	$\dfrac{27}{8}$	$\dfrac{27}{8}$	3
-1	$\dfrac{2}{3}$	$\dfrac{2}{3}$	-1
-2	$\dfrac{4}{9}$	$\dfrac{4}{9}$	-2

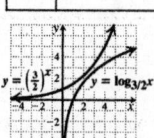

103. Graph: $y = \log_3 |x+1|$

x	y
0	0
2	1
8	2
−2	0
−4	1
−9	2

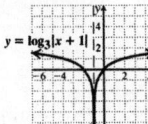

105. $\log_4(3x-2) = 2$
$$4^2 = 3x - 2$$
$$16 = 3x - 2$$
$$18 = 3x$$
$$6 = x$$

107. $\log_{10}(x^2 + 21x) = 2$
$$10^2 = x^2 + 21x$$
$$0 = x^2 + 21x - 100$$
$$0 = (x+25)(x-4)$$

$x = -25$ or $x = 4$

109. Let $\log_{1/5} 25 = x$. Then
$$\left(\frac{1}{5}\right)^x = 25$$
$$\left(5^{-1}\right)^x = 25$$
$$5^{-x} = 5^2$$
$$-x = 2$$
$$x = -2.$$
Thus, $\log_{1/5} 25 = -2$.

111. $\log_{10}\left(\log_4\left(\log_3 81\right)\right)$
$$= \log_{10}\left(\log_4 4\right) \qquad (\log_3 81 = 4)$$
$$= \log_{10} 1 \qquad\qquad (\log_4 4 = 1)$$
$$= 0$$

113. Let $b = 0$, $x = 1$, and $y = 2$. Then $0^1 = 0^2$, but $1 \neq 2$.

Let $b = 1$, $x = 1$, and $y = 2$. Then $1^1 = 1^2$, but $1 \neq 2$.

Exercise Set 9.4

---

1. Use the product rule for logarithms.

$\log_7 20 = \log_7(5 \cdot 4) = \log_7 5 + \log_7 4$; choice (e) is correct.

3. Use the quotient rule for logarithms.

$\log_7 \frac{5}{4} = \log_7 5 - \log_7 4$; choice (a) is correct.

5. The exponent to which we raise 7 to get 1 is 0, so choice (c) is correct.

7. $\log_3(81 \cdot 27) = \log_3 81 + \log_3 27$ Using the product rule

9. $\log_4(64 \cdot 16) = \log_4 64 + \log_4 16$ Using the product rule

11. $\log_c rst = \log_c r + \log_c s + \log_c t$ Using the product rule

13. $\log_a 2 + \log_a 10 = \log_a(2 \cdot 10)$ Using the product rule
The result can also be expressed as $\log_a 20$.

15. $\log_c t + \log_c y = \log_c(t \cdot y)$ Using the product rule

17. $\log_a r^8 = 8 \log_a r$ Using the power rule

19. $\log_2 y^{1/3} = \frac{1}{3}\log_2 y$ Using the power rule

21. $\log_b C^{-3} = -3\log_b C$ Using the power rule

23. $\log_2 \frac{5}{11} = \log_2 5 - \log_2 11$ Using the quotient rule

25. $\log_b \frac{m}{n} = \log_b m - \log_b n$ Using the quotient rule

27. $\log_a 19 - \log_a 2 = \log_a \frac{19}{2}$ Using the quotient rule

29. $\log_b 36 - \log_b 4 = \log_b \frac{36}{4}$, Using the quotient rule
or $\log_b 9$

31. $\log_a x - \log_a y = \log_a \frac{x}{y}$ Using the quotient rule

33. $\log_a(xyz)$
$= \log_a x + \log_a y + \log_a z$ Using the product rule

35. $\log_a\left(x^3 z^4\right)$
$= \log_a x^3 + \log_a z^4$ Using the product rule
$= 3\log_a x + 4\log_a z$ Using the power rule

37. $\log_a\left(w^2 x^{-2} y\right)$
$= \log_a w^2 + \log_a x^{-2} + \log_a y$ Using the product rule
$= 2\log_a w - 2\log_a x + \log_a y$ Using the power rule

39. $\log_a \frac{x^5}{y^3 z}$
$= \log_a x^5 - \log_a y^3 z$ Using the quotient rule
$= \log_a x^5 - \left(\log_a y^3 + \log_a z\right)$ Using the product rule
$= \log_a x^5 - \log_a y^3 - \log_a z$ Removing the parentheses
$= 5\log_a x - 3\log_a y - \log_a z$ Using the power rule

41. $\log_b \dfrac{xy^2}{wz^3}$

$= \log_b xy^2 - \log_b wz^3 \qquad \text{Using the quotient rule}$

$= \log_b x + \log_b y^2 - \left(\log_b w + \log_b z^3\right)$
$\qquad\qquad\qquad\qquad \text{Using the product rule}$

$= \log_b x + \log_b y^2 - \log_b w - \log_b z^3$
$\qquad\qquad\qquad\qquad \text{Removing parentheses}$

$= \log_b x + 2\log_b y - \log_b w - 3\log_b z$
$\qquad\qquad\qquad\qquad \text{Using the power rule}$

43. $\log_a \sqrt{\dfrac{x^7}{y^5 z^8}}$

$= \log_a \left(\dfrac{x^7}{y^5 z^8}\right)^{1/2}$

$= \dfrac{1}{2}\log_a \dfrac{x^7}{y^5 z^8} \qquad \text{Using the power rule}$

$= \dfrac{1}{2}\left(\log_a x^7 - \log_a y^5 z^8\right) \qquad \text{Using the quotient rule}$

$= \dfrac{1}{2}\left[\log_a x^7 - (\log_a y^5 + \log_a z^8)\right] \quad \text{Using the}$
$\qquad\qquad\qquad\qquad\qquad \text{product rule}$

$= \dfrac{1}{2}\left(\log_a x^7 - \log_a y^5 - \log_a z^8\right) \quad \text{Removing parentheses}$

$= \dfrac{1}{2}\left(7\log_a x - 5\log_a y - 8\log_a z\right) \quad \text{Using the power rule}$

45. $\log_a \sqrt[3]{\dfrac{x^6 y^3}{a^2 z^7}}$

$= \log_a \left(\dfrac{x^6 y^3}{a^2 z^7}\right)^{1/3}$

$= \dfrac{1}{3}\log_a \dfrac{x^6 y^3}{a^2 z^7} \qquad \text{Using the power rule}$

$= \dfrac{1}{3}\left(\log_a x^6 y^3 - \log_a a^2 z^7\right) \qquad \text{Using the quotient rule}$

$= \dfrac{1}{3}\left[\log_a x^6 + \log_a y^3 - \left(\log_a a^2 + \log_a z^7\right)\right] \quad \text{Using the}$
$\qquad\qquad\qquad\qquad\qquad\qquad \text{product rule}$

$= \dfrac{1}{3}\left(\log_a x^6 + \log_a y^3 - \log_a a^2 - \log_a z^7\right)$
$\qquad\qquad\qquad\qquad\qquad \text{Removing parentheses}$

$= \dfrac{1}{3}\left(\log_a x^6 + \log_a y^3 - 2 - \log_a z^7\right)$
$\qquad\qquad\qquad 2 \text{ is the number to which}$
$\qquad\qquad\qquad \text{we raise } a \text{ to get } a^2.$

$= \dfrac{1}{3}\left(6\log_a x + 3\log_a y - 2 - 7\log_a z\right)$
$\qquad\qquad\qquad\qquad \text{Using the power rule}$

47. $8\log_a x + 3\log_a z$

$= \log_a x^8 + \log_a z^3 \qquad \text{Using the power rule}$

$= \log_a \left(x^8 z^3\right) \qquad \text{Using the product rule}$

49. $\log_a x^2 - 2\log_a \sqrt{x}$

$= \log_a x^2 - \log_a \left(\sqrt{x}\right)^2 \qquad \text{Using the power rule}$

$= \log_a x^2 - \log_a \left(\sqrt{x}\right)^2 = x$

$= \log_a \dfrac{x^2}{x} \qquad \text{Using the quotient rule}$

$= \log_a x \qquad \text{Simplifying}$

51. $\dfrac{1}{2}\log_a x + 5\log_a y - 2\log_a x$

$= \log_a x^{1/2} + \log_a y^5 - \log_a x^2 \qquad \text{Using the power rule}$

$= \log_a x^{1/2} y^5 - \log_a x^2 \qquad \text{Using the product rule}$

$= \log_a \dfrac{x^{1/2} y^5}{x^2} \qquad \text{Using the quotient rule}$

The result can also be expressed as

$\log_a \dfrac{\sqrt{x}\,y^5}{x^2} \quad \text{or as} \quad \log_a \dfrac{y^5}{x^{3/2}}.$

53. $\log_a (x^2 - 9) - \log_a (x + 3)$

$= \log_a \dfrac{x^2 - 9}{x + 3} \qquad \text{Using the quotient rule}$

$= \log_a \dfrac{(x+3)(x-3)}{x+3}$

$= \log_a \dfrac{\cancel{(x+3)}(x-3)}{\cancel{x+3}} \qquad \text{Simplifying}$

$= \log_a (x - 3)$

55. $\log_b 15 = \log_b (3 \cdot 5)$
$\qquad\quad = \log_b 3 + \log_b 5 \qquad \text{Using the product rule}$
$\qquad\quad = 0.792 + 1.161$
$\qquad\quad = 1.953$

57. $\log_b \dfrac{3}{5} = \log_b 3 - \log_b 5 \qquad \text{Using the quotient rule}$
$\qquad\quad = 0.792 - 1.161$
$\qquad\quad = -0.369$

59. $\log_b \dfrac{1}{5} = \log_b 1 - \log_b 5 \qquad \text{Using the quotient rule}$
$\qquad\quad = 0 - 1.161 \qquad (\log_b 1 = 0)$
$\qquad\quad = -1.161$

61. $\log_b \sqrt{b^3} = \log_b b^{3/2} = \dfrac{3}{2} \qquad 3/2 \text{ is the number to which}$
$\qquad\qquad\qquad\qquad\qquad\qquad\qquad \text{we raise } b \text{ to get } b^{3/2}.$

63. $\log_b 8$

Since 8 cannot be expressed using the numbers 1, 3, and 5, we cannot find $\log_b 8$ using the given information.

65. $\log_t t^{10} = 10 \qquad 10 \text{ is the exponent to which}$
$\qquad\qquad\qquad\qquad\qquad \text{we raise } t \text{ to get } t^{10}.$

67. $\log_e e^m = m \qquad m \text{ is the exponent to which}$
$\qquad\qquad\qquad\qquad\qquad \text{we raise } e \text{ to get } e^m.$

69. *Writing Exercise.* The logarithm of a quotient is an expression like $\log_a \dfrac{x}{y}$ which can be simplified, using the quotient rule, to $\log_a x - \log_a y$. A quotient of logarithms is an expression like $\dfrac{\log_a x}{\log_a y}$, which cannot be simplified.

71. Graph $f(x) = \sqrt{x} - 3$.

We construct a table of values, plot points, and connect them with a smooth curve. Note that we must choose nonnegative values of x in order for $\sqrt{x}$ to be a real number.

x	$f(x)$
0	-3
1	-2
4	-1
9	0

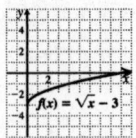

73. Graph: $g(x) = x^3 + 2$.

We construct a table of values, plot points, and connect them with a smooth curve.

x	$g(x)$
-2	-6
-1	1
0	2
1	3
2	10

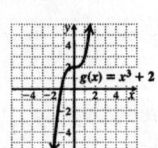

75. For $f(x) = \dfrac{x-3}{x+7}$,

$$x + 7 = 0$$
$$x = -7$$

The domain is $\{x \mid x \text{ is a real number and } x \neq -7\}$ or $(-\infty, -7) \cup (-7, \infty)$.

77. For $g(x) = \sqrt{10 - x}$

$$10 - x \geq 0$$
$$-x \geq -10$$
$$x \leq 10$$

The domain is $\{x \mid x \leq 10\}$ or $(-\infty, 10]$.

79. *Writing Exercise.* The student didn't subtract the logarithm of the entire denominator after using the quotient rule. The correct procedure is as follows:

$$\log_b \frac{1}{x} = \log_b \frac{x}{xx}$$
$$= \log_b x - \log_b xx$$
$$= \log_b x - (\log_b x + \log_b x)$$
$$= \log_b x - \log_b x - \log_b x$$
$$= -\log_b x$$

(Note that $-\log_b x$ is equivalent to $\log_b 1 - \log_b x$.)

81. $\log_a \left(x^8 - y^8\right) - \log_a \left(x^2 + y^2\right)$

$$= \log_a \frac{x^8 - y^8}{x^2 + y^2}$$
$$= \log_a \frac{\left(x^4 + y^4\right)\left(x^2 + y^2\right)(x+y)(x-y)}{x^2 + y^2}$$
$$= \log_a \left[\left(x^4 + y^4\right)\left(x^2 - y^2\right)\right] \quad \text{Simplifying}$$
$$= \log_a \left(x^6 - x^4 y^2 + x^2 y^4 - y^6\right)$$

83. $\log_a \sqrt{1 - s^2}$

$$= \log_a \left(1 - s^2\right)^{1/2}$$
$$= \frac{1}{2}\log_a \left(1 - s^2\right)$$
$$= \frac{1}{2}\log_a [(1-s)(1+s)]$$
$$= \frac{1}{2}\log_a (1-s) + \frac{1}{2}\log_a (1+s)$$

85. $\log_a \dfrac{\sqrt[3]{x^2 z}}{\sqrt[3]{y^2 z^{-2}}}$

$$= \log_a \left(\frac{x^2 z^3}{y^2}\right)^{1/3}$$
$$= \frac{1}{3}\left(\log_a x^2 z^3 - \log_a y^2\right)$$
$$= \frac{1}{3}(2\log_a x + 3\log_a z - 2\log_a y)$$
$$= \frac{1}{3}[2 \cdot 2 + 3 \cdot 4 - 2 \cdot 3]$$
$$= \frac{1}{3}(10)$$
$$= \frac{10}{3}$$

87. $\log_a x = 2$, so $a^2 = x$.

Let $\log_{1/a} x = n$ and solve for n.

$$\log_{1/a} a^2 = n \quad \text{Substituting } a^2 \text{ for } x$$
$$\left(\frac{1}{a}\right)^n = a^2$$
$$\left(a^{-1}\right)^n = a^2$$
$$a^{-n} = a^2$$
$$-n = 2$$
$$n = -2$$

Thus, $\log_{1/a} x = -2$ when $\log_a x = 2$.

89. $\log_2 80 + \log_2 x = 5$

$$\log_2 80x = 5$$
$$2^5 = 80x$$
$$\frac{32}{80} = x$$
$$\frac{2}{5} = x$$

91. True; $\log_a \left(Q + Q^2\right) = \log_a [Q(1+Q)]$
$$= \log_a Q + \log_a (1+Q) = \log_a Q + \log_a (Q+1)$$

Exercise Set 9.5

1. True; see page 613 in the text.

3. True; see page 614 in the text.

5. True; $\log 18 - \log 2 = \log \dfrac{18}{2} = \log 9$

7. True; $\ln 81 = \ln 9^2 = 2 \ln 9$

9. True; see Example 7.

11. 0.8451

13. 1.1367

15. Since $10^3 = 1000$, $\log 1000 = 3$.

17. -0.1249

19. 13.0014

21. 50.1187

23. 0.0011

25. 2.1972

27. -5.0832

29. 96.7583

31. 15.0293

33. 0.0305

35. We will use common logarithms for the conversion.
Let $a = 10$, $b = 3$, and $M = 28$ and substitute in the change-of-base formula.

$$\log_b M = \frac{\log_a M}{\log_a b}$$

$$\log_3 28 = \frac{\log_{10} 28}{\log_{10} 3}$$

$$\approx \frac{1.447158031}{0.477121254}$$

$$\approx 3.0331$$

37. We will use common logarithms for the conversion. Let $a = 10$, $b = 2$, and $M = 100$ and substitute in the change-of-base formula.

$$\log_2 100 = \frac{\log_{10} 100}{\log_{10} 2} \approx \frac{2}{0.3010} \approx 6.6439$$

39. We will use natural logarithms for the conversion.
Let $a = e$, $b = 4$, and $M = 5$ and substitute in the change-of-base formula.

$$\log_4 5 = \frac{\ln 5}{\ln 4} \approx \frac{1.6094}{1.3863} \approx 1.1610$$

41. We will use natural logarithms for the conversion.
Let $a = e$, $b = 0.1$, and $M = 2$ and substitute in the change-of-base formula.

$$\log_{0.1} 2 = \frac{\ln 2}{\ln 0.1} \approx \frac{0.6931}{-2.3026} \approx -0.3010$$

43. We will use common logarithms for the conversion.
Let $a = 10$, $b = 2$, and $M = 0.1$ and substitute in the change-of-base formula.

$$\log_2 0.1 = \frac{\log_{10} 0.1}{\log_{10} 2} \approx \frac{-1}{0.3010} \approx -3.3220$$

45. We will use natural logarithms for the conversion.
Let $a = e$, $b = \pi$, and $M = 10$ and substitute in the change-of-base formula.

$$\log_\pi 10 = \frac{\ln 10}{\ln \pi} \approx \frac{2.3026}{1.1447} \approx 2.0115$$

47. Graph: $f(x) = e^x$

We find some function values with a calculator. We use these values to plot points and draw the graph.

x	e^x
1	2.7
2	7.4
3	20.1
-1	0.4
-2	0.1

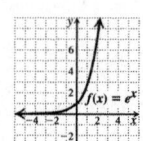

The domain is the set of real numbers and the range is $(0, \infty)$.

49. Graph: $f(x) = e^x + 3$

We find some function values, plot points, and draw the graph.

x	$e^x + 3$
0	4
1	5.72
2	10.39
-1	3.37
-2	3.14

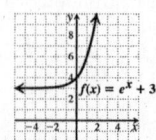

The domain is the set of real numbers and the range is $(3, \infty)$.

51. Graph: $f(x) = e^x - 2$

We find some function values, plot points, and draw the graph.

x	$e^x - 2$
0	-1
1	0.72
2	5.4
-1	-1.6
-2	-1.9

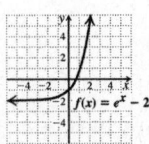

The domain is the set of real numbers and the range is $(-2, \infty)$.

53. Graph: $f(x) = 0.5e^x$

We find some function values, plot points, and draw the graph.

x	$0.5e^x$
0	0.5
1	1.36
2	3.69
-1	0.18
-2	0.07

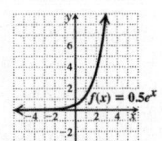

The domain is the set of real numbers and the range is $(0, \infty)$.

55. Graph: $f(x) = 0.5e^{2x}$

We find some function values, plot points, and draw the graph.

x	$0.5e^{2x}$
0	0.5
1	3.69
2	27.30
-1	0.07
-2	0.01

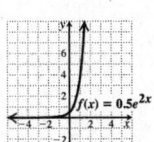

The domain is the set of real numbers and the range is $(0, \infty)$.

57. Graph: $f(x) = e^{x-3}$

We find some function values, plot points, and draw the graph.

x	e^{x-3}
2	0.37
3	1
4	2.72
-2	0.01

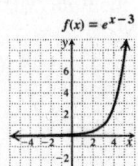

The domain is the set of real numbers and the range is $(0, \infty)$.

59. Graph: $f(x) = e^{x+2}$

We find some function values, plot points, and draw the graph.

x	e^{x+2}
-1	2.72
-2	1
-3	0.37
-4	0.14

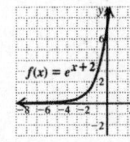

The domain is the set of real numbers and the range is $(0, \infty)$.

61. Graph: $f(x) = -e^x$

We find some function values, plot points, and draw the graph.

x	$-e^x$
0	-1
1	-2.72
2	-7.39
-1	-0.37
-3	-0.05

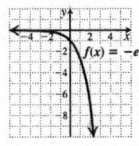

The domain is the set of real numbers and the range is $(-\infty, 0)$.

63. Graph: $g(x) = \ln x + 1$

We find some function values, plot points, and draw the graph.

x	$\ln x + 1$
1	1
3	2.10
5	2.61
7	2.95

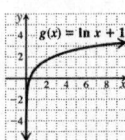

The domain is $(0, \infty)$ and the range is the set of real numbers.

65. Graph: $g(x) = \ln x - 2$

x	$\ln x + 1$
1	1
3	-0.90
5	-0.39
7	-0.05

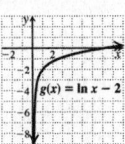

The domain is $(0, \infty)$ and the range is the set of real numbers.

67. Graph: $f(x) = 2 \ln x$

x	$2 \ln x$
0.5	-1.4
1	0
2	1.4
3	2.2
4	2.8
5	3.2
6	3.6

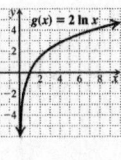

The domain is $(0, \infty)$ and the range is the set of real numbers.

69. Graph: $g(x) = -2 \ln x$

x	$-2\ln x$
0.5	1.4
1	0
2	-1.4
3	-2.2
4	-2.8
5	-3.2
6	-3.6

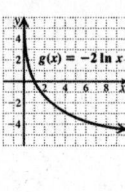

The domain is $(0, \infty)$ and the range is the set of real numbers.

71. Graph: $f(x) = \ln(x+2)$

We find some function values, plot points, and draw the graph.

x	$\ln(x+2)$
1	1.10
3	1.61
5	1.95
-1	0
-2	Undefined

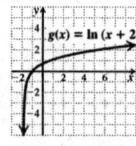

The domain is $(-2, \infty)$ and the range is the set of real numbers.

73. Graph: $g(x) = \ln(x-1)$

We find some function values, plot points, and draw the graph.

x	$\ln(x-1)$
2	0
3	0.69
4	1.10
6	1.61

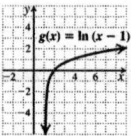

The domain is $(1, \infty)$ and the range is the set of real numbers.

75. *Writing Exercise.* $\log 10 < \log 79 < \log 100$, or

$1 < \log 79 < 2$

77. $x^2 - 3x - 28 = 0$
$(x+4)(x-7) = 0$
$x+4 = 0 \quad or \quad x-7 = 0$
$\quad x = -4 \quad or \quad \quad x = 7$

The solutions are -4 and 7.

79. $17x - 15 = 0$
$\qquad 17x = 15$
$\qquad x = \dfrac{15}{17}$

The solution is $\dfrac{15}{17}$.

81. $(x-5) \cdot 9 = 11$
$\qquad x - 5 = \dfrac{11}{9}$
$\qquad x = \dfrac{56}{9}$

The solution is $\dfrac{56}{9}$.

83. $x^{1/2} - 6x^{1/4} + 8 = 0$
Let $u = x^{1/4}$.

$u^2 - 6u + 8 = 0 \quad$ Substituting
$(u-4)(u-2) = 0$

$u = 4 \qquad or \qquad u = 2$
$x^{1/4} = 4 \qquad or \qquad x^{1/4} = 2$
$x = 256 \quad or \qquad x = 16 \quad$ Raising both sides
$\qquad\qquad\qquad\qquad\qquad\qquad$ to the fourth power

Both numbers check. The solutions are 256 and 16.

85. *Writing Exercise.* Reflect the graph of $f(x) = e^x$ across the line $y = x$ and then translate it up one unit.

87. We use the change-of-base formula.

$$\log_6 81 = \frac{\log 81}{\log 6}$$
$$= \frac{\log 3^4}{\log(2 \cdot 3)}$$
$$= \frac{4\log 3}{\log 2 + \log 3}$$
$$\approx \frac{4(0.477)}{0.301 + 0.477}$$
$$\approx 2.452$$

89. We use the change-of-base formula.

$$\log_{12} 36 = \frac{\log 36}{\log 12}$$
$$= \frac{\log(2 \cdot 3)^2}{\log(2^2 \cdot 3)}$$
$$= \frac{2\log(2 \cdot 3)}{\log 2^2 + \log 3}$$
$$= \frac{2(\log 2 + \log 3)}{2\log 2 + \log 3}$$
$$\approx \frac{2(0.301 + 0.477)}{2(0.301) + 0.477}$$
$$\approx 1.442$$

91. Use the change-of-base formula with $a = e$ and $b = 10$.
We obtain

$$\log M = \frac{\ln M}{\ln 10}.$$

93. $\log(492x) = 5.728$
$\qquad 10^{5.728} = 492x$
$\qquad \dfrac{10^{5.728}}{492} = x$
$\qquad 1086.5129 \approx x$

95. $\log 692 + \log x = \log 3450$
$$\log x = \log 3450 - \log 692$$
$$\log x = \log \frac{3450}{692}$$
$$x = \frac{3450}{692}$$
$$x \approx 4.9855$$

97. a) Domain: $\{x \mid x > 0\}$, or $(0, \infty)$; range:
$\{y \mid y < 0.5135\}$, or $(-\infty, 0.5135)$;

b) $[-1, 5, -10, 5]$;

c)

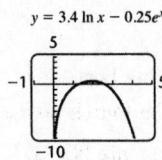
$y = 3.4 \ln x - 0.25e^x$

99. a) Domain $\{x \mid x > 0\}$, or $(0, \infty)$;

range: $\{y \mid y > -0.2453\}$, or $(-0.2453, \infty)$

b) $[-1, 5, -1, 10]$;

c)

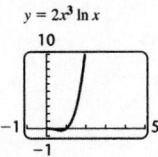
$y = 2x^3 \ln x$

101. *Graphing Calculator Exercise*

Connecting the Concepts

1. $\log_4 16 = \log_4 4^2$
$$= 2 \log_4 4 \qquad \text{Using the power rule}$$
$$= 2 \qquad\qquad \log_4 4 = 1$$

3. $\log_{100} 10 = \log_{100} 100^{1/2}$
$$= \frac{1}{2} \log_{100} 100$$
$$= \frac{1}{2}$$

5. $\log 10 = \log_{10} 10 = 1$

7. $\log 10^4 = 4 \log_{10} 10 = 4$

9. $e^{\ln 7} = 7$

11. $\log_x 3 = m$
$$x^m = 3$$

13. $e^t = x$
$$\ln x = t$$

15. $\log_x 64 = 3$
$$x^3 = 64$$
$$x^3 = 4^3$$
$$x = 4$$

17. $\log \sqrt{\dfrac{x^2}{yz^3}} = \log \left(\dfrac{x^2}{yz^3} \right)^{1/2} = \dfrac{1}{2} \log \dfrac{x^2}{yz^3}$
$$= \frac{1}{2} \left(\log x^2 - \log y - \log z^3 \right)$$
$$= \frac{1}{2} \left(2 \log x - \log y - 3 \log z \right)$$
$$= \log x - \frac{1}{2} \log y - \frac{3}{2} \log z$$

19. $\log_4 8 = \dfrac{\log 8}{\log 4} \approx \dfrac{0.9031}{0.6021} \approx 1.5$

Exercise Set 9.6

1. If we take the common logarithm on both sides, we see that choice (e) is correct.

3. $\ln x = 3$ means that 3 is the exponent to which we raise e to get x, so choice (f) is correct.

5. By the product rule for logarithms, $\log_5 x + \log_5 (x - 2)$
$$= \log_5 [x(x - 2)] = \log_5 (x^2 - 2x), \text{ so choice (b) is correct.}$$

7. By the quotient rule for logarithms,

$\ln x - \ln(x - 2) = \ln \dfrac{x}{x - 2}$, so choice (g) is correct.

9. $3^{2x} = 81$
$$3^{2x} = 3^4$$
$$2x = 4 \qquad \text{Equating the exponents}$$
$$x = 2$$

11. $4^x = 32$
$$2^{2x} = 2^5$$
$$2x = 5 \qquad \text{Equating the exponents}$$
$$x = \frac{5}{2}$$

13. $2^x = 10$
$$\log 2^x = \log 10 \qquad \text{Taking log on both sides}$$
$$\log 2^x = 1 \qquad\qquad \log 10 = 1$$
$$x \log 2 = 1$$
$$x = \frac{1}{\log 2}$$
$$x \approx 3.322 \qquad \text{Using a calculator}$$

15. $2^{x+5} = 16$
$$2^{x+5} = 2^4$$
$$x + 5 = 4 \qquad \text{Equating the exponents}$$
$$x = -1$$

17.
$$8^{x-3} = 19$$
$$\log 8^{x-3} = \log 19 \qquad \text{Taking log on both sides}$$
$$(x-3)\log 8 = \log 19$$
$$x - 3 = \frac{\log 19}{\log 8}$$
$$x = \frac{\log 19}{\log 8} + 3$$
$$x \approx 4.416 \qquad \text{Using a calculator}$$

19.
$$e^t = 50$$
$$\ln e^t = \ln 50 \qquad \text{Taking ln on both sides}$$
$$t = \ln 50 \qquad \ln e^t = t \ln e = t \cdot 1 = t$$
$$t \approx 3.912$$

21.
$$e^{-0.02t} = 8$$
$$\ln e^{-0.02t} = \ln 8 \qquad \text{Taking ln on both sides}$$
$$-0.02t = \ln 8$$
$$t = \frac{\ln 8}{-0.02}$$
$$t \approx -103.972$$

23.
$$4.9^x - 87 = 0$$
$$4.9^x = 87$$
$$\log 4.9^x = \log 87$$
$$x \log 4.9 = \log 87$$
$$x = \frac{\log 87}{\log 4.9}$$
$$x \approx 2.810$$

25.
$$19 = 2e^{4x}$$
$$\frac{19}{2} = e^{4x}$$
$$\ln\left(\frac{19}{2}\right) = \ln e^{4x}$$
$$\ln\left(\frac{19}{2}\right) = 4x$$
$$\frac{\ln\left(\frac{19}{2}\right)}{4} = x$$
$$0.563 \approx x$$

27.
$$7 + 3e^{-x} = 13$$
$$3e^{-x} = 6$$
$$e^{-x} = 2$$
$$\ln e^{-x} = \ln 2$$
$$-x = \ln 2$$
$$x = -\ln 2$$
$$x \approx -0.693$$

29.
$$\log_3 x = 4$$
$$x = 3^4 \qquad \text{Writing an equivalent exponential equation}$$
$$x = 81$$

31.
$$\log_4 x = -2$$
$$x = 4^{-2} \qquad \text{Writing an equivalent exponential equation}$$
$$x = \frac{1}{4^2}, \text{ or } \frac{1}{16}$$

33.
$$\ln x = 5$$
$$x = e^5 \qquad \text{Writing an equivalent exponential equation}$$
$$x \approx 148.413$$

35.
$$\ln 4x = 3$$
$$4x = e^3$$
$$x = \frac{e^3}{4} \approx 5.021$$

37.
$$\log x = 1.2 \qquad \text{The base is 10.}$$
$$x = 10^{1.2}$$
$$x \approx 15.849$$

39.
$$\ln(2x + 1) = 4$$
$$2x + 1 = e^4$$
$$2x = e^4 - 1$$
$$x = \frac{e^4 - 1}{2} \approx 26.799$$

41.
$$\ln x = 1$$
$$x = e \approx 2.718$$

43.
$$5 \ln x = -15$$
$$\ln x = -3$$
$$x = e^{-3} \approx 0.050$$

45.
$$\log_2(8 - 6x) = 5$$
$$8 - 6x = 2^5$$
$$8 - 6x = 32$$
$$-6x = 24$$
$$x = -4$$
The answer checks. The solution is –4.

47.
$$\log(x - 9) + \log x = 1 \qquad \text{The base is 10.}$$
$$\log_{10}\left[(x-9)(x)\right] = 1 \qquad \text{Using the product rule}$$
$$x(x - 9) = 10^1$$
$$x^2 - 9x = 10$$
$$x^2 - 9x - 10 = 0$$
$$(x + 1)(x - 10) = 0$$
$$x = -1 \ \text{or} \ x = 10$$

Check: For -1:
$$\log(x - 9) + \log x = 1$$
$$\log(-1 - 9) + \log(-1) = 1 \qquad \text{FALSE}$$

For 10:
$$\log(x - 9) + \log x = 1$$
$$\begin{array}{c|c} \log(10 - 9) + \log(10) & 1 \\ \log 1 + \log 10 & \\ 0 + 1 & \\ \hline 1 = 1 & \text{TRUE} \end{array}$$

The number –1 does not check, because negative numbers do not have logarithms. The solution is 10.

49. $\log x - \log(x+3) = 1$ The base is 10.

$$\log_{10} \frac{x}{x+3} = 1 \quad \text{Using the quotient rule}$$
$$\frac{x}{x+3} = 10^1$$
$$x = 10(x+3)$$
$$x = 10x + 30$$
$$-9x = 30$$
$$x = -\frac{10}{3}$$

The number $-\frac{10}{3}$ does not check. The equation has no solution.

51. We observe that since $\log(2x+1) = \log 5$, then
$$2x + 1 = 5$$
$$2x = 4$$
$$x = 2$$

53.
$$\log_4(x+3) = 2 + \log_4(x-5)$$
$$\log_4(x+3) - \log_4(x-5) = 2$$
$$\log_4 \frac{x+3}{x-5} = 2 \quad \text{Using the quotient rule}$$
$$\frac{x+3}{x-5} = 4^2$$
$$\frac{x+3}{x-5} = 16$$
$$x+3 = 16(x-5)$$
$$x+3 = 16x - 80$$
$$83 = 15x$$
$$\frac{83}{15} = x$$

The number $\frac{83}{15}$ checks. It is the solution.

55. $\log_7(x+1) + \log_7(x+2) = \log_7 6$
$$\log_7[(x+1)(x+2)] = \log_7 6 \quad \text{Using the product rule}$$
$$\log_7(x^2 + 3x + 2) = \log_7 6$$
$$x^2 + 3x + 2 = 6 \quad \text{Using the property of logarithmic equality}$$
$$x^2 + 3x - 4 = 0$$
$$(x+4)(x-1) = 0$$

$x = -4 \quad or \quad x = 1$

The number 1 checks, but -4 does not.

The solution is 1.

57. $\log_5(x+4) + \log_5(x-4) = \log_5 20$
$$\log_5[(x+4)(x-4)] = \log_5 20 \quad \text{Using the product rule}$$
$$\log_5(x^2 - 16) = \log_5 20$$
$$x^2 - 16 = 20 \quad \text{Using the property of logarithmic equality}$$
$$x^2 = 36$$
$$x = \pm 6$$

The number 6 checks, but -6 does not.

The solution is 6.

59. $\ln(x+5) + \ln(x+1) = \ln 12$
$$\ln[(x+5)(x+1)] = \ln 12$$
$$\ln(x^2 + 6x + 5) = \ln 12$$
$$x^2 + 6x + 5 = 12$$
$$x^2 + 6x - 7 = 0$$
$$(x+7)(x-1) = 0$$

$x = -7 \quad or \quad x = 1$

The number -7 does not check, but 1 does.

The solution is 1.

61. $\log_2(x-3) + \log_2(x+3) = 4$
$$\log_2[(x-3)(x+3)] = 4$$
$$(x-3)(x+3) = 2^4$$
$$x^2 - 9 = 16$$
$$x^2 = 25$$
$$x = \pm 5$$

The number 5 checks, but -5 does not.

The solution is 5.

63. $\log_{12}(x+5) - \log_{12}(x-4) = \log_{12} 3$
$$\log_{12} \frac{x+5}{x-4} = \log_{12} 3$$
$$\frac{x+5}{x-4} = 3 \quad \text{Using the property of logarithmic equality}$$
$$x+5 = 3(x-4)$$
$$x+5 = 3x - 12$$
$$17 = 2x$$
$$\frac{17}{2} = x$$

The number $\frac{17}{2}$ checks and is the solution.

65. $\log_2(x-2) + \log_2 x = 3$
$$\log_2[(x-2)(x)] = 3$$
$$x(x-2) = 2^3$$
$$x^2 - 2x = 8$$
$$x^2 - 2x - 8 = 0$$
$$(x-4)(x+2) = 0$$

$x = 4 \quad or \quad x = -2$

The number 4 checks, but -2 does not.

The solution is 4.

67. *Writing Exercise.* Madison mistakenly thinks that, because negative numbers do not have logarithms, negative numbers cannot be solutions of logarithmic equations.

69. *Familiarize.* Let w represent the width of the rectangle, in ft. Then $w + 6$ represents the length. Recall that the formula for the perimeter P of a rectangle with length l and width w is $P = 2l + 2w$.

Translate.

Perimeter	is	26 ft,
$\downarrow$	$\downarrow$	$\downarrow$
$2 \cdot (w + 6) + 2 \cdot w$	$=$	26

Carry out. We solve the equation.

$$2 \cdot (w + 6) + 2 \cdot w = 26$$
$$2w + 12 + 2w = 26$$
$$4w + 12 = 26$$
$$4w = 14$$
$$w = 3.5$$

When $w = 3.5$, then $w + 6 = 3.5 + 6 = 9.5$.

Check. If the length is 9.5 ft and the width is 3.5 ft, then the length is 6 ft longer than the width. Also $P = 2 \cdot 9.5 + 2 \cdot 3.5 = 19 + 7 = 26$ ft. The answer checks.

State. The length of the rectangle is 9.5 ft, and the width is 3.5 ft.

71. ***Familiarize.*** Let $x =$ the number of pounds of Golden Days and $y =$ the number of pounds of Snowy Friends to be used in the mixture. The amount of sunflower seeds in the mixture is 33%(50 lb), or 16.5.

Translate. We organize the information in a table.

	Golden Days	Snowy Friends	Mixture
Number of pounds	x	y	50
Percent of sunflower seeds	25%	40%	33%
Amount of sunflower seeds	$0.25x$	$0.40y$	16.5 lb

We get one equation from the "Number of pounds" row of the table:

$$x + y = 50$$

The last row of the table yields a second equation:

$$0.25x + 0.4y = 16.5$$

After clearing decimals, we have the problem translated to a system of equations:

$$x + y = 50, \quad (1)$$
$$25x + 40y = 1650 \quad (2)$$

Carry out. We use the elimination method to solve the system of equations.

$$\begin{array}{r} -25x - 25y = -1250 \\ 25x + 40y = 1650 \\ \hline 15y = 400 \\ y = 26\frac{2}{3} \end{array} \quad \text{Multiplying (1) by } -25$$

Substitute $26\frac{2}{3}$ for y in (1) and solve for x.

$$x + 26\frac{2}{3} = 50$$
$$x = 23\frac{1}{3}$$

Check. The amount of the mixture is $26\frac{2}{3}$ lb $+ 23\frac{1}{3}$ lb or 50 lb. The amount of sunflower seeds in the mixture is $0.25\left(23\frac{1}{3}\text{ lb}\right) + 0.4\left(26\frac{2}{3}\text{ lb}\right) = 5\frac{5}{6}\text{ lb} + 10\frac{2}{3}\text{ lb} = 16.5\text{ lb}$. The answer checks.

State. $23\frac{1}{3}$ lb of Golden Days and $26\frac{2}{3}$ lb of Snowy Friends should be mixed.

73. ***Familiarize.*** Let $t =$ the time, in hours, it takes Max and Miles to key in the score, working together. Then in t hours Max does $\frac{t}{2}$ of the job, Miles does $\frac{t}{3}$, and together they do 1 entire job.

Translate.

$$\frac{t}{2} + \frac{t}{3} = 1$$

Carry out. We solve the equation. First we multiply by the LCD, 6.

$$6\left(\frac{t}{2} + \frac{t}{3}\right) = 6 \cdot 1$$
$$6 \cdot \frac{t}{2} + 6 \cdot \frac{t}{3} = 6$$
$$3t + 2t = 6$$
$$5t = 6$$
$$t = \frac{6}{5}$$

Check. In $\frac{6}{5}$ hr Max does $\frac{6/5}{2}$, or $\frac{3}{5}$ of the job, and Miles does $\frac{6/5}{2}$, or $\frac{2}{5}$ of the job. Together they do $\frac{3}{5} + \frac{2}{5}$ or 1 entire job. The answer checks.

State. It takes Max and Miles $\frac{6}{5}$ hr, or $1\frac{1}{5}$ hr, to do the job, working together.

75. ***Writing Exercise.*** No; let $m = 2$, $n = -2$, and $f(x) = x^2$. Then $f(2) = f(-2)$, but $2 \neq -2$.

77.
$$8^x = 16^{3x+9}$$
$$(2^3)^x = (2^4)^{3x+9}$$
$$2^{3x} = 2^{12x+36}$$
$$3x = 12x + 36$$
$$-36 = 9x$$
$$-4 = x$$

The solution is –4.

79.
$$\log_6(\log_2 x) = 0$$
$$\log_2 x = 6^0$$
$$\log_2 x = 1$$
$$x = 2^1$$
$$x = 2$$

The solution is 2.

81. $\log_5 \sqrt{x^2 - 9} = 1$

$\sqrt{x^2 - 9} = 5^1$

$\left(\sqrt{x^2 - 9}\right)^2 = 5^2$

$x^2 - 9 = 25$

$x^2 = 34$

$x = \pm\sqrt{34}$

The solutions are $\pm\sqrt{34}$.

83. $2^{x^2 + 4x} = \dfrac{1}{8}$

$2^{x^2 + 4x} = \dfrac{1}{2^3}$

$2^{x^2 + 4x} = 2^{-3}$

$x^2 + 4x = -3$

$x^2 + 4x + 3 = 0$

$(x + 3)(x + 1) = 0$

$x = -3 \ \ or \ \ x = -1$

The solutions are -3 and -1.

85. $\log_5 |x| = 4$

$|x| = 5^4$

$|x| = 625$

$x = 625 \ \ or \ \ x = -625$

The solutions are 625 and -625.

87. $\log \sqrt{2x} = \sqrt{\log 2x}$

$\log(2x)^{1/2} = \sqrt{\log 2x}$

$\dfrac{1}{2}\log 2x = \sqrt{\log 2x}$

$\dfrac{1}{4}(\log 2x)^2 = \log 2x$ Squaring both sides

$\dfrac{1}{4}(\log 2x)^2 - \log 2x = 0$

Let $u = \log 2x$.

$\dfrac{1}{4}u^2 - u = 0$

$u\left(\dfrac{1}{4}u - 1\right) = 0$

$u = 0 \quad or \quad \dfrac{1}{4}u - 1 = 0$

$u = 0 \quad or \quad \dfrac{1}{4}u = 1$

$u = 0 \quad or \quad u = 4$

$\log 2x = 0 \quad or \quad \log 2x = 4$ Replacing u with $\log 2$

$2x = 10^0 \quad or \quad 2x = 10^4$

$2x = 1 \quad or \quad 2x = 10{,}000$

$x = \dfrac{1}{2} \quad or \quad x = 5000$

Both numbers check. The solutions are $\dfrac{1}{2}$ and 5000.

89. $3^{x^2} \cdot 3^{4x} = \dfrac{1}{27}$

$3^{x^2 + 4x} = 3^{-3}$

$x^2 + 4x = -3$ The exponents must be equal.

$x^2 + 4x + 3 = 0$

$(x + 1)(x + 3) = 0$

$x = -1 \ \ or \ \ x = -3$

Both numbers check. The solutions are -1 and -3.

91. $\log x^{\log x} = 25$

$\log x (\log x) = 25$ Using the power rule

$(\log x)^2 = 25$

$\log x = \pm 5$

$x = 10^5 \quad or \quad x = 10^{-5}$

$x = 100{,}000 \quad or \quad x = \dfrac{1}{100{,}000}$

Both numbers check. The solutions are $100{,}000$ and

$\dfrac{1}{100{,}000}$.

93. $\left(81^{x-2}\right)\left(27^{x+1}\right) = 9^{2x-3}$

$\left[\left(3^4\right)^{x-2}\right]\left[\left(3^3\right)^{x+1}\right] = \left(3^2\right)^{2x-3}$

$\left(3^{4x-8}\right)\left(3^{3x+3}\right) = 3^{4x-6}$

$3^{7x-5} = 3^{4x-6}$

$7x - 5 = 4x - 6$

$3x = -1$

$x = -\dfrac{1}{3}$

The solution is $-\dfrac{1}{3}$.

95. $2^y = 16^{x-3} \quad and \quad 3^{y+2} = 27^x$

$2^y = \left(2^4\right)^{x-3} \quad and \quad 3^{y+2} = \left(3^3\right)^x$

$y = 4x - 12 \quad and \quad y + 2 = 3x$

$12 = 4x - y \quad and \quad 2 = 3x - y$

Solving this system of equations we get $x = 10$ and $y = 28$. Then $x + y = 10 + 28 = 38$.

97. Find the first coordinate of the point of intersection of $y_1 = \ln x$ and $y_2 = \log x$. The value of x for which the natural logarithm of x is the same as the common logarithm of x is 1.

Exercise Set 9.7

1. a) Replace $A(t)$ with 4000 and solve for t.

$$A(t) = 77(1.283)^t$$
$$4000 = 77(1.283)^t$$
$$\frac{4000}{77} = 1.283^t$$
$$\ln\frac{4000}{77} = \ln 1.283^t$$
$$\ln\frac{4000}{77} = t\ln 1.283$$
$$\frac{\ln\frac{4000}{77}}{\ln 1.283} = t$$
$$16 \approx t$$

The number of known asteroids first reached 4000 about 16 yr after 1990, or in 2006.

b) $A(0) = 77(1.283)^0 = 77 \cdot 1 = 77$, so to find the doubling time, we replace $A(t)$ with $2(77)$, or 154 and solve for t.

$$154 = 77(1.283)^t$$
$$2 = 1.283^t$$
$$\ln 2 = \ln 1.283^t$$
$$\ln 2 = t\ln 1.283$$
$$\frac{\ln 2}{\ln 1.283} = t$$
$$2.8 \approx t$$

The doubling time is about 2.8 years.

3. a) Replace $S(t)$ with 100 and solve for t.

$$S(t) = 180(0.97)^t$$
$$100 = 180(0.97)^t$$
$$\frac{100}{180} = (0.97)^t$$
$$\ln\frac{100}{180} = \ln(0.97)^t$$
$$\ln\frac{100}{180} = t\ln 0.97$$
$$\frac{\ln\frac{100}{180}}{\ln 0.97} = t$$
$$19 \approx t$$

The death rate reached 100 about 19 yr after 1960, or in 1979.

b) Replace $S(t)$ with 25 and solve for t.

$$S(t) = 180(0.97)^t$$
$$25 = 180(0.97)^t$$
$$\frac{25}{180} = (0.97)^t$$
$$\ln\frac{25}{180} = \ln(0.97)^t$$
$$\ln\frac{25}{180} = t\ln 0.97$$
$$\frac{\ln\frac{25}{180}}{\ln 0.97} = t$$
$$65 \approx t$$

The death rate will reach 25 about 65 yr after 1960, or in 2025.

5. a) Replace $A(t)$ with 35,000 and solve for t.

$$A(t) = 29,000(1.03)^t$$
$$35,000 = 29,000(1.03)^t$$
$$1.207 \approx (1.03)^t$$
$$\log 1.207 \approx \log(1.03)^t$$
$$\log 1.207 \approx t\log 1.03$$
$$\frac{\log 1.207}{\log 1.03} \approx t$$
$$6.4 \approx t$$

The amount due will reach \$35,000 after about 6.4 years.

b) Replace $A(t)$ with $2(29,000)$, or 58,000, and solve for t.

$$58,000 = 29,000(1.03)^t$$
$$2 = (1.03)^t$$
$$\log 2 = \log(1.03)^t$$
$$\log 2 = t\log 1.03$$
$$\frac{\log 2}{\log 1.03} = t$$
$$23.4 \approx t$$

The doubling time is about 23.4 years.

7. a) Substitute 50 for $W(t)$ and solve for t.

$$W(t) = 89(0.837)^t$$
$$50 = 89(0.837)^t$$
$$\frac{50}{89} = (0.837)^t$$
$$\log\frac{50}{89} = \log(0.837)^t$$
$$\log\frac{50}{89} = t\log 0.837$$
$$\frac{\log\frac{50}{89}}{\log 0.837} = t$$
$$3 \approx t$$

The number dropped below 50% 3 yr after 1988, or 1991.

b) Substitute 1 for $W(t)$ and solve for t.

$$W(t) = 89(0.837)^t$$
$$1 = 89(0.837)^t$$
$$\frac{1}{89} = (0.837)^t$$
$$\log\frac{1}{89} = \log(0.837)^t$$
$$-\log 89 = t\log 0.837 \qquad \log\frac{1}{89} = -\log 89$$
$$\frac{-\log 89}{\log 0.837} = t$$
$$25 \approx t$$

The number will drop below 1% 25 yr

after 1988, or 2013.

9. a) $P(t)$ is given in thousands, so we substitute 30 for

$P(t)$ and solve for t.

$$P(t) = 5.5(1.047)^t$$
$$30 = 5.5(1.047)^t$$
$$5.455 \approx 1.047^t$$
$$\log 5.455 \approx \log 1.047^t$$
$$\log 5.455 \approx t\log 1.047$$
$$\frac{\log 5.455}{\log 1.047} \approx t$$
$$36.9 \approx t$$

The humpback whale population will reach 30,000

about 36.9 yr after 1982, or in 2018.

b) $P(0) = 5.5(1.047)^0 = 5.5(1) = 5.5$ and $2(5.5) = 11$,

so we substitute 11 for $P(t)$ and solve for t.

$$11 = 5.5(1.047)^t$$
$$2 = 1.047^t$$
$$\log 2 = \log 1.047^t$$
$$\log 2 = t\log 1.047$$
$$\frac{\log 2}{\log 1.047} = t$$
$$15.1 \approx t$$

The doubling time is about 15.1 yr.

11. $\mathrm{pH} = -\log[H^+]$
$$= -\log(1.3 \times 10^{-5})$$
$$\approx -(-4.886057) \quad \text{Using a calculator}$$
$$\approx 4.9$$

The pH of fresh-brewed coffee is about 4.9.

13. $\mathrm{pH} = -\log[H^+]$
$$7.0 = -\log[H^+]$$
$$-7.0 = \log[H^+]$$
$$10^{-7.0} = [H^+] \quad \text{Converting to an}$$
$$\text{exponential equation}$$

The hydrogen ion concentration is 10^{-7} moles per liter.

15. $L = 10\cdot\log\dfrac{I}{I_0}$
$$= 10\cdot\log\frac{10}{10^{-12}}$$
$$= 10\cdot\log 10^{13}$$
$$= 10\cdot 13\cdot\log 10$$
$$= 130$$

The sound level is 130 dB.

17. $L = 10\cdot\log\dfrac{I}{I_0}$
$$128.8 = 10\cdot\log\frac{I}{10^{-12}}$$
$$12.88 = \log\frac{I}{10^{-12}}$$
$$12.88 = \log I - \log 10^{-12} \quad \text{Using the quotient rule}$$
$$12.88 = \log I - (-12)$$
$$12.88 = \log I + 12$$
$$0.88 = \log I$$
$$10^{0.88} = I \quad \text{Converting to an exponential equation}$$
$$7.6 \approx I$$

The intensity of the sound is $10^{0.88}$ W/m^2, or about

7.6 W/m^2.

19. $$M = \log\frac{v}{1.34}$$
$$7.5 = \log\frac{v}{1.34}$$
$$10^{7.5} = \frac{v}{1.34}$$
$$1.34(10^{7.5}) = v$$
$$42{,}400{,}000 \approx v$$

Approximately 42.4 million messages per day are sent by

that network.

21. a) Substitute 0.025 for k:

$$P(t) = P_0\, e^{0.025t}$$

b) To find the balance after one year, replace

P_0 with 5000 and t with 1. We find $P(1)$:

$$P(1) = 5000\, e^{0.025(1)} = 5000\, e^{0.025} \approx \$5126.58$$

To find the balance after 2 years, replace P_0 with

5000 and t with 2. We find $P(2)$:

$$P(2) = 5000\, e^{0.025(2)} = 5000\, e^{0.05} \approx \$5256.36$$

c) To find the doubling time, replace P_0 with 5000 and

$P(t)$ with 10,000 and solve for t.

$$10{,}000 = 5000\, e^{0.025t}$$
$$2 = e^{0.025t}$$
$$\ln 2 = \ln e^{0.025t} \quad \text{Taking the natural}$$
$$\text{logarithm on both sides}$$
$$\ln 2 = 0.025t \quad \text{Finding the logarithm of}$$
$$\frac{\ln 2}{0.025} = t \qquad \qquad \text{the base to a power}$$
$$27.7 \approx t$$

The investment will double in about 27.7 years.

23. a) $P(t) = 304e^{0.009t}$, where $P(t)$ is in millions and t is the number of years after 2008.

b) In 2012, $t = 2012 - 2008 = 4$. Find $P(4)$.

$P(4) = 304e^{0.009(4)} = 304e^{0.036} \approx 315$

The U.S. population will be about 315 million in 2012.

c) Substitute 325 for $P(t)$ and solve for t.

$$325 = 304e^{0.009t}$$
$$\frac{325}{304} = e^{0.009t}$$
$$\ln\frac{325}{304} = \ln e^{0.009t}$$
$$\ln\frac{325}{304} = 0.009t$$
$$\frac{\ln\frac{325}{304}}{0.009} = t$$
$$7 \approx t$$

The U.S. population will reach 325 million about 7 yr after 2008, or in 2015.

25. The exponential growth function is $S(t) = S_0 e^{3.40t}$. We replace $S(t)$ with $2S_0$ and solve for t.

$$2S_0 = S_0 e^{3.40t}$$
$$2 = e^{3.40t}$$
$$\ln 2 = \ln e^{3.40t}$$
$$\ln 2 = 3.40t$$
$$\frac{\ln 2}{3.40} = t$$
$$0.2 \approx t$$

The doubling time for the zebra mussels is about 0.2 yr.

27. $Y(x) = 71.41\ln\frac{x}{4.6}$

a) $Y(10) = 71.41\ln\frac{10}{4.6} \approx 55$

The world population will reach 10 billion about 55 yr after 2000, or in 2055.

b) $Y(12) = 71.41\ln\frac{12}{4.6} \approx 68$

The world population will reach 12 billion about 68 yr after 2000, or in 2068.

c) Plot the points found in parts (a) and (b) and others as necessary and draw the graph.

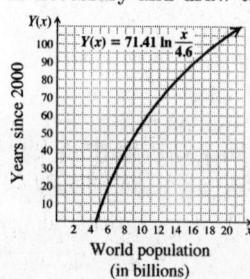

29. a) $S(0) = 68 - 20\log(0 + 1) = 68 - 20\log 1$
$= 68 - 20(0) = 68\%$

b) $S(4) = 68 - 20\log(4 + 1) = 68 - 20\log 5$
$\approx 68 - 20(0.69897) \approx 54\%$

$S(24) = 68 - 20\log(24 + 1)$
$= 68 - 20\log 25 \approx 68 - 20(1.39794) \approx 40\%$

c) Using the values we computed in parts (a) and (b) and any others we wish to calculate, we sketch the graph:

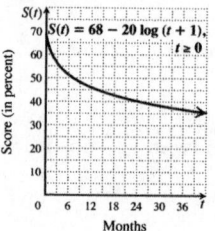

d) $$50 = 68 - 20\log(t + 1)$$
$$-18 = -20\log(t + 1)$$
$$0.9 = \log(t + 1)$$
$$10^{0.9} = t + 1$$
$$7.9 \approx t + 1$$
$$6.9 \approx t$$

After about 6.9 months, the average score was 50.

31. a) We start with the exponential growth equation

$$P(t) = P_0 e^{kt}.$$

Substituting 2000 for P_0, we have $P(t) = 2000e^{kt}$. To find the exponential growth rate k, observe that the wind-power capacity was 17,000 mW in 2007 or 17 years after 1990. We substitute and solve for k.

$$P(t) = 2000e^{k \cdot 17}$$
$$17,000 = 2000e^{17k}$$
$$8.5 = e^{17k}$$
$$\ln 8.5 = \ln e^{17k}$$
$$\ln 8.5 = 17k$$
$$\frac{\ln 8.5}{17} = k$$
$$0.126 \approx k$$

Thus, the exponential growth function is

$P(t) = 2000e^{0.126t}$, where t is the number of years after 1990.

b) Substitute 50,000 for $P(t)$ and solve for t.

$$50,000 = 2000e^{0.126t}$$
$$25 = e^{0.126t}$$
$$\ln 25 = \ln e^{0.126t}$$
$$\ln 25 = 0.126t$$
$$\frac{\ln 25}{0.126} = t$$
$$25 \approx t$$

The wind-power capacity will reach 50,000 mW about 25 yr after 1990, or in 2015.

33. a) We start with the exponential growth equation

$$P(t) = P_0 e^{kt}, \text{ where } t \text{ is the number of}$$
$$\text{years after 1997.}$$

Substituting 8200 for P_0, we have

$$P(t) = 8200 e^{kt}.$$

To find the exponential growth rate k, observe that the cost was \$500 in 2007 or 10 years after 1997. We substitute and solve for k.

$$P(t) = 8200 e^{k \cdot 10}$$
$$500 = 8200 e^{10k}$$
$$\frac{5}{82} = e^{10k}$$
$$\ln \frac{5}{82} = \ln e^{10k}$$
$$\ln \frac{5}{82} = 10k$$
$$\frac{\ln \frac{5}{82}}{10} = k$$
$$-0.280 \approx k$$

Thus, the exponential growth function is

$$P(t) = 8200 e^{-0.280t}, \text{ where } t \text{ is the number}$$
$$\text{of years after 1997.}$$

b) In 2010, $t = 2010 - 1997 = 13$

$$P(t) = 8200 e^{-0.280(15)} \approx \$215$$

The cost in 2010 will be about \$215 per gigabit per second per mile.

c) Substitute 1 for $P(t)$ and solve for t.

$$1 = 8200 e^{-0.280t}$$
$$\frac{1}{8200} = e^{-0.280t}$$
$$\ln \frac{1}{8200} = \ln e^{-0.280t}$$
$$\ln \frac{1}{8200} = -0.280t$$
$$\frac{\ln \frac{1}{8200}}{-0.280} = t$$
$$32 \approx t$$

The cost will be \$1 per gigabit per second per mile about 32 yr after 1997, or in 2029.

35. We will use the function derived in Example 7:

$$P(t) = P_0 e^{-0.00012t}$$

If the seed had lost 21% of its carbon-14 from the initial amount P_0, then $79\%(P_0)$ is the amount present. To find the age t of the seed, we substitute $79\%(P_0)$, or $0.79P_0$ for $P(t)$ in the function above and solve for t.

$$0.79P_0 = P_0 e^{-0.00012t}$$
$$0.79 = e^{-0.00012t}$$
$$\ln 0.79 = \ln e^{-0.00012t}$$
$$\ln 0.79 = -0.00012t$$
$$\frac{\ln 0.79}{-0.00012} = t$$
$$1964 \approx t$$

The seed is about 1964 yr old.

37. The function $P(t) = P_0 e^{-kt}$, $k > 0$, can be used to model decay. For iodine-131, $k = 9.6\%$, or 0.096. To find the half-life we substitute 0.096 for k and $\frac{1}{2}P_0$ for $P(t)$, and solve for t.

$$\frac{1}{2}P_0 = P_0 e^{-0.096t}, \text{ or } \frac{1}{2} = e^{-0.096t}$$
$$\ln \frac{1}{2} = \ln e^{-0.096t} = -0.096t$$
$$t = \frac{\ln 0.5}{-0.096} \approx \frac{-0.6931}{-0.096} \approx 7.2 \text{ days}$$

39. a) The function $P(t) = P_0 e^{-kt}$, $k > 0$, can be used to model decay. We substitute $\frac{1}{2}P_0$ for $P(t)$ and 5 for t and solve for the decay rate k.

$$\frac{1}{2}P_0 = P_0 e^{-k \cdot 5}$$
$$\frac{1}{2} = e^{-5k}$$
$$\ln \frac{1}{2} = \ln e^{-5k}$$
$$-\ln 2 = -5k$$
$$\frac{\ln 2}{5} = k$$
$$0.139 \approx k$$

The decay rate is 0.139, or 13.9% per hour.

b) 95% consumed = 5% remains

$$0.05P_0 = P_0 e^{-0.139t}$$
$$0.05 = e^{-0.139t}$$
$$\ln 0.05 = \ln e^{-0.139t}$$
$$\ln 0.05 = -0.139t$$
$$\frac{\ln 0.05}{-0.139} = t$$
$$21.6 \approx t$$

How will take approximately 21.6 hr for 95% of the caffeine to leave the body.

41. a) We start with the exponential growth equation

$$V(t) = V_0 e^{kt}, \text{ where } t \text{ is the number of}$$
$$\text{years after 1991.}$$

Substituting 451,000 for V_0, we have

$$V(t) = 451,000 e^{kt}.$$

To find the exponential growth rate k, observe that the card sold for \$2.8 million, or \$2,800,000 in 2007, or 16 years after 1991. We substitute and solve for k.

$$V(16) = 451{,}000e^{k\cdot 16}$$
$$2{,}800{,}000 = 451{,}000e^{16k}$$
$$\frac{2800}{451} = e^{16k}$$
$$\ln\frac{2800}{451} = \ln e^{16k}$$
$$\ln\frac{2800}{451} = 16k$$
$$\frac{\ln\frac{2800}{451}}{16} = k$$
$$0.114 \approx k$$

Thus, the exponential growth function is

$$V(t) = 451{,}000e^{0.114t}, \text{ where } t \text{ is the number}$$
$$\text{of years after 1991.}$$

b) In 2012, $t = 2012 - 1991 = 21$

$$V(21) = 451{,}000e^{0.114(21)} \approx 4{,}900{,}000$$

The card's value in 2012 will be about $4.9 million.

c) Substitute 2($451,000), or $902,000 for $V(t)$ and
solve for t.

$$902{,}000 = 451{,}000e^{0.114t}$$
$$2 = e^{0.114t}$$
$$\ln 2 = \ln e^{0.114t}$$
$$\ln 2 = 0.114t$$
$$\frac{\ln 2}{0.114} = t$$
$$6.1 \approx t$$

The doubling time is about 6.1 years.

d) Substitute $4,000,000 for $V(t)$ and solve for t.

$$4{,}000{,}000 = 451{,}000e^{0.114t}$$
$$\frac{4000}{451} = e^{0.114t}$$
$$\ln\frac{4000}{451} = \ln e^{0.114t}$$
$$\ln\frac{4000}{451} = 0.114t$$
$$\frac{\ln\frac{4000}{451}}{0.114} = t$$
$$19 \approx t$$

The value of the card will first exceed $4,000,000
about 19 years after 1991, or in 2010.

43. *Writing Exercise.* Answers will vary. The problem could
be modeled after Exercises 19-22, 29, 30, 39, or 40.

45. Using the distance formula $d = \sqrt{(x_2 - x_1)^2 + (y_2 - y_1)^2}$
for the points (–3, 7) and (–2, 6),

$$d = \sqrt{[-2-(-3)]^2 + (6-7)^2}$$
$$= \sqrt{1^2 + (-1)^2} = \sqrt{1+1}$$
$$= \sqrt{2}$$

47. Using the midpoint formula $\left(\dfrac{x_1 + x_2}{2},\ \dfrac{y_1 + y_2}{2}\right)$ for the

points (3, –8) and (5, –6),

$$\left(\frac{3+5}{2},\ \frac{-8+(-6)}{2}\right), \text{ or } (4, -7)$$

49.
$$x^2 + 8x = 1$$
$$x^2 + 8x + 16 = 1 + 16$$
$$(x+4)^2 = 17$$
$$x + 4 = \pm\sqrt{17}$$
$$x = -4 \pm \sqrt{17}$$

51. Graph $y = x^2 - 5x - 6$

First we find the vertex.

$$-\frac{b}{2a} = -\frac{-5}{2\cdot 1} = \frac{5}{2}$$

When $x = \dfrac{5}{2},\ y = \left(\dfrac{5}{2}\right)^2 - 5\left(\dfrac{5}{2}\right) - 6 = -12\dfrac{1}{4}$

The vertex is $\left(\dfrac{5}{2}, -12\dfrac{1}{4}\right)$ and the axis of symmetry is

$x = \dfrac{5}{2}$.

x	y
–1	0
0	–6
1	–10
4	–10
5	–6
6	0

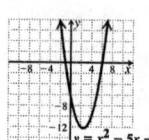

53. *Writing Exercise.* No; the model predicts that the number
of text messages will be 1.4×10^{15} in 2030. This is not
realistic.

55. We will use the exponential growth function $V(t) = V_0 e^{kt}$,
where t is the number of years after 2008 and $V(t)$ is in
millions of dollars. Substitute 20 for $V(t)$, 0.04 for k, and
8 for t and solve for V_0.

$$V(t) = V_0 e^{kt}$$
$$20 = V_0 e^{0.04(8)}$$
$$\frac{20}{e^{0.32}} = V_0$$
$$14.5 \approx V_0$$

About $14.5 million would need to be invested.

57. a) Substitute 1390 for I and solve for m.

$$m(I) = -(19 + 2.5\cdot\log I)$$
$$m = -(19 + 2.5\cdot\log 1390)$$
$$m \approx -26.9$$

The apparent stellar magnitude is about –26.9.

b) Substitute 23 for m and solve for I.

$$m(I) = -(19 + 2.5 \cdot \log I)$$
$$23 = -(19 + 2.5 \cdot \log I)$$
$$-23 = 19 + 2.5 \cdot \log I$$
$$-42 = 2.5 \cdot \log I$$
$$-16.8 = \log I$$
$$10^{-16.8} = I$$
$$1.58 \times 10^{-17} \approx I$$

The intensity is about 1.58×10^{-17} W/m^2.

59. Consider an exponential growth function $P(t) = P_0 e^{kt}$.

Suppose that at time T, $P(T) = 2P_0$.

Solve for T:

$$2P_0 = P_0 e^{kT}$$ **61.**   *Writing Exercise and Graphing*
$$2 = e^{kT}$$
$$\ln 2 = \ln e^{kT}$$
$$\ln 2 = kT$$
$$\frac{\ln 2}{k} = T$$

Calculator Exercise.

Answers may vary.

a) Using a graphing calculator to graph the number of applications data, it appears that an exponential function might be a better fit for this data.

b) Using a graphing calculator to graph the number of approvals data, it appears that a linear function might be a better fit for this data.

c) For the first set of data, we find

$f(x) = 241(1.24)^x$ as the regression model.

For the second set of data, we find

$g(x) = 41x + 266$ as the regression model.

d) To find x when $\frac{1}{2}f(x) = g(x)$, use the graphing calculator to make an approximation. The solution is about 7 yr. In approximately 2001+7, or 2008, there will be only half as many approvals as applications.

Chapter 9 Review

1. True.

3. True.

5. False; log, which has base 10, is not the same as ln, which has base e. In addition, the power rule states

$\log x^a = a \log x$.

7. False; the domain of $f(x) = 3^x$ is all real numbers.

9. True; if $F(-2) = F(-5)$, then the function has the same output for two different inputs, and therefore is not one-to-one.

11. $(f \circ g)(x) = f(g(x)) = f(2x - 3)$
$$= (2x - 3)^2 + 1$$
$$= 4x^2 - 12x + 9 + 1$$
$$= 4x^2 - 12x + 10$$
$(g \circ f)(x) = g(f(x)) = g(x^2 + 1)$
$$= 2(x^2 + 1) - 3$$
$$= 2x^2 + 2 - 3$$
$$= 2x^2 - 1$$

13. $f(x) = 4 - x^2$

The graph of this function is a parabola that opens down. Thus, there are many horizontal lines that cross the graph more than once. In particular, the line $y = -4$ crosses the graph more than once. The function is not one-to-one.

15. $y = \frac{3x + 1}{2}$ Replace $g(x)$.

$x = \frac{3y + 1}{2}$ Interchange variables.

$\frac{2x - 1}{3} = y$ Solve for y.

$g^{-1}(x) = \frac{2x - 1}{3}$ Replace y.

17. Graph $f(x) = 3^x + 1$

We compute some function values, thinking of y as $f(x)$, and keep the results in a table.

$$f(-2) = 3^{-2} + 1 = \frac{1}{9} + 1 = \frac{10}{9}$$
$$f(-1) = 3^{-1} + 1 = \frac{1}{3} + 1 = \frac{4}{3}$$
$$f(0) = 3^0 + 1 = 1$$
$$f(1) = 3^1 + 1 = 4$$
$$f(2) = 3^2 + 1 = 10$$

x	y, or $f(x)$
-2	$\frac{10}{9}$
-1	$\frac{4}{3}$
0	1
1	4
2	10

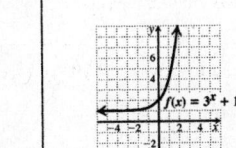

19. Graph: $f(x) = \log_5 x$

Think of $f(x)$ as y. Then $y = \log_5 x$ is equivalent to $5^y = x$. We find ordered pairs by choosing values for y and computing the corresponding x-values. Then we plot the points and connect them with a smooth curve.

For $y = 0$, $x = 5^0 = 1$
For $y = 1$, $x = 5^1 = 5$
For $y = 2$, $x = 5^2 = 25$
For $y = -1$, $x = 5^{-1} = \dfrac{1}{5}$
For $y = -2$, $x = 5^{-2} = \dfrac{1}{25}$

x, or 5^y	y
1	0
5	1
25	2
$\frac{1}{5}$	-1
$\frac{1}{25}$	-2

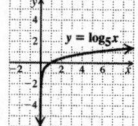

21. $\log_3 \dfrac{1}{9} = \log_3 3^{-2} = -2\log_3 3 = -2$

23. $\log_{16} 4 = \log_{16} 16^{1/2} = \dfrac{1}{2}\log_{16} 16 = \dfrac{1}{2}$

25. $25^{1/2} = 5$ is equivalent to $\dfrac{1}{2} = \log_{25} 5$.

27. $\log_8 1 = 0$ is equivlant to $1 = 8^0$.

29. $\log_a \dfrac{x^5}{yz^2} = \log_a x^5 - \log_a y - \log_a z^2$
$$= 5\log_a x - \log_a y - 2\log_a z$$

31. $\log_a 5 + \log_a 8 = \log_a (5 \cdot 8) = \log_a 40$

33. $\dfrac{1}{2}\log a - \log b - 2\log c = \log a^{1/2} - \log b - \log c^2$
$$= \log \dfrac{a^{1/2}}{bc^2}$$

35. $\log_m m = 1$

37. $\log_m m^{17} = 17\log_m m = 17 \cdot 1 = 17$

39. $\log_a \dfrac{2}{7} = \log_a 2 - \log_a 7$
$$= 1.8301 - 5.0999$$
$$= -3.2698$$

41. $\log_a 3.5 = \log_a \dfrac{7}{2}$
$$= \log_a 7 - \log_a 2$$
$$= 5.0999 - 1.8301$$
$$= 3.2698$$

43. $\log_a \dfrac{1}{4} = \log_a 1 - \log_a 4$
$$= 0 - \log_a 2^2$$
$$= -2\log_a 2$$
$$= -2(1.8301)$$
$$= -3.6602$$

45. 61.5177

47. 0.3753

49. We will use common logarithms for the conversion. Let $a = 10$, $b = 6$, and $M = 5$ and substitute in the change-of-base formula.

$$\log_6 5 = \dfrac{\log 5}{\log 6} \approx \dfrac{0.6890}{0.7782} \approx 0.8982$$

51. Graph $g(x) = 0.6\ln x$.

We find some function values, plot points, and draw the graph.

x	$0.6\ln x$
0.5	-0.42
1	0
2	0.42
3	0.66

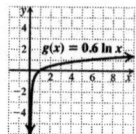

The domain is $(0, \infty)$ and the range is all real numbers.

53. $3^{2x} = \dfrac{1}{9}$
$3^{2x} = 3^{-2}$
$2x = -2$
$x = -1$

The solution is -1.

55. $\log_x 16 = 4$
$x^4 = 16$
$x = \pm\sqrt[4]{16} = \pm 2$

Since x cannot be negative, we use only the positive solution. The solution is 2.

57. $6\ln x = 18$
$\ln x = 3$
$x = e^3 \approx 20.0855$

The solution is e^3 or approximately 20.0855.

59. $2^x = 12$
$$x = \log_2 12 = \dfrac{\log 12}{\log 2} \approx 3.5850$$

61. $2\ln x = -6$
$\ln x = -3$
$x = e^{-3} \approx 0.0498$

63. $\log_4 x - \log_4 (x - 15) = 2$
$$\log_4 \dfrac{x}{x - 15} = 2$$
$$\dfrac{x}{x - 15} = 4^2$$
$$\dfrac{x}{x - 15} = 16$$
$$x = 16(x - 15)$$
$$x = 16x - 240$$
$$240 = 15x$$
$$16 = x$$

65. $S(t) = 82 - 18 \log(t+1)$

a) $S(0) = 82 - 18 \log(0+1) = 82 - 18 \log 1 = 82$

b) $S(6) = 82 - 18 \log(6+1) = 82 - 18 \log 7 \approx 66.8$

c) Substitute 54 for $S(t)$ and solve for t.

$$54 = 82 - 18 \log(t+1)$$
$$-28 = -18 \log(t+1)$$
$$\frac{14}{9} = \log(t+1)$$
$$t+1 = 10^{14/9}$$
$$t = 10^{14/9} - 1 \approx 35$$

The average score will be 54 after about 35 months.

67. a) We start with the exponential growth equation

$$A(t) = A_0 e^{kt}, \text{ where } t \text{ is the number of}$$
$$\text{years after 2005.}$$

Substituting $885 for A_0, we have

$$A(t) = 885 e^{kt}.$$

Substitute 5 for t, $1100 for $A(t)$ and solve for k.

$$A(5) = 885 e^{k(5)}$$
$$1100 = 885 e^{5k}$$
$$\frac{1100}{885} = e^{5k}$$
$$\ln \frac{1100}{885} = \ln e^{5k}$$
$$\ln \frac{1100}{885} = 5k$$
$$\frac{\ln \frac{1100}{885}}{5} = k$$
$$0.043 \approx k$$

Thus, the exponential growth function is

$$A(t) = 885 e^{0.043t}, \text{ where } t \text{ is the number}$$
$$\text{of years after 2005.}$$

b) In 2008, $t = 2008 - 2005 = 3$

$$A(3) = 885 e^{0.043(3)} \approx 1000$$

The amount spent in 2008 will be about $1.0 billion.

c) Substitute $2000 for $A(t)$ and solve for t.

$$2000 = 885 e^{0.043t}$$
$$\frac{2000}{885} = e^{0.043t}$$
$$\ln \frac{2000}{885} = \ln e^{0.043t}$$
$$\ln \frac{2000}{885} = 0.043t$$
$$\frac{\ln \frac{2000}{885}}{0.043} = t$$
$$18 \approx t$$

The $2 billion will be spent on advertising about 18 years after 2005, or in 2023.

d) Substitute 2($885), or $1170 for $A(t)$ and solve for t.

$$1770 = 885 e^{0.043t}$$
$$2 = e^{0.043t}$$
$$\ln 2 = \ln e^{0.043t}$$
$$\ln 2 = 0.043t$$
$$\frac{\ln 2}{0.043} = t$$
$$16.1 \approx t$$

The doubling time is about 16.1 years.

69. The doubling time of the initial investment P_0 would be $2P_0$ and $t = 6$ years. Substitute this information into the exponential growth formula and solve for k.

$$2P_0 = P_0 e^{k(6)}$$
$$2 = e^{6k}$$
$$\ln 2 = \ln e^{6k}$$
$$\ln 2 = 6k$$
$$\frac{\ln 2}{6} = k$$
$$0.11553 \approx k$$

The rate is 11.553% per year.

71. We will use the function $P(t) = P_0 e^{-0.00012t}$

If the skull had lost 34% of its carbon-14 from the initial amount P_0, then $66\%(P_0)$ is the amount present. To find the age t of the skull, we substitute $66\%(P_0)$, or $0.66P_0$ for $P(t)$ in the function above and solve for t.

$$0.66P_0 = P_0 e^{-0.00012t}$$
$$0.66 = e^{-0.00012t}$$
$$\ln 0.66 = \ln e^{-0.00012t}$$
$$\ln 0.66 = -0.00012t$$
$$\frac{\ln 0.66}{-0.00012} = t$$
$$3463 \approx t$$

The skull is about 3463 yr old.

73. $L = 10 \cdot \log \frac{I}{I_0}$

$$= 10 \cdot \log \frac{2.5 \times 10^{-1}}{10^{-12}}$$
$$= 10 \log (2.5 \times 10^{11})$$
$$\approx 114$$

The sound is about 114 dB.

75. *Writing Exercise.* If $f(x) = e^x$, then to find the inverse function, we let $y = e^x$ and interchange x and y: $x = e^y$. If $x = e^y$, then $\log_e x = y$ by the definition of logarithms. Since $\log_e x = \ln x$, we have $y = \ln x$ or $f^{-1}(x) = \ln x$. Thus, $g(x) = \ln x$ is the inverse of $f(x) = e^x$. Another approach is to find $(f \circ g)(x)$ and $(g \circ f)(x)$:

$$(f \circ g)(x) = e^{\ln x} = x, \text{ and}$$
$$(g \circ f)(x) = \ln e^x = x.$$

Thus, g and f are inverse functions.

77. $2^{x^2+4x} = \dfrac{1}{8}$ can be written as $2^{x^2+4x} = 2^{-3}$.

So the exponents must be equal.

$$x^2 + 4x = -3$$
$$x^2 + 4x + 3 = 0$$
$$(x+3)(x+1) = 0$$
$$x + 3 = 0 \quad or \quad x + 1 = 0$$
$$x = -3 \quad or \quad x = -1$$

The solutions are -3 and -1.

Chapter 9 Test

1. $(f \circ g)(x) = f(g(x)) = f(2x+1)$
$$= (2x+1) + (2x+1)^2$$
$$= 2x + 1 + 4x^2 + 4x + 1$$
$$= 4x^2 + 6x + 2$$
$(g \circ f)(x) = g(f(x)) = g(x + x^2)$
$$= 2(x + x^2) + 1$$
$$= 2x + 2x^2 + 1$$
$$= 2x^2 + 2x + 1$$

3. $f(x) = x^2 + 3$

Observe that the graph of this function is a parabola that opens up. Thus, there are many horizontal lines that cross the graph more than once. In particular, the line $y = 4$ crosses the graph more than once. The function is not one-to-one.

5.
$$y = (x+1)^3 \qquad \text{Replace } g(x).$$
$$x = (y+1)^3 \qquad \text{Interchange variables.}$$
$$\sqrt[3]{x} - 1 = y \qquad \text{Solve for } y.$$
$$g^{-1}(x) = \sqrt[3]{x} - 1 \qquad \text{Replace } y.$$

7. Graph: $f(x) = \log_7 x$

Think of $f(x)$ as y. Then $y = \log_7 x$ is equivalent to $7^y = x$. We find ordered pairs by choosing values for y and computing the corresponding x-values. Then we plot the points and connect them with a smooth curve.

For $y = 0$, $x = 7^0 = 1$	x, or 7^y	y
For $y = 1$, $x = 7^1 = 7$	1	0
For $y = 2$, $x = 7^2 = 49$	7	1
For $y = -1$, $x = 7^{-1} = \dfrac{1}{7}$	49	2
For $y = -2$, $x = 7^{-2} = \dfrac{1}{49}$	$\dfrac{1}{7}$	-1
	$\dfrac{1}{49}$	-2

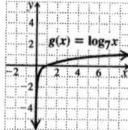

9. $\log_{100} 10 = \log_{100} 100^{1/2} = \dfrac{1}{2}\log_{100} 100 = \dfrac{1}{2}\cdot 1 = \dfrac{1}{2}$

11. $\log_n n = 1$

13. $\log_a a^{19} = 19\log_a a = 19 \cdot 1 = 19$

15. $m = \log_2 \dfrac{1}{2}$ is equivalent to $2^m = \dfrac{1}{2}$.

17. $\dfrac{1}{3}\log_a x + 2\log_a z = \log_a x^{1/3} + \log_a z^2$
$$= \log_a \sqrt[3]{x} + \log_a z^2$$
$$= \log_a \left(z^2 \sqrt[3]{x}\right)$$

19. $\log_a 3 = \log_a \dfrac{6}{2}$
$$= \log_a 6 - \log_a 2$$
$$= 0.778 - 0.301$$
$$= 0.477$$

21. 1.3979

23. -0.9163

25. We will use common logarithms for the conversion.

Let $a = 10$, $b = 3$, and $M = 14$ and substitute in the change-of-base formula.

$$\log_3 14 = \dfrac{\log 14}{\log 3} \approx \dfrac{1.1461}{0.4771} \approx 2.4022$$

27. Graph $g(x) = \ln(x-4)$.

We find some function values, plot points, and draw the graph.

x	$\ln(x-4)$
4.5	-0.69
5	0
6	0.69
7	1.10

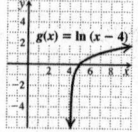

The domain is $(4, \infty)$ and the range is all real numbers.

29. $\log_4 x = \dfrac{1}{2}$
$$x = 4^{1/2}$$
$$x = 2$$

31. $5^{4-3x} = 87$

$4 - 3x = \log_5 87$

$-3x = \log_5 87 - 4$

$x = -\dfrac{1}{3}\left(\dfrac{\log 87}{\log 5} - 4\right) \approx 0.4084$

33. $\ln x = 3$

$x = e^3 \approx 20.0855$

35. $R = 0.37 \ln P + 0.05$

a) Substitute 383 for P and solve for R.

$R = 0.37 \ln 383 + 0.05 = 2.25$

The average walking speed is about 2.25 ft/sec.

b) Substitute 3 for R and solve for P.

$3 = 0.37 \ln P + 0.05$

$2.95 = 0.37 \ln P$

$\dfrac{2.95}{0.37} = \ln P$

$P = e^{2.95/0.37} \approx 2{,}901$

The population is approximately 2,901,000.

37. a) We start with the exponential growth equation

$C(t) = C_0 e^{kt}$, where t is the number of years after 2001.

Substituting \$21,855 for C_0, we have

$C(t) = 21{,}855 e^{kt}$.

Substitute 5 for t, \$27,317 for $C(t)$ and solve for k.

$27{,}317 = 21{,}855 e^{k(5)}$

$\dfrac{27{,}317}{21{,}855} = e^{5k}$

$\ln \dfrac{27{,}317}{21{,}855} = \ln e^{5k}$

$\ln \dfrac{27{,}317}{21{,}855} = 5k$

$\dfrac{\ln \dfrac{27{,}317}{21{,}855}}{5} = k$

$0.045 \approx k$

Thus, the exponential growth function is

$C(t) = 21{,}855 e^{0.045t}$, where t is the number of years after 2001.

b) In 2012, $t = 2012 - 2001 = 11$

$C(11) = 21{,}855 e^{0.045(11)} \approx 35{,}853$

The cost in 2012 will be about \$35,853.

c) Substitute \$50,000 for $C(t)$ and solve for t.

$50{,}000 = 21{,}855 e^{0.045t}$

$\dfrac{50{,}000}{21{,}855} = e^{0.045t}$

$\ln \dfrac{50{,}000}{21{,}855} = \ln e^{0.045t}$

$\ln \dfrac{50{,}000}{21{,}855} = 0.045t$

$\dfrac{\ln \dfrac{50{,}000}{21{,}855}}{0.045} = t$

$18 \approx t$

Cost for college will be \$50,000 about 18 years after 2001, or in 2019.

39. We will use the function $P(t) = P_0 e^{-0.00012t}$

If the bone had lost 43% of its carbon-14 from the initial amount P_0, then $57\%(P_0)$ is the amount present. To find the age t of the bone, we substitute $57\%(P_0)$, or $0.57 P_0$ for $P(t)$ in the function above and solve for t.

$0.57 P_0 = P_0 e^{-0.00012t}$

$0.57 = e^{-0.00012t}$

$\ln 0.57 = \ln e^{-0.00012t}$

$\ln 0.57 = -0.00012t$

$\dfrac{\ln 0.57}{-0.00012} = t$

$4684 \approx t$

The bone is about 4684 yr old.

41. $\text{pH} = -\log[H^+]$

$= -\log(1.0 \times 10^{-7})$

$= 7.0$

The pH of water is 7.0.

43. $\log_a \dfrac{\sqrt[3]{x^2 z}}{\sqrt[3]{y^2 z^{-1}}}$

$= \log_a \sqrt[3]{\dfrac{x^2 z}{y^2 z^{-1}}}$

$= \log_a \sqrt[3]{\dfrac{x^2 z^2}{y^2}}$

$= \log_a \left(\dfrac{x^2 z^2}{y^2}\right)^{1/3}$

$= \dfrac{1}{3} \log_a \dfrac{x^2 z^2}{y^2}$

$= \dfrac{1}{3}\left(\log_a x^2 + \log_a z^2 - \log_a y^2\right)$

$= \dfrac{1}{3}\left(2\log_a x + 2\log_a z - 2\log_a y\right)$

$= \dfrac{1}{3}(2 \cdot 2 + 2 \cdot 4 - 2 \cdot 3)$

$= 2$

Chapter 10

Conic Sections

Exercise Set 10.1

1. $(x-2)^2+(y+5)^2=9$, or $(x-2)^2+[y-(-5)]^2=3^2$, is the equation of a circle with center $(2,-5)$ and radius 3, so choice (f) is correct.

3. $(x-5)^2+(y+2)^2=9$, or $(x-5)^2+[y-(-2)]^2=3^2$, is the equation of a circle with center $(5,-2)$ and radius 3, so choice (g) is correct.

5. $y=(x-2)^2-5$ is the equation of a parabola with vertex $(2,-5)$ that opens upward, so choice (c) is correct.

7. $x=(y-2)^2-5$ is the equation of a parabola with vertex $(-5,\ 2)$ that opens to the right, so choice (d) is correct.

9. $y=-x^2$

This is equivalent to $y=-(x-0)^2+0$. The vertex is $(0,\ 0)$.

We choose some x-values on both sides of the vertex and compute the corresponding values of y. The graph opens down, because the coefficient of x^2, -1, is negative.

x	y
0	0
1	-1
2	-4
-1	-1
-2	-4

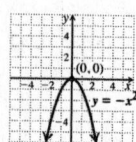

11. $y=-x^2+4x-5$

We can find the vertex by computing the first coordinate, $x=-b/2a$, and then substituting to find the second coordinate:

$$x=-\frac{b}{2a}=-\frac{4}{2(-1)}=2$$
$$y=-x^2+4x-5=-(2)^2+4(2)-5=-1$$

The vertex is $(2,-1)$.

We choose some x-values and compute the corresponding values for y. The graph opens downward because the

coefficient of x^2, -1, is negative.

x	y
2	-1
3	-2
4	-5
1	-2
0	-5

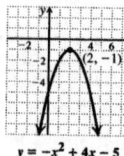

13. $x=y^2-4y+2$

We find the vertex by completing the square.

$$x=(y^2-4y+4)+2-4$$
$$x=(y-2)^2-2$$

The vertex is $(-2,2)$.

To find ordered pairs, we choose values for y and compute the corresponding values of x. The graph opens to the right, because the coefficient of y^2, 1, is positive.

x	y
7	-1
2	0
-1	1
-2	2
-1	3

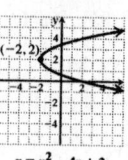

15. $x=y^2+3$

$$x=(y-0)^2+3$$

The vertex is $(3,0)$.

To find the ordered pairs, we choose y-values and compute the corresponding values for x. The graph opens to the right, because the coefficient of y^2, 1, is positive.

x	y
3	0
4	1
7	2
4	-1
7	-2

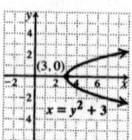

17. $x=2y^2$

$$x=2(y-0)^2+0$$

The vertex is $(0,\ 0)$.

We choose y-values and compute the corresponding values for x. The graph opens to the right, because the coefficient of y^2, 2, is positive.

x	y
0	0
2	1
2	-1
8	2
8	-2

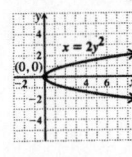

19. $x = -y^2 - 4y$

We find the vertex by computing the second coordinate, $y = -b/2a$, and then substituting to find the first coordinate:

$$y = -\frac{b}{2a} = -\frac{-4}{2(-1)} = -2$$
$$x = -y^2 - 4y = -(-2)^2 - 4(-2) = 4$$

The vertex is $(4, -2)$.

We choose y-values and compute the corresponding values for x. The graph opens to the left, because the coefficient of y^2, -1, is negative.

x	y
4	-2
-5	1
0	0
3	-1
3	-3

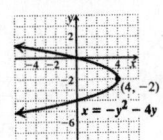

21. $y = x^2 - 2x + 1$

$y = (x-1)^2 + 0$

The vertex is $(1, 0)$.

We choose x-values and compute the corresponding values for y. The graph opens upward, because the coefficient of x^2, 1, is positive.

x	y
1	0
0	1
-1	4
2	1
3	4

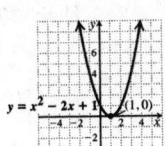

23. $x = -\frac{1}{2}y^2$

$x = -\frac{1}{2}(y-0)^2 + 0$

The vertex is $(0, 0)$.

We choose y-values and compute the corresponding values for x. The graph opens to the left, because the coefficient of y^2, $-\frac{1}{2}$, is negative.

x	y
0	0
-2	2
-8	4
-2	-2
-8	-4

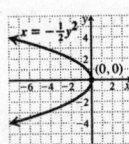

25. $x = -y^2 + 2y - 1$

We find the vertex by computing the second coordinate, $y = -b/2a$, and then substituting to find the first coordinate.

$$y = -\frac{b}{2a} = -\frac{2}{2(-1)} = 1$$
$$x = -y^2 + 2y - 1 = -(1)^2 + 2(1) - 1 = 0$$

The vertex is $(0, 1)$.

We choose y-values and compute the corresponding values for x. The graph opens to the left, because the coefficient of y^2, -1, is negative.

x	y
-4	3
-1	2
-1	0
-4	-1
-4	3

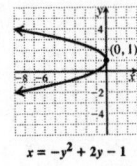

27. $x = -2y^2 - 4y + 1$

We find the vertex by completing the square.

$$x = -2(y^2 + 2y) + 1$$
$$x = -2(y^2 + 2y + 1) + 1 + 2$$
$$x = -2(y+1)^2 + 3$$

The vertex is $(3, -1)$.

We choose y-values and compute the corresponding values for x. The graph opens to the left, because the coefficient of y^2, -2, is negative.

x	y
3	-1
1	-2
-5	-3
1	0
-5	1

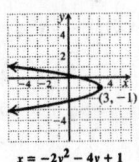

29. $(x-h)^2 + (y-k)^2 = r^2$ Standard form
$(x-0)^2 + (y-0)^2 = 8^2$ Substituting
$x^2 + y^2 = 64$ Simplifying

31. $(x-h)^2 + (y-k)^2 = r^2$ Standard form
$(x-7)^2 + (y-3)^2 = (\sqrt{6})^2$ Substituting
$(x-7)^2 + (y-3)^2 = 6$

33. $(x-h)^2 + (y-k)^2 = r^2$
$[x-(-4)]^2 + (y-3)^2 = (3\sqrt{2})^2$
$(x+4)^2 + (y-3)^2 = 18$

35. $(x-h)^2 + (y-k)^2 = r^2$
$[x-(-5)]^2 + [y-(-8)]^2 = (10\sqrt{3})^2$
$(x+5)^2 + (y+8)^2 = 300$

37. Since the center is $(0, 0)$, we have
$$(x-0)^2 + (y-0)^2 = r^2 \text{ or } x^2 + y^2 = r^2$$
The circle passes through $(-3, 4)$. We find r^2 by substituting -3 for x and 4 for y.
$$(-3)^2 + 4^2 = r^2$$
$$9 + 16 = r^2$$
$$25 = r^2$$
Then $x^2 + y^2 = 25$ is an equation of the circle.

39. Since the center is $(-4, 1)$, we have
$$[x-(-4)]^2 + (y-1)^2 = r^2, \text{ or}$$
$$(x+4)^2 + (y-1)^2 = r^2.$$
The circle passes through $(-2, 5)$. We find r^2 by substituting -2 for x and 5 for y.
$$(-2+4)^2 + (5-1)^2 = r^2$$
$$4 + 16 = r^2$$
$$20 = r^2$$
Then $(x+4)^2 + (y-1)^2 = 20$ is an equation of the circle.

41. We write standard form.
$$(x-0)^2 + (y-0)^2 = 1^2$$
The center is $(0, 0)$, and the radius is 1.

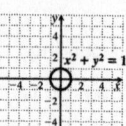

43.
$$(x+1)^2 + (y+3)^2 = 49$$
$$[x-(-1)]^2 + [y-(-3)]^2 = 7^2 \quad \text{Standard form}$$
The center is $(-1, -3)$, and the radius is 7.

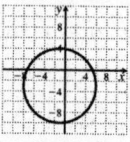

45.
$$(x-4)^2 + (y+3)^2 = 10$$
$$(x-4)^2 + [y-(-3)]^2 = \left(\sqrt{10}\right)^2$$
The center is $(4, -3)$, and the radius is $\sqrt{10}$.

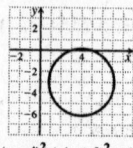

47.
$$x^2 + y^2 = 8$$
$$(x-0)^2 + (y-0)^2 = \left(\sqrt{8}\right)^2 \quad \text{Standard form}$$
The center is $(0, 0)$, and the radius is $\sqrt{8}$, or $2\sqrt{2}$.

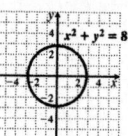

49.
$$(x-5)^2 + y^2 = \frac{1}{4}$$
$$(x-5)^2 + (y-0)^2 = \left(\frac{1}{2}\right)^2 \quad \text{Standard form}$$
The center is $(5, 0)$, and the radius is $\frac{1}{2}$.

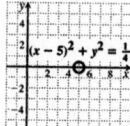

51.
$$x^2 + y^2 + 8x - 6y - 15 = 0$$
$$x^2 + 8x + y^2 - 6y = 15$$
$$(x^2 + 8x + 16) + (y^2 - 6y + 9) = 15 + 16 + 9 \quad \text{Completing the square twice}$$
$$(x+4)^2 + (y-3)^2 = 40$$
$$[x-(-4)]^2 + (y-3)^2 = \left(\sqrt{40}\right)^2$$
$$\text{Standard form}$$
The center is $(-4, 3)$, and the radius is $\sqrt{40}$, or $2\sqrt{10}$.

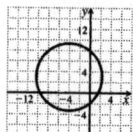

$$x^2 + y^2 + 8x - 6y - 15 = 0$$

53.
$$x^2 + y^2 - 8x + 2y + 13 = 0$$
$$x^2 - 8x + y^2 + 2y = -13$$
$$(x^2 - 8x + 16) + (y^2 + 2y + 1) = -13 + 16 + 1 \quad \text{Completing the square twice}$$
$$(x-4)^2 + (y+1)^2 = 4$$
$$(x-4)^2 + [y-(-1)]^2 = 2^2$$
$$\text{Standard form}$$
The center is $(4, -1)$, and the radius is 2.

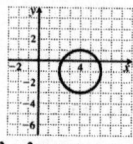

$$x^2 + y^2 - 8x + 2y + 13 = 0$$

55.
$$x^2 + y^2 + 10y - 75 = 0$$
$$x^2 + y^2 + 10y = 75$$
$$x^2 + (y^2 + 10y + 25) = 75 + 25$$
$$(x - 0)^2 + (y + 5)^2 = 100$$
$$(x - 0)^2 + [y - (-5)]^2 = 10^2$$

The center is $(0, -5)$, and the radius is 10.

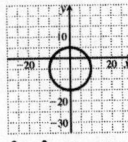

$x^2 + y^2 + 10y - 75 = 0$

57.
$$x^2 + y^2 + 7x - 3y - 10 = 0$$
$$x^2 + 7x + y^2 - 3y = 10$$
$$\left(x^2 + 7x + \frac{49}{4}\right) + \left(y^2 - 3y + \frac{9}{4}\right) = 10 + \frac{49}{4} + \frac{9}{4}$$
$$\left(x + \frac{7}{2}\right)^2 + \left(y - \frac{3}{2}\right)^2 = \frac{98}{4}$$
$$\left[x - \left(-\frac{7}{2}\right)\right]^2 + \left(y - \frac{3}{2}\right)^2 = \left(\sqrt{\frac{98}{4}}\right)^2$$

The center is $\left(-\frac{7}{2}, \frac{3}{2}\right)$, and the radius is $\sqrt{\frac{98}{4}}$, or $\frac{\sqrt{98}}{2}$,

or $\frac{7\sqrt{2}}{2}$.

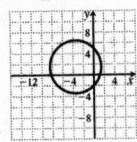

$x^2 + y^2 + 7x - 3y - 10 = 0$

59.
$$36x^2 + 36y^2 = 1$$
$$x^2 + y^2 = \frac{1}{36} \quad \text{Multiplying by } \frac{1}{36}$$
$$\text{on both sides}$$
$$(x - 0)^2 + (y - 0)^2 = \left(\frac{1}{6}\right)^2$$

The center is $(0, 0)$, and the radius is $\frac{1}{6}$.

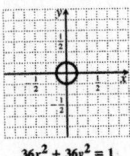

$36x^2 + 36y^2 = 1$

61. *Writing Exercise.* No; a circle is defined to be the set of points in a plane that are a fixed distance from the center. Thus, unless $r = 0$ and the circle is one point, the center is not part of the circle.

63.
$$\frac{y^2}{16} = 1$$
$$y^2 = 16$$
$$y = 4 \text{ or } y = -4 \quad \text{Using the principle of square roots}$$
The solutions are ± 4.

65.
$$\frac{(x - 1)^2}{25} = 1$$
$$(x - 1)^2 = 25$$
$$x - 1 = 5 \quad or \quad x - 1 = -5$$
$$x = 6 \quad or \qquad x = -4$$

The solutions are -4 and 6.

67.
$$\frac{1}{4} + \frac{(y + 3)^2}{36} = 1$$
$$\frac{(y + 3)^2}{36} = \frac{3}{4}$$
$$(y + 3)^2 = 27$$
$$y + 3 = \sqrt{27} \qquad or \quad y + 3 = -\sqrt{27}$$
$$y = -3 + 3\sqrt{3} \quad or \qquad y = -3 - 3\sqrt{3}$$

The solutions are $-3 \pm 3\sqrt{3}$.

69. *Writing Exercise.* The points appear to form a parabola. A circle is formed from a fixed distance, the radius, from a point. The result is a circle. This set of points was formed from a fixed distance from a point *and* a line. The result is a U-shaped figure.

71. We make a drawing of the circle with center $(3, -5)$ and tangent to the y-axis.

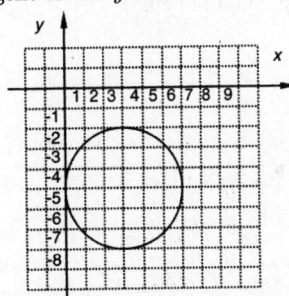

We see that the circle touches the y-axis at $(0, -5)$. Hence the radius is the distance between $(0, -5)$ and $(3, -5)$, or $\sqrt{(3 - 0)^2 + [-5 - (-5)]^2}$, or 3. Now we write the equation of the circle.
$$(x - h)^2 + (y - k)^2 = r^2$$
$$(x - 3)^2 + [y - (-5)]^2 = 3^2$$
$$(x - 3)^2 + (y + 5)^2 = 9$$

73. First we use the midpoint formula to find the center:
$$\left(\frac{7 + (-1)}{2}, \frac{3 + (-3)}{2}\right), \text{ or } \left(\frac{6}{2}, \frac{0}{2}\right), \text{ or } (3, 0)$$
The length of the radius is the distance between the center $(3, 0)$ and either endpoint of a diameter. We will use endpoint $(7, 3)$ in the distance formula:
$$r = \sqrt{(7 - 3)^2 + (3 - 0)^2} = \sqrt{25} = 5$$

Now we write the equation of the circle:

$$(x-h)^2 + (y-k)^2 = r^2$$
$$(x-3)^2 + (y-0)^2 = 5^2$$
$$(x-3)^2 + y^2 = 25$$

75. Let $(0, y)$ be the point on the y-axis that is equidistant from $(2, 10)$ and $(6, 2)$. Then the distance between $(2, 10)$ and $(0, y)$ is the same as the distance between $(6, 2)$ and $(0, y)$.

$$\sqrt{(0-2)^2 + (y-10)^2} = \sqrt{(0-6)^2 + (y-2)^2}$$
$$(-2)^2 + (y-10)^2 = (-6)^2 + (y-2)^2 \quad \text{Squaring}$$
$$\text{both sides}$$
$$4 + y^2 - 20y + 100 = 36 + y^2 - 4y + 4$$
$$64 = 16y$$
$$4 = y$$

This number checks. The point is $(0, 4)$.

77. For the outer circle, $r^2 = \dfrac{81}{4}$. For the inner circle, $r^2 = 16$. The area of the red zone is the difference between the areas of the outer and inner circles. Recall that the area A of a circle with radius r is given by the formula $A = \pi r^2$.

$$\pi \cdot \frac{81}{4} - \pi \cdot 16 = \frac{81}{4}\pi - \frac{64}{4}\pi = \frac{17}{4}\pi$$

The area of the red zone is $\dfrac{17}{4}\pi$ m^2, or about 13.4 m^2.

79. Superimposing a coordinate system on the snowboard as in Exercise 78, and observing that $1160/2 = 580$, we know that three points on the circle are $(-580, 0)$, $(0, 23.5)$ and $(580, 0)$. Let $(0, k)$ represent the center of the circle. Use the fact that $(0, k)$ is equidistant from $(-580, 0)$ and $(0, 23.5)$.

$$\sqrt{(-580-0)^2 + (0-k)^2} = \sqrt{(0-0)^2 + (23.5-k)^2}$$
$$\sqrt{336,400 + k^2} = \sqrt{552.25 - 47k + k^2}$$
$$336,400 + k^2 = 552.25 - 47k + k^2$$
$$335,847.75 = -47k$$
$$-7145.7 \approx k$$

Then to find the radius, we find the distance from the center $(0, -7145.7)$ to any one of the three known points on the circle. We use $(0, 23.5)$.

$$r = \sqrt{(0-0)^2 + (-7145.7 - 23.5)^2} \approx 7169 \text{ mm}$$

81. a) When the circle is positioned on a coordinate system as shown in the text, the center lies on the y-axis. To find the center, we will find the point on the y-axis that is equidistant from $(-4, 0)$ and $(0, 2)$. Let $(0, y)$ be this point.

$$\sqrt{[0-(-4)]^2 + (y-0)^2} = \sqrt{(0-0)^2 + (y-2)^2}$$
$$4^2 + y^2 = 0^2 + (y-2)^2$$
$$\text{Squaring both sides}$$
$$16 + y^2 = y^2 - 4y + 4$$
$$12 = -4y$$
$$-3 = y$$

The center of the circle is $(0, -3)$.

b) We find the radius of the circle.

$$(x-0)^2 + [y-(-3)]^2 = r^2 \quad \text{Standard form}$$
$$x^2 + (y+3)^2 = r^2$$
$$(-4)^2 + (0+3)^2 = r^2 \quad \text{Substituting}$$
$$16 + 9 = r^2 \quad (-4, 0) \text{ for } (x, y)$$
$$25 = r^2$$
$$5 = r$$

The radius is 5 ft.

83. We write the equation of a circle with center $(0, 30.6)$ and radius 24.3:

$$x^2 + (y-30.6)^2 = 590.49$$

85. Substitute 6 for N.

$$H = \frac{D^2 N}{2.5} = \frac{D^2 \cdot 6}{2.5} = 2.4D^2$$

Find some ordered pairs for $2.5 \le D \le 8$ and draw the graph.

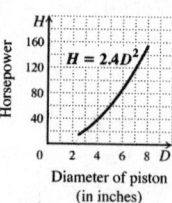

Using the graph, a horse power of 120, on the vertical axis, relates to a diameter of 7 in., on the horizontal axis.

87. *Writing Exercise.* One method is to enter $y_1 = ax^2 + bx + c$, deselect it, and then use the Draw Inverse operation to graph the inverse of y_1. This inverse is the graph of $x_1 = ay^2 + by + c$.

Exercise Set 10.2

1. True; see page 660 in the text.

3. False; see page 660 in the text.

5. True; see page 660 in the text.

7. True; see page 662 in the text.

9. $\dfrac{x^2}{1} + \dfrac{y^2}{4} = 1$

$\dfrac{x^2}{1^2} + \dfrac{y^2}{2^2} = 1$

The x-intercepts are $(1, 0)$ and $(-1, 0)$, and the y-intercepts are $(0, 2)$ and $(0, -2)$. We plot these points and connect them with an oval-shaped curve.

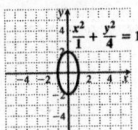

11. $\dfrac{x^2}{25} + \dfrac{y^2}{9} = 1$

$\dfrac{x^2}{5^2} + \dfrac{y^2}{3^2} = 1$

The x-intercepts are $(5, 0)$ and $(-5, 0)$, and the y-intercepts are $(0, 3)$ and $(0, -3)$. We plot these points and connect them with an oval-shaped curve.

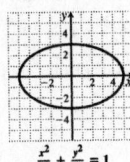

13. $4x^2 + 9y^2 = 36$

$\dfrac{1}{36}(4x^2 + 9y^2) = \dfrac{1}{36}(36)$ Multiplying by $\dfrac{1}{36}$

$\dfrac{x^2}{9} + \dfrac{y^2}{4} = 1$

$\dfrac{x^2}{3^2} + \dfrac{y^2}{2^2} = 1$

The x-intercepts are $(-3, 0)$ and $(3, 0)$, and the y-intercepts are $(0, -2)$ and $(0, 2)$. We plot these points and connect them with an oval-shaped curve.

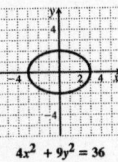

15. $16x^2 + 9y^2 = 144$

$\dfrac{x^2}{9} + \dfrac{y^2}{16} = 1$ Multiplying by $\dfrac{1}{144}$

$\dfrac{x^2}{3^2} + \dfrac{y^2}{4^2} = 1$

The x-intercepts are $(3, 0)$ and $(-3, 0)$, and the y-intercepts are $(0, 4)$ and $(0, -4)$. We plot these points and connect them with an oval-shaped curve.

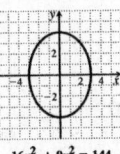

17. $2x^2 + 3y^2 = 6$

$\dfrac{x^2}{3} + \dfrac{y^2}{2} = 1$ Multiplying by $\dfrac{1}{6}$

$\dfrac{x^2}{\left(\sqrt{3}\right)^2} + \dfrac{y^2}{\left(\sqrt{2}\right)^2} = 1$

The x-intercepts are $\left(\sqrt{3},\, 0\right)$ and $\left(-\sqrt{3},\, 0\right)$, and the y-intercepts are $\left(0,\, \sqrt{2}\right)$ and $\left(0, -\sqrt{2}\right)$. We plot these points and connect them with an oval-shaped curve.

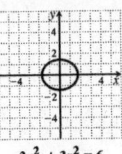

19. $5x^2 + 5y^2 = 125$

Observe that the x^2- and y^2-terms have the same coefficient. We divide both sides of the equation by 5 to obtain $x^2 + y^2 = 25$. This is the equation of a circle with center $(0, 0)$ and radius 5.

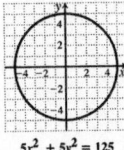

21. $3x^2 + 7y^2 - 63 = 0$

$3x^2 + 7y^2 = 63$

$\dfrac{x^2}{21} + \dfrac{y^2}{9} = 1$ Multiplying by $\dfrac{1}{63}$

$\dfrac{x^2}{\left(\sqrt{21}\right)^2} + \dfrac{y^2}{3^2} = 1$

The x-intercepts are $\left(\sqrt{21},\, 0\right)$ and $\left(-\sqrt{21},\, 0\right)$, or about $(4.583, 0)$ and $(-4.583, 0)$. The y-intercepts are $(0, 3)$ and $(0, -3)$. We plot these points and connect them with an oval-shaped curve.

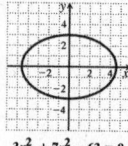

$3x^2 + 7y^2 - 63 = 0$

23. $16x^2 = 16 - y^2$

$16x^2 + y^2 = 16$

$\dfrac{x^2}{1} + \dfrac{y^2}{16} = 1$

The x-intercepts are $(1, 0)$ and $(-1, 0)$, and the y-intercepts are $(0, 4)$ and $(0, -4)$. We plot these points and connect them with an oval-shaped curve.

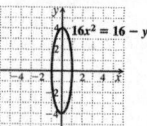

25. $16x^2 + 25y^2 = 1$

Note that $16 = \dfrac{1}{\frac{1}{16}}$ and $25 = \dfrac{1}{\frac{1}{25}}$. Thus, we can rewrite the equation:

$$\dfrac{x^2}{\frac{1}{16}} + \dfrac{y^2}{\frac{1}{25}} = 1$$

$$\dfrac{x^2}{\left(\frac{1}{4}\right)^2} + \dfrac{y^2}{\left(\frac{1}{5}\right)^2} = 1$$

The x-intercepts are $\left(\frac{1}{4}, 0\right)$ and $\left(-\frac{1}{4}, 0\right)$, and the y-intercepts are $\left(0, \frac{1}{5}\right)$ and $\left(0, -\frac{1}{5}\right)$. We plot these points and connect them with an oval-shaped curve.

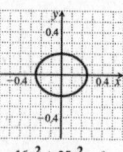

$16x^2 + 25y^2 = 1$

27. $\dfrac{(x-3)^2}{9} + \dfrac{(y-2)^2}{25} = 1$

$\dfrac{(x-3)^2}{3^2} + \dfrac{(y-2)^2}{5^2} = 1$

The center of the ellipse is $(3, 2)$. Note that $a = 3$ and $b = 5$. We locate the center and then plot the points $(3+3, 2)$ $(3-3, 2)$, $(3, 2+5)$, and $(3, 2-5)$, or $(6, 2)$, $(0, 2)$, $(3, 7)$, and $(3, -3)$. Connect these points with an

oval-shaped curve.

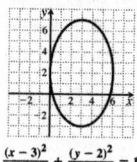

$\dfrac{(x-3)^2}{9} + \dfrac{(y-2)^2}{25} = 1$

29. $\dfrac{(x+4)^2}{16} + \dfrac{(y-3)^2}{49} = 1$

$\dfrac{(x-(-4))^2}{4^2} + \dfrac{(y-3)^2}{7^2} = 1$

The center of the ellipse is $(-4, 3)$. Note that $a = 4$ and $b = 7$. We locate the center and then plot the points $(-4+4, 3)$, $(-4-4, 3)$, $(-4, 3+7)$, and $(-4, 3-7)$, or $(0, 3)$, $(-8, 3)$, $(-4, 10)$, and $(-4, -4)$. Connect these points with an oval-shaped curve.

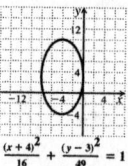

$\dfrac{(x+4)^2}{16} + \dfrac{(y-3)^2}{49} = 1$

31. $12(x-1)^2 + 3(y+4)^2 = 48$

$\dfrac{(x-1)^2}{4} + \dfrac{(y+4)^2}{16} = 1$

$\dfrac{(x-1)^2}{2^2} + \dfrac{(y-(-4))^2}{4^2} = 1$

The center of the ellipse is $(1, -4)$. Note that $a = 2$ and $b = 4$. We locate the center and then plot the points $(1+2, -4)$, $(1-2, -4)$, $(1, -4+4)$, and $(1, -4-4)$, or $(3, -4)$, $(-1, -4)$, $(1, 0)$, and $(1, -8)$. Connect these points with an oval-shaped curve.

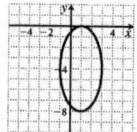

$12(x-1)^2 + 3(y+4)^2 = 48$

33. $4(x+3)^2 + 4(y+1)^2 - 10 = 90$

$4(x+3)^2 + 4(y+1)^2 = 100$

Observe that the x^2- and y^2-terms have the same coefficient. Dividing both sides by 4, we have

$(x+3)^2 + (y+1)^2 = 25$.

This is the equation of a circle with center $(-3, -1)$ and radius 5.

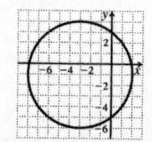

$4(x+3)^2 + 4(y+1)^2 - 10 = 90$

35. *Writing Exercise.* Write the equation of an ellipse in standard form, $\frac{x^2}{a^2} + \frac{y^2}{b^2} = 1$. If $a^2 > b^2$, then the ellipse is horizontal. If $b^2 > a^2$, then the ellipse is vertical.

37. $x^2 - 5x + 3 = 0$

$x = \dfrac{5 \pm \sqrt{25 - 4 \cdot 1 \cdot 3}}{2 \cdot 1}$ Using the quadratic formula

$= \dfrac{5 \pm \sqrt{13}}{2}$

39. $\dfrac{4}{x+2} + \dfrac{3}{2x-1} = 2$

$4(2x-1) + 3(x+2) = 2(x+2)(2x-1)$

$8x - 4 + 3x + 6 = 4x^2 + 6x - 4$

$4x^2 - 5x - 6 = 0$

$(4x + 3)(x - 2) = 0$

$4x + 3 = 0 \quad or \quad x - 2 = 0$

$x = -\frac{3}{4} \quad or \qquad x = 2$

41. $x^2 = 11$

$x = \pm\sqrt{11}$

43. *Writing Exercise.* Since the coefficients of the squared terms $9x^2$ and y^2 have the same sign, then the equation is an ellipse.

45. Plot the given points.

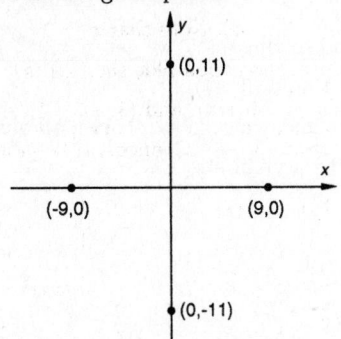

From the location of these points, we see that the ellipse that contains them is centered at the origin with $a = 9$ and $b = 11$. We write the equation of the ellipse:

$\dfrac{x^2}{9^2} + \dfrac{y^2}{11^2} = 1$

$\dfrac{x^2}{81} + \dfrac{y^2}{121} = 1$

47. Plot the given points.

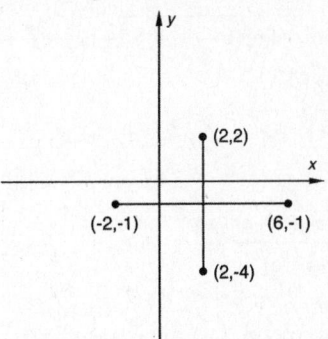

The midpoint of the segment from $(-2, -1)$ to $(6, -1)$ is $\left(\dfrac{-2+6}{2}, \dfrac{-1-1}{2}\right)$, or $(2, -1)$. The midpoint of the segment from $(2, -4)$ to $(2, 2)$ is $\left(\dfrac{2+2}{2}, \dfrac{-4+2}{2}\right)$, or $(2, -1)$. Thus, we can conclude that $(2, -1)$ is the center of the ellipse. The distance from $(-2, -1)$ to $(2, -1)$ is

$\sqrt{[2-(-2)]^2 + [-1-(-1)]^2} = \sqrt{16} = 4$, so $a = 4$. The distance from $(2, 2)$ to $(2, -1)$ is $\sqrt{(2-2)^2 + (-1-2)^2}$

$= \sqrt{9} = 3$, so $b = 3$. We write the equation of the ellipse.

$\dfrac{(x-2)^2}{4^2} + \dfrac{(y-(-1))^2}{3^2} = 1$

$\dfrac{(x-2)^2}{16} + \dfrac{(y+1)^2}{9} = 1$

49. We make a drawing.

The distance between vertex $(a, 0)$ and the sun is the same as the distance between vertex $(-a, 0)$ and the other focus. Then

$d = 2.48 \times 10^8 - 3.46 \times 10^7$

$= 2.48 \times 10^8 - 0.346 \times 10^8 = 2.134 \times 10^8$ mi.

51. a)   Let $F_1 = (-c, 0)$ and $F_2 = (c, 0)$. Then the sum of the distances from the foci to P is $2a$. By the distance formula,

$\sqrt{(x+c)^2 + y^2} + \sqrt{(x-c)^2 + y^2} = 2a$, or

$\sqrt{(x+c)^2 + y^2} = 2a - \sqrt{(x-c)^2 + y^2}$.

Squaring, we get

$$(x+c)^2 + y^2 = 4a^2 - 4a\sqrt{(x-c)^2 + y^2} + (x-c)^2 + y^2 ,$$

or $x^2 + 2cx + c^2 + y^2$

$$= 4a^2 - 4a\sqrt{(x-c)^2 + y^2} + x^2 - 2cx + c^2 + y^2 .$$

Thus

$$-4a^2 + 4cx = -4a\sqrt{(x-c)^2 + y^2}$$
$$a^2 - cx = a\sqrt{(x-c)^2 + y^2} .$$

Squaring again, we get

$$a^4 - 2a^2 cx + c^2 x^2 = a^2 (x^2 - 2cx + c^2 + y^2)$$
$$a^4 - 2a^2 cx + c^2 x^2 = a^2 x^2 - 2a^2 cx + a^2 c^2 + a^2 y^2 ,$$

or

$$x^2(a^2 - c^2) + a^2 y^2 = a^2(a^2 - c^2)$$
$$\frac{x^2}{a^2} + \frac{y^2}{a^2 - c^2} = 1 .$$

b)   When P is at $(0, b)$, it follows that $b^2 = a^2 - c^2$.

Substituting, we have $\dfrac{x^2}{a^2} + \dfrac{y^2}{b^2} = 1$.

53. For the given ellipse, $a = 6/2$, or 3, and $b = 2/2$, or 1. The patient's mouth should be at a distance 2c from the light source, where the coordinates of the foci of the ellipse are $(-c, 0)$ and $(c, 0)$. From Exercise 51(b), we know $b^2 = a^2 - c^2$. We use this to find c.

$$b^2 = a^2 - c^2$$
$$1^2 = 3^2 - c^2 \quad \text{Substituting}$$
$$c^2 = 8$$
$$c = \sqrt{8}$$

Then $2c = 2\sqrt{8} \approx 5.66$. The patient's mouth should be about 5.66 ft from the light source.

55.
$$x^2 - 4x + 4y^2 + 8y - 8 = 0$$
$$x^2 - 4x + 4y^2 + 8y = 8$$
$$x^2 - 4x + 4(y^2 + 2y) = 8$$
$$(x^2 - 4x + 4 - 4) + 4(y^2 + 2y + 1 - 1) = 8$$
$$(x^2 - 4x + 4) + 4(y^2 + 2y + 1) = 8 + 4 + 4 \cdot 1$$
$$(x - 2)^2 + 4(y + 1)^2 = 16$$
$$\frac{(x - 2)^2}{16} + \frac{(y + 1)^2}{4} = 1$$
$$\frac{(x - 2)^2}{4^2} + \frac{(y - (-1))^2}{2^2} = 1$$

The center of the ellipse is $(2, -1)$. Note that $a = 4$ and $b = 2$. We locate the center and then plot the points $(2+4, -1)$, $(2-4, -1)$, $(2, -1+2)$, $(2, -1-2)$, or $(6, -1)$, $(-2, -1)$, $(2, 1)$, and $(2, -3)$. Connect these

points with an oval-shaped curve.

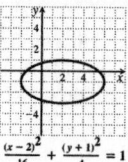

$$\frac{(x-2)^2}{16} + \frac{(y+1)^2}{4} = 1$$

57. *Graphing Calculator Exercise*

Exercise Set 10.3

1. d; see page 666 in the text.

3. h; see page 670 in the text.

5. g; see page 670 in the text.

7. c; see page 670 in the text.

9. $\dfrac{y^2}{16} - \dfrac{x^2}{16} = 1$

$$\frac{y^2}{4^2} - \frac{x^2}{4^2} = 1$$

$a = 4$ and $b = 4$, so the asymptotes are $y = \dfrac{4}{4}x$ and $y = -\dfrac{4}{4}x$, or $y = x$ and $y = -x$. We sketch them.

Replacing x with 0 and solving for y, we get $y = \pm 4$, so the intercepts are $(0, 4)$ and $(0, -4)$.

We plot the intercepts and draw smooth curves through them that approach the asymptotes.

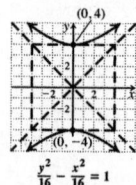

$$\frac{y^2}{16} - \frac{x^2}{16} = 1$$

11. $\dfrac{x^2}{4} - \dfrac{y^2}{25} = 1$

$$\frac{x^2}{2^2} - \frac{y^2}{5^2} = 1$$

$a = 2$ and $b = 5$, so the asymptotes are $y = \dfrac{5}{2}x$ and $y = -\dfrac{5}{2}x$. We sketch them.

Replacing y with 0 and solving for x, we get $x = \pm 2$, so the intercepts are $(2, 0)$ and $(-2, 0)$.

We plot the intercepts and draw smooth curves through

them that approach the asymptotes.

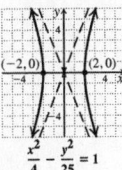

$$\frac{x^2}{4} - \frac{y^2}{25} = 1$$

13. $\dfrac{y^2}{36} - \dfrac{x^2}{9} = 1$

$$\frac{y^2}{6^2} - \frac{x^2}{3^2} = 1$$

$a = 3$ and $b = 6$, so the asymptotes are $y = \dfrac{6}{3}x$ and

$y = -\dfrac{6}{3}x$, or $y = 2x$ and $y = -2x$. We sketch them.

Replacing x with 0 and solving for y, we get $y = \pm 6$, so the intercepts are $(0, 6)$ and $(0, -6)$.

We plot the intercepts and draw smooth curves through them that approach the asymptotes.

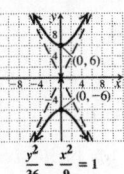

$$\frac{y^2}{36} - \frac{x^2}{9} = 1$$

15. $y^2 - x^2 = 25$

$$\frac{y^2}{25} - \frac{x^2}{25} = 1$$

$$\frac{y^2}{5^2} - \frac{x^2}{5^2} = 1$$

$a = 5$ and $b = 5$, so the asymptotes are $y = \dfrac{5}{5}x$ and

$y = -\dfrac{5}{5}x$, or $y = x$ and $y = -x$. We sketch them.

Replacing x with 0 and solving for y, we get $y = \pm 5$, so the intercepts are $(0, 5)$ and $(0, -5)$.

We plot the intercepts and draw smooth curves through them that approach the asymptotes.

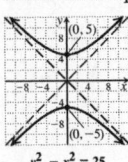

$$y^2 - x^2 = 25$$

17. $25x^2 - 16y^2 = 400$

$$\frac{x^2}{16} - \frac{y^2}{25} = 1 \quad \text{Multiplying by } \frac{1}{400}$$

$$\frac{x^2}{4^2} - \frac{y^2}{5^2} = 1$$

$a = 4$ and $b = 5$, so the asymptotes are $y = \dfrac{5}{4}x$ and

$y = -\dfrac{5}{4}x$. We sketch them.

Replacing y with 0 and solving for x, we get $x = \pm 4$, so

the intercepts are $(4, 0)$ and $(-4, 0)$.

We plot the intercepts and draw smooth curves through them that approach the asymptotes.

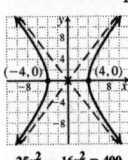

$$25x^2 - 16y^2 = 400$$

19. $xy = -6$

$$y = -\frac{6}{x} \qquad \text{Solving for } y$$

We find some solutions, keeping the results in a table.

x	y
$\frac{1}{6}$	36
1	-6
6	-1
12	$-\frac{1}{2}$
$-\frac{1}{6}$	36
-1	6
-6	1
-12	$\frac{1}{2}$

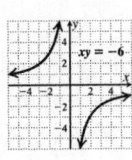

Note that we cannot use 0 for x. The x-axis and the y-axis are the asymptotes.

21. $xy = 4$

$$y = \frac{4}{x} \quad \text{Solving for } y$$

We find some solutions, keeping the results in a table.

x	y
$\frac{1}{2}$	8
1	4
4	1
8	$\frac{1}{2}$
$-\frac{1}{2}$	-8
-1	-4
-2	-2
-4	-1

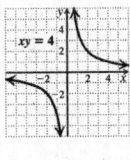

Note that we cannot use 0 for x. The x-axis and the y-axis are the asymptotes.

23. $xy = -2$

$$y = -\frac{2}{x} \quad \text{Solving for } y$$

x	y
$\frac{1}{2}$	-4
1	-2
2	-1
4	$-\frac{1}{2}$
$-\frac{1}{2}$	4
-1	2
-2	1
-4	$\frac{1}{2}$

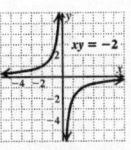

Note that we cannot use 0 for x. The x-axis and the y-axis are the asymptotes.

25. $xy = 1$

$y = \dfrac{1}{x}$ Solving for y

x	y
$\frac{1}{4}$	4
$\frac{1}{2}$	2
1	1
2	$\frac{1}{2}$
4	$\frac{1}{4}$
$-\frac{1}{4}$	-4
$-\frac{1}{2}$	-2
-1	-1
-2	$-\frac{1}{2}$
-4	$-\frac{1}{4}$

Note that we cannot use 0 for x. The x-axis and the y-axis are the asymptotes.

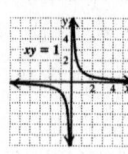

27. $x^2 + y^2 - 6x + 10y - 40 = 0$

Completing the square twice, we obtain an equivalent equation:

$$\left(x^2 - 6x\right) + \left(y^2 + 10y\right) = 40$$
$$\left(x^2 - 6x + 9\right) + \left(y^2 + 10y + 25\right) = 40 + 9 + 25$$
$$(x - 3)^2 + (y + 5)^2 = 74$$

The graph is a circle.

29. $9x^2 + 4y^2 - 36 = 0$

$$9x^2 + 4y^2 = 36$$
$$\frac{x^2}{4} + \frac{y^2}{9} = 1$$

The graph is an ellipse.

31. $4x^2 - 9y^2 - 72 = 0$

$$4x^2 - 9y^2 = 72$$
$$\frac{x^2}{18} - \frac{y^2}{8} = 1$$

The graph is a hyperbola.

33. $y^2 = 20 - x^2$

$$x^2 + y^2 = 20$$

The graph is a circle.

35. $x - 10 = y^2 - 6y$

$$x - 10 + 9 = y^2 - 6y + 9$$
$$x - 1 = (y - 3)^2$$

The graph is a parabola.

37. $x - \dfrac{3}{y} = 0$

$$x = \frac{3}{y}$$
$$xy = 3$$

The graph is a hyperbola.

39. $y + 6x = x^2 + 5$

$$y = x^2 - 6x + 5$$

The graph is a parabola

41. $25y^2 = 100 + 4x^2$

$$25y^2 - 4x^2 = 100$$
$$\frac{y^2}{4} - \frac{x^2}{25} = 1$$

The graph is a hyperbola.

43. $3x^2 + y^2 - x = 2x^2 - 9x + 10y + 40$

$$x^2 + y^2 + 8x - 10y = 40$$

Both variables are squared, so the graph is not a parabola. The plus sign between x^2 and y^2 indicates that we have either a circle or an ellipse. Since the coefficients of x^2 and y^2 are the same, the graph is a circle.

45. $16x^2 + 5y^2 - 12x^2 + 8y^2 - 3x + 4y = 568$

$$4x^2 + 13y^2 - 3x + 4y = 568$$

Both variables are squared, so the graph is not a parabola. The plus sign between x^2 and y^2 indicates that we have either a circle or an ellipse. Since the coefficients of x^2 and y^2 are different, the graph is an ellipse.

47. *Writing Exercise.* The equation of the ellipse in standard form is the sum of the squared terms. The equation of the hyperbola in standard form is the difference of the squared terms.

49. $5x + 2y = -3$ (1)

$2x + 3y = 12$ (2)

We multiply twice to make two terms become additive inverses.

From (1)	$15x + 6y = -9$	Multiplying by 3
From (2)	$-4x - 6y = -24$	Multiplying by -2
	$11x \qquad = -33$	
	$x = -3$	

Substitute -3 for x in Equation (2) and solve for y.

$$2x + 3y = 12$$
$$2(-3) + 3y = 12$$
$$-6 + 3y = 12$$
$$3y = 18$$
$$y = 6$$

The solution is $(-3, 6)$.

51.
$$\frac{3}{4}x^2 + x^2 = 7$$
$$\frac{7}{4}x^2 = 7$$
$$x^2 = 4$$
$$x^2 - 4 = 0$$
$$(x+2)(x-2) = 0$$
$$x + 2 = 0 \quad or \quad x - 2 = 0$$
$$x = -2 \quad or \quad x = 2$$

53. $x^2 - 3x - 1 = 0$
$$x = \frac{3 \pm \sqrt{(-3)^2 - 4 \cdot 1 \cdot (-1)}}{2 \cdot 1}$$
$$x = \frac{3 \pm \sqrt{9 + 4}}{2}$$
$$x = \frac{3 \pm \sqrt{13}}{2}$$

55. *Writing Exercise.* The ratio b/a controls how wide open the branches of a hyperbola are. The larger this ratio, the steeper the slant of the asymptotes. For a hyperbola with a horizontal axis, the steeper the slant the more wide open the branches. For a hyperbola with a vertical axis, the steeper the slant the less wide open the branches.

57. Since the intercepts are $(0, 6)$ and $(0, -6)$, we know that

the hyperbola is of the form $\dfrac{y^2}{b^2} - \dfrac{x^2}{a^2} = 1$ and that $b = 6$.

The equations of the asymptotes tell us that $b/a = 3$, so
$$\frac{6}{a} = 3$$
$$a = 2.$$

The equation is $\dfrac{y^2}{6^2} - \dfrac{x^2}{2^2} = 1$, or $\dfrac{y^2}{36} - \dfrac{x^2}{4} = 1$.

59.
$$\frac{(x-5)^2}{36} - \frac{(y-2)^2}{25} = 1$$
$$\frac{(x-5)^2}{6^2} - \frac{(y-2)^2}{5^2} = 1$$

$h = 5$, $k = 2$, $a = 6$, $b = 5$

Center: $(5, 2)$

Vertices: $(5 - 6, 2)$ and $(5 + 6, 2)$, or $(-1, 2)$ and $(11, 2)$

Asymptotes: $y - 2 = \frac{5}{6}(x-5)$ and $y - 2 = -\frac{5}{6}(x-5)$

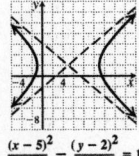

$$\frac{(x-5)^2}{36} - \frac{(y-2)^2}{25} = 1$$

61.
$$8(y+3)^2 - 2(x-4)^2 = 32$$
$$\frac{(y+3)^2}{4} - \frac{(x-4)^2}{16} = 1$$
$$\frac{(y-(-3))^2}{2^2} - \frac{(x-4)^2}{4^2} = 1$$

$h = 4$, $k = -3$, $a = 4$, $b = 2$

Center: $(4, -3)$

Vertices: $(4, -3 + 2)$ and $(4, -3 - 2)$, or $(4, -1)$ and $(4, -5)$

Asymptotes: $y - (-3) = \frac{2}{4}(x-4)$ and
$y - (-3) = -\frac{2}{4}(x-4)$, or $y + 3 = \frac{1}{2}(x-4)$ and
$y + 3 = -\frac{1}{2}(x-4)$

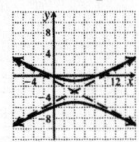

$8(y+3)^2 - 2(x-4)^2 = 32$

63.
$$4x^2 - y^2 + 24x + 4y + 28 = 0$$
$$4(x^2 + 6x) - (y^2 - 4y) = -28$$
$$4(x^2 + 6x + 9 - 9) - (y^2 - 4y + 4 - 4) = -28$$
$$4(x^2 + 6x + 9) - (y^2 - 4y + 4) = -28 + 4 \cdot 9 - 4$$
$$4(x+3)^2 - (y-2)^2 = 4$$
$$\frac{(x+3)^2}{1} - \frac{(y-2)^2}{4} = 1$$
$$\frac{(x-(-3))^2}{1^2} - \frac{(y-2)^2}{2^2} = 1$$

$h = -3$, $k = 2$, $a = 1$, $b = 2$

Center: $(-3, 2)$

Vertices: $(-3 - 1, 2)$, and $(-3 + 1, 2)$, or $(-4, 2)$ and $(-2, 2)$

Asymptotes: $y - 2 = \frac{2}{1}(x-(-3))$ and
$y - 2 = -\frac{2}{1}(x-(-3))$, or $y - 2 = 2(x+3)$ and
$y - 2 = -2(x+3)$

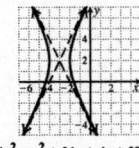

$4x^2 - y^2 + 24x + 4y + 28 = 0$

65. *Graphing Calculator Exercise*

Connecting the Concepts

1. $y = 3(x-4)^2 + 1$ parabola

Vertex: $(4, 1)$

Axis of symmetry: $x = 4$

3. $(x-3)^2 + (y-2)^2 = 5$ circle

Center: $(3, 2)$

5. $\dfrac{x^2}{144} + \dfrac{y^2}{81} = 1$ ellipse

x-intercepts: $(-12, 0)$ and $(12, 0)$

y-intercepts: $(0, 9)$ and $(0, -9)$

7. $4y^2 - x^2 = 4$ hyperbola

$\dfrac{y^2}{1} - \dfrac{x^2}{4} = 1$

Vertices: $(0, -1)$ and $(0, 1)$

9. $x^2 + y^2 = 36$ is a circle.

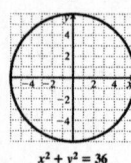

11. $\dfrac{x^2}{25} + \dfrac{y^2}{49} = 1$ is an ellipse.

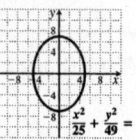

13. $x = (y+3)^2 + 2$ is a parabola.

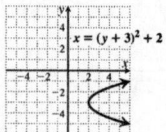

15. $xy = -4$ is a hyperbola.

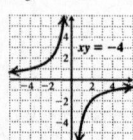

17. $x^2 + y^2 - 8y - 20 = 0$

$x^2 + y^2 - 8y + 16 = 20 + 16$

$x^2 + (y-4)^2 = 36$ is a circle.

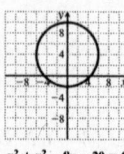

19. $16y^2 - x^2 = 16$

$\dfrac{y^2}{1} - \dfrac{x^2}{16} = 1$ is a hyperbola.

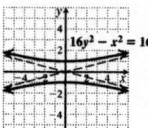

Exercise Set 10.4

1. True.

3. False; see page 677 in the text.

5. True; see page 675 in the text.

7. $x^2 + y^2 = 41,$ (1)

    $y - x = 1$ (2)

First solve Equation (2) for y.

    $y = x + 1$ (3)

Then substitute $x + 1$ for y in Equation (1) and solve for x.

$$x^2 + y^2 = 41$$
$$x^2 + (x+1)^2 = 41$$
$$x^2 + x^2 + 2x + 1 = 41$$
$$2x^2 + 2x - 40 = 0$$
$$x^2 + x - 20 = 0 \quad \text{Multiplying by } \tfrac{1}{2}$$
$$(x+5)(x-4) = 0$$

$x + 5 = 0$ *or* $x - 4 = 0$ Principle of zero products

   $x = -5$ *or* $x = 4$

Now substitute these numbers in Equation (3) and solve for y.

    For $x = -5,$ $y = -5 + 1 = -4$

    For $x = 4,$ $y = 4 + 1 = 5$

The pairs $(-5, -4)$ and $(4, 5)$ check, so they are the solutions.

9. $4x^2 + 9y^2 = 36,$ (1)

    $3y + 2x = 6$ (2)

First solve Equation (2) for y.

    $3y = -2x + 6$

    $y = -\tfrac{2}{3}x + 2$ (3)

Then substitute $-\tfrac{2}{3}x + 2$ for y in Equation (1) and solve for x.

$$4x^2 + 9y^2 = 36$$
$$4x^2 + 9\left(-\tfrac{2}{3}x + 2\right)^2 = 36$$
$$4x^2 + 9\left(\tfrac{4}{9}x^2 - \tfrac{8}{3}x + 4\right) = 36$$
$$4x^2 + 4x^2 - 24x + 36 = 36$$
$$8x^2 - 24x = 0$$
$$x^2 - 3x = 0$$
$$x(x-3) = 0$$
$$x = 0 \ \ or \ \ x = 3$$

Now substitute these numbers in Equation (3) and solve for y.

$$\text{For } x = 0, \quad y = -\tfrac{2}{3}\cdot 0 + 2 = 2$$
$$\text{For } x = 3, \quad y = -\tfrac{2}{3}\cdot 3 + 2 = 0$$

The pairs $(0, 2)$ and $(3, 0)$ check, so they are the solutions.

11. $y^2 = x + 3, \quad (1)$
$\ \ \ 2y = x + 4 \quad (2)$

First solve Equation (2) for x.

$$2y - 4 = x \quad (3)$$

Then substitute $2y - 4$ for x in Equation (1) and solve for y.

$$y^2 = x + 3$$
$$y^2 = (2y - 4) + 3$$
$$y^2 = 2y - 1$$
$$y^2 - 2y + 1 = 0$$
$$(y-1)(y-1) = 0$$
$$y - 1 = 0 \ \ or \ \ y - 1 = 0$$
$$y = 1 \ \ or \ \ \ \ \ \ y = 1$$

Now substitute 1 for y in Equation (3) and solve for x.
$$2\cdot 1 - 4 = x$$
$$-2 = x$$

The pair $(-2, 1)$ checks. It is the solution.

13. $x^2 - xy + 3y^2 = 27, \quad (1)$
$\ \ \ \ \ \ \ \ \ x - y = 2 \quad (2)$

First solve Equation (2) for y.

$$x - 2 = y \quad (3)$$

Then substitute x - 2 for y in Equation (1) and solve for x.

$$x^2 - xy + 3y^2 = 27$$
$$x^2 - x(x-2) + 3(x-2)^2 = 27$$
$$x^2 - x^2 + 2x + 3x^2 - 12x + 12 = 27$$
$$3x^2 - 10x - 15 = 0$$
$$x = \frac{-(-10) \pm \sqrt{(-10)^2 - 4(3)(-15)}}{2\cdot 3}$$
$$x = \frac{10 \pm \sqrt{100 + 180}}{6} = \frac{10 \pm \sqrt{280}}{6}$$
$$x = \frac{10 \pm 2\sqrt{70}}{6} = \frac{5 \pm \sqrt{70}}{3}$$

Now substitute these numbers in Equation (3) and solve

for y.

$$\text{For } x = \frac{5 + \sqrt{70}}{3}, \quad y = \frac{5 + \sqrt{70}}{3} - 2 = \frac{-1 + \sqrt{70}}{3}$$
$$\text{For } x = \frac{5 - \sqrt{70}}{3}, \quad y = \frac{5 - \sqrt{70}}{3} - 2 = \frac{-1 - \sqrt{70}}{3}$$

The pairs $\left(\dfrac{5 + \sqrt{70}}{3}, \dfrac{-1 + \sqrt{70}}{3}\right)$ and $\left(\dfrac{5 - \sqrt{70}}{3}, \dfrac{-1 - \sqrt{70}}{3}\right)$ check, so they are the solutions.

15. $x^2 + 4y^2 = 25, \quad (1)$
$\ \ \ \ \ \ x + 2y = 7 \quad (2)$

First solve Equation (2) for x.

$$x = -2y + 7 \quad (3)$$

Then substitute $-2y + 7$ for x in Equation (1) and solve for y.

$$x^2 + 4y^2 = 25$$
$$(-2y + 7)^2 + 4y^2 = 25$$
$$4y^2 - 28y + 49 + 4y^2 = 25$$
$$8y^2 - 28y + 24 = 0$$
$$2y^2 - 7y + 6 = 0$$
$$(2y - 3)(y - 2) = 0$$
$$y = \frac{3}{2} \ \ or \ \ y = 2$$

Now substitute these numbers in Equation (3) and solve for x.

$$\text{For } y = \frac{3}{2}, \quad x = -2\cdot\frac{3}{2} + 7 = 4$$
$$\text{For } y = 2, \quad x = -2\cdot 2 + 7 = 3$$

The pairs $\left(4, \dfrac{3}{2}\right)$ and $(3, 2)$ check, so they are the solutions.

17. $x^2 - xy + 3y^2 = 5, \quad (1)$
$\ \ \ \ \ \ \ \ \ x - y = 2 \quad (2)$

First solve Equation (2) for y.

$$x - 2 = y \quad (3)$$

Then substitute $x - 2$ for y in Equation (1) and solve for x.

$$x^2 - xy + 3y^2 = 5$$
$$x^2 - x(x-2) + 3(x-2)^2 = 5$$
$$x^2 - x^2 + 2x + 3x^2 - 12x + 12 = 5$$
$$3x^2 - 10x + 7 = 0$$
$$(3x - 7)(x - 1) = 0$$
$$x = \frac{7}{3} \ \ or \ \ x = 1$$

Now substitute these numbers in Equation (3) and solve for y.

$$\text{For } x = \frac{7}{3}, \quad y = \frac{7}{3} - 2 = \frac{1}{3}$$
$$\text{For } x = 1, \quad y = 1 - 2 = -1$$

The pairs $\left(\dfrac{7}{3}, \dfrac{1}{3}\right)$ and $(1, -1)$ check, so they are the solutions.

19. $3x + y = 7$,   (1)
     $4x^2 + 5y = 24$ (2)

First solve Equation (1) for y.

$$y = 7 - 3x \quad (3)$$

Then substitute $7 - 3x$ for y in Equation (2) and solve for x.

$$4x^2 + 5y = 24$$
$$4x^2 + 5(7 - 3x) = 24$$
$$4x^2 + 35 - 15x = 24$$
$$4x^2 - 15x + 11 = 0$$
$$(4x - 11)(x - 1) = 0$$

$$x = \frac{11}{4} \quad or \quad x = 1$$

Now substitute these numbers into Equation (3) and solve for y.

For $x = \frac{11}{4}$, $y = 7 - 3 \cdot \frac{11}{4} = -\frac{5}{4}$
For $x = 1$, $y = 7 - 3 \cdot 1 = 4$

The pairs $\left(\frac{11}{4}, -\frac{5}{4}\right)$ and $(1, 4)$ check, so they are the solutions.

21. $a + b = 6$,   (1)
     $ab = 8$ (2)

First solve Equation (1) for a.

$a = -b + 6$ (3)

Then substitute $-b + 6$ for a in Equation (2) and solve for b.

$$(-b + 6)b = 8$$
$$-b^2 + 6b = 8$$
$$0 = b^2 - 6b + 8$$
$$0 = (b - 2)(b - 4)$$
$$b - 2 = 0 \quad or \quad b - 4 = 0$$
$$b = 2 \quad or \quad\quad b = 4$$

Now substitute these numbers in Equation (3) and solve for a.

For $b = 2$, $a = -2 + 6 = 4$
For $b = 4$, $a = -4 + 6 = 2$

The pairs $(4, 2)$ and $(2, 4)$ check, so they are the solutions.

23. $2a + b = 1$,   (1)
     $b = 4 - a^2$ (2)

Equation (2) is already solved for b. Substitute $4 - a^2$ for b in Equation (1) and solve for a.

$$2a + 4 - a^2 = 1$$
$$0 = a^2 - 2a - 3$$
$$0 = (a - 3)(a + 1)$$
$$a = 3 \quad or \quad a = -1$$

Substitute these numbers in Equation (2) and solve for b.

For $a = 3$, $b = 4 - 3^2 = -5$
For $a = -1$, $b = 4 - (-1)^2 = 3$

The pairs $(3, -5)$ and $(-1, 3)$ check, so they are the solutions.

25. $a^2 + b^2 = 89$,   (1)
     $a - b = 3$ (2)

First solve Equation (2) for a.

$a = b + 3$ (3)

Then substitute $b + 3$ for a in Equation (1) and solve for b.

$$(b + 3)^2 + b^2 = 89$$
$$b^2 + 6b + 9 + b^2 = 89$$
$$2b^2 + 6b - 80 = 0$$
$$b^2 + 3b - 40 = 0$$
$$(b + 8)(b - 5) = 0$$

$$b = -8 \quad or \quad b = 5$$

Substitute these numbers in Equation (3) and solve for a.

For $b = -8$, $a = -8 + 3 = -5$
For $b = 5$, $a = 5 + 3 = 8$

The pairs $(-5, -8)$ and $(8, 5)$ check, so they are the solutions.

27. $y = x^2$,   (1)
     $x = y^2$ (2)

Equation (1) is already solved for y. Substitute x^2 for y in Equation (2) and solve for x.

$$x = y^2$$
$$x = \left(x^2\right)^2$$
$$x = x^4$$
$$0 = x^4 - x$$
$$0 = x(x^3 - 1)$$
$$0 = x(x - 1)(x^2 + x + 1)$$

$$x = 0 \quad or \quad x = 1 \quad or \quad x = \frac{-1 \pm \sqrt{1^2 - 4 \cdot 1 \cdot 1}}{2}$$

$$x = 0 \quad or \quad x = 1 \quad or \quad x = -\frac{1}{2} \pm \frac{\sqrt{3}}{2}i$$

Substitute these numbers in Equation (1) and solve for y.

For $x = 0$, $y = 0^2 = 0$
For $x = 1$, $y = 1^2 = 1$
For $x = -\frac{1}{2} + \frac{\sqrt{3}}{2}i$, $y = \left(-\frac{1}{2} + \frac{\sqrt{3}}{2}i\right)^2 = -\frac{1}{2} - \frac{\sqrt{3}}{2}i$
For $x = -\frac{1}{2} - \frac{\sqrt{3}}{2}i$, $y = \left(-\frac{1}{2} - \frac{\sqrt{3}}{2}i\right)^2 = -\frac{1}{2} + \frac{\sqrt{3}}{2}i$

The pairs $(0, 0)$, $(1, 1)$, $\left(-\frac{1}{2} + \frac{\sqrt{3}}{2}i, \ -\frac{1}{2} - \frac{\sqrt{3}}{2}i\right)$,

and $\left(-\frac{1}{2} - \frac{\sqrt{3}}{2}i, \ -\frac{1}{2} + \frac{\sqrt{3}}{2}i\right)$ check, so they are the solutions.

29. $x^2 + y^2 = 16,$ (1)
$x^2 - y^2 = 16$ (2)

Here we use the elimination method.

$$\begin{array}{ll} x^2 + y^2 = 16 & (1) \\ \underline{x^2 - y^2 = 16} & (2) \\ 2x^2 \qquad\;\; = 32 & \text{Adding} \\ \quad\; x^2 = 16 \\ \quad\;\; x = \pm 4 \end{array}$$

If $x = 4$, $x^2 = 16$, and if $x = -4$, $x^2 = 16$, so substituting 4 or -4 in Equation (2) gives us

$$\begin{aligned} x^2 + y^2 &= 16 \\ 16 + y^2 &= 16 \\ y^2 &= 0 \\ y &= 0 \end{aligned}$$

The pairs $(4, 0)$ and $(-4, 0)$ check. They are the solutions.

31. $x^2 + y^2 = 25,$ (1)
$xy = 12$ (2)

First we solve Equation (2) for y.

$$\begin{aligned} xy &= 12 \\ y &= \frac{12}{x} \end{aligned}$$

Then we substitute $\frac{12}{x}$ for y in Equation (1) and solve for x.

$$\begin{aligned} x^2 + y^2 &= 25 \\ x^2 + \left(\frac{12}{x}\right)^2 &= 25 \\ x^2 + \frac{144}{x^2} &= 25 \\ x^4 + 144 &= 25x^2 \quad \text{Multiplying by } x^2 \\ x^4 - 25x^2 + 144 &= 0 \\ u^2 - 25u + 144 &= 0 \qquad \text{Letting } u = x^2 \\ (u - 9)(u - 16) &= 0 \\ u = 9 \;\; &or \;\; u = 16 \end{aligned}$$

We now substitute x^2 for u and solve for x.

$$\begin{array}{ccc} x^2 = 9 & or & x^2 = 16 \\ x = \pm 3 & or & x = \pm 4 \end{array}$$

Since $y = 12/x$, if $x = 3$, $y = 4$; if $x = -3$, $y = -4$; if $x = 4$, $y = 3$; and if $x = -4$, $y = -3$. The pairs $(3, 4)$, $(-3, -4)$, $(4, 3)$, and $(-4, -3)$ check. They are the solutions.

33. $x^2 + y^2 = 9,$ (1)
$25x^2 + 16y^2 = 400$ (2)

$$\begin{array}{ll} -16x^2 - 16y^2 = -144 & \text{Multiplying (1) by } -16 \\ \underline{25x^2 + 16y^2 = 400} \\ 9x^2 \qquad\;\;\; = 256 & \text{Adding} \\ \qquad x = \pm \dfrac{16}{3} \end{array}$$

$$\frac{256}{9} + y^2 = 9 \quad \text{Substituting in (1)}$$
$$y^2 = 9 - \frac{256}{9}$$
$$y^2 = -\frac{175}{9}$$
$$y = \pm\sqrt{-\frac{175}{9}} = \pm\frac{5\sqrt{7}}{3}i$$

The pairs $\left(\dfrac{16}{3},\ \dfrac{5\sqrt{7}}{3}i\right)$, $\left(\dfrac{16}{3},\ -\dfrac{5\sqrt{7}}{3}i\right)$, $\left(-\dfrac{16}{3},\ \dfrac{5\sqrt{7}}{3}i\right)$, and $\left(-\dfrac{16}{3},\ -\dfrac{5\sqrt{7}}{3}i\right)$ check. They are the solutions.

35. $x^2 + y^2 = 14,$ (1)
$x^2 - y^2 = 4$ (2)

$$\begin{array}{ll} 2x^2 \quad\;\; = 18 & \text{Adding} \\ \quad x^2 = 9 \\ \quad\; x = \pm 3 \end{array}$$

$$\begin{aligned} 9 + y^2 &= 14 \quad \text{Substituting in Eq. (1)} \\ y^2 &= 5 \\ y &= \pm\sqrt{5} \end{aligned}$$

The pairs $(-3, -\sqrt{5})$, $(-3, \sqrt{5})$, $(3, -\sqrt{5})$, and $(3, \sqrt{5})$ check. They are the solutions.

37. $x^2 + y^2 = 10,$ (1)
$xy = 3$ (2)

First we solve Equation (2) for y.

$$\begin{aligned} xy &= 3 \\ y &= \frac{3}{x} \end{aligned}$$

Then we substitute $\frac{3}{x}$ for y in Equation (1) and solve for x.

$$\begin{aligned} x^2 + y^2 &= 10 \\ x^2 + \left(\frac{3}{x}\right)^2 &= 10 \\ x^2 + \frac{9}{x^2} &= 10 \\ x^4 + 9 &= 10x^2 \quad \text{Multiplying by } x^2 \\ x^4 - 10x^2 + 9 &= 0 \\ u^2 - 10u + 9 &= 0 \qquad \text{Letting } u = x^2 \\ (u - 1)(u - 9) &= 0 \\ u = 1 \quad &or \quad u = 9 \\ x^2 = 1 \quad &or \quad x^2 = 9 \quad \text{Substitute } x^2 \text{ for } u \\ x = \pm 1 \quad &or \quad x = \pm 3 \qquad \text{and solve for } x. \end{aligned}$$

$y = 3/x$, so if $x = 1$, $y = 3$; if $x = -1$, $y = -3$; if $x = 3$, $y = 1$; if $x = -3$, $y = -1$. The pairs $(1, 3)$, $(-1, -3)$, $(3, 1)$, and $(-3, -1)$ check. They are the solutions.

39. $x^2 + 4y^2 = 20,$ (1)
$xy = 4$ (2)

First we solve Equation (2) for y.

$$y = \frac{4}{x}$$

Then we substitute $\frac{4}{x}$ for y in Equation (1) and solve for x.

$$x^2 + 4\left(\frac{4}{x}\right)^2 = 20$$
$$x^2 + \frac{64}{x^2} = 20$$
$$x^4 + 64 = 20x^2$$
$$x^4 - 20x^2 + 64 = 0$$
$$u^2 - 20u + 64 = 0 \quad \text{Letting } u = x^2$$
$$(u - 16)(u - 4) = 0$$
$$u = 16 \quad or \quad u = 4$$
$$x^2 = 16 \quad or \quad x^2 = 4$$
$$x = \pm 4 \quad or \quad x = \pm 2$$

$y = 4 / x$, so if $x = 4$, $y = 1$; if $x = -4$, $y = -1$; if $x = 2$, $y = 2$; and if $x = -2$, $y = -2$. The pairs $(4, 1)$, $(-4, -1)$, $(2, 2)$, and $(-2, -2)$ check. They are the solutions.

41. $2xy + 3y^2 = 7$, (1)
$3xy - 2y^2 = 4$ (2)

$$6xy + 9y^2 = 21 \quad \text{Multiplying (1) by 3}$$
$$\underline{-6xy + 4y^2 = -8} \quad \text{Multiplying (2) by} -2$$
$$13y^2 = 13$$
$$y^2 = 1$$
$$y = \pm 1$$

Substitute for y in Equation (1) and solve for x.

When $y = 1$: $\quad 2 \cdot x \cdot 1 + 3 \cdot 1^2 = 7$
$$2x = 4$$
$$x = 2$$

When $y = -1$: $\quad 2 \cdot x \cdot (-1) + 3(-1)^2 = 7$
$$-2x = 4$$
$$x = -2$$

The pairs $(2, 1)$ and $(-2, -1)$ check. They are the solutions.

43. $4a^2 - 25b^2 = 0$, (1)
$2a^2 - 10b^2 = 3b + 4$ (2)

$$4a^2 - 25b^2 = 0$$
$$\underline{-4a^2 + 20b^2 = -6b - 8} \quad \text{Multiplying (2) by} -2$$
$$-5b^2 = -6b - 8$$

$$0 = 5b^2 - 6b - 8$$
$$0 = (5b + 4)(b - 2)$$

$$b = -\frac{4}{5} \quad or \quad b = 2$$

Substitute for b in Equation (1) and solve for a.

When $b = -\frac{4}{5}$: $\quad 4a^2 - 25\left(-\frac{4}{5}\right)^2 = 0$
$$4a^2 = 16$$
$$a^2 = 4$$
$$a = \pm 2$$

When $b = 2$: $\quad 4a^2 - 25(2)^2 = 0$
$$4a^2 = 100$$
$$a^2 = 25$$
$$a = \pm 5$$

The pairs $\left(2, -\frac{4}{5}\right)$, $\left(-2, -\frac{4}{5}\right)$, $(5, 2)$ and $(-5, 2)$ check. They are the solutions.

45. $ab - b^2 = -4$, (1)
$ab - 2b^2 = -6$ (2)

$$ab - b^2 = -4$$
$$\underline{-ab + 2b^2 = 6} \quad \text{Multiplying (2) by} -1$$
$$b^2 = 2$$
$$b = \pm\sqrt{2}$$

Substitute for b in Equation (1) and solve for a.

When $b = \sqrt{2}$: $\quad a(\sqrt{2}) - (\sqrt{2})^2 = -4$
$$a\sqrt{2} = -2$$
$$a = -\frac{2}{\sqrt{2}} = -\sqrt{2}$$

When $b = -\sqrt{2}$: $\quad a(-\sqrt{2}) - (-\sqrt{2})^2 = -4$
$$-a\sqrt{2} = -2$$
$$a = \frac{-2}{-\sqrt{2}} = \sqrt{2}$$

The pairs $(-\sqrt{2}, \sqrt{2})$ and $(\sqrt{2}, -\sqrt{2})$ check. They are the solutions.

47. *Familiarize*. We first make a drawing. We let l and w represent the length and width, respectively.

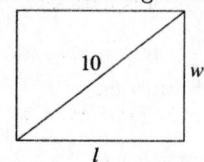

Translate. The perimeter is 28 cm.

$$2l + 2w = 28, \text{ or } l + w = 14$$

Using the Pythagorean theorem we have another equation.

$$l^2 + w^2 = 10^2, \text{ or } l^2 + w^2 = 100$$

Carry out. We solve the system:

$$l + w = 14, \quad (1)$$
$$l^2 + w^2 = 100 \quad (2)$$

First solve Equation (1) for w.

$$w = 14 - l \quad (3)$$

Then substitute $14 - l$ for w in Equation (2) and solve for l.

$$l^2 + w^2 = 100$$
$$l^2 + (14 - l)^2 = 100$$
$$l^2 + 196 - 28l + l^2 = 100$$
$$2l^2 - 28l + 96 = 0$$
$$l^2 - 14l + 48 = 0$$
$$(l - 8)(l - 6) = 0$$

$l = 8$ or $l = 6$

If $l = 8$, then $w = 14 - 8$, or 6. If $l = 6$, then $w = 14 - 6$, or 8. Since the length is usually considered to be longer than the width, we have the solution $l = 8$ and $w = 6$, or $(8, 6)$.

Check. If $l = 8$ and $w = 6$, then the perimeter is $2 \cdot 8 + 2 \cdot 6$, or 28. The length of a diagonal is $\sqrt{8^2 + 6^2}$, or $\sqrt{100}$, or 10. The numbers check.

State. The length is 8 cm, and the width is 6 cm.

49. Familiarize. Let $l =$ the length and $w =$ the width of the rectangle.

Translate. The perimeter is 6 in., so we have one equation:

$$2l + 2w = 6 \text{, or } l + w = 3$$

Using the Pythagorean theorem we have another equation.

$$l^2 + w^2 = \left(\sqrt{5}\right)^2 \text{, or } l^2 + w^2 = 5$$

Carry out. We solve the system of equations:

$$l + w = 3, \quad (1)$$
$$l^2 + w^2 = 5 \quad (2)$$

Solve Equation (1) for l: $l = 3 - w$. Substitute $3 - w$ for l in Equation (2) and solve for w.

$$l^2 + w^2 = 5$$
$$(3 - w)^2 + w^2 = 5$$
$$9 - 6w + w^2 + w^2 = 5$$
$$2w^2 - 6w + 4 = 0$$
$$2\left(w^2 - 3w + 2\right) = 0$$
$$2(w - 2)(w - 1) = 0$$

$w = 2$ or $w = 1$

For $w = 2$, $l = 3 - 2 = 1$.
For $w = 1$, $l = 3 - 1 = 2$.

Check. The solutions are $(1, 2)$ and $(2, 1)$. We choose the larger number for the length. $2 + 1 = 3$, and $2^2 + 1^2 = 5$. The solution checks.

State. The length is 2 in. and the width is 1 in.

51. Familiarize. We first make a drawing. Let $l =$ the length and $w =$ the width of the cargo area, in feet.

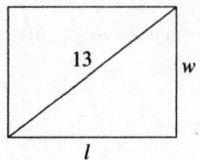

Translate. The cargo area must be 60 ft^2, so we have one equation:

$$lw = 60$$

The Pythagorean equation gives us another equation:

$$l^2 + w^2 = 13^2 \text{, or } l^2 + w^2 = 169$$

Carry out. We solve the system of equations.

$$lw = 60, \quad (1)$$
$$l^2 + w^2 = 169 \quad (2)$$

First solve Equation (1) for w:

$$lw = 60$$
$$w = \frac{60}{l} \quad (3)$$

Then substitute $60/l$ for w in Equation (2) and solve for l.

$$l^2 + w^2 = 169$$
$$l^2 + \left(\frac{60}{l}\right)^2 = 169$$
$$l^2 + \frac{3600}{l^2} = 169$$
$$l^4 + 3600 = 169l^2$$
$$l^4 - 169l^2 + 3600 = 0$$

Let $u = l^2$ and $u^2 = l^4$ and substitute.

$$u^2 - 169u + 3600 = 0$$
$$(u - 144)(u - 25) = 0$$
$$u = 144 \quad \text{or} \quad u = 25$$
$$l^2 = 144 \quad \text{or} \quad l^2 = 25 \quad \text{Replacing } u \text{ with } l^2$$
$$l = \pm 12 \quad \text{or} \quad l = \pm 5$$

Since the length cannot be negative, we consider only 12 and 5. We substitute in Equation (3) to find w. When $l = 12$, $w = 60/12 = 5$; when $l = 5$, $w = 60/5 = 12$. Since we usually consider length to be longer than width, we check the pair $(12, 5)$.

Check. If the length is 12 ft and the width is 5 ft, then the area is $12 \cdot 5$, or 60 ft^2. Also $12^2 + 5^2 = 144 + 25 = 169 = 13^2$. The answer checks.

State. The length is 12 ft and the width is 5 ft.

53. Familiarize. Let x and y represent the numbers.

Translate. The product of the numbers is 90, so we have

$$xy = 90 \quad (1)$$

The sum of the squares of the numbers is 261, so we have

$$x^2 + y^2 = 261 \quad (2)$$

Carry out. We solve the system of equations.

$$xy = 90, \quad (1)$$
$$x^2 + y^2 = 261 \quad (2)$$

First solve Equation (1) for y:

$$xy = 90$$
$$y = \frac{90}{x} \quad (3)$$

Then substitute $90/x$ for y in Equation (2) and solve for x.

$$x^2 + y^2 = 261$$
$$x^2 + \left(\frac{90}{x}\right)^2 = 261$$
$$x^2 + \frac{8100}{x^2} = 261$$
$$x^4 + 8100 = 261x^2$$
$$x^4 - 261x^2 + 8100 = 0$$

Let $u = x^2$ and $u^2 = x^4$ and substitute.

$$u^2 - 261u + 8100 = 0$$
$$(u - 36)(u - 225) = 0$$
$$u = 36 \quad or \quad u = 225$$
$$x^2 = 36 \quad or \quad x^2 = 225 \quad \text{Replacing } u \text{ with } x^2$$
$$x = \pm 6 \quad or \quad x = \pm 15$$

We use Equation (3) to find y.

When $x = 6$, $y = 90/6 = 15$;

when $x = -6$, $y = 90/(-6) = -15$;

when $x = 15$, $y = 90/15 = 6$;

when $x = -15$, $y = 90/(-15) = -6$. We see that the numbers can be 6 and 15 or –6 and –15.

Check. $6 \cdot 15 = 90$ and $6^2 + 15^2 = 261$; also

$-6(-15) - 90$ and $(-6)^2 + (-15)^2 = 261$. The solutions check.

State. The numbers are 6 and 15 or –6 and –15.

55. Familiarize. We let $x =$ the length of a side of one flower bed, in feet, and $y =$ the length of a side of the other flower bed. Make a drawing.

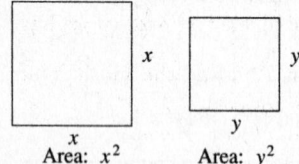

Area: x^2 Area: y^2

Translate. The sum of the areas is 832 ft^2, so we have

$$x^2 + y^2 = 832 . \quad (1)$$

The difference of the areas is 320 ft^2, so we have

$$x^2 - y^2 = 320 . \quad (2)$$

Carry out. We solve the system of equations.

$$x^2 + y^2 = 832 \quad (1)$$
$$\underline{x^2 - y^2 = 320} \quad (2)$$
$$2x^2 \qquad = 1152 \quad \text{Adding}$$
$$x^2 = 576$$
$$x = \pm 24$$

Since the length cannot be negative we consider only 24. We substitute 24 for x in Equation (1) and solve for y.

$$24^2 + y^2 = 832$$
$$576 + y^2 = 832$$
$$y^2 = 256$$
$$y = \pm 16$$

Again we consider only the positive number.

Check. If the lengths of the sides of the beds are 24 ft and 16 ft, the areas of the beds are 24^2, or 576 ft^2, and 16^2, or 256 ft^2, respectively. Then

$576 \text{ ft}^2 + 256 \text{ ft}^2 = 832 \text{ ft}^2$, and $576 \text{ ft}^2 - 256 \text{ ft}^2 = 320 \text{ ft}^2$, so the answer checks.

State. The lengths of the sides of the beds are 24 ft and 16 ft.

57. Familiarize. Let $l =$ the length and $w =$ the width of the rectangle area, in meters.

Translate. The area must be $\sqrt{3}$ m^2, so we have one equation: $lw = \sqrt{3}$

The Pythagorean equation gives us another equation:

$$l^2 + w^2 = 2^2, \text{ or } l^2 + w^2 = 4$$

Carry out. We solve the system of equations.

$$lw = \sqrt{3}, \quad (1)$$
$$l^2 + w^2 = 4 \quad (2)$$

First solve Equation (1) for l:

$$l = \frac{\sqrt{3}}{w} \quad (3)$$

Then substitute $\sqrt{3}/w$ for l in Equation (2) and solve for w.

$$l^2 + w^2 = 4$$
$$\left(\frac{\sqrt{3}}{w}\right)^2 + w^2 = 4$$
$$\frac{3}{w^2} + w^2 = 4$$
$$3 + w^4 = 4w^2$$
$$w^4 - 4w^2 + 3 = 0$$

Let $u = w^2$ and $u^2 = w^4$ and substitute.

$$u^2 - 4u + 3 = 0$$
$$(u - 1)(u - 3) = 0$$
$$u = 1 \quad or \quad u = 3$$
$$w^2 = 1 \quad or \quad w^2 = 3 \quad \text{Replacing } u \text{ with } w^2$$
$$w = \pm 1 \quad or \quad w = \pm\sqrt{3}$$

Since the length cannot be negative, we consider only 1 and $\sqrt{3}$. We substitute in Equation (3) to find l. When $w = \sqrt{3}$, $l = \sqrt{3}/\sqrt{3} = 1$; when $w = 1$, $l = \sqrt{3}/1 = \sqrt{3}$. Since we usually consider length to be longer than width, we check the pair $\left(\sqrt{3},\ 1\right)$.

Check. If the length is $\sqrt{3}$ m and the width is 1 m, then the area is $\sqrt{3}\cdot 1$, or $\sqrt{3}$ m^2. Also $\sqrt{3}^2 + 1^2 = 3 + 1 = 4 = 2^2$. The answer checks.

State. The length is $\sqrt{3}$ m and the width is 1 m.

59. *Writing Exercise*. When we can visualize the graphs of the equations in the system, we can determine how many real number solutions the system has and whether the solutions found algebraically seem reasonable.

61. $(-1)^9 (-3)^2 = -1 \cdot 9 = -9$

63. $\dfrac{(-1)^k}{k-6} = \dfrac{(-1)^7}{7-6} = \dfrac{-1}{1} = -1$

65. $\dfrac{n}{2}(3+n) = \dfrac{11}{2}(3+11) = \dfrac{11}{2}(14) = 77$

67. *Writing Exercise*. Answers may vary. One possibility is given. A rectangular banner has a diagonal of 2 ft and an area of 5 ft^2. Find its dimensions.

69. Let (h, k) represent the point on the line $5x + 8y = -2$ which is the center of a circle that passes through the points $(-2, 3)$ and $(-4, 1)$. The distance between (h, k) and $(-2, 3)$ is the same as the distance between (h, k) and $(-4, 1)$. This gives us one equation:

$$\sqrt{[h-(-2)]^2 + (k-3)^2} = \sqrt{[h-(-4)]^2 + (k-1)^2}$$
$$(h+2)^2 + (k-3)^2 = (h+4)^2 + (k-1)^2$$
$$h^2 + 4h + 4 + k^2 - 6k + 9 = h^2 + 8h + 16 + k^2 - 2k + 1$$
$$4h - 6k + 13 = 8h - 2k + 17$$
$$-4h - 4k = 4$$
$$h + k = -1$$

We get a second equation by substituting (h, k) in $5x + 8y = -2$.

$$5h + 8k = -2$$

We now solve the following system:
$$h + k = -1,$$
$$5h + 8k = -2$$

The solution, which is the center of the circle, is $(-2, 1)$. Next we find the length of the radius. We can find the distance between either $(-2, 3)$ or $(-4, 1)$ and the center $(-2, 1)$. We use $(-2, 3)$.

$$r = \sqrt{[-2-(-2)]^2 + (1-3)^2}$$
$$r = \sqrt{0^2 + (-2)^2}$$
$$r = \sqrt{4} = 2$$

We can write the equation of the circle with center $(-2, 1)$ and radius 2.

$$(x-h)^2 + (y-k)^2 = r^2$$
$$[x-(-2)]^2 + (y-1)^2 = 2^2$$
$$(x+2)^2 + (y-1)^2 = 4$$

71. $p^2 + q^2 = 13, \quad$ (1)
$\dfrac{1}{pq} = -\dfrac{1}{6} \quad$ (2)

Solve Equation (2) for p.

$$\frac{1}{q} = -\frac{p}{6}$$
$$-\frac{6}{q} = p$$

Substitute $-6/q$ for p in Equation (1) and solve for q.

$$\left(-\frac{6}{q}\right)^2 + q^2 = 13$$
$$\frac{36}{q^2} + q^2 = 13$$
$$36 + q^4 = 13q^2$$
$$q^4 - 13q^2 + 36 = 0$$
$$u^2 - 13u + 36 = 0 \quad \text{Letting } u = q^2$$
$$(u-9)(u-4) = 0$$
$$u = 9 \quad \text{or} \quad u = 4$$
$$x^2 = 9 \quad \text{or} \quad x^2 = 4$$
$$x = \pm 3 \quad \text{or} \quad x = \pm 2$$

Since $p = -6/q$, if $q = 3$, $p = -2$; if $q = -3$, $p = 2$; if $q = 2$, $p = -3$; and if $q = -2$, $p = 3$. The pairs $(-2, 3)$, $(2, -3)$, $(-3, 2)$, and $(3, -2)$ check. They are the solutions.

73. *Familiarize*. Let $l =$ the length of the rectangle, in feet, and let $w =$ the width.

Translate. 100 ft of fencing is used, so we have
$$l + w = 100. \quad (1)$$
The area is 2475 ft^2, so we have
$$lw = 2475. \quad (2)$$

Carry out. Solving the system of equations, we get $(55, 45)$ and $(45, 55)$. Since length is usually considered to be longer than width, we have $l = 55$ and $w = 45$.

Check. If the length is 55 ft and the width is 45 ft, then $55 + 45$, or 100 ft, of fencing is used. The area is $55 \cdot 45$, or 2475 ft^2. The answer checks.

State. The length of the rectangle is 55 ft, and the width is 45 ft.

75. *Familiarize*. We let x and y represent the length and width of the base of the box, in inches, respectively. Make a drawing.

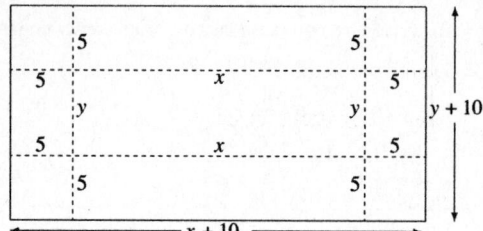

The dimensions of the metal sheet are $x + 10$ and $y + 10$.

Translate. The area of the sheet of metal is 340 in^2, so we have

$$(x+10)(y+10) = 340 . \quad (1)$$

The volume of the box is 350 in^3, so we have

$$x \cdot y \cdot 5 = 350 . \quad (2)$$

Carry out. Solving the system of equations, we get $(10, 7)$ and $(7, 10)$. Since length is usually considered to be longer than width, we have $l = 10$ and $w = 7$.

Check. The dimensions of the metal sheet are $10 + 10$, or 20, and $7 + 10$, or 17, so the area is $20 \cdot 17$, or 340 in^2. The volume of the box is $7 \cdot 10 \cdot 5$, or 350 in^3. The answer checks.

State. The dimensions of the box are 10 in. by 7 in. by 5 in.

77. *Familiarize*. Let $l =$ the length, and $h =$ the height, in inches.

Translate. Since the ratio of the length to height is 16 to 9, we have one equation:

$$\frac{l}{h} = \frac{16}{9}$$

The Pythagorean equation gives us a second equation:

$$l^2 + h^2 = 73^2$$

We have a system of equations.

$$\frac{l}{h} = \frac{16}{9}$$
$$l^2 + h^2 = 5329$$

Carry out. Solving the system of equations, we get $(35.8, 63.6)$ and $(-35.8, -63.6)$. Since the dimensions cannot be negative, we consider only $(35.8, 63.6)$.

Check. The ratio of 63.6 to 35.8 is $\frac{63.6}{35.8} \approx 1.777 \approx \frac{16}{9}$. Also, $(63.6)^2 + (35.8)^2 = 5326.6 \approx 5329$. The answer checks.

State. The length is about 63.6 in., and the height is about 35.8 in.

79. *Graphing Calculator Exercise*

Chapter 10 Review

---

1. True; see page 652 in the text.

3. False; see page 659 in the text.

5. True; see page 670 in the text.

7. False; see page 677 in the text.

9.
$$(x+3)^2 + (y-2)^2 = 16$$
$$[x-(-3)]^2 + (y-2)^2 = 16 \quad \text{Standard form}$$

The center is $(-3, 2)$ and the radius is 4.

11.
$$x^2 + y^2 - 6x - 2y + 1 = 0$$
$$x^2 - 6x + y^2 - 2y = -1$$
$$(x^2 - 6x + 9) + (y^2 - 2y + 1) = -1 + 9 + 1$$
$$(x-3)^2 + (y-1)^2 = 9$$
$$(x-3)^2 + (y-1)^2 = 3^2$$

The center is $(3, 1)$ and the radius is 3.

13.
$$(x-h)^2 + (y-k)^2 = r^2$$
$$[x-(-4)]^2 + (y-3)^2 = 4^2$$
$$(x+4)^2 + (y-3)^2 = 16$$

15. Circle
$$5x^2 + 5y^2 = 80$$
$$x^2 + y^2 = 16$$

Center: $(0, 0)$; radius: 4

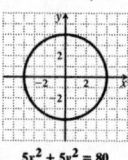

$5x^2 + 5y^2 = 80$

17. Parabola
$$y = -x^2 + 2x - 3$$
$$x = -\frac{b}{2a} = -\frac{2}{2(-1)} = 1$$
$$y = -x^2 + 2x - 3 = -1^2 + 2 \cdot 1 - 3 = -2$$

The vertex is $(1, -2)$. The graph opens downward because the coefficient of x^2, -1 is negative.

x	y
-1	-6
0	-3
1	-2
2	-3
3	-6

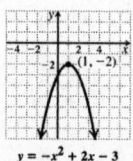

$y = -x^2 + 2x - 3$

19. Hyperbola

$xy = 9$

$y = \dfrac{9}{x}$ Solving for y

x	y
1	9
3	3
9	1
$\frac{1}{2}$	18
−1	−9
−3	−3
−9	−1
−$\frac{1}{2}$	−18

Note that we cannot use 0 for x. The x-axis and the y-axis are the asymptotes.

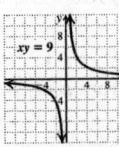

21. Ellipse

$$\dfrac{(x+1)^2}{3} + (y-3)^2 = 1 \quad 3 > 1 \text{ ellipse is horizontal}$$

The center is $(-1, 3)$. Note $a = \sqrt{3}$ and $b = 1$.

Vertices: $(-1, 2)$, $(-1, 4)$, $\left(-1 - \sqrt{3},\ 3\right)$, $\left(-1 + \sqrt{3},\ 3\right)$

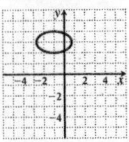

$$\dfrac{(x+1)^2}{3} + (y-3)^2 = 1$$

23. $x^2 - y^2 = 21$,   (1)

  $x + y = 3$ (2)

First we solve Equation (2) for y.

$$y = -x + 3 \quad (3)$$

Then we substitute $-x + 3$ for y in Equation (1) and solve for x.

$$x^2 - (-x+3)^2 = 21$$
$$x^2 - x^2 + 6x - 9 = 21$$
$$6x = 30$$
$$x = 5$$

Substitute 5 for x in Equation (3).

$$y = -5 + 3 = -2$$

The solution is $(5, -2)$.

25. $x^2 - y = 5$,   (1)

  $2x - y = 5$ (2)

First we solve Equation (2) for y.

$$y = 2x - 5 \quad (3)$$

Here we multiply Equation (2) by −1 and then add.

$$x^2 - y = 5$$
$$\underline{-2x + y = -5} \quad \text{Multiplying by } -1$$
$$x^2 - 2x = 0 \quad \text{Adding}$$
$$x(x - 2) = 0$$
$$x = 0 \text{ or } x = 2$$

Now we substitute these numbers for x in Equation (3).

If $x = 0,\ y = 2 \cdot 0 - 5 = -5$

If $x = 2,\ y = 2 \cdot 2 - 5 = -1$

The solutions are $(0, -5)$ and $(2, -1)$.

27. $x^2 - y^2 = 3$,   (1)

  $y = x^2 - 3$ (2)

First we solve Equation (2) for x.

$$y + 3 = x^2$$
$$\pm\sqrt{y + 3} = x \quad (3)$$

Adding Equations (1) and (2).

$$x^2 - y^2 = 3 \quad (1)$$
$$\underline{-x^2 + y = -3} \quad (2)$$
$$-y^2 + y = 0 \quad \text{Adding}$$
$$-y(y - 1) = 0$$
$$y = 0 \text{ or } y = 1$$

Now we substitute these numbers for y in Equation (3).

For $y = 0,\ x = \pm\sqrt{0 + 3} = \pm\sqrt{3}$

For $y = 1,\ x = \pm\sqrt{1 + 3} = \pm\sqrt{4} = \pm 2$

The solutions are $\left(\sqrt{3},\ 0\right)$, $\left(-\sqrt{3},\ 0\right)$, $(-2, 1)$ and $(2, 1)$.

29. $x^2 + y^2 = 100$,   (1)

  $2x^2 - 3y^2 = -120$ (2)

We use the elimination method.

$$-2x^2 - 2y^2 = -200 \quad \text{Multiplying (1) by } -2$$
$$\underline{2x^2 - 3y^2 = -120} \quad (2)$$
$$-5y^2 = -320 \quad \text{Adding}$$
$$y^2 = 64$$
$$y = \pm 8$$

Solve Equation (1) for x.

$$x^2 + y^2 = 100$$
$$x^2 = 100 - y^2$$
$$x = \pm\sqrt{100 - y^2} \quad (3)$$

Since x is solved in terms of y^2, we need only substitute once in Equation (3).

For $y^2 = 64,\ x = \pm\sqrt{100 - 64} = \pm\sqrt{36} = \pm 6$

The solutions are $(6, 8)$, $(6, -8)$, $(-6, 8)$ and $(-6, -8)$.

31. *Familiarize*. Let $l =$ the length and $w =$ the width of the bandstand.

Translate.

Perimeter: $2l + 2w = 38$, or $l + w = 19$

Area: $lw = 84$

Carry out. We solve the system:

Solve the first equation for l: $l = 19 - w$.

Substitute $19 - w$ for l in the second equation and solve for w.

$$(19-w)w = 84$$
$$19w - w^2 = 84$$
$$0 = w^2 - 19w + 84$$
$$0 = (w-7)(w-12)$$
$$w = 7 \quad or \quad w = 12$$

If $w = 7$, then $l = 19 - 7$, or 12. If $w = 12$, then $l = 19 - 12$, or 7. Since length is usually considered to be longer than width, we have the solution $l = 12$ and $w = 7$, or (12, 7)

Check. If $l = 12$ and $w = 7$, the area is $12 \cdot 7$, or 84. The perimeter is $2 \cdot 12 + 2 \cdot 7$, or 38. The numbers check.

State. The length is 12 m and the width is 7 m.

33. Familiarize. Let x represent the length of a side of one square mirror and y represent the length of a side of the other square mirror.

Translate.
$$4x = 4y + 12, \text{ or } x = y + 3$$
$$x^2 = y^2 + 39$$

Carry out. We solve the system of equations.
$$x = y + 3 \qquad (1)$$
$$x^2 = y^2 + 39 \qquad (2)$$

We substitute $y + 3$ for x into Equation (2).
$$(y+3)^2 = y^2 + 39$$
$$y^2 + 6y + 9 = y^2 + 39$$
$$6y + 9 = 39$$
$$6y = 30$$
$$y = 5$$

Then $x = 5 + 3 = 8$.

Check. If $x = 8$ and $y = 5$, then $4 \cdot 5 + 12 = 32 = 4 \cdot 8$, and $5^2 + 39 = 64 = 8^2$. The numbers check. The perimeter of the first mirror is $4 \cdot 8$, or 32, and the perimeter of the second mirror is $4 \cdot 5$, or 20.

State. The perimeter of each mirror is 32 cm and 20 cm, respectively.

35. Writing Exercise. The graph of a parabola has one branch whereas the graph of a hyperbola has two branches. A hyperbola has asymptotes, but a parabola does not.

37. $\quad 4x^2 - x - 3y^2 = 9, \quad (1)$
$\quad -x^2 + x + y^2 = 2 \quad (2)$

We use the elimination method.
$$4x^2 - x - 3y^2 = 9 \qquad (1)$$
$$\underline{-3x^2 + 3x + 3y^2 = 6} \qquad \text{Multiplying (2) by 3}$$
$$x^2 + 2x = 15 \qquad \text{Adding}$$
$$x^2 + 2x - 15 = 0$$
$$(x+5)(x-3) = 0$$

$$x = -5 \quad or \quad x = 3$$

Solving Equation (2) for y:
$$-x^2 + x + y^2 = 2$$
$$y^2 = x^2 - x + 2$$
$$y = \pm\sqrt{x^2 - x + 2}$$

For $x = -5$, $y = \pm\sqrt{(-5)^2 - (-5) + 2} = \pm\sqrt{32} = \pm 4\sqrt{2}$

For $x = 3$, $y = \pm\sqrt{3^2 - 3 + 2} = \pm\sqrt{8} = \pm 2\sqrt{2}$

The solutions are $\left(-5, -4\sqrt{2}\right)$, $\left(-5,\ 4\sqrt{2}\right)$, $\left(3, -2\sqrt{2}\right)$, $\left(3,\ 2\sqrt{2}\right)$.

39. The three points are equidistant from the center of the circle, (h, k). Using each of the three points in the equation of a circle, we get three different equations.

For $(-2, -4)$, $\quad [x-(-2)]^2 + [y-(-4)]^2 = r^2$
$$(x+2)^2 + (y+4)^2 = r^2$$
$$x^2 + 4x + 4 + y^2 + 8y + 16 = r^2$$
$$x^2 + 4x + y^2 + 8y + 20 = r^2 \qquad (1)$$

For $(5, -5)$, $\qquad (x-5)^2 + [y-(-5)]^2 = r^2$
$$(x-5)^2 + (y+5)^2 = r^2$$
$$x^2 - 10x + 25 + y^2 + 10y + 25 = r^2$$
$$x^2 - 10x + y^2 + 10y + 50 = r^2 \qquad (2)$$

For $(6, 2)$, $\qquad (x-6)^2 + (y-2)^2 = r^2$
$$x^2 - 12x + 36 + y^2 - 4y + 4 = r^2$$
$$x^2 - 12x + y^2 - 4y + 40 = r^2 \qquad (3)$$

Since the radius is equal, we can set Equation (1) equal to Equation (2) and simplify.
$$x^2 + 4x + y^2 + 8y + 20 = x^2 - 10x + y^2 + 10y + 50$$
$$14x - 2y = 30$$
$$7x - y = 15 \qquad (4)$$

Next, we set Equation (1) equal to Equation (3).
$$x^2 + 4x + y^2 + 8y + 20 = x^2 - 12x + y^2 - 4y + 40$$
$$16x + 12y = 20$$
$$4x + 3y = 5 \qquad (5)$$

We solve the system of Equations (4) and (5) using the elimination method.
$$21x - 3y = 45 \qquad \text{Multiplying (4) by 3}$$
$$\underline{4x + 3y = 5 \qquad (5)}$$
$$25x \qquad = 50 \qquad \text{Adding}$$
$$x = 2$$
$$y = -1$$

The center of the circle is $(2, -1)$. Thus the equation is
$$(x-2)^2 + (y+1)^2 = r^2.$$

We may choose any of the three points on the circle to determine r^2.

$$(6-2)^2 + (2+1)^2 = r^2$$
$$4^2 + 3^2 = r^2$$
$$25 = r^2$$

The equation of the circle is $(x-2)^2 + (y+1)^2 = 25$.

41. Let $(x, 0)$ be the point on the x-axis that is equidistant from $(-3, 4)$ and $(5, 6)$.

$$\sqrt{[x-(-3)]^2 + (0-4)^2} = \sqrt{(x-5)^2 + (0-6)^2}$$
$$x^2 + 6x + 9 + 16 = x^2 - 10x + 25 + 36$$
$$16x = 36$$
$$x = \frac{9}{4}$$

The point is $\left(\frac{9}{4}, 0\right)$.

Chapter 10 Test

1. For circle with center $(3, -4)$ and radius $2\sqrt{3}$,

$$(x-3)^2 + [y-(-4)]^2 = (2\sqrt{3})^2$$
$$(x-3)^2 + (y+4)^2 = 12$$

3.
$$x^2 + y^2 + 4x - 6y + 4 = 0$$
$$x^2 + 4x + y^2 - 6y = -4$$
$$(x^2 + 4x + 4) + (y^2 - 6y + 9) = -4 + 4 + 9$$
$$(x+2)^2 + (y-3)^2 = 3^2$$

The center is $(-2, 3)$ and the radius is 3.

5. Circle
$$x^2 + y^2 + 2x + 6y + 6 = 0$$
$$x^2 + 2x + y^2 + 6y = -6$$
$$(x^2 + 2x + 1) + (y^2 + 6y + 9) = -6 + 1 + 9$$
$$(x+1)^2 + (y+3)^2 = 4$$

The center is $(-1, -3)$ and the radius is 2.

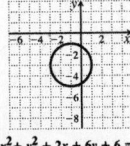

$x^2 + y^2 + 2x + 6y + 6 = 0$

7. Ellipse
$$16x^2 + 4y^2 = 64 \quad 4 < 16 \text{ ellipse is vertical}$$
$$\frac{x^2}{4} + \frac{y^2}{16} = 1$$

The center is $(0, 0)$. Note $a = 2$ and $b = 4$.

Vertices: $(0, 2)$, $(0, -2)$, $(4, 0)$, and $(-4, 0)$

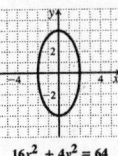

$16x^2 + 4y^2 = 64$

9. Parabola
$$x = -y^2 + 4y$$
$$y = -\frac{b}{2a} = -\frac{4}{2(-1)} = 2$$
$$x = -y^2 + 4y = -2^2 + 4 \cdot 2 = 4$$

The vertex is $(4, 2)$. The graph opens to the left because the coefficient of y^2, -1 is negative.

x	y
0	0
3	1
4	2
3	3
0	4

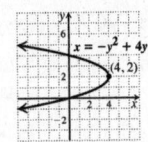

$x = -y^2 + 4y$

11. $x^2 - y = 3, \quad (1)$
$2x + y = 5 \quad (2)$

Use the elimination method.

$$\begin{aligned} x^2 - y &= 3 \quad (1) \\ 2x + y &= 5 \quad (2) \\ \hline x^2 + 2x &= 8 \quad \text{Adding} \end{aligned}$$
$$x^2 + 2x - 8 = 0$$
$$(x+4)(x-2) = 0$$
$$x = -4 \quad \text{or} \quad x = 2$$

Substitute for x in Equation (2).

$$y = 5 - 2x \quad \text{Solving Eq. (2) for } y$$

For $x = -4$, $y = 5 - 2(-4) = 13$
For $x = 2$, $y = 5 - 2(2) = 1$

The pairs $(-4, 13)$ and $(2, 1)$ check. They are the solutions.

13. $x^2 + y^2 = 10, \quad (1)$
$x^2 = y^2 + 2 \quad (2)$

Substitute $y^2 + 2$ for x^2 in Equation (1) and solve for y.

$$x^2 + y^2 = 10$$
$$y^2 + 2 + y^2 = 10$$
$$2y^2 - 8 = 0$$
$$2(y^2 - 4) = 0$$
$$2(y+2)(y-2) = 0$$
$$y = -2 \quad \text{or} \quad y = 2$$

Now substitute these numbers in Equation (2) and solve for x.

For $y = -2$, $\quad x^2 = (-2)^2 + 2 = 6$, so $x = \pm\sqrt{6}$
For $y = 2$, $\quad\quad x^2 = 2^2 + 2 = 6$, so $x = \pm\sqrt{6}$

The pairs $\left(-\sqrt{6}, -2\right)$, $\left(-\sqrt{6},\ 2\right)$, $\left(-\sqrt{6},\ 2\right)$, $\left(\sqrt{6},\ 2\right)$ check, so they are the solutions.

15. *Familiarize.* We let x = the length of a side of one square, in meters, and y = the length of a side of the other square. Make a drawing.

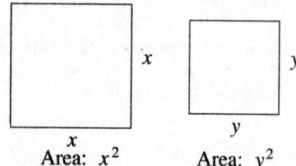

Area: x^2 $\quad\quad\quad$ Area: y^2

Translate. The sum of the areas is 8 m^2, so we have

$$x^2 + y^2 = 8. \quad (1)$$

The difference of the areas is 2 m^2, so we have

$$x^2 - y^2 = 2. \quad (2)$$

Carry out. We solve the system of equations.

$$
\begin{array}{rl}
x^2 + y^2 = 8 & (1) \\
\underline{x^2 - y^2 = 2} & (2) \\
2x^2 \quad = 10 & \text{Adding} \\
x^2 = 5 & \\
x = \pm\sqrt{5} &
\end{array}
$$

Since the length cannot be negative we consider only $\sqrt{5}$. We substitute $\sqrt{5}$ for x in Equation (1) and solve for y.

$$
\begin{aligned}
\left(\sqrt{5}\right)^2 + y^2 &= 8 \\
5 + y^2 &= 8 \\
y^2 &= 3 \\
y &= \pm\sqrt{3}
\end{aligned}
$$

Again we consider only the positive number.

Check. If the lengths of the sides of the squares are $\sqrt{5}$ m and $\sqrt{3}$ m, the areas of the squares are $\left(\sqrt{5}\right)^2$, or 5 m^2, and $\left(\sqrt{3}\right)^2$, or 3 ft^2, respectively. Then $5\text{ m}^2 + 3\text{ m}^2 = 8\text{ m}^2$, and $5\text{ m}^2 - 3\text{ m}^2 = 2\text{ m}^2$, so the answer checks.

State. The lengths of the sides of the squares are $\sqrt{5}$ m and $\sqrt{3}$ m.

17. *Familiarize.* Let p = the principal and r = the interest rate. We recall the formula $I = prt$. Since the time is one year, $t = 1$, and the formula simplifies to $I = pr$.

Translate.

Brett invested p dollars at interest rate r with \$72 in interest.

Erin invested $p + 240$ dollars at $\dfrac{5}{6}r$ interest rate with \$72 in interest. We have two equations.

$$
\begin{array}{ll}
72 = pr & (1) \\
72 = (p + 240)\dfrac{5}{6}r & (2)
\end{array}
$$

Carry out. We solve the system of equations using substitution.

We solve Equation (1) for r: $r = \dfrac{72}{p}$.

We substitute $\dfrac{72}{p}$ for r in Equation (2) and solve for p.

$$
\begin{aligned}
72 &= (p + 240)\frac{5}{6}\left(\frac{72}{p}\right) \\
72 &= (p + 240)\frac{60}{p} \\
72p &= 60p + 14{,}400 \\
12p &= 14{,}400 \\
p &= 1200 \\
r &= \frac{72}{1200} = 0.06 \text{ or } 6\%
\end{aligned}
$$

Check. $1200 \cdot 0.06 = 72$ and $(1200 + 240)\dfrac{5}{6} \cdot 0.06 = 72$, so the numbers check.

State. The principal was \$1200 and the interest rate was 6%.

19. Let $(0, y)$ be the point on the y-axis which is equidistant from $(-3, -5)$ and $(4, -7)$. We equate their distances and solve

$$
\begin{aligned}
\sqrt{[0 - (-3)]^2 + [y - (-5)]^2} &= \sqrt{(0 - 4)^2 + [y - (-7)]^2} \\
\sqrt{9 + y^2 + 10y + 25} &= \sqrt{16 + y^2 + 14y + 49} \\
y^2 + 10y + 34 &= y^2 + 14y + 65 \\
-4y &= 31 \\
y &= -\frac{31}{4}
\end{aligned}
$$

The point on the y-axis is $\left(0, -\dfrac{31}{4}\right)$.

21. Let the actor be in the center at $(0, 0)$. Using the information, we have $(-4, 0)$ and $(4, 0)$ and $(0, -7)$ and $(0, 7)$. Thus, $a = 4$ and $b = 7$. We write the equation of the ellipse.

$$
\begin{aligned}
\frac{x^2}{4^2} + \frac{y^2}{7^2} &= 1 \\
\frac{x^2}{16} + \frac{y^2}{49} &= 1
\end{aligned}
$$

Chapter 11

Sequences, Series, and the Binomial Theorem

Exercise Set 11.1

1. f

3. d

5. c

7. $a_n = 5n + 3$
$a_8 = 5 \cdot 8 + 3 = 40 + 3 = 43$

9. $a_n = (3n+1)(2n-5)$
$a_9 = (3 \cdot 9 + 1)(2 \cdot 9 - 5) = 28 \cdot 13 = 364$

11. $a_n = (-1)^{n-1}(3.4n - 17.3)$
$a_{12} = (-1)^{12-1}[3.4(12) - 17.3] = -23.5$

13. $a_n = 3n^2(9n - 100)$
$a_{11} = 3 \cdot 11^2(9 \cdot 11 - 100) = 3 \cdot 121(-1) = -363$

15. $a_n = \left(1 + \dfrac{1}{n}\right)^2$
$a_{20} = \left(1 + \dfrac{1}{20}\right)^2 = \left(\dfrac{21}{20}\right)^2 = \dfrac{441}{400}$

17. $a_n = 3n - 1$
$a_1 = 3 \cdot 1 - 1 = 2$
$a_2 = 3 \cdot 2 - 1 = 5$
$a_3 = 3 \cdot 3 - 1 = 8$
$a_4 = 3 \cdot 4 - 1 = 11$
$a_{10} = 3 \cdot 10 - 1 = 29$
$a_{15} = 3 \cdot 15 - 1 = 44$

19. $a_n = n^2 + 2$
$a_1 = 1^2 + 2 = 3$
$a_2 = 2^2 + 2 = 6$
$a_3 = 3^2 + 2 = 11$
$a_4 = 4^2 + 2 = 18$
$a_{10} = 10^2 + 2 = 102$
$a_{15} = 15^2 + 2 = 227$

21. $a_n = \dfrac{n}{n+1}$
$a_1 = \dfrac{1}{1+1} = \dfrac{1}{2}$
$a_2 = \dfrac{2}{2+1} = \dfrac{2}{3}$
$a_3 = \dfrac{3}{3+1} = \dfrac{3}{4}$
$a_4 = \dfrac{4}{4+1} = \dfrac{4}{5}$
$a_{10} = \dfrac{10}{10+1} = \dfrac{10}{11}$
$a_{15} = \dfrac{15}{15+1} = \dfrac{15}{16}$

23. $a_n = \left(-\dfrac{1}{2}\right)^{n-1}$
$a_1 = \left(-\dfrac{1}{2}\right)^{1-1} = 1$
$a_2 = \left(-\dfrac{1}{2}\right)^{2-1} = -\dfrac{1}{2}$
$a_3 = \left(-\dfrac{1}{2}\right)^{3-1} = \dfrac{1}{4}$
$a_4 = \left(-\dfrac{1}{2}\right)^{4-1} = -\dfrac{1}{8}$
$a_{10} = \left(-\dfrac{1}{2}\right)^{10-1} = -\dfrac{1}{512}$
$a_{15} = \left(-\dfrac{1}{2}\right)^{15-1} = \dfrac{1}{16,384}$

25. $a_n = \dfrac{(-1)^n}{n}$
$a_1 = \dfrac{(-1)^1}{1} = -1$
$a_2 = \dfrac{(-1)^2}{2} = \dfrac{1}{2}$
$a_3 = \dfrac{(-1)^3}{3} = -\dfrac{1}{3}$
$a_4 = \dfrac{(-1)^4}{4} = \dfrac{1}{4}$
$a_{10} = \dfrac{(-1)^{10}}{10} = \dfrac{1}{10}$
$a_{15} = \dfrac{(-1)^{15}}{15} = -\dfrac{1}{15}$

27. $a_n = (-1)^n (n^3 - 1)$

$a_1 = (-1)^1 (1^3 - 1) = 0$

$a_2 = (-1)^2 (2^3 - 1) = 7$

$a_3 = (-1)^3 (3^3 - 1) = -26$

$a_4 = (-1)^4 (4^3 - 1) = 63$

$a_{10} = (-1)^{10} (10^3 - 1) = 999$

$a_{15} = (-1)^{15} (15^3 - 1) = -3374$

29. 2, 4, 6, 8, 10,...

These are even integers beginning with 2, so the general term could be $2n$.

31. $-1, 1, -1, 1,...$

-1 and 1 alternate, beginning with -1, so the general term could be $(-1)^n$.

33. $1, -2, 3, -4,...$

These are the first four natural numbers, but with alternating signs, beginning with a positive number. The general term could be $(-1)^{n+1} \cdot n$.

35. 3, 5, 7, 9,...

These are odd integers beginning with 3, so the general term could be $2n + 1$.

37. 0, 3, 8, 15, 24,...

We can see a pattern if we write the sequence $1^2 - 1$, $2^2 - 1$, $3^2 - 1$, $4^2 - 1$, $5^2 - 1$,... The general term could be $n^2 - 1$, or $(n+1)(n-1)$.

39. $\dfrac{1}{2}, \dfrac{2}{3}, \dfrac{3}{4}, \dfrac{4}{5}, \dfrac{5}{6},...$

These are fractions in which the denominator is 1 greater than the numerator. Also, each numerator is 1 greater than the preceding numerator. The general term could be $\dfrac{n}{n+1}$.

41. 0.1, 0.01, 0.001, 0.0001,...

This is negative powers of 10, or positive powers of 0.1, so the general term is 10^{-n}, or $(0.1)^n$.

43. $-1, 4, -9, 16,...$

This is the squares of the first four natural numbers, but with alternating signs, beginning with a negative number. The general terms could be $(-1)^n \cdot n^2$.

45. $-1, 2, -3, 4, -5, 6,...$

$S_{10} = -1 + 2 - 3 + 4 - 5 + 6 - 7 + 8 - 9 + 10 = 5$

47. $1, \dfrac{1}{10}, \dfrac{1}{100}, \dfrac{1}{1000},...$

$S_6 = 1 + \dfrac{1}{10} + \dfrac{1}{100} + \dfrac{1}{1000} + \dfrac{1}{10,000} + \dfrac{1}{100,000} = 1.11111$

49. $\displaystyle\sum_{k=1}^{5} \dfrac{1}{2k} = \dfrac{1}{2 \cdot 1} + \dfrac{1}{2 \cdot 2} + \dfrac{1}{2 \cdot 3} + \dfrac{1}{2 \cdot 4} + \dfrac{1}{2 \cdot 5}$

$= \dfrac{1}{2} + \dfrac{1}{4} + \dfrac{1}{6} + \dfrac{1}{8} + \dfrac{1}{10}$

$= \dfrac{60}{120} + \dfrac{30}{120} + \dfrac{20}{120} + \dfrac{15}{120} + \dfrac{12}{120}$

$= \dfrac{137}{120}$

51. $\displaystyle\sum_{k=0}^{4} 10^k = 10^0 + 10^1 + 10^2 + 10^3 + 10^4$

$= 1 + 10 + 100 + 1000 + 10,000$

$= 11,111$

53. $\displaystyle\sum_{k=2}^{8} \dfrac{k}{k-1}$

$= \dfrac{2}{2-1} + \dfrac{3}{3-1} + \dfrac{4}{4-1} + \dfrac{5}{5-1} + \dfrac{6}{6-1} + \dfrac{7}{7-1} + \dfrac{8}{8-1}$

$= \dfrac{2}{1} + \dfrac{3}{2} + \dfrac{4}{3} + \dfrac{5}{4} + \dfrac{6}{5} + \dfrac{7}{6} + \dfrac{8}{7}$

$= \dfrac{1343}{140}$

55. $\displaystyle\sum_{k=1}^{8} (-1)^{k+1} 2^k$

$= (-1)^{1+1} 2^1 + (-1)^{2+1} 2^2 + (-1)^{3+1} 2^3 + (-1)^{4+1} 2^4$

$\quad + (-1)^{5+1} 2^5 + (-1)^{6+1} 2^6 + (-1)^{7+1} 2^7 + (-1)^{8+1} 2^8$

$= 2 - 4 + 8 - 16 + 32 - 64 + 128 - 256$

$= -170$

57. $\displaystyle\sum_{k=0}^{5} (k^2 - 2k + 3)$

$= (0^2 - 2 \cdot 0 + 3) + (1^2 - 2 \cdot 1 + 3) + (2^2 - 2 \cdot 2 + 3)$

$\quad + (3^2 - 2 \cdot 3 + 3) + (4^2 - 2 \cdot 4 + 3) + (5^2 - 2 \cdot 5 + 3)$

$= 3 + 2 + 3 + 6 + 11 + 18$

$= 43$

59. $\displaystyle\sum_{k=3}^{5} \dfrac{(-1)^k}{k(k+1)} = \dfrac{(-1)^3}{3(3+1)} + \dfrac{(-1)^4}{4(4+1)} + \dfrac{(-1)^5}{5(5+1)}$

$= \dfrac{-1}{3 \cdot 4} + \dfrac{1}{4 \cdot 5} + \dfrac{-1}{5 \cdot 6}$

$= -\dfrac{1}{12} + \dfrac{1}{20} - \dfrac{1}{30}$

$= -\dfrac{4}{60} = -\dfrac{1}{15}$

61. $\dfrac{2}{3} + \dfrac{3}{4} + \dfrac{4}{5} + \dfrac{5}{6} + \dfrac{6}{7}$

This is a sum of fractions in which the denominator is one greater than the numerator. Also, each numerator is 1 greater than the preceding numerator. Sigma notation is

$\displaystyle\sum_{k=1}^{5} \dfrac{k+1}{k+2}$.

63. $1+4+9+16+25+36$

This is the sum of the squares of the first six natural numbers. Sigma notation is

$$\sum_{k=1}^{6} k^2.$$

65. $4-9+16-25+...+(-1)^n n^2$

This is a sum of terms of the form $(-1)^k k^2$, beginning with $k=2$ and continuing through $k=n$. Sigma notation is

$$\sum_{k=2}^{n} (-1)^k k^2.$$

67. $6+12+18+24+...$

This is the sum of all the positive multiples of 6. It is an infinite series. Sigma notation is

$$\sum_{k=1}^{\infty} 6k.$$

69. $\dfrac{1}{1\cdot 2}+\dfrac{1}{2\cdot 3}+\dfrac{1}{3\cdot 4}+\dfrac{1}{4\cdot 5}+...$

This is a sum of fractions in which the numerator is 1 and the denominator is a product of two consecutive integers. The larger integer in each product is the smaller integer in the succeeding product. It is an infinite series. Sigma notation is

$$\sum_{k=1}^{\infty} \dfrac{1}{k(k+1)}.$$

71. *Writing Exercise.* The graph of f is a set of points $\left(x,\ x^2\right)$ where x is a natural number. The graph of $y=x^2$ for $x>0$ is formed by connecting these points with a smooth curve. The graph of $y=x^2$ also contains the points $\left(x,\ x^2\right)$ for $x\le 0$.

73. $\dfrac{7}{2}(a_1+a_7)=\dfrac{7}{2}(8+20)=\dfrac{7}{2}(28)=98$

75. $\left(a_1+3d\right)+d=a_1+3d+d=a_1+4d$

77. $\left(a_1+a_n\right)+\left(a_1+a_n\right)+\left(a_1+a_n\right)$
$=3\left(a_1+a_n\right)=3a_1+3a_n$

79. *Writing Exercise.*

$$\sum_{k=1}^{n}\left(a_k+b_k\right)=\left(a_1+b_1\right)+\left(a_2+b_2\right)+...+\left(a_n+b_n\right)$$
$$=\left(a_1+a_2+...+a_n\right)+\left(b_1+b_2+...+b_n\right)$$

Using the commutative and associative laws of addition

$$=\sum_{k=1}^{n} a_k+\sum_{k=1}^{n} b_k$$

81. $a_1=1,\ a_{n+1}=5a_n-2$
$a_1=1$
$a_2=5\cdot 1-2=3$
$a_3=5\cdot 3-2=13$
$a_4=5\cdot 13-2=63$
$a_5=5\cdot 63-2=313$
$a_6=5\cdot 313-2=1563$

83. Find each term by multiplying the preceding term by 0.80: \$2500, \$2000, \$1600, \$1280, \$1024, \$819.20, \$655.36, \$524.29, \$419.43, \$335.54.

85. $a_n=(-1)^n$

This sequence is of the form $-1, 1, -1, 1, ...$ Each pair of terms adds to 0. S_{100} has 50 such pairs, so $S_{100}=0$. S_{101} consists of the 50 pairs in S_{100} that add to 0 as well as a_{101}, or -1, so $S_{101}=-1$.

87. $a_n=i^n$
$a_1=i^1=1$
$a_2=i^2=-1$
$a_3=i^3=i^2\cdot i=-1\cdot i=-i$
$a_4=i^4=\left(i^2\right)^2=(-1)^2=1$
$a_5=i^5=\left(i^2\right)^2\cdot i=(-1)^2\cdot i=1\cdot i=i$
$S_5=i-1-i+1+i=i$

89. Enter $y_1=x^5-14x^4+6x^3+416x^2-655x-1050$. Then scroll through a table of values. We see that $y_1=6144$ when $x=11$, so the 11th term of the sequence is 6144.

Exercise Set 11.2

1. True; see page 699 in the text.

3. False; see page 699 in the text.

5. True; see page 701 in the text.

7. False; $S_5=a_1+a_2+a_3+a_4+a_5$.

9. 8, 13, 18, 23, ...

$a_1 = 8$

$d = 5$ $(13 - 8 = 5, 18 - 13 = 5, 23 - 18 = 5)$

11. 7, 3, -1, -5, ...

$a_1 = 7$

$d = -4$ $(3 - 7 = -4, -1 - 3 = -4, -5 - (-1) = -4)$

13. $\dfrac{3}{2}, \dfrac{9}{4}, 3, \dfrac{15}{4}, ...$

$a_1 = \dfrac{3}{2}$

$d = \dfrac{3}{4}$ $\left(\dfrac{9}{4} - \dfrac{3}{2} = \dfrac{3}{4}, \; 3 - \dfrac{9}{4} = \dfrac{3}{4}\right)$

15. $8.16, $8.46, $8.76, $9.06, ...

$a_1 = 8.16

$d = 0.30 $($8.46 - $8.16 = $0.30,$

$8.76 - $8.46 = $0.30,$

$9.06 - $8.76 = $0.30)$

17. 10, 18, 26, ...

$a_1 = 10, \; d = 8, \text{ and } n = 19$

$a_n = a_1 + (n-1)d$

$a_{19} = 10 + (19-1)8 = 10 + 18 \cdot 8 = 10 + 144 = 154$

19. 8, 2, -4, ...

$a_1 = 8, \; d = -6, \text{ and } n = 18$

$a_n = a_1 + (n-1)d$

$a_{18} = 8 + (18-1)(-6) = 8 + 17(-6) = 8 - 102 = -94$

21. $1200, $964.32, $728.64, ...

$a_1 = $1200, \; d = $964.32 - $1200 = -$235.68$

and $n = 13$

$a_n = a_1 + (n-1)d$

$a_{13} = $1200 + (13-1)(-$235.68)$

$= $1200 + 12(-$235.68) = $1200 - 2828.16

$= -$1628.16$

23. $a_1 = 10, \; d = 8$

$a_n = a_1 + (n-1)d$

Let $a_n = 210,$ and solve for n.

$210 = 10 + (n-1)8$

$210 = 10 + 8n - 8$

$210 = 2 + 8n$

$208 = 8n$

$26 = n$

The 26th term is 210.

25. $a_1 = 8, \; d = -6$

$a_n = a_1 + (n-1)d$

$-328 = 8 + (n-1)(-6)$

$-328 = 8 - 6n + 6$

$-328 = 14 - 6n$

$-342 = -6n$

$57 = n$

The 57th term is -328.

27. $a_n = a_1 + (n-1)d$

$a_{18} = 8 + (18-1)10$ Substituting 18 for n,

8 for a_1, and 10 for d

$= 8 + 17 \cdot 10$

$= 8 + 170$

$= 178$

29. $a_n = a_1 + (n-1)d$

$33 = a_1 + (8-1)4$ Substituting 33 for a_8,

8 for n, and 4 for d

$33 = a_1 + 28$

$5 = a_1$

31. $a_n = a_1 + (n-1)d$

$-76 = 5 + (n-1)(-3)$ Substituting -76 for a_n,

5 for a_1, and -3 for d

$-76 = 5 - 3n + 3$

$-76 = 8 - 3n$

$-84 = -3n$

$28 = n$

33. We know that $a_{17} = -40$ and $a_{28} = -73$. We would have to add d eleven times to get from a_{17} to a_{28}. That is,

$-40 + 11d = -73$

$11d = -33$

$d = -3.$

Since $a_{17} = -40,$ we subtract d sixteen times to get a_1.

$a_1 = -40 - 16(-3) = -40 + 48 = 8$

We write the first five terms of the sequence:

8, 5, 2, -1, -4.

35. $a_{13} = 13$ and $a_{54} = 54$

Observe that for this to be true, $a_1 = 1$ and $d = 1$.

37. $1 + 5 + 9 + 13 + ...$

Note that $a_1 = 1, \; d = 4, \text{ and } n = 20$. Before using the formula for S_n, we find a_{20}:

$a_{20} = 1 + (20-1)4$ Substituting into

the formula for a_n

$= 1 + 19 \cdot 4$

$= 77$

Then using the formula for S_n,

$S_{20} = \dfrac{20}{2}(1 + 77) = 10(78) = 780.$

39. The sum is $1+2+3+\ldots+249+250$. This is the sum of the arithmetic sequence for which $a_1=1$, $a_n=250$, and $n=250$. We use the formula for S_n.

$$S_n = \frac{n}{2}(a_1+a_n)$$
$$S_{250} = \frac{250}{2}(1+250) = 125(251) = 31{,}375$$

41. The sum is $2+4+6+\ldots+98+100$. This is the sum of the arithmetic sequence for which $a_1=2$, $a_n=100$, and $n=50$. We use the formula for S_n.

$$S_n = \frac{n}{2}(a_1+a_n)$$
$$S_{50} = \frac{50}{2}(2+100) = 25(102) = 2550$$

43. The sum is $6+12+18+\ldots+96+102$. This is the sum of the arithmetic sequence for which $a_1=6$, $a_n=102$, and $n=17$. We use the formula for S_n.

$$S_n = \frac{n}{2}(a_1+a_n)$$
$$S_{17} = \frac{17}{2}(6+102) = \frac{17}{2}(108) = 918$$

45. Before using the formula for S_n, we find a_{20}:

$$a_{20} = 4+(20-1)5 \qquad \text{Substituting into the formula for } a_n$$
$$= 4+19\cdot5$$
$$= 99$$

Then using the formula for S_n,

$$S_{20} = \frac{20}{2}(4+99) = 10(103) = 1030.$$

47. *Familiarize*. We want to find the fifteenth term and the sum of an arithmetic sequence with $a_1=7$, $d=2$, and $n=15$. We will first use the formula for a_n to find a_{15}. This result is the number of musicians in the last row. Then we will use the formula for S_n to find S_{15}. This is the total number of musicians .

Translate. Substituting into the formula for a_n, we have

$$a_{15} = 7+(15-1)2.$$

Carry out. We first find a_{15}.

$$a_{15} = 7+14\cdot2 = 35$$

Then use the formula for S_n to find S_{15}.

$$S_{15} = \frac{15}{2}(7+35) = \frac{15}{2}(42) = 315$$

Check. We can do the calculations again. We can also do the entire addition.

$$7+9+11+\ldots+35.$$

State. There are 35 musicians in the last row, and there are 315 musicians altogether.

49. *Familiarize*. We want to find the sum of the arithmetic sequence $36+32+\ldots+4$. Note that $a_1=36$, and $d=-4$. We will first use the formula for a_n to find n. Then we will use the formula for S_n.

Translate. Substituting into the formula for a_n, we have

$$4 = 36+(n-1)(-4).$$

Carry out. We solve for n.

$$4 = 36+(n-1)(-4)$$
$$4 = 36-4n+4$$
$$4 = 40-4n$$
$$-36 = -4n$$
$$9 = n$$

Now we find S_9.

$$S_9 = \frac{9}{2}(36+4) = \frac{9}{2}(40) = 180$$

Check. We can do the calculations again. We can also do the entire addition.

$$36+32+\ldots+4.$$

State. There are 180 stones in the pyramid.

51. *Familiarize*. We want to find the sum of the arithmetic sequence with $a_1=10¢$, $d=10¢$, and $n=31$. First we will find a_{31} and then we will find S_{31}.

Translate. Substituting in the formula for a_n, we have

$$a_{31} = 10+(31-1)(10).$$

Carry out. First we find a_{31}.

$$a_{31} = 10+30\cdot10 = 10+300 = 310$$

Then we use the formula for S_n to find S_{31}.

$$S_{31} = \frac{31}{2}(10+310) = \frac{31}{2}(320) = 4960$$

Check. We can do the calculations again.

State. The amount saved is 4960¢, or \$49.60.

53. *Familiarize*. We want to find the sum of an arithmetic sequence with $a_1=20$, $d=2$, and $n=16$. We will use the formula for a_n to find a_{16}, and then we will use the formula for S_n to find S_{16}.

Translate. Substituting into the formula for a_n, we have $a_{16} = 20+(16-1)2$.

Carry out. We find a_{16}.

$$a_{16} = 20+15\cdot2 = 50$$

Then we use the formula for S_n to find S_{16}.

$$S_{16} = \frac{16}{2}(20+50) = 560$$

Check. We do the calculations again.

State. There are 560 seats.

55. *Writing Exercise.*

$1 + 2 + 3 + ... + 100$
$= (1 + 100) + (2 + 99) + (3 + 98) + ... + (50 + 51)$
$= \underbrace{101 + 101 + 101 + ... + 101}_{\text{50 addends of 101}}$
$= 50 \cdot 101$
$= 5050$

57. Using the slope-intercept form, where $m = \frac{1}{3}$ and $b = 10$,

we have $y = \frac{1}{3}x + 10$.

59. Rewrite the equation.

$2x + y = 8$
$\qquad y = -2x + 8$

The slope of the parallel line is -2.

Use point-slope form.

$y - y_1 = m(x - x_1)$
$y - 0 = -2(x - 5)$
$\qquad y = -2x + 10$

61. A circle with center $(0, 0)$ and radius 4 is $x^2 + y^2 = 16$.

63. *Writing Exercise.* To explain why S_n is always an integer, we recall how S_n was first developed.

$S_n = a_1 + (a_1 + d) + (a_1 + 2d) + ... + (a_n - 2d) + (a_n - d) + a_n$

Since a_1, d and a_n are all integers, the sum is also an integer.

65. The frog climbs $4 - 1$, or 3 ft, with each jump. Then the total distance the frog has jumped with each successive jump is given by the arithmetic sequence 3, 6, 9, ..., 96. When the frog has climbed 96 ft, it will reach the top of the hole on the next jump because it will have climbed $96 + 4$, or 100 ft with that jump. Then the total number of jumps is the number of terms of the sequence above plus the final jump. We find n for the sequence with $a_1 = 3$, $d = 3$, and $a_n = 96$:

$a_n = a_1 + (n - 1)d$
$96 = 3 + (n - 1)3$
$96 = 3 + 3n - 3$
$96 = 3n$
$32 = n$

The total number of jumps is $32 + 1$, or 33 jumps.

67. Let $d = $ the common difference. Since p, m, and q form an arithmetic sequence, $m = p + d$, and $q = p + 2d$. Then

$\dfrac{p + q}{2} = \dfrac{p + (p + 2d)}{2} = p + d = m.$

69. Each integer from 501 through 750 is 500 more than the corresponding integer from 1 through 250. There are 250 integers from 501 through 750, so their sum is the sum of the integers from 1 to 250 plus $250 \cdot 500$. From Exercise 39, we know that the sum of the integers from 1 through 250 is 31,375. Thus, we have

$31,375 + 250 \cdot 500$, or $156,375.$

Exercise Set 11.3

1. $\dfrac{a_{n+1}}{a_n} = 2$, so this is a geometric sequence.

3. $a_{n+1} = a_n - 3$, so this is a arithmetic sequence.

5. $\dfrac{a_{n+1}}{a_n} = 5$, so this is a geometric series.

7. $\dfrac{a_{n+1}}{a_n} = -\dfrac{1}{2}$, so this is a geometric series.

9. 10, 20, 40, 80, ...

$\dfrac{20}{10} = 2$, $\dfrac{40}{20} = 2$, $\dfrac{80}{40} = 2$

$r = 2$

11. 6, −0.6, 0.06, −0.006, ...

$-\dfrac{0.6}{6} = -0.1$, $\dfrac{0.06}{-0.6} = -0.1$, $\dfrac{-0.006}{0.06} = -0.1$

$r = -0.1$

13. $\dfrac{1}{2}$, $-\dfrac{1}{4}$, $\dfrac{1}{8}$, $-\dfrac{1}{16}$, ...

$\dfrac{-\frac{1}{4}}{\frac{1}{2}} = -\dfrac{1}{4} \cdot \dfrac{2}{1} = -\dfrac{2}{4} = -\dfrac{1}{2}$, $\dfrac{\frac{1}{8}}{-\frac{1}{4}} = \dfrac{1}{8} \cdot \left(-\dfrac{4}{1}\right) = -\dfrac{4}{8} = -\dfrac{1}{2}$,

$\dfrac{-\frac{1}{16}}{\frac{1}{8}} = -\dfrac{1}{16} \cdot \dfrac{8}{1} = -\dfrac{8}{16} = -\dfrac{1}{2}$

$r = -\dfrac{1}{2}$

15. 75, 15, 3, $\dfrac{3}{5}$, ...

$\dfrac{15}{75} = \dfrac{1}{5}$, $\dfrac{3}{15} = \dfrac{1}{5}$, $\dfrac{\frac{3}{5}}{3} = \dfrac{3}{5} \cdot \dfrac{1}{3} = \dfrac{1}{5}$

$r = \dfrac{1}{5}$

17. $\dfrac{1}{m}, \dfrac{6}{m^2}, \dfrac{36}{m^3}, \dfrac{216}{m^4}, \ldots$

$$\dfrac{\frac{6}{m^2}}{\frac{1}{m}} = \dfrac{6}{m^2} \cdot \dfrac{m}{1} = \dfrac{6}{m}, \quad \dfrac{\frac{36}{m^3}}{\frac{6}{m^2}} = \dfrac{36}{m^3} \cdot \dfrac{m^2}{6} = \dfrac{6}{m}$$

$$\dfrac{\frac{216}{m^4}}{\frac{36}{m^3}} = \dfrac{216}{m^4} \cdot \dfrac{m^3}{36} = \dfrac{6}{m}$$

$$r = \dfrac{6}{m}$$

19. $2, 6, 18, \ldots$

$a_1 = 2$, $n = 7$, and $r = \dfrac{6}{2} = 3$

We use the formula $a_n = a_1 r^{n-1}$.

$a_7 = 2 \cdot 3^{7-1} = 2 \cdot 3^6 = 2 \cdot 729 = 1458$

21. $\sqrt{3}, 3, 3\sqrt{3}, \ldots$

$a_1 = \sqrt{3}$, $n = 10$, and $r = \dfrac{3\sqrt{3}}{3} = \sqrt{3}$

$a_n = a_1 r^{n-1}$

$a_{10} = \sqrt{3}\left(\sqrt{3}\right)^{10-1} = \sqrt{3}\left(\sqrt{3}\right)^9 = \left(\sqrt{3}\right)^{10} = 243$

23. $-\dfrac{8}{243}, \dfrac{8}{81}, -\dfrac{8}{27}, \ldots$

$a_1 = -\dfrac{8}{243}$, $n = 14$, and $r = \dfrac{\frac{8}{81}}{-\frac{8}{243}} = \dfrac{8}{81}\left(-\dfrac{243}{8}\right) = -3$

$a_n = a_1 r^{n-1}$

$a_{14} = -\dfrac{8}{243}(-3)^{14-1} = -\dfrac{8}{243}(-3)^{13}$

$\qquad = -\dfrac{8}{243}(-1{,}594{,}323) = 52{,}488$

25. $\$1000, \$1040, \$1081.60, \ldots$

$a_1 = \$1000$, $n = 10$, and $r = \dfrac{1040}{1000} = 1.04$

$a_n = a_1 r^{n-1}$

$a_{10} = \$1000(1.04)^{10-1} \approx \$1000(1.423311812) \approx \$1423.31$

27. $1, 5, 25, 125, \ldots$

$a_1 = 1$, and $r = \dfrac{5}{1} = 5$

$a_n = a_1 r^{n-1}$

$a_n = 1 \cdot 5^{n-1} = 5^{n-1}$

29. $1, -1, 1, -1, \ldots$

$a_1 = 1$, and $r = \dfrac{-1}{1} = -1$

$a_n = a_1 r^{n-1}$

$a_n = 1(-1)^{n-1} = (-1)^{n-1}$

31. $\dfrac{1}{x}, \dfrac{1}{x^2}, \dfrac{1}{x^3}, \ldots$

$a_1 = \dfrac{1}{x}$, and $r = \dfrac{\frac{1}{x^2}}{\frac{1}{x}} = \dfrac{1}{x^2} \cdot \dfrac{x}{1} = \dfrac{1}{x}$

$a_n = a_1 r^{n-1}$

$a_n = \dfrac{1}{x}\left(\dfrac{1}{x}\right)^{n-1} = \dfrac{1}{x} \cdot \dfrac{1}{x^{n-1}} = \dfrac{1}{x^{1+n-1}} = \dfrac{1}{x^n}$, or x^{-n}

33. $6 + 12 + 24 + \ldots$

$a_1 = 6$, $n = 9$, and $r = \dfrac{12}{6} = 2$

$S_n = \dfrac{a_1\left(1 - r^n\right)}{1 - r}$

$S_9 = \dfrac{6\left(1 - 2^9\right)}{1 - 2} = \dfrac{6(1 - 512)}{-1} = \dfrac{6(-511)}{-1} = 3066$

35. $\dfrac{1}{18} - \dfrac{1}{6} + \dfrac{1}{2} - \ldots$

$a_1 = \dfrac{1}{18}$, $n = 7$, and $r = \dfrac{-\frac{1}{6}}{\frac{1}{18}} = -\dfrac{1}{6} \cdot \dfrac{18}{1} = -3$

$S_n = \dfrac{a_1\left(1 - r^n\right)}{1 - r}$

$S_7 = \dfrac{\frac{1}{18}\left[1 - (-3)^7\right]}{1 - (-3)} = \dfrac{\frac{1}{18}(1 + 2187)}{4} = \dfrac{\frac{1}{18}(2188)}{4}$

$\qquad = \dfrac{1}{18}(2188)\left(\dfrac{1}{4}\right) = \dfrac{547}{18}$

37. $1 + x + x^2 + x^3 + \ldots$

$a_1 = 1$, $n = 8$, and $r = \dfrac{x}{1}$, or x

$S_n = \dfrac{a_1\left(1 - r^n\right)}{1 - r}$

$S_8 = \dfrac{1\left(1 - x^8\right)}{1 - x} = \dfrac{\left(1 + x^4\right)\left(1 - x^4\right)}{1 - x}$

$\quad = \dfrac{\left(1 + x^4\right)\left(1 + x^2\right)\left(1 - x^2\right)}{1 - x}$

$\quad = \dfrac{\left(1 + x^4\right)\left(1 + x^2\right)(1 + x)(1 - x)}{1 - x}$

$\quad = \left(1 + x^4\right)\left(1 + x^2\right)(1 + x)$

39. $\$200 + \$200(1.06) + \$200(1.06)^2 + \ldots$

$a_1 = \$200$, $n = 16$, and $r = \dfrac{\$200(1.06)}{\$200} = 1.06$

$S_n = \dfrac{a_1\left(1 - r^n\right)}{1 - r}$

$S_{16} = \dfrac{\$200\left[1 - (1.06)^{16}\right]}{1 - 1.06} \approx \dfrac{\$200(1 - 2.540351685)}{-0.06}$

$\qquad \approx \$5134.51$

41. $18 + 6 + 2 + \ldots$

$|r| = \left|\dfrac{6}{18}\right| = \left|\dfrac{1}{3}\right| = \dfrac{1}{3}$, and since $|r| < 1$, the series does have a limit.

$S_\infty = \dfrac{a_1}{1-r} = \dfrac{18}{1 - \dfrac{1}{3}} = \dfrac{18}{\dfrac{2}{3}} = 18 \cdot \dfrac{3}{2} = 27$

43. $7 + 3 + \dfrac{9}{7} + \ldots$

$|r| = \left|\dfrac{3}{7}\right| = \dfrac{3}{7}$, and since $|r| < 1$, the series does have a limit.

$S_\infty = \dfrac{a_1}{1-r} = \dfrac{7}{1 - \dfrac{3}{7}} = \dfrac{7}{\dfrac{4}{7}} = 7 \cdot \dfrac{7}{4} = \dfrac{49}{4}$

45. $3 + 15 + 75 + \ldots$

$|r| = \left|\dfrac{15}{3}\right| = |5| = 5$, and since $|r| \not< 1$, the series does not have a limit.

47. $4 - 6 + 9 - \dfrac{27}{2} + \ldots$

$|r| = \left|\dfrac{-6}{4}\right| = \left|-\dfrac{3}{2}\right| = \dfrac{3}{2}$, and since $|r| \not< 1$, the series does not have a limit.

49. $0.43 + 0.0043 + 0.000043 + \ldots$

$|r| = \left|\dfrac{0.0043}{0.43}\right| = |0.01| = 0.01$, and since $|r| < 1$, the series does have a limit.

$S_\infty = \dfrac{a_1}{1-r} = \dfrac{0.43}{1 - 0.01} = \dfrac{0.43}{0.99} = \dfrac{43}{99}$

51. $\$500(1.02)^{-1} + \$500(1.02)^{-2} + \$500(1.02)^{-3} + \ldots$

$|r| = \left|\dfrac{\$500(1.02)^{-2}}{\$500(1.02)^{-1}}\right| = |(1.02)^{-1}| = (1.02)^{-1}$, or $\dfrac{1}{1.02}$, and since $|r| < 1$, the series does have a limit.

$S_\infty = \dfrac{a_1}{1-r} = \dfrac{\$500(1.02)^{-1}}{1 - \left(\dfrac{1}{1.02}\right)} = \dfrac{\dfrac{\$500}{1.02}}{\dfrac{0.02}{1.02}} = \dfrac{\$500}{1.02} \cdot \dfrac{1.02}{0.02}$

$= \$25{,}000$

53. $0.5555\ldots = 0.5 + 0.05 + 0.005 + 0.0005 + \ldots$

This is an infinite geometric series with $a_1 = 0.5$.

$|r| = \left|\dfrac{0.05}{0.5}\right| = |0.1| = 0.1 < 1$, so the series has a limit.

$S_\infty = \dfrac{a_1}{1-r} = \dfrac{0.5}{1 - 0.1} = \dfrac{0.5}{0.9} = \dfrac{5}{9}$

Fractional notation for $0.5555\ldots$ is $\dfrac{5}{9}$.

55. $3.4646\ldots = 3 + 0.4646\ldots$

$0.464646\ldots = 0.46 + 0.0046 + 0.000046 + \ldots$

$|r| = \left|\dfrac{0.0046}{0.46}\right| = |0.01| = 0.01 < 1$, so the series has a limit.

$S_\infty = \dfrac{a_1}{1-r} = \dfrac{0.46}{1 - 0.01} = \dfrac{0.46}{0.99} = \dfrac{46}{99}$

Fractional notation for $0.4646\ldots$ is $\dfrac{46}{99}$.

Fractional notation for $3.4646\ldots$ is $3 + \dfrac{46}{99} = \dfrac{343}{99}$.

57. $0.15151515\ldots = 0.15 + 0.0015 + 0.000015 + \ldots$

This is an infinite geometric series with $a_1 = 0.15$.

$|r| = \left|\dfrac{0.0015}{0.15}\right| = |0.01| = 0.01 < 1$, so the series has a limit.

$S_\infty = \dfrac{a_1}{1-r} = \dfrac{0.15}{1 - 0.01} = \dfrac{0.15}{0.99} = \dfrac{15}{99} = \dfrac{5}{33}$

Fractional notation for $0.15151515\ldots$ is $\dfrac{5}{33}$.

59. *Familiarize.* The rebound distances form a geometric sequence:

$$\dfrac{1}{4} \times 20, \quad \left(\dfrac{1}{4}\right)^2 \times 20, \quad \left(\dfrac{1}{4}\right)^3 \times 20, \ldots,$$

or $5, \quad \dfrac{1}{4} \times 5, \quad \left(\dfrac{1}{4}\right)^2 \times 5, \ldots$

The height of the 6th rebound is the 6th term of the sequence.

Translate. We will use the formula $a_n = a_1 r^{n-1}$, with $a_1 = 5$, $r = \dfrac{1}{4}$, and $n = 6$:

$$a_6 = 5\left(\dfrac{1}{4}\right)^{6-1}$$

Carry out. We calculate to obtain $a_6 = \dfrac{5}{1024}$.

Check. We can do the calculation again.

State. It rebounds $\dfrac{5}{1024}$ ft the 6th time.

61. *Familiarize.* In one year, the population will be $100{,}000 + 0.03(100{,}000)$, or $(1.03)100{,}000$. In two years, the population will be $(1.03)100{,}000 + 0.03(1.03)100{,}000$, or $(1.03)^2 100{,}000$. Thus, the populations form a geometric sequence:

$100{,}000, \quad (1.03)100{,}000, \quad (1.03)^2 100{,}000, \ldots$

The population in 15 years will be the 16th term of the sequence.

Translate. We will use the formula $a_n = a_1 r^{n-1}$ with $a_1 = 100{,}000$, $r = 1.03$, and $n = 16$:

$$a_{16} = 100{,}000(1.03)^{16-1}$$

Carry out. We calculate to obtain $a_{16} \approx 155{,}797$.

Check. We can do the calculation again.

State. In 15 years the population will be about $155{,}797$.

63. Familiarize. At the end of each minute the population is 96% of the previous population.

We have a geometric sequence:

$$5000,\ 5000(0.96),\ 5000(0.96)^2, \ldots$$

The number of fruit flies remaining alive after 15 minutes is given by the 16th term of the sequence.

Translate. We use the formula $a_n = a_1 r^{n-1}$ with $a_1 = 5000$, $r = 0.96$, and $n = 16$:

$$a_{16} = 5000(0.96)^{16-1}$$

Carry out. We calculate to obtain $a_{16} \approx 2710$.

Check. We can do the calculation again.

State. About 2710 flies will be alive after 15 min.

65. Familiarize. Each year the number of espresso-based coffees sold in the U.S. is 104% of the number sold the previous year. These numbers form a geometric sequence:

$$17,\ 17(1.04),\ 17(1.04)^2, \ldots$$

The number of espresso-based coffees sold from 2007 to 2015 is the sum of the first 9 terms of this sequence.

Translate. We use the formula $S_n = \dfrac{a_1(1 - r^n)}{1 - r}$ with $a_1 = 17$, $r = 1.04$ and $n = 9$.

$$S_9 = \frac{17(1 - 1.04^9)}{1 - 1.04}$$

Carry out. We use a calculator to obtain

$$S_9 \approx 179.9 \text{ billion}$$

Check. We can do the calculation again.

State. About 179.9 billion espresso-based coffees were sold from 2007 to 2015.

67. Familiarize. The lengths of the falls form a geometric sequence:

$$556,\ 556\left(\frac{3}{4}\right),\ 556\left(\frac{3}{4}\right)^2,\ 556\left(\frac{3}{4}\right)^3, \ldots$$

The total length of the first 6 falls is the sum of the first six terms of this sequence. The heights of the rebounds also form a geometric sequence:

$$556\left(\frac{3}{4}\right),\ 556\left(\frac{3}{4}\right)^2,\ 556\left(\frac{3}{4}\right)^3, \ldots \text{ or }$$

$$417,\ 417\left(\frac{3}{4}\right),\ 417\left(\frac{3}{4}\right)^2, \ldots$$

When the ball hits the ground for the 6th time, it will

have rebounded 5 times. Thus the total length of the rebounds is the sum of the first five terms of this sequence.

Translate. We use the formula $S_n = \dfrac{a_1(1 - r^n)}{1 - r}$ twice, once with $a_1 = 556$, $r = \dfrac{3}{4}$, and $n = 6$ and a second time with $a_1 = 417$, $r = \dfrac{3}{4}$, and $n = 5$.

$D = $ Length of falls + length of rebounds

$$= \frac{556\left[1 - \left(\frac{3}{4}\right)^6\right]}{1 - \frac{3}{4}} + \frac{417\left[1 - \left(\frac{3}{4}\right)^5\right]}{1 - \frac{3}{4}}$$

Carry out. We use a calculator to obtain $D \approx 3100.35$.

Check. We can do the calculations again.

State. The ball will have traveled about 3100.35 ft.

69. Familiarize. The heights of the stack form a geometric sequence:

$$0.02,\ 0.02(2),\ 0.02(2)^2,\ \ldots$$

The height of the stack after it is doubled 10 times is given by the 11th term of this sequence.

Translate. We have a geometric sequence with $a_1 = 0.02$, $r = 2$, and $n = 11$. We use the formula $a_n = a_1 r^{n-1}$.

Carry out. We substitute and calculate.

$$a_{11} = 0.02\left(2^{11-1}\right)$$
$$a_{11} = 0.02(1024) = 20.48$$

Check. We can do the calculations again.

State. The final stack will be 20.48 in. high.

71. Writing Exercise. One circumstance in which this situation occurs is in an alternating sequence, with $a_1 > 0$ and $r > 1$. One example is the sequence

$$1,\ -2,\ 4,\ -8,\ 16,\ -32,\ 64, \ldots$$

73. $(x + y)^2 = (x + y)(x + y) = x^2 + 2xy + y^2$

75. $(x - y)^3 = (x - y)(x - y)(x - y)$
$$= \left(x^2 - 2xy + y^2\right)(x - y)$$
$$= x^3 - 3x^2 y + 3xy^2 - y^3$$

77. $(2x + y)^3 = (2x + y)(2x + y)(2x + y)$
$$= \left(4x^2 + 4xy + y^2\right)(2x + y)$$
$$= 8x^3 + 12x^2 y + 6xy^2 + y^3$$

79. *Writing Exercise.* Answers may vary. One possibility is given.

Casey invests $900 at 8% interest, compounded annually. How much will be in the account at the end of 40 years?

81. $\displaystyle\sum_{k=1}^{\infty} 6(0.9)^k = 6(0.9) + 6(0.9)^2 + 6(0.9)^3 + \ldots$

$|r| = \left|\dfrac{6(0.9)^2}{6(0.9)}\right| = |0.9| = 0.9 < 1,$ so the series has a limit.

$S_\infty = \dfrac{a_1}{1-r} = \dfrac{6(0.9)}{1-0.9} = \dfrac{5.4}{0.1} = 54$

83. $x^2 - x^3 + x^4 - x^5 + \ldots$

This is a geometric series with $a_1 = x^2$ and $r = -x$.

$S_n = \dfrac{a_1\left(1-r^n\right)}{1-r} = \dfrac{x^2\left[1-(-x)^n\right]}{1-(-x)} = \dfrac{x^2\left[1-(-x)^n\right]}{1+x}$

85. The length of a side of the first square is 16 cm. The length of a side of the next square is the length of the hypotenuse of a right triangle with legs 8 cm and 8 cm, or $8\sqrt{2}$ cm. The length of a side of the next square is the length of the hypotenuse of a right triangle with legs $4\sqrt{2}$ cm and $4\sqrt{2}$ cm, or 8 cm. The areas of the squares form a sequence:

$(16)^2, \; \left(8\sqrt{2}\right)^2, \; (8)^2, \; \ldots,$ or
$256, \; 128, \; 64, \; \ldots$

This is a geometric series with $a_1 = 256$ and $r = \dfrac{1}{2}$.

We find the sum of the infinite geometric series

$256 + 128 + 64 + \ldots$

$S_\infty = \dfrac{a_1}{1-r} = \dfrac{256}{1-\dfrac{1}{2}} = \dfrac{256}{\dfrac{1}{2}} = 512 \text{ cm}^2$

87. *Writing Exercise.* If the graph shows that the points (n, a_n) approach a horizontal line as n increases, then the geometric series has a limit. If this does not occur, then the series does not have a limit.

Connecting the Concepts

1. $a_n = n^2 - 5n$
$a_{20} = 20^2 - 5\cdot 20 = 300$

3. 1, 2, 3, 4, ...

Note that $a_1 = 1$, $d = 1$, and $n = 12$. Before using the formula to find S_{12}, we find a_{12}.

$a_{12} = 1 + (12-1)1 = 12$

Then using the formula for S_n,

$S_{12} = \dfrac{12}{2}(1+12) = 6(13) = 78.$

5. $1 - 2 + 3 - 4 + 5 - 6 = \displaystyle\sum_{k=1}^{6}(-1)^{k+1}\cdot k$

7. 10, 15, 20, 25, ...

Note that $a_1 = 10$, $d = 5$, and $n = 21$.

$a_{21} = 10 + (21-1)5 = 10 + 100 = 110$

9. $a_n = a_1 + (n-1)d$
$a_{25} = 9 + (25-1)(-2) = 9 + (24)(-2) = -39$

11. $a_n = a_1 + (n-1)d$

$0 = 5 + (n-1)\left(-\dfrac{1}{2}\right)$

$-5 = -\dfrac{n}{2} + \dfrac{1}{2}$

$10 = n-1$

$11 = n$

13. $\dfrac{1}{3}, \; -\dfrac{1}{6}, \; \dfrac{1}{12}, \; -\dfrac{1}{24}, \; \ldots$

$r = \dfrac{-\dfrac{1}{6}}{\dfrac{1}{3}} = -\dfrac{1}{6}\cdot\dfrac{3}{1} = -\dfrac{1}{2}$

15. 2, -2, 2, -2, ...

$a_1 = 2$, and $r = \dfrac{-2}{2} = -1$

$a_n = a_1 r^{n-1}$

$a_n = 2(-1)^{n-1}$ or $2(-1)^{n+1}$

17. $0.9 + 0.09 + 0.009 + \ldots$

$|r| = \left|\dfrac{0.09}{0.9}\right| = |0.1| = 0.1 < 1,$ so the series has a limit.

$S_\infty = \dfrac{0.9}{1-0.1} = \dfrac{0.9}{0.9} = 1$

Thus, $0.9 + 0.09 + 0.009 + \ldots = 1$.

19. $\$1 + \$2 + \$3 + \$4 + \ldots$

This is an arithmetic sequence $a_1 = 1$, $d = 1$, and $n = 30$. Before using the formula to find S_{30}, we find a_{30}.

$a_{30} = 1 + (30-1)1 = 30$

Then using the formula for S_n,

$S_{30} = \dfrac{30}{2}(1+30) = 15(31) = 465.$

She earns $465.

Exercise Set 11.4

1. 2^5, or 32

3. 9

5. $\binom{8}{5}$, or $\binom{8}{3}$

7. 1

9. $4! = 4 \cdot 3 \cdot 2 \cdot 1 = 24$

11. $10! = 10 \cdot 9 \cdot 8 \cdot 7 \cdot 6 \cdot 5 \cdot 4 \cdot 3 \cdot 2 \cdot 1 = 3{,}628{,}800$

13. $\dfrac{10!}{8!} = \dfrac{10 \cdot 9 \cdot 8!}{8!} = 10 \cdot 9 = 90$

15. $\dfrac{9!}{4!5!} = \dfrac{9 \cdot 8 \cdot 7 \cdot 6 \cdot 5!}{4!5!} = \dfrac{9 \cdot 8 \cdot 7 \cdot 6}{4 \cdot 3 \cdot 2 \cdot 1} = 3 \cdot 7 \cdot 6 = 126$

17. $\binom{10}{4} = \dfrac{10!}{6!4!} = \dfrac{10 \cdot 9 \cdot 8 \cdot 7 \cdot 6!}{6!4!} = \dfrac{10 \cdot 9 \cdot 8 \cdot 7}{4 \cdot 3 \cdot 2 \cdot 1} = 10 \cdot 3 \cdot 7 = 210$

19. $\binom{9}{9} = \dfrac{9!}{0!9!} = \dfrac{9!}{1 \cdot 9!} = \dfrac{9!}{9!} = 1$

21. $\binom{30}{2} = \dfrac{30!}{28!2!} = \dfrac{30 \cdot 29 \cdot 28!}{28!2!} = \dfrac{30 \cdot 29}{2 \cdot 1} = 15 \cdot 29 = 435$

23. $\binom{40}{38} = \dfrac{40!}{2!38!} = \dfrac{40 \cdot 39 \cdot 38!}{2!38!} = \dfrac{40 \cdot 39}{2 \cdot 1} = 20 \cdot 39 = 780$

25. Expand $(a-b)^4$.

We have $a = a$, $b = -b$, and $n = 4$.

Form 1: We use the fifth row of Pascal's triangle:

$$1 \quad 4 \quad 6 \quad 4 \quad 1$$

$$(a-b)^4 = 1 \cdot a^4 + 4a^3(-b) + 6a^2(-b)^2 + 4a(-b)^3 + 1 \cdot (-b)^4$$
$$= a^4 - 4a^3b + 6a^2b^2 - 4ab^3 + b^4$$

Form 2:

$$(a-b)^4 = \binom{4}{0}a^4 + \binom{4}{1}a^3(-b) + \binom{4}{2}a^2(-b)^2$$
$$+ \binom{4}{3}a(-b)^3 + \binom{4}{4}(-b)^4$$
$$= \frac{4!}{4!0!}a^4 + \frac{4!}{3!1!}a^3(-b) + \frac{4!}{2!2!}a^2(-b)^2$$
$$+ \frac{4!}{1!3!}a(-b)^3 + \frac{4!}{0!4!}(-b)^4$$
$$= a^4 - 4a^3b + 6a^2b^2 - 4ab^3 + b^4$$

27. Expand $(p+q)^7$.

We have $a = p$, $b = q$, and $n = 7$.

Form 1: We use the 8th row of Pascal's triangle:

$$1 \quad 7 \quad 21 \quad 35 \quad 35 \quad 21 \quad 7 \quad 1$$

$$(p+q)^7$$
$$= p^7 + 7p^6q^1 + 21p^5q^2 + 35p^4q^3$$
$$+ 35p^3q^4 + 21p^2q^5 + 7pq^6 + q^7$$

Form 2:

$$(p+q)^7$$
$$= \binom{7}{0}p^7 + \binom{7}{1}p^6q^1 + \binom{7}{2}p^5q^2 + \binom{7}{3}p^4q^3$$
$$+ \binom{7}{4}p^3q^4 + \binom{7}{5}p^2q^5 + \binom{7}{6}pq^6 + \binom{7}{7}q^7$$
$$= \frac{7!}{7!0!}p^7 + \frac{7!}{6!1!}p^6q^1 + \frac{7!}{5!2!}p^5q^2 + \frac{7!}{4!3!}p^4q^3$$
$$+ \frac{7!}{3!4!}p^3q^4 + \frac{7!}{2!5!}p^2q^5 + \frac{7!}{1!6!}pq^6 + \frac{7!}{0!7!}q^7$$
$$= p^7 + 7p^6q^1 + 21p^5q^2 + 35p^4q^3$$
$$+ 35p^3q^4 + 21p^2q^5 + 7pq^6 + q^7$$

29. Expand $(3c-d)^7$.

We have $a = 3c$, $b = -d$, and $n = 7$.

Form 1: We use the 8th row of Pascal's triangle:

$$1 \quad 7 \quad 21 \quad 35 \quad 35 \quad 21 \quad 7 \quad 1$$

$$(3c-d)^7$$
$$= (3c)^7 + 7(3c)^6(-d)^1 + 21(3c)^5(-d)^2 + 35(3c)^4(-d)^3$$
$$+ 35(3c)^3(-d)^4 + 21(3c)^2(-d)^5 + 7(3c)(-d)^6 + (-d)^7$$
$$= 2187c^7 - 5103c^6d + 5103c^5d^2 - 2835c^4d^3$$
$$+ 945c^3d^4 - 189c^2d^5 + 21cd^6 - d^7$$

Form 2:

$$(3c-d)^7$$
$$= \binom{7}{0}(3c)^7 + \binom{7}{1}(3c)^6(-d)^1 + \binom{7}{2}(3c)^5(-d)^2$$
$$+ \binom{7}{3}(3c)^4(-d)^3 + \binom{7}{4}(3c)^3(-d)^4 + \binom{7}{5}(3c)^2(-d)^5$$
$$+ \binom{7}{6}(3c)(-d)^6 + \binom{7}{7}(-d)^7$$
$$= \frac{7!}{7!0!}(3c)^7 + \frac{7!}{6!1!}(3c)^6(-d)^1 + \frac{7!}{5!2!}(3c)^5(-d)^2$$
$$+ \frac{7!}{4!3!}(3c)^4(-d)^3 + \frac{7!}{3!4!}(3c)^3(-d)^4 + \frac{7!}{2!5!}(3c)^2(-d)^5$$
$$+ \frac{7!}{1!6!}(3c)(-d)^6 + \frac{7!}{0!7!}(-d)^7$$
$$= 2187c^7 - 5103c^6d + 5103c^5d^2 - 2835c^4d^3$$
$$+ 945c^3d^4 - 189c^2d^5 + 21cd^6 - d^7$$

31. Expand $(t^{-2}+2)^6$.

We have $a = t^{-2}$, $b = 2$, and $n = 6$.

Form 1: We use the 7th row of Pascal's triangle:

$$1 \quad 6 \quad 15 \quad 20 \quad 15 \quad 6 \quad 1$$

$$(t^{-2}+2)^6$$
$$= 1 \cdot (t^{-2})^6 + 6(t^{-2})^5(2)^1 + 15(t^{-2})^4(2)^2 + 20(t^{-2})^3(2)^3$$
$$+ 15(t^{-2})^2(2)^4 + 6(t^{-2})^1(2)^5 + 1 \cdot (2)^6$$
$$= t^{-12} + 12t^{-10} + 60t^{-8} + 160t^{-6} + 240t^{-4} + 192t^{-2} + 64$$

Form 2:

$$\left(t^{-2}+2\right)^6$$
$$=\binom{6}{0}\left(t^{-2}\right)^6+\binom{6}{1}\left(t^{-2}\right)^5(2)^1+\binom{6}{2}\left(t^{-2}\right)^4(2)^2$$
$$\quad+\binom{6}{3}\left(t^{-2}\right)^3(2)^3+\binom{6}{4}\left(t^{-2}\right)^2(2)^4$$
$$\quad+\binom{6}{5}\left(t^{-2}\right)^1(2)^5+\binom{6}{6}(2)^6$$
$$=\frac{6!}{6!0!}\left(t^{-2}\right)^6+\frac{6!}{5!1!}\left(t^{-2}\right)^5(2)^1+\frac{6!}{4!2!}\left(t^{-2}\right)^4(2)^2$$
$$\quad+\frac{6!}{3!3!}\left(t^{-2}\right)^3(2)^3+\frac{6!}{2!4!}\left(t^{-2}\right)^2(2)^4$$
$$\quad+\frac{6!}{1!5!}\left(t^{-2}\right)^1(2)^5+\frac{6!}{0!6!}(2)^6$$
$$=t^{-12}+12t^{-10}+60t^{-8}+160t^{-6}+240t^{-4}+192t^{-2}+64$$

33. Expand $(x-y)^5$.

We have $a=x$, $b=-y$, and $n=5$.

Form 1: We use the 6th row of Pascal's triangle:

$$1\quad 5\quad 10\quad 10\quad 5\quad 1$$

$$(x-y)^5$$
$$=1\cdot x^5+5x^4(-y)^1+10x^3(-y)^2$$
$$\quad+10x^2(-y)^3+5x^1(-y)^4+1\cdot(-y)^5$$
$$=x^5-5x^4y+10x^3y^2-10x^2y^3+5xy^4-y^5$$

Form 2:

$$(x-y)^5=\binom{5}{0}x^5+\binom{5}{1}x^4(-y)+\binom{5}{2}x^3(-y)^2$$
$$\quad+\binom{5}{3}x^2(-y)^3+\binom{5}{4}x(-y)^4+\binom{5}{5}(-y)^5$$
$$=\frac{5!}{5!0!}x^5+\frac{5!}{4!1!}x^4(-y)+\frac{5!}{3!2!}x^3(-y)^2$$
$$\quad+\frac{5!}{2!3!}x^2(-y)^3+\frac{5!}{1!4!}x(-y)^4+\frac{5!}{0!5!}(-y)^5$$
$$=x^5-5x^4y+10x^3y^2-10x^2y^3+5xy^4-y^5$$

35. Expand $\left(3s+\frac{1}{t}\right)^9$.

We have $a=3s$, $b=\frac{1}{t}$, and $n=9$.

Form 1: We use the tenth row of Pascal's triangle:

$$1\quad 9\quad 36\quad 84\quad 126\quad 126\quad 84\quad 36\quad 9\quad 1$$

$$\left(3s+\frac{1}{t}\right)^9$$
$$=1\cdot(3s)^9+9(3s)^8\left(\frac{1}{t}\right)^1+36(3s)^7\left(\frac{1}{t}\right)^2+84(3s)^6\left(\frac{1}{t}\right)^3$$
$$\quad+126(3s)^5\left(\frac{1}{t}\right)^4+126(3s)^4\left(\frac{1}{t}\right)^5+84(3s)^3\left(\frac{1}{t}\right)^6$$
$$\quad+36(3s)^2\left(\frac{1}{t}\right)^7+9(3s)^1\left(\frac{1}{t}\right)^8+1\cdot\left(\frac{1}{t}\right)^9$$
$$=19{,}683s^9+\frac{59{,}049s^8}{t}+\frac{78{,}732s^7}{t^2}+\frac{61{,}236s^6}{t^3}$$
$$\quad+\frac{30{,}618s^5}{t^4}+\frac{10{,}206s^4}{t^5}+\frac{2268s^3}{t^6}+\frac{324s^2}{t^7}+\frac{27s}{t^8}+\frac{1}{t^9}$$

Form 2:

$$\left(3s+\frac{1}{t}\right)^9$$
$$=\binom{9}{0}(3s)^9+\binom{9}{1}(3s)^8\left(\frac{1}{t}\right)^1+\binom{9}{2}(3s)^7\left(\frac{1}{t}\right)^2+\binom{9}{3}(3s)^6\left(\frac{1}{t}\right)^3$$
$$\quad+\binom{9}{4}(3s)^5\left(\frac{1}{t}\right)^4+\binom{9}{5}(3s)^4\left(\frac{1}{t}\right)^5+\binom{9}{6}(3s)^3\left(\frac{1}{t}\right)^6$$
$$\quad+\binom{9}{7}(3s)^2\left(\frac{1}{t}\right)^7+\binom{9}{8}(3s)^1\left(\frac{1}{t}\right)^8+\binom{9}{9}\left(\frac{1}{t}\right)^9$$
$$=\frac{9!}{9!0!}(3s)^9+\frac{9!}{8!1!}(3s)^8\left(\frac{1}{t}\right)^1+\frac{9!}{7!2!}(3s)^7\left(\frac{1}{t}\right)^2+\frac{9!}{6!3!}(3s)^6\left(\frac{1}{t}\right)^3$$
$$\quad+\frac{9!}{5!4!}(3s)^5\left(\frac{1}{t}\right)^4+\frac{9!}{4!5!}(3s)^4\left(\frac{1}{t}\right)^5+\frac{9!}{3!6!}(3s)^3\left(\frac{1}{t}\right)^6$$
$$\quad+\frac{9!}{2!7!}(3s)^2\left(\frac{1}{t}\right)^7+\frac{9!}{1!8!}(3s)^1\left(\frac{1}{t}\right)^8+\frac{9!}{0!9!}\left(\frac{1}{t}\right)^9$$
$$=19{,}683s^9+\frac{59{,}049s^8}{t}+\frac{78{,}732s^7}{t^2}+\frac{61{,}236s^6}{t^3}$$
$$\quad+\frac{30{,}618s^5}{t^4}+\frac{10{,}206s^4}{t^5}+\frac{2268s^3}{t^6}+\frac{324s^2}{t^7}+\frac{27s}{t^8}+\frac{1}{t^9}$$

37. Expand $\left(x^3-2y\right)^5$.

We have $a=x^3$, $b=-2y$, and $n=5$.

Form 1: We use the 6th row of Pascal's triangle:

$$1\quad 5\quad 10\quad 10\quad 5\quad 1$$

$$\left(x^3-2y\right)^5$$
$$=1\cdot\left(x^3\right)^5+5\left(x^3\right)^4(-2y)^1+10\left(x^3\right)^3(-2y)^2$$
$$\quad+10\left(x^3\right)^2(-2y)^3+5\left(x^3\right)^1(-2y)^4+1\cdot(-2y)^5$$
$$=x^{15}-10x^{12}y+40x^9y^2-80x^6y^3+80x^3y^4-32y^5$$

Form 2:

$$\left(x^3-2y\right)^5$$
$$=\binom{5}{0}\left(x^3\right)^5+\binom{5}{1}\left(x^3\right)^4(-2y)+\binom{5}{2}\left(x^3\right)^3(-2y)^2$$
$$\quad+\binom{5}{3}\left(x^3\right)^2(-2y)^3+\binom{5}{4}\left(x^3\right)(-2y)^4+\binom{5}{5}(-2y)^5$$
$$=\frac{5!}{5!0!}\left(x^3\right)^5+\frac{5!}{4!1!}\left(x^3\right)^4(-2y)+\frac{5!}{3!2!}\left(x^3\right)^3(-2y)^2$$
$$\quad+\frac{5!}{2!3!}\left(x^3\right)^2(-2y)^3+\frac{5!}{1!4!}\left(x^3\right)(-2y)^4+\frac{5!}{0!5!}(-2y)^5$$
$$=x^{15}-10x^{12}y+40x^9y^2-80x^6y^3+80x^3y^4-32y^5$$

39. Expand $\left(\sqrt{5}+t\right)^6$.

We have $a=\sqrt{5}$, $b=t$, and $n=6$.

Form 1: We use the 7th row of Pascal's triangle:

$$1\quad 6\quad 15\quad 20\quad 15\quad 6\quad 1$$

$$\left(\sqrt{5}+t\right)^6$$
$$=1\cdot\left(\sqrt{5}\right)^6+6\left(\sqrt{5}\right)^5t^1+15\left(\sqrt{5}\right)^4t^2+20\left(\sqrt{5}\right)^3t^3$$
$$\quad+15\left(\sqrt{5}\right)^2t^4+6\left(\sqrt{5}\right)^1t^5+1\cdot t^6$$
$$=125+125\sqrt{5}\,t+375t^2+100\sqrt{5}\,t^3+75t^4+6\sqrt{5}\,t^5+t^6$$

Form 2:

$$\left(\sqrt{5}+t\right)^6$$

$$=\binom{6}{0}\left(\sqrt{5}\right)^6+\binom{6}{1}\left(\sqrt{5}\right)^5 t^1+\binom{6}{2}\left(\sqrt{5}\right)^4 t^2+\binom{6}{3}\left(\sqrt{5}\right)^3 t^3$$

$$+\binom{6}{4}\left(\sqrt{5}\right)^2 t^4+\binom{6}{5}\left(\sqrt{5}\right)^1 t^5+\binom{6}{6}t^6$$

$$=\frac{6!}{6!0!}\left(\sqrt{5}\right)^6+\frac{6!}{5!1!}\left(\sqrt{5}\right)^5 t^1+\frac{6!}{4!2!}\left(\sqrt{5}\right)^4 t^2+\frac{6!}{3!3!}\left(\sqrt{5}\right)^3 t^3$$

$$+\frac{6!}{2!4!}\left(\sqrt{5}\right)^2 t^4+\frac{6!}{1!5!}\left(\sqrt{5}\right)^1 t^5+\frac{6!}{0!6!}t^6$$

$$=125+125\sqrt{5}\,t+375t^2+100\sqrt{5}\,t^3+75t^4+6\sqrt{5}\,t^5+t^6$$

41. Expand $\left(\dfrac{1}{\sqrt{x}}-\sqrt{x}\right)^6$.

We have $a=\dfrac{1}{\sqrt{x}}$, $b=-\sqrt{x}$, and $n=6$.

Form 1: We use the 7th row of Pascal's triangle:

$$1\quad 6\quad 15\quad 20\quad 15\quad 6\quad 1$$

$$\left(\frac{1}{\sqrt{x}}-\sqrt{x}\right)^6$$

$$=1\cdot\left(\frac{1}{\sqrt{x}}\right)^6+6\left(\frac{1}{\sqrt{x}}\right)^5\left(-\sqrt{x}\right)^1+15\left(\frac{1}{\sqrt{x}}\right)^4\left(-\sqrt{x}\right)^2$$

$$+20\left(\frac{1}{\sqrt{x}}\right)^3\left(-\sqrt{x}\right)^3+15\left(\frac{1}{\sqrt{x}}\right)^2\left(-\sqrt{x}\right)^4$$

$$+6\left(\frac{1}{\sqrt{x}}\right)^1\left(-\sqrt{x}\right)^5+1\cdot\left(-\sqrt{x}\right)^6$$

$$=x^{-3}-6x^{-2}+15x^{-1}-20+15x-6x^2+x^3$$

Form 2:

$$\left(\frac{1}{\sqrt{x}}-\sqrt{x}\right)^6$$

$$=\binom{6}{0}\left(\frac{1}{\sqrt{x}}\right)^6+\binom{6}{1}\left(\frac{1}{\sqrt{x}}\right)^5\left(-\sqrt{x}\right)^1+\binom{6}{2}\left(\frac{1}{\sqrt{x}}\right)^4\left(-\sqrt{x}\right)^2$$

$$+\binom{6}{3}\left(\frac{1}{\sqrt{x}}\right)^3\left(-\sqrt{x}\right)^3+\binom{6}{4}\left(\frac{1}{\sqrt{x}}\right)^2\left(-\sqrt{x}\right)^4$$

$$+\binom{6}{5}\left(\frac{1}{\sqrt{x}}\right)^1\left(-\sqrt{x}\right)^5+\binom{6}{6}\left(-\sqrt{x}\right)^6$$

$$=\frac{6!}{6!0!}\left(\frac{1}{\sqrt{x}}\right)^6+\frac{6!}{5!1!}\left(\frac{1}{\sqrt{x}}\right)^5\left(-\sqrt{x}\right)^1+\frac{6!}{4!2!}\left(\frac{1}{\sqrt{x}}\right)^4\left(-\sqrt{x}\right)^2$$

$$+\frac{6!}{3!3!}\left(\frac{1}{\sqrt{x}}\right)^3\left(-\sqrt{x}\right)^3+\frac{6!}{2!4!}\left(\frac{1}{\sqrt{x}}\right)^2\left(-\sqrt{x}\right)^4$$

$$+\frac{6!}{1!5!}\left(\frac{1}{\sqrt{x}}\right)^1\left(-\sqrt{x}\right)^5+\frac{6!}{0!6!}\left(-\sqrt{x}\right)^6$$

$$=x^{-3}-6x^{-2}+15x^{-1}-20+15x-6x^2+x^3$$

43. Find the 3rd term of $(a+b)^6$.

First we note that $3=2+1$, $a=a$, $b=b$, and $n=6$.

Then the 3rd term of the expansion of $(a+b)^6$ is

$$\binom{6}{2}a^{6-2}b^2,\text{ or }\frac{6!}{4!2!}a^4b^2,\text{ or }15a^4b^2.$$

45. Find the 12th term of $(a-3)^{14}$.

First we note that $12=11+1$, $a=a$, $b=-3$, and $n=14$.

Then the 12th term of the expansion of $(a-3)^{14}$ is

$$\binom{14}{11}a^{14-11}\cdot(-3)^{11}=\frac{14!}{3!11!}a^3\,(-177,147)$$

$$=364a^3\,(-177,147)$$

$$=-64,481,508a^3$$

47. Find the 5th term of $\left(2x^3+\sqrt{y}\right)^8$.

First we note that $5=4+1$, $a=2x^3$, $b=\sqrt{y}$, and $n=8$.

Then the 5th term of the expansion of $\left(2x^3+\sqrt{y}\right)^8$ is

$$\binom{8}{4}\left(2x^3\right)^{8-4}\left(\sqrt{y}\right)^4=\frac{8!}{4!4!}\left(2x^3\right)^4\left(\sqrt{y}\right)$$

$$=70\left(16x^{12}\right)\left(y^2\right)$$

$$=1120x^{12}y^2$$

49. The expansion of $\left(2u+3v^2\right)^{10}$ has 11 terms so the 6th term is the middle term. Note that $6=5+1$, $a=2u$, $b=-3v^2$, and $n=10$. Then the 6th term of the expansion of $\left(2u+3v^2\right)^{10}$ is

$$\binom{10}{5}(2u)^{10-5}\left(3v^2\right)^5=\frac{10!}{5!5!}(2u)^5\left(3v^2\right)^5$$

$$=252\left(32u^5\right)\left(243v^{10}\right)$$

$$=1,959,552u^5v^{10}$$

51. The 9th term of $(x-y)^8$ is the last term, y^8.

53. *Writing Exercise.* The binomial coefficients of the first and last terms are always 1, so Maya needs only to find a^n and b^n to find these two terms. The binomial coefficients of the second term and the next-to-the-last term are always n, so to find these two terms Maya has only to compute na^{n-1} and nb^{n-1}.

55. Graph $y=x^2-5$.

This is a parabola with vertex $(0,-5)$.

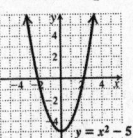

57. Graph $y\geq x-5$.

Use a solid line to form the line $y=x-5$. Since the test point $(0,0)$ is a solution, shade this side of the line.

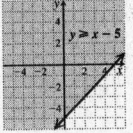

59. Graph $f(x) = \log_5 x$.

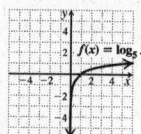

61. *Writing Exercise.* The $(r+1)$st term of $\left(x - \dfrac{3}{x}\right)^{10}$ is

$\dbinom{10}{r} x^{10-r}\left(-\dfrac{3}{x}\right)^r$. In the x^2-term the exponent $10 - r$ is 2

more than the exponent r, so we have:

$$10 - r = 2 + r$$
$$8 = 2r$$
$$4 = r$$

Thus, we would find the fifth term of the expansion

$\dbinom{10}{4} x^6 \left(-\dfrac{3}{x}\right)^4$.

63. Consider the set of 5 elements $\{a, b, c, d, e\}$. List all the

subsets of size 3:

$\{a, b, c\}$, $\{a, b, d\}$, $\{a, b, e\}$, $\{a, c, d\}$, $\{a, c, e\}$, $\{a, d, e\}$,
$\{b, c, d\}$, $\{b, c, e\}$, $\{b, d, e\}$, $\{c, d, e\}$.

There are exactly 10 subsets of size 3 and $\dbinom{5}{3} = 10$, so there

are exactly $\dbinom{5}{3}$ ways of forming a subset of size 3 from a

set of 5 elements.

65. Find the sixth term of $(0.15 + 0.85)^8$.

$\dbinom{8}{5}(0.15)^{8-5}(0.85)^5 = \dfrac{8!}{3!5!}(0.15)^3(0.85)^5 \approx 0.084$

67. Find and add the 7th through 9th terms of

$(0.15 + 0.85)^9$.

$\dbinom{8}{6}(0.15)^2(0.85)^6 + \dbinom{8}{7}(0.15)(0.85)^7 + \dbinom{8}{8}(0.85)^8 \approx 0.89$

69. $\dbinom{n}{n-r} = \dfrac{n!}{[n-(n-r)]!(n-r)!} = \dfrac{n!}{r!(n-r)!} = \dbinom{n}{r}$

71. $\dfrac{\dbinom{5}{3}(p^2)^2\left(-\dfrac{1}{2}p\sqrt[3]{q}\right)^3}{\dbinom{5}{2}(p^2)^3\left(-\dfrac{1}{2}p\sqrt[3]{q}\right)^2} = \dfrac{-\dfrac{1}{8}p^7 q}{\dfrac{1}{4}p^8\sqrt[3]{q^2}} = -\dfrac{\dfrac{1}{8}p^7 q}{\dfrac{1}{4}p^8 q^{2/3}}$

$= -\dfrac{1}{8} \cdot \dfrac{4}{1} \cdot p^{7-8} \cdot q^{1-2/3} = -\dfrac{1}{2}p^{-1}q^{1/3} = -\dfrac{\sqrt[3]{q}}{2p}$

73. $\left(x^2 + 2xy + y^2\right)\left(x^2 + 2xy + y^2\right)^2(x+y)$

$= (x+y)^2\left[(x+y)^2\right]^2(x+y) = (x+y)^7$

We can find the given product by finding the binomial

expansion of $(x+y)^7$. It is   (See Exercise 27.)

$x^7 + 7x^6 y + 21x^5 y^2 + 35x^4 y^3 + 35x^3 y^4 + 21x^2 y^5 + 7xy^6 + y^7$.

Chapter 11 Review

1. False; the next term of the arithmetic sequence

10, 15, 20, ... is $20 + 5$, or 25.

3. True.

5. False; a geometric sequence has a common *ratio*.

7. False; $n! = n \cdot (n-1) \cdot (n-2) \cdot \ldots \cdot 3 \cdot 2 \cdot 1$.

9. $a_n = 10n - 9$
$a_1 = 10 \cdot 1 - 9 = 1$
$a_2 = 10 \cdot 2 - 9 = 11$
$a_3 = 10 \cdot 3 - 9 = 21$
$a_4 = 10 \cdot 4 - 9 = 31$
$a_8 = 10 \cdot 8 - 9 = 71$
$a_{12} = 10 \cdot 12 - 9 = 111$

11. $-5, -10, -15, -20, \ldots$

These are negative multiples of 5 beginning with -5, so

the general term could be $-5n$.

13. $\displaystyle\sum_{k=1}^{5}(-2)^k = (-2)^1 + (-2)^2 + (-2)^3 + (-2)^4 + (-2)^5$

$= -2 + 4 + (-8) + 16 + (-32) = -22$

15. $7 + 14 + 21 + 28 + 35 + 42$

This is the sum of the first six positive multiples of 7. It

is an finite series. Sigma notation is

$\displaystyle\sum_{k=1}^{6} 7k$.

17. $-3, -7, -11, \ldots$

$a_1 = -3$, $d = -4$, and $n = 14$
$a_n = a_1 + (n-1)d$
$a_{14} = -3 + (14-1)(-4) = -3 + 13(-4) = -55$

19. We know that $a_8 = 20$ and $a_{24} = 100$. We would have to add d sixteen times to get from a_8 to a_{24}. That is,

$$20 + 16d = 100$$
$$16d = 80$$
$$d = 5.$$

Since $a_8 = 20$, we subtract d seven times to get a_1.

$$a_1 = 20 - 7(5) = 20 - 35 = -15$$

21. The sum is $5 + 10 + 15 + \ldots + 495 + 500$. This is the sum of the arithmetic sequence for which $a_1 = 5$, $a_n = 500$, and $n = 100$. We use the formula for S_n.

$$S_n = \frac{n}{2}(a_1 + a_n)$$
$$S_{100} = \frac{100}{2}(5 + 500) = 50(505) = 25{,}250$$

23. $r = \dfrac{30}{40} = \dfrac{3}{4}$

25. $3, \dfrac{3}{4}x, \dfrac{3}{16}x^2, \ldots$

$a_1 = 3$, and $r = \dfrac{\frac{3}{4}x}{3} = \dfrac{3x}{4} \cdot \dfrac{1}{3} = \dfrac{x}{4}$

$a_n = a_1 r^{n-1}$

$a_n = 3\left(\dfrac{x}{4}\right)^{n-1}$

27. $3x - 6x + 12x - \ldots$

$a_1 = 3x$, $n = 12$, and $r = \dfrac{-6x}{3x} = -2$

$$S_n = \frac{a_1(1 - r^n)}{1 - r}$$
$$S_{12} = \frac{3x(1 - (-2)^{12})}{1 - (-2)} = \frac{3x(1 - 4096)}{3} = x(-4095) = -4095x$$

29. $7 - 4 + \dfrac{16}{7} - \ldots$

$|r| = \left|\dfrac{-4}{7}\right| = \dfrac{4}{7} < 1$, the series has a limit.

$$S_\infty = \frac{a_1}{1 - r} = \frac{7}{1 - \left(-\frac{4}{7}\right)} = \frac{7}{\frac{11}{7}} = 7 \cdot \frac{7}{11} = \frac{49}{11}$$

31. $0.04 + 0.08 + 0.16 + 0.32 + \ldots$

$|r| = \left|\dfrac{0.08}{0.04}\right| = |2| = 2$, and since $|r| \not< 1$, the series does not have a limit.

33. $0.5555\ldots = 0.5 + 0.05 + 0.005 + 0.0005 + \ldots$

This is an infinite geometric series with $a_1 = 0.5$.

$|r| = \left|\dfrac{0.05}{0.5}\right| = |0.1| = 0.1 < 1$, so the series has a limit.

$$S_\infty = \frac{a_1}{1 - r} = \frac{0.5}{1 - 0.1} = \frac{0.5}{0.9} = \frac{5}{9}$$

Fractional notation for $0.5555\ldots$ is $\dfrac{5}{9}$.

35. *Familiarize.* A \$0.40 raise every 3 months for 8 years, is 32 raises. We want to find the 33th term of an arithmetic sequence with $a_1 = \$11.50$, $d = \$0.40$, and $n = 33$. We will first use the formula for a_n to find a_{33}. This result is the hourly wage after 8 years.

Translate. Substituting into the formula for a_n, we have

$$a_{33} = \$11.50 + (33 - 1)(\$0.40).$$

Carry out. We first find a_{33}.

$$a_{33} = \$11.50 + 32(\$0.40) = \$24.30$$

Check. We can do the calculations again.

State. After 8 years, Tyrone earns \$24.30 an hour.

37. *Familiarize.* At the end of each year, the interest at 4% will be added to the previous year's amount. We have a geometric sequence:

$\$12{,}000$, $\$12{,}000(1.04)$, $\$12{,}000(1.04)^2, \ldots$

We are looking for the amount after 7 years.

Translate. We use the formula $a_n = a_1 r^{n-1}$ with $a_1 = \$12{,}000$, $r = 1.04$, and $n = 8$:

$$a_8 = \$12{,}000(1.04)^{8-1}$$

Carry out. We calculate to obtain $a_8 \approx \$15{,}791.18$.

Check. We can do the calculation again.

State. After 7 years, the amount of the loan will be about \$15,791.18.

39. $7! = 7 \cdot 6 \cdot 5 \cdot 4 \cdot 3 \cdot 2 \cdot 1 = 5040$

41. $\dbinom{20}{2}a^{20-2}b^2 = 190a^{18}b^2$

43. *Writing Exercise.* For a geometric sequence with $|r| < 1$, as n gets larger, the absolute value of the terms gets smaller, since $|r^n|$ gets smaller.

45. $1 - x + x^2 - x^3 + \ldots$

$a_1 = 1$, $n = n$, and $r = \dfrac{-x}{1} = -x$

$$S_n = \frac{1(1 - (-x)^n)}{1 - (-x)} = \frac{1 - (-x)^n}{1 + x}$$

Chapter 11 Test

1. $a_n = \dfrac{1}{n^2+1}$

$a_1 = \dfrac{1}{1^2+1} = \dfrac{1}{2}$

$a_2 = \dfrac{1}{2^2+1} = \dfrac{1}{5}$

$a_3 = \dfrac{1}{3^2+1} = \dfrac{1}{10}$

$a_4 = \dfrac{1}{4^2+1} = \dfrac{1}{17}$

$a_5 = \dfrac{1}{5^2+1} = \dfrac{1}{26}$

$a_{12} = \dfrac{1}{12^2+1} = \dfrac{1}{145}$

3. $\displaystyle\sum_{k=2}^{5}(1-2^k) = (1-2^2)+(1-2^3)+(1-2^4)+(1-2^5)$
$$= -3+(-7)+(-15)+(-31) = -56$$

5. $\dfrac{1}{2},\ 1,\ \dfrac{3}{2},\ 2,\ ...$

$a_1 = \dfrac{1}{2},\ d = \dfrac{1}{2}$, and $n = 13$

$a_n = a_1+(n-1)d$

$a_{13} = \dfrac{1}{2}+(13-1)\left(\dfrac{1}{2}\right) = \dfrac{1}{2}+12\left(\dfrac{1}{2}\right) = \dfrac{13}{2}$

7. We know that $a_5 = 16$ and $a_{10} = -3$. We would have to add d five times to get from a_5 to a_{10}. That is,

$16+5d = -3$
$5d = -19$
$d = -3.8.$

Since $a_5 = 16$, we subtract d four times to get a_1.

$a_1 = 16-4(-3.8) = 16+15.2 = 31.2$

9. $-3,\ 6,\ -12,...$

$a_1 = -3,\ n = 10$, and $r = \dfrac{6}{-3} = -2$

$a_n = a_1 r^{n-1}$

$a_{10} = -3(-2)^{10-1} = -3(-2)^9 = -3(-512) = 1536$

11. $3,\ 9,\ 27,...$

$a_1 = 3$, and $r = \dfrac{9}{3} = 3$

$a_n = a_1 r^{n-1}$

$a_n = 3(3)^{n-1} = 3^n$

13. $0.5+0.25+0.125+...$

$|r| = \left|\dfrac{0.25}{0.5}\right| = |0.5| = 0.5 < 1$, the series has a limit.

$S_\infty = \dfrac{a_1}{1-r} = \dfrac{0.5}{1-0.5} = \dfrac{0.5}{0.5} = 1$

15. $\$1000+\$80+\$6.40+...$

$|r| = \left|\dfrac{\$80}{\$1000}\right| = |0.08| = 0.08 < 1$, the series has a limit.

$S_\infty = \dfrac{a_1}{1-r} = \dfrac{\$1000}{1-0.08} = \dfrac{\$1000}{0.92} = \dfrac{\$25,000}{23} \approx \$1086.96$

17. *Familiarize.* We want to find the seventeenth term of an arithmetic sequence with $a_1 = 31$, $d = 2$, and $n = 17$. We will first use the formula for a_n to find a_{17}. This result is the number of seats in the 17th row.

Translate. Substituting into the formula for a_n, we have

$a_{17} = 31+(17-1)2.$

Carry out. We first find a_{17}.

$a_{17} = 31+16\cdot2 = 63$

Check. We can do the calculations again.

State. There are 63 seats in the 17th row.

19. *Familiarize.* At the end of each week, the price will be 95% of the previous week's price.

We have a geometric sequence:

$\$10,000,\ \$10,000(0.95),\ \$10,000(0.95)^2,...$

We are looking for the price after 10 weeks.

Translate. We use the formula $a_n = a_1 r^{n-1}$ with $a_1 = \$10,000$, $r = 0.95$, and $n = 11$:

$a_{10} = \$10,000(0.95)^{11-1}$

Carry out. We calculate to obtain $a_{11} \approx \$5987.37$.

Check. We can do the calculation again.

State. After 10 weeks, the price of the boat will be about $\$5987.37$.

21. $\dbinom{12}{9} = \dfrac{12!}{3!9!} = \dfrac{12\cdot11\cdot10\cdot9!}{3!9!} = \dfrac{12\cdot11\cdot10}{3\cdot2\cdot1} = 4\cdot11\cdot5 = 220$

23. $\dbinom{12}{3}a^{12-3}x^3 = 220a^9x^3$

25. $1+\dfrac{1}{x}+\dfrac{1}{x^2}+\dfrac{1}{x^3}+...$

$a_1 = 1,\ n = n$, and $r = \dfrac{\frac{1}{x}}{1} = \dfrac{1}{x}$

$S_n = \dfrac{1\left(1-\left(\frac{1}{x}\right)^n\right)}{1-\frac{1}{x}} = \dfrac{1-\left(\frac{1}{x}\right)^n}{1-\frac{1}{x}}$, or $\dfrac{x^n-1}{x^{n-1}(x-1)}$